Advances in Experimental Medicine and Biology

Volume 1457

Advances in Experimental Medicine and Biology provides a platform for scientific contributions in the main disciplines of the biomedicine and the life sciences. This series publishes thematic volumes on contemporary research in the areas of microbiology, immunology, neurosciences, biochemistry, biomedical engineering, genetics, physiology, and cancer research. Covering emerging topics and techniques in basic and clinical science, it brings together clinicians and researchers from various fields.

Advances in Experimental Medicine and Biology has been publishing exceptional works in the field for over 40 years, and is indexed in SCOPUS, Medline (PubMed), EMBASE, Reaxys, EMBiology, the Chemical Abstracts Service (CAS), and Pathway Studio.

2023 CiteScore: 5.9

Nima Rezaei
Editor

The COVID-19 Aftermath

Volume I: Ongoing Challenges

Springer

Editor
Nima Rezaei
Network of Immunity in Infection, Autoimmunity and Malignancy (NIIMA)
Universal Scientific Education and Research Network (USERN)
Stockholm, Sweden

Research Center for Immunodeficiencies, Children's Medical Center
Tehran University of Medical Sciences
Tehran, Iran

ISSN 0065-2598 ISSN 2214-8019 (electronic)
Advances in Experimental Medicine and Biology
ISBN 978-3-031-61938-0 ISBN 978-3-031-61939-7 (eBook)
https://doi.org/10.1007/978-3-031-61939-7

This Springer imprint is published by the registered company Springer Nature Switzerland AG
The registered company address is: Gewerbestrasse 11, 6330 Cham, Switzerland

Preface

Coronavirus disease 2019 (COVID-19) emerged in December 2019 and continued to distribute worldwide; so, it became a global epidemic. Despite many efforts launched to limit the epidemic's spread and reduce its burden, COVID has been the cause of death for more than seven million people. It reflects the programs have failures. However, the recovery of about 700 million people from the disease is evidence that those programs have also had successes. Therefore, the epidemic has taught us many lessons in a relatively short period, though it was a very strict teacher; we paid a heavy price-human suffering and death-to learn, making it urgent to document these lessons in detail for the next time and even the next times no matter we are alive or dead at that time or times, the matter is that humanity is there.

After successful publication of the book *Coronavirus Disease 2019 (COVID-19)* at the start of the pandemic https://link.springer.com/book/10.1007/978-3-030-63761-3, which was very welcomed by the readers, we decided to publish two further volumes on COVID-19 Aftermath, to explain the ongoing challenges and the lessons learned from a pandemic.

In *The COVID-19 Aftermath—Volume I: Ongoing Challenges*, different challenges imposed to the globe following the COVID-19 pandemic are discussed by experts in this field. It offers lessons learned from COVID related to numerous domains, from transmission containment, diagnostics, therapeutics, and vaccines, to research in different conditions.

The book starts with a general chapter with the title of Chap. 1 to highlight the lessons learned from the pandemic. Structure of coronavirus and its targeting is discussed in Chaps. 2 and 3. Clinical outcomes of the disease, especially respiratory involvement, failure, and support are the subjects of Chaps. 4 and 5. COVID-19-related extrapyramidal disorders, acute encephalopathies, and neurocovid are explained in Chaps. 6–8, respectively. Chapters 9–16 focus on the COVID-19 vaccination and therapeutics and challenges. A few regional challenges are also discussed in Chaps. 17–21. Finally, challenges in different conditions, such as children and adolescents with chronic illness, ophthalmology, obesity, and the world of sports are the main discussion of Chaps. 22–25.

I would like to hereby acknowledge the expertise of all contributors, for generously devoting their time and considerable effort in preparing their respective chapters. I would also like to express my gratitude to the Springer Nature publication for providing me the opportunity to publish this new volume as well. Finally, I hope that this timely book will be comprehensible, cogent, and of special value for researchers and clinicians who wish to extend their knowledge on ongoing challenges from a pandemic.

Stockholm, Sweden/Tehran, Iran Nima Rezaei MD, Ph.D.

Acknowledgments I would like to express my gratitude to the Editorial Assistant of this book series, Dr. Amene Saghazadeh.

Contents

About the Editor

Prof. Nima Rezaei, MD, Ph.D. gained his medical degree (MD) from Tehran University of Medical Sciences and subsequently obtained an M.Sc. in Molecular and Genetic Medicine and a Ph.D. in Clinical Immunology and Human Genetics from the University of Sheffield, UK. He also spent a short-term fellowship of Pediatric Clinical Immunology and Bone Marrow Transplantation in the Newcastle General Hospital. Professor Rezaei is now the full professor of Immunology and vice dean of Research and Technologies at the School of Medicine, Tehran University of Medical Sciences, and the Co-founder and head of the Research Center for Immunodeficiencies. He is also the founder of the Universal Scientific Education and Research Network (USERN). Professor Rezaei has already been the director of more than 200 research projects and has designed and participated in several international collaborative projects. Professor Rezaei is the editor, editorial assistant, or editorial board member of more than 50 international journals. He has edited more than 100 international books, has presented more than a 1000 lectures/posters in congresses/meetings, and has published more than 1500 scientific papers in the international journals.

Global Challenges After a Global Challenge: Lessons Learned from the COVID-19 Pandemic

1

Niloufar Yazdanpanah, Constantine Sedikides, Hans D. Ochs, Carlos A. Camargo Jr., Gary L. Darmstadt, Artemi Cerda, Valentina Cauda, Godefridus J. Peters, Frank Sellke, Nathan D. Wong, Elisabetta Comini, Alberto Ruiz Jimeno, Vivette Glover, Nikos Hatziargyriou, Christian E. Vincenot, Stéphane P. A. Bordas, Idupulapati M. Rao, Hassan Abolhassani, Gevork B. Gharehpetian, Ralf Weiskirchen, Manoj Gupta, Shyam Singh Chandel, Bolajoko O. Olusanya, Bruce Cheson, Alessio Pomponio, Michael Tanzer, Paul S. Myles, Wen-Xiu Ma, Federico Bella, Saeid Ghavami, S. Moein Moghimi, Domenico Pratico, Alfredo M. Hernandez, Maria Martinez-Urbistondo, Diego Martinez Urbistondo, Seyed-Mohammad Fereshtehnejad, Imran Ali, Shinya Kimura, A. Wallace Hayes, Wenju Cai, Chua K. J. Ernest, Sabu Thomas, Kazem Rahimi, Armin Sorooshian, Michael Schreiber, Koichi Kato, John H. T. Luong, Stefano Pluchino, Andres M. Lozano, John F. Seymour, Kenneth S. Kosik, Stefan G. Hofmann, Roger S. McIntyre, Matjaz Perc, Alexander Leemans, Robyn S. Klein, Shuji Ogino, Christopher Wlezien, George Perry, Juan J. Nieto, Lisa Levin, Daniel J. Klionsky, Bahram Mobasher, Tommaso Dorigo, Nima Rezaei, and USERN Advisory Board

---

N. Yazdanpanah · C. Sedikides · H. D. Ochs · C. A. Camargo Jr. · G. L. Darmstadt · A. Cerda · V. Cauda · G. J. Peters · F. Sellke · N. D. Wong · E. Comini · A. R. Jimeno · V. Glover · N. Hatziargyriou · C. E. Vincenot · S. P. A. Bordas · I. M. Rao · H. Abolhassani · G. B. Gharehpetian · R. Weiskirchen · M. Gupta · S. S. Chandel · B. O. Olusanya · B. Cheson · A. Pomponio · M. Tanzer · P. S. Myles · W.-X. Ma · F. Bella · S. Ghavami · S. Moein Moghimi · D. Pratico · A. M. Hernandez · M. Martinez-Urbistondo · D. M. Urbistondo · S.-M. Fereshtehnejad · I. Ali · S. Kimura · A. Wallace Hayes · W. Cai · C. K. J. Ernest · S. Thomas · K. Rahimi · A. Sorooshian · M. Schreiber · K. Kato · J. H. T. Luong · S. Pluchino · A. M. Lozano · J. F. Seymour · K. S. Kosik · S. G. Hofmann · R. S. McIntyre · M. Perc · A. Leemans · R. S. Klein · S. Ogino · C. Wlezien · G. Perry · J. J. Nieto · L. Levin · D. J. Klionsky · B. Mobasher · T. Dorigo · N. Rezaei (✉) · USERN Advisory Board
Universal Scientific Education and Research Network (USERN), URL: https://usern.org/
e-mail: rezaei_nima@yahoo.com

N. Rezaei (ed.), *The COVID-19 Aftermath*, Advances in Experimental Medicine and Biology 1457,
https://doi.org/10.1007/978-3-031-61939-7_1

N. Yazdanpanah · N. Rezaei
School of Medicine, Tehran University of Medical Sciences, Tehran, Iran

N. Yazdanpanah · H. Abolhassani · N. Rezaei
Research Center for Immunodeficiencies, Children's Medical Center, Tehran University of Medical Sciences, Tehran, Iran

C. Sedikides
School of Psychology, University of Southampton, Southampton, UK

H. D. Ochs
Department of Pediatrics, Seattle Children's Research Institute, University of Washington School of Medicine, Seattle, WA, USA

C. A. Camargo Jr.
Department of Emergency Medicine, Massachusetts General Hospital, Harvard Medical School, Boston, MA, USA

G. L. Darmstadt
Department of Pediatrics, Stanford University School of Medicine, Stanford, CA, USA

A. Cerda
Soil Erosion and Degradation Research Group, Department of Geography, Valencia University, Blasco Ibàñez, Valencia, Spain

V. Cauda · F. Bella
Department of Applied Science and Technology, Politecnico di Torino, Corso Duca degli Abruzzi, Turin, Italy

G. J. Peters
Laboratory Medical Oncology, Amsterdam University Medical Centers, Location VUMC, Amsterdam, the Netherlands

Department of Biochemistry, Medical University of Gdańsk, Gdańsk, Poland

F. Sellke
Warren Alpert Medical School, Brown University, Providence, RI, USA

Division of Cardiothoracic Surgery, Rhode Island Hospital, Providence, RI, USA

N. D. Wong
Heart Disease Prevention Program, Division of Cardiology, University of California Irvine, C-240 Medical Sciences, Irvine, CA, USA

E. Comini
SENSOR Laboratory, University of Brescia, Brescia, Italy

A. R. Jimeno
Instituto de Física de Cantabria (IFCA), CSIC-Universidad de Cantabria, Santander, Spain

V. Glover
Department of Metabolism, Digestion and Reproduction Hammersmith Hospital Campus, Institute of Reproductive and Developmental Biology, Imperial College London, London, UK

N. Hatziargyriou
School of Electrical and Computer Engineering, National Technical University of Athens (NTUA), Athens, Greece

C. E. Vincenot · S. P. A. Bordas
Faculty of Science, Technology and Medicine (FSTM), University of Luxembourg, Esch-sur-Alzette, Grand Duchy of Luxembourg

I. M. Rao
Alliance of Bioversity International, International Center for Tropical Agriculture, Cali, Colombia

International Centre of Insect Physiology and Ecology (ICIPE), Nairobi, Kenya

H. Abolhassani
Division of Immunology, Department of Medical Biochemistry and Biophysics, Karolinska Institute, Stockholm, Sweden

G. B. Gharehpetian
Amirkabir University of Technology, Tehran, Iran

R. Weiskirchen
Institute of Molecular Pathobiochemistry, Experimental Gene Therapy and Clinical Chemistry (IFMPEGKC), RWTH University Hospital Aachen, Aachen, Germany

M. Gupta · C. K. J. Ernest
Department of Mechanical Engineering, National University of Singapore, Singapore, Singapore

S. S. Chandel
Photovoltaics Research Group, Centre of Excellence in Energy Science and Technology, Shoolini University, Solan, Himachal Pradesh 173212, India

B. O. Olusanya
Centre for Healthy Start Initiative, Ikoyi, Lagos, Nigeria

B. Cheson
Center for Cancer and Blood Disorders, Bethesda, MD, USA

A. Pomponio
Dipartimento di Meccanica, Matematica e Management, Politecnico di Bari, Bari, Italy

M. Tanzer
Division of Orthopedic Surgery, McGill University, Montreal, QC, Canada

P. S. Myles
Alfred Hospital and Monash University, Melbourne, Australia

W.-X. Ma
Department of Mathematics and Statistics, University of South Florida, Tampa, FL, USA

Department of Mathematics, King Abdulaziz University, Jeddah 21589, Saudi Arabia

Material Science Innovation and Modelling, North-West University, Mafikeng Campus, Mmabatho 2735, South Africa

S. Ghavami
Human Anatomy and Cell Science, University of Manitoba, Winnipeg, MB, Canada

S. Moein Moghimi
School of Pharmacy, Newcastle University, Newcastle upon Tyne NE1 7RU, UK

Faculty of Health and Medical Sciences, Translational and Clinical Research Institute, Newcastle University, Newcastle upon Tyne NE2 4HH, UK

Colorado Center for Nanomedicine and Nanosafety, University of Colorado Anschutz Medical Center, Aurora, CO, USA

D. Pratico
Alzheimer's Center at Temple, Lewis Katz School of Medicine, Temple University, Philadelphia, PA, USA

A. M. Hernandez
Medicine and Endocrinology Department, Universidad de Valladolid and IMDEA, Madrid, Spain

M. Martinez-Urbistondo
Internal Medicine Department, Hospital Universitario Puerta de Hierro, Madrid, Spain

D. M. Urbistondo
Hospital Universitario HM Sanchinarro, Madrid, Spain

S.-M. Fereshtehnejad
Division of Neurology, Faculty of Medicine, University of Ottawa, Ottawa, ON, Canada

Division of Clinical Geriatrics, NVS Department, Karolinska Institutet, Stockholm, Sweden

I. Ali
Department of Chemistry, Jamia Millia Islamia (Central University), New Delhi, India

S. Kimura
Division of Hematology, Respiratory Medicine and Oncology, Department of Internal Medicine, Faculty of Medicine, Saga University, Saga, Japan

A. Wallace Hayes
Center for Environmental/Occupational Risk Analysis and Management, College of Public Health, University of South Florida, Tampa, FL 33612, USA

Michigan State University, East Lansing, MI, USA

W. Cai
CSIRO Environment, Hobart, TAS, Australia

S. Thomas
School of Energy Materials, Mahatma Gandhi University, Kottayam, Kerala, India

K. Rahimi
Deep Medicine, Nuffield Department of Women's and Reproductive Health, University of Oxford, Oxford, UK

A. Sorooshian
Department of Chemical and Environmental Engineering, University of Arizona, Tucson, AZ, USA

M. Schreiber
Institut für Physik, Technische Universität Chemnitz, 09107 Chemnitz, Germany

K. Kato
Exploratory Research Center on Life and Living Systems (ExCELLS) and Institute for Molecular Science (IMS), National Institutes of Natural Sciences, Okazaki, Aichi, Japan

J. H. T. Luong
School of Chemistry, University College Cork, Cork T12 YN60, Ireland

S. Pluchino
Department of Clinical Neurosciences, University of Cambridge, Cambridge, UK

A. M. Lozano
Division of Neurosurgery, Toronto Western Hospital, University Health Network, University of Toronto, Toronto, ON, Canada

Krembil Research Institute, Toronto, ON, Canada

J. F. Seymour
Clinical Haematology, Peter MacCallum Cancer Centre and The Royal Melbourne Hospital, Melbourne, VIC, Australia

Faculty of Medicine, Dentistry and Health Sciences, University of Melbourne, Melbourne, VIC, Australia

K. S. Kosik
Department of Molecular Cellular Developmental Biology, Neuroscience Research Institute, University of California, Santa Barbara, CA, USA

S. G. Hofmann
Department of Psychology, Philipps-University Marburg, Marburg, Germany

R. S. McIntyre
Department of Pharmacology and Toxicology, University of Toronto, Toronto, ON, Canada

Department of Psychiatry, University of Toronto, Toronto, ON, Canada

M. Perc
Faculty of Natural Sciences and Mathematics, University of Maribor, Koroška cesta 160, Maribor, Slovenia

Department of Medical Research, China Medical University Hospital, China Medical University, Taichung 404332, Taiwan

Alma Mater Europaea, Slovenska ulica 17, 2000 Maribor, Slovenia

Complexity Science Hub Vienna, Josefstädterstraße 39, 1080 Vienna, Austria

Department of Physics, Kyung Hee University, 26 Kyungheedae-ro, Dongdaemun-gu, Seoul, Republic of Korea

A. Leemans
Image Sciences Institute, University Medical Center Utrecht, Utrecht, the Netherlands

R. S. Klein
Center for Neuroimmunology and Neuroinfectious Diseases, St. Louis, MO, USA

Departments of Medicine, Pathology and Immunology, and Neuroscience, Washington University School of Medicine, St. Louis, MO, USA

S. Ogino
Department of Epidemiology, Harvard T.H. Chan School of Public Health, Boston, MA, USA

Program in MPE Molecular Pathological Epidemiology, Department of Pathology, Brigham and Women's Hospital, and Harvard Medical School, Boston, MA, USA

Broad Institute of MIT and Harvard, Cambridge, MA, USA

C. Wlezien
Department of Government, University of Texas at Austin, Austin, TX, USA

G. Perry
Department of Neuroscience, Developmental and Regenerative Biology, University of Texas at San Antonio, San Antonio, TX, USA

J. J. Nieto
CITMAga, University of Santiago de Compostela, A Coruña, Spain

L. Levin
Center for Marine Biodiversity and Conservation, Integrative Oceanography Division, Scripps Institution of Oceanography, University of California, San Diego, San Diego, CA, USA

D. J. Klionsky
Life Sciences Institute, University of Michigan, Ann Arbor, MI, USA

Department of Molecular, Cellular and Developmental Biology, University of Michigan, Ann Arbor, MI, USA

B. Mobasher
Department of Physics and Astronomy, University of California, Riverside, CA, USA

T. Dorigo
Lulea University of Technology, Laboratorievagen 14, Lulea, Sweden

Istituto Nazionale di Fisica Nucleare (INFN), Via Francesco Marzolo, Sezione di Padova, Italy

Abstract

Coronavirus disease 2019 (COVID-19) has affected not only individual lives but also the world and global systems, both natural and human-made. Besides millions of deaths and environmental challenges, the rapid spread of the infection and its very high socioeconomic impact have affected healthcare, economic status and wealth, and mental health across the globe. To better appreciate the pandemic's influence, multidisciplinary and interdisciplinary approaches are needed. In this chapter, world-leading scientists from different backgrounds share collectively their views about the pandemic's footprint and discuss challenges that face the international community.

Keywords

COVID-19 · Post-pandemic · Physiological · Mental health · Environment · Economics · Scientific · Infodemic · Long COVID-19 · Complications

Abbreviations

AI	Artificial intelligence
COVID-19	Coronavirus disease 2019
HIV	Human immunodeficiency virus
RdRP	RNA-dependent RNA polymerase
SARS-CoV-2	Severe acute respiratory syndrome coronavirus 2
SDGs	Sustainable development goals
WHO	World Health Organization
PHEIC	Public Health Emergency of International Concern

1.1 Introduction

Coronavirus disease 2019 (COVID-19) has affected not only individual lives but also the world and global systems, both natural and human-made. The latter includes adverse consequences to the environment, global trade, and supply chains, as well as politics. It has been more than three years since the World Health Organization (WHO) declared the newly emerged epidemic viral infection as a Public Health Emergency of International Concern (PHEIC); since then, the world has been facing, and will be facing, many pandemic-related challenges.

Although the COVID-19 pandemic put a remarkable and perhaps unprecedented strain on healthcare infrastructure and care delivery, its impact was also felt on education systems, physiological and psychological health, economics, and the environment. Some of the pandemic's effects may only become fully apparent over time. To better appreciate the pandemic's influence, multidisciplinary and interdisciplinary approaches are needed. In this chapter, world-leading scientists from different backgrounds share collectively their views about the pandemic's footprint and discuss challenges that face the international community.

1.2 Interdisciplinary Approach to Combat the Virus

1.2.1 Modeling of COVID-19

The prediction of COVID-19's evolution was crucial. Mathematical models, computational simulations, and epidemiological simulations play a pivotal role in anticipating and controlling the pandemic and potential future epidemics (Moradian et al. 2020).

To model effectively the dynamics of a specific infection, researchers need to consider the influence of different sub-population compartments. Many variables ranging from micro-host–pathogen interactions to host-to-host encounters as well as the prevailing cultural, social, economic, and local customs worldwide need to be taken into account. Mathematical modeling may help to minimize the effects of the epidemic by understanding the virus transmission mechanisms, thus providing the opportunity to predict the spread of the virus and controlling the pandemic. As an example, the relevant compartments of the asymptomatic persons and, most important, the super-spreaders have been used to model the COVID-19 transmission dynamics (Ndaïrou et al. 2020).

Most of the mathematical models contain ordinary or partial differential equations. These mathematical tools, when combined with large quantities of data along with machine learning and artificial intelligence, can create new technologies to facilitate prevention and combat viral epidemics (Palaniappan et al. 2022), highlighting the potential of interdisciplinary studies to address complex problems.

1.2.2 Virus Detection

Early detection of COVID-19 is crucial for health policy. As such, sensitive, robust, and reliable sensors are needed. Researchers have been proposing and testing new techniques and kits necessary for virus detection but not adequate sensitivity, selectivity, and scalability have been achieved so far. Lately, technologies involving nanomaterial-based sensors have been developed, characterized by the potential for a rapid coronavirus detection (Naikoo et al. 2022). These technologies are not invasive and may afford real-time results at low fabrication costs. We highlight surface plasmon resonance and localized surface plasmon resonance sensors (Lee et al. 2018) that detect the viral nucleic-acid particles and nucleocapsid antibodies against coronavirus in undiluted human serum, exhaled breath sensors based on inorganic or organic nanomaterials that rely on a change in the electrical resistance caused by the interaction with the analyte (Shan et al. 2020; Naikoo et al. 2021), graphene-based field-effect transistor (FET)-based biosensors (Seo et al. 2020), electrochemical sensors (Mujawar et al. 2020; Maduraiveeran et al. 2018), and colorimetric sensors that are based on gold nanoparticles (AuNPs) linked to thiol modified antisense oligonucleotides (Moitra et al. 2020). Yet, nanomaterial-based sensors for COVID-19 detection did not fully achieve the requirements for commercialization. Accuracy needs to improve together with reduction of false positives. On better news, artificial intelligence (AI) hold promises for addressing these challenges (Jain et al. 2021). The achievement of reliable and wide-spreadable sensors may allow mass COVID-19 detection along with its large area monitoring that in turn will provide a clear view on the rapid progression of viral infections.

Finally, sensors may be important for reducing the risks of infections in a different way. Given that the virus is present in aerosols in the exhaled breath of infected persons, and the inhalation of high doses of these aerosols may infect healthy persons, it is pivotal to monitor the presence of exhaled breath in closed environments in order to alert or tart a ventilation process. Forcing ventilation can decrease the presence of aerosols indoors and decrease the risk of spreading infection. This can be achieved via air quality sensors (Bosch Sensortec 2019).

1.2.3 Development of New Vaccines

The pandemic has pushed scientists to rapidly develop, validate, and clinically translate next-generation vaccines. The successful approval of vaccines against COVID-19 has led to the administration of billions of doses worldwide, protected people from severe acute respiratory syndrome coronavirus 2 (SARS-CoV-2), and saved many lives (Zheng et al. 2022). The existing types of COVID-19 vaccines include at least one of the following four approaches: (i) nucleic-acid vaccines (mainly mRNA encapsulated in lipid nanoparticles) able to instruct cells to produce the SARS-CoV-2 spike protein; (ii) inactivated vaccines derived from the coronavirus itself; (iii) protein vaccines based on the spike protein or its receptor-binding domain; and (iv) viral-vector vaccines, which adopt modified viruses to shuttle genetic instructions for synthesizing the spike protein into cells (Chavda et al. 2021; Ndwandwe and Wiysonge 2021; Deng et al. 2022).

However, challenges still exist for the COVID-19 vaccines. As the viral variants can evade the immunity provided by the original vaccination, the upgrade and improvement of COVID-19 vaccines are ongoing (Callaway 2023). Consequently, vaccines are currently developed as an upgrade of previous versions, such as the bivalent vaccines that comprise portions of the original virus and a variant (e.g., Omicron). In addition, new vaccines are currently developed relying on new technologies and nanosized platforms. These include mosaic vaccines based on nanoparticles supporting coronavirus proteins, and their RBDs derived from multiple coronavirus variants. The common objective of these new vaccines is to provide a long-lasting protection to the patient and overcome viral (present and future) variants. Furthermore, these new vaccines will become cheaper and even more potent, while minimizing side effects (Yamamoto 2022).

1.2.4 Looking for Treatments

The pandemic propelled drug development and repurposing of existing anticancer and anti-inflammatory drugs. Given that the SARS-CoV-2 virus infection led to COVID-19, a serious inflammatory condition, many existing drugs were used to relieve symptoms, such as corticosteroids (e.g., dexamethasone). These drugs relieved symptoms but did not control or reduce the viral infection. Several drugs, which were registered for other viral infections (e.g., human immunodeficiency virus [HIV]) or were in development by various pharmaceutical companies and research institutes, were immediately investigated for their antiviral SARS-CoV-2 activity. Antiviral drugs are targeted against either host factors or viral factors. The best known host factor is angiotensin-converting enzyme 2 (ACE2), a receptor responsible for the uptake of the virus. The major disadvantage of ACE2 targeting is its normal cellular function (Jackson et al. 2022). Targeting the viral life cycle (i.e., replication) is more likely to be specific, as it addresses a viral-encoded component of the reproduction machinery in the cell. This is the backbone of the successful treatment of HIV infections. Such practice led to rapid development and registration of the nucleoside analogs remdesivir and molnupiriavir, drugs inhibiting RNA-dependent RNA polymerase (RdRP) (Lee et al. 2022a; Beigel et al. 2020; von Delft et al. 2023). Viral proteases are essential in the viral life cycle, and nirmatrelvir is an example of a drug inhibiting SARS-CoV-2 proteases (Hammond et al. 2022). This drug inhibits Mpro (3CL protease) and is combined with ritonavir, which inhibits CYP3A4/CYP450 to increase its half-life. However, there is a problem with these drugs. As has been observed with HIV infection, the virus mutates rapidly, leading to drug resistance. This can be solved effectively by developing combinations of drugs directed to different viral targets, such as RdRP and Mpro. It can also be solved by repurposing already registered anticancer drugs, which are likely or proven to be effective; examples are antimetabolites such as 6-mercaptopurine or antifolates such as methotrexate (Stegmann et al. 2021; Bayoumy et al. 2020; Mohamed et al. 2021a). For instance, methotrexate enhances the efficacy of remdesivir, whereas inhibition of dihydroorotate dehydrogenase potentiates the effect of molnupiravir (Stegmann et al. 2022).

The development of these combinations is rapidly proceeding. With experience in designing combinations for HIV treatment and cancer, more effective combination treatments will likely be available both for SARS-CoV-2 and other viral infections.

1.3 Medical and Health Challenges

1.3.1 Long COVID-19

Many patients experience difficult, even life-threatening conditions, days and weeks following infection with SARS-CoV-2, and for high-risk patients immediate mortality remains a danger (Rezaei et al. 2021). Nevertheless, most infected people will survive the acute episode, although the health challenges may not end quickly. A good number of patients face a range of ongoing complications after COVID-19 recovery, a condition named long COVID-19. It is estimated that more than 10% of patients develop long COVID-19, and more than 65 million individuals worldwide suffer from it. Multiple organs and systems are affected by long COVID-19 with diverse clinical manifestations, including those that resemble chronic fatigue syndrome/myalgic encephalomyelitis and postural orthostatic tachycardia syndrome (Davis et al. 2021; Ceban et al. 2022; Larsen et al. 2022). Incidences of long COVID-19 are most frequent in hospitalized patients, followed by non-hospitalized and vaccinated individuals (Davis et al. 2023). Patients present with clinical manifestations of the organ system and tissues involved, including neurological and cognitive system (memory loss, cognitive impairment, paresthesia, dizziness, and balance issues, sensitivity to light and noise, loss of smell or taste, autonomic dysfunction, tinnitus, hearing loss, vertigo), reproductive system (menstrual irregularities), respiratory system (shortness of breath, cough), gastrointestinal system (nausea, abdominal pain, loss of appetite, heartburn, constipation), dysautonomia, chronic fatigue syndrome/myalgic encephalomyelitis (postexertional malaise, unrefreshing sleep, cognitive impairment, and/or orthostatic intolerance), vascular tissue and organ damage (increased risk of deep vein thrombosis, pulmonary embolism, bleeding), and endocrine disorders (low cortisol) (Nalbandian et al. 2021; Han et al. 2022).

Various mechanisms have been proposed to explain long COVID-19 symptoms. ACE2, which is well recognized as the COVID-19 entry portal into cells, is also expressed in different non-respiratory tissues. Therefore, viral replication in ACE2-expressing cells in different tissues may induce a local inflammatory response, which has been suggested to contribute to prolonged COVID-19 pathophysiology (Khazaal et al. 2022). In addition, the host immune system may be a culprit. Activation of the innate immune system can contribute to disease pathology (Lucas et al. 2020). For instance, SARS-CoV-2 nucleocapsid (N)-protein and spike proteins can trigger complement activation in different ways, which might induce tissue damage and thrombosis (Flude et al. 2021; Yu et al. 2020). These findings have prompted clinical studies on the use of complement inhibitors, such as AMY-101, with successful therapeutic outcomes in COVID-19 patients (Soung et al. 2022; Mastaglio et al. 2020).

However, the misdirected immune response is not limited to innate immunity, in view of the frequently present autoantibodies that target immunomodulatory proteins (such as anti-type I interferon neutralizing antibodies) and the considerable increase in autoantibody reactivities frequently detected in COVID-19 patients (Wang et al. 2021a). As an example, a type of autoantibody targeting the orexin/hypocretin receptor HCRTR2 impedes orexin signaling, which plays a crucial role in wakefulness and appetite regulation (Nixon et al. 2015). Many cytokines and chemokines are elevated, and several autoantibodies have been observed in affected individuals, including autoantibodies against ACE2, ADRB2/β2-adrenoceptor, CHRM2/muscarinic M2 receptor, AGTR1/angiotensin II AT1 receptor, and the angiotensin 1–7-MAS1 receptor (Davis et al. 2023).

In addition, prolonged inflammatory state due to the persistence of the virus in some body tissues (Li et al. 2020a; Wu et al. 2020; Yong 2021) can lead to endovascular and tissue fibrosis. These may be responsible for some lung injuries and cardiac problems post-COVID-19, which, in turn, may trigger more complications in other organs (Hama Amin et al. 2022; Liu et al. 2020). Autoimmunity is also a contributing mechanism to long COVID-19 (Rojas et al. 2022; Chang et al. 2023). The evidence for long COVID is supported by the long-term complications that occur following earlier outbreaks of SARS-CoV and Middle East respiratory syndrome coronavirus (MERS-CoV) (Ahmed et al. 2020; Hui et al. 2005; Lee et al. 2019). Although some post-COVID-19 health complications are recognized, others may reveal over time. Regular follow-up of patients who suffer from post-COVID-19 complications is helpful in identifying long-term complications.

1.3.2 Chronic Diseases

Disrupted follow-up of patients with chronic diseases has attracted the medical community's attention during the pandemic. Autoimmune diseases (e.g., inflammatory bowel disease, multiple sclerosis), rheumatological diseases (e.g., rheumatoid arthritis, systemic lupus erythematosus), congenital disorders, primary immunodeficiency diseases, neurological diseases (e.g., Alzheimer's disease, dementia), cardiometabolic diseases (e.g., heart failure, obesity, diabetes mellitus), cancers, and chronic diseases (e.g., kidney disease, asthma) place a heavy burden on patients' families and the healthcare system. However, the pandemic, mandatory lockdown and home confinement, as well as fear of infection, caused a remarkable disruption in medical follow-up of chronic diseases (Barach et al. 2020). In addition, patients with chronic diseases may be more susceptible to COVID-19, and patients infected with COVID-19 may be at risk of developing chronic conditions (Davis et al. 2021; Han et al. 2022; Papadopoulou et al. 2021; Yazdanpanah and Rezaei 2022; Kunutsor and Laukkanen 2020). Next, we consider relevant examples.

1.3.2.1 Alzheimer's Disease

Reports of the impact of COVID-19 on nursing home residents early in the pandemic raised the alarm for Alzheimer's disease (AD) (Perry 2020). With nearly two-thirds of nursing home residents suffering from dementia, mostly from AD, and estimates of over 60% of COVID-19 deaths in nursing homes (Gaugler et al. 2014; Neumann 2020), the alarm was justified. Epidemiological studies demonstrated that AD patients are more susceptible to COVID-19 (Reyes-Bueno et al. 2020; Wang et al. 2021b), and, conversely, those recovered from COVID-19 are more susceptible to AD onset (Wang et al. 2022a). As long COVID evolves, it is critical to evaluate the coincidence of AD for understanding and reducing the underlying risk factors.

1.3.2.2 Cancer

Delays in screening and early symptom assessment brought on by COVID-19 impacted the capacity of the health system and will likely lead to a surge in cancer cases in the near future, although the short-term effect is a decline in reported cancer cases due to disrupted diagnosis and screening programs. However, the expected increase in cancer cases and cancer-related mortality could be explained by the disrupted follow-up and missed or delayed regular screening programs due to changes in healthcare system priorities during the pandemic, lockdown, and fear of visiting health-related facilities (Oakes et al. 2023; Siegel et al. 2022, 2023). The role of COVID-19 itself in triggering cancer formation is still under debate. Multiple mechanisms are being considered to explain a potential effect of COVID-19 on carcinogenesis and cancer progression. In particular, increased production and prolonged presence of pro-inflammatory cytokines, T-cell depletion leading to host immune dysfunction, sustained activation of oncogenic signaling pathways such as JAK-STAT, MAPK, and NFKB/NF-κB, hypoxia-induced oxidative stress, and a prolonged inflammatory environment have been proposed as explanations of the link between COVID-19 and tumor formation and progression (Del Valle et al. 2020; André et al. 2022; Li et al. 2020b; Satarker et al. 2021; Goel et al. 2021; Abassi et al. 2020; Muz et al. 2015; Greten and Grivennikov 2019).

During SARS-CoV infection, non-structural protein 3 (Nsp3) and endoribonuclease Nsp15 provoke degradation of tumor suppressor protein TP53 and RB transcriptional corepressor 1 (RB1), in preclinical models, respectively (Ma-Lauer et al. 2016; Bhardwaj et al. 2012; Fujiyoshi et al. 2021). Moreover, in silico studies on SARS-CoV-2 proteins revealed that the S2 subunit interacts with TP53 as well as BRCA1 and BRCA2 proteins (Gorodetska et al. 2019; Mantovani et al. 2019; Singh and Bharara 2020). Taken together, the SARS-CoV viral family has the potential functional capacity to induce genomic instability by impairing the function of tumor suppressor proteins.

One of the most important mechanisms associated with cancer is macroautophagy (hereafter autophagy) (Gaebler et al. 2021). Autophagy is responsible for removing damaged organelles and misfolded proteins, and defects in autophagy are associated with several pathological conditions including cancer (Levine and Kroemer 2019; Alizadeh et al. 2023). As a homeostatic mechanism, autophagy crosstalks with other such pathways including the unfolded protein response, the induction of which also could be an initiator for cancer (Dastghaib et al. 2020; Høyer-Hansen and Jäättelä 2007). In general, autophagy is considered a double-edged sword: It could be a mechanism against cancer or a mechanism to promote cancer progression (White 2015). Both SARS-CoV-2 receptors, angiotensin-converting enzyme 2 (ACE2) and transmembrane serine protease 2 (TMPRSS2) are highly expressed in the human gastrointestinal tract; thus, SARS-CoV-2 is able to infect and replicate in human enterocytes (Zang et al. 2020). In a recent investigation on the gastrointestinal biopsies of 14 patients, researchers located virus nucleic acid and antigens in 50% of the cases (Gaebler et al. 2021). These results were repeated in the nasopharyngeal specimens obtained from individuals with previous SARS-CoV-2 infection (Habibzadeh et al. 2020, 2021). Therefore, continuous expression of viral nucleic acids could have significant long-term effects in COVID-19 survivors. Importantly, SARS-COV-2 proteins could interact with the autophagy machinery and induce changes in autophagy flux in mammalian cells. As an example, ORF3a protein interacts with VPS39 and induces autophagy flux inhibition via targeting autophagosome and lysosome interactions (Zhang et al. 2021). Nsp15 blocks the initial stage of autophagy (induction), whereas ORF7a protein decreases lysosomal acidification and slows down the autophagy flux (Hayn et al. 2021). Therefore, SARS-COV-2 infection could reduce autophagy flux via targeting both the initiation and degradation steps of autophagy.

The effects of COVID-19 on cancer formation and progression are not simply attributable to a single direct mechanism but are likely a cumulative consequence of multiple mechanisms. Of note, the currently available statistics may reflect only the tip of the iceberg, as the effect of COVID-19 on cancer trends will take years to ascertain.

Furthermore, the effect of the COVID-19 pandemic is not limited to cancer pathophysiology; indeed, treatment of cancers has been severely affected during the pandemic. The development of cancer therapy has suffered greatly, especially in blood cancers. Many of the available most effective therapeutics are B-cell-depleting agents, leading to a great susceptibility to COVID-19. Recently, highly effective drugs have failed to receive regulatory approval, because of the incidence of COVID-19 in clinical trials. Physicians need to balance treatment efficacy with COVID-19 risk (Cheson 2022).

1.3.2.3 Obesity

According to a recent report by the WHO, more than one billion individuals are affected by obesity and the number is growing (World Health Organization 2022). The pandemic worsened this trend. Lockdown or mandatory home confinement has increased the risk of obesity through multiple mechanisms (Ghanemi et al. 2020; Tan et al. 2020). Fear of getting infected with COVID-19 by visiting outdoor public places, such as food markets, caused many individuals to choose highly processed food products containing large amounts of calories (especially in the form of saturated fats) and extra salt, with long shelf lives but low nutritional value, rather than fresh products. Moreover, the pandemic had a negative impact on mental health (Hettich et al. 2022). Mental health problems are mutually associated with disruption of normal sleep cycles and sleep deprivation (Zhao et al. 2022). The combination of sleep disturbance and psychological problems such as stress, anxiety, and depression, may result in increased food consumption, which, when associated with a lack of physical activity (due to home confinement), considerably increases the risk of obesity. Mental health problems are also associated with increased alcohol intake (Grossman et al. 2020), which is another source of nutritionally empty calories. From a socioeconomic point of view, as numerous businesses were closed during the pandemic, many employees faced unemployment and financial difficulties. As a result, some families could not afford fresh healthy food, and instead used more junk foods or preserved foods, which cost less. In addition, both fear of visiting public places and financial problems led to a noteworthy decline in the use of public gyms and healthcare facilities for weight control, which further intensified the obesity trend (Ghanemi et al. 2020; Tan et al. 2020). Obesity is strongly linked to cardiovascular diseases, diabetes mellitus, and metabolic complications, as well as musculoskeletal disorders such as osteoarthritis; moreover, it is a risk factor for some types of cancer. As such, when the obesity trend is accelerated due to the pandemic, more complications and a heavier burden are to be expected in the near future. This merits a multidisciplinary call for global action.

1.3.2.4 Cardiovascular Diseases

Various cardiovascular complications are associated with long COVID. These include fatigue, chest pain, and shortness of breath, as well as autonomic manifestations such as postural orthostatic tachycardia (i.e., a low volume of blood returning to the heart upon standing). All these symptoms can lead to greater anxiety and disability. Key cardiovascular issues that have been reported beyond the acute phase of COVID-19 include myocardial inflammation, myocardial infarction, right ventricular dysfunction, and arrhythmias (Al-Aly et al. 2021; Raman et al. 2022; Lorente-Ros et al. 2023; Dale et al. 2023). There is also some evidence of myocardial edema, active inflammation, as well as both left and right ventricular dysfunction. Overall, there is an increased risk of cardiovascular outcomes, as evidenced by large-scale observational studies, such as coronary artery disease, cardiomyopathy, and arrhythmias, compared to the general population (Abbasi 2022). Mechanisms of how various cardiovascular outcomes are linked to COVID-19 are not adequately understood, but COVID-19 likely modulates the long-term risk of these, and other cardiovascular complications often prevalent among individuals suffering greatly from COVID-19. Although the acute cardiovascular consequences of COVID-19 were more prominent in the hospitalized patients compared to those enduring a milder illness, the heightened risk of long-term cardiovascular events was observed even in patients who endured mild forms of the illness and were not hospitalized (Al-Aly et al. 2021).

During the pandemic, cardiovascular surgery was all but halted except for emergency operations such as for aortic dissection. A recent study examined (Ad et al. 2021) the effects of the COVID-19 pandemic in 67 adult cardiac surgery institutions in Canada and the USA. Non-urgent surgery markedly dwindled during the month of March 2020 in nearly all institutions, resulting in a decline to 45% of baseline. Hospitals with a high burden of patients hospitalized with COVID-19 showed decline trends in total volume similar to institutions in areas with fewer cases of COVID-19. As a proportion of total surgical volume, there was a relative increase in coronary artery bypass grafting surgery, heart transplantation, and the use of extracorporeal membrane oxygenation but a decrease in valvular operations. With the end of the pandemic, cardiac surgical volumes normalized, but the hospital staffing including nurses and respiratory therapies showed high turnover, markedly affecting patient care. Staff shortages continue to be a problem to the present day and probably to the future.

1.3.3 Outbreaks of Vaccine-Preventable Diseases

A large proportion of healthcare resources and facilities were devoted to the pandemic. Even where resources were available, fear of visiting healthcare centers resulted in considerable disruption of routine childhood vaccination programs in parts of the world. In addition, some authorities dictated vaccination against COVID-19, likely prompting opposition and increased fear for other vaccinations. Failure to receive life-saving childhood vaccines increases the risk of future large outbreaks of infectious diseases. Indeed, during the first two months of 2022, globally reported measles cases showed a 79% increase compared to the same period in 2021, which raises concerns of possible outbreaks of vaccine-preventable diseases with potential fatal outcomes (World Health Organization JNR 2022). Due to the highly contagious nature of measles, as social distancing and strict health-related protocols weaken and vaccination rates decline, the disease could spread quickly. The same is true for other vaccine-preventable infectious diseases, perhaps at a more gradual pace. Besides its acute impact on general health, measles temporarily impairs the immune system and predisposes the patient to a variety of secondary bacterial and viral infections, which could be devastating for children. The outbreak of vaccine-preventable diseases is a worldwide serious threat requiring global action.

More generally, a post-COVID rebound has been acknowledged also in non-preventable infectious diseases such as hand-foot-and-mouth disease (HFMD) and some infectious respiratory diseases as a result of the loss of immunity following COVID-19 social distancing measures (Farahat et al. 2022; Lee et al. 2022b). Although the situation may normalize in the years to come, it is unclear how long this "immunity debt" will last and what its consequences will be.

1.3.4 Repercussion for Young Children

Many people suffered increases in anxiety, depression, and stress during COVID-19 (Muñoz-Vela et al. 2023). Domestic violence also increased (Huldani et al. 2022). If a pregnant woman is exposed to such increased stress, her child is at greater risk of a range of neurodevelopmental problems, including attention deficit hyperactivity disorder (ADHD), conduct disorder, anxiety, and depression, as well as autism spectrum disorders (Glover et al. 2018). Prenatal infection independently predicts child social and communication problems (O'Connor et al. 2022) and the children are also at increased risk of some physical problems, such as asthma. All these conditions will likely be more frequent among the children of women who were pregnant during the COVID-19 pandemic. There should be extra screening programs for these children and early intervention to help.

1.4 Psychological and Mental Health Challenges

1.4.1 Stress, Grief, and Mental Disorders

The pandemic disrupted life and brought many coping challenges to individuals and families. Lockdowns, distancing, and strict health protocols affected mental health (Xiong et al. 2020). In particular, isolation, home confinement, deprivation from social interactions, persistent fear of being infected, loss of friends and family members, and long-term hospitalization worsened mental health (Gruber et al. 2021). Isolation had particularly adverse consequences when family members or friends were not allowed to visit dying family members in hospitals or residential care homes for the elderly; these consequences included guilt for being unable to express an appropriate "goodbye" to loved ones (Szuhany et al. 2021).

In addition, many businesses were closed, and the workers faced unemployment and financial difficulties. Relatedly, mental health deteriorated due to the unavailability of mental healthcare services, especially during the first weeks of the pandemic when online services were just about to be established (Li et al. 2020c). The psychological repercussions of the pandemic were most prevalent in vulnerable populations, such as children and older adults. Moreover, healthcare personnel experienced long periods of isolation from family besides long hours of working in hospitals and healthcare centers, which put considerable psychological strain on them. Anxiety, depression, post-traumatic stress disorder (PTSD), obsessive-compulsive disorder (OCD), and sleep disturbances were commonly reported during the pandemic (Zeng et al. 2023; Peng et al. 2023; Ghahramani et al. 2022; Theberath et al. 2022).

Maladaptive behaviors such as self-harm ideation, unprotected sex, aggressive behavior, hoarding, binge eating episodes, drug abuse, and alcohol consumption were also increasingly observed (Pedrini et al. 2022; Nowak et al. 2020). A multidisciplinary, long-term action is needed to prevent serious physical and psychosocial complications (Holmes et al. 2020).

1.4.2 Information Overload and Misinformation

Infodemic, defined as the combination of excessively mostly correct and relevant information with incorrect or unreliable information that is typically widespread, has become a challenge since the early days of the pandemic (Naeem and Bhatti 2020). Infodemic leads to the accumulation of false or distorted news, causes uncertainty, provokes emotional distress, and negatively affects mental health. When prolonged, infodemic may foster alternative coping strategies, such as neglecting pandemic-related information, which may deprive people of receiving effective health-related information to combat the pandemic (Soroya et al. 2021; Zarocostas 2020). Infodemic, information overload, and misinformation also challenged the acceptance of the COVID-19 vaccination program in many susceptible communities, with harmful consequences (Singh et al. 2022; Chen et al. 2022). Infodemic and misinformation resulted in a significant social pathology during the peak of the infection and beyond, as they created a divide between abiding and non-complying citizens, and generated friction in otherwise socially cohesive groups (Mohamed et al. 2021b).

1.4.3 COVID-19 Vaccine Hesitancy

Under the COVID-19 campaign worldwide, many people manifested vaccine refusal or hesitancy. A study analyzing COVID-19 vaccine hesitancy in 23 countries showed that this is the result of misperceptions about the safety, efficacy, and risks posed by the vaccine, as well as mistrust in the science and clinical validation of vaccine development or in the institutions responsible for vaccination campaigns (Lazarus et al. 2022). Other factors associated with vaccine hesitancy varied by country and included personal experience with COVID-19 (e.g., loss of a family member, sickness) and demographic characteristics (e.g., gender, education, income). For example, one study (Lazarus et al. 2022) showed that low education levels and low household income are correlated with high levels of hesitancy in 19 and 11 countries, respectively, among the 23 included. In contrast, a loss of income due to the pandemic and loss of a family member were factors associated with vaccine acceptance in some countries. Likely, people who lost their jobs, lost their socioeconomic status, or were sick desired to quickly return to normality and were more prone to accept the COVID-19 vaccination. In the same study, although these cases of vaccine hesitancy accounted for 24.8% in June 2021, the reported vaccine acceptance was as high as 75.2% among the 23 countries (Lazarus et al. 2022). However, this percentage of acceptance is lower than the estimation needed to control the pandemic's progress. The misperception of vaccine safety and efficacy has been also associated with wrong or inaccurate communication and dissemination, often politicized (Aliasin et al. 2022; Larson et al. 2022). Finally, negative vaccine respondents may be more willing to accept a vaccine if recommended by their trusted doctor.

The challenges for the ongoing vaccination campaigns are thus to succeed in overcoming such misperception and resistance and expanding vaccine access in low- and middle-income countries. Such campaigns would need to include the importance of the correct information and communication by trusted sources throughout the entire population, which could improve perceptions of vaccine safety and efficacy and increase vaccine acceptance (Shakeel et al. 2022).

1.5 Digital Transformation Challenges

1.5.1 Education

Due to closure of schools, universities, and other educational centers, educational authorities started considering alternative routes for mandated education programs. Digitalization in education and e-learning had been expected to become part of routine educational programs, particularly as a supplemental tool, but the pandemic pushed the plans rapidly in practice and quickly added virtual learning methods to established curriculums during the middle of a semester. Although neither the education system nor the students were ready for the unprecedented shift, they had to adapt quickly. Post-pandemic educational models have become topics of scholarly interest (Peimani and Kamalipour 2021; O'Dea and Stern 2022). Despite some concern that digital technologies could change the conventional role played by universities and educational institutions in modifying social life, digitalization of university education is inevitable and the process is well underway (Peimani and Kamalipour 2021). It has been generally accepted that online education cannot replace face-to-face education and in-person classes; yet, digitalization and remote learning methods are helpful in redesigning and developing novel learning and teaching programs. Indeed, the pandemic-induced virtual learning methods could be considered a model for evaluating institutional, local, national, and global efforts to introduce digitalization into education, and for providing an opportunity to assess innovative pedagogical approaches, students' perception and expectations, and the psychological effects of online learning and teaching (O'Dea and Stern 2022; Zhao and Watterston 2021). The role of online education in primary school requires special mention. Engaging children to behind-the-computer learning is challenging, given that interindividual learning capabilities are different and remote education does not support the social growth of children to the same extend as in-person education (Alsubaie 2022). It is important to address how the shift in the education model during the pandemic has affected the overall student's ability to participate in life-long learning and to develop a sense of professionalism.

Closure of schools and universities had a different impact in developing versus developed countries, due to the great gap in access to the internet (and, sometimes, also to electricity) and in information and communications technology infrastructures. Indeed, there is a massive regional variation in internet penetration globally. In the poorest (vs. wealthiest) countries, the percentage of people with internet connectivity is still very low. Improving the situation would require a considerable investment, which is unlikely to occur. As such, it is worth considering low-cost solutions that could bridge the digital divide such as the community communication center (C3) model, massive open online courses (MOOCs), use of social media as a learning platform, and use of digital platforms (Osman 2022).

To be more concrete with regard to higher education, financial strain, and research disruption are key challenges. Many universities faced financial difficulties as a result of the pandemic, due to a decline in enrollment and government funding, which reduced their revenue. These financial difficulties will likely continue to impact on higher education institutions for years to come (Dubey and Pandey 2020). The pandemic has disrupted research in many fields, with laboratories and research facilities having to close temporarily. These disruptions are likely to have a lasting influence and may change the way research is conducted in the future. In fact, pandemic-related effects on the supply chain for many reagents, consumable products, and equipment items are still being experienced by researchers globally.

The technology for virtualizing training, meetings between research teams, and international collaborations and supervision has undergone an extraordinary but uneven evolution in a very short time, depending on a university's technical capacities. Developments included the installation of digitized techniques both in the training process and the defense phase of doctoral theses, leading to a better economic return. The possibility of holding a greater number of virtual meetings of higher quality makes the training process more flexible and increases the capacity for internationalization. Regarding the defense of doctoral theses, digital technology has worked well, making examination committees more international and effective. The protocol and solemnity of the process may have been affected, but the economic cost has been lowered (Mullen 2021; Raju 2021).

1.5.2 Careers

A change in careers occurred, starting at the early days of the pandemic and lockdown. Many non-essential employees switched to working remotely from home, and many companies deployed automatic systems and used robots to continue work with the lowest risk of spreading the infection. Permanent career changes are likely in the post-pandemic era.

Some companies have considered flexible workplace arrangements (in-person and remote), as they realized that remote working during the pandemic could be more efficient and often save physical space, financially benefiting the company. In addition, employees' transportation and meal expenses may decrease. The possibility of working remotely and the widespread application of virtual meetings limits business travel, which reduces companies' expenses and saves time (Hanaei et al. 2021, 2022). However, online conferences and online meetings may provide reduced opportunities for informal social interactions, which could otherwise culminate in stimulating ideas. Before the pandemic, informal contacts at conferences were considered by many as the most important features or opportunities.

Conversely, whereas business travels may be reduced, tourism and leisure-related businesses are expected to experience a surge in demand post-pandemic (Lund et al. 2021). Furthermore, both businesses and consumers have found e-commerce more convenient and likely will continue to extend it post-pandemic. Many companies have started to use automation and artificial intelligence (AI), initiating a quick adoption of these technologies in work categories with a high level of physical interaction (direct human-to-human contact) such as in warehouses, call centers, and grocery stores. Application of automation and AI is predicted to continue and even expand to other work categories in the post-pandemic era. The demand for healthcare-related workers may increase, as the pandemic has alerted attention to health. Overall, low-wage occupations as well as those of part-time and minority workers are predicted to be vulnerable jobs facing cuts post-pandemic (McKinsey Global Institute 2021; Battisti et al. 2022).

1.5.3 Tele-Healthcare

Tele-healthcare is the application of distant communication and information technologies to deliver healthcare as well as promote and monitor patient's self-management via constant monitoring and personalized feedback. During the pandemic, a considerable burden was imposed on healthcare staff, which triggered accelerated in-practice utilization of robotic and AI technologies in healthcare settings. For instance, disinfection of the hospitals' environment, patients' risk assessment, and delivery and supply chain activities were conducted using robotic and AI technologies (Sarker et al. 2021). However, due to the high expenses for developing robotic and AI-aiding technologies and lack of proper infrastructure, healthcare settings in some regions of the world did not benefit from these technologies. The COVID-19 pandemic sped up the tele-healthcare revolution. Although many patients and physicians prefer in-person clinical visit and care, tele-healthcare and "home visits" are predicted to become popular and perhaps the norm. A strong surveillance system and proper infrastructure are essential for continuous and extensive use of tele-healthcare; otherwise, serious health issues may occur, particularly in regions with underdeveloped infrastructure (Sarker et al. 2021; Hollander and Carr 2020). Different challenges are associated with widespread application of tele-health. For example, lack of face-to-face physical contact may negatively affect the quality of visits as emotional information is more likely to be missed. Another challenge will be the overuse of available services when they are unnecessary, which may result in mismanagement of the patient's chief health complaint. Similar to digitalization in other fields, such as education, poor internet access in some regions is a big challenge for tele-healthcare (Anaya et al. 2022; Omboni et al.

2022). Although tele-healthcare has been a great resource during the pandemic crisis, it is arguable to what extent the widespread application of robotics, AI, and advanced technologies in a vital situation, such as human health, would be safe and useful in the post-pandemic era.

1.6 Scientific Challenges

1.6.1 Data Disclosure Process

The health toll caused by the COVID-19 crisis called for the collection of a large volume of data and a speed-up in the flow of information at an unprecedented scale and magnitude. The scientific community adapted rather promptly by facilitating the exchange of information between research groups, creating rules governing publishing in such cases, and ultimately through a sharp rise in the use of preprint archives. The perceived value of preprint manuscripts increased considerably, as they became a source of information for the media (Watson 2022). This trend has drawbacks: the pertinent manuscript has not undergone peer review and hence may entail mistakes.

Similarly, public repositories of raw data, especially of genetic sequences, boomed to facilitate storage and shared analysis. Although this was largely a salutary measure, particular cases invited caution by the scientific community, as some repositories (the Global Initiative on Sharing All Influenza Data [GISAID], in particular) have operated based on terms and policies contrary to the best interest and ethical needs (Enserink 2023).

1.7 Economic Challenges

The pandemic had an enormous on global economy. However, the pandemic's consequences were most severe among emerging economies, exacerbating their fragility. With the lockdown and closure of many businesses, over 50% of households (not only in emerging but also in advanced economies) were unable to keep up basic consumption for more than 90 days in the face of income loss. In addition, an average business was unable to cover expenses for more than 55 days with cash reserves. Overall, the pandemic had a huge influence on global poverty, which increased at a rate unobserved in a generation. Disproportionate unemployment and income loss resulted in a remarkable rise in fiscal inequality nationally and globally (World Bank 2022a).

The World Bank estimated that recovery from the pandemic-induced crisis would be similarly unequal to its initial economic effects. In the post-pandemic era, high-income countries are expected to recover at a faster pace than low-income countries. As the interest rates rise worldwide, the low-income countries' currencies will depreciate, making it difficult to pay back their debts (which they incurred by borrowing in foreign currency during the pandemic) and hindering their economic recovery (World Bank 2022b). Inequality in economic development has been a global challenge but was exacerbated by the pandemic. The swift trend of digitalization contributed to an increase in the use of robots and automated systems for work that was previously carried out by low/middle-skilled workers, thus raising unemployment and inequality (World Bank 2022a; Park et al. 2022).

Interestingly, the economic crisis caused by the pandemic has provided an opportunity to facilitate transformation to a more efficient and sustainable global economy. Climate change has always been a global challenge, and different measures, including economic practices, have been suggested to tackle global warming. Revising the cost of carbon emissions and supporting green economy approaches and the application of sustainable technologies are important practices to slow down climate change (Drews et al. 2022; Savin et al. 2022). The pandemic led to an inevitable economic slowdown, which varied between countries, offering governments the opportunity to adopt green economic approaches (World Bank 2022b; Sandbrook et al. 2022).

Furthermore, the increased healthcare expenses caused by the pandemic resulted in reallocation of funds. Also, large amounts of money were spent on vaccine development programs. Consequently, the allocation of large sums of money to health-related sectors has inevitably reduced the budget for other sectors.

On a related topic, additional funding needs to be committed to worldwide health monitoring programs. For example, the USA ended funding for the Predict program that was run by the US Agency for International Development (USAID) and was acting as an early warning system for pandemics, just months before the outbreak of the current pandemic. Similarly, the Centers for Disease Control and Prevention (CDC), the leading public health agency in the USA, is chronically underfunded, and there has been a consistent decline in independent funding for the WHO. Overall, public health is receiving less funding than that devoted to individualized medicine, undermining preparedness for coping with new pandemics.

1.8 Environmental Challenges

1.8.1 Climate Change, Recycling, and Air Pollution

Deforestation to build farms or pastures results in migration of wild species to new habitats, which leads to contact among different species (wild-wild, wild-domestic, and wild-human contacts) that would not have previously interacted when living in separate habitats. Furthermore, as shown in the case of bats, which face multiple and synergistic threats (Kingston et al. 2021), environmental stress is responsible for increased viral shedding (Eby et al. 2023). These processes facilitate the spread of emerging infectious diseases (Wolfe et al. 2007; Rohr et al. 2020).

Biodiversity affects human health; it is reported that about 75% of anti-infection medications are derived from natural products (Romanelli et al. 2015). Moreover, the checks and balances of natural systems help to prevent contagious pathogens from worldwide spread and becoming a pandemic (Everard et al. 2020).

Climate change is considered the biggest global environmental threat, which compounds the toll on mental health brought on by the pandemic (Kornei 2023). It is estimated that a further 1.5 °C increase in global temperature can happen by 2040 (McNeely 2021). The global temperature rise causes a sea level rise, and extreme climatic events such as droughts, floods, heat waves, extensive forest fires, and more dust emissions. Hence, population movement and changes in animals' habitat raises the possibility of spreading new infections. Also, prolonged wildfire seasons induced by climate change produce a huge amount of carbon dioxide (CO_2) that, in addition to regular anthropogenic inputs, exacerbates global warming and climate change (Landrum and Holland 2020; Hugelius et al. 2020) resulting in a negative downward cycle. Emissions of particulate matter from such fires have a deleterious impact on society (e.g., destroyed communities) and public health.

During the pandemic, environmental changes were accelerated. As the pandemic inflicted serious economic problems and priorities changed in budget allocation, environmental protection agencies weakened, allowing easier exploitation of natural resources. Moreover, in some regions, particularly tropical areas, people started to attack wildlife and protected species and to cut trees due to unemployment and loss of income (McNeely 2021; Badola 2020). Bats and to a lesser extent pangolins have been stigmatized and increasingly persecuted (Gupta et al. 2022; Lu et al. 2021).

Increased plastic wastes also pose an environmental challenge, as the use of personal protection equipment, plastic-packaged food, and disposable utensils increased considerably during the pandemic. A safe process to eliminate plastic waste is necessary on both national and global scales to preserve the environment from greater damage (Vanapalli et al. 2021).

The pandemic has also raised concerns about global recycling efforts to mitigate environmental challenges, and in particular materials that have been exposed to the virus. New guidelines and regulations have been formulated regarding waste management (Das et al. 2021). These include specific requirements for the handling of potentially contaminated materials such as medical masks, gloves, and suits. Overall, the pandemic has presented a new set of challenges for waste recycling, and adherence to new guidelines can minimize the spread of the virus while sustaining the continued global recycling efforts (Das et al. 2021; Sarkodie and Owusu 2021).

Air pollution decreased in some cities following lockdowns and home confinement regulations. Reduced rates of transportation and limited activity of large industries resulted in lower global consumption of fossil fuels, which, in turn, slightly lessened the emission of greenhouse gas (Forster et al. 2020; International Energy Agency 2020). This experience has encouraged some governments to speed up the development of renewable infrastructure, such as wind and solar power generation.

During lockdowns, the world learned to still be connected without much physical presence or international travel. Although the latter has rebounded substantially (Lund et al. 2021), a certain level of reduced travel to and from offices has been achieved, as flexible workplace arrangement is maintained. This may contribute to a reduction in emissions of greenhouse gases in the long run.

1.8.2 Food Systems

The pandemic increased global food insecurity in almost every country by slashing incomes and disrupting food supply chains. The conditions during the pandemic were worsened by Russia's invasion of Ukraine. According to the World Food Program (WFP), due to the compounding effects of social, political, and economic crises, the total number of food insecure people rose from 276 to 323 million (World Food Programme 2023). The pandemic continues to wreak havoc on global hunger and poverty—especially on the poorest and most vulnerable populations. In particular, the pandemic had an impact upon the agri-food systems and their economy, stressing critical factors like food production, demand, price hikes, security, and supply chain resilience (Sridhar et al. 2023). Also, the pandemic posed huge barriers in providing affordable and nutritious food to poor people in developing countries. A lesson learned is the need to build a resilient food system. This can be achieved by adopting sustainable agricultural practices with a prime focus on techniques like urban agriculture, crop rotation, hydroponics, and family farming. Scientific advances in using modern digital tools such as artificial intelligence, machine learning, deep learning, and blockchain technology can contribute toward developing sustainable agri-food systems for a self-reliant society (Sridhar et al. 2023).

1.8.3 Energy

The spread of SARS-CoV-2 had a negative impact upon the economic growth of many countries. For example, according to the International Energy Agency (IEA) report published in 2020, the US gross domestic product (GDP) decreased by 7.4% in 2020. As a result of the economic slowdown, energy consumption saw a worldwide reduction. Of note, the pattern of energy consumption differed across countries. Travel restrictions and lockdowns had a considerable influence on energy demand and consequently on the need for oil and other forms of energy. By mid-April 2020, weekly energy demand had fallen by 25% for countries in complete lockdown and 18% for those in partial lockdown. The fall in demand led to a sharp fall in the price of natural gas and oil, reaching a decade low in April 2021 as the virus spread around the world. Renewable energy proved to be more resilient than other forms of energy. Electricity systems increased the shares of renewables in their operation, as they operated effectively, albeit under reduced load conditions (International Energy Agency 2022; Guida et al. 2020).

Besides its drastic effects on demand and energy prices, the pandemic has revealed the crucial role of energy in current society. Without energy, vital functions of the health system and the critical telecommunication system would collapse. It is suggested that given the surplus of energy supplies, the importance of efficient, resilient energy networks for the transmission and distribution of electrical energy in such extreme conditions cannot be overstated. During the pandemic, network operators have taken successful mitigation measures to ensure continuity and maximization of operations both for field and office activities. In some cases, grid maintenance works have been postponed or canceled, due to travel restrictions without risking system operation in short or medium terms. Moreover, hygiene and sanitary measures, especially for operational staff and travel restriction or cancelation of arrangements to prevent virus contamination, have been imposed. The transmission and distribution supply industry has been able to continue its essential operations to provide network operators with the necessary equipment and services, although its activity has been affected by reduced market demand and supply chain disruptions (International Renewable Energy Agency 2020).

The global economy recovered at record speed in 2021. As energy intensity improvements were postponed, global energy demand increased by 5.4%, and the demand for all fossil fuels grew by at least 5% in 2021. The historic plunge in global energy consumption in the early months of 2020 that drove the price of many fossil fuels to their lowest level in decades has rebounded fast since mid-2021. Oil prices have returned to around or above USD 100/barrel, and coal and natural gas prices have reached record levels (International Energy Agency 2022).

Surging energy demand was in part met by increased use of coal, resulting in a jump in emissions in 2021, the largest annual increase in global CO_2 emissions from the energy sector ever recorded. This contrasted with a fall of about 8% in CO_2 emissions growth in 2020 compared to 2019, the largest annual drop since the Second World War. The rebound of global economy among other factors, like the impact of weather-related events on demand and electricity generation, has led to a global energy crisis with energy prices touching record levels in many markets. The global energy crisis has hurt households, industries, and entire economies around the world, with the poorest and most vulnerable suffering particular hardship. One of the striking results of the combination of the COVID-19 pandemic and the energy crisis is a reversal of global progress on electricity access for the first time in more than a decade (International Energy Agency 2022).

The increasing trend of energy prices was exacerbated in early 2022 by the Ukraine conflict, pointing to the need for increased energy security by reducing dependence on imported fuels (International Energy Agency 2022). Moreover, concerns about climate change have led several countries to accelerate their plans to reach net zero emissions through faster deployment of renewable energy sources, both in the power sector and beyond (Guida et al. 2020).

During the pandemic, forced stays at home increased, and therefore, the management of energy consumption became crucial (Karami et al. 2014). As a solution for solving this challenge, Smart Homes provide a solution in the framework of Smart Grid and Smart Cities (El-Dessouki and Saeed 2021). Microgrids, as novel network structures, provide the most effective way to control local energy resources and supply critical loads increasing the system resilience in case of rare events, like the pandemic and natural disasters (Schwaegerl and Tao 2013). In a Smart Grid framework, it is possible to realize the e-health infrastructure, distance working, and e-learning, while also tracing the ill people in the city (a very important consideration during the outbreak of infectious diseases). Investment in infrastructures for the development of Smart Grids and Smart Cities will play a key role in future outbreaks of infectious diseases.

In some parts of the world, political institutions have invested a large portion of public savings to carry out an energy and ecological transition. For example, Next-Generation EU (NGEU) is the European Union economic recovery package to support the EU member states in recovering from the COVID-19 pandemic, especially those that have been particularly hard hit. The instrument is worth €750 billion and, among various initiatives, the complete substitution of combustion vehicles with electric ones has been decided and will be fully adopted starting in 2035. This mandate has pushed researchers to find new rechargeable battery solutions, in terms of safety, recycling, raw materials, and engineering (Fagiolari et al. 2022; Siccardi et al. 2022; Gandolfo et al. 2022).

1.8.4 Ocean and Marine Studies

Humans have become increasingly reliant on the ocean, which cover 70% of the planet. Thus, it is not surprising to find ocean interactions with SARS-CoV-2 in multiple ways—both positive and negative. Marine scientists have made numerous contributions regarding COVID-19 transmission, detection, and the search for drug treatment. Early work by scientists at the University of California San Diego identified the airborne transmission of COVID-19 that confirms the effectiveness of masking (Prather et al. 2020) and pioneered the surveillance and early detection of COVID-19 using campus wastewater (Karthikeyan et al. 2021). Active compounds derived from marine organisms, including flavonoids, tannins, alkaloids, terpenoids, peptides, lectins, polysaccharides, and lipids, have been found to influence COVID-19 entry into cells, replication, and release from cells. Marine sponges and algae in particular produce RdRp inhibitors or exhibit general antiviral properties (Geahchan et al. 2021; Rahman et al. 2022).

Ocean-based livelihoods were affected by the pandemic with reductions in fishing, recreation, and tourism. This hit particularly hard for tropical island countries reliant on tourism as a primary source of income, and recovery is just now happening. Marine science observation contributes to the knowledge of weather, climate, and ocean hazards as well as basic science. For open ocean scientific observing, a COVID-19 impacts survey in April 2020 revealed that the cancelation of ship activity meant that existing moored equipment could not be serviced, limited operations of autonomous underwater vehicles (AUVs) and gliders, and no reseeding of fundamental drifter arrays like ARGO (Heslop et al. 2020).

A direct negative effect of the pandemic on the ocean was the generation of medical waste including 129 billion masks and 65 billion gloves every month (Prata et al. 2020), leading to significant plastic excess in the ocean (Wang et al. 2022b). It is estimated that 25,000 tons of single-use plastic entered the global ocean, creating danger to marine ecosystems (Peng et al. 2021). In contrast, the limited activities of humans during 2020 and parts of 2021 reduced other human impacts on the ocean from tropical waters to Antarctica. There was a 70.2% decrease in marine traffic within Exclusive Economic Zones associated with restrictions on human mobility and in consumer supply and demand (March et al. 2021). Reduced noise, ship strikes of mammals, less commercial and recreational fish catching, and less direct human contact all offer some respite to marine ecosystems.

For marine science projects engaging stakeholders, engagement was more difficult, and the move to an internet-based, digital mode of operation created distinct disadvantages for specific regions and groups (e.g., artisanal fishers, aquaculture farmers) (Köpsel et al. 2021). Marine science is still catching up but with some changed practices. Funding agencies work harder to support early career scientists in transition between jobs, conferences are now routinely hybrid, and ships are back in action but with the COVID-19 precautions in place.

1.8.5 Geographical Studies

Time and space are researched by geographers, who traced the spread of the pandemic via maps and graphs and by studying travel behavior (Rose-Redwood et al. 2020; Bissell 2021; Franch-Pardo et al. 2020; Xi et al. 2023). Several research teams contributed to the accumulated knowledge, spanning countries such as Spain (Maza and Hierro 2022), Sweden (Florida and Mellander 2022), Denmark (Holmager et al. 2021), and Italy (Ascani et al. 2021). This knowledge helped to prevent the pandemic's impact on human health (Brinkman and Mangum 2022; Gaskin et al. 2021). Also, geographers analyzed the role of climate and weather on the spread of the coronavirus, which is highly affected by atmospheric conditions (Yang et al. 2021). Geographical studies were useful for developing the understanding and knowledge of the pandemic (Franch-Pardo et al. 2021; Maiti et al. 2021). As discussed in previous sections, the pandemic opened a new challenge for the environment (e.g., increased plastic waste), biodiversity (e.g., execution of some species and wildlife overuse due to economic problems in some areas), and soil security (e.g., soil security being affected as it is closely linked to water, food, energy; environment security being affected) (Geraghty and Kerski 2020; El-Ramady et al. 2021). Hence, geographical studies can help to tackle environmental and perhaps long-lasting, challenges.

1.9 Sustainable Development Goals

Sustainable development goals (SDGs) were adopted by the UN General Assembly on September 25, 2015, and agreed upon by 193 countries under the 2030 Agenda for Sustainable Development. The SDGs were formulated to align the world to act toward achieving a better quality of life and sustainable growth, as well as to mitigate the effect of climate change. There are 17 SDGs classifiable into five groups:

- *People*: no poverty, zero hunger, good health and well-being, quality education, and gender equality
- *Planet*: drinking water and sanitation, climate action, responsible consumption and production, life below water, and life on land
- *Prosperity*: affordable and clean energy for all, reduced inequality, sustainable cities and communities, decent work and economic growth, and industry, innovation and infrastructure
- *Peace*: peace, justice and strong institutions
- *Partnership*: partnerships for achieving the targets of goals.

With the emergence of the COVID-19 pandemic, progress toward achieving SDGs is likely to be hindered. Hence, the pandemic may be a distraction from achieving SDGs. Yet, the pandemic could be viewed as an opportunity to identify pitfalls in the route toward achieving the SDGs.

1.10 Future Perspectives and Conclusions

The spreading of COVID-19 represented an unprecedented challenge for humanity. This is not only because of the complexity of protecting public health and the need to quickly develop and administer effective vaccines but also because of immediate repercussions to the global economy. Besides millions of deaths, the rapid diffusion of the infection and its very high socioeconomic impact have affected healthcare, economic status and wealth, and mental health across the globe. Yet, COVID-19 also constituted an opportunity, as it tested individuals' and societies' capability to react to an existential threat. The lessons learned may pay off in preparing for another virus outbreak. Science can play a critical role in shaping the future agenda and political decisions.

In most healthy people, the SARS-CoV-2 infection is asymptomatic and mild, although it can occasionally result in severe or life-threatening illness, or permanent disabling symptoms. The pandemic proved that several characteristics mainly affecting the immune response of humans have a significant impact on the consequence of infection-host interaction. Therefore cases with higher susceptibility should be identified and protected or vaccinated compared to other immunocompetent cases to avoid the same adverse consequences observed during the early phase of the pandemic. The human immune function, the key factor in determining the severity of COVID-19 illness, is substantially diminished by age, with the over 65-year-olds having the highest likelihood of needing intensive care. Comorbidities such as obesity, metabolic syndrome, and chronic liver and kidney diseases associated with immune dysregulation are associated with severe COVID-19 illness. Also, men are more vulnerable than women due to the impact of X-chromosome-located immune genes on the final response against the virus, as many genetic defects have been identified in X-linked patterns (e.g., TLR7 deficiency). The presence of neutralizing autoantibodies against immune components (e.g., type I interferons) was strongly over-represented among individuals who developed life-threatening COVID-19. Young children appear to be less severely affected compared to people with other respiratory viral infections, but those with inborn errors of immunity could present severe COVID-19 mainly with defects in interferon pathways. Therefore, active identification of cases with immune susceptibility to other viral infections, mainly zoonotic respiratory pathogens, should be conducted at high speed and in advance to prevent the impact of the next potential pandemic on this vulnerable group of patients and manage better the prioritization of treatment and vaccination. However, given the relatively small group of patients with such defects, it is necessary to consider the available resources and think of the best strategy to protect both vulnerable populations and the meet the healthcare-related needs of the overall population.

Long COVID-19 is associated with more than 200 symptoms that affect multiple organ systems. Consequently, there will be a great number of new conditions requiring novel treatment strategies. Therefore, from the medical point of view, new recommendations and guidelines are essential for each medical discipline. Moreover, it will be a great challenge to define standards for diagnostics. The question will arise "What is long COVID?." There will be a risk that doctors classify patients with a rare, unknown, or even unrecognized disease as a patient with COVID-19 complications.

The pandemic has opened up new research priorities. For example, in medical sciences, it will be necessary to identify the pathobiochemical mechanisms by which the SARS-CoV-2 mediates its pleiotropic effect on different organs. Receptor and downstream signaling must be investigated. It would be ideal to pinpoint biomarkers with high sensitivity and specificity to identify patients suffering from long COVID-19. In addition, there are many patients that are assumed to have developed diseases after vaccination. Therefore, there is a need to establish strategies to define what lesions were induced by the vaccination. In addition, there is an urgent need for strategies to cure or improve post-vaccination sequelae. Finally, mental health issues (psychological distress, depression, anxiety, stress symptoms, suicidal thoughts) associated with COVID-19 are more prominent in women than men. The reasons need to be understood.

Some recommended treatments for COVID-19 are challenging. IgG antibodies formed specifically against enveloped viruses can undergo a unique form of glycosylation (afucosylation); although these antibodies can be effective in eliminating cancer cells, they can also amplify cytokine storms and immune-mediated pathologies, which could contribute to the severity of COVID-19 (Larsen et al. 2021). Further research is essential to design proper medications for COVID-19. In biological, chemical, and physical sciences, the importance of interdisciplinary work to develop preventive vaccines, diagnostic tools, and therapeutic agents has been thoroughly highlighted by the pandemic. Collaborative studies between scientists from different disciplines such as environmental, biological, and chemical sciences are crucial in studying the origin of SARS-CoV-2 and how to prevent future viral spreads. Scientists from environmental sciences, economics, and politics need to work together to develop protocols about how to protect nature, biodiversity, and the ecosystem. Taken together, as scientific inquiry becomes increasingly interdisciplinary, research that integrates insights from multiple fields will only become more important in the years ahead.

During the pandemic, social media have been used as a platform for spreading various conspiracy theories about the source and consequences of COVID-19. Better ways to communicate health information and prevent the spread of misinformation are needed.

Inflation and the disruption of supply chains lead to families making difficult financial decisions regarding basic needs. Consumer prices rose 7.1% from November 2021 to November 2022, with food costs up 10.6%, prompting the 2023 World Economic Forum to name the cost-of-living crisis as the most significant short-term global risk. Strategies for alleviating poverty through the widespread promotion of medium-term practices of food storage on a yearly or bi-yearly basis should also be encouraged to avoid the 55–90 days individual support limit observed during COVID-19. Such practices have the potential to alleviate inequality and promote economic recovery in future pandemics.

The pandemic has had a notable political impact, with many governments around the world facing criticism for their handling of the crisis. It has also led to increased nationalism and racism in some countries. In addition, it has highlighted existing health disparities and inequities, particularly in vulnerable populations.

Anthropogenic ecological disequilibria could and did generate a global health crisis. Yet, it is still something that people and particularly, to a lesser extent, the scientific community, have not yet fully come to terms with. Yet, the COVID-19 crisis apparently started due to habitat clearance putting pressure on wildlife, and wildlife trade markets, which act as potential spillover sites. We have now entered the paradigms of "One Earth" and "One Health." The creation of international networks to prevent outbreaks and pandemics is another task that was started but remains to be finalized. In particular, it is important to remember that the COVID pandemic is of zoonotic nature. Hence, given the environmental factors that will make spillovers more frequent in the future, the fostering of networks, such as Preventing Zoonotic Disease Emergence/PREZODE, focused on the prevention of such events is crucial.

There are also positive lessons learned. The pandemic has increased awareness of public health and resulted in increased investment and focus on healthcare and research. The development and use of mRNA vaccines have opened the door to create vaccines for numerous other viruses that were previously not amenable to preventative measures. The development of mRNA-loaded nanoparticle-based vaccines, like the ones introduced by Pfizer-BioNTech and Moderna, has showcased the unprecedented potential of mRNA technology and has opened the field for the development of next-generation lipid-based nanoparticle vaccines for other viral and bacterial infections to improve preparation for future threats, and for other types of disease, such as solid tumors. Indeed, encouraging results from early clinical trials with mRNA-loaded nanoparticle-based vaccines as a monotherapy or in combination with immunotherapy have been obtained in patients with resected solid tumors and gastrointestinal cancer (Lorentzen et al. 2022). The ongoing studies on next-generation mRNA cancer vaccines are more likely to succeed in combination with immunotherapeutic treatments such as adoptive cell therapy and immune checkpoint inhibitors. Nevertheless, several challenges remain and include identification of individual cancer neoantigens, validation of the most feasible vaccine administration methods, and regulatory hurdles.

The pandemic period has accelerated the introduction of digital tools and new teaching methods. These will also be important in the future, in the case of teacher unavailability, public transport strikes, student indisposition, or development of online courses. In engineering sciences, many countries have moved to spend a share of public money to accelerate environmental and energy policies, paving the way to a more sustainable world. In this context, Smart Grid technologies have proved to increase the system resilience by supplying critical loads in case of rare events, like pandemics and natural disasters.

The role of science and academia during global crises has also come to the fore. First, the safeguarding of freedom of expression, academic freedom, debate, and protection of international whistleblowers are crucial for science to operate properly. The COVID crisis has demonstrated shortcomings in these aspects. Second, science needs to have voices. Structures (e.g., venues, communication media, and official speakers) are necessary to ensure that vocal minorities do not take over the public dialog and that scientists can express the majority of the community's opinion and thereby are in a position to counter any potential political agendas that are contrary to the public interest.

Pandemics have threatened the health and well-being of societies, and the global order, in the past. The crucial roles of science and public health, vaccination programs, the WHO, and all other mutually beneficial international collaborations must continue to play a central role in informing healthcare policy and delivery. Now that we are at what is considered the end of the pandemic, we should look back, and by assessing the toll we as humans paid in each of these areas, we can learn how to be better prepared for the next pandemic. It will help if scientists understand issues outside their immediate areas of expertise/interest and see how the different challenges are interrelated. In addition to tackling the many challenges in this article, we need to consider what will happen in the next pandemic. What have we done to try to prevent or at least mitigate the impact of that next threat to global health?

Acknowledgements We appreciate the comments and contribution of the USERN Advisory Board members.

Disclosures Dr. Matjz Perc was supported by the Slovenian Research Agency (Javna agencija za raziskovalno dejavnost RS) (Grant Nos. P1-0403).

Dr. Roger McIntyre has received research grant support from CIHR/GACD/National Natural Science Foundation of China (NSFC) and the Milken Institute; speaker/consultation fees from Lundbeck, Janssen, Alkermes, Neumora Therapeutics, Boehringer Ingelheim, Sage, Biogen, Mitsubishi Tanabe, Purdue, Pfizer, Otsuka, Takeda, Neurocrine, Neurawell, Sunovion, Bausch Health, Axsome, Novo Nordisk, Kris, Sanofi, Eisai, Intra-Cellular, NewBridge Pharmaceuticals, Viatris, Abbvie, Atai Life Sciences. Dr. Roger McIntyre is the CEO of Braxia Scientific Corp.

Dr. Hofmann receives financial support by the Alexander von Humboldt Foundation (as part of the Alexander von Humboldt Professur), the Hessische Ministerium für Wissenschaft und Kunst (as part of the LOEWE Spitzenprofessur), and the DYNAMIC center, funded by the LOEWE program of the Hessian Ministry of Science and Arts (Grant Number: LOEWE 1/16/519/03/09.001(0009)/98). He also receives compensation for his work as editor from SpringerNature and royalties and payments for his work from various publishers.

References

Abbasi J (2022) The COVID heart—one year after SARS-CoV-2 infection, patients have an array of increased cardiovascular risks. JAMA 327(12):1113–1114

Abassi Z, Higazi AAR, Kinaneh S, Armaly Z, Skorecki K, Heyman SN (2020) ACE2, COVID-19 infection, inflammation, and coagulopathy: missing pieces in the puzzle. Front Physiol 11:574753

Ad N, Luc JG, Nguyen TC, Arora RC, Balkhy HH, Bender EM et al (2021) Cardiac surgery in North America and coronavirus disease 2019 (COVID-19): regional variability in burden and impact. J Thorac Cardiovasc Surg 162(3):893–903.e4

Ahmed H, Patel K, Greenwood DC, Halpin S, Lewthwaite P, Salawu A et al (2020) Long-term clinical outcomes in survivors of severe acute respiratory syndrome and Middle East respiratory syndrome coronavirus outbreaks after hospitalisation or ICU admission: a systematic review and meta-analysis. J Rehabil Med 52(5):jrm00063

Al-Aly Z, Xie Y, Bowe B (2021) High-dimensional characterization of post-acute sequelae of COVID-19. Nature 594(7862):259–264

Aliasin MM, Yazdanpanah N, Rezaei N (2022) Neurological and neuropsychological adverse effects of SARS-CoV-2 vaccines—where do we stand? Rev Neurosci 33(7):721–743

Alizadeh J, Kavoosi M, Singh N, Lorzadeh S, Ravandi A, Kidane B et al (2023) Regulation of autophagy via carbohydrate and lipid metabolism in cancer. Cancers 15(8)

Alsubaie MA (2022) Distance education and the social literacy of elementary school students during the COVID-19 pandemic. Heliyon 8(7):e09811

Anaya YB-M, Bañuelos Mota A, Hernandez GD, Osorio A, Hayes-Bautista DE (2022) Post-pandemic telehealth policy for primary care: an equity perspective. J Am Board Fam Med 35(3):588

André S, Picard M, Cezar R, Roux-Dalvai F, Alleaume-Butaux A, Soundaramourty C et al (2022) T cell apoptosis characterizes severe COVID-19 disease. Cell Death Differ 29(8):1486–1499

Ascani A, Faggian A, Montresor S (2021) The geography of COVID-19 and the structure of local economies: the case of Italy. J Reg Sci 61(2):407–441

Badola S (2020) Indian wildlife amidst the COVID-19 crisis: an analysis of status of poaching and illegal wildlife trade. TRAFFIC India office

Barach P, Fisher SD, Adams MJ, Burstein GR, Brophy PD, Kuo DZ et al (2020) Disruption of healthcare: will the COVID pandemic worsen non-COVID outcomes and disease outbreaks? Prog Pediatr Cardiol 59:101254

Battisti E, Alfiero S, Leonidou E (2022) Remote working and digital transformation during the COVID-19 pandemic: economic–financial impacts and psychological drivers for employees. J Bus Res 150:38–50

Bayoumy AB, Ansari AR, De Abreu RA, Peters GJ, Mulder CJ (2020) Multi-drug therapy schedules for SARS-COV-2: smart repurposing of old drugs. J Explor Res Pharmacol 5:29–30

Beigel JH, Tomashek KM, Dodd LE, Mehta AK, Zingman BS, Kalil AC et al (2020) Remdesivir for the treatment of COVID-19—final report. N Engl J Med 383(19):1813–1826

Bhardwaj K, Liu P, Leibowitz JL, Kao CC (2012) The coronavirus endoribonuclease Nsp15 interacts with retinoblastoma tumor suppressor protein. J Virol 86(8):4294–4304

Bissell D (2021) A changing sense of place: geography and COVID-19. Geogr Res 59(2):150–159

Bosch Sensortec. How sensing solutions help to fight against the spread of COVID-19. Available from: https://www.bosch-sensortec.com/news-and-stories/stories/how-sensing-solutions-help-to-fight-against-the-spread-of-covid-19.html

Brinkman J, Mangum K (2022) JUE insight: the geography of travel behavior in the early phase of the COVID-19 pandemic. J Urban Econ 127:103384

Callaway E (2023) The next generation of coronavirus vaccines. Nature 614:22–25

Ceban F, Ling S, Lui LM, Lee Y, Gill H, Teopiz KM et al (2022) Fatigue and cognitive impairment in post-COVID-19 syndrome: a systematic review and meta-analysis. Brain Behav Immunity 101:93–135

Chang R, Chen TY-T, Wang SI, Hung YM, Chen HY, Wei CJ (2023) Risk of autoimmune diseases in patients with COVID-19: a retrospective cohort study. eClinicalMedicine 56:101783

Chavda VP, Hossain MK, Beladiya J, Apostolopoulos V (2021) Nucleic acid vaccines for COVID-19: a paradigm shift in the vaccine development arena. Biologics 1(3):337–356

Chen X, Lee W, Lin F (2022) Infodemic, institutional trust, and COVID-19 vaccine hesitancy: a cross-national survey. Int J Environ Res Public Health 19(13)

Cheson BD (2022) Death and clinical trials in the plague years. The ASCO Post. Available from: https://ascopost.com/issues/july-10-2022/death-and-clinical-trials-in-the-plague-years/

Dale CE, Takhar R, Carragher R, Katsoulis M, Torabi F, Duffield S et al (2023) The impact of the COVID-19 pandemic on cardiovascular disease prevention and management. Nat Med 29(1):219–225

Das AK, Islam MN, Billah MM, Sarker A (2021) COVID-19 pandemic and healthcare solid waste management strategy—a mini-review. Sci Total Environ 778:146220

Dastghaib S, Shojaei S, Mostafavi-Pour Z, Sharma P, Patterson JB, Samali A et al (2020) Simvastatin induces unfolded protein response and enhances temozolomide-induced cell death in glioblastoma cells. Cells 9(11)

Davis HE, Assaf GS, McCorkell L, Wei H, Low RJ, Re'em Y et al (2021) Characterizing long COVID in an international cohort: 7 months of symptoms and their impact. eClinicalMedicine 38:101019

Davis HE, McCorkell L, Vogel JM, Topol EJ (2023) Long COVID: major findings, mechanisms and recommendations. Nat Rev Microbiol 1–14

Del Valle DM, Kim-Schulze S, Huang H-H, Beckmann ND, Nirenberg S, Wang B et al (2020) An inflammatory cytokine signature predicts COVID-19 severity and survival. Nat Med 26(10):1636–1643

Deng S, Liang H, Chen P, Li Y, Li Z, Fan S et al (2022) Viral vector vaccine development and application during the COVID-19 pandemic. Microorganisms 10(7):1450

Drews S, Savin I, Van Den Bergh JC, Villamayor-Tomás S (2022) Climate concern and policy acceptance before and after COVID-19. Ecol Econ 199:107507

Dubey P, Pandey D (2020) Distance learning in higher education during pandemic: challenges and opportunities. Int J Indian Psychol 8(2):43–46

Eby P, Peel AJ, Hoegh A, Madden W, Giles JR, Hudson PJ et al (2023) Pathogen spillover driven by rapid changes in bat ecology. Nature 613(7943):340–344

El-Dessouki I, Saeed N (eds) (2021) Smart grid integration into smart cities. In: 2021 IEEE international smart cities conference (ISC2), 7–10 Sept 2021

El-Ramady HR, Singh A, Rajput VD, Amer MM, Omara AE-D, Elsakhawy T et al (2021) Environment, biodiversity and soil security: a new dimension in the era of COVID-19. Environ Biodivers Soil Secur 5(2021):1–14

Enserink M (2023) Dispute simmers over who first shared SARS-CoV-2's genome. Science. Available from: https://www.science.org/content/article/dispute-simmers-over-who-first-shared-sars-cov-2-s-genome

Everard M, Johnston P, Santillo D, Staddon C (2020) The role of ecosystems in mitigation and management of COVID-19 and other zoonoses. Environ Sci Policy 111:7–17

Fagiolari L, Versaci D, Di Berardino F, Amici J, Francia C, Bodoardo S et al (2022) An exploratory study of MoS_2 as anode material for potassium batteries. Batteries 8(11)

Farahat RA, Shaheen N, Kundu M, Shaheen A, Abdelaal A (2022) The resurfacing of hand, foot, and mouth disease: are we on the verge of another epidemic? Ann Med Surg 81:104419

Florida R, Mellander C (2022) The geography of COVID-19 in Sweden. Ann Reg Sci 68(1):125–150
Flude BM, Nannetti G, Mitchell P, Compton N, Richards C, Heurich M et al (2021) Targeting the complement serine protease MASP-2 as a therapeutic strategy for coronavirus infections. Viruses 13(2):312
Forster PM, Forster HI, Evans MJ, Gidden MJ, Jones CD, Keller CA et al (2020) Current and future global climate impacts resulting from COVID-19. Nat Clim Change 10(10):913–919
Franch-Pardo I, Napoletano BM, Rosete-Verges F, Billa L (2020) Spatial analysis and GIS in the study of COVID-19. A review. Sci Total Environ 739:140033
Franch-Pardo I, Desjardins MR, Barea-Navarro I, Cerdà A (2021) A review of GIS methodologies to analyze the dynamics of COVID-19 in the second half of 2020. Trans GIS 25(5):2191–2239
Fujiyoshi K, Bruford EA, Mroz P, Sims CL, O'Leary TJ, Lo AWI et al (2021) Opinion: standardizing gene product nomenclature—a call to action. Proc Natl Acad Sci USA 118(3)
Gaebler C, Wang Z, Lorenzi JCC, Muecksch F, Finkin S, Tokuyama M et al (2021) Evolution of antibody immunity to SARS-CoV-2. Nature 591(7851):639–644
Gandolfo M, Amici J, Fagiolari L, Francia C, Bodoardo S, Bella F (2022) Designing photocured macromolecular matrices for stable potassium batteries. Sustain Mater Technol 34:e00504
Gaskin DJ, Zare H, Delarmente BA (2021) Geographic disparities in COVID-19 infections and deaths: the role of transportation. Transp Policy 102:35–46
Gaugler JE, Yu F, Davila HW, Shippee T (2014) Alzheimer's disease and nursing homes. Health Aff (Proj Hope) 33(4):650–657
Geahchan S, Ehrlich H, Rahman MA (2021) The antiviral applications of marine resources for COVID-19 treatment: an overview. Mar Drugs 19(8)
Geraghty E, Kerski J (2020) The impact of COVID-19 on geography, GIS, and education. J Read J Res Didact Geogr 2
Ghahramani S, Kasraei H, Hayati R, Tabrizi R, Marzaleh MA (2022) Health care workers' mental health in the face of COVID-19: a systematic review and meta-analysis. Int J Psychiatry Clin Pract 1–10
Ghanemi A, Yoshioka M, St-Amand J (2020) Will an obesity pandemic replace the coronavirus disease-2019 (COVID-19) pandemic? Med Hypotheses 144:110042
Glover V, O'Donnell KJ, O'Connor TG, Fisher J (2018) Prenatal maternal stress, fetal programming, and mechanisms underlying later psychopathology—a global perspective. Dev Psychopathol 30(3):843–854
Goel S, Saheb Sharif-Askari F, Saheb Sharif Askari N, Madkhana B, Alwaa AM, Mahboub B et al (2021) SARS-CoV-2 switches 'on' MAPK and NFκB signaling via the reduction of nuclear DUSP1 and DUSP5 expression. Front Pharmacol 12:631879
Gorodetska I, Kozeretska I, Dubrovska A (2019) BRCA genes: the role in genome stability, cancer stemness and therapy resistance. J Cancer 10(9):2109–2127
Greten FR, Grivennikov SI (2019) Inflammation and cancer: triggers, mechanisms, and consequences. Immunity 51(1):27–41
Grossman ER, Benjamin-Neelon SE, Sonnenschein S (2020) Alcohol consumption during the COVID-19 pandemic: a cross-sectional survey of US adults. Int J Environ Res Public Health 17(24)
Gruber J, Prinstein MJ, Clark LA, Rottenberg J, Abramowitz JS, Albano AM et al (2021) Mental health and clinical psychological science in the time of COVID-19: challenges, opportunities, and a call to action. Am Psychol 76:409–426
Guida G, Hatziargyriou N, Russo E (2020) ETIP SNET views on post COVID 19 recovery. Available from: https://smart-networks-energy-transition.ec.europa.eu/news/etip-snet-views-post-covid-19-recovery
Gupta SK, Minocha R, Thapa PJ, Srivastava M, Dandekar T (2022) Role of the pangolin in origin of SARS-CoV-2: an evolutionary perspective. Int J Mol Sci 23(16)
Habibzadeh P, Sajadi MM, Emami A, Karimi MH, Yadollahie M, Kucheki M et al (2020) Rate of re-positive RT-PCR test among patients recovered from COVID-19. Biochem Med 30(3):030401
Habibzadeh P, Mofatteh M, Silawi M, Ghavami S, Faghihi MA (2021) Molecular diagnostic assays for COVID-19: an overview. Crit Rev Clin Lab Sci 58(6):385–398
Hama Amin BJ, Kakamad FH, Ahmed GS, Ahmed SF, Abdulla BA, Mohammed SH et al (2022) Post COVID-19 pulmonary fibrosis; a meta-analysis study. Ann Med Surg 77:103590
Hammond J, Leister-Tebbe H, Gardner A, Abreu P, Bao W, Wisemandle W et al (2022) Oral nirmatrelvir for high-risk, nonhospitalized adults with COVID-19. N Engl J Med 386(15):1397–1408
Han Q, Zheng B, Daines L, Sheikh A (2022) Long-term sequelae of COVID-19: a systematic review and meta-analysis of one-year follow-up studies on post-COVID symptoms. Pathogens (Basel, Switzerland) 11(2)
Hanaei S, Sarzaeim M, Yazdanpanah N, Pirkoohi ZR, Ziaei H, Mohamed K et al (2021) The hybrid USERN 2020 congress: new standards for events in practice. Acta Med Iran
Hanaei S, Takian A, Majdzadeh R, Maboloc CR, Grossmann I, Gomes O et al (2022) Emerging standards and the hybrid model for organizing scientific events during and after the COVID-19 pandemic. Disaster Med Public Health Prep 16(3):1172–1177
Hayn M, Hirschenberger M, Koepke L, Nchioua R, Straub JH, Klute S et al (2021) Systematic functional analysis of SARS-CoV-2 proteins uncovers viral innate immune antagonists and remaining vulnerabilities. Cell Rep 35(7):109126
Heslop A, Fischer A, Tanhua T, Legler D, Belbeoch M, Kramp M et al (2020) COVID-19's impact on the ocean observing system and our ability to forecast weather and predict climate change. Glob Ocean Observ Syst

Hettich N, Entringer TM, Kroeger H, Schmidt P, Tibubos AN, Braehler E et al (2022) Impact of the COVID-19 pandemic on depression, anxiety, loneliness, and satisfaction in the German general population: a longitudinal analysis. Soc Psychiatry Psychiatr Epidemiol 57(12):2481–2490

Hollander JE, Carr BG (2020) Virtually perfect? Telemedicine for COVID-19. N Engl J Med 382(18):1679–1681

Holmager TL, Lynge E, Kann CE, St-Martin G (2021) Geography of COVID-19 in Denmark. Scand J Public Health 49(1):88–95

Holmes EA, O'Connor RC, Perry VH, Tracey I, Wessely S, Arseneault L et al (2020) Multidisciplinary research priorities for the COVID-19 pandemic: a call for action for mental health science. Lancet Psychiatry 7(6):547–560

Høyer-Hansen M, Jäättelä M (2007) Connecting endoplasmic reticulum stress to autophagy by unfolded protein response and calcium. Cell Death Differ 14(9):1576–1582

Hugelius G, Loisel J, Chadburn S, Jackson RB, Jones M, MacDonald G et al (2020) Large stocks of peatland carbon and nitrogen are vulnerable to permafrost thaw. Proc Natl Acad Sci 117(34):20438–20446

Hui DS, Joynt GM, Wong KT, Gomersall CD, Li TS, Antonio G et al (2005) Impact of severe acute respiratory syndrome (SARS) on pulmonary function, functional capacity and quality of life in a cohort of survivors. Thorax 60(5):401–409

Huldani H, Kamal Abdelbasset W, Abdalkareem Jasim S, Suksatan W, Turki Jalil A, Thangavelu L et al (2022) Intimate partner violence against pregnant women during the COVID-19 pandemic: a systematic review and meta-analysis. Women Health 62(6):556–564

International Energy Agency (2020) Global energy review. Available from: https://www.iea.org/reports/global-energy-review-2020

International Energy Agency (2022) World energy outlook. Available from: https://www.iea.org/reports/world-energy-outlook-2022

International Renewable Energy Agency (2020) Post-COVID recovery: an agenda for resilience, development and equality. Available from: https://www.irena.org/publications/2020/Jun/Post-COVID-Recovery

Jackson CB, Farzan M, Chen B, Choe H (2022) Mechanisms of SARS-CoV-2 entry into cells. Nat Rev Mol Cell Biol 23(1):3–20

Jain S, Nehra M, Kumar R, Dilbaghi N, Hu T, Kumar S et al (2021) Internet of medical things (IoMT)-integrated biosensors for point-of-care testing of infectious diseases. Biosens Bioelectron 179:113074

Karami H, Sanjari MJ, Hosseinian SH, Gharehpetian GB (2014) An optimal dispatch algorithm for managing residential distributed energy resources. IEEE Trans Smart Grid 5(5):2360–2367

Karthikeyan S, Nguyen A, McDonald D, Zong Y, Ronquillo N, Ren J et al (2021) Rapid, large-scale wastewater surveillance and automated reporting system enable early detection of nearly 85% of COVID-19 cases on a university campus. mSystems 6(4):e0079321

Khazaal S, Harb J, Rima M, Annweiler C, Wu Y, Cao Z et al (2022) The pathophysiology of long COVID throughout the renin-angiotensin system. Molecules (Basel, Switzerland) 27(9)

Kingston T, Florens FV, Vincenot CE (2021) Crumbling island keystones: threat diversity and intensification on islands push large island fruit bats to the brink. EcoEvoRxiv. Preprint publication

Köpsel V, de Moura Kiipper G, Peck MA (2021) Stakeholder engagement vs. social distancing—how does the COVID-19 pandemic affect participatory research in EU marine science projects? Marit Stud 20(2):189–205

Kornei K (2023) The mental toll of climate change. Eos. Available from: https://eos.org/features/the-mental-toll-of-climate-change

Kunutsor SK, Laukkanen JA (2020) Renal complications in COVID-19: a systematic review and meta-analysis. Ann Med 52(7):345–353

Landrum L, Holland MM (2020) Extremes become routine in an emerging new Arctic. Nat Clim Change 10(12):1108–1115

Larsen MD, de Graaf EL, Sonneveld ME, Plomp HR, Nouta J, Hoepel W et al (2021) Afucosylated IgG characterizes enveloped viral responses and correlates with COVID-19 severity. Science (New York, NY) 371(6532):eabc8378

Larsen NW, Stiles LE, Shaik R, Schneider L, Muppidi S, Tsui CT et al (2022) Characterization of autonomic symptom burden in long COVID: a global survey of 2,314 adults. Medrxiv

Larson HJ, Gakidou E, Murray CJ (2022) The vaccine-hesitant moment. N Engl J Med 387(1):58–65

Lazarus JV, Wyka K, White TM, Picchio CA, Rabin K, Ratzan SC et al (2022) Revisiting COVID-19 vaccine hesitancy around the world using data from 23 countries in 2021. Nat Commun 13(1):3801

Lee T, Ahn J-H, Park SY, Kim G-H, Kim J, Kim T-H et al (2018) Recent advances in AIV biosensors composed of nanobio hybrid material. Micromachines 9(12)

Lee SH, Shin HS, Park HY, Kim JL, Lee JJ, Lee H et al (2019) Depression as a mediator of chronic fatigue and post-traumatic stress symptoms in Middle East respiratory syndrome survivors. Psychiatry Investig 16(1):59–64

Lee TC, Murthy S, Del Corpo O, Senécal J, Butler-Laporte G, Sohani ZN et al (2022a) Remdesivir for the treatment of COVID-19: a systematic review and meta-analysis. Clin Microbiol Infect Off Publ Eur Soc Clin Microbiol Infect Dis 28(9):1203–1210

Lee SS, Viboud C, Petersen E (2022b) Understanding the rebound of influenza in the post COVID-19 pandemic period holds important clues for epidemiology and control. Int J Infect Dis IJID Off Publ Int Soc Infect Dis 122:1002–1004

Levine B, Kroemer G (2019) Biological functions of autophagy genes: a disease perspective. Cell 176(1–2):11–42

Li Q, Zheng XS, Shen XR, Si HR, Wang X, Wang Q et al (2020a) Prolonged shedding of severe acute respiratory syndrome coronavirus 2 in patients with COVID-19. Emerg Microbes Infect 9(1):2571–2577

Li G, Fan Y, Lai Y, Han T, Li Z, Zhou P et al (2020b) Coronavirus infections and immune responses. J Med Virol 92(4):424–432

Li W, Yang Y, Liu ZH, Zhao YJ, Zhang Q, Zhang L et al (2020c) Progression of mental health services during the COVID-19 outbreak in China. Int J Biol Sci 16(10):1732–1738

Liu PP, Blet A, Smyth D, Li H (2020) The science underlying COVID-19: implications for the cardiovascular system. Circulation 142(1):68–78

Lorente-Ros M, Das S, Elias J, Frishman WH, Aronow WS (2023) Cardiovascular manifestations of the long COVID syndrome. Cardiol Rev

Lorentzen CL, Haanen JB, Met Ö, Svane IM (2022) Clinical advances and ongoing trials on mRNA vaccines for cancer treatment. Lancet Oncol 23(10):e450–e458

Lu M, Wang X, Ye H, Wang H, Qiu S, Zhang H et al (2021) Does public fear that bats spread COVID-19 jeopardize bat conservation? Biol Conserv 254:108952

Lucas C, Wong P, Klein J, Castro TBR, Silva J, Sundaram M et al (2020) Longitudinal analyses reveal immunological misfiring in severe COVID-19. Nature 584(7821):463–469

Lund SAM, Manyika J, Smit S, Ellingrud K, Robinson O (2021) The future of work after COVID-19. McKinsey Global Institute. Available from: https://www.mckinsey.com/featured-insights/future-of-work/the-future-of-work-after-covid-19

Maduraiveeran G, Sasidharan M, Ganesan V (2018) Electrochemical sensor and biosensor platforms based on advanced nanomaterials for biological and biomedical applications. Biosens Bioelectron 103:113–129

Maiti A, Zhang Q, Sannigrahi S, Pramanik S, Chakraborti S, Cerda A et al (2021) Exploring spatiotemporal effects of the driving factors on COVID-19 incidences in the contiguous United States. Sustain Cities Soc 68:102784

Ma-Lauer Y, Carbajo-Lozoya J, Hein MY, Müller MA, Deng W, Lei J et al (2016) p53 down-regulates SARS coronavirus replication and is targeted by the SARS-unique domain and PLpro via E3 ubiquitin ligase RCHY1. Proc Natl Acad Sci USA 113(35):E5192–E5201

Mantovani F, Collavin L, Del Sal G (2019) Mutant p53 as a guardian of the cancer cell. Cell Death Differ 26(2):199–212

March D, Metcalfe K, Tintoré J, Godley BJ (2021) Tracking the global reduction of marine traffic during the COVID-19 pandemic. Nat Commun 12(1):2415

Mastaglio S, Ruggeri A, Risitano AM, Angelillo P, Yancopoulou D, Mastellos DC et al (2020) The first case of COVID-19 treated with the complement C3 inhibitor AMY-101. Clin Immunol 215:108450

Maza A, Hierro M (2022) Modelling changing patterns in the COVID-19 geographical distribution: Madrid's case. Geogr Res 60(2):218–231

McKinsey Global Institute (2021) The future of work after COVID-19. Available from: https://www.mckinsey.com/featured-insights/future-of-work/the-future-of-work-after-covid-19

McNeely JA (2021) Nature and COVID-19: the pandemic, the environment, and the way ahead. Ambio 50(4):767–781

Mohamed K, Yazdanpanah N, Saghazadeh A, Rezaei N (2021a) Computational drug discovery and repurposing for the treatment of COVID-19: a systematic review. Bioorg Chem 106:104490

Mohamed K, Yazdanpanah N, Saghazadeh A, Rezaei N (2021b) Cognitive biases affecting the maintenance of COVID-19 pandemic. Acta Bio-Med Atenei Parmensis 92(2):e2021102

Moitra P, Alafeef M, Dighe K, Frieman MB, Pan D (2020) Selective naked-eye detection of SARS-CoV-2 mediated by N gene targeted antisense oligonucleotide capped plasmonic nanoparticles. ACS Nano 14(6):7617–7627

Moradian N, Ochs HD, Sedikies C, Hamblin MR, Camargo CA Jr, Martinez JA et al (2020) The urgent need for integrated science to fight COVID-19 pandemic and beyond. J Transl Med 18(1):205

Mujawar MA, Gohel H, Bhardwaj SK, Srinivasan S, Hickman N, Kaushik A (2020) Nano-enabled biosensing systems for intelligent healthcare: towards COVID-19 management. Mater Today Chem 17:100306

Mullen CA (2021) Online doctoral mentoring in a pandemic: help or hindrance to academic progress on dissertations? Int J Mentor Coach Educ 10(2):139–157

Muñoz-Vela FJ, Rodríguez-Díaz L, Gómez-Salgado J, Fernández-Carrasco FJ, Allande-Cussó R, Vázquez-Lara JM et al (2023) Fear and anxiety in pregnant women during the COVID-19 pandemic: a systematic review. Int J Public Health 41

Muz B, de la Puente P, Azab F, Azab AK (2015) The role of hypoxia in cancer progression, angiogenesis, metastasis, and resistance to therapy. Hypoxia (Auckland, NZ) 3:83–92

Naeem SB, Bhatti R (2020) The COVID-19 'infodemic': a new front for information professionals. Health Info Libr J 37(3):233–239

Naikoo GA, Awan T, Hassan IU, Salim H, Arshad F, Ahmed W et al (2021) Nanomaterials-based sensors for respiratory viral detection: a review. IEEE Sens J 21(16):17643–17656

Naikoo GA, Arshad F, Hassan IU, Awan T, Salim H, Pedram MZ et al (2022) Nanomaterials-based sensors for the detection of COVID-19: a review. Bioeng Transl Med 7(3):e10305

Nalbandian A, Sehgal K, Gupta A, Madhavan MV, McGroder C, Stevens JS et al (2021) Post-acute COVID-19 syndrome. Nat Med 27(4):601–615

Ndaïrou F, Area I, Nieto JJ, Torres DFM (2020) Mathematical modeling of COVID-19 transmission dynamics with a case study of Wuhan. Chaos Solitons Fractals 135:109846

Ndwandwe D, Wiysonge CS (2021) COVID-19 vaccines. Curr Opin Immunol 71:111–116

Neumann A (2020) It has the highest death rate of any nursing home in the US. The Guardian. Available from: https://www.theguardian.com/us-news/2020/oct/28/nursing-homes-america-new-jersey-coronavirus

Nixon JP, Mavanji V, Butterick TA, Billington CJ, Kotz CM, Teske JA (2015) Sleep disorders, obesity, and aging: the role of orexin. Ageing Res Rev 20:63–73

Nowak B, Brzóska P, Piotrowski J, Sedikides C, Żemojtel-Piotrowska M, Jonason PK (2020) Adaptive and maladaptive behavior during the COVID-19 pandemic: the roles of dark triad traits, collective narcissism, and health beliefs. Personality Individ Differ 167:110232

O'Connor TG, Ciesla AA, Sefair AV, Thornburg LL, Brown AS, Glover V et al (2022) Maternal prenatal infection and anxiety predict neurodevelopmental outcomes in middle childhood. J Psychopathol Clin Sci 131(4):422

O'Dea XC, Stern J (2022) Virtually the same?: Online higher education in the post COVID-19 era. Br J Educ Technol J Counc Educ Technol 53(3):437–442

Oakes AH, Boyce K, Patton C, Jain S (2023) Rates of routine cancer screening and diagnosis before vs after the COVID-19 pandemic. JAMA Oncol 9(1):145–146

Omboni S, Padwal RS, Alessa T, Benczúr B, Green BB, Hubbard I et al (2022) The worldwide impact of telemedicine during COVID-19: current evidence and recommendations for the future. Connect Health 1:7–35

Osman AJK (2022) The impact of COVID-19 on education systems in the commonwealth. The Commonwealth iLibrary. Available from: https://www.thecommonwealth-ilibrary.org/index.php/comsec/catalog/book/950

Palaniappan S, Ragavi V, David B, Pathur Nisha S (2022) Prediction of epidemic disease dynamics on the infection risk using machine learning algorithms. SN Comput Sci 3(1):47

Papadopoulou A, Musa H, Sivaganesan M, McCoy D, Deloukas P, Marouli E (2021) COVID-19 susceptibility variants associate with blood clots, thrombophlebitis and circulatory diseases. PLoS ONE 16(9):e0256988

Park YC, Beck T, Obstfeld M, Acemoğlu D (2022) Prospects of the global economy after COVID-19. CEPR Press. Available from: https://cepr.org/voxeu/columns/prospects-global-economy-after-covid-19

Pedrini L, Meloni S, Lanfredi M, Ferrari C, Geviti A, Cattaneo A et al (2022) Adolescents' mental health and maladaptive behaviors before the COVID-19 pandemic and 1-year after: analysis of trajectories over time and associated factors. Child Adolesc Psychiatry Ment Health 16(1):42

Peimani N, Kamalipour H (2021) Online education in the post COVID-19 era: students' perception and learning experience. Educ Sci [Internet] 11(10)

Peng Y, Wu P, Schartup AT, Zhang Y (2021) Plastic waste release caused by COVID-19 and its fate in the global ocean. Proc Natl Acad Sci 118(47):e2111530118

Peng P, Hao Y, Liu Y, Chen S, Wang Y, Yang Q et al (2023) The prevalence and risk factors of mental problems in medical students during COVID-19 pandemic: a systematic review and meta-analysis. J Affect Disord 321:167–181

Perry G (2020) Alzheimer's disease patients in the crosshairs of COVID-19. J Alzheimer's Dis JAD 76(1):1

Prata JC, Silva ALP, Walker TR, Duarte AC, Rocha-Santos T (2020) COVID-19 pandemic repercussions on the use and management of plastics. Environ Sci Technol 54(13):7760–7765

Prather KA, Marr LC, Schooley RT, McDiarmid MA, Wilson ME, Milton DK (2020) Airborne transmission of SARS-CoV-2. Science (New York, NY) 370(6514):303–304

Rahman MM, Islam MR, Shohag S, Hossain ME, Shah M, Shuvo S et al (2022) Multifaceted role of natural sources for COVID-19 pandemic as marine drugs. Environ Sci Pollut Res 29(31):46527–46550

Raju V (2021) Implementing flexible systems in doctoral viva defense through virtual mechanism. Glob J Flex Syst Manag 22(2):127–139

Raman B, Bluemke DA, Lüscher TF, Neubauer S (2022) Long COVID: post-acute sequelae of COVID-19 with a cardiovascular focus. Eur Heart J 43(11):1157–1172

Reyes-Bueno JA, Mena-Vázquez N, Ojea-Ortega T, Gonzalez-Sotomayor MM, Cabezudo-Garcia P, Ciano-Petersen NL et al (2020) Case fatality of COVID-19 in patients with neurodegenerative dementia. Neurología (Engl Ed) 35(9):639–645

Rezaei N, Ashkevarian S, Fathi MK, Hanaei S, Kolahchi Z, Ladi Seyedian S-S et al (2021) Introduction on coronavirus disease (COVID-19) pandemic: the global challenge. In: Rezaei N (ed) Coronavirus disease—COVID-19. Springer International Publishing, Cham, pp 1–22

Rohr JR, Civitello DJ, Halliday FW, Hudson PJ, Lafferty KD, Wood CL et al (2020) Towards common ground in the biodiversity–disease debate. Nat Ecol Evol 4(1):24–33

Rojas M, Rodríguez Y, Acosta-Ampudia Y, Monsalve DM, Zhu C, Li Q-Z et al (2022) Autoimmunity is a hallmark of post-COVID syndrome. J Transl Med 20(1):129

Romanelli C, Cooper D, Campbell-Lendrum D, Maiero M, Karesh WB, Hunter D et al (eds) (2015) Connecting global priorities: biodiversity and human health: a state of knowledge review. World Health Organisation/Secretariat of the UN Convention on Biological Diversity

Rose-Redwood R, Kitchin R, Apostolopoulou E, Rickards L, Blackman T, Crampton J et al (2020) Geographies of the COVID-19 pandemic. Dialog Hum Geogr 10(2):97–106

Sandbrook C, Gómez-Baggethun E, Adams WM (2022) Biodiversity conservation in a post-COVID-19 economy. Oryx 56(2):277–283

Sarker S, Jamal L, Ahmed SF, Irtisam N (2021) Robotics and artificial intelligence in healthcare during COVID-19 pandemic: a systematic review. Robot Auton Syst 146:103902

Sarkodie SA, Owusu PA (2021) Impact of COVID-19 pandemic on waste management. Environ Dev Sustain 23(5):7951–7960

Satarker S, Tom AA, Shaji RA, Alosious A, Luvis M, Nampoothiri M (2021) JAK-STAT pathway inhibition and their implications in COVID-19 therapy. Postgrad Med 133(5):489–507

Savin I, Drews S, Van Den Bergh J, Villamayor-Tomas S (2022) Public expectations about the impact of COVID-19 on climate action by citizens and government. PLoS ONE 17(6):e0266979

Schwaegerl C, Tao L (2013) The microgrids concept. In: Hatziargyriou N (ed) Microgrids: architectures and control. Wiley-IEEE Press, pp 1–24

Seo G, Lee G, Kim MJ, Baek S-H, Choi M, Ku KB et al (2020) Rapid detection of COVID-19 causative virus (SARS-CoV-2) in human nasopharyngeal swab specimens using field-effect transistor-based biosensor. ACS Nano 14(4):5135–5142

Shakeel CS, Mujeeb AA, Mirza MS, Chaudhry B, Khan SJ (2022) Global COVID-19 vaccine acceptance: a systematic review of associated social and behavioral factors. Vaccines 10(1)

Shan B, Broza YY, Li W, Wang Y, Wu S, Liu Z et al (2020) Multiplexed nanomaterial-based sensor array for detection of COVID-19 in exhaled breath. ACS Nano 14(9):12125–12132

Siccardi S, Amici J, Colombi S, Carvalho JT, Versaci D, Quartarone E et al (2022) UV-cured self-healing gel polymer electrolyte toward safer room temperature lithium metal batteries. Electrochim Acta 433:141265

Siegel RL, Miller KD, Fuchs HE, Jemal A (2022) Cancer statistics, 2022. CA Cancer J Clin 72(1):7–33

Siegel RL, Miller KD, Wagle NS, Jemal A (2023) Cancer statistics, 2023. CA Cancer J Clin 73(1):17–48

Singh N, Bharara SA (2020) S2 subunit of SARS-nCoV-2 interacts with tumor suppressor protein p53 and BRCA: an in silico study. Transl Oncol 13(10):100814

Singh K, Lima G, Cha M, Cha C, Kulshrestha J, Ahn YY et al (2022) Misinformation, believability, and vaccine acceptance over 40 countries: takeaways from the initial phase of the COVID-19 infodemic. PLoS ONE 17(2):e0263381

Soroya SH, Farooq A, Mahmood K, Isoaho J, Zara S-e (2021) From information seeking to information avoidance: understanding the health information behavior during a global health crisis. Inf Process Manag 58(2):102440

Soung AL, Vanderheiden A, Nordvig AS, Sissoko CA, Canoll P, Mariani MB et al (2022) COVID-19 induces CNS cytokine expression and loss of hippocampal neurogenesis. Brain 145(12):4193–4201

Sridhar A, Balakrishnan A, Jacob MM, Sillanpää M, Dayanandan N (2023) Global impact of COVID-19 on agriculture: role of sustainable agriculture and digital farming. Environ Sci Pollut Res 30(15):42509–42525

Stegmann KM, Dickmanns A, Gerber S, Nikolova V, Klemke L, Manzini V et al (2021) The folate antagonist methotrexate diminishes replication of the coronavirus SARS-CoV-2 and enhances the antiviral efficacy of remdesivir in cell culture models. Virus Res 302:198469

Stegmann KM, Dickmanns A, Heinen N, Blaurock C, Karrasch T, Breithaupt A et al (2022) Inhibitors of dihydroorotate dehydrogenase cooperate with molnupiravir and N4-hydroxycytidine to suppress SARS-CoV-2 replication. Iscience 25(5):104293

Szuhany KL, Malgaroli M, Miron CD, Simon NM (2021) Prolonged grief disorder: course, diagnosis, assessment, and treatment. Focus (Am Psychiatr Publ) 19(2):161–172

Tan M, He FJ, MacGregor GA (2020) Obesity and COVID-19: the role of the food industry. BMJ 369:m2237

Theberath M, Bauer D, Chen W, Salinas M, Mohabbat AB, Yang J et al (2022) Effects of COVID-19 pandemic on mental health of children and adolescents: a systematic review of survey studies. SAGE Open Med 10:20503121221086710

Vanapalli KR, Sharma HB, Ranjan VP, Samal B, Bhattacharya J, Dubey BK et al (2021) Challenges and strategies for effective plastic waste management during and post COVID-19 pandemic. Sci Total Environ 750:141514

von Delft A, Hall MD, Kwong AD, Purcell LA, Saikatendu KS, Schmitz U et al (2023) Accelerating antiviral drug discovery: lessons from COVID-19. Nat Rev Drug Discov 1–19

Wang EY, Mao T, Klein J, Dai Y, Huck JD, Jaycox JR et al (2021a) Diverse functional autoantibodies in patients with COVID-19. Nature 595(7866):283–288

Wang Q, Xu R, Volkow ND (2021b) Increased risk of COVID-19 infection and mortality in people with mental disorders: analysis from electronic health records in the United States. World Psychiatry Off J World Psychiatr Assoc (WPA) 20(1):124–130

Wang L, Davis PB, Volkow ND, Berger NA, Kaelber DC, Xu R (2022a) Association of COVID-19 with new-onset Alzheimer's disease. J Alzheimer's Dis JAD 89(2):411–414

Wang Q, Huang R, Li R (2022b) Impact of the COVID-19 pandemic on research on marine plastic pollution—a bibliometric-based assessment. Mar Policy 146:105285

Watson C (2022) Rise of the preprint: how rapid data sharing during COVID-19 has changed science forever. Nat Med 28(1):2–5

White E (2015) The role for autophagy in cancer. J Clin Investig 125(1):42–46

Wolfe ND, Dunavan CP, Diamond J (2007) Origins of major human infectious diseases. Nature 447(7142):279–283

World Bank (2022a) World development report 2022. Chapter 1. The economic impacts of the COVID-19 crisis 2022. Available from: https://www.worldbank.org/en/publication/wdr2022/brief/chapter-1-introduction-the-economic-impacts-of-the-covid-19-crisis

World Bank (2022b) World development report 2022. Chapter 6. Policy priorities for the recovery. Available from: https://www.worldbank.org/en/publication/wdr2022/brief/chapter-6-policy-priorities-for-the-recovery

World Food Programme (2023) Food insecurity. Available from: https://www.wfp.org/

World Health Organization (2022) World obesity day 2022—accelerating action to stop obesity 2022. Available from: https://www.who.int/news/item/04-03-2022-world-obesity-day-2022-accelerating-action-to-stop-obesity#:~:text=More%20than%201%20billion%20people,they%20are%20overweight%20or%20obese

World Health Organization JNR (2022) UNICEF and WHO warn of perfect storm of conditions for measles outbreaks, affecting children. Available from: https://www.who.int/news/item/27-04-2022-unicef-and-who-warn-of--perfect-storm--of-conditions-for-measles-outbreaks--affecting-children

Wu Y, Guo C, Tang L, Hong Z, Zhou J, Dong X et al (2020) Prolonged presence of SARS-CoV-2 viral RNA in faecal samples. Lancet Gastroenterol Hepatol 5(5):434–435

Xi J, Liu X, Wang J, Yao L, Zhou C (2023) A systematic review of COVID-19 geographical research: machine learning and bibliometric approach. Ann Am Assoc Geogr 113(3):581–598

Xiong J, Lipsitz O, Nasri F, Lui LMW, Gill H, Phan L et al (2020) Impact of COVID-19 pandemic on mental health in the general population: a systematic review. J Affect Disord 277:55–64

Yamamoto K (2022) Adverse effects of COVID-19 vaccines and measures to prevent them. Virol J 19(1):1–3

Yang X-D, Li H-L, Cao Y-E (2021) Influence of meteorological factors on the COVID-19 transmission with season and geographic location. Int J Environ Res Public Health 18(2):484

Yazdanpanah N, Rezaei N (2022) Autoimmune complications of COVID-19. J Med Virol 94(1):54–62

Yong SJ (2021) Long COVID or post-COVID-19 syndrome: putative pathophysiology, risk factors, and treatments. Infect Dis (Lond Engl) 53(10):737–754

Yu J, Yuan X, Chen H, Chaturvedi S, Braunstein EM, Brodsky RA (2020) Direct activation of the alternative complement pathway by SARS-CoV-2 spike proteins is blocked by factor D inhibition. Blood 136(18):2080–2089

Zang R, Gomez Castro MF, McCune BT, Zeng Q, Rothlauf PW, Sonnek NM et al (2020) TMPRSS2 and TMPRSS4 promote SARS-CoV-2 infection of human small intestinal enterocytes. Sci Immunol 5(47)

Zarocostas J (2020) How to fight an infodemic. The Lancet 395(10225):676

Zeng N, Zhao Y-M, Yan W, Li C, Lu Q-D, Liu L et al (2023) A systematic review and meta-analysis of long term physical and mental sequelae of COVID-19 pandemic: call for research priority and action. Mol Psychiatry 28(1):423–433

Zhang Y, Sun H, Pei R, Mao B, Zhao Z, Li H et al (2021) The SARS-CoV-2 protein ORF3a inhibits fusion of autophagosomes with lysosomes. Cell Discov 7(1):31

Zhao Y, Watterston J (2021) The changes we need: education post COVID-19. J Educ Change 22(1):3–12

Zhao J, Xu J, He Y, Xiang M (2022) Children and adolescents' sleep patterns and their associations with mental health during the COVID-19 pandemic in Shanghai, China. J Affect Disord 301:337–344

Zheng C, Shao W, Chen X, Zhang B, Wang G, Zhang W (2022) Real-world effectiveness of COVID-19 vaccines: a literature review and meta-analysis. Int J Infect Dis 114:252–260

Coronavirus Anatomy and Its Analytical Approaches for Targeting COVID-19

2

Santheraleka Ramanathan, M. B. Malarvili, Subash C. B. Gopinath, and Tjok Gde Tirta Nindhia

Abstract

The whole world has lived in fear of the Coronavirus 2019 (COVID-19) for more than one-fourth of a year in 2020. Since then, numerous studies have been done and others are still being conducted to identify the coronavirus causal agent and determine the ideal antiviral therapies for treating this illness. Despite the fact of significant global efforts, which are being made to develop vaccines and identify therapeutic medicines, the emergence of multiple variants has rumbled the research for developing an ideal diagnostic and therapeutic tool for targeting COVID-19. A thorough anatomy of the coronavirus virus and its global dissemination is essential for healthcare initiatives, as well as for predicting and preventing new epidemics. Testing for active COVID-19 infections is a critical public health strategy for tracking the epidemic. The widespread adoption of polymerase chain reaction (PCR) and antigen rapid test assays has increased the supply of test kits. In this chapter, the significance of molecular technique to develop prognostic analytical techniques for targeting COVID-19 along with the advantages and limitations of available techniques is discussed.

Keywords

Coronaviruses · Molecular · Detection method

2.1 Introduction

The coronavirus family of positive-sense, enveloped RNA viruses infects both mammals and birds and is responsible for several diseases. The Nidovirales order, which includes the families

S. Ramanathan · M. B. Malarvili (✉)
School of Biomedical Engineering, Universiti Teknologi Malaysia, UTM, 81310 Skudai, Johor, Malaysia
e-mail: malarvili@utm.my

S. Ramanathan
e-mail: santheraleka@utm.my

S. C. B. Gopinath
Institute of Nano Electronic Engineering, Universiti Malaysia Perlis, 01000 Kangar, Perlis, Malaysia
e-mail: subash@unimap.edu.my

Faculty of Chemical Engineering Technology, Universiti Malaysia Perlis, 02600 Arau, Perlis, Malaysia

Faculty of Applied Sciences, Centre of Excellence for Nanobiotechnology and Nanomedicine (CoExNano), AIMST University, 08100 Semeling, Kedah, Malaysia

T. G. T. Nindhia
Faculty of Engineering, Udayana University, Jl. Raya Kampus UNUD, 80361 Bukit Jimbaran, Kuta Selatan, Badung-Bali, Malaysia
e-mail: tirta.nindhia@me.unud.ac.id

N. Rezaei (ed.), *The COVID-19 Aftermath*, Advances in Experimental Medicine and Biology 1457,
https://doi.org/10.1007/978-3-031-61939-7_2

Arteriviridae, Coronaviridae, Roniviridae, and Mesoniviridae, comprises the largest number of viruses. Coronaviruses are distinguished structurally by their club-like glycoprotein spikes that emerge from the exterior of their envelope and their exceptionally large RNA genome. The term "coronavirus" originated because of these spikes, which produce a distinctive ultrastructure appearance mimicking the solar corona (Zhu et al. 2020; Firouzabadi et al. 2023). Human coronaviruses, which were first discovered in 1965, have lately become well-known due to their involvement in widespread epidemics.

The early stages of COVID-19 infection include anosmia, tiredness, fever, and a dry cough. Most cases (about 80%) are moderate and self-limited, affecting mainly the upper airway with minimal lung involvement. In 15% of the cases, there is a severe infection that is characterized by dyspnea, tachypnea, hypoxemia, cardiovascular symptoms, and severe lung illness. The rest are critical infections characterized by multisystem organ dysfunction, acute respiratory distress syndrome (ARDS), and respiratory failure (Kaye et al. 2021; Pushparajah et al. 2021). It is also frequently worsened by disseminated intravascular coagulopathy (DIC). The projected case fatality rate is 49% in patients who are critically ill and 2% overall. Individuals with concomitant illnesses are more likely to experience negative effects with COVID-19, including those with hypertension, cardiovascular disease, lung disease, and cerebrovascular disease. The COVID-19 virus also causes an increase in morbidity and mortality in elderly people (Guo et al. 2021).

In this overview, we first go through the molecular biology of coronaviruses, covering the genome structure of the virus. The pathogenesis of coronavirus infection and SARS-CoV-2 is then covered, with special emphasis on the significance of the ACE2 receptor in viral entrance into host cells. The chapter outlines the evolution of targeting COVID-19 diseases and the current approaches in managing the pandemic toward the endemic era of the virus infection. The advantages, limitations, and the molecular theory involved in the detection techniques are discussed.

2.2 Structure of SARS-CoV-2

The structure of SARS-CoV-2 allows the virus to infect host cells. The outermost layer of the virus is the viral envelope, which is a lipid bilayer that is studded with spike proteins, membrane proteins, and envelope proteins. The envelope provides protection for the virus and helps it to enter host cells. The spike protein, which gives the virus its name, is the most distinguishing feature of SARS-CoV-2 (coronavirus, meaning "crown-like"). The S1 and S2 subunits make up the spike protein. Angiotensin-converting enzyme 2 (ACE2) is the host cell receptor that binds to the S1 subunit's receptor-binding domain (RBD). The fusion peptide found in the S2 subunit enables the virus to join the host cell membrane and penetrate the cell. Since the spike protein is essential for viral entrance into host cells, it is the focus of numerous vaccines and treatments (Aydın et al. 2021; Lybarger et al. 2021). Important elements of the viral envelope include the membrane protein and envelope protein. The membrane protein aids in the formation of the viral envelope and is involved in viral assembly. It is believed that the envelope protein contributes to viral release from host cells. The positive-sense single-stranded RNA virus SARS-CoV-2 is an enveloped member of the Corona viridine family. The virus is spherical in shape and ranges in diameter from 60 to 140 nm. Spike proteins that are embedded in the virus's envelope give it its distinctive crown-like look. The spike proteins are essential for virus penetration into the host cells. Membrane and envelope proteins are also present in the envelope and are necessary for viral assembly and release (Huang et al. 2020; Taleghani and Taghipour 2021). Figure 2.1 shows the schematic illustration of SARS-CoV-2 genome structure and the structural proteins, which are mainly focused on developing techniques to target virus infection.

2.3 Proteins of SARS-CoV-2

SARS-CoV-2 proteins are required for viral replication, assembly, and infection. The spike protein, which is the virus's most extensively researched protein, oversees facilitating viral entry into host cells. The receptor-binding domain (RBD) of the S1 subunit of the spike protein, which consists of two subunits, allows it to attach to the host cell receptor, ACE2 (Mauer et al. 2021). Another important target for vaccinations and treatments is the spike protein. The RNA-dependent RNA polymerase, which oversees reproducing the viral genome, and the nucleocapsid protein, which binds to the viral RNA and aids in viral assembly, are two more crucial SARS-CoV-2 proteins (Lukose et al. 2021). The viral proteases, which include the primary protease and the papain-like protease, are crucial for viral replication and could serve as potential therapeutic targets (Basin et al. 2022).

2.4 Genetic Sequencing and Its Role in Studying SARS-CoV-2 Virus

The genome of SARS-CoV-2 is approximately 29.9 kilobases in length and encodes for 29 proteins. The genome is organized into several open reading frames (ORFs), which are responsible for the production of viral proteins. The ORFs are flanked by untranslated regions (UTRs), which regulate viral replication and translation. The genome also contains several non-coding regions, which are thought to play a role in viral replication and assembly (Long et al. 2020; Xu et al. 2021). Genetic sequencing has played a crucial role in understanding the biology of the SARS-CoV-2 virus, the causative agent of COVID-19. This technology has allowed researchers to study the genome of the virus and identify important genetic features that contribute to its infectivity, transmissibility, and pathogenicity. In this report, we will discuss

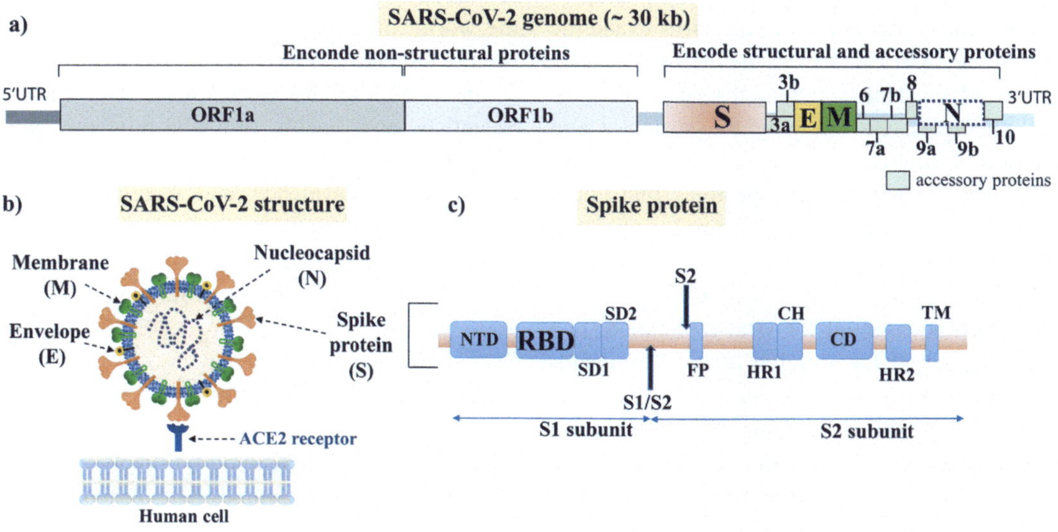

Fig. 2.1 Illustration of SARS-CoV-2 genome and structural proteins. **a** Open reading frames, which encode structural and non-structural proteins. **b** Four major structural proteins of SARS-CoV-2. **c** S1 and S2 subunits of the spike protein, which plays role in the receptor-binding between ACE2 and host cell. Reproduced with permission (Pushparajah et al. 2021) Copyright 2021 Elsevier

the different aspects of genetic sequencing related to SARS-CoV-2 and their implications for COVID-19 research. Genome sequencing and phylogenetic analysis have been essential tools for studying the evolution and spread of SARS-CoV-2 (Shan et al. 2020). By comparing the genetic sequences of different strains of the virus, researchers have been able to trace the origins of the pandemic and monitor the emergence of new viral variants. Phylogenetic analyses have also shown that the virus has evolved into several distinct lineages, some of which are associated with increased transmissibility or severity of disease. One of the most important applications of genetic sequencing in COVID-19 research has been the identification of viral mutations. Mutations in the SARS-CoV-2 genome have been shown to affect the structure and function of viral proteins, potentially altering the virus's behavior and pathogenicity. For instance, the B.1.1.7 variant, first identified in the UK, contains several mutations in the spike protein that may increase its transmissibility. Similarly, the B.1.351 variant, identified in South Africa, induces spike protein mutations that may help the virus evade the immune system. Genetic sequencing has also been used to identify potential targets for antiviral therapies. By studying the structure and function of viral proteins, researchers have identified several enzymes and proteins that could be targeted by drugs to inhibit viral replication or block viral entry into host cells (Syed et al. 2021). For example, the viral RNA polymerase, which is essential for viral replication, is a promising target for antiviral drugs. The development of COVID-19 vaccines has also relied heavily on genetic sequencing. By studying the genetic sequence of the virus, researchers were able to identify the spike protein as a potential target for vaccine development (Galgali et al. 2023; Reece et al. 2023). Several COVID-19 vaccines, including those developed by Pfizer-BioNTech, Moderna, and Johnson & Johnson, target the spike protein to induce an immune response. Ongoing sequencing efforts will be critical for monitoring the evolution of the virus and identifying any mutations that may reduce the efficacy of existing vaccines. Finally, genetic sequencing has also been used to monitor the efficacy of COVID-19 vaccines (Fleming-Dutra et al. 2023; Skoy et al. 2023). By sequencing viral isolates from vaccinated individuals who later test positive for the virus, researchers can determine whether any mutations have emerged that could reduce the effectiveness of the vaccine. This information helps to develop new vaccine formulations or booster shots to address emerging viral variants. Genetic sequencing has been an essential tool for capturing the biology of the SARS-CoV-2 virus and developing strategies for controlling the spread of COVID-19 (Islam et al. 2023). Ongoing sequencing efforts will be critical for monitoring the evolution of the virus and developing effective treatments and vaccines to combat the ongoing pandemic.

2.5 COVID-19 Sampling Techniques-Analytical Content

As the COVID-19 pandemic wreaks havoc on the world, there is a rapidly increasing need to sample human populations in a bid to better comprehend the disease's spread and impact. Scientists and health officials rely on sampling techniques to gather data on the virus's transmission, prevalence, and severity. We will explore some of the human sampling techniques that have been used in COVID-19 research, including their strengths and limitations.

2.5.1 Random Sampling

COVID-19 sampling techniques involve the collection of biological samples from individuals in a population to estimate the virus's prevalence. This technique's biological aspects include the detection and analysis of viral genetic material or antigens in collected samples (Seo et al. 2020). Various equipment and tests are used to medically recognize and test for COVID-19. The reverse transcription polymerase chain reaction (RT-PCR) test, which detects virus genetic material in respiratory samples, is the gold standard for testing. This test is highly sensitive and specific but requires specialized equipment and trained personnel to perform (Tan et al. 2021; Kudhail et al. 2022). Rapid antigen tests, which provide results in minutes, are also used for COVID-19 testing. These tests detect specific proteins on the surface of the virus and are less sensitive than PCR tests, but they are faster and less expensive. In addition to these tests, serological assays that detect antibodies against the virus in blood samples are used to estimate the proportion of the population that has been infected with the virus. Depending on the purpose of the study, available resources, and the population being tested, random sampling techniques for COVID-19 testing may use any of these tests. The sensitivity, specificity, and accuracy of the equipment and tests used should be considered. Overall, random human sampling techniques for COVID-19 are critical for understanding the virus's biological aspects and estimating its prevalence in a population. It is critical to use appropriate testing equipment and techniques to ensure accurate results and inform public health responses (Al Ghobain et al. 2022; Sciavicco et al. 2022).

2.5.2 Convenience Sampling

Convenience sampling is a technique used in research studies to select participants based on their availability and willingness to participate. This sampling method can also be used for COVID-19 testing, especially when randomized sampling is not feasible or practical. Convenience sampling involves testing people who are easily accessible, such as those who go to testing clinics, seek medical attention, or work in high-risk environments. However, because individuals who are more likely to seek testing may differ from the general population, this sampling technique may introduce bias into the study (Huang et al. 2009; Ghodake et al. 2021). Various equipment and tests are used to medically recognize and test for COVID-19 in convenience sampling. The most used tests for COVID-19 are RT-PCR and antigen tests, which detect the genetic material or specific proteins of the virus in respiratory samples such as nasal swabs. The sensitivity, specificity, and accuracy of the equipment and tests used in convenience sampling for COVID-19 testing should be considered (Park et al. 2020). Rapid antigen tests yield results in minutes and can be performed in point-of-care settings such as clinics, whereas RT-PCR tests are more accurate but necessitate specialized equipment and trained personnel. Sum up, convenience sampling can provide useful information about the prevalence of COVID-19 in specific populations or settings, but it must be used with caution to avoid bias and ensure accurate testing. It is critical to use appropriate testing equipment and techniques to inform public health responses and control measures (Kalaiyarasi et al. 2020; Ramanathan et al. 2022).

2.5.3 Stratified Sampling

Stratified sampling is a probability sampling technique used in research studies to ensure that different subgroups of a population are represented. This sampling technique can also be used for COVID-19 testing, especially when certain subgroups are at a higher risk of infection. Various equipment and tests are used in stratified sampling to medically recognize and test for COVID-19 (Lam et al. 2023; Pezzullo et al. 2023). The most used tests for COVID-19 are RT-PCR and antigen tests, which detect the genetic material or specific proteins of the virus in respiratory samples such as nasal swabs. These tests are carried out in clinics, hospitals, and other healthcare facilities by trained medical professionals or technicians. Stratified sampling for COVID-19 testing can be used to ensure that different subgroups of the population are represented, such as healthcare workers, the elderly, or people with underlying health conditions. This sampling method can provide a more accurate estimate of the virus's prevalence in different subgroups, which can inform public health responses and control measures. The sensitivity, specificity, and accuracy of the equipment and tests used in stratified sampling for COVID-19 testing should be considered (Plank et al. 2022; Santopolo et al. 2022). Rapid antigen tests yield results in minutes and can be performed in point-of-care settings such as clinics, whereas RT-PCR tests are more accurate but necessitate specialized equipment and trained personnel. Overall, stratified sampling of human COVID-19 sampling techniques can provide valuable information about the virus's epidemiology in different subgroups of the population. It is critical to use appropriate testing equipment and techniques to ensure accurate results and inform public health responses (Chen et al. 2022).

2.5.4 Cluster Sampling

Cluster sampling is a probability sampling technique used in research studies to choose groups of people, or "clusters," rather than individuals. This sampling technique can also be used for COVID-19 testing, especially when large populations must be tested. Various equipment and tests are used in cluster sampling to medically recognize and test COVID-19 (Zoha et al. 2021; Ilu and Prasad 2023). The most used COVID-19 tests are RT-PCR and antigen tests, which detect the virus's genetic material or specific proteins in respiratory samples such as nasal swabs. These tests are carried out in clinics, hospitals, and other healthcare facilities by trained medical professionals or technicians. Cluster sampling for COVID-19 testing can be used to pinpoint specific geographic areas or communities where the virus is likely to be spreading rapidly. It can also be used in settings where many people are at risk of infection, such as schools, universities, or workplaces. The sensitivity, specificity, and accuracy of the equipment and tests used in cluster sampling for COVID-19 testing should be considered. Rapid antigen tests yield results in minutes and can be performed in point-of-care settings such as clinics, whereas RT-PCR tests are more accurate but necessitate specialized equipment and trained personnel (Jana et al. 2023; Luo et al. 2023). Overall, cluster sampling of human COVID-19 sampling techniques can provide valuable information about the virus's prevalence in specific geographic locations or communities. It can assist public health officials in identifying and containing outbreaks, as well as allocating resources more effectively. However, it is essential to ensure that cluster sampling is conducted in a random and unbiased manner to avoid potential selection biases. Furthermore, ethical issues related to COVID-19 testing in cluster sampling must be considered, such as obtaining informed consent from participants, ensuring the privacy and confidentiality of participants' health information, and providing appropriate counseling and support for those who test positive for the virus. Cluster sampling can be an effective tool for COVID-19 testing in a variety of settings. To ensure that the results obtained are accurate and meaningful, careful consideration should be given to the choice of tests and equipment used, as well as ethical considerations (De Sola et al. 2022).

2.6 COVID-19 Detection Techniques-Current Approaches

2.6.1 Polymerase Chain Reaction (PCR)

The COVID-19 pandemic has caused significant disruptions around the world, affecting millions of people's lives. The use of polymerase chain reaction (PCR) to detect the SARS-CoV-2 virus that causes the disease has been one of the most important tools in the fight against COVID-19. In this chapter, we will look at the principles, benefits, and limitations of PCR in COVID-19 testing.

2.6.1.1 Principles

PCR is a technique that amplifies DNA sequences, allowing specific genetic material in a sample to be detected. The method involves heating and cooling a reaction mixture containing DNA polymerase, primers, and nucleotides multiple times. The DNA double-stranded molecule is denatured into single-stranded molecules during the heating step (Nappier et al. 2019; Kapoor et al. 2021). The primers anneal to the single-stranded DNA during the cooling step, providing a starting point for DNA polymerase to copy and amplify the specific sequence (Fu-Ming et al. 2017). In COVID-19 testing, PCR is used to detect the SARS-CoV-2 virus by targeting specific regions of the viral genome. The viral RNA is extracted from patient samples like nasal swabs or saliva and then reverse-transcribed into cDNA, which serves as a template for PCR. The PCR reaction is then carried out using specific primers that target regions of the viral genome, allowing the virus to be amplified and detected (Bhat and Emslie 2016).

2.6.1.2 Advantages

Because of its high sensitivity and specificity, PCR has become a cornerstone of COVID-19 testing. PCR tests can detect even trace amounts of the virus in patient samples, allowing for the early identification and isolation of infected people. PCR testing is also highly specific, which means it can differentiate between the SARS-CoV-2 virus and other viruses that are similar, such as the common cold (Farasani 2021). PCR testing is also relatively simple and can be automated, enabling high-throughput testing of large numbers of patient samples. This was especially important during the COVID-19 pandemic, as testing was a critical component of public health strategies to control disease spread (Shahi et al. 2018; Caméléna et al. 2021).

2.6.1.3 Limitations

While PCR testing has proven to be useful in the fight against COVID-19, it is not without limitations. One of the main limitations of PCR testing is that it can produce false negatives, which means that infected people may test negative for the virus despite being infected. This can happen if the viral load in the sample is too low or if the sample was not properly collected. False negatives can cause infected people to unknowingly spread the virus to others, contributing to the disease's spread (Mimmack et al. 2004). Another limitation of PCR testing is that it can detect the virus's presence even in people who are no longer infectious. Individuals who have recovered from COVID-19 and are no longer contagious may still test positive for the virus for some time. This may result in unnecessary isolation and quarantine for people who are no longer at risk of spreading the disease. The sensitive and specific detection of the SARS-CoV-2 virus in patient samples has made PCR a critical tool in the fight against COVID-19. While PCR testing has limitations, it has been useful in identifying infected individuals and assisting in disease control. As the pandemic continues, PCR testing is likely to play an important role in public health strategies to combat COVID-19 (Yagci et al. 2020; Hwang et al. 2023).

2.6.2 Antigen Rapid Test Kit (RTK)

Rapid antigen tests, also known as Rapid Test Kit (RTK) COVID-19 tests, have grown in popularity as a method of detecting the SARS-CoV-2 virus, which causes COVID-19. These tests have a few advantages over other COVID-19 testing methods, including their speed and simplicity (Rahman et al. 2021).

2.6.2.1 Principles

RTK COVID-19 tests look for SARS-CoV-2 antigens in patient samples like nasal swabs or saliva. These antigens are proteins found on the surface of the virus that can cause an immune response in infected people. To detect the presence of the virus in patient samples, RTK COVID-19 tests employ antibodies that bind to these antigens. RTK COVID-19 tests are intended to produce results quickly, usually within 15–30 min. This is accomplished using lateral flow technology, which involves the addition of a patient sample to a test strip containing antibodies that bind to the SARS-CoV-2 antigen. If the virus is present in the sample, it will bind to the antibodies on the test strip, resulting in the appearance of a visible line (Manchikalapati et al. 2021; Majumder et al. 2023).

2.6.2.2 Advantages

The speed of RTK COVID-19 tests is one of their main advantages. These tests can produce results in minutes, making them an invaluable tool in situations where quick diagnosis is critical, such as hospitals, clinics, and airports (Krüttgen et al. 2021). The rapidity of these tests also allows for rapid contact tracing and isolation of infected individuals, which can aid in virus control. RTK COVID-19 tests are also relatively simple to use, requiring no specialized equipment or trained personnel. This makes them useful in situations where other testing methods may be impractical, such as remote or low-resource areas. Furthermore, the lower cost of RTK COVID-19 tests versus PCR tests makes them an appealing option for widespread screening programs (Zhou et al. 2021; Soh et al. 2022).

2.6.2.3 Limitations

While RTK COVID-19 tests have several benefits, they also have some drawbacks. One of the main limitations of these tests is that they are less sensitive than PCR tests. If the viral load in the sample is low or the test is not performed correctly, RTK COVID-19 tests may produce false negative results. This means that infected people may test negative for the virus despite being infected, increasing the risk of the virus spreading to others (Fernandes et al. 2022). Another limitation of RTK COVID-19 tests is that they are less specific than PCR tests. If the antibodies in the test strip cross-react with other viral antigens or non-specific proteins in the sample, RTK COVID-19 tests may produce false positive results. This may result in unnecessary isolation and quarantine measures for individuals who are not infected. RTK COVID-19 tests are an important tool in the fight against the COVID-19 pandemic, allowing for quick and easy testing for the SARS-CoV-2 virus. While these tests have many advantages, they also have some drawbacks that must be considered. As the pandemic continues, RTK COVID-19 tests are likely to play an important role in public health strategies to control disease spread (Pavia and Plummer 2021; Filchakova et al. 2022).

2.6.3 Serological Test

COVID-19 serological tests, also known as antibody tests, detect the antibodies developed in the blood in response to a SARS-CoV-2 infection.

2.6.3.1 Principles

Serological tests look for antibodies in the blood that are produced by the immune system in response to an infection. In response to a SARS-CoV-2 infection, two types of antibodies are produced: IgM and IgG antibodies. The first antibodies produced by the immune system in response to an infection are IgM antibodies. They can be detected in the blood within a few days of infection and usually go away within a few weeks. IgG antibodies are produced later in the infection's course and can remain in the blood for months (Sidiq et al. 2020; Devi et al. 2022).

2.6.3.2 Advantages

One of the primary benefits of serological tests for COVID-19 diagnosis is their ability to detect previous infections. In contrast to PCR and antigen tests, which can only detect current infections, serological tests can detect infections that occurred weeks or months ago. Serological tests are also relatively simple to administer and can yield results quickly. Many serological tests can be performed on a small blood sample from a finger prick and provide results in as little as 10–15 min (Moura et al. 2020; Chan et al. 2023). Furthermore, serological tests can be used to estimate COVID-19 prevalence in a population. Public health officials can gain a better understanding of how many people have been infected with SARS-CoV-2 and the extent of the virus's spread in a specific region by testing a representative sample of the population (Tran et al. 2021).

2.6.3.3 Limitations

While serological tests have several advantages, there are some limitations to be aware of. One of the main drawbacks is that they are less sensitive than PCR tests. Serological tests can detect the presence of antibodies in the blood for several days or weeks, which means they may miss recent infections (Ong et al. 2021; de Souza Medeiros et al. 2022). Furthermore, the timing of the test can affect the accuracy of serological tests. In the early stages of infection, antibody levels may be too low to detect, and antibody levels can decrease over time, resulting in false negative results. Another limitation of serological tests is their inability to distinguish between previous SARS-CoV-2 infections and other coronavirus infections. This can lead to false positive results, especially in areas where other methods are used (Nishizawa et al. 2023; Shah et al. 2023). Serological tests are an important tool in the diagnosis and management of COVID-19, as they provide information on previous infections and estimate the disease's prevalence in a population. While serological tests have several advantages, there are some limitations to be aware of. As the pandemic continues, serological tests are likely to play an important role in the diagnosis and management of COVID-19, especially in settings where other testing methods may be impractical. To achieve accurate and timely diagnoses, serological tests must be used in conjunction with other diagnostic tests (La Marca et al. 2020; Guo et al. 2023).

2.6.4 Computed Tomography (CT) Scan

CT scans have been used to diagnose COVID-19 since the pandemic's early stages. CT scans are a non-invasive imaging method that can detect lung abnormalities associated with COVID-19, such as ground-glass opacities (GGOs) and consolidation (Golikhatir et al. 2022).

2.6.4.1 Principles

CT scans create cross-sectional images of the body using X-rays. The images are created by rotating an X-ray tube around the body and using a detector to detect the X-rays that pass through the body. The detector's data is used to generate a three-dimensional image of the body. CT scans are used in the context of COVID-19 diagnosis to detect lung abnormalities associated with the disease (Abuladze et al. 2023). Ground-glass opacities (GGOs), which appear as hazy areas in the lungs, and consolidation, which appear as solid areas, are examples of these abnormalities. These abnormalities may indicate COVID-19 or other lung diseases such as pneumonia (van Bakel et al. 2023).

2.6.4.2 Advantages

The ability of CT scans to provide high-resolution images of the lungs is one of the main advantages of using them for COVID-19 diagnosis. This enables radiologists to detect even minor abnormalities that would otherwise go undetected on a chest X-ray (Alsharif and Qurashi 2021). Furthermore, CT scans can provide information on the extent and severity of lung abnormalities, assisting clinicians in making treatment decisions. CT scans have the added benefit of being a non-invasive imaging method (Gupta and Bajaj 2023). This means that patients will not have to undergo any invasive procedures or be exposed to harmful radiation levels. The procedure is also relatively quick, usually only taking a few minutes to complete.

2.6.4.3 Limitations

While CT scans have many advantages, they do have some drawbacks that must be considered. One of the main drawbacks is that they are less specific than PCR tests. CT scans can detect COVID-19-related lung abnormalities, but they cannot definitively diagnose the disease (Dong et al. 2023). Furthermore, CT scans can detect lung abnormalities other than COVID-19, such as those associated with other lung diseases. Another disadvantage of CT scans for COVID-19 diagnosis is their higher cost in comparison to other testing methods such as RT-PCR tests (Moezzi et al. 2021). CT scans also necessitate specialized equipment and trained personnel, both of which are not always available in all settings (Hassan et al. 2022).

2.7 Conclusion

Understanding the anatomy of the SARS-CoV-2 virus and genetic sequencing is essential for controlling the COVID-19 pandemic. The spike protein and other components of the virus are targets for developing vaccines and therapeutics. Genetic sequencing provides important insights into the virus's evolution and spread, which is essential for developing effective strategies to control the pandemic. Genetic sequencing has played a crucial role in our understanding of the biology of the SARS-CoV-2 virus and its impact on the ongoing COVID-19 pandemic. By conducting genome sequencing and phylogenetic analyses, researchers have identified distinct lineages of the virus and tracked its spread in different regions of the world. Additionally, genetic sequencing has been used to identify viral mutations and develop potential targets for antiviral therapies. It has also been critical for monitoring the efficacy of COVID-19 vaccines. Ongoing sequencing efforts will be essential for understanding the biology of the virus and developing effective treatments and vaccines to combat the ongoing pandemic. The application of genetic sequencing to COVID-19 research has highlighted the importance of genomic surveillance and will likely lead to significant advances in our ability to diagnose, treat, and prevent viral infections in the future.

References

Abuladze LR et al (2023) CT imaging of HIV-associated pulmonary disorders in COVID-19 pandemic. Clin Imaging

Al Ghobain M et al (2022) The efficacy of Zafirlukast as a SARS-CoV-2 helicase inhibitor in adult patients with moderate COVID-19 Pneumonia (pilot randomized clinical trial). J Infect Public Health 15(12):1546–1550

Alsharif W, Qurashi A (2021) Effectiveness of COVID-19 diagnosis and management tools: a review. Radiography 27(2):682–687

Aydın EB et al (2021) Highly selective and sensitive sandwich immunosensor platform modified with MUA-capped GNPs for detection of spike Receptor Binding Domain protein: a precious marker of COVID 19 infection. Sens Actuators B Chem 345:130355

Basin S et al (2022) Progression to a severe form of COVID-19 among patients with chronic respiratory diseases. Respir Med Res 81:100880

Bhat S, Emslie KR (2016) Digital polymerase chain reaction for characterisation of DNA reference materials. Biomol Detect Quantif 10:47–49

Caméléna F et al (2021) Performance of a multiplex polymerase chain reaction panel for identifying bacterial pathogens causing pneumonia in critically ill patients with COVID-19. Diagn Microbiol Infect Dis 99(1):115183

Chan A et al (2023) Evaluation and comparison of four quantitative SARS-CoV-2 serological assays in COVID-19 patients and immunized healthy individuals, cancer patients, and patients with immunosuppressive therapy. Clin Biochem 116:1–6

Chen X et al (2022) Stratified changes in emotional distress during the COVID-19 pandemic: evidence from a longitudinal survey of residents in Hubei province, China. J Psychosom Res 160:110959

De Sola H et al (2022) Understanding the impact of the COVID-19-related lockdown in university workers. Identifying groups through cluster analysis. Heliyon 8(12):e12483

de Souza Medeiros EJ et al (2022) Genetic basis underlying the serological affinity of leptospiral serovars from serogroups Sejroe, Mini and Hebdomadis. Infect Genet Evol 103:105345

Devi MJ et al (2022) A review post-vaccination SARS-CoV-2 serological test: method and antibody titer response. Anal Biochem 114902

Dong X et al (2023) Macro-porosity and skeleton of a cement-gravel-treated granite residual soil for subgrade from CT scanning. Constr Build Mater 371:130703

Farasani A (2021) Genetic analysis of the 2019 coronavirus pandemic with from real-time reverse transcriptase polymerase chain reaction. Saudi J Biol Sci 28(1):911–916

Fernandes RS et al (2022) Recent advances in point of care testing for COVID-19 detection. Biomed Pharmacother 113538

Filchakova O et al (2022) Review of COVID-19 testing and diagnostic methods. Talanta 123409

Firouzabadi N et al (2023) Update on the effectiveness of COVID-19 vaccines on different variants of SARS-CoV-2. Int Immunopharmacol 109968

Fleming-Dutra KE et al (2023) Safety and effectiveness of maternal COVID-19 vaccines among pregnant people and infants. Obstet Gynecol Clin North Am

Fu-Ming S et al (2017) Development of nano-polymerase chain reaction and its application. Chin J Anal Chem 45(11):1745–1753

Galgali MS et al (2023) Feeling too isolated to be vaccinated? The contributing role of subjective interpersonal isolation factors towards COVID-19 vaccine hesitancy and resistance. Soc Sci Med 115865

Ghodake GS et al (2021) Biological characteristics and biomarkers of novel SARS-CoV-2 facilitated rapid development and implementation of diagnostic tools

and surveillance measures. Biosens Bioelectron 177:112969
Golikhatir I et al (2022) Comparison of the diagnostic accuracy of CT scan with oral and intravenous contrast versus CT scan with intravenous contrast alone in the diagnosis of blunt abdominal traumas. Chin J Traumatol
Guo J et al (2021) 5G-enabled ultra-sensitive fluorescence sensor for proactive prognosis of COVID-19. Biosens Bioelectron 181:113160
Guo R et al (2023) When COVID-19 sits on people's laps: a systematic review of SARS-CoV-2 infection prevalence in household dogs and cats. One Health 100497
Gupta K, Bajaj V (2023) Deep learning models-based CT-scan image classification for automated screening of COVID-19. Biomed Signal Process Control 80:104268
Hassan H et al (2022). Supervised and weakly supervised deep learning models for COVID-19 CT diagnosis: a systematic review. Comput Methods Programs Biomed 106731
Huang JC et al (2009) Detection of severe acute respiratory syndrome (SARS) coronavirus nucleocapsid protein in human serum using a localized surface plasmon coupled fluorescence fiber-optic biosensor. Biosens Bioelectron 25(2):320–325
Huang Y et al (2020) Structural and functional properties of SARS-CoV-2 spike protein: potential antivirus drug development for COVID-19. Acta Pharmacol Sin 41(9):1141–1149
Hwang J et al (2023) Real-time transcription polymerase chain reaction cycle threshold values as criteria for utilization of incidental COVID-19 positive lung donors. J Heart Lung Transplant 42(3):301–304
Ilu SY, Prasad R (2023) Improved autoregressive integrated moving average model for COVID-19 prediction by using statistical significance and clustering techniques. Heliyon 9(2)
Islam M et al (2023) Evidence of sample selectivity in highway injury-severity models: the case of risky driving during COVID-19. Anal Methods Accid Res 38:100263
Jana A et al (2023) Spatial shifting of COVID-19 clusters and disease association with environmental parameters, India. Environ Res 115288
Kalaiyarasi J et al (2020) Graphitic carbon nitride/graphene nanoflakes hybrid system for electrochemical sensing of DNA bases in meat samples. Sci Rep 10(1):1–16
Kapoor M et al (2021) Cycle threshold values versus reverse transcription-polymerase chain reaction positivity in COVID-19 de-isolation. Indian J Med Microbiol 39(1):133–135
Kaye AD et al (2021) Biology of COVID-19 and related viruses: epidemiology, signs, symptoms, diagnosis, and treatment. Best Pract Res Clin Anaesthesiol 35(3):269–292
Krüttgen A et al (2021) Comparison of the SARS-CoV-2 rapid antigen test to the real star SARS-CoV-2 RT PCR kit. J Virol Methods 288:114024
Kudhail K et al (2022) Randomised controlled trials in COVID-19 patients: a systematic review and critical appraisal. Int J Infect Dis
La Marca A et al (2020) Testing for SARS-CoV-2 (COVID-19): a systematic review and clinical guide to molecular and serological in-vitro diagnostic assays. Reprod Biomed Online 41(3):483–499
Lam MTY et al (2023) Dynamic activity in cis-regulatory elements of leukocytes identifies transcription factor activation and stratifies COVID-19 severity in ICU patients. Cell Rep Med 4(2)
Long C et al (2020) Diagnosis of the coronavirus disease (COVID-19): rRT-PCR or CT? Eur J Radiol 126:108961
Lukose J et al (2021) Optical technologies for the detection of viruses like COVID-19: progress and prospects. Biosens Bioelectron 178:113004
Luo Z et al (2023) Time series clustering of COVID-19 pandemic-related data. Data Sci Manage
Lybarger K et al (2021) Extracting COVID-19 diagnoses and symptoms from clinical text: a new annotated corpus and neural event extraction framework. J Biomed Inform 117:103761
Majumder S et al (2023) Comparison of SARS-CoV-2 diagnosis by rapid antigen detection kit with RT-qPCR in a tertiary care setup in North Eastern India. Indian J Med Microbiol 42:12–16
Manchikalapati Y et al (2021) Comparative evaluation of commercially available Dengue NS1 Antigen ELISA & Rapid diagnostic test kits to detect early Dengue viral infection. Indian J Med Microbiol 39:S129
Mauer E et al (2021) A predictive model of clinical deterioration among hospitalized COVID-19 patients by harnessing hospital course trajectories. J Biomed Inform 118:103794
Mimmack ML et al (2004) Quantitative polymerase chain reaction: validation of microarray results from postmortem brain studies. Biol Psychiatry 55(4):337–345
Moezzi M et al (2021) The diagnostic accuracy of artificial intelligence-assisted CT imaging in COVID-19 disease: a systematic review and meta-analysis. Inform Med Unlocked 24:100591
Moura DTHD et al (2020) Diagnostic characteristics of serological-based COVID-19 testing: a systematic review and meta-analysis. Clinics 75
Nappier SP et al (2019) Advancements in mitigating interference in quantitative polymerase chain reaction (qPCR) for microbial water quality monitoring. Sci Total Environ 671:732–740
Nishizawa T et al (2023) Pocket structures of surface protrusions shared among serologically distinct nervous necrosis viruses (NNVs) were predicted in silico to bind to sialylated N-glycans, a host cellular receptor. Aquaculture 565:739157

Ong DS et al (2021) How to interpret and use COVID-19 serology and immunology tests. Clin Microbiol Infect 27(7):981–986

Park M et al (2020) Optimization of primer sets and detection protocols for SARS-CoV-2 of coronavirus disease 2019 (COVID-19) using PCR and real-time PCR. Exp Mol Med 52(6):963–977

Pavia CS, Plummer MM (2021) The evolution of rapid antigen detection systems and their application for COVID-19 and other serious respiratory infectious diseases. J Microbiol Immunol Infect 54(5):776–786

Pezzullo AM et al (2023) Age-stratified infection fatality rate of COVID-19 in the non-elderly population. Environ Res 216:114655

Plank K et al (2022) School meal access and changes in meal participation during COVID-19 school closures: a stratified random sample of CalFresh Healthy Living Eligible school districts in California. Prev Med Rep 28:101794

Pushparajah D et al (2021) Advances in gene-based vaccine platforms to address the COVID-19 pandemic. Adv Drug Deliv Rev 170:113–141

Rahman MM et al (2021) Clinical evaluation of SARS-CoV-2 antigen-based rapid diagnostic test kit for detection of COVID-19 cases in Bangladesh. Heliyon 7(11):e08455

Ramanathan S et al (2022) Nanodiamond conjugated SARS-CoV-2 spike protein: electrochemical impedance immunosensing on a gold microelectrode. Microchim Acta 189(6):226

Reece S et al (2023) Hesitant adopters: COVID-19 vaccine hesitancy among diverse vaccinated adults in the United States. Infect Med

Santopolo G et al (2022) Plasma-induced nanoparticle aggregation for stratifying COVID-19 patients according to disease severity. Sens Actuators B Chem 373:132638

Sciavicco G et al (2022) The voice of COVID19: breath and cough recording classification with temporal decision trees and random forests

Seo G et al (2020) Rapid detection of COVID-19 causative virus (SARS-CoV-2) in human nasopharyngeal swab specimens using field-effect transistor-based biosensor. ACS Nano 14(4):5135–5142

Shah AA et al (2023) Limited diagnostic utility of serologic testing for neurologic manifestations of systemic disease in the evaluation of suspected multiple sclerosis: a single-center observational study. Mult Scler Relat Disord 69:104443

Shahi S et al (2018) Polymerase chain reaction (PCR)-based methods: promising molecular tools in dentistry. Int J Biol Macromol 117:983–992

Shan B et al (2020) Multiplexed nanomaterial-based sensor array for detection of COVID-19 in exhaled breath. ACS Nano 14(9):12125–12132

Sidiq Z et al (2020) Benefits and limitations of serological assays in COVID-19 infection. Indian J Tuberc 67(4):S163–S166

Skoy E et al (2023) Preparedness for a pandemic: independent community pharmacists' experiences delivering COVID-19 vaccines. J Am Pharm Assoc

Soh JH et al (2022) CRISPR-based systems for sensitive and rapid on-site COVID-19 diagnostics. Trends Biotechnol

Syed MN et al (2021) Effect of anti-tumor necrosis factor therapy on the risk of respiratory tract infections and related symptoms in patients with psoriasis—a meta-estimate of pivotal phase 3 trials relevant to decision making during the COVID-19 pandemic. J Am Acad Dermatol 84(1):161–163

Taleghani N, Taghipour F (2021) Diagnosis of COVID-19 for controlling the pandemic: a review of the state-of-the-art. Biosens Bioelectron 174:112830

Tan J-S et al (2021) Genetic predisposition to COVID-19 may increase the risk of hypertension disorders in pregnancy: a two-sample Mendelian randomization study. Pregnancy Hypertens 26:17–23

Tran S et al (2021) Evaluation of COVID-19 vaccine response in patients with cancer: an interim analysis. Eur J Cancer 159:259–274

van Bakel S et al (2023) CT-derived muscle but not fat area independently predicts mortality in COVID-19. Chest

Xu R et al (2021) Progressive deterioration of the upper respiratory tract and the gut microbiomes in children during the early infection stages of COVID-19. J Genet Genomics 48(9):803–814

Yagci AK et al (2020) Relationship of the cycle threshold values of SARS-CoV-2 polymerase chain reaction and total severity score of computerized tomography in patients with COVID 19. Int J Infect Dis 101:160–166

Zhou Y et al (2021) Point-of-care COVID-19 diagnostics powered by lateral flow assay. TrAC Trends Anal Chem 145:116452

Zhu W et al (2020) RNA-dependent RNA polymerase as a target for COVID-19 drug discovery. SLAS Discov 25(10):1141–1151

Zoha N et al (2021) A numerical approach to maximize the number of testing of COVID-19 using conditional cluster sampling method. Inform Med Unlocked 23:100532

Ethanol Inhalation as a Method to Denature the Spike Protein of SARS-CoV-2

3

Thomas J. Manning, Teighlor Livingston, Capri Persaud, Akshil Patel, Madelyn Adair, Taylor Taylor, and Paige Bland

Abstract

The SARS-CoV-2 virus caused the 2019 COVID pandemic by infecting almost eight hundred million people worldwide. Because it was a new viral infection, there were no vaccines or small molecule medications that could prevent or treat the disease. This chapter provides some details for an obscure treatment for COVID-19, that has decades of anti-viral activity data both in vitro and in vivo in the literature. The medicinal molecules are compared to other small molecules that were identified as possible medications for COVID-19. We developed a computational method that ranks small molecules and their ability to penetrate mucus in the lungs of a COVID-19 patient. Our focus is ethanol as a COVID-19 treatment. The results discussed here are based on Lipinski Rules and QSAR computational methods as well as in vitro and in vivo data. These parameters indicate that ethanol should be a strong candidate for future evaluations.

T. J. Manning (✉) · C. Persaud · A. Patel · M. Adair · T. Taylor · P. Bland
Chemistry Department, Valdosta State University, Valdosta, GA 31698, USA
e-mail: tmanning@valdosta.edu

T. Livingston
Philadelphia College of Osteopathic Medicine, Moultrie, GA, USA

Keywords

Corona virus · COVID-19 · Spike protein · Denature · Ethanol · Inhalation · SARS-CoV-2

3.1 Introduction

The highly transmissible SARS-CoV-2 virus is a positive-sense ss-RNA virus that is in the coronaviridae family. It is the virus responsible for COVID-19, which was identified in over 230 countries. SARS-CoV-2 encodes at least 29 proteins, including nonstructural proteins, structural proteins, auxiliary proteins and the spike protein. Spike proteins are a target for small molecule pharmaceuticals because of their role in helping the virus enter the cell. This chapter compares a total of four in vivo applications, two of which utilized ethanol inhalation to treat patients with COVID-19. The other two trials, including the Solidarity trial, utilized Remdesivir. An equation is developed that utilizes calculated chemical and physical values for the molecules to determine which is most likely to penetrate the mucus in a patient's lungs. The data, in vivo, in vitro and computational, suggests that ethanol is highly effective as a treatment for COVID-19. Ethanol has a significant body of scientific and medicinal data, such as pharmacokinetics,

N. Rezaei (ed.), *The COVID-19 Aftermath*, Advances in Experimental Medicine and Biology 1457,
https://doi.org/10.1007/978-3-031-61939-7_3

pharmacodynamics, mechanism of action (MOA) and in vitro studies. The background data provides a significant advantage for drug development over new molecules. Parameters such as dosing, allergies and side effects, degradation and excretion routes and a reliable production and distribution route for a medicinal grade product are well known for ethanol.

During the COVID-19 pandemic, the Centers for Disease Control recommended the use of hand sanitizers that contained more than 60% ethanol as a preferred method of disinfection (National institute of Health 2020). The National Institutes of Health showed that N95 respirators could be decontaminated by ethanol (Fischer et al. 2020). The World Health Organization also made a number of recommendations involving the use of ethanol as a disinfecting agent for COVID-19 (World Health Organization 2022). Ethanol is ubiquitous in modern times as a disinfectant against a variety of viral infections in a health care setting (Kampf et al. 2020).

The scientific community tested various methods to denature the spike protein on the SARS-CoV-2 virus. The spike protein is the critical entity that is needed to infect a cell. Its receptor is the angiotensin-converting enzyme 2, ACE2. Once the SARS-CoV-2 virus enters the cell, its replication process begins, resulting in the eventual lysis of the cell and the release of the new viral particles (Jackson et al. 2022). A concern with any specific pharmaceutical treatment or vaccine is the ability of the virus to mutate. For vaccines, this is recognized as a legitimate problem in their long-term ability to prevent infection (Arora et al. 2021; Rath and Kumar 2020; Malik et al. 2022).

The development of small molecules that interfere with the protein–protein interactions (PPI) has been an ongoing endeavor against MERS for SARS (Buchwald 2022; Chien et al. 2020). Considering the relatively small number of drugs synthesized and used in vitro and in vivo, and the much larger number that have been developed via computational methods, the number of valid candidates and their performance in vivo is fairly low. The IUPAC definition of protein denaturation is "*The process of partial or total alteration of the native structure of a macromolecule resulting from the loss of tertiary or tertiary and secondary structure that is a consequence of the disruption of stabilizing weak bonds. Denaturation can occur when proteins and nucleic acids are subjected to elevated temperature or to extremes of pH, or to non-physiological concentrations of salt, organic solvents, urea or other chemical agents.*" (Pure and Applied Chemistry 1992, 1994).

Two common organic chemicals that are used to denature proteins are urea and guanidinium chloride (aka guanidine hydrochloride). The Gibbs free energy (ΔG) of the protein will change as its geometry changes. This change is directly related to the concentration of the molecule that induces the denaturing.

$$\Delta G = \Delta G^{\circ} - m[\text{denaturant}] \qquad (3.1)$$

ΔG is the Gibbs free energy (kJ/mol) before the protein undergoes denaturing, for zero concentration of the denaturant, $m = 0$. ΔG is the value as the denaturant concentration changes (Greene and Pace 1974; Freire et al. 2013).

$$\Delta G = \Delta S - T\Delta S \qquad (3.1\text{A})$$

In the case of a protein in thermodynamic equilibrium with its surrounding (Zeroth Law of Thermodynamics), a primary component of the ΔG value is the increase in entropy or disorder involved in the protein structure being disrupted by the denaturing agent. The interaction of ethanol with the protein will result in hydrogen bonds, and weaker dipole–dipole interactions, forming and breaking within the protein structure. The dipole moment, logP, TPSA and molecular volume values of ethanol (2.37 D, 0.06, 20.23 A^2, 54.02 A^3) are different than water (1.84 D, −0.29, 29.27 A^2, 19.33 A^3). This impacts how and if the molecule can competitively occupy sites on the exterior of the protein and/or wedge itself inside the structure and cause the strains to be separated and the protein denatured.

Common disinfectants include ethanol, propanol, chlorine and chlorine compounds (i.e., Clorox), formaldehyde, glutaraldehyde, hydrogen peroxide, iodophors, peracetic acid, quaternary ammonium compounds, metals as microbicides (i.e., Ag(I)), ultraviolet light and pasteurization. These common disinfecting approaches are not specific for a target, so they can cause indiscriminate and nonselective damage at a molecular level. Several can also leave an undesired chemical residue (i.e., Clorox leaving a chlorine-based residue). Many of these same substances can denature proteins, rendering them unable to function. Because of the large-scale use of ethanol as a disinfectant, a drug carrier and consumed in recreational beverages, its impact on tissues, specific organs, red blood cells, etc., is well known and understood.

This paper outlines the results of an inhaled ethanol treatment that was utilized for treating patients inflicted with COVID-19 and compares it to other medications. QSAR calculations are also used for fifty-one medications and supplements at various levels of development. A review paper by Kamphf et al. provides disinfection data for ethanol, 1-propanol, 2-propanol, benzalkonium chloride, didecyldimethyl ammonium chloride, chlorhexidine digluconate, sodium hypochlorite, hydrogen peroxide, formaldehyde, povidone iodine and glutardialdehyde against viruses such as Middle East Respiratory Syndrome (MERS); human coronavirus (HCoV); transmissible gastroenteritis virus (TGEV); mouse hepatitis virus (MHV); and Severe Acute Respiratory Syndrome (SARS). Other sources provide additional supporting data (Martins et al. 2022; Xiling et al. 2021; Huang et al. 2022; Henwood 2020). For ethanol applied to the corona virus on various surfaces, relatively high concentrations (i.e., 70–90%) resulted in rapid disinfections (10 to 15 s) reducing the viral load up to 10,000-fold. For example, the corona virus on a ceramic surface showed a 10^4 fold decrease in numbers in 15 s when treated with 70% ethanol (Kampf 2020). Most COVID-19 patients have the bulk of the viral load in their lungs, which can be exacerbated by cytokine or bradykinin storms, increasing the phlegm or sputum (mucus, microbial and organic matter), resulting in lower levels of oxygen reaching serum. Inhaling ethanol with the goal of denaturing the spike and/or ACE2 proteins before the replication process starts is the proposed MOA (see Figs. 3.1, and 3.2) (Tanford 1968; Shimizu and Shimizu 1999). Given the favorable medicinal, chemical and physical qualities of ethanol, perhaps the biggest uncertainty before any patient was treated was if the bulk of the viral load could be reached within the mucus by the alcohol.

Pharmacokinetics of Ethanol

When a new pharmaceutical is evaluated, one of the top priorities is understanding the pharmacokinetics (PK) parameters. In a patient, ethanol pharmacokinetics are defined by absorption, distribution, metabolism, excretion and toxicity (ADMET) and have been studied since the advent of modern clinical trials (Jones 2019; Wilkinson 1980). The reaction of ethanol with acetaldehyde to produce acetic acid, then the intermediate acetyl coenzyme A, results in the production of water and carbon dioxide (Reaction 3.3). This is the primary mechanism (95%) that the body uses to break down ethanol.

$$C_2H_6O \rightarrow C_2H_4O \rightarrow C_2H_4O_2 \rightarrow \text{acetyl} \\ -\text{CoA} \rightarrow 3H_2O + 2CO_2 \quad (3.3)$$

The human body naturally produces ethanol by fermentation, with an average value for an adult being between two and three milliliters per day. Alcohol Dehydrogenase (ADH) metabolizes ethanol to acetaldehyde. When the quantity of ethanol becomes high, typically through consumption, the enzyme Cytochrome P450 (CYP2E1) metabolizes ethanol to acetaldehyde (Peter Guengerich and Avadhani 2018; Cederbaum 2012). The inhalation of ethanol for the treatment of various conditions has been ongoing for decades. For example, the treatment of pulmonary edema demonstrated that ethanol vapor by inhalation was successfully implemented by using a nasal catheter, and the implantation of a positive pressure mask (Luisada et al. 1952). Ethanol inhalation has also been employed to aid patients with alcohol withdrawal syndrome (Zhang et al. 2011).

Ethanol's ability to denature proteins has been studied for decades and is well understood (Lui et al. 2022; Andreadis and Moschakis 2023). Its mechanism of action is to interfere with the

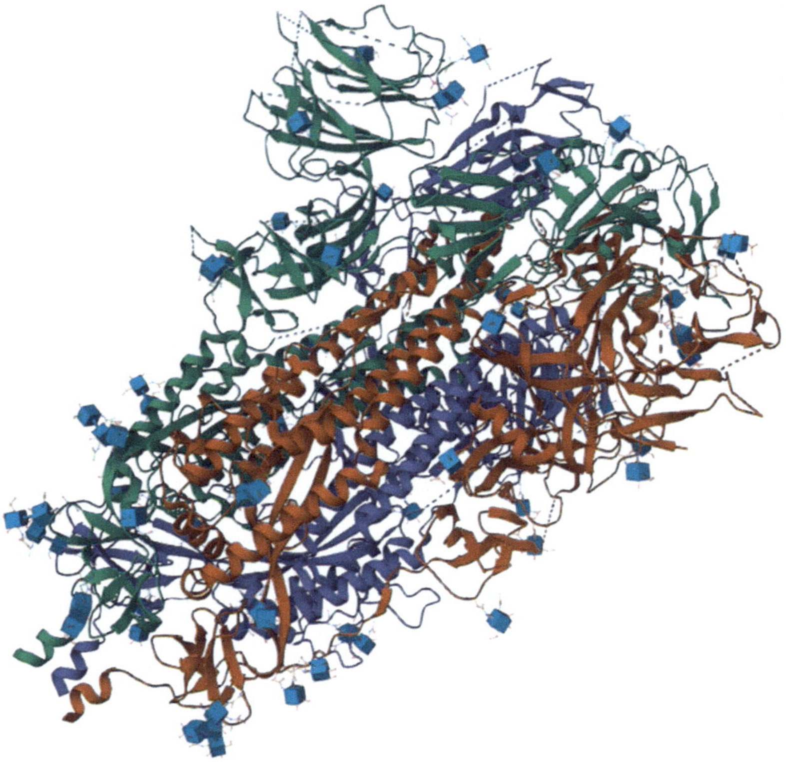

Fig. 3.1 Structure of the SARS-CoV-2 spike glycoprotein, total structure molar mass: 438.26 kDa, atom count: 23,694 (6VXX PDB, closed state)

intramolecular hydrogen bonding between the proteins' side chains. These bonds are critical to keep the tertiary protein structure intact. With the –OH group on ethanol, it is competitive with different functional groups within the protein to be a hydrogen bond donor or a hydrogen bond acceptor. Other alcohols (i.e., methanol, propanol) could also do this but have additional toxicity rendering them useless for in vivo applications. The levels of ethanol and the impact are an approximation and not an exact determination;

(i) Below 4 mmol/L asymptomatic.
(ii) Up to 10 mmol/l behavioral cognitive changes.
(iii) Levels over 10 mmol/L can be considered toxic levels.
(iv) Above 60 mmol/L can result in a coma or death (Lui et al. 2022; Andreadis and Moschakis 2023). The values utilized in inhalation result in serum concentrations significantly lower than are correlated with toxicity and death (Peng et al. 2016).

To illustrate this denaturation process on a small scale, two sets of amino acid sequences that are aligned parallel to each other in the structure of the SARS-CoV-2 spike glycoprotein (closed state, 6VXX, PDB) are utilized for demonstration. They are hydrogen bonded to each other (see Fig. 3.3a). The two parallel molecular structures have a length of 40.2 Angstroms. The ten shortest hydrogen bonds linking the two structures are between 2.79 and 3.56 Angstroms. A nano-droplet that is composed of 256 ethanol molecules is constructed and its energy minimized (see Fig. 3.3b). The primary bonding mechanism that holds the droplet together are the hydrogen bonds involving the R-OH groups in ethanol. The hydrogen bonded peptide chains are inserted into the ethanol nano-droplet, and the energy is minimized. The shortest distance involving the same pairs in

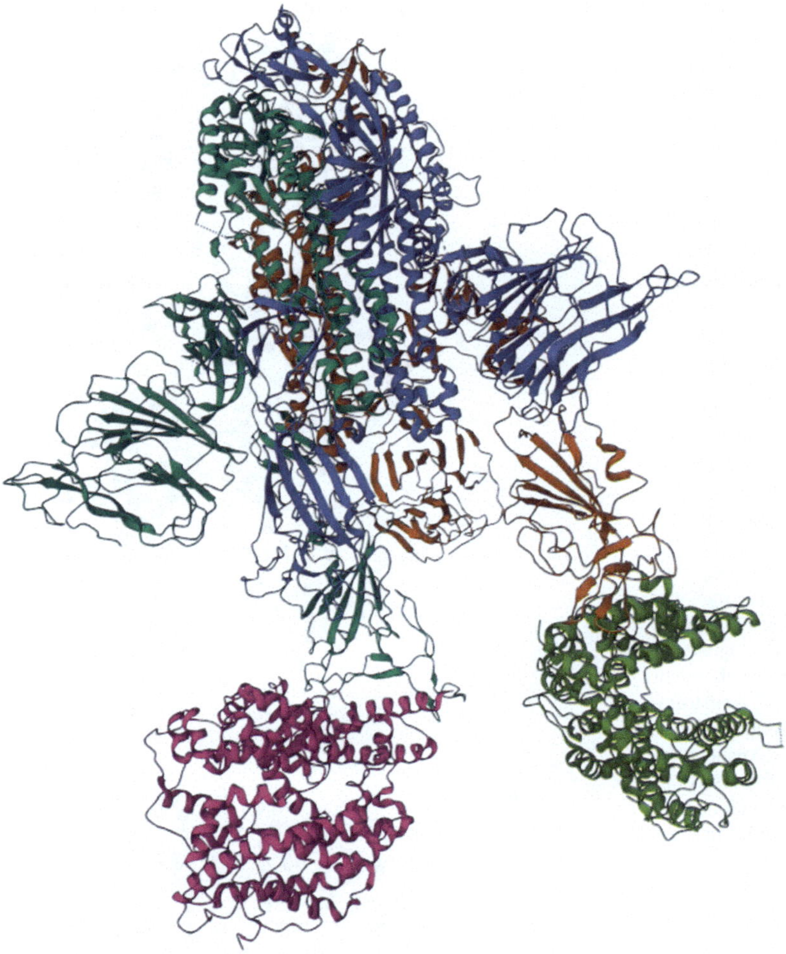

Fig. 3.2 3D image of the SARS-CoV-2 spike glycoprotein with 2 ACE2 bound, total structure weight: 577.91 kDa; atom count: 34,626; (PDB: 7A97)

Fig. 3.3 is approximately 4.72 Angstroms, with several of them extending to over 8 Angstroms, well beyond the distance considered for constituting a hydrogen bond. Hydrogen bonds play a critical role in proteins, maintaining their geometry and subsequently their function (Pace et al. 2014; Derewenda et al. 1995). Other approaches to denaturing the spike protein range from the use of microwaves (2.45 GHz electromagnetic radiation) (Afaghi et al. 2021) to the development of new pharmaceutical agents (Choudhary et al. 2021).

Viral Load

A critical component in evaluating medications for COVID-19 is the primary location of the viral load in patients. Studies indicate that viral loads in the lungs can be correlated with COVID-19 disease severity (Derewenda et al. 1995; Afaghi et al. 2021; Choudhary et al. 2021). The results also indicate that measurements involving SARS-CoV-2 at the time of ICU admission can be correlated with a patient's COVID-19 severity and their final outcome. A numerical presentation by Sender et al. (Afaghi et al. 2021) indicates that an infected human carries between 1 and 100 billion virions during their infection period. This results in a total mass of between one and one hundred micrograms of virion material. They also make estimates that indicate the (i) lungs, (ii) Tonsils, (iii) lymph node, (iv)

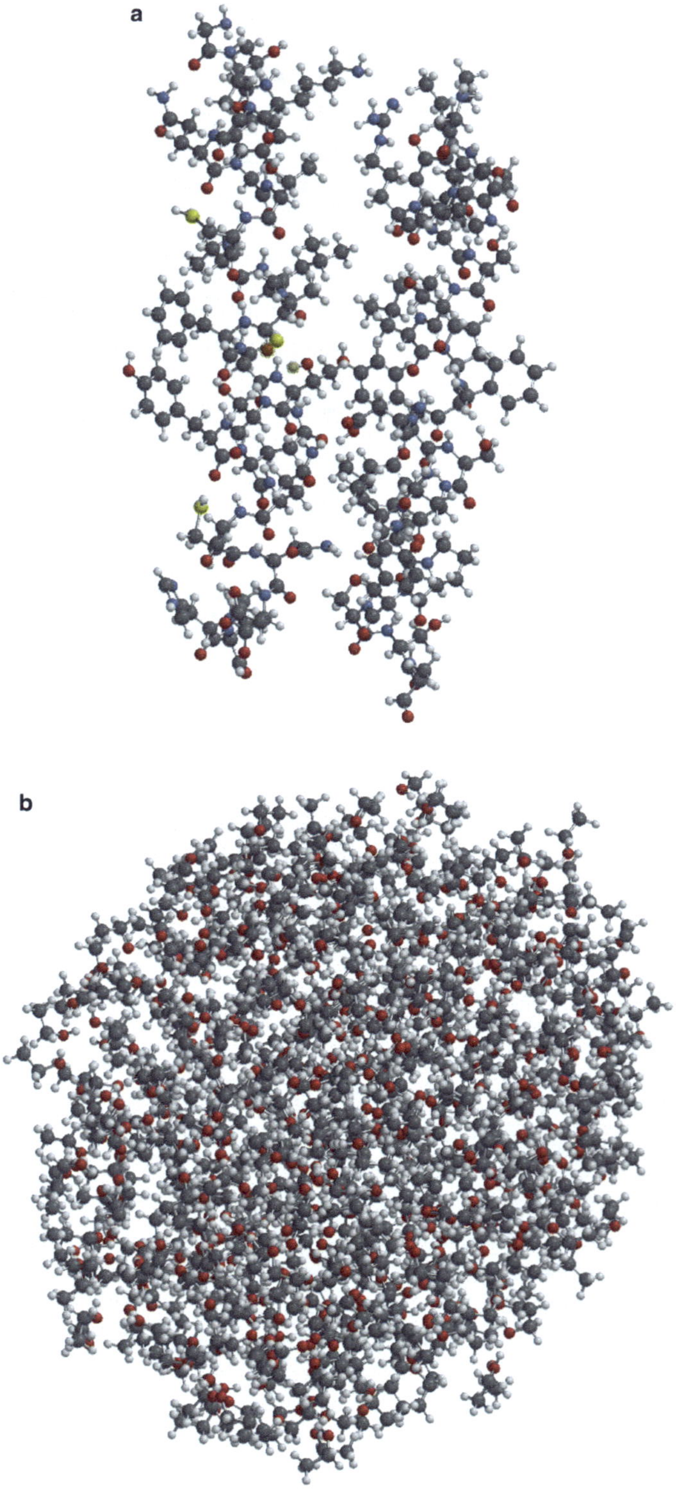

Fig. 3.3 **a** two subsections of the SARS virus spike protein are held together by hydrogen bonding. It has a length of 40.2 Angstroms. (Red = oxygen, gray = carbon; white = hydrogen; purple = nitrogen; yellow = sulfur), **b** A nano-droplet composed of 256 ethanol molecules. The nano-droplet has an average diameter of 41.7 Angstroms,
c This simulation illustrates the protein denaturing process. The two hydrogen bonded peptide chains are inserted into the ethanol nano-droplet and the energy of the complex is minimized. The average diameter is approximately 48.5 Angstroms, **d** Hydrogen bonds, which have a typical length in the 3–3.5 Angstrom range, were frequently increased to a point where the bond no longer existed (red arrow; (example above; the bond above was 2.7 Angstrom in the peptide-peptide structure (Fig. 3.3a), and it expanded to 11.4 Angstrom, after insertion and energy minimization, into 250 ethanol molecules in a nano-droplet

Fig. 3.3 (continued)

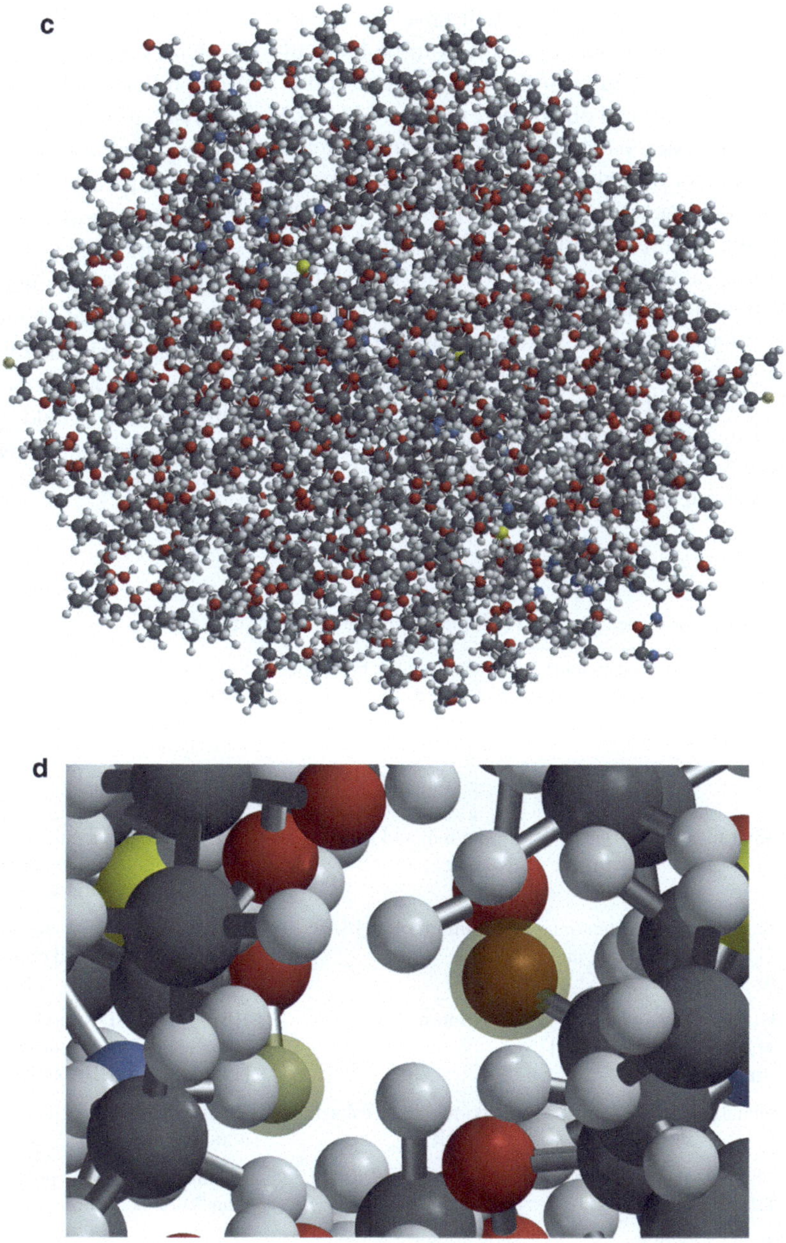

trachea, (v) bronchus, (vi) nasal mucosa, (vii) Pharynx, (viii) digestive tract have the highest to lowest viral loads. The first seven (Tonsils to pharynx) would also be treated by ethanol inhalation.

In an infected individual, the number of infected cells to healthy cells ratio is between 1/100,000 and 1/10,000,000. This would indicate that a direct treatment of the infection, from the mouth to the alveoli, would be the most efficient and effective.

Lipinski Rules and QSAR

Lipinski's rules are a molecular level system applied to small molecules being considered as pharmaceutical agents (Abdullaev et al. 2021; Sender et al. 2021). It is an aid to help in the drug development process to evaluate if a biological active structure will have the needed physical and chemical properties to be bioavailable. Lipinski's rules of five relate to absorption, distribution, metabolism and excretion (ADME). This includes parameters such as molar mass, the number of hydrogen bond donors and hydrogen bond acceptors, water solubility evaluated by the partition coefficient (logP), total polar surface area (TPSA or PSA) and the number of rotatable bonds (single bonds not linked to a terminal atom). The general parameters applied are a partition coefficient (logP) not greater than 5; a TPSA less than 140 A^2; no more than 5 hydrogen bond donors (HBD); no more than 10 hydrogen bond acceptors (HBA); and a molecular mass less than 500 Daltons (see Table 3.1).

The BCS or Biopharmaceutics Classification System divides a molecule into four groups. Lipophilicity indicates the ability of a molecule to dissolve in fats, lipids and non-polar solvents. For our ranking, we believe that more parameters must be considered if the molecule is going to penetrate mucus and still interact with the virus. A molecule (i.e., a fatty acid) could form a structure such as a micelle and have little mobility and minimal chance of interacting with the virus particle in the mucus environment.

There are additional parameters that could be imposed and would provide additional insight into the drug's potential. For example, Ghose filter suggests utilizing molar refractivity values

Table 3.1 Parameters related to Lipinski rules adjusted for inhalation

logP (partition coefficient)	+ 1.6 and lower is passing	Sucrose Water Penicillin G[a] 1-Octanol Hexane Taxol[b]	− 3.71 − 0.29 + 1.82 + 3.18 + 3.66 + 4.95
Molar mass (g/mol)	400 g/mol and lower is passing	Many pharmaceuticals with higher MW's (> 500 Amu's) also have low water solubility and need a special delivery agent to be solubilized	
Hydrogen bond acceptors: –H bonded to N or O, and attracted to N or O	Maximum of 4 HBA's	Vascular endothelial and pulmonary epithelium, among other bio-layers, need to be penetrated by medication to inactivate viral load	
Hydrogen bond donors: –H bonded to N or O, and attracted to N or O	Maximum of 4 HBD's	HBA's and HBD's can increase the polarity (logP) of a molecule but also result in unwanted H-bonds. A balance between HBD/HBA and logP needs to be considered	
Rotatable bonds	Maximum of 4 rotatable bonds (RB)	RB is a single bond (not in ring), linked to non-terminal, non-H atom[c]	
Total polar surface area (TPSA)[d]	TPSA $0 < Å^2 < 140$	If denaturing SARS spike and/or ACE2 proteins, less than 90 $Å^2$ is needed	
Molar refractivity	It is proportional to the polarization of a molecule	Not utilized in this study	
Number of atoms	Most drugs incorporated CNHO and occasionally SFP (all fairly low mass atoms), we used molar mass	The range from 2 to 70 is considered optimal by some researchers. This would include a range of reactive oxidation species (ROS)	

[a] Penicillin G's solubility in water is described as slightly soluble (210 mg/L)

[b] Taxol (Paclitaxel) is injected with polyoxyethylated castor oil to increase water solubility

[c] Acetylsalicylic acid, active ingredient in aspirin, has 3 RB's

[d] Not incorporated in this study, the TPSA/(# atoms in molecule) would likely give a value with predictive power. TPSA values in the 10–110 Å2 range with logP values in the − 1 to 4 range are typically needed for good intestinal absorption, blood–brain barrier (BBB) Solidarity Trial

Table 3.2 This table provides values used in the calculation of the Mucus Penetration Score (MPS). Hexane is an example for a nonpolar molecule, and water is an example for a polar molecule. Some common medicinal agents that are not used to treat COVID, are provided as additional examples

Drug candidate	Structure (2D)	Molar mass (g/mol)	# of atoms	Empirical formula, molar mass (g/mol)	# For Cl atoms	# DB, TP not in terminal group + #aromatic structure	Protease inhibitor values	logP	TPSA (Å²)	HBA	HBD	nrotb	Mechanism of action	Pass/Fail Lipinski/Veter rules	Pass/Fail	Physical-chemical (penetration score)
Hexane (rep. nonpolar)		86.18	20	C_6H_{14}	0	0	– 3.62	3.66	0	0	0	3	• Used as reference for low water solubility (LogP) and inert activity standard	Molar mass: Pass LogP: Pass TPSA: Pass HBA: Pass HBD: Pass nrotb: Pass	Pass	14.58
Water—rep polar	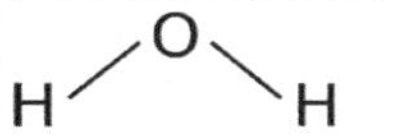	18.01	3	H_2O	0	0	– 3.87	– 0.29	29.27	1	2	0	• Represents solubility value logP for serum and represents inactive medicinal activity	Molar mass: Pass LogP: Pass TPSA: Pass HBA: Pass HBD: Pass nrotb: Pass	Pass	5.99
Remdesivir	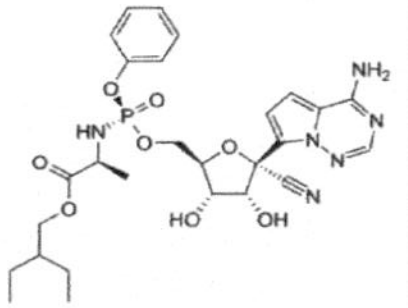	602.6	77	$C_{27}H_{35}N_6O_8P$	0	7	0.49	2.82	204	14	5	14	• Inhibits the SARS-CoV-2 RNA-dependent RNA polymerase (RdRp)	Molar mass: Fail LogP: Pass HBA: Fail HBD: Pass TPSA: Fail nrotb: Fail	Fail	92.21
Favipiravir	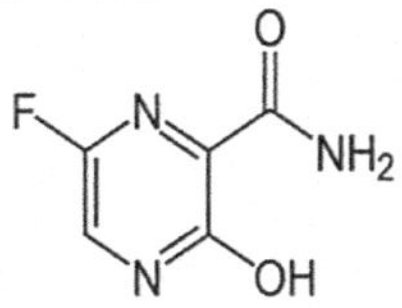	157.1	15	$C_5H_4FN_3O_2$	1	3	– 0.91	– 0.98	88.9	5	3	1	• Targets RNA-dependent RNA polymerase (RdRp)	Molar mass: Pass LogP: Pass HBA: Pass HBD: Pass TPSA: Pass nrotb: Pass	Pass	21.74

(continued)

Table 3.2 (continued)

Drug candidate	Structure (2D)	Molar mass (g/mol)	# of atoms	Empirical formula, molar mass (g/mol)	# For Cl atoms	# DB, TP not in terminal group + #aromatic structure	Protease inhibitor values	logP	TPSA (Å^2)	HBA	HBD	nrotb	Mechanism of action	Pass/Fail Lipinski/Veter rules	Pass/Fail	Physical-chemical (penetration score)
ASC-09 (AKA TMC-310911)	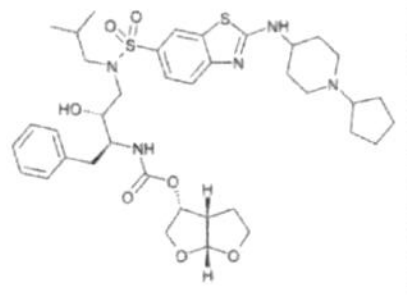	756	105	$C_{38}H_{53}N_5O_7S_2$	0	6	0.18	7.46	143	12	3	15	• Activates $NLRP_3$ inflammasome-mediated IL-1β production • Inhibits protease	Molar mass: Fail LogP: Fail TPSA: Pass HBA: Fail HBD: Pass nrotb: Fail	Fail	106.13
Apabetalone	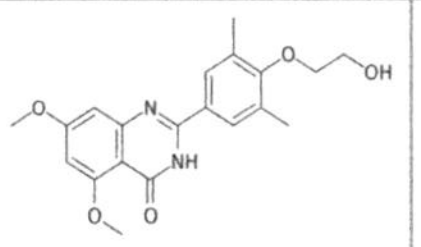	370.4	49	$C_{20}H_{22}N_2O_5$	0	9	− 0.28	3.3	93.7	7	2	6	• Interferes with viral entry • Downregulates ACE2 expression reducing protein levels on host cells	Molar mass: Pass LogP: Pass TPSA: Pass HBA: Pass HBD: Pass nrotb: Pass	Pass	46.23
Lopinavir	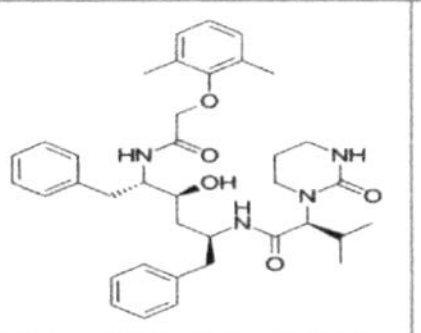	628.8	94	$C_{37}H_{48}N_4O_5$	0	12	0.42	5.69	120	9	4	15	• Inhibits polyprotein cleavage into functional proteins • Binds to site of viral protease activity	Molar mass: Fail LogP: Fail TPSA: Pass HBA: Pass HBD: Pass nrotb: Fail	Fail	86.57
Chloroquine	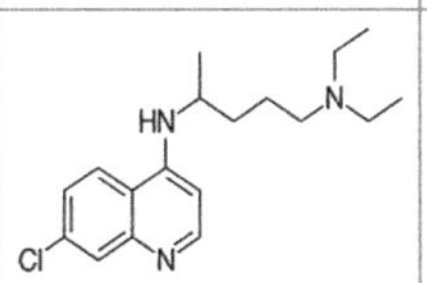	319.9	48	$C_{18}H_{26}ClN_3$	1	6	0.05	5	28.2	3	1	8	• Inhibits terminal glyosylation of ACE2 receptors	Molar mass: Pass LogP: Fail TPSA: Pass HBA: Pass HBD: Pass nrotb: Pass	Fail	39.90
Hydroxychloroqu ine	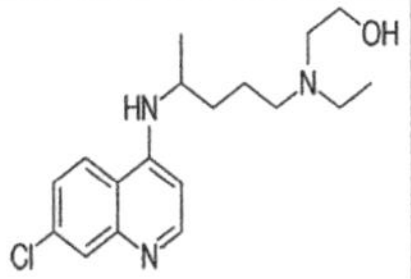	335.9	49	$C_{18}H_{26}C1N_3O$	1	6	0.12	4	48.4	4	2	9	• Restricts pH mediated spike protein cleavage at ACE2 binding site • Interferes with endocytic pathways	Molar mass: Pass LogP: Pass TPSA: Pass HBA: Pass HBD: Pass nrotb: Pass	Pass	44.79

(continued)

Table 3.2 (continued)

Drug candidate	Structure (2D)	Molar mass (g/mol)	# of atoms	Empirical formula, molar mass (g/mol)	# For Cl atoms	# DB, TP not in terminal group + #aromatic structure	Protease inhibitor values	logP	TPSA (Å^2)	HBA	HBD	nrotb	Mechanism of action	Pass/Fail Lipinski/Veter rules	Pass/Fail	Physical-chemical (penetration score)
Ethanol	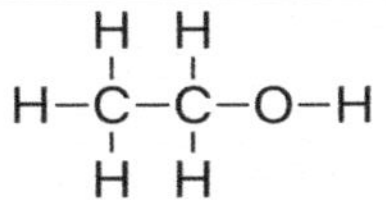	46.07	9	C_2H_6O	0	0	− 3.77	0.06	20.2	1	1	0	• Antiviral due to action of solvent on lipids (pericapsid) and denaturation of proteins (capsid)	Molar mass: Pass LogP: Pass TPSA: Pass HBA: Pass HBD: Pass nrotb: Pass	Pass	8.25
Dexamethasone	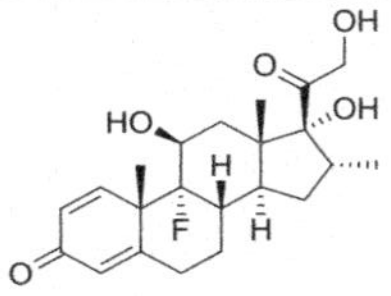	392.5	57	$C_{22}H_{29}FO_5$	0	2	0.76	2.06	94.8	5	3	2	• Inhibits chemokines/pro-inflammatory cytokine release/action	Molar mass: Pass LogP: Pass TPSA: Pass HBA: Pass HBD: Pass nrotb: Pass	Pass	52.12
Ivermectin (B1a)	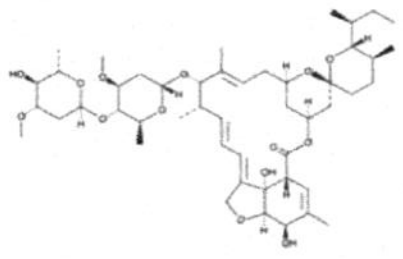	875.1	136	$C_{48}H_{74}O_{14}$	0	4	− 1.89	4.58	170	14	3	8	• Inhibits/disrupts binding of spike protein at ACE-2 receptors	Molar mass: Fail LogP: Pass TPSA: Fail HBA: Fail HBD: Pass nrotb: Pass	No	120.93
Ivermectin (B1b)	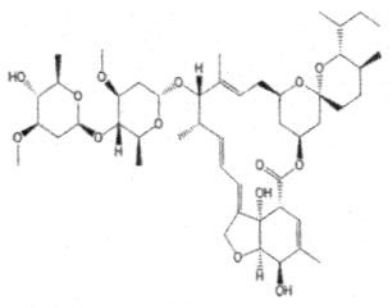	861	133	$C_{47}H_{72}O_{14}$	0	4	− 1.68	4.08	170	14	3	7	• Inhibits/disrupts binding of spike protein at ACE-2 receptors	Molar mass: Fail LogP: Pass TPSA: Fail HBA: Fail HBD: Pass nrotb: Pass	No	118.13

(continued)

Table 3.2 (continued)

Drug candidate	Structure (2D)	Molar mass (g/mol)	# of atoms	Empirical formula, molar mass (g/mol)	# For Cl atoms	# DB, TP not in terminal group + #aromatic structure	Protease inhibitor values	logP	TPSA (Å^2)	HBA	HBD	nrotb	Mechanism of action	Pass/Fail Lipinski/Veter rules	Pass/Fail	Physical-chemical (penetration score)
Prednisone	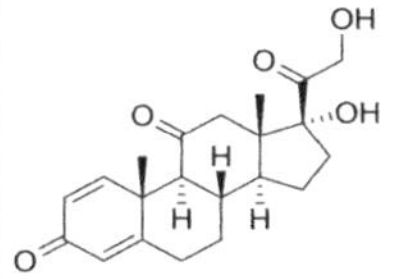	358.4	52	$C_{21}H_{26}O_5$	0	2	0.06	1.41	91.7	5	2	2	• Decrease inflammation and increase capillary permeability	Molar mass: Pass LogP: Pass TPSA: Pass HBA: Pass HBD: Pass nrotb: Pass	Pass	47.43
Methylprednisolone	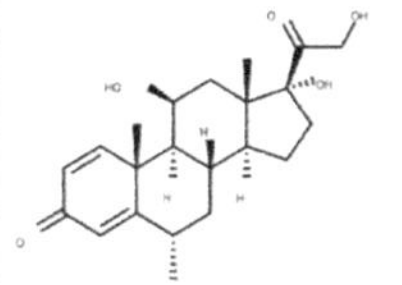	374.5	57	$C_{22}H_{30}O_5$	0	2	0.3	2.07	94.8	5	3	2	• Inhibits transcription	Molar mass: Pass LogP: Pass HBA: Pass HBD: Pass TPSA: Pass nrotb: Pass	Pass	51.12
Hydrocortisone	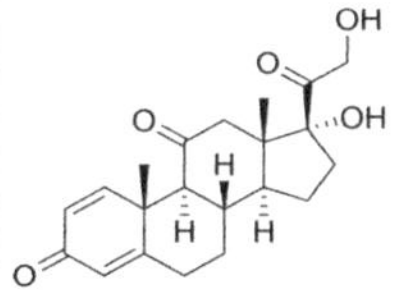	362.5	56	$C_{21}H_{30}O_5$	0	2	0.09	1.62	94.8	5	3	2	• Inhibits phospholipase	Molar mass: Pass LogP: Pass TPSA: Pass HBA: Pass HBD: Pass nrotb: Pass	Pass	50.11
Ritonavir	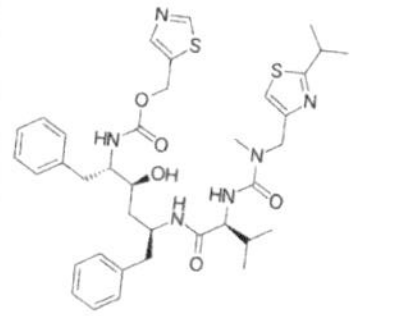	720.9	98	$C_{37}H_{48}N6O_5S_2$	0	10	0.35	7.51	146	11	4	18	• Inhibits protease in the body	Molar mass: Fail LogP: Fail TPSA: Fail HBA: Fail HBD: Pass nrotb: Fail	Fail	100.96

(continued)

Table 3.2 (continued)

Drug candidate	Structure (2D)	Molar mass (g/mol)	# of atoms	Empirical formula, molar mass (g/mol)	# For Cl atoms	# DB, TP not in terminal group + #aromatic structure	Protease inhibitor values	logP	TPSA (Å^2)	HBA	HBD	nrotb	Mechanism of action	Pass/Fail Lipinski/Veter rules	Pass/Fail	Physical-chemical (penetration score)
Metformin	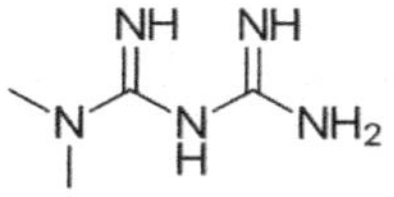	129.2	20	$C_4H_{11}N_5$	0	2	− 1.39	− 1.26	91.5	5	5	2	• Inhibits macrophage activation and contributes to COVID-19 morbidity	Molar mass: Pass LogP: Pass TPSA: Pass HBA: Pass HBD: Pass nrotb: Pass	Pass	26.93
Colchicine	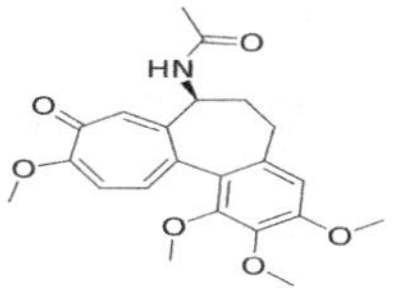	399.4	54	$C_{22}H_{25}NO_6$	0	7	− 0.09	1.1	83.1	7	1	5	• Inhibits inflammasome signaling and cytokine production	Molar mass: Pass LogP: Pass TPSA: Pass HBA: Fail HBD: Pass nrotb: Pass	Fail	49.07
Fluvoxamine	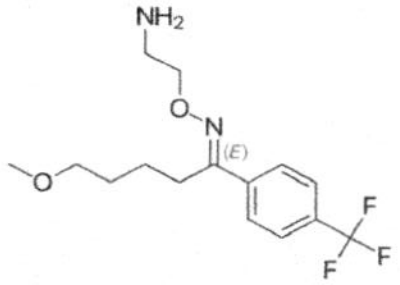	318.3	43	$C_{15}H_{21}F_3N_2O_2$	3	5	0.26	2.53	56.9	4	2	10	• Inhibits viral entry and replication	Molar mass: Pass LogP: Pass TPSA: Pass HBA: Pass HBD: Pass nrotb: Pass	Pass	42.05
Vitamin D (supplement)	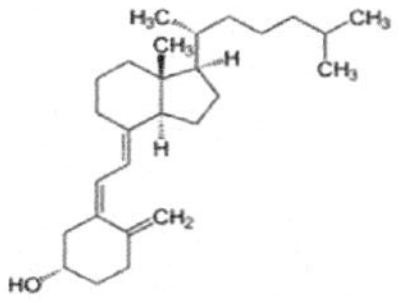	384.6	71	$C_{27}H_{44}O$	0	2	0.12	7.68	20.2	1	1	6	• Blocks differentiation of B cells and immunoglobulin secretion • Inhibits proliferation	Molar mass: Pass LogP: Fail TSPA: Pass HBA: Pass HBD: Pass Nrotb: Pass	Fail	51.98

(continued)

Table 3.2 (continued)

Drug candidate	Structure (2D)	Molar mass (g/mol)	# of atoms	Empirical formula, molar mass (g/mol)	# For Cl atoms	# DB, TP not in terminal group + #aromatic structure	Protease inhibitor values	logP	TPSA (Å^2)	HBA	HBD	nrotb	Mechanism of action	Pass/Fail Lipinski/Veter rules	Pass/Fail	Physical-chemical (penetration score)
Vitamin C (supplement)	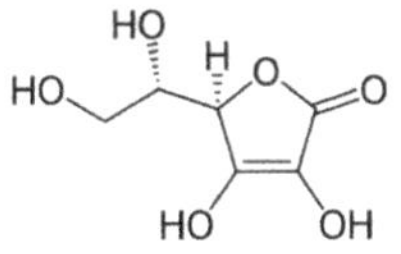	176.1	20	$C_6H_8O_6$	0	1	− 0.81	− 1.4	107	6	4	2	• Stimulates neutrophil migration to the site of infection	Molar mass: Pass LogP: Pass TSPA: Pass HBA: Pass HBD: Pass Nrotb: Pass	Pass	31.08
Quercetin	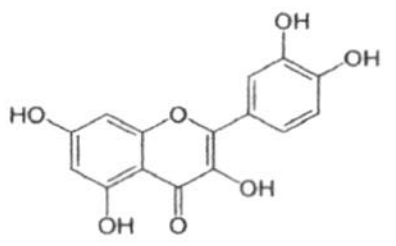	302.2	32	$C_{15}H_{10}O_7$	0	9	− 0.25	1.68	131	7	5	1	• Hinders SARS-CoV-2 protease by producing hydrogen bonds	Molar mass: Pass LogP: Fail TPSA: Pass HBA: Pass HBD: Pass Nrotb: Pass	Fail	36.01
Silymarin	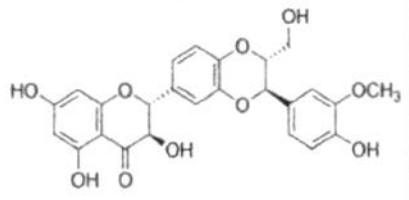	482.4	57	$C_{25}H_{22}O_{10}$	0	12	0.02	1.47	155	10	5	4	• Inhibits the expression of TMPRSS2	Molar mass: Pass LogP: Pass TPSA: Fail HBA: Pass HBD: Pass Nrotb: Pass	Fail	58.17
Luteolin	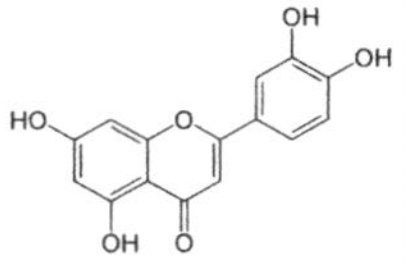	286.2	31	$C_{15}H_{10}O_6$	0	9	− 0.22	1.97	111	6	4	1	• Inhibit the expression of SARS-CoV 3CL protease	Molar mass: Pass LogP: Fail TPSA: Pass HBA: Pass HBD: Pass Nrotb: Pass	Fail	32.11
Baicalin	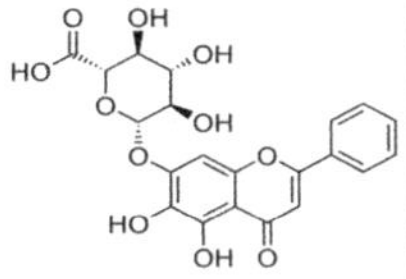	446.4	50	$C_{21}H_{18}O_{11}$	0	9	− 0.05	0.55	187	11	6	4	• Inhibits replication of SARS-CoV-2	Molar mass: Pass LogP: Pass TPSA: Fail HBA: Fail HBD: Fail Nrotb: Pass	Fail	59.89

(continued)

Table 3.2 (continued)

Drug candidate	Structure (2D)	Molar mass (g/mol)	# of atoms	Empirical formula, molar mass (g/mol)	# For Cl atoms	# DB, TP not in terminal group + #aromatic structure	Protease inhibitor values	logP	TPSA (Å^2)	HBA	HBD	nrotb	Mechanism of action	Pass/Fail Lipinski/Veter rules	Pass/Fail	Physical-chemical (penetration score)
EGCG	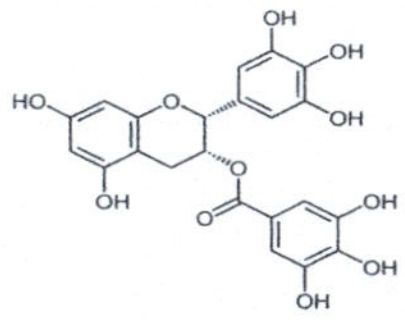	458.4	51	$C_{22}H_{18}O_{11}$	9	3	0.13	2.25	197	11	8	4	• Blocks the primary low-affinity attachment of virions to cells	Molar mass: Pass LogP: Fail TPSA: Fail HBA: Fail HBD: Fail Nrotb: Pass	Fail	60.30
Molnupiravir	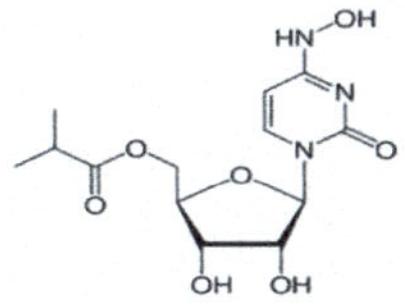	329.3	42	$C_{13}H_{19}N_3O_7$	0	2	0.28	– 0.26	143	10	4	6	• Induces viral RNA mutagenesis	Molar mass: Pass LogP: Pass TPSA: Fail HBA: Pass HBD: Pass nortb: Pass	Fail	55.19
Thioguanosine	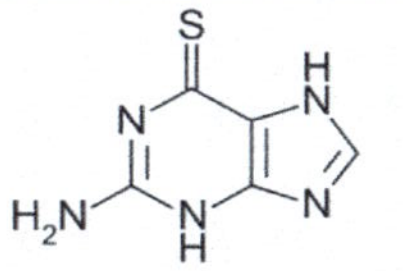	299.3	33	$C_{10}H_{13}N_5O_4S$	0	3	– 0.14	– 1.67	142	6	5	2	• Inhibits SARS-CoV-2 PLpro-catalyzed viral poly protein cleavage • Inhibits replication of SARS-CoV-2 in Vero-E6 cells	Molar mass: Pass LogP: Pass TPSA: Fail HBA: Pass HBD: Pass nortb: Pass	Fail	42.45
Sofosbuvir	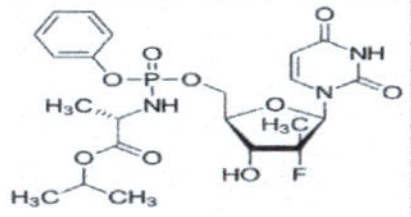	529.5	65	$C_{22}H_{29}FN_3O_9P$	1	5	0.53	0.64	158	12	3	11	• Directs acting antivirals that bind to HCV RdRp • Is incorporated in nascent strand of HCV, preventing addition of next nucleotide	Molar mass: Fail LogP: Pass TPSA: Fail HBA: Fail HBD: Pass nortb: Fail	Fail	76.52

(continued)

Table 3.2 (continued)

Drug candidate	Structure (2D)	Molar mass (g/mol)	# of atoms	Empirical formula, molar mass (g/mol)	# For Cl atoms	# DB, TP not in terminal group + #aromatic structure	Protease inhibitor values	logP	TPSA (Å^2)	HBA	HBD	nrotb	Mechanism of action	Pass/Fail Lipinski/Veter rules	Pass/Fail	Physical-chemical (penetration score)
Baricitinib	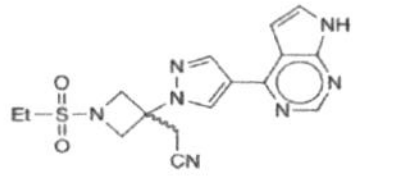	371.4	43	$C_{16}H_{17}N_7O_2S$	0	7	− 0.03	− 0.24	121	9	1	5	• Inhibits signaling of cytokines	Molar mass: Pass LogP: Pass TPSA: Pass HBA: Pass HBD: Pass nrotb: Pass	Pass	47.09
Caffeic acid	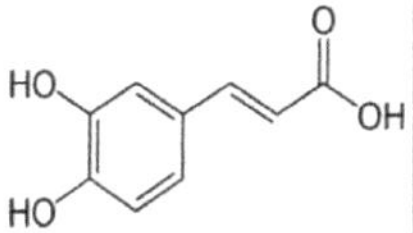	180.2	21	$C_9H_8O_4$	0	4	− 0.79	0.94	77.8	4	3	2	• Potent modulators of COVID-19 and potent binder of SARS-CoV-2	Molar mass: Pass LogP: Pass TPSA: Pass HBA: Pass HBD: Pass nrotb: Pass	Pass	24.70
Bergamottin	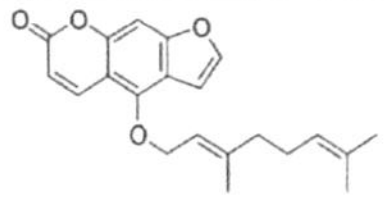	338.4	47	$C_{21}H_{22}O_4$	0	8	− 0.43	5.8	52.6	4	0	6	• Blocks SARS-CoV-2 spike-membrane fusion and reduces viral synthesis	Molar mass: Pass LogP: Fail TPSA: Pass HBA: Pass HBD: Pass nrotb: Pass	Fail	38.46
Phillyrin	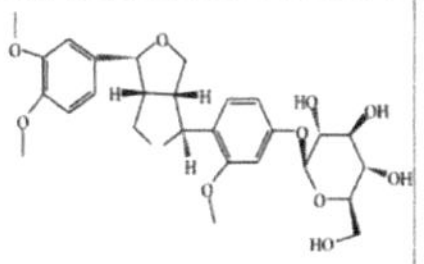	534.6	72	$C_{27}H_{34}O_{11}$	0	8	− 0.05	0.88	146	11	4	8	• Helps regulate immune and anti-inflammatory effects	Molar mass: Fail LogP: Pass TPSA: Fail HBA: Fail HBD: Pass nrotb: Pass	Fail	73.71
Nirmatrelvir	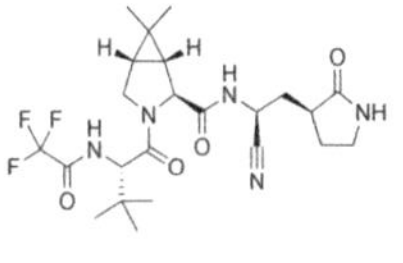	499.5	67	$C_{23}H_{32}F_3N_5O_4$	3	0	1.23	1.66	131	9	3	8	• Viral processing prevention for precursors required for replication	Molar mass: Pass LogP: Pass TPSA: Pass HBA: Pass HBD: Pass nrotb: Pass	Pass	71.64

(continued)

Table 3.2 (continued)

Drug candidate	Structure (2D)	Molar mass (g/mol)	# of atoms	Empirical formula, molar mass (g/mol)	# For Cl atoms	# DB, TP not in terminal group + #aromatic structure	Protease inhibitor values	logP	TPSA (Å^2)	HBA	HBD	nrotb	Mechanism of action	Pass/Fail Lipinski/Veter rules	Pass/Fail	Physical-chemical (penetration score)
Dexamethasone	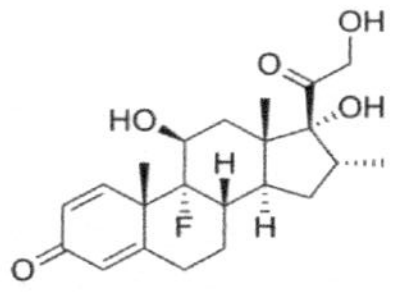	392.5	57	$C_{22}H_{29}FO_5$	1	2	0.76	2.06	94.8	5	3	2	• Soothes hyperinflammatory state caused by SARS-CoV-2	Molar mass: Pass LogP: Pass TPSA: Pass HBA: Pass HBD: Pass nrotb: Pass	Pass	51.12
Artemisinin	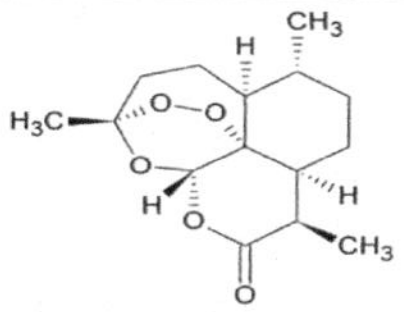	282.3	42	$C_{15}H_{22}O_5$	0	0	– 0.19	3.32	54	5	0	0	• Blocks NF-κB signaling by inhibiting IκB kinase	Molar mass: Pass LogP: Pass TPSA: Pass HBA: Pass HBD: Pass nrotb: Pass	Pass	36.65
Andrographolide	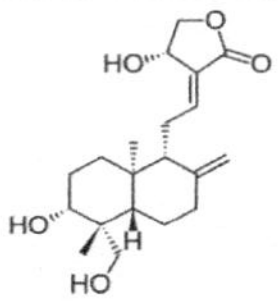	350.4	55	$C_{20}H_{30}O_5$	0	1	0.26	1.05	87	5	3	3	• Blocks TNF-induced cytokine storm	Molar mass: Pass LogP: Pass TPSA: Pass HBA: Pass HBD: Pass nrotb: Pass	Pass	50.81
Glycyrrhizic acid	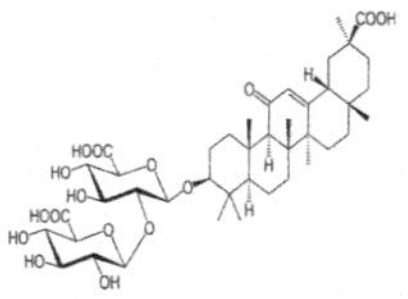	822.9	120	$C_{42}H_{62}O_{16}$	0	1	– 1.26	1.97	267	16	8	7	• Causes disorganization important for SARS-CoV-2 to gain entry into a cell	Molar mass: Fail LogP: Pass TPSA: Fail HBA: Fail HBD: Fail nrotb: Pass	Fail	124.93
Nitazoxanide	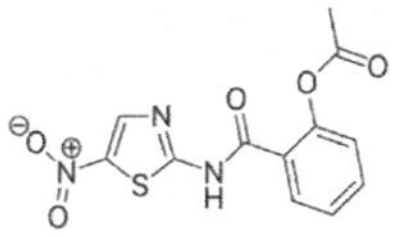	307.3	30	$C_{12}H_9N_3O_5S$	0	6	– 0.66	2.02	114	8	1	5	• Reduces inflammatory reaction associated with Covid-19	Molar mass: Pass LogP: Pass TPSA: Pass HBA: Pass HBD: Pass nrotb: Pass	Pass	39.05

(continued)

Table 3.2 (continued)

Drug candidate	Structure (2D)	Molar mass (g/mol)	# of atoms	Empirical formula, molar mass (g/mol)	# For Cl atoms	# DB, TP not in terminal group + #aromatic structure	Protease inhibitor values	logP	TPSA (Å²)	HBA	HBD	nrotb	Mechanism of action	Pass/Fail Lipinski/Veter rules	Pass/Fail	Physical-chemical (penetration score)
Asapiprant	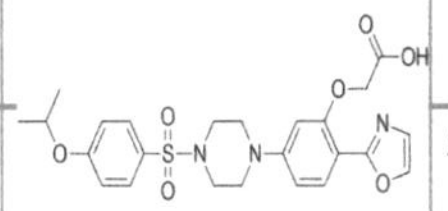	501.6	61	$C_{24}H_{27}N_3O_7S$	0	10	0.13	3.07	122	10	1	9	• Reduces inflammatory responses associated with morbidity	Molar mass: Fail LogP: Pass TPSA: Pass HBA: Pass HBD: Pass nrotb: Pass	Fail	62.50
Itraconazole	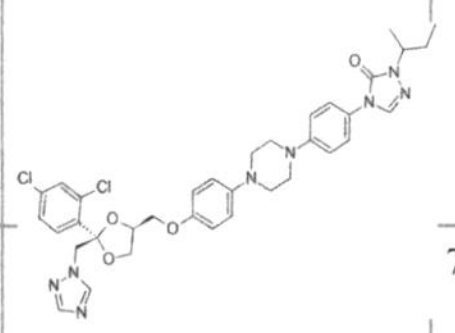	705.7	87	$C_{35}H_{38}Cl_2N_8O_4$	2	15	− 0.66	5.32	105	12	0	11	• Inhibits cytochrome P450-dependent ergosterol	Molar mass: Fail LogP: Fail TPSA: Pass HBA: Fail HBD: Pass nrotb: Fail	Fail	77.97
Chloroquine	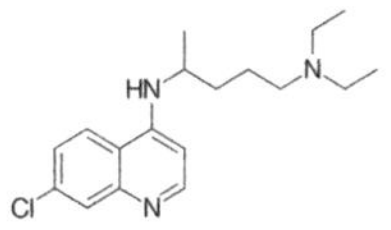	319.9	48	$C_{18}H_{26}ClN_3$	1	3	0.05	5.00	28.2	3	1	8	• Down regulates cytokine production by monocyte-macrophages • Increases pH of acidic organelles, hindering virion transport and assembly	Molar mass: Pass LogP: Pass TPSA: Pass HBA: Pass HBD: Pass nrotb: Pass	Pass	42.90
Emetine	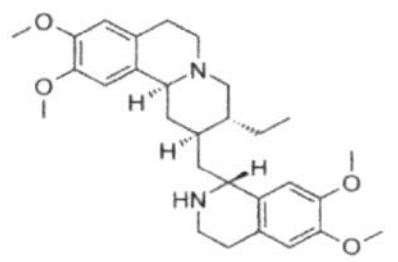	480.7	75	$C_{29}H_{40}N_2O_4$	0	8	0.08	3.64	52.20	6	1	7	• Prevents SARS-CoV-2 mRNA from binding to eIF4E by binding to eIF4E's cap-binding pocket	Molar mass: Pass LogP: Pass TPSA: Pass HBA: Pass HBD: Pass nrotb: Pass	Pass	59.61

(continued)

Table 3.2 (continued)

Drug candidate	Structure (2D)	Molar mass (g/mol)	# of atoms	Empirical formula, molar mass (g/mol)	# For Cl atoms	# DB, TP not in terminal group + #aromatic structure	Protease inhibitor values	logP	TPSA (Å²)	HBA	HBD	nrotb	Mechanism of action	Pass/Fail Lipinski/Veter rules	Pass/Fail	Physical-chemical (penetration score)
Melatonin	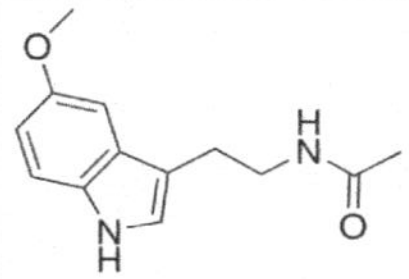	232.3	33	$C_{13}H_{16}N_2O_2$	0	5	– 0.24	1.45	54.1	4	2	4	• Downregulates NF-κB activation in T cells and lung tissue	Molar mass: Pass LogP: Pass TPSA: Pass HBA: Pass HBD: Pass nrotb: Pass	Pass	30.81
GS-441524	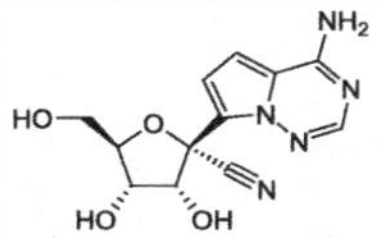	291.3	34	$C_{12}H_{13}N_5O_4$	0	4	0.07	– 0.88	150	9	5	2	• Terminates RNA chain and acts as a alternative substrate	Molar mass: Pass LogP: Fail TPSA: Fail HBA: Pass HBD: Pass nortb: Pass	Fail	44.62
Remedesivir triphosphate	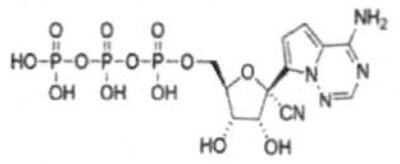	531.2	49	$C_{12}H_{16}N_5O_{13}P_3$	0	4	0.66	– 3.56	290	18	8	8	• Efficiently inhibits RNA-dependent RNA polymerase	Molar mass: Fail LogP: Fail TPSA: Fail HBA: Fail HBD: Fail nrotb: Pass	Fail	85.64
Imatinib	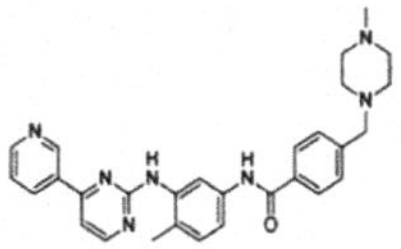	493.6	40	$C_{29}H_{31}N_7O$	0	16	– 0.08	3.89	86.28	8	2	7	• Inhibits tyrosine kinase enzymes	Molar mass: Pass LogP: Fail TPSA: Pass HBA: Fail HBD: Pass nrotb: Fail	Fail	44.84

(continued)

Table 3.2 (continued)

Drug candidate	Structure (2D)	Molar mass (g/mol)	# of atoms	Empirical formula, molar mass (g/mol)	# For Cl atoms	# DB, TP not in terminal group + #aromatic structure	Protease inhibitor values	logP	TPSA (Å^2)	HBA	HBD	nrotb	Mechanism of action	Pass/Fail Lipinski/Veter rules	Pass/Fail	Physical-chemical (penetration score)
Dapagliflozin (Farxiga)	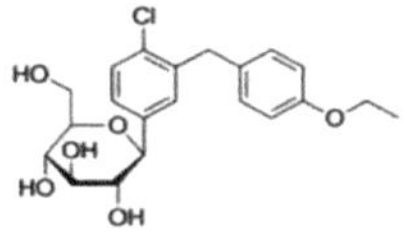	408.8	53	$C_{21}H_{25}ClO_6$	1	8	0.06	2.6	99.4	6	4	6	• Prevents decrease in cytosolic pH and reduces viral load	Molar mass: Pass LogP: Fail TPSA: Pass HBA: Fail HBD: Pass nrotb: Fail	Fail	50.87
Ensitrelvir (Xocova)	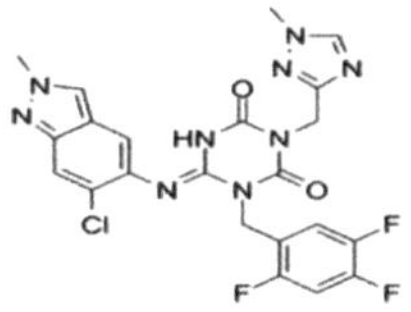	531.8	54	$C_{22}H_{17}ClF_3N_9O_2$	4	10	− 0.17	2.54	117	11	1	6	• Targets 3C-like protease of SARS-CoV-2	Molar mass: Fail LogP: Fail TPSA: Pass HBA: Fail HBD: Pass nrotb: Fail	Fail	55.65
Sabizabulin	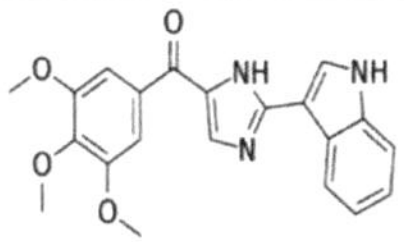	377.4	47	$C_{21}H_{19}N_3O_4$	0	11	− 0.07	3.52	89.3	7	2	6	• Disrupts cytoskeleton	Molar mass: Pass LogP: Fail TPSA: Pass HBA: Fail HBD: Pass nrotb: Fail	Fail	43.81

from 40 to 130 $m^3\ mol^{-1}$. Molecular weights from 80 to 480 g/mol; and the maximum number atoms is 70. In some cases, it is thought that the TPSA (< 140 $Å^2$) and the number of rotatable bonds ($\leq$ 10) are better indicators than molecular weight boundaries. Table 3.2 provides a comparison of various pharmaceuticals and supplements either utilized to treat COVID-19 patients or developed for this application. We developed a relationship that is based on the Lipinski Rules and is adapted to delivering a drug or treatment via inhalation to treat the viral load in the lungs. In some cases, such as bradykinin or cytokine storms, a patient can have a mucus coating in the lung that limits the amount of oxygen entering the serum. This mucus is where the majority of the viral is located.

We propose a physical–chemical-based equation which results in a Penetration Score (PS) to estimate or rank if a molecule might penetrate a mucus (phlegm, sputum) layer. This is an important value for the direct treatment of a COVID-19 patient, particularly one with mucus in their lungs. This value (PS) is calculated for each drug relative to water, which is the dominant molecular species in vivo. For a polar molecule, water is considered to have the best parameters to serve as a model for a molecule to penetrate mucus and surround a virus. Water not only has the ability to penetrate mucus but to also penetrate a protein. For a nonpolar molecule, which has a low solubility in water, hexane is considered as an example (sol. < 9.56 mg/L). Given that mucus can have significant nonpolar components (i.e., long chain lipids and fatty acids), hexane or a small nonpolar molecule could penetrate the mucus. Neither water nor hexane provide any direct medicinal value for selectively treating a viral infection but are used as standards for penetrating the mucus and potentially the protein. Equation 3.4 represents the calculation and parameters utilized in our evaluation.

The physical/chemical values of the pharmaceutical are represented with the subscript "p" (i.e., $\#HBD_p$), while the values for water are represented with subscript "w" ($\#HBD_w$). Because water is the solvent for testing or treatment in vitro and in vivo, all values are normalized against water's QSAR parameters.

Equation 3.4 Physical–Chemical parameters are used to rate the ability of a small molecule to penetrate mucus and interact with a viral particle. It is defined as a Mucus Penetration Score (MPS).

$$\begin{aligned}&(\#HBA/1)^a + (\#HBD/2)^b + (\text{Molar mass}/18)^c \\ &+ (\log P/29.7)^d \\ &+ (\text{nrotb} + 1)/(H_2O + 1)^e + (\text{TPSA}/29.2)^f \\ &+ (\#\text{atoms}/3)^g - (\#F + \#Cl)^h - (\#DB + \#TB)^i \\ &= MPS^k\end{aligned} \tag{3.4}$$

a. #HBA is hydrogen bond acceptor and is divided by 1, the value for water.
b. #HBD is hydrogen bond donors, and is divided by 2, the value for water.
c. Molar mass (g/mol)is the value for the pharmaceutical agent and is divided by 18 g/mol, the value for water. The units g/mol are canceled.
d. logP is the calculated partition coefficient for the pharmaceutical agent, and is divided by 29.7.
e. nrotb is the number of rotatable bonds for the pharmaceutical agent, and since H_2O's value is 0, both numerator and the denominator have a + 1 included to avoid a value of zero.
f. TPSA is the total polar surface area (A^2) of the pharmaceutical agent, and 29.2 A^2 is the value for a water molecule.
g. # of atoms is the total number of atoms in the pharmaceutical agent and does not include species included as salts or any excipients. It is divided by 3, the value for water.
h. #F and #Cl are the number of fluorine and chlorine atoms covalently bound to pharmaceutical agent. Chloride or fluoride in an excipient or salt are not included. The negative value − (#F + #Cl) indicates they are lowering the score and potentially beneficial to the structure.
i. # DB is the number of double bonds in the molecule but excludes those in an aromatic (C_6CH_6) structure, TP is the # of triple bonds not in terminal group, the (#aromatic structure) is C_6H_6 and is considered as a value of 4. The negative value indicates it results in reducing the penetration score to a more favorable value.

j. Protease inhibitor is a calculated value, with water and hexane having values in the −3.7 range and considered to be inert, while a common drug such penicillin G has a value of 0.9 and is considered to have some medicinal activity for specific situations. Some medications are labeled as protease inhibitors, many have other mechanisms of action, including protein denaturation, so this value is not utilized in the equation.
k. The calculated Mucus Penetration score (Physical–Chemical values) provides a relative value that ranks the ability of the molecule to infiltrate the mucus (sputum, phlegm) in a patient's lungs. It is a combination of parameters used individually in the Lipinski and related models. Typically, ethanol applied concentrations lower than 30–40% are considered ineffective, but once it penetrates the infected region, its effective concentration should be lower.

Table 3.1 provides parameters that are applied to small molecules being delivered by an inhalation method to reduce the viral load in the lungs. Equation 3.4 utilizes the chemical and physical parameters of a molecule that model the potential molecular level interactions the pharmaceutical agent can make with the molecular and cellular components of mucus. The lowest values for the Penetration Score (PS) are considered the best candidates. The calculated value identifies molecules that can dissolve in, and penetrate the matrix composed of small and large molecules, cellular debris, minerals, water, large and small strings of molecules that form polymeric strings, etc. The molecule penetrating this matrix should have a low number of hydrogen bond donors and acceptors, a low number of atoms, a low TPSA value that correlates to with a smaller molecule with a low number of electronegative HBA and HBD to minimize interactions with unwanted species, yet at the same time have a logP value that emphasizes water solubility and lipophilic parameters. These parameters only identify the physical and chemical parameters that are desired but do not identify pharmaceutical actives.

Waters' favorable value (5.5) indicates it would penetrate the mucus easily, while the nonpolar hexane (14.2) would also penetrate efficiently. While hexane has some potential advantage (no HBD or HBA, low TPSA, low molar mass), its poor logP and larger size and mass are overriding factors and indicates it would likely aggregate in nonpolar regions such as extracellular lipids and fatty acids.

In some cases, such as bradykinin or cytokine storms, a patient can have a mucus coating in the lung that limits the amount of oxygen entering the serum (Lipinski et al. 1997). This mucus is where the majority of the viral load is located. Research conducted identified three long, stringy species that are in the mucus of COVID-19 patients and determined it impacted their ability to breathe on their own (Lipinski 2000). A significant component of the sputum is high-molecular mass polysaccharides found outside of the cells. Inhalation potentially provides a direct route to deliver the medication with a significantly higher efficiency than a tablet or an IV, considering the viral load in the lungs is key to treatment (Mehta et al. 2017; Liu et al. 2020; Kratochvil et al. 2022).

The physical and chemical parameters needed to penetrate the sputum (Eq. 3.4) coupled with the antiviral activity required for the medication to be effective outline an almost unrealistic requirement for a treatment:

i. Slightly exfoliate, penetrate and dissolve the sputum
ii. Have minimal interactions with the chemical composition of the sputum
iii. Not irritate lung tissue, particularly for patients receiving oxygen (tube or mask) or is intubated
iv. Denature and inactivate the spike protein of viruses in sputum, some entering cells; others involved in the replication process in the cytoplasm.

The first two entries on the table are water and hexane and serve as benchmarks for polar and nonpolar logP values, calculated high and low TPSA, and active and inert for calculated protease inhibitors. The two primary molecules being compared are ethanol and Remdesivir, and the other species have been aligned with COVID-19 as a potential treatment, or tested as a treatment and subsequently ruled ineffective. Ethanol was projected early in the COVID-19 pandemic as a possible treatment due to its in vitro data (Mehta et al. 2017; Liu et al. 2020; Kratochvil et al. 2022). Ethanol's value (5.99) is the lowest of the molecular species that have efficacy against the SARS infection. Its ability to penetrate the mucus coupled with its well-known ability to denature proteins suggests it could be treatment to reduce SARS levels in the lung. Supplements (Zinc(II), Vitamins C) have favorable values and, if needed, would be among the most effective at penetrating the mucus. Given Zinc does not contain many of the parameters (HBA, HBD, rotatable bonds, TPSA), it is not considered.

Other molecular species with favorable chemical and physical characteristics are; Favipiravir, Prednisone, Hydrocortisone, Molnupiravir, Metformin, Thioguanosine, Baricitinib, Caffeic Acid, Artemisinin, Andrographolide, Melatonin, Remdesivir Triphosphate, and GS-441524. Remdesivir (49.54) value suggests it would have difficulty being delivered to a viral load that is contained within mucus.

The small molecules were assigned a *pass* or *fail* grade based on passing all of the parameters outlined in Table 3.1. The values should be examined and the pass/fail not considered a binary yes/no evaluation. For example, Molnupiravir has a TPSA of 143.2 Å slightly above the 140 Å limit so is listed as failing despite having an overall favorable penetration score. If we consider the drug on the surface of the mucus layer and the amount penetrated, there would be an equilibrium ratio that is proportional penetration score established:

$$\text{Penetration Score } \alpha \frac{\text{Molnupiravir (surface)}}{\text{Molnupiravir(mucus)}}$$

The molecules in Table 3.2 with the lowest Penetration Scores are: Water, Ethanol, Hexane, Favipiravir, Caffeic Acid, Metaformin, Melatonin, Vitamin C, Luteolin, and Quercetin. The molecules with the largest or least favorable Penetration Scores (PS) are; Glycyrrhizic acid, Ivermectin B1a, Ivermectin B1b, ASC-09, Ritonavir, Remdesivir, Lopinavir, Remdesivir Triphosphate, Iitraconazole, and Sofosbuvir.

3.2 In Vivo Results

Table 3.3 is the summaries of four medical interventions in which either ethanol was utilized alone or with another medication, or Remdesivir' efficacy was the primary objective of the clinical trial. It is impossible to make a direct and accurate assessment between a large-scale international endeavor such as the solidarity trial and a single physician using a treatment off label simply because nothing else helped his most critical patients (Manning et al. 2020a). The fact that a direct comparison of data cannot be made should not imply that all sets of data should not be judged based on their results, coupled with data (i.e., in vitro) that resulted in its medicinal application. As cited earlier, there is a large body of published data, studies that quantitate ethanol's tremendous efficacy against a range of viral infections, as well as its favorably pharmacokinetic (ADMET) data (Manning et al. 2020b).

The worldwide Solidarity clinical trial was organized by the World Health Organization (WHO) to evaluate various treatments (Manning et al. 2020a). The WHO Solidarity trial enrolled 14,221 consenting adults (≥ 18 years old) hospitalized with COVID-19. The trial ran from March, 2020 to January, 2021. The patients were distributed over 454 hospitals in thirty countries. The trials were implemented at a Phase III–IV level and the selected patients had no contraindication to any of the selected medications. At that time COVID-19 had confirmed cases in Africa, Antarctica, Asia, Europe, North America, Oceania and South America. In addition it was confirmed on cruise ships, maritime vessels and navy ships worldwide. The four medications selected for Solidarity were:

Table 3.3 A summary of applications of Remdesivir and/or ethanol with COVID-19 patients and controls

Drug(s) tested	Dexamethasone + Remdesivir + Nebulized ethanol	Nebulized ethanol (EtOH)	IV Remdesivir (RDV)	Remdesivir (RDV), hydroxychloroquine (HCQ), Lopinavir (LPV) and Interferon-β1a (IFN)
Study type/location	– Randomized, double-blind – Usabn-e-Maryam Hospital, Medical University of Isfahan, Iran	– Randomized controlled trials – Observational; case series – Forrest General Hospital (Harrisburg, Mississippi)	– Randomized, open-label trial – 105 hospital throughout the United States, Asia, and Europe	– Randomized, open-label trial and meta-analyzes – 454 hospitals in 35 countries within all six WHO regions
Name	Efficacy and safety of inhalation of nebulized ethanol in COVID-19 Treatment: a randomized clinical trial	A novel treatment for a novel virus: nebulized 96% ethyl alcohol for COVID-19	Effect of Remdesivir versus Standard care on clinical status at 11 days in patients with moderate COVID-19: a randomized clinical trial	Remdesivir and three other drugs for hospitalized patients with COVID-19: final results of the WHO Solidarity randomized trial and updated meta-analyzes
Dates	September 2021	April 2020–August 2020	March 15, 2020–April 18, 2020	March 22, 2020–January 29, 2021
# of patients	99	66	584	14,221
# of controls	44	44 (decreased to 36)	200	RDV control: 4129 HCQ control: 900 LPV control: 1368 LFN control: 2147
# males in control	18	21	125	RDV: 2639 HCQ: 532 LPV: 800 LFN: 1331
# females in control	37	23	75	RDV: 1490 HCQ: 368 LPV: 568 LFN: 816
Inclusion criteria	– PCR-positive SARS-CoV-2 patients – Oxygen saturation = 90–94% or lung involvement – Informed consent – 12 years of age and older – No pregnancy – No history of asthma, alcoholism or epilepsy – No contraindications to ethanol – No drug use that would interact with ethanol	– Patients with confirmed COVID-19 who required non-intensive care treatment – Consent form – No allergies or serious adverse reaction to alcohol – No history of alcoholism or drug addiction – No liver failure, cirrhosis, infectious hepatitis – No history of pancreatitis, uncontrolled diabetes mellitus, psychosis	– Written informed consent for those 18 and older – Assent for those 12–18 years old – Confirmed SARS-CoV2 via PCR test 4 days prior to randomization – Currently hospitalized for COVID-19 – Radiographic evidence of pulmonary infiltrates	– 18 years or older – Positive for COVID-19 – Not received any of the study drugs – Not expected to be transferred within 72 h

(continued)

Table 3.3 (continued)

Drug(s) tested	Dexamethasone + Remdesivir + Nebulized ethanol	Nebulized ethanol (EtOH)	IV Remdesivir (RDV)	Remdesivir (RDV), hydroxychloroquine (HCQ), Lopinavir (LPV) and Interferon-β1a (IFN)
Exclusion criteria	– Intolerance to inhaled ethanol – Partial/incomplete treatment	– Inability to give informed consent – Refused 2nd ETOH Tx due to cough and/or "burning throat" ($n=3$) – Did not complete 1st ETOH due to extremis, requiring incubation ($n=1$)	– Participation in other experimental COVID-19 trials – Concurrent treatment that contain actual/potential antiviral properties against SARS-CoV-2 – Those needing mechanical ventilation at screening	– Contraindications including CLD, heart disease, or pregnancy – Patients with a refuted COVID-19 diagnosis – No consent encrypted into database
Treatment (control and intervention groups)	– Treated with Intramuscular dexamethasone and Remdesivir	– Received one or more of the following – Remdesivir – Tocilizumab – Dexamethasone – Hydroxychloroquine – Azithromycin – Convalescent plasma – Vitamins C – D3 – Zinc	– All groups continued SOC (standard of care) therapy during trial, though SOC methodology not clear	– Local SOC (standard of care) therapy
Control group (CG) specific parameters	– Sprayed a mask with distilled water three times a day	– Local SCC (standard of care) therapy	– Local SOC (standard of care) therapy	– SOC only, no placebo given
Intervention group (IG) specific parameters	– Sprayed a mask with 35% ethanol three times a day	– One daily nebulized weight-based treatment, over sixty to seventy-five minutes, for a total of three treatments	– 5-day RDV: SOC therapy with IV	– RDV: IV infusion – HCQ: Oral tablet – LPV: Oral tablet – IFN: SQ injection
Health indicators monitored	– Global symptomatic score (GSS): calculates scores of clinical signs and symptoms – Oxygenation status: Monitored using a pulse oximeter – Inflammatory status: variation in serum C-reactive protein (CPR) levels – Seven-point ordinal scale: used to assess clinical condition on day seven (1) Death (2) Hospitalized on an invasive mechanical ventilation (3) Hospitalized on non-invasive ventilation (4) Hospitalized, requiring low-flow oxygen (5) Hospitalized, requiring medical care or supplemental oxygen at home (6) Symptoms of COVID-19 without requiring supplemental oxygen (7) Complete recovery	– Tolerance of treatment – SpO_2 levels – Vital signs before, during, and after treatment – Severity of hypoxemia was measured using the ratio of SpO_2 to FiO_2 – Hypoxemia measured after initiation of treatment and ninety-six hours later	– Seven-point ordinal scale (clinical status scale; CSS): Assessed daily – Adverse events exploratory endpoints: Assessed on days 5, 7, and 11 – Other exploratory endpoints: Duration of hospitalization, Duration of different modes of respiratory support, All-cause mortality	– In-hospital mortality – Disease severity – Initiation of ventilation – Duration of hospital stay – Risk

(continued)

Table 3.3 (continued)

Drug(s) tested	Dexamethasone + Remdesivir + Nebulized ethanol	Nebulized ethanol (EtOH)	IV Remdesivir (RDV)	Remdesivir (RDV), hydroxychloroquine (HCQ), Lopinavir (LPV) and Interferon-β1a (IFN)
Outcomes	– GSS was same at the start of study for both groups – GSS for IG group decreased more rapidly than CG – No significant difference in blood oxygen saturation between the CG and IG – Inflammatory status (CRP) demonstrated IG having an increased rate of reduction – The IG had a 5.7 times higher odds of having a better CSS than CG	– At 96 h after the last treatment, average SER in treatment group was 3.31 – Patients receiving only one or part of one ETOH treatment had a 3% increase in SFR 96 h post treatment – Inpatient mortality was less in the treatment group than in the controls (11% v. 36%) – Average length of stay in the treatment group was five days – Average length of stay in control group was 14 days – All patients (22%) who received ETOH improved, and were discharged – Less need for supplemental oxygen – Decreased progression of disease requiring transfer to ICU – Decreased need for intubation – Decreased length of stay – Improved survival – Decreased need for post-acute hospital care – Treatments were safe since no patient became noticeable intoxicated – Some patients complained about burning of the nasopharynx and throat	– Median length of treatment for those in 5-day RDV group: 5 days – Median length of treatment for those in 10-day RDV group: 6 days – Day 11: patients in 5-day RDV group had statistical higher odds of better clinical status distribution than those receiving SOC – Day 11: no significant difference between the 10-day RDV and SOC groups – Day 28: 9 patients died – Nausea, hypokalemia, and headache were more frequent among RDV-treated versus SOC patients – Patients with moderate COVID-19 in the 10-day RDV course did not have a statistically significant difference in clinical status compared with SOC at 11 days after initiation of treatment – Patients with moderate COVID-19 in the 5-day RDV course had a statistically significant difference in clinical status versus SOC, but it was of uncertain clinical importance	– Hydroxychloroquine, lopinavi, and IFN were discontinued due to futility **Remdesivir results** – 14.5% patients assigned to RDV died versus 15.6% in control died – Ventilated patients: 42.1% in RDV died versus 38.6% in control died – Non-ventilated patients (on oxygen): 14.6% patients in RDV died versus 16.3% in control died – No oxygen patients: 2.9% patients in RDV died versus 3.8% in control died – Non-ventilated initially: 11.9% in RDV died versus 13.5% in control died **Conclusions** – Remdesivir has no effect on already ventilated patients with COVID-19 – Remdesivir has small effect against death or progression to ventilation in other patients

(continued)

Table 3.3 (continued)

Drug(s) tested	Dexamethasone + Remdesivir + Nebulized ethanol	Nebulized ethanol (EtOH)	IV Remdesivir (RDV)	Remdesivir (RDV), hydroxychloroquine (HCQ), Lopinavir (LPV) and Interferon-β1a (IFN)
References	Amoushahi A, Moazam E, Tabatabaei AR, Ghasimi G, Grant-Whyte I, Salvatori P, Ezz AR. Efficacy and Safety of Inhalation of Nebulized Ethanol in COVID-19 Treatment: A Randomized Clinical Trial. Cureus. 2022 Dec 5;14(12):e32218. https://doi.org/10.7759/cureus.32218. PMID: 36505954; PMCID: PMC9728981	Results summarized in: Application to U.S. FDA Division of Pulmonology, Allergy, and Critical Care. FDA Review: Dr. Ladan Jafari; CPMS Chief, Project Management Staff DATE: January 26, 2022Submission by Dr. S. Stogner; Prepared by: Dr. Thomas Manning PIND 156307; Emergency Use Application (EUA)	Spinner CD, Gottlieb RL, Criner GJ, et al. Effect of Remdesivir versus Standard Care on Clinical Status at 11 Days in Patients With Moderate COVID-19: A Randomized Clinical Trial. JAMA. 2020;324(11):1048–1057. https://doi.org/10.1001/jama.2020.16349	Remdesivir and three other drugs for hospitalized patients with COVID-19: Final results of the WHO solidarity randomized trial and updated meta-analyzes. The Lancet (British Edition) 2022 May 21,;399(10339):1941–53

Abbreviations meaning at bottom of table

Abbreviations ALT—alanine aminotransferase, AST—aminotransferase, CG—control group, CPR—C-reactive protein, CSS—clinical status scale, EtOH—ethanol, GSS—global symptomatic score, HCQ—hydroxychloroquine, IFN—interferon-β1a, IG—intervention group, LPV—Lopinavir, PCR—polymerase chain reaction, RDV—IV Remdesivir, RR—mortality rate ratio, SFR—SpO_2 to FiO_2 ratio, SOC—standard of care, ULN—upper limit of normal

Lopinavir: A small molecule treatment that has been used against HIV. It is in the protease inhibitor. It would be combined with another protease inhibitor, ritonavir. It was also found to have efficacy against MERS and HPV.

- Lopinavir patients were treated with two tablets, administrated twice daily for two weeks (56 tablets total). The tablets were composed of 200 mg of Lopinavir and 50 mg of ritonavir.
- Lopinavir trials were discontinued for futility, but trials for Remdesivir continued.

Hydroxychloroquine: A small molecule treatment that has been used for decades to treat malaria. It emerged as resistance to other medications such as quinine and chloroquine emerged worldwide. It has been used to rheumatoid arthritis and lupus.

- Hydroxychloroquine patients were given four tablets containing 155 mg of hydroxychloroquine. The regiment included, at hour 0, four tablets at hour 6, and, beginning at the twenty-second hour, two tablets every twelve hours for ten days. Most tablets used sulfate but some used the chloride salt.
- Hydroxychloroquine trials were discontinued for futility, but trials for Remdesivir continued.

IFN-β1a: Interferon beta-1a is a cytokine in the interferon family used to treat multiple sclerosis. It is a cytokines, which are a group of small proteins (< 25 kDa).

- Patients assigned to IFN-β1a mainly received, via subcutaneous injection, 44 μg on days 0, 3, and 6. A little-used alternative was daily intravenous infusion of 10 μg for

6 days. Where intravenous IFN was available, it was restricted to patients receiving high-flow oxygen, ventilation, or extracorporeal membrane oxygenation.
- IFN-β1a trials were discontinued for futility, but trials for Remdesivir continued.

Remdesivir: It is a small molecule prodrug. It is a broad-spectrum antiviral medication developed to treat hepatitis, and was later tested for the treatment of the Ebola virus and Marburg virus infections (Manning et al. 2020b, 2021; Rusz et al. 2021; Salvatori 2021; Santoro and Carafoli 2021).

- Patients received daily infusions.
- 602 (14.5%) patients given Remdesivir died (4146 patients total for Remdesivir treatment).
- 643 (15.6%) patients in control died. (4129 total patients for control).
- For ventilated patients; 42.1% of the 359 patients died that were given Remdesivir.
- For the ventilated patients; 38.6% assigned to the control.
- For those administrated Remdesivir that were on oxygen but not on ventilation; 14.6% died.
- For those administrated oxygen but no Remdesivir (control), 16.3% died.
- A combined outcome analysis showed that death or ventilation occurred in 19.6% for patients given Remdesivir.
- A combined outcome analysis showed that death or ventilation occurred in 22.5% for patients given the control.

It is recommended that Remdesivir should be administered where hypersensitivity reactions can be managed by trained medical professionals. It can also result in an increase in prothrombin, gastrointestinal symptoms and an elevated transaminase levels (Salvatori 2021; Santoro and Carafoli 2021).

Ethanol: access to the viral load in the lungs involves two steps; the inhalation and the ability to penetrate the mucus. → Inhalation → penetrate mucus. When it interacts with the virus, its primary target is the spike protein on the surface, which eliminates any inefficiency that would be encountered by a drug that has to penetrate the virus to interact with genetic material, or if it has to enter a cell to interfere with the replication processes. This abbreviated route coupled with ethanol's PC properties, makes the efficiency, on a per molecule basis, very high.

Remdesivir: an injection is given into a vein by a health care professional. This process includes distribution into the entire circulatory system, diluting the drug and resulting in some being excreted in bile, via kidney function and being found in feces (Wang et al. 2020). In addition the drug has to penetrate the Pulmonary Epithelial Lining Fluid, which will lower the quantity reaching the lungs. At that point, its penetration score will introduce another inefficiency, lowering the efficacy of the drug. While many drugs overcome barriers such as these by floating the patient with excess drugs which results in more side effects such as back pain, chest tightness, chills, cough, dark-colored urine, swallowing, fast heartbeat, fever, headache, hives, itching, light-colored stools, nausea and vomiting, puffiness\swelling of the eyes, face, lips, or tongue, stomach pain, trouble breathing, unusual tiredness or weakness and/or yellow eyes or skin. Remdesivir in vitro data against coronaviruses (MERS-CoV, SARS-CoV, and SARS-CoV-2) is very encouraging (Veklury 2022), however because of some of the reasons outlined above, it is less effective in vivo (human patients).

Remdesivir (Veklury) is a small molecule broad-spectrum antiviral developed by Gilead Sciences (Takashita et al. 2022; Veklury 2022). It was originally developed to treat hepatitis, followed by testing against Ebola. It does have in vitro activiy against Filoviridae, Ebola, Marburg, Paramyxoviridae, Nipah, Hendra, Parainfluenza type 3, Measles, Pneumoviridae Respiratory syncytial virus, Human metapneumovirus, Coronaviridae, HCoV-OC43, HCoV-229E, HCoV-NL63, MERS-CoV, SARS-CoV and SARS-CoV-2 (Institute and of Diabetes and Digestive and Kidney Diseases 2012). Remdesivir is a nucleoside analog used to inhibit the action of RNA polymerase and would also have to penetrate the cell membrane to interfere with the viral replication process, again lowering its efficacy but reducing the number of molecules present during the viral replication process. Remdesivir is also a prodrug that transforms into GS-441524 triphosphate. It is not clear if the additional interactions occurring during the serum-lung barrier and the mucus composition can interfere with the molecule to prodrug transition. Remdesivir was the primary pharmaceutical agent in the Solitary Trials organized by the World Health Organization.

While other repurposed pharmaceuticals were tested, some were quickly removed. The World Health Organization did release a statement entitled: *COVID-19: Remdesivir has little or no impact on survival, WHO trial shows* (Institute and of Diabetes and Digestive and Kidney Diseases 2012). It would later be published that there were some small statistical successes with Remdesivir data such as 14.5% of patients died (602/4146) that were on the medication, while 15.6% (643/4129) that were controls died.

Table 3.3 provides a summary of four trials related to the evaluation of Remdesivir and/or ethanol. The Solidarity trial involve the greatest number of participants and countries. It had two goals; (i) too eliminate specific medications that had no efficacy in vivo. Hydroxychloroquine, a malaria drug, was considered controversial but its medicinal logic was an immunomodulatory effect that would lower or eliminate the production of cytokines. In the solidarity plus trial Artesunate, Imatinib and Infliximab were tested. When cytokines are at significant levels, it can produce serious medical conditions due to the synthesis for proinflammatory cytokines, such as IFN-γ, TNF-α. IL-1, IL-6 and IL-12. In vitro antiviral activity vis-à-vis SARS-CoV-2, HCQ is more potent and less toxic (Institute and of Diabetes and Digestive and Kidney Diseases 2012; Deb et al. 2021; Frediansyah et al. 2021).

Table 3.3: A summary of key points for four different applications involving Remdesivir and/or ethanol. At the bottom of the table is a reference to location and reference.

3.3 Conclusion

As of March of 2023, worldwide there have been almost 7 million deaths and almost 700 million verified cases. While new vaccines were released, new strains emerged. The development of new molecules, with the goal being a cure for the viral infection, emerged and none truly succeeded (Dyer 2020; Eastman et al. 2020; Khuroo 2020; Ynga-Durand et al. 2022).

The evaluation of forty-seven molecules included QSAR type calculations that evaluated physical and chemical parameters of the small molecule pharmaceuticals that had some connection, from Phase III results in a clinical trial down to a preliminary synthesis and some preclinical data. For ethanol, the in vitro and in vivo data strongly suggest its application as a treatment for COVID-19, via an inhalation approach, should be pursued. Remdesevir had very encouraging pre-clinical results, but its clinical results were not as encouraging. Remdesevir's Penetration Score clearly indicates its ability to penetrate membranes, a mucus layer and perhaps lung tissue, negatively impacts its ability to reduce the viral load in the lungs of the COVID patient.

Figure 3.4 provides a visualization of lung cells covered in mucus. The penetration score equation outlined in this work not only explains why some molecules with encouraging in vitro trials did not perform well in COVID-19 patients, it can also provide insight to how those scores can be improved making the medication more effective.

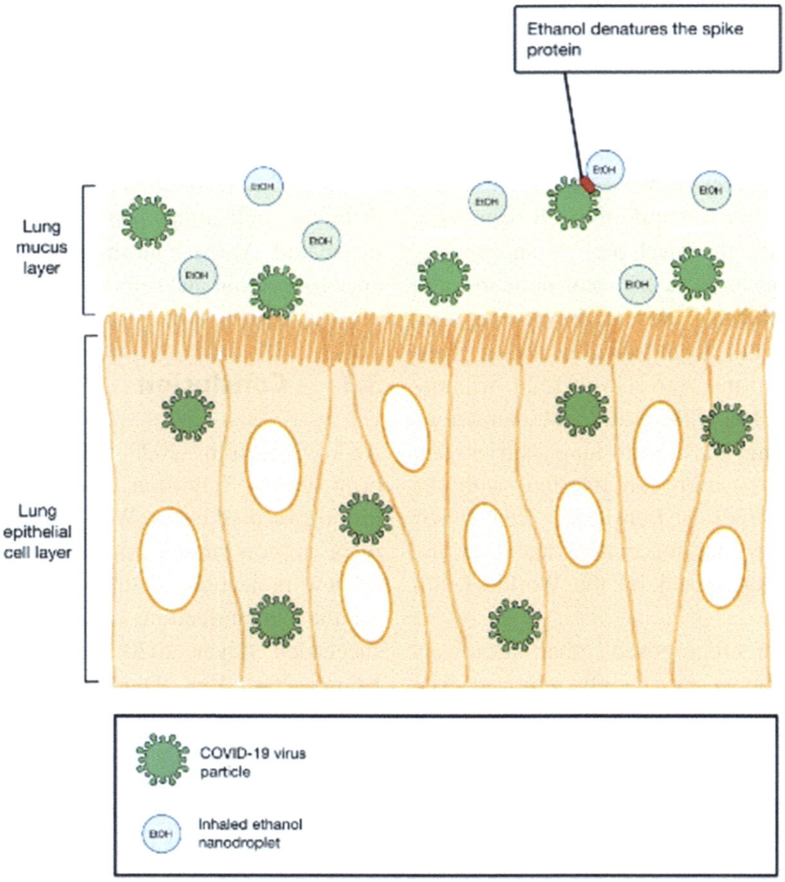

Fig. 3.4 The images outlines the path a medication must take if it is to inactivate virus particles in mucus, at the surface of cells and inside cells where viral replication will take place. Ethanol can be inhaled as a nano-droplet but disperse in the mucus as individual molecules

Acknowledgements We would like to thank our many interactions with Dr. Stephen Stogner who utilized ethanol with his patients in Mississippi. The chemistry department at Valdosta State University is thanked for the use of facilities. Dr. Jose Vasquez of the Medical College of Georgia (Augusta) is thanked for the many discussions related to COVID-19%.

References

Abdullaev A, Odilov A, Ershler M, Volkov A, Lipina T, Gasanova T, Lebedin Y, Babichenko I, Sudarikov A (2021) Viral load and patterns of SARS-CoV-2 dissemination to the lungs, mediastinal lymph nodes, and spleen of patients with COVID-19 associated lymphopenia. Viruses 13:1410. https://doi.org/10.3390/v13071410

Afaghi P, Lapolla MA, Ghandi K (2021) Denaturation of the SARS-CoV-2 spike protein under non-thermal microwave radiation. Sci Rep 11:23373. https://doi.org/10.1038/s41598-021-02753-7

Andreadis M, Moschakis T (2023) Effect of ethanol on gelation and microstructure of whey protein gels in the presence of NaCl. Food Hydrocolloids 134:107985

Arora P, Rocha C, Kempf A, Nehlmeier I, Graichen L, Winkler MS, Lier M, Schulz S, Jäck HM, Cossmann A, Stankov MV (2021) The spike protein of SARS-CoV-2 variant A. 30 is heavily mutated and evades vaccine-induced antibodies with high efficiency. Cell Mol Immunol 18:2673–2675. https://doi.org/10.1038/s41423-021-00779-5

Buchwald P (2022) Developing small-molecule inhibitors of protein-protein interactions involved in viral entry as potential antivirals for COVID-19. Front Drug Discov https://doi.org/10.3389/fddsv.2022.898035

Cederbaum AI (2012) Alcohol metabolism. Clin Liver Dis 16:667–685. https://doi.org/10.1016/j.cld.2012.08.002

Chien M, Anderson TK, Jockusch S, Tao C, Li X, Kumar S, Russo JJ, Kirchdoerfer RN, Ju J (2020) Nucleotide analogues as inhibitors of SARS-CoV-2 polymerase, a key drug target for COVID-19. J Proteome Res 19(11):4690–4697

Choudhary V, Gupta A, Sharma R, Parmar HS (2021) Therapeutically effective covalent spike protein inhibitors in treatment of SARS-CoV-2. J Proteins Proteom 12:257–270. https://doi.org/10.1007/s42485-021-00074-x

Deb S, Reeves AA, Hopefl R, Bejusca R (2021) ADME and pharmacokinetic properties of Remdesivir: its drug interaction potential. Pharmaceuticals (basel) 14:655. https://doi.org/10.3390/ph14070655

Derewenda ZS, Lee L, Derewenda U (1995) The occurrence of C-H...O hydrogen bonds in proteins. J Mol Biol 252:248–262. https://doi.org/10.1006/jmbi.1995.0492

Dyer O (2020) Covid-19: Remdesivir has little or no impact on survival, WHO trial shows. BMJ 371:m4057. https://doi.org/10.1136/bmj.m4057

Eastman RT, Roth JS, Brimacombe KR, Simeonov A, Shen M, Patnaik S, Hall MD (2020) Remdesivir: a review of its discovery and development leading to emergency use authorization for treatment of COVID-19. ACS Cent Sci 6:672–683. https://doi.org/10.1021/acscentsci.0c00489

Fischer RJ, Morris DH, van Doremalen N, Sarchette S, Matson MJ, Bushmaker T, Yinda CK, Seifert SN, Gamble A, Williamson BN, Judson SD (2020) Assessment of N95 respirator decontamination and re-use for SARS-CoV-2. medrxiv. https://doi.org/10.1101/2020.04.11.20062018

Frediansyah A, Nainu F, Dhama K, Mudatsir M, Harapan H (2021) Remdesivir and its antiviral activity against COVID-19: a systematic review. Clin Epidemiol Glob Health 9:123–127. https://doi.org/10.1016/j.cegh.2020.07.011

Freire E, Schön A, Hutchins BM, Brown RK (2013) Chemical denaturation as a tool in the formulation optimization of biologics. Drug Discov Today 18(19–20):1007–1013. https://doi.org/10.1016/j.drudis.2013.06.005

Greene RF, Pace CN (1974) Urea and guanidine hydrochloride denaturation of ribonuclease, lysozyme, alpha-chymotrypsin, and beta-lactoglobulin. J Biol Chem 249:5388–5393

Henwood AF (2020) Coronavirus disinfection in histopathology. J Histotechnol 43:102–104. https://doi.org/10.1080/01478885.2020.1734718

Huang Y, Xiao S, Song D, Yuan, Z (2022) Evaluating the virucidal activity of four disinfectants against SARS-CoV-2. Am J Infect Control 50:319–324. https://doi.org/10.1016/j.ajic.2021.10.035

Jackson CB, Farzan M, Chen B, Choe H (2022) Mechanisms of SARS-CoV-2 entry into cells. Nat Rev Mol Cell Biol 23(1):3–20

Jones AW (2019) Alcohol, its absorption, distribution, metabolism, and excretion in the body and pharmacokinetic calculations. Wires Forensic Sci 1:e1340. https://doi.org/10.1002/wfs2.1340

Kampf G (2020) Potential role of inanimate surfaces for the spread of coronaviruses and their inactivation with disinfectant agents. Infect Prev Pract 2:100044. https://doi.org/10.1016/j.infpip.2020.100044

Kampf G, Todt D, Pfaender S, Steinmann E (2020) Persistence of coronaviruses on inanimate surfaces and their inactivation with biocidal agents. J Hosp Infect 104:246–251. https://doi.org/10.1016/j.jhin.2020.01.022

Khuroo MS (2020) Chloroquine and hydroxychloroquine in coronavirus disease 2019 (COVID-19). Facts, fiction and the hype: a critical appraisal. Int J Antimicrob Agents 56:106101. https://doi.org/10.1016/j.ijantimicag.2020.106101

Kratochvil MJ, Kaber G, Demirdjian S, Cai PC, Burgener EB, Nagy N, Barlow GL, Popescu M, Nicolls MR, Ozawa MG, Regula DP, Pacheco-Navarro AE, Yang S, de Jesus Perez VA, Karmouty-Quintana H, Peters AM, Zhao B, Buja ML, Johnson PY, Vernon RB, Wight TN, Stanford COVID-19 Biobank Study Group, Milla CE, Rogers AJ, Spakowitz AJ, Heilshorn SC, Bollyky PL (2022) Biochemical, biophysical, and immunological characterization of respiratory secretions in severe SARS-CoV-2 infections. JCI Insight 7:e152629. https://doi.org/10.1172/jci.insight.152629

Lipinski CA (2000) Drug-like properties and the causes of poor solubility and poor permeability. J Pharm Tox Meth 44(1):235–249

Lipinski CA, Lombardo F, Dominy BW, Feeney PJ (1997) Experimental and computational approaches to estimate solubility and permeability in drug discovery and development settings. Adv Drug Deliv Rev 23(1–3):3–25

Liu R, Yi S, Zhang J, Lv Z, Zhu C, Zhang Y (2020) Viral load dynamics in sputum and nasopharyngeal swab in patients with COVID-19. J Dent 99:1239–1244. https://doi.org/10.1177/0022034520946251

Lui LH, Egbu R, Graver T, Williams GR, Brocchini S, Velayudhan A (2022) Computational and experimental evaluation of the stability of a GLP-1-like peptide in ethanol-water mixtures. Pharmaceutics 14:1462. https://doi.org/10.3390/pharmaceutics14071462

Luisada AA, Goldmann MA, Weyl R (1952) Alcohol vapor by inhalation in the treatment of acute pulmonary edema. Circulation 5(3):363–369. https://doi.org/10.1161/01.cir.5.3.363

Malik JA, Ahmed S, Mir A, Shinde M, Bende O, Alshammari F, Ansari M, Anwar S (2022) The SARS-CoV-2 mutation versus vaccine effectiveness: new opportunities to new challenges. J Infect Public Health. https://doi.org/10.1016/j.jiph.2021.12.014

Manning TJ, Thomas-Richardson J, Cowan M, Beard T (2020a) Vaporization, bioactive formulations and a marine natural product: different perspectives on antivirals. Drug Discovery Today 25(6):956–958

Manning TJ, Thomas-Richardson J, Cowan M, Thomas-Richardson G (2020b) Should ethanol be considered a treatment for COVID-19? Rev Assoc Med Bras 66(9):1169–1171
Manning T, Thomas-Richardson J, Stogner SM (2021) A three-component solvent for the treatment for COVID-19: is the data in place for clinical trial? https://doi.org/10.25159/UnisaRxiv/000003.v1
Martins CV, Xavier CS, Cobrado L (2022) Disinfection methods against SARS-CoV-2: a systematic review. J Hosp Infect 119:84–117. https://doi.org/10.1016/j.jhin.2021.07.014
Mehta MU, Uppoor RS, Conner DP, Seo P, Vaidyanathan J, Volpe DA, Stier E, Chilukuri D, Dorantes A, Ghosh T, Mandula H (2017) Impact of the US FDA "biopharmaceutics classification system" (BCS) guidance on global drug development. Mol Pharm 14(12):4334–4338
National Institute of Diabetes and Digestive and Kidney Diseases (2012) LiverTox: clinical and research information on drug-induced liver injury. National Institute of Diabetes and Digestive and Kidney Diseases
National institute of Health (2020) NIH safety guidance for return to physical workplace. https://oamp.od.nih.gov/sites/default/files/DGS/Return-to-Work-Guidance.pdf. Accessed 31 Jan 2023
Pace CN, Fu H, Lee Fryar K, Landua J, Trevino SR, Schell D, Thurlkill RL, Imura S, Scholtz JM, Gajiwala K, Sevcik J, Urbanikova L, Myers JK, Takano K, Hebert EJ, Shirley BA, Grimsley GR (2014) Contribution of hydrogen bonds to protein stability. Protein Sci 23:652–661. https://doi.org/10.1002/pro.2449
Peng SH, Hsu SY, Kuo PJ, Rau CS, Cheng YA, Hsieh CH (2016) Influence of alcohol use on mortality and expenditure during hospital admission: a cross-sectional study. BMJ Open 6(11):e013176
Peter Guengerich F, Avadhani NG (2018) Roles of cytochrome P450 in metabolism of ethanol and carcinogens. Adv Exp Med Biol 1032:15–35. https://doi.org/10.1007/978-3-319-98788-0_2
Pure and Applied Chemistry (1992) Glossary for chemists of terms used in biotechnology. https://publications.iupac.org/pac/64/1/0143/index.html. Accessed 31 Jan 2023
Pure and Applied Chemistry (1994) Glossary of bioanalytical nomenclature—part 1: general terminology, body fluids, enzymology, immunology. https://publications.iupac.org/pac/66/12/2587/index.html. Accessed 31 Jan 2023
Rath SL, Kumar K (2020) Investigation of the effect of temperature on the structure of SARS-CoV-2 spike protein by molecular dynamics simulations. Front Mol Biosci 7:583523. https://doi.org/10.3389/fmolb.2020.583523
Rusz CM, Ősz BE, Jîtcă G, Miklos A, Bătrînu MG, Imre S (2021) Off-label medication: from a simple concept to complex practical aspects. Int J Environ Res Public Health 18:10447. https://doi.org/10.3390/ijerph181910447
Salvatori P (2021) The rationale of ethanol inhalation for disinfection of the respiratory tract in SARS-CoV-2-positive asymptomatic subjects. Pan Afr Med J 40:201. https://doi.org/10.11604/pamj.2021.40.201.31211
Santoro MG, Carafoli E (2021) Remdesivir: from Ebola to COVID-19. Biochem Biophys Res Commun 538:145–150. https://doi.org/10.1016/j.bbrc.2020.11.043
Sender R, Bar-On YM, Gleizer S, Bernshtein B, Flamholz A, Phillips R, Milo R (2021) The total number and mass of SARS-CoV-2 virions. Proc Natl Acad Sci 118:e2024815118. https://doi.org/10.1073/pnas.2024815118
Shimizu S, Shimizu K (1999) Alcohol denaturation: thermodynamic theory of peptide unit solvation. J Am Chem Soc 121:2387–2394. https://doi.org/10.1021/ja982560s
Takashita E, Kinoshita N, Yamayoshi S, Sakai-Tagawa Y, Fujisaki S, Ito M, Iwatsuki-Horimoto K, Halfmann P, Watanabe S, Maeda K, Imai M, Mitsuya H, Ohmagari N, Takeda M, Hasegawa H, Kawaoka Y (2022) Efficacy of antiviral agents against the SARS-CoV-2 Omicron subvariant BA.2. N Engl J Med 386:1475–1477. https://doi.org/10.1056/NEJMc2201933
Tanford C (1968) Protein denaturation. Adv Protein Chem 23:121–282. https://doi.org/10.1016/S0065-3233(08)60401-5
(a) Remdesivir (Veklury) [package insert]. Gilead Sciences. 2022. Available at: https://www.gilead.com/-/media/files/pdfs/medicines/covid-19/veklury/veklury_pi.pdf. (b) Veklury (2023) https://www.vekluryhcp.com/dosing-and-admin/
Wang Y, Zhang D, Du G, Du R, Zhao J, Jin Y, Fu S, Gao L, Cheng Z, Lu Q, Hu Y, Luo G, Wang K, Lu Y, Li H, Wang S, Ruan S, Yang C, Mei C, Wang Y, Ding D, Wu F, Tang X, Ye X, Ye Y, Liu B, Yang J, Yin W, Wang A, Fan G, Zhou F, Liu Z, Gu X, Xu J, Shang L, Zhang Y, Cao L, Guo T, Wan Y, Qin H, Jiang Y, Jaki T, Hayden FG, Horby PW, Cao B, Wang C (2020) Remdesivir in adults with severe COVID-19: a randomized, double-blind, placebo-controlled, multicentre trial. The Lancet 395:1596–1578. https://doi.org/10.1016/S0140-6736(20)31022-9
Wilkinson P (1980) Pharmacokinetics of ethanol: a review. Alcohol: Clin Exp Res 4(1):6–21
World Health Organization (2022) Coronavirus disease (COVID-19) advice for the public: Mythbusters. https://www.who.int/emergencies/diseases/novel-coronavirus-2019/advice-for-public/mythbusters#alcohol. Accessed 31 Jan 2023
Xiling G, Yin C, Ling W, Xiaosong W, Jingjing F, Fang L, Xiaoyan Z, Yiyue G, Ying C, Lunbiao C, Liubo Z (2021) In vitro inactivation of SARS-CoV-2 by commonly used disinfection products and methods. Sci Rep 11:1–9. https://doi.org/10.1038/s41598-021-82148-w

Ynga-Durand M, Maaß H, Milošević M, Krstanović F, Pribanić Matešić M, Jonjić S, Protić A, Brizić I, Šustić A, Čičin-Šain L (2022) SARS-CoV-2 viral load in the pulmonary compartment of critically Ill COVID-19 patients correlates with viral serum load and fatal outcomes. Viruses 14:1292. https://doi.org/10.3390/v14061292

Zhang P, Yang Z, Zhao Y, Liu Y, Zhang L, Shao G (2011) Inhalation of alcohol vapor driven by oxygen is a useful therapeutic method for postoperative alcohol withdrawal syndrome in a patient with esophageal cancer: a case report. Alcohol Alcohol 46(4):424–426

Frailty in Critically Ill Patients with COVID-19: The Lessons Learned

4

Ashwin Subramaniam, Ryan Ruiyang Ling, and Kiran Shekar

Abstract

The coronavirus disease 2019 (COVID-19) pandemic has unquestionably transformed the field of intensive care medicine. Never have we witnessed millions of patients develop acute respiratory failure in such a short span of time. This led to extensive resource constraints and difficulty in treating patients. However, this also gave rise to several innovations that have spurred the development and progress of intensive care medicine as a specialty. In this chapter, we explore an overview of frailty, the impact of frailty in patients with severe COVID-19 respiratory failure, and the available supports, by summarising the current literature. This chapter also discusses the lessons learnt from each of the sections that can be applied to daily clinical practice. The chapter also proposes insights into future research.

Keywords

Frailty · Critical care · COVID-19 · SARS-CoV-2 · Pandemic · Lessons learned

4.1 Introduction

4.1.1 Overview of Frailty

Frailty is a multidimensional geriatric syndrome that is characterised by a decline in physical, physiological, and cognitive reserves (Clegg

A. Subramaniam (✉)
Department of Intensive Care Medicine, Monash Health, Dandenong Hospital, Dandenong, VIC, Australia
e-mail: ashwin.subramaniam@monash.edu

Department of Intensive Care Medicine, Peninsula Health, Frankston, VIC, Australia

Department of Acute and Critical Care, Department of Epidemiology and Preventive Medicine, Monash University, Melbourne, VIC, Australia

Peninsula Clinical School, Monash University, Frankston, VIC, Australia

R. R. Ling
Yong Loo Lin School of Medicine, National University of Singapore, National University Health System, Singapore, Singapore
e-mail: ryan.ling@u.nus.edu

K. Shekar
Adult Intensive Care Services, The Prince Charles Hospital, Brisbane, QLD, Australia
e-mail: Kiran.Shekar@health.qld.gov.au

Faculty of Medicine, University of Queensland, Brisbane, QLD, Australia

Faculty of Health, Queensland University of Technology, Brisbane, QLD, Australia

Faculty of Health Sciences and Medicine, Bond University, Gold Coast, QLD, Australia

N. Rezaei (ed.), *The COVID-19 Aftermath*, Advances in Experimental Medicine and Biology 1457,
https://doi.org/10.1007/978-3-031-61939-7_4

et al. 2013). The risk of frailty rises with age and is characterised by a varied combination of decreased mobility, weakness, reduced muscle mass, poor nutritional status, and diminished cognitive function (Clegg et al. 2013). Patients with frailty are at a higher risk of requiring help in daily activities, being vulnerable to adverse events, and of death (Joseph et al. 2014; Muscedere et al. 2017).

Rather than universally accepted symptoms or specific biomarkers (Biasio et al. 2020), the five core components that describe frailty are (a) vulnerability; (b) multifactorial nature; (c) heterogeneous presentation; (d) clinical measurability; and (e) predisposition to adverse outcomes (Junius-Walker et al. 2018). These components reflect the variability and complexity in the presentation of frailty and summarise its inherent ambiguity, some of which stems from a complex interplay between biological foundations with intrinsic and environmental factors (Biasio et al. 2020).

There are two accepted paradigms of frailty: phenotypic construct (Fried et al. 2001) and deficit accumulation model (Rockwood et al. 2005). The phenotype construct is based on a cluster of signs and symptoms such as self-reported exhaustion, slowed performance (by walking speed), weakness (by grip strength), unintentional weight loss (4.5 kg in the past year), and low physical activity (Fried et al. 2001). The deficit accumulation model other the other hand is quantified based on the number rather than the nature of health problems (Rockwood et al. 2005; Mitnitski et al. 2013), along with biochemical and physiological impairments. An overlap exists between the two constructs (Fried et al. 2001; Rockwood et al. 2005), their sum contributing to a risk state (Theou et al. 2015).

4.1.2 Evaluation of Frailty

Patients with frailty have higher rates of adverse outcomes and need care adaptations, personalised interventions, and modifications of standard protocols (Clegg et al. 2013; Theou and Rockwood 2015). It is vital to routinely consider frailty when managing older patients and identifying them early (Theou et al. 2018). While various instruments have been developed to estimate frailty, there is no consensus on which is feasible in acute settings. A scoping review in 2018 of 617 eligible studies found 89 frailty measures and found that the Clinical Frailty Scale (CFS), the Frailty Index (FI), and the Frailty Phenotype (FP) were most commonly used (Theou et al. 2018). These tools differ in their utility, while some focus on identifying individuals at high risk for adverse outcomes, and others focus on clinical interventions to prevent or treat frailty. Some are highly complex and time-consuming to operate, which has consequently reduced their acceptance for use in clinical settings. Additionally, the agreement between these tools varies greatly and the optimal frailty measurement is unclear. Comprehensive geriatric assessment (CGA) considered a gold standard frailty measure (Parker et al. 2018) recapitulates both the phenotype and deficit accumulation models, however, it is time-consuming and impractical for day-to-day use in ICU.

4.1.3 The Impact of Frailty on Critically Ill Patients

Frailty in patients with critical illness is distinctly vulnerable and generally portends poorer outcomes including increased short-term and long-term mortality (Muscedere et al. 2017; Darvall et al. 2019a, 2021a). In addition, patients with frailty are more likely to have falls, longer hospital stays, delayed recovery, and be readmitted to the ICU (Bagshaw et al. 2014). These patients are also more likely to lose independence following hospital discharge (Muscedere et al. 2017) and have poorer quality of life related to both physical (Bagshaw et al. 2014; Brummel et al. 2017) and mental health (Bagshaw et al. 2015). With more older and frailer patients being admitted to ICU (Nguyen et al. 2011), more intensive care services are needed for older adults, which will impact intensive care services globally. As a result, assessing frailty is often recommended upon ICU admission (Robert et al. 2017). It is crucial to how care is best tailored to the unique needs of this cohort (Brummel and Ferrante 2018; Falvey and Ferrante 2019).

4.1.4 Quantifying Frailty in the Critically Ill

Using all three models (CFS, FI, and FP) to operationalise and identify frailty may prompt more cautious and considerate care (Falvey and Ferrante 2019). Regrettably, frailty assessments are difficult to routinely perform in the ICU (Pugh et al. 2018). The phenotype construct is not easily diagnosed in the ICU setting given the measurements that are needed. The deficit accumulation model has several challenges in this setting. A well-constructed model with many elements is difficult to identify in the ICU setting (Searle et al. 2008). However, FI has been successfully used with ICU patients (Darvall et al. 2021b; Mueller et al. 2016; Zeng et al. 2015; Heyland et al. 2015; Kara et al. 2018). FI is one of the most used frailty screening tools in the ICU environment (Theou et al. 2018) and may be difficult to deliver on a routine basis by the critical care team (Pugh et al. 2018). Shorter modified versions (5 or 11 conditions), using administrative datasets, have been explored (Rockwood and Mitnitski 2012). Although the multidimensional CGA is considered the gold standard, it is an impractical tool in the ICU setting; frailty is, therefore, most commonly assessed (Muscedere et al. 2017) using the CFS (Rockwood et al. 2005), and the FI (Mitnitski et al. 2001).

The most widely used frailty screening tool in the ICU setting is the CFS (Rockwood et al. 2005), which was originally developed to summarise a CGA and provide a care plan (Theou et al. 2021). This tool, using visual and written analogues, has been designed to holistically screen for frailty according to physical activity, functional status, chronic illness burden, and cognition (Rockwood et al. 2005). Patients are categorised as being non-frail (CFS score ≤ 3), vulnerable (CFS score 4), mildly frail (CFS score 5); moderately frail (CFS score 6) to severely/very severely frail (CFS score 7/8); or terminally ill (CFS score of 9) (Fig. 4.1) (Bagshaw et al. 2014). The scale was originally validated as a predictor for mortality in the primary care setting and correlated with the deficit accumulation FI (Rockwood et al. 2005). The convenience of CFS has led it to be used very commonly in the ICU (Muscedere et al. 2017; Biasio et al. 2020; Pugh et al. 2018; Fisher et al. 2015). Despite its subjectivity, the CFS is validated to stratify older adults based on vulnerability (Rockwood et al. 2005) and reliably predicts outcomes in critically ill patients (Bagshaw et al. 2014, 2015; Fisher et al. 2015; Maguet et al. 2014; Subramaniam et al. 2022a). Darvall et al. (2019a) and Bagshaw et al. (2014) and correlates well with other frailty scales (Subramaniam et al. 2022a, b; Tipping et al. 2019; Darvall et al. 2019b). An updated version of the CFS at the start of the pandemic to aid with scarce resource allocation, where they revised level names (e.g. CFS 4 which was previously "vulnerable" was revised to "living with very mild frailty") (Fig. 4.1) (Rockwood and Theou 2020). A recent study provided brief aids to guide CFS scoring that could either be assessed by clinicians (Fig. 4.2, top panel: CFS-ED) or the patients themselves (Fig. 4.2, bottom panel: CFS-Self) (Chong et al. 2021). In another study, the authors provided a frailty classification tree that could reliably quantify frailty even by an inexperienced user (Fig. 4.3) (Theou et al. 2021).

4.1.5 Clinical Course of Critically Ill Patients with Frailty

A large systematic review (Muscedere et al. 2017) and a recent registry-based study (Darvall et al. 2022) found that frailty was increased with in-hospital and long-term mortality, and fewer discharges to home and persistent critical illness; combining frailty with the APACHE-IIIj score significantly improved risk prediction for mortality (Darvall et al. 2021b). Additionally, patients with frailty had a higher chance of developing and dying from persistent

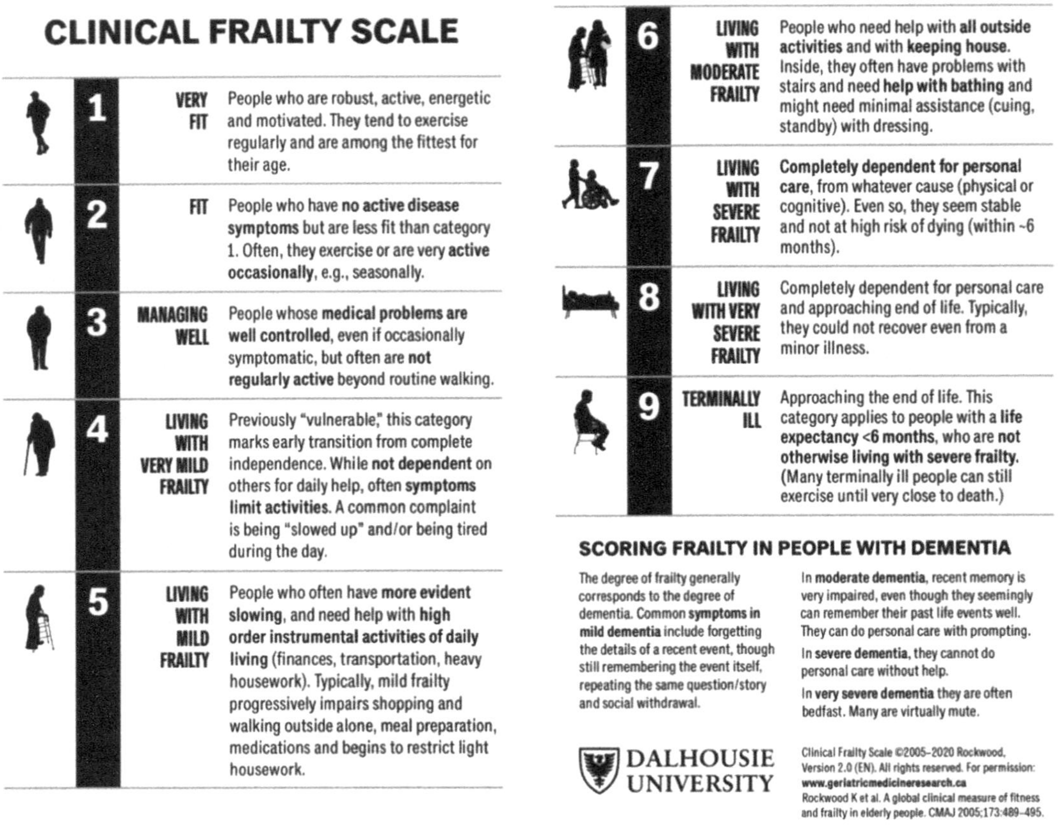

Fig. 4.1 The clinical frailty scale [copyright and used with permission (Rockwood and Theou 2020)]. Rockwood and Theou (2020). The content is a subject to copyright

critical illness (Darvall et al. 2022). Patients with frailty admitted to ICU following a rapid-response team's decision were more likely to die in the hospital (So et al. 2018) as well as have a longer length of stay (Fisher et al. 2015). Another large registry-based study from Australia and New Zealand found that a higher frailty score was independently associated with a longer duration of mechanical ventilation. A higher frailty score also contributed to identifying patients who were less likely to be liberated from mechanical ventilation (Okahara et al. 2022). The patients with a CFS score ≥ 6 had between 4 and10 times greater adjusted risk of new discharge to chronic care (Darvall et al. 2021b).

4.1.6 Critically Ill Patients with Severe COVID-19

A novel coronavirus was identified in December 2019 and quickly spread, resulting in a global pandemic that came to be known as coronavirus disease 2019 (COVID-19) (World Health Organization 2020). As of January 2023, over 660 million confirmed cases of COVID-19 with 6,706,735 deaths have been reported globally (Medicine JHUa 2023). Several variants of SARS-CoV-2 have emerged that are notable because of their increased transmissibility.

The clinical spectrum ranges widely from asymptomatic to severe respiratory and multi-organ failure, and death (Zhou et al. 2020; He

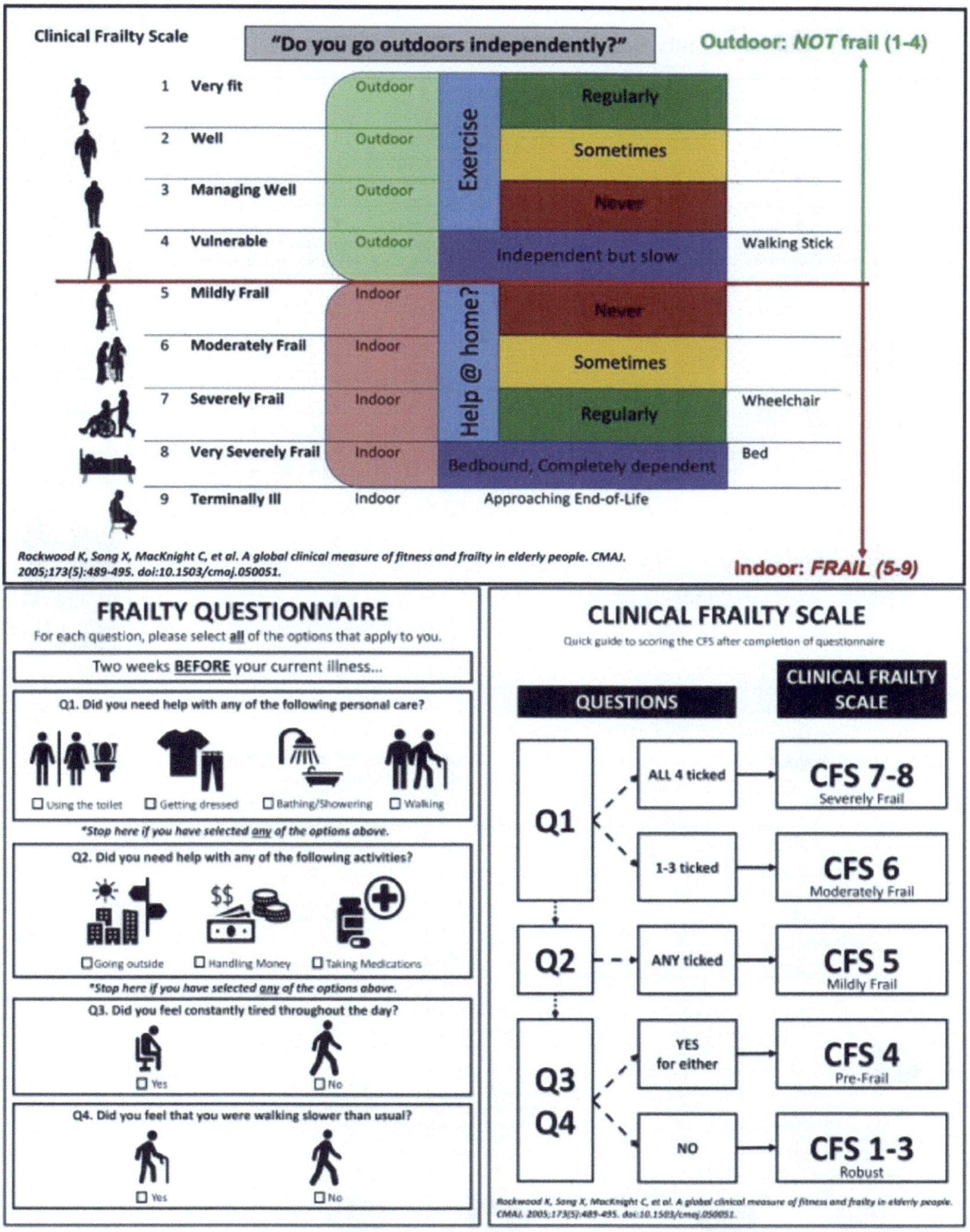

Fig. 4.2 Clinical frailty scale emergency department (CFS-ED, top panel) and self-questionnaire (CFS-Self, bottom panel) [permission obtained (Chong et al. 2021)]. Copyright © 2021 Copyright © 2021 The Authors. Published by Elsevier, Inc. on behalf of the Society of JAMDA

et al. 2020). Reports during the initial phase of the pandemic suggest that 25% of patients hospitalised with COVID-19 received ICU admission (Yang et al. 2020; Wang et al. 2020; Wu and McGoogan 2020; Richardson et al. 2020; Guan et al. 2020; Grasselli et al. 2020; Arentz et al.

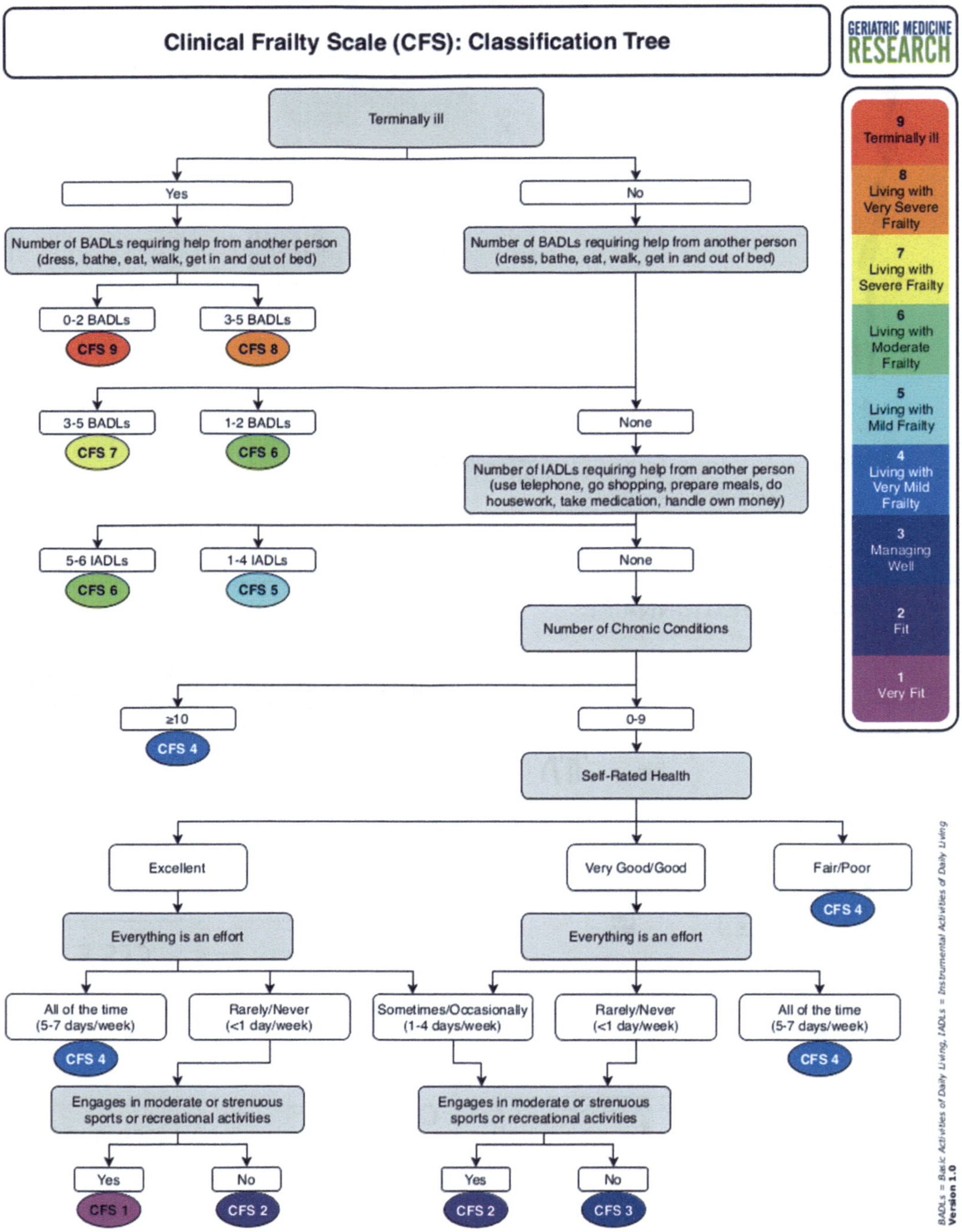

Fig. 4.3 The clinical frailty scale classification tree [permission obtained (Theou et al. 2021)]. Theou et al. (2021). The content is a subject to copyright

2020; Livingston and Bucher 2020), but progressively decreased over time (Vahidy et al. 2020; Kurtz et al. 2021; Garg et al. 2021). A surge in critically ill patients with respiratory failure has overwhelmed ICU capacity in many healthcare systems across the world. A systematic review and meta-analysis, earlier in the pandemic, reported a case fatality rate of 45% in patients with COVID-19 who needed invasive mechanical ventilation, with mortality rates exponentially increasing with age (Lim et al. 2021).

Vaccination has also helped in substantially reducing disease severity and ICU admissions (Polack et al. 2020; Baden et al. 2021), but death can occur from several other conditions including cardiac arrhythmia, cardiac arrest, and pulmonary embolism. The highest rates of death occur in those ≥ 64 years. As the pandemic progresses, mortality may be decreasing for several reasons. The length of stay appears to be longer than that in patients with non-COVID-19-related ARDS. Long-term sequelae may also be higher in this population.

4.2 Impact of COVID-19 on Critically Ill Patients with Frailty

4.2.1 Resource Allocation

Achieving a balance between patients with COVID-19 requiring more intensive monitoring/support and the number of available ICU beds mandated the solicitation of an ethical approach (Sekulić et al. 2020). Resource allocation in intensive care medicine is a decision-making process, and deliberations of autonomy are secondary to distributive justice (Truog et al. 2020). The principles to respect human dignity and non-abandonment must also be endorsed, particularly in those who do not receive ICU care, who should receive appropriate palliative care (Arya et al. 2020). There are subsequent ethical considerations as to how resources are allocated. As such, it is relevant we understand how frailty has impacted clinical decisions which may then drive improvements in future practice. Frailty has sustained relevance in a critical care environment with a continual scarcity of resources (Pugh et al. 2021). During the COVID-19 pandemic, two competing ethical methods have emerged: medical triage (focussing on the most severely ill patient) and a priori selection of patients (targeting a group of patients meeting certain criteria) (Sekulić et al. 2020). The American Geriatric Society recommends that the emergency resource allocation strategies during COVID-19 and future pandemics must not disproportionately disfavour older adults and that healthcare providers should not be expected to make rationing decisions in isolation (Farrell et al. 2020a, b). With healthcare resources worldwide being overstretched, compounded by the scarcity of intensive care resources faced by many areas, critical care triage becomes ethically complex and challenging. This led to countries rapidly preparing triage guidelines to aid doctors in deciding the appropriateness and potential benefits of admitting patients, especially older adults, to critical care (Phua et al. 2020). It was insufficient for triage committees and principles to ration decisions based on reliable prognosis and utilitarian principles when resources including beds and staff were scarce on an unprecedented scale (Supady et al. 2021). Triage committees can have a valuable role in setting standards to guide and support rationing decisions (Azoulay et al. 2020). Many frameworks were designed to include considerations that did not unethically disadvantage older patients, those living with disability, and those with chronic health conditions. Consequently, the goal is to maximise hospital survival rather than life-years saved; a triage team is involved in the care of the patient who is critically ill (Valiani et al. 2020). Frailty has become a factor in clinical decision-making for patients with COVID-19 in some settings (Institute and for Health and Care Excellence: Clinical Guidelines 2022). In patients diagnosed with COVID-19, the CFS was also found to be useful to predict adverse health outcomes and mortality. In patients ≥ 65 years with serious comorbidities, the proposed CFS threshold for ICU admission of patients with COVID-19 ranged from ≥ 5 to 6 (Institute and for Health and Care Excellence: Clinical Guidelines 2022; Swiss Academy of

Medical Sciences 2020), though these guidelines were criticised for extrapolating recommendations from pre-pandemic data (Azoulay et al. 2020). The seminal COPE study (Hewitt et al. 2020) found that frailty independently predicted the time from hospital admission to death. The authors concluded their results supported the use of CFS to inform medical decision-making in adults with COVID-19 admitted to the hospital. Nonetheless, this study did not adjust for disease severity and did not report on patients in the ICU or those receiving mechanical ventilation (Darvall et al. 2020a).

4.2.2 Patient Demographics and Hospital Mortality

The demographic of patients admitted to the ICU continues to drift towards a younger, fitter, and less co-morbid population (Hubbard et al. 2020; O'Caoimh et al. 2020). Yet, people of older age groups and with frailty are disproportionately affected by COVID-19 and are more likely to die from COVID-19 (Hubbard et al. 2020; Hagg et al. 2020). Frailty is a useful tool to highlight the risk of poor outcomes in COVID-19 patients (Kow and Hasan 2021). About 20% of the patients admitted to ICU are frail (Jung et al. 2021; Subramaniam et al. 2022c, d; Taniguchi et al. 2022a); frailty is associated with advanced age (Jung et al. 2021; Subramaniam et al. 2022c, d). The evidence on the prognosis of patients of older age groups remains equivocal. While some report that frailty was associated with mortality only among middle-aged patients (Geer et al. 2022), and others suggest that outcomes were poor for patients with frailty regardless of age group (Jung et al. 2021). A systematic review observed that the patients who died were older than those who survived (76.7 ± 9.6 vs. 69.2 ± 13.4 years) (Subramaniam et al. 2022e). Patients with frailty are usually male (Jung et al. 2021; Subramaniam et al. 2022d, e) and compared to females, had higher short-term mortality risk (RR = 1.08; 95%-CI 1.06–1.11) (Subramaniam et al. 2022e). A high degree of frailty and cumulative comorbidities (such as diabetes mellitus, chronic kidney disease, cardiovascular disease, cerebrovascular accident, dementia, and hypertension) have been associated with higher mortality rates in patients with COVID-19 (Jung et al. 2021; Subramaniam et al. 2022e; Davis et al. 2020; Wang et al. 2020; Guan et al. 2020; Chen et al. 2020). Patients with frailty also had higher illness severity scores (Jung et al. 2021; Subramaniam et al. 2022c, d; Taniguchi et al. 2022a).

Similar to the established literature (Hussien et al. 2021; Rottler et al. 2022), a systematic review found that frailty was strongly associated with mortality in COVID-19, and it concluded that frailty-based patient management guides the treatment of COVID-19 (Rottler et al. 2022), which another recent review echoed (Fernandes and Pereira 2022). A large Australian and New Zealand study found that frailty similarly affected hospitals in patients with and without COVID-19 pneumonitis but had low discriminatory capacity (Fig. 4.4) (Subramaniam et al. 2022d). However, the COVIP study found that the outcome in terms of mortality in patients with a frailty level ≥ 5 is similar in patients across all age groups ≥ 70 years, however, frailty was independently associated with lower survival (Jung et al. 2021). Many studies used frailty, measured with the CFS as a predictor for mortality (Wernly et al. 2023).

4.2.3 ICU Supports

It may be that patients with frailty have a poor immune response to SARS-CoV-2, resulting in higher short-term mortality, slower recovery, and further functional decline (Pan et al. 2020). While they are more likely to develop severe illnesses requiring intensive medical care, they are often unable to endure these invasive treatments or medical side effects, resulting in a greater likelihood of death during treatment. A meta-analysis pre-pandemic demonstrated that patients with frailty have a 1.71-fold risk of mortality if they are critically ill (Muscedere et al. 2017). Thus, patients with frailty with COVID-19 can develop a vicious sequence of impairment (Zhang et al. 2021).

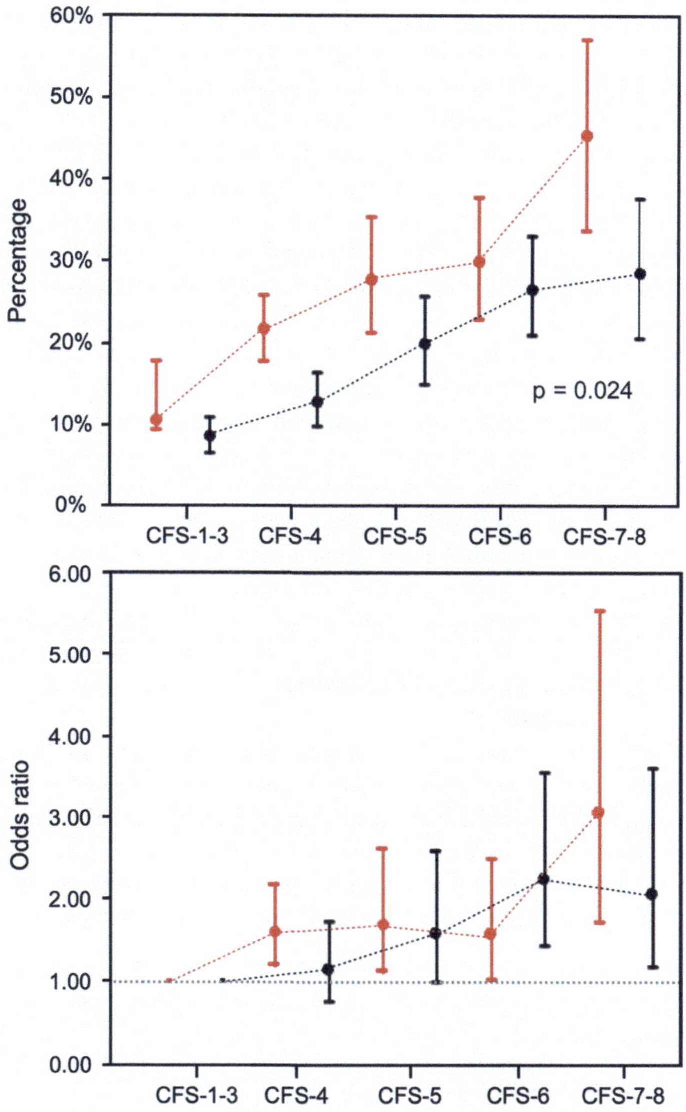

Fig. 4.4 Hospital mortality according to clinical frailty scale (CFS) score for all patients with (red lines) and without (black lines) COVID-19. The top panel is unadjusted for hospital mortality, while the bottom panel is adjusted for ANZROD and sex [permission obtained (Subramaniam et al. 2022d)]. Subramaniam et al. (2022d). The content is a subject to copyright

4.2.3.1 Invasive Mechanical Ventilation

Early invasive mechanical ventilation was employed during the initial stages of the pandemic due to concerns about an aerosol generation from non-invasive oxygenation therapies facilitating nosocomial viral transmission (Phua et al. 2020; McGain et al. 2020). Patients with frailty and older age groups were less likely to receive mechanical ventilation (Jung et al. 2021; Subramaniam et al. 2022c), received invasive mechanical ventilation for shorter periods of time, and were less likely to be admitted to the ICU (Ecarnot et al. 2022). However, an individual patient data meta-analysis found that there was no difference in mortality between frail and non-frail patients receiving invasive mechanical ventilation (68.2% vs. 77.5%; $p=0.21$) (Subramaniam et al. 2022c). Similarly, a study from Brazil also found that most patients required mechanical ventilation during their ICU stay, regardless of frailty status (Taniguchi et al. 2022a). The unadjusted mortality rates were higher in patients requiring mechanical ventilation than those who did not (Jung et al. 2021; Subramaniam et al. 2022c). However, among those admitted to the ICU, almost two-thirds of patients with frailty with COVID-19 died in the hospital, with a greater risk of death for those receiving mechanical ventilation, when compared with patients without frailty (Subramaniam et al. 2022c). Hospital mortality was relatively lower in Australian and New Zealand patients receiving mechanical ventilation (Subramaniam et al. 2022d) when compared to the published literature (Subramaniam et al. 2022c, e; Jung et al. 2021; Lim et al. 2020; Guidet et al. 2022). However, Fig. 4.5 describes the cumulative risk of death over time among patients who received mechanical ventilation, which demonstrated that the cumulative risk of death decreases with more days on mechanical ventilation (Subramaniam et al. 2022c).

4.2.3.2 Non-invasive Ventilatory Support

Although there were concerns about aerosol generation from non-invasive ventilation (NIV) early in the pandemic, NIV has become a promising alternative to invasive mechanical ventilation with particular importance amidst the shortage of ICU beds during the COVID-19 pandemic. NIV may be a particularly appealing therapeutic option in elderly frail patients with COVID-19, including those with the "do not intubate" order. Patients with frailty more commonly had treatment limitations at ICU admission and were more likely to receive non-invasive ventilatory support (Darvall et al. 2020b). A sub-study of the COVIP study demonstrated that the use of NIV was highly variable in Europe and was associated with high failure and mortality rates (Polok et al. 2022). The authors observed an association between higher SOFA and CFS scores and an increased risk of NIV failure and also found observed a long time to intubation in patients requiring escalation from NIV to mechanical ventilation was associated with increased 30-day mortality (Polok et al. 2022).

4.2.3.3 Other ICU Supports

The need for vasoactive and inotropic support among differed patients with frailty in Australia (35–40%) (Subramaniam et al. 2022d) and in Brazil (75%) (Taniguchi et al. 2022a). Fewer patients with frailty were receipt of renal replacement therapy (Subramaniam et al. 2022d). Patients with frailty less commonly received extracorporeal membrane oxygenation therapy (Subramaniam et al. 2022d).

Fig. 4.5 Graphical representation of cumulative incidence of death for frailty [copyright and used with permission (Subramaniam et al. 2022c)]. Subramaniam et al. (2022c). Copyright © 2023 Copyright © 2022 The Authors. Published by Wolters Kluwer Health, Inc. on behalf of the Society of Critical Care Medicine

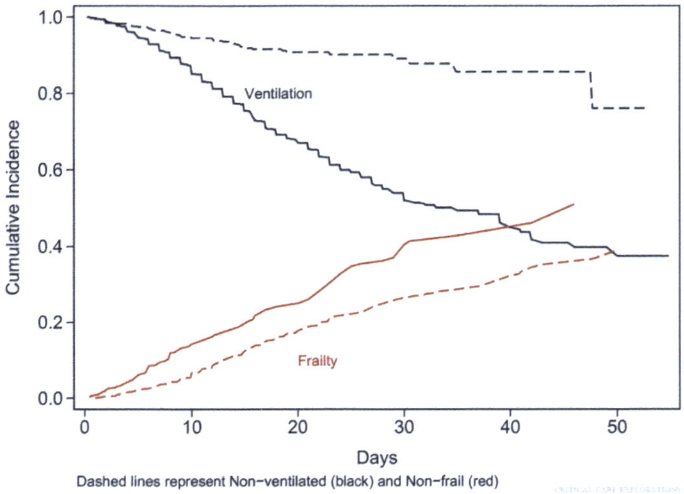

4.2.4 Long-Term Survival in Critically Ill Patients with Frailty with COVID-19

Although vaccination has diminished the impact of the severity and mortality, many patients are still affected by the consequences of COVID-19. Symptoms can persist in the longer term ("long-COVID") (Crook et al. 2021), and potentially affect the cardiovascular, musculoskeletal, and neurologic systems (Crook et al. 2021; Huang et al. 2021; Peñuelas et al. 2021). The first studies translating "long-COVID" into patient-centred outcomes (including frailty) found that long-term survival in patients with frailty with critical COVID-19 is poor (Taniguchi et al. 2022a, b; Geer et al. 2022). Even after hospital discharge, frailty increases the risk of patient-centred adverse outcomes in COVID-19 ICU survivors (Taniguchi et al. 2022b). Frailty and disability were more frequent 3–6 months after hospital discharge compared with baseline in COVID-19 patients admitted to the ICU; patients with CFS ≥ 4 had a lower six-month survival (Taniguchi et al. 2022a, b). A third of the ICU patients were frail, had worse health status, and developed new disabilities at 3 months follow-up (Fig. 4.6) (Taniguchi et al. 2022b), however, significant improvements can occur at 1-year follow-up (Taniguchi et al. 2023). This highlights the importance of longitudinal assessment of this population (Auriemma and Ferrante 2023). Providers should be aware that assessing frailty can bring valuable information to their clinical decisions and treatment plans before, during, and even after critical illness in COVID-19. Moreover, future studies should collect detailed

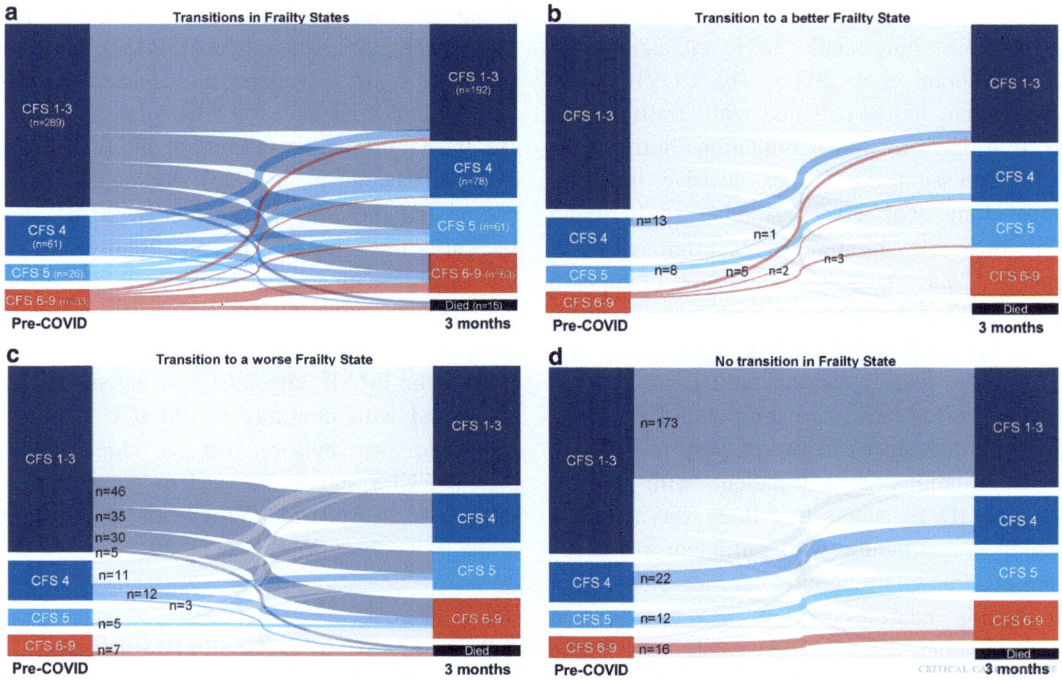

Fig. 4.6 Transition of frailty states from Pre-COVID to ICU Admission at and 90 days after hospital discharge. **a** Changes in frailty states from pre-COVID-19 (2–4 weeks before infection) to 3-mo follow-up among survivors of critical COVID-19 [copyright and used with permission (Taniguchi et al. 2022b)]. **b** The number of patients who transitioned to better frailty states. **c** The number of patients who transitioned to worse frailty states. **d** The number of patients who had no transition in a frailty state. CFS = clinical frailty scale Taniguchi et al. (2022b). Copyright © 2023 by the Society of Critical Care Medicine and Wolters Kluwer Health, Inc. All rights reserved

information about treatment characteristics (mechanical ventilation, paralytics and sedatives, physical therapy), and important potential mediators (delirium) to enhance our understanding of how frailty, disability, and cognitive impairment are related to both underlying disease and ICU management (Auriemma and Ferrante 2023).

4.2.5 Goals of Care

Early in the pandemic, there was significant emphasis on advanced care planning and treatment limitations in older patients (Chase 2020; O'Mara et al. 2020; Straw et al. 2021; Lee et al. 2020). Despite that, a recent study identified that timely goals of care documentation occurred less frequently in patients with frailty during the COVID-19 pandemic (Subramaniam et al. 2021). There is good evidence that the presence of treatment limitations independently predicts mortality among patients with COVID-19 (Jung et al. 2021; Alhatem et al. 2020; Tehrani et al. 2021). The COVIP study found that more patients with frailty were likely to have treatment limitations during their ICU admission than other invasive interventions (Jung et al. 2021). Another study found that treatment limitations occurred approximately 12 days after ICU admission (Taniguchi et al. 2022a). In Australia and New Zealand, more than a third of patients with frailty and COVID-19 had treatment limitations at ICU admission (Subramaniam et al. 2023), and there were no differences in the proportion of treatment limitations between patients with or without COVID-19, suggesting there was a robust health infrastructure with sufficient resources and economic strength to ensure strict public health measures before mass vaccination (Subramaniam et al. 2023). The study also found that patients with frailty were more likely to die in the ICU or hospital, require mechanical ventilatory and vasopressor support, and were less likely to be discharged home (Subramaniam et al. 2023). Despite vaccinations and public health measures to mitigate this pandemic, COVID-19 may continue to severely impact frail older and vulnerable patients. Further improvements in the outcomes in these patients are likely with increased uptake of vaccinations. As the ethical approach in managing critically ill patients is to facilitate early goals of care conversations (Chase 2020; O'Mara et al. 2020; Alhatem et al. 2020; Kramer et al. 2020), all patients must receive timely goal-concordant care, which may involve avoiding burdensome treatment (O'Mara et al. 2020; Subramaniam et al. 2021; Noortgate and Block 2022).

The concept of time-limited trial has been proposed as a practical approach to establishing the goals of ICU care when the long-term outcomes and overall benefits of aggressive interventions are uncertain (Vink et al. 2018). The "F-R-A-I-L" screening checklist (Functional impairment, Recurrent hospitalisations, Advanced malignancy or chronic diseases, Irreversible organ failure, and Long hospital stay) can be used as a prompt to admit critically ill patients who may benefit from a time-limited trial in the ICU (Cheung et al. 2022). Patients with one of these criteria are considered vulnerable and frail and were postulated to benefit from upfront discussions about limitations of care (Cheung et al. 2022). Although it cannot predict mortality (Wernly et al. 2023), this screening tool may improve the quality of care for patients with critical COVID-19. However, the ability to predict mortality is inherent in any critically ill patient evaluation tool. The CFS but not the FRAIL checklist was independently associated with mortality in old ICU patients. Therefore, we believe that in elderly ICU patients, CFS should be used to assess frailty because it also provides prognostic information in the frail cohort (Wernly et al. 2023).

4.2.6 Frailty in Critically Ill Patients with COVID-19: Lessons Learned

What are some lessons that we can learn from the past three years in managing patients with frailty with severe COVID-19 needing ICU care? The COVID-19 pandemic has certainly led to an increasingly rapid understanding of managing acute respiratory failure and ARDS in patients with frailty. First, about 20% of the patients admitted to ICU with severe COVID-19 were frail, intensive care teams either followed stringent triaging guidelines or were innately more selective in admitting these patients into the ICU (Subramaniam et al. 2022c, d, e). Patients with frailty are generally older, more likely to be male, had higher comorbidities, and had higher illness severity scores. Second, although many frailty measures were used, the CFS was the most used screening tool during the COVID-19 pandemic. Although hospital mortality varied globally, many patients with frailty died in-hospital, and frailty was associated with mortality and unfavourable patient-centred. Third, the impact of frailty on the use of mechanical ventilation was age-dependent: older patients with frailty were less likely to receive mechanical ventilation, and if they did, the duration of mechanical ventilation was shorter. In contrast, NIV was more commonly used in older patients with frailty, particularly in those with treatment limitations. There was an association between higher illness severity scores and CFS scores with NIV failure. If patients with frailty are failing NIV, it is important to decide early if they need intubation and mechanical ventilation, if appropriate. Fourth, most COVID-19 critical care survivors transition to poorer health status, highlighting the importance of long-term medical follow-up for this population. Fifth, patients with frailty with treatment limitations were commonly admitted to ICU. More patients with frailty were likely to have treatment limitations during their ICU admission than other invasive interventions. Patients with frailty with treatment limitations were less likely to be discharged from hospital. It is therefore essential to ensure that frail and older patients receive timely goal-concordant care, which may avoid burdensome treatment. Finally, frailty should be assessed in day-to-day clinical practice in all older ICU patients. Frailty in intensive care medicine must be measured, interpreted, and considered to improve individual care and resource allocation (Jung et al. 2023). Even as our daily practice in critical care becomes increasingly data-driven, it is important to take a step back and evaluate our patients holistically. Reflecting on the past three years, "less medications, more Medicine" may be the most important lesson for clinicians in the ICU. Rather than treating the numbers and data that are presented to us, it would be more appropriate to treat the patient present before us, taking into account not just their disease severity and comorbidities, but also their goals of care.

References

Alhatem A, Spruijt O, Heller DS, Chokshi RJ, Schwartz RA, Lambert WC, (2020) "Do-not-resuscitate (DNR)" status determines mortality in patients with COVID-19. Clin Dermatol

Arentz M, Yim E, Klaff L, Lokhandwala S, Riedo FX, Chong M, Lee M (2020) Characteristics and outcomes of 21 critically ill patients with COVID-19 in Washington State. JAMA 323:1612–1614

Arya A, Buchman S, Gagnon B, Downar J (2020) Pandemic palliative care: beyond ventilators and saving lives. CMAJ 192:E400–E404

Auriemma CL, Ferrante LE (2023) Better but not well: disability, frailty, and cognitive impairment one year after COVID-19 critical illness. Ann Am Thorac Soc 20:202–203

Azoulay É, Beloucif S, Guidet B, Pateron D, Vivien B, Le Dorze M (2020) Admission decisions to intensive care units in the context of the major COVID-19 outbreak: local guidance from the COVID-19 Paris-region area. Crit Care 24:293

Baden LR, El Sahly HM, Essink B, Kotloff K, Frey S, Novak R, Diemert D, Spector SA, Rouphael N, Creech CB, McGettigan J, Khetan S, Segall N, Solis J, Brosz A, Fierro C, Schwartz H, Neuzil K, Corey L, Gilbert P, Janes H, Follmann D, Marovich M, Mascola J, Polakowski L, Ledgerwood J, Graham BS, Bennett H, Pajon R, Knightly C, Leav B, Deng W, Zhou H, Han S, Ivarsson M, Miller J, Zaks T (2021) Efficacy and safety of the mRNA-1273 SARS-CoV-2 vaccine. N Engl J Med 384:403–416

Bagshaw SM, Stelfox HT, McDermid RC, Rolfson DB, Tsuyuki RT, Baig N, Artiuch B, Ibrahim Q, Stollery DE, Rokosh E, Majumdar SR (2014) Association between frailty and short- and long-term outcomes among critically ill patients: a multicentre prospective cohort study. CMAJ 186:E95-102

Bagshaw SM, Stelfox HT, Johnson JA, McDermid RC, Rolfson DB, Tsuyuki RT, Ibrahim Q, Majumdar SR (2015) Long-term association between frailty and health-related quality of life among survivors of

critical illness: a prospective multicenter cohort study. Crit Care Med 43:973–982
Brummel NE, Ferrante LE (2018) Integrating geriatric principles into critical care medicine: the time is now. Ann Am Thorac Soc 15:518–522
Brummel NE, Bell SP, Girard TD, Pandharipande PP, Jackson JC, Morandi A, Thompson JL, Chandrasekhar R, Bernard GR, Dittus RS, Gill TM, Ely EW (2017) Frailty and subsequent disability and mortality among patients with critical illness. Am J Respir Crit Care Med 196:64–72
Chase J (2020) Caring for frail older adults during COVID-19: integrating public health ethics into clinical practice. J Am Geriatr Soc 68:1666–1670
Chen N, Zhou M, Dong X, Qu J, Gong F, Han Y, Qiu Y, Wang J, Liu Y, Wei Y, Xia J, Yu T, Zhang X, Zhang L (2020) Epidemiological and clinical characteristics of 99 cases of 2019 novel coronavirus pneumonia in Wuhan, China: a descriptive study. Lancet 395:507–513
Cheung EH-L, Cheung JC-H, Yip Y-Y (2022) Raising awareness for time-limited trial discussion upon ICU triage and admission. Intensive Care Med 48:240–241
Chong E, Tham A, Chew J, Lim WS, Tan HN, Ang H, Chan M (2021) Brief aids to guide clinical frailty scale scoring at the front door of acute hospitals. J Am Med Dir Assoc 22:1116-1117.e1112
Clegg A, Young J, Iliffe S, Rikkert MO, Rockwood K (2013) Frailty in elderly people. Lancet 381:752–762
Crook H, Raza S, Nowell J, Young M, Edison P (2021) Long covid-mechanisms, risk factors, and management. BMJ 374:n1648
Darvall JN, Bellomo R, Paul E, Subramaniam A, Santamaria JD, Bagshaw SM, Rai S, Hubbard RE, Pilcher D (2019a) Frailty in very old critically ill patients in Australia and New Zealand: a population-based cohort study. Med J Aust 211:318–323
Darvall JN, Greentree K, Braat MS, Story DA, Lim WK (2019b) Contributors to frailty in critical illness: multi-dimensional analysis of the clinical frailty scale. J Crit Care 52:193–199
Darvall JN, Bellomo R, Young PJ, Rockwood K, Pilcher D (2020a) Frailty and mortality in patients with COVID-19. Lancet Public Health 5:e580
Darvall JN, Bellomo R, Bailey M, Paul E, Young PJ, Rockwood K, Pilcher D (2020b) Frailty and outcomes from pneumonia in critical illness: a population-based cohort study. Br J Anaesth 125:730–738
Darvall JN, Bellomo R, Paul E, Bailey M, Young PJ, Reid A, Rockwood K, Pilcher D (2021a) Routine frailty screening in critical illness: a population-based cohort study in Australia and New Zealand. Chest
Darvall JN, Bellomo R, Paul E, Bailey M, Young PJ, Reid A, Rockwood K, Pilcher D (2021b) Routine frailty screening in critical illness: a population-based cohort study in Australia and New Zealand. Chest 160:1292–1303
Darvall JN, Bellomo R, Bailey M, Young PJ, Rockwood K, Pilcher D (2022) Impact of frailty on persistent critical illness: a population-based cohort study. Intensive Care Med 48:343–351
Davis P, Gibson R, Wright E, Bryan A, Ingram J, Lee RP, Godwin J, Evans T, Burleigh E, Wishart S, Capek E, Mitchell L (2020) Atypical presentations in the hospitalised older adult testing positive for SARS-CoV-2: a retrospective observational study in Glasgow, Scotland. Scott Med J 36933020962891
De Biasio JC, Mittel AM, Mueller AL, Ferrante LE, Kim DH, Shaefi S (2020) Frailty in critical care medicine: a review. Anesth Analg 130:1462–1473
De Geer L, Fredrikson M, Chew MS (2022) Frailty is a stronger predictor of death in younger intensive care patients than in older patients: a prospective observational study. Ann Intensive Care 12:120
Ecarnot F, Rebora P, Focà E, Zucchelli A, Citerio G, Valsecchi MG, Marengoni A, Bellelli G (2022) Mechanical ventilation in COVID-19 patients: insights into the role of age and frailty from a multicentre observational study. Aging Dis 13:340–343
Falvey JR, Ferrante LE (2019) Frailty assessment in the ICU: translation to 'real-world' clinical practice. Anaesthesia 74:700–703
Farrell TW, Ferrante LE, Brown T, Francis L, Widera E, Rhodes R, Rosen T, Hwang U, Witt LJ, Thothala N, Liu SW, Vitale CA, Braun UK, Stephens C, Saliba D (2020a) AGS position statement: resource allocation strategies and age-related considerations in the COVID-19 era and beyond. J Am Geriatr Soc 68:1136–1142
Farrell TW, Francis L, Brown T, Ferrante LE, Widera E, Rhodes R, Rosen T, Hwang U, Witt LJ, Thothala N, Liu SW, Vitale CA, Braun UK, Stephens C, Saliba D (2020b) Rationing limited healthcare resources in the COVID-19 era and beyond: ethical considerations regarding older adults. J Am Geriatr Soc 68:1143–1149
Fernandes AL, Pereira RMR (2022) Frailty in the context of COVID-19 pandemic: a life-threatening condition. Front Med (Lausanne) 9:965562
Fisher C, Karalapillai DK, Bailey M, Glassford NG, Bellomo R, Jones D (2015a) Predicting intensive care and hospital outcome with the Dalhousie clinical frailty scale: a pilot assessment. Anaesth Intensive Care 43:361–368
Fried LP, Tangen CM, Walston J, Newman AB, Hirsch C, Gottdiener J, Seeman T, Tracy R, Kop WJ, Burke G, McBurnie MA, Cardiovascular Health Study Collaborative Research G (2001) Frailty in older adults: evidence for a phenotype. J Gerontol A Biol Sci Med Sci 56:M146–M156
Garg S, Patel K, Pham H, Whitaker M, O'Halloran A, Milucky J, Anglin O, Kirley PD, Reingold A, Kawasaki B, Herlihy R, Yousey-Hindes K, Maslar A, Anderson EJ, Openo KP, Weigel A, Teno K, Ryan PA, Monroe ML, Reeg L, Kim S, Como-Sabetti K, Bye E, Shrum Davis S, Eisenberg N, Muse A, Barney G, Bennett NM, Felsen CB, Billing L, Shiltz J, Sutton M, Abdullah N, Talbot HK, Schaffner W, Hill

M, Chatelain R, Wortham J, Taylor C, Hall A, Fry AM, Kim L, Havers FP (2021) Clinical trends among U.S. adults hospitalized with COVID-19, March to December 2020: a cross-sectional study. Ann Intern Med 174:1409–1419

Grasselli G, Pesenti A, Cecconi M (2020) Critical care utilization for the COVID-19 outbreak in Lombardy, Italy: early experience and forecast during an emergency response. JAMA 323:1545–1546

Guan WJ, Ni ZY, Hu Y, Liang WH, Ou CQ, He JX, Liu L, Shan H, Lei CL, Hui DSC, Du B, Li LJ, Zeng G, Yuen KY, Chen RC, Tang CL, Wang T, Chen PY, Xiang J, Li SY, Wang JL, Liang ZJ, Peng YX, Wei L, Liu Y, Hu YH, Peng P, Wang JM, Liu JY, Chen Z, Li G, Zheng ZJ, Qiu SQ, Luo J, Ye CJ, Zhu SY, Zhong NS, China Medical Treatment Expert Group for Covid-19 (2020a) Clinical characteristics of coronavirus disease 2019 in China. N Engl J Med 382:1708–1720

Guidet B, Jung C, Flaatten H, Fjolner J, Artigas A, Pinto BB, Schefold JC, Beil M, Sigal S, van Heerden PV, Szczeklik W, Joannidis M, Oeyen S, Kondili E, Marsh B, Andersen FH, Moreno R, Cecconi M, Leaver S, De Lange DW, Boumendil A, VIP2 and COVIP Study Groups (2022) Increased 30-day mortality in very old ICU patients with COVID-19 compared to patients with respiratory failure without COVID-19. Intensive Care Med

Hagg S, Jylhava J, Wang Y, Xu H, Metzner C, Annetorp M, Garcia-Ptacek S, Khedri M, Bostrom AM, Kadir A, Johansson A, Kivipelto M, Eriksdotter M, Cederholm T, Religa D (2020) Age, frailty, and comorbidity as prognostic factors for short-term outcomes in patients with coronavirus disease 2019 in geriatric care. J Am Med Dir Assoc 21(1555–1559):e1552

He F, Deng Y, Li W (2020) Coronavirus disease 2019: what we know? J Med Virol 92:719–725

Hewitt J, Carter B, Vilches-Moraga A, Quinn TJ, Braude P, Verduri A, Pearce L, Stechman M, Short R, Price A, Collins JT, Bruce E, Einarsson A, Rickard F, Mitchell E, Holloway M, Hesford J, Barlow-Pay F, Clini E, Myint PK, Moug SJ, McCarthy K, Collaborators CS (2020) The effect of frailty on survival in patients with COVID-19 (COPE): a multicentre, European, observational cohort study. Lancet Public Health 5:e444–e451

Heyland DK, Garland A, Bagshaw SM, Cook D, Rockwood K, Stelfox HT, Dodek P, Fowler RA, Turgeon AF, Burns K, Muscedere J, Kutsogiannis J, Albert M, Mehta S, Jiang X, Day AG (2015) Recovery after critical illness in patients aged 80 years or older: a multi-centre prospective observational cohort study. Intensive Care Med 41:1911–1920

Huang L, Yao Q, Gu X, Wang Q, Ren L, Wang Y, Hu P, Guo L, Liu M, Xu J, Zhang X, Qu Y, Fan Y, Li X, Li C, Yu T, Xia J, Wei M, Chen L, Li Y, Xiao F, Liu D, Wang J, Wang X, Cao B (2021) 1-year outcomes in hospital survivors with COVID-19: a longitudinal cohort study. Lancet 398:747–758

Hubbard RE, Maier AB, Hilmer SN, Naganathan V, Etherton-Beer C, Rockwood K (2020) Frailty in the face of COVID-19. Age Ageing 49:499–500

Hussien H, Nastasa A, Apetrii M, Nistor I, Petrovic M, Covic A (2021) Different aspects of frailty and COVID-19: points to consider in the current pandemic and future ones. BMC Geriatr 21:389

Joseph B, Pandit V, Zangbar B, Kulvatunyou N, Hashmi A, Green DJ, O'Keeffe T, Tang A, Vercruysse G, Fain MJ, Friese RS, Rhee P (2014) Superiority of frailty over age in predicting outcomes among geriatric trauma patients: a prospective analysis. JAMA Surg 149:766–772

Jung C, Flaatten H, Fjølner J, Bruno RR, Wernly B, Artigas A, Bollen Pinto B, Schefold JC, Wolff G, Kelm M, Beil M, Sviri S, van Heerden PV, Szczeklik W, Czuczwar M, Elhadi M, Joannidis M, Oeyen S, Zafeiridis T, Marsh B, Andersen FH, Moreno R, Cecconi M, Leaver S, Boumendil A, De Lange DW, Guidet B, Flaatten H, Wernly B, Artigas A, Beil M, Sviri S, van Heerden PV, Szczeklik W, Elhadi M, Zafeiridis T, Moreno R, Cecconi M, Boumendil A, Abosheaishaa HM, Abualqumboz EMY, Ahmed AK, Ahmed H, Aidoni Z, Aldecoa C, Alexandru N, Ali YKNM, Al-Sadawi M, Andersen K, Andersen FH, Assis R, Azab MA, Azzam AY, Badawy MR, Balleby IR, Barth E, Barth E, Ben-Hamouda N, Besch G, Besset S, Bjerregaard AT, Brix H, Bruno RR, Brushoej J, Bundgaard H, Burtin P, Caillard A, Canas-Perez I, Charron C, Chrisanthopoulou E, Comellini V, Cornet A, Cubero PJ, Czuczwar M, Dauger S, Diaz-Rodriguez C, Dieperink W, Dindane Z, Djibré M, Dormans T, Dullenkopf A, Dumas G, Elgazzar YA, Eller P, Elsaka A, Evers M, Faltlhauser A, Ferreira AF, Fjølner J, Fleury Y, Galbois A, Garcon P, Garnier M, Gawda R, Ghannam A, Goebel U, Gomà G, Goncalves B, Gordinho A, Groenendijk M, Guerot E, Guidet B, Gurjar M, Haake H, Haas L, Habib AA, Hahn M, Hansen MA, Hilles MMY, Hussein AARM, Iglesias D, Joannidis M, Jung C, Jurcisin I, Kabitz H-J, Kelm M, Kindgen-Milles D, Klimkiewicz J, Kuhn KF, Kunstein A, Kurt M, De Lange DW, Leaver S, Lutz M, Mahmoodpoor A, Maizel J, Marin N, Marsh B, Megarbane B, Mesotten D, Meybohm P, Meyer C, Mira AP, Namendys-Silva Sa, Nedergaard HK, Nseir S, Oeyen S, Olasveengen T, Oliveira AIP, Oziel J, Papadogoulas A, Perez-Torres D, Bollen Pinto B, Piton G, Plantefeve G, Poerner T, Priego J, Rabha A, Randerath W, Raphaelen J-h, Reper P, Rigaud J-P, Rivera SA, Roberti A, Romundstad L, Rovina N, Salah R, Saleh M, Sancho S, de Lurdes Campos Santos M, Santos H, Schaller S, Schefold JC, Schuster M, Shala G, Sjøbø B, Steiner S, Strietzel HF, Sviri S, Swinnen W, Tamayo-Lomas L, Tharwat S, Tomasa T, Uhrenholt S, Vaissiere M, Valent A, Valette X, Vanderlinden T, Vázquez EM, Villamayor MI, Villefrance M, Voigt

I, Wassim K, Welte M, Wolff G, Wollborn J, Zalba-Etayo B, Zegers M, COVIP Study Group (2021) The impact of frailty on survival in elderly intensive care patients with COVID-19: the COVIP study. Crit Care 25:149

Jung C, Guidet B, Flaatten H, VIP Study Group (2023) Frailty in intensive care medicine must be measured, interpreted and taken into account! Intensive Care Med 49:87–90

Junius-Walker U, Onder G, Soleymani D, Wiese B, Albaina O, Bernabei R, Marzetti E, ADVANTAGE JA WP4 Group (2018) The essence of frailty: A systematic review and qualitative synthesis on frailty concepts and definitions. Eur J Intern Med 56:3–10

Kara I, Aydogdu M, Gursel G (2018) The impact of frailty on noninvasive mechanical ventilation in elderly medical intensive care unit patients. Aging Clin Exp Res 30:683

Kow CS, Hasan SS (2021) Letter to the editor: prevalence of frailty in patients with COVID-19: a meta-analysis. J Frailty Aging 10:189–190

Kramer DB, Lo B, Dickert NW (2020) CPR in the Covid-19 era—an ethical framework. N Engl J Med 383:e6

Kurtz P, Bastos LSL, Dantas LF, Zampieri FG, Soares M, Hamacher S, Salluh JIF, Bozza FA (2021) Evolving changes in mortality of 13,301 critically ill adult patients with COVID-19 over 8 months. Intensive Care Med 47:538–548

Le Maguet P, Roquilly A, Lasocki S, Asehnoune K, Carise E, Saint Martin M, Mimoz O, Le Gac G, Somme D, Cattenoz C, Feuillet F, Malledant Y, Seguin P (2014) Prevalence and impact of frailty on mortality in elderly ICU patients: a prospective, multicenter, observational study. Intensive Care Med 40:674–682

Lee J, Abrukin L, Flores S, Gavin N, Romney ML, Blinderman CD, Nakagawa S (2020) Early intervention of palliative care in the emergency department during the COVID-19 pandemic. JAMA Intern Med 180:1252–1254

Lim ZJ, Subramaniam A, Reddy MP, Blecher G, Kadam U, Afroz A, Billah B, Ashwin S, Kubicki M, Bilotta F, Curtis JR, Rubulotta F (2020) Case fatality rates for COVID-19 patients requiring invasive mechanical ventilation: a meta-analysis. Am J Respir Crit Care Med

Lim ZJ, Subramaniam A, Ponnapa Reddy M, Blecher G, Kadam U, Afroz A, Billah B, Ashwin S, Kubicki M, Bilotta F, Curtis JR, Rubulotta F (2021) Case fatality rates for patients with COVID-19 requiring invasive mechanical ventilation. A meta-analysis. Am J Respir Crit Care Med 203:54–66

Livingston E, Bucher K (2020) Coronavirus disease 2019 (COVID-19) in Italy. JAMA 323:1335

McGain F, Humphries RS, Lee JH, Schofield R, French C, Keywood MD, Irving L, Kevin K, Patel J, Monty J (2020) Aerosol generation related to respiratory interventions and the effectiveness of a personal ventilation hood. Crit Care Resusc

Medicine JHUa (2023) COVID-19 dashboard. In: COVID-19 Dashboard

Mitnitski AB, Mogilner AJ, Rockwood K (2001) Accumulation of deficits as a proxy measure of aging. Sci World J 1:323–336

Mitnitski A, Song X, Rockwood K (2013) Assessing biological aging: the origin of deficit accumulation. Biogerontology 14:709–717

Mueller N, Murthy S, Tainter CR, Lee J, Riddell K, Fintelmann FJ, Grabitz SD, Timm FP, Levi B, Kurth T, Eikermann M (2016) Can sarcopenia quantified by ultrasound of the rectus femoris muscle predict adverse outcome of surgical intensive care unit patients as well as frailty? A prospective, observational Cohort study. Ann Surg 264:1116–1124

Muscedere J, Waters B, Varambally A, Bagshaw SM, Boyd JG, Maslove D, Sibley S, Rockwood K (2017) The impact of frailty on intensive care unit outcomes: a systematic review and meta-analysis. Intensive Care Med 43:1105–1122

National Institute for Health and Care Excellence: Clinical Guidelines (2022) COVID-19 rapid guideline: managing COVID-19. National Institute for Health and Care Excellence (NICE). Copyright © NICE 2021, London

Nguyen YL, Angus DC, Boumendil A, Guidet B (2011) The challenge of admitting the very elderly to intensive care. Ann Intensive Care 1:29

O'Caoimh R, Kennelly S, Ahern E, O'Keeffe S, Ortuno RR (2020) Letter to the editor: Covid-19 and the challenges of frailty screening in older adults. J Frailty Aging 9:185–186

Okahara S, Subramaniam A, Darvall JN, Ueno R, Bailey M, Pilcher DV (2022) The relationship between frailty and mechanical ventilation: a population-based cohort study. Ann Am Thorac Soc 19:264–271

O'Mara L SS, Orkaby AR, Ouchi K, Bernacki R (2020) A framework to triage older adults with Covid-19 to provide patient-centered care. NEJM Catal 1–11

Pan D, Sze S, Minhas JS, Squire IB, Pareek M (2020) Frailty and mortality in patients with COVID-19. Lancet Public Health 5:e581

Parker SG, McCue P, Phelps K, McCleod A, Arora S, Nockels K, Kennedy S, Roberts H, Conroy S (2018) What is comprehensive geriatric assessment (CGA)? An umbrella review. Age Ageing 47:149–155

Peñuelas O, Del Campo-Albendea L, de Aledo ALG, Añón JM, Rodríguez-Solís C, Mancebo J, Vera P, Ballesteros D, Jiménez J, Maseda E, Figueira JC, Franco N, Algaba Á, Avilés JP, Díaz R, Abad B, Canabal A, Abella A, Gordo F, García J, Suarez JG, Cedeño J, Martínez-Palacios B, Manteiga E, Martínez Ó, Blancas R, Bardi T, Pestaña D, Lorente J, Muriel A, Esteban A, Frutos-Vivar F (2021) Long-term survival of mechanically ventilated patients with severe COVID-19: an observational cohort study. Ann Intensive Care 11:143

Phua J, Weng L, Ling L, Egi M, Lim CM, Divatia JV, Shrestha BR, Arabi YM, Ng J, Gomersall CD, Nishimura M, Koh Y, Du B, Asian Critical Care Clinical Trials Group (2020) Intensive care management of coronavirus disease 2019 (COVID-19): challenges and recommendations. Lancet Respir Med 8:506–517

Polack FP, Thomas SJ, Kitchin N, Absalon J, Gurtman A, Lockhart S, Perez JL, Pérez Marc G, Moreira ED, Zerbini C, Bailey R, Swanson KA, Roychoudhury S, Koury K, Li P, Kalina WV, Cooper D, Frenck RW Jr, Hammitt LL, Türeci Ö, Nell H, Schaefer A, Ünal S, Tresnan DB, Mather S, Dormitzer PR, Şahin U, Jansen KU, Gruber WC (2020) Safety and efficacy of the BNT162b2 mRNA Covid-19 vaccine. N Engl J Med 383:2603–2615

Polok K, Fronczek J, Artigas A, Flaatten H, Guidet B, De Lange DW, Fjølner J, Leaver S, Beil M, Sviri S, Bruno RR, Wernly B, Bollen Pinto B, Schefold JC, Studzińska D, Joannidis M, Oeyen S, Marsh B, Andersen FH, Moreno R, Cecconi M, Jung C, Szczeklik W (2022) Noninvasive ventilation in COVID-19 patients aged ≥ 70 years-a prospective multicentre cohort study. Crit Care 26:224

Pugh RJ, Ellison A, Pye K, Subbe CP, Thorpe CM, Lone NI, Clegg A (2018) Feasibility and reliability of frailty assessment in the critically ill: a systematic review. Crit Care 22:49

Pugh RJ, Bailey R, Szakmany T, Al Sallakh M, Hollinghurst J, Akbari A, Griffiths R, Battle C, Thorpe C, Subbe CP, Lyons RA (2021) Long-term trends in critical care admissions in Wales*. Anaesthesia 76:1316–1325

Richardson S, Hirsch JS, Narasimhan M, Crawford JM, McGinn T, Davidson KW, Barnaby DP, Becker LB, Chelico JD, Cohen SL, Cookingham J, Coppa K, Diefenbach MA, Dominello AJ, Duer-Hefele J, Falzon L, Gitlin J, Hajizadeh N, Harvin TG, Hirschwerk DA, Kim EJ, Kozel ZM, Marrast LM, Mogavero JN, Osorio GA, Qiu M, Zanos TP (2020) Presenting characteristics, comorbidities, and outcomes among 5700 patients hospitalized with COVID-19 in the New York City Area. JAMA 323:2052–2059

Robert R, Skrifvars MB, Ranzani OT (2017) Is this critically ill patient elderly or too old? Intensive Care Med 43:1884–1886

Rockwood K, Mitnitski A (2012) How might deficit accumulation give rise to frailty? J Frailty Aging 1:8–12

Rockwood K, Theou O (2020) Using the clinical frailty scale in allocating scarce health care resources. Can Geriatr J 23:210–215

Rockwood K, Song X, MacKnight C, Bergman H, Hogan DB, McDowell I, Mitnitski A (2005) A global clinical measure of fitness and frailty in elderly people. CMAJ 173:489–495

Rottler M, Ocskay K, Sipos Z, Görbe A, Virág M, Hegyi P, Molnár T, Erőss B, Leiner T, Molnár Z (2022) Clinical frailty scale (CFS) indicated frailty is associated with increased in-hospital and 30-day mortality in COVID-19 patients: a systematic review and meta-analysis. Ann Intensive Care 12:17

Searle SD, Mitnitski A, Gahbauer EA, Gill TM, Rockwood K (2008) A standard procedure for creating a frailty index. BMC Geriatr 8:24

Sekulić A, Likić R, Matas M (2020) How to allocate intensive care resources during the COVID-19 pandemic: medical triage or a priori selection? Croat Med J 61:276–278

So RKL, Bannard-Smith J, Subbe CP, Jones DA, van Rosmalen J, Lighthall GK (2018) The association of clinical frailty with outcomes of patients reviewed by rapid response teams: an international prospective observational cohort study. Crit Care 22:227

Straw S, McGinlay M, Drozd M, Slater TA, Cowley A, Kamalathasan S, Maxwell N, Bird RA, Koshy AO, Prica M, Patel PA, Relton SD, Gierula J, Cubbon RM, Kearney MT, Witte KK (2021) Advanced care planning during the COVID-19 pandemic: ceiling of care decisions and their implications for observational data. BMC Palliat Care 20:10

Subramaniam A, Pilcher D, Tiruvoipati R, Wilson J, Mitchell H, Xu D, Bailey M (2021) Timely goals of care documentation in patients with frailty in the COVID-19 era: a retrospective multi-site study. Intern Med J

Subramaniam A, Ueno R, Tiruvoipati R, Srikanth V, Bailey M, Pilcher D (2022a) Comparison of the predictive ability of clinical frailty scale and hospital frailty risk score to determine long-term survival in critically ill patients: a multicentre retrospective cohort study. Crit Care 26:121

Subramaniam A, Ueno R, Tiruvoipati R, Darvall J, Srikanth V, Bailey M, Pilcher D, Bellomo R (2022b) Comparing the clinical frailty scale and an international classification of diseases-10 modified frailty index in predicting long-term survival in critically ill patients. Crit Care Explor 4:e0777

Subramaniam A, Anstey C, Curtis JR, Ashwin S, Ponnapa Reddy M, Aliberti MJR, Avelino-Silva TJ, Welch C, Koduri G, Prowle JR, Wan YI, Laurent MR, Marengoni A, Lim JP, Pilcher D, Shekar K (2022c) Characteristics and outcomes of patients with frailty admitted to ICU with coronavirus disease 2019: an individual patient data meta-analysis. Crit Care Explor 4:e0616

Subramaniam A, Shekar K, Anstey C, Tiruvoipati R, Pilcher D (2022d) Impact of frailty on clinical outcomes in patients with and without COVID-19 pneumonitis admitted to intensive care units in Australia and New Zealand: a retrospective registry data analysis. Crit Care 26:301. https://doi.org/10.1186/s13054-022-04177-9

Subramaniam A, Shekar K, Afroz A, Ashwin S, Billah B, Brown H, Kundi H, Lim ZJ, Ponnapa Reddy M, Curtis JR (2022e) Frailty and mortality associations in patients with COVID-19: a systematic review and meta-analysis. Intern Med J 52:724–739

Subramaniam A, Tiruvoipati R, Pilcher D, Bailey M (2023) Treatment limitations and clinical outcomes in critically ill frail patients with and without COVID-19 pneumonitis. J Am Geriatr Soc 71:145–156

Supady A, Curtis JR, Abrams D, Lorusso R, Bein T, Boldt J, Brown CE, Duerschmied D, Metaxa V, Brodie D (2021) Allocating scarce intensive care resources during the COVID-19 pandemic: practical challenges to theoretical frameworks. Lancet Respir Med 9:430–434

Swiss Academy of Medical Sciences (2020) COVID-19 pandemic: triage for intensive-care treatment under resource scarcity (3rd, updated version). Swiss Med Wkly 150:w20401

Taniguchi LU, Avelino-Silva TJ, Dias MB, Jacob-Filho W, Aliberti MJR (2022a) Association of frailty, organ support, and long-term survival in critically ill patients with COVID-19. Crit Care Explor 4:e0712

Taniguchi LU, Avelino-Silva TJ, Dias MB, Jacob-Filho W, Aliberti MJR (2022b) Patient-centered outcomes following COVID-19: frailty and disability transitions in critical care survivors. Crit Care Med 50:955–963

Taniguchi LU, Aliberti MJR, Dias MB, Jacob-Filho W, Avelino-Silva TJ (2023) Twelve months and counting: following clinical outcomes in critical COVID-19 survivors. Ann Am Thorac Soc 20:289–295

Tehrani S, Killander A, Astrand P, Jakobsson J, Gille-Johnson P (2021) Risk factors for death in adult COVID-19 patients: frailty predicts fatal outcome in older patients. Int J Infect Dis 102:415–421

Theou O, Rockwood K (2015) Comparison and clinical applications of the frailty phenotype and frailty index approaches. Interdiscip Top Gerontol Geriatr 41:74–84

Theou O, Walston J, Rockwood K (2015) Operationalizing frailty using the frailty phenotype and deficit accumulation approaches. Interdiscip Top Gerontol Geriatr 41:66–73

Theou O, Squires E, Mallery K, Lee JS, Fay S, Goldstein J, Armstrong JJ, Rockwood K (2018) What do we know about frailty in the acute care setting? A scoping review. BMC Geriatr 18:139

Theou O, Pérez-Zepeda MU, van der Valk AM, Searle SD, Howlett SE, Rockwood K (2021) A classification tree to assist with routine scoring of the clinical frailty scale. Age Ageing 50(4):1406–1411. https://doi.org/10.1093/ageing/afab006

Tipping CJ, Hodgson CL, Harrold M, Chan T, Holland AE (2019) Frailty in patients with trauma who are critically Ill: a prospective observational study to determine feasibility, concordance, and construct and predictive validity of 2 frailty measures. Phys Ther 99:1089–1097

Truog RD, Mitchell C, Daley GQ (2020) The toughest triage—allocating ventilators in a pandemic. N Engl J Med 382:1973–1975

Vahidy FS, Drews AL, Masud FN, Schwartz RL, Askary BB, Boom ML, Phillips RA (2020) Characteristics and outcomes of COVID-19 patients during initial peak and resurgence in the Houston Metropolitan Area. JAMA 324:998–1000

Valiani S, Terrett L, Gebhardt C, Prokopchuk-Gauk O, Isinger M (2020) Development of a framework for critical care resource allocation for the COVID-19 pandemic in Saskatchewan. CMAJ 192:E1067–E1073

Van Den Noortgate NJ, Van den Block L (2022) End-of-life care for older people: the way forward. Age Ageing 51

Vink EE, Azoulay E, Caplan A, Kompanje EJO, Bakker J (2018) Time-limited trial of intensive care treatment: an overview of current literature. Intensive Care Med 44:1369–1377

Wang D, Hu B, Hu C, Zhu F, Liu X, Zhang J, Wang B, Xiang H, Cheng Z, Xiong Y, Zhao Y, Li Y, Wang X, Peng Z (2020) Clinical characteristics of 138 hospitalized patients with 2019 novel coronavirus-infected pneumonia in Wuhan, China. JAMA 323:1061–1069

Wernly B, Flaatten H, Leaver S, Guidet B, Jung C, Fjølner J, Beil M, Oeyen S, Szczeklik W, Elhadi M, Sviri S, deLange D, Moreno R, Artigas A, Dudzinski D, Serck N, Nedergaard HK, Shaat I, Hussein AM, Zanaty M, Hassanin E, Hussein AM, Mohamed N, Omar M, El-wahed GAAA, Touny S, Aujayeb A, Nseir S, Urbina T, Garcon P, Rigaud J-P, Vanderlinden T, Valette X, Megarbane B, Baron E, Nigeon O, Plantefeve G, Foucault C, Monchi M, Fuest K, Bruno R, Kelm M, Kabitz H-J, Schaller S, Abusalama A, Embarek H, Anaiba M, Taher A, Alkaseek A, Evers M, Dieperink W, Cornet AD, Brochado F, Lopez-Cuenca S, Aldiabat M, Al-Sadawi M, the COVIP Investigators (2023) The clinical frailty scale, but not the FRAIL checklist is associated with mortality in old critically ill patients with COVID-19. Crit Care 27:101

World Health Organization (2020) Director-general's remarks at the media briefing on 2019-nCoV on 11 February 2020. World Health Organization

Wu Z, McGoogan JM (2020) Characteristics of and important lessons from the coronavirus disease 2019 (COVID-19) outbreak in China: summary of a report of 72 314 cases from the Chinese Center for Disease Control and Prevention. JAMA 323:1239–1242

Yang X, Yu Y, Xu J, Shu H, Xia J, Liu H, Wu Y, Zhang L, Yu Z, Fang M, Yu T, Wang Y, Pan S, Zou X, Yuan S, Shang Y (2020) Clinical course and outcomes of critically ill patients with SARS-CoV-2 pneumonia in Wuhan, China: a single-centered, retrospective, observational study. Lancet Respir Med 8:475–481

Zeng A, Song X, Dong J, Mitnitski A, Liu J, Guo Z, Rockwood K (2015) Mortality in relation to frailty in patients admitted to a specialized geriatric intensive care unit. J Gerontol A Biol Sci Med Sci 70:1586–1594

Zhang XM, Jiao J, Cao J, Huo XP, Zhu C, Wu XJ, Xie XH (2021) Frailty as a predictor of mortality among patients with COVID-19: a systematic review and meta-analysis. BMC Geriatr 21:186

Zhou F, Yu T, Du R, Fan G, Liu Y, Liu Z, Xiang J, Wang Y, Song B, Gu X, Guan L, Wei Y, Li H, Wu X, Xu J, Tu S, Zhang Y, Chen H, Cao B (2020) Clinical course and risk factors for mortality of adult inpatients with COVID-19 in Wuhan, China: a retrospective cohort study. Lancet 395:1054–1062

Respiratory Support in COVID-19-Related Respiratory Failure: Lessons Learnt

5

Ryan Ruiyang Ling, Kollengode Ramanathan, Ashwin Subramaniam, and Kiran Shekar

Abstract

The COVID-19 pandemic has undeniably changed the way intensivists manage acute hypoxaemic respiratory failure. Paradigms had evolved particularly in the way we support patients with respiratory failure, and the adjunctive therapies which can be used. Many questions have been answered, and many more generated, from the last few years. For example, is COVID-19 acute respiratory failure and acute respiratory distress syndrome similar to non-COVID-19? How can we personalize therapy in patients with COVID-19, and what are some new statistical tools that we can use to aid in this approach? Is intubation and invasive mechanical ventilation the only way to support patients with acute respiratory failure, or can we turn to other modalities of respiratory support? And what about patients with the most severe form of respiratory failure, how can we support them? In this chapter, we explore the lessons learnt, identifying gaps and advances in knowledge in terms of the pathophysiology of acute respiratory failure, its prognostic factors, oxygen supports, and other therapies. We also touch on how physicians treating patients can tap on international networks to create a "whole that is more than the sum of its parts", and impart clinical insights on the management of acute respiratory failure. Finally, we highlight the importance of a

R. R. Ling · K. Ramanathan
Yong Loo Lin School of Medicine, National University of Singapore, National University Health System, Singapore, Singapore

R. R. Ling
Australian and New Zealand Intensive Care Research Centre, School of Preventive Medicine and Public Health, Monash University, Melbourne, VIC, Australia

K. Ramanathan
Cardiothoracic Intensive Care Unit, National University Heart Centre Singapore, National University Health System, Singapore, Singapore

A. Subramaniam
Department of Intensive Care Medicine, Peninsula Health, Frankston, VIC, Australia

Department of Intensive Care Medicine, Dandenong Hospital, Monash Health, Melbourne, VIC, Australia

Peninsula Clinical School, Monash University, Clayton, VIC, Australia

K. Shekar (✉)
Adult Intensive Care Services, The Prince Charles Hospital, Brisbane, QLD, Australia
e-mail: kiran.shekar@health.qld.gov.au

Faculty of Medicine, University of Queensland, Brisbane, QLD, Australia

Institute of Health and Biomedical Innovation, Queensland University of Technology, Brisbane, QLD, Australia

Faculty of Health Sciences and Medicine, Bond University, Gold Coast, QLD, Australia

N. Rezaei (ed.), *The COVID-19 Aftermath*, Advances in Experimental Medicine and Biology 1457,
https://doi.org/10.1007/978-3-031-61939-7_5

cautious skepticism in our approach to both clinical medicine and evidence-based medicine, highlighting how evidence in a pandemic can rapidly evolve both within an ICU, and longitudinally around the world.

Keywords

COVID-19 · Acute respiratory failure · Intensive care medicine · Invasive mechanical ventilation · Extracorporeal membrane oxygenation

The coronavirus disease 2019 (COVID-19) pandemic has undoubtedly changed the field of intensive care medicine, just like the polio epidemics that laid the foundations for mechanical ventilation and organized intensive care. Never have we seen hundreds of millions of patients develop acute respiratory failure in such a short span of time, which led to extensive resource constraints and difficulty in treating patients. However, this also gave rise to several innovations that have spurred the development and progress of intensive care medicine as a specialty. In this chapter, we explore the pathophysiology of COVID-19 respiratory failure, the available respiratory supports in patients with COVID-19, and lessons learnt from that can be applied to daily clinical practice.

5.1 Pathophysiology of COVID-19-Related Acute Respiratory Failure

Patients with COVID-19-related acute respiratory failure fall under the National Institutes of Health (2022) and World Health Organisation (2022) definition of severe-to-critical COVID-19. These patients typically require oxygen supplementation to maintain physiological levels of oxygen saturation. These patients are likely to fulfil the Berlin criteria for acute respiratory distress syndrome (ARDS) (The ARDS Definition Task Force 2012), and mortality rates can reach in excess of 40% (Lim et al. 2021).

The exact mechanism by which COVID-19-related respiratory failure occurs is not completely understood. Radiological investigations and post-mortem autopsies have shown that patients with COVID-19 display features of "typical" ARDS disease pathophysiology, including diffuse alveolar damage with hyaline membrane formation and an increase in lung weight (Sofizan et al. 2022). This has been attributed to two mechanisms: direct alveolar injury by viral particles, and the hyperinflammatory cytokine cascade following the alveolar injury (Ball et al. 2022). Yet, certain features of COVID-19 respiratory failure do not align with the "typical" ARDS. First, the immune milieu in COVID-19 may not mirror the cytokine "storm" observed in other hyperinflammatory syndromes (Leisman et al. 2020). Second, the extent of hypoxemia and shunt in COVID-19-ARDS appeared out-of-proportion to the reduction in the respiratory system mechanics and static respiratory compliance (Gattinoni et al. 2020a). Finally, the rates of thromboembolic and haemorrhagic complications (Sofizan et al. 2022), and coagulopathy (Mitra et al. 2021) following COVID-19 appear higher than non-COVID-19-ARDS. Some experts have hypothesized COVID-19 as being primarily a "vascular" disease (Cherian 2021) which increases dead space, compared to the 'typical' ARDS where the increased shunt is caused due to diffuse alveolar infiltrates and injury.

What can we learn from the current discussion about the mechanism of COVID-19 ARDS? A better understanding of the underlying disease pathophysiology will guide better management. It is imperative for the debate to continue healthily for a deeper understanding of how COVID-19-ARDS occurs, and what pathways we can target to treat such patients. As the pandemic progressed, it became increasingly clear that the underlying disease pathology may not be identical for each patient and as such experts have recommended a more personalized approach to managing COVID-19-ARDS.

5.2 Phenotypes and Precision Medicine in COVID-19 Respiratory Failure

The breadth of the Berlin consensus definition of ARDS allowed for efficient patient enrolment within randomized clinical trials (RCTs). However, there are concerns if we are approaching the ceiling of detectable benefit in ARDS, and observational studies show that mortality rates have not decreased substantially with time (Bellani et al. 2016). It is increasingly becoming recognized that ARDS is a syndrome that has many possible aetiologies, and is not a diagnosis in and of itself (Gattinoni and Marini 2021). As such, it is difficult to establish whether current trials investigating one therapy for a heterogeneous group are targeting the appropriate population. To design meaningful clinical trials, there is a need to identify appropriate subgroups of patients in which these interventions will be beneficial. Central to this problem lies the concepts of prognostic enrichment (selecting for the population where event rates are higher for more clinically meaningful benefits and larger effect estimates) and predictive enrichment (selecting patients who are more likely to respond to the therapy based on a biological mechanism, which can include serum biomarkers) (Stanski and Wong 2020). Outside of COVID-19-ARDS, several studies have used clustering approaches (*k*-means clustering, latent class analyses) to hypothesize the existence of ARDS phenotypes—most famously the "hyperinflammatory" and "hypoinflammatory" phenotypes. This has been applied to more recent trials, including the Statins for Acutely Injured Lungs from Sepsis (SAILS) study, and the ARDSnet trials on tidal volume (ARMA), positive end-expiratory pressure (ALVEOLI,) and fluid management strategies (FACTT) (Duggal et al. 2022; Sinha et al. 2018; Famous et al. 2017; Calfee et al. 2014).

These methods have been extended to patients with COVID-19-related respiratory failure and ARDS. The initial hypothesis of potential phenotypes was proposed by Gattinoni and colleagues (Gattinoni et al. 2020a), who found divergent phenotypes of COVID-19: the type "L" phenotype with low elastance, ventilation-perfusion ratio, lung weight, and recruitability, and the type "H" phenotype with high elastance, right-to-left shunt, lung weight, and recruitability (Gattinoni et al. 2020b). There is still an intense debate about whether these phenotypes exist, and if they are of prognostic significance. A meta-analysis found that the static respiratory compliance of patients receiving invasive mechanical ventilation for COVID-19 ARDS was not different from the normal distribution (Reddy et al. 2022). Another large observational study utilizing latent class analysis and group-based trajectory modelling found that patients receiving invasive mechanical ventilation (IMV) for COVID-19-ARDS developed sub-phenotypes not at baseline, but over time. These sub-phenotypes were not defined based on respiratory system compliance, but dead space ventilation, mechanical power and ventilatory ratio (Bos et al. 2021).

Apart from sub-phenotypes based on disease factors and prognosis, studies have also found sub-phenotypes in patients receiving IMV for COVID-19 based on differential responses to interventions. A latent class analysis found steroid-resistant and steroid-responsive sub-phenotypes in COVID-19 and highlighted significant differences in mortality outcomes. These phenotypes interestingly appeared to mirror the hypoinflammatory and hyperinflammatory sub-phenotypes in non-COVID-19 ARDS (Sinha et al. 2021). Unsupervised clustering analyses of the randomized, embedded, multifactorial, adaptive platform trial for community-acquired pneumonia (REMAP-CAP) found three unique sub-phenotypes with different biomarker patterns, response to therapy, and mortality outcomes (Fish et al. 2022). Finally, in patients with ARDS refractory to conventional treatment, a single-centre study has shown three different sub-phenotypes of patients with divergent outcomes: survival to ICU discharge was 96% in one phenotype, and 50% in others. The phenotypes also displayed variable degrees of organ failure and organ support requirements, as well as inflammation (Zhang et al. 2021). Further studies on phenotypes in patients with COVID-19-ARDS, and non-COVID-19-ARDS are currently underway, and the results are eagerly awaited. Bearing in mind the current understanding of COVID-19 acute respiratory failure and ARDS, we further explore the utility of various interventions that have been investigated in the last few years of the pandemic.

5.3 Oxygen Supplementation and Trends in the Escalation of Respiratory Support in COVID-19

Early on in the pandemic, it was unclear if the virus could potentially be aerosolized (Hamilton 2021). To prioritize the safety of healthcare professionals and reduce the nosocomial spread of the virus, many centres elected for early endotracheal intubation. As such, the main form of respiratory support was IMV. However, as the pandemic progressed and newer waves of COVID-19 overwhelmed healthcare systems, more physicians avoided endotracheal intubation and IMV (Shekar and Ling 2022). This led to the increased uptake of non-invasive positive pressure ventilation (NIV) as well as high-flow nasal cannula (HFNC). It is important to note that the optimal method of respiratory support in patients with COVID-19-related respiratory failure is still a subject of controversy. In the next few sections, we discuss the evidence base about these respiratory supports.

5.4 Awake-Prone Positioning

While extensive research has gone into prone positioning in patients who received mechanical ventilation (covered later in this chapter), data on patients who were not invasively ventilated, or receiving ECMO remained sparse before COVID-19. In this section, we explore the use of prone positioning as an adjunct in COVID-19-related respiratory failure and focus our attention on these populations where there was a relatively smaller volume of literature being published.

Drawing from its physiological benefits in non-COVID-19 ARDS and partly due to resource constraints, prone positioning was used in patients who did not receive mechanical ventilation. A meta-analysis in 2020 based on observational studies found significant improvements in oxygenation, but not intubation and mortality (Ponnapa Reddy et al. 2021). A large, multinational meta-trial found a significant reduction in intubation, but not mortality (Ehrmann et al. 2021), which was echoed by a second meta-analysis incorporating both observational and randomized data in 2022 (Li et al. 2022*).* Yet, following this, three new trials published with similar durations of prone positioning (4–6 h daily) did not show a significant reduction in intubation or mortality (Alhazzani et al. 2022; Fralick et al. 2022; Qian et al. 2022).

What then do we make of the current evidence in awake prone positioning? Awake prone positioning is likely more appropriate in patients with more severe respiratory failure (admission to ICU, receiving advanced respiratory support, or PaO_2/FiO_2 ratio of < 150), and were able to receive awake prone positioning for a prolonged period of time (Ponnapa Reddy et al. 2021; Li et al. 2022). Incidentally, patients with more severe respiratory failure were more likely to have more intensive monitoring, lower nurse-to-patient ratio, and as such, better adherence to prone positioning (Fralick et al. 2022). Several studies have shown a dose–response relationship between prone positioning and its beneficial effects on oxygenation (Shekar and Ling 2022). As such, while awake prone positioning can be beneficial in patients with acute respiratory failure, it must be conducted appropriately in a setting that optimizes adherence and encourages patient participation in the intervention.

5.5 Oxygen Supplementation and Non-invasive Supplementation

The two main modalities of NIV for COVID-19 comprise continuous positive airway pressure (CPAP) and bilevel positive airway pressure (BiPAP). A network meta-analysis in 2020 investigating non-invasive respiratory support in RCTs conducted before the pandemic found that the use of a helmet or face mask NIV, and HFNC were associated with a reduction in endotracheal intubation and mortality when compared to conventional oxygen therapy (Ferreyro et al. 2020).

Numerous RCTs have been conducted, and the results from the overall literature appear conflicting. These results are briefly summarized in Table 5.1.

An updated network meta-analysis in 2022 including trials conducted in patients with and without COVID-19 found that HFNC and CPAP reduce mortality, while HFNC and mask BiPAP reduce the risk of mechanical ventilation (Zeraatkar et al. 2022). The analysis also found that all non-invasive respiratory supports reduce hospitalization length of stay. Nonetheless, in the face of conflicting evidence from the RCTs, more direct evidence is required to confirm this finding of the network meta-analysis.

5.6 Invasive Mechanical Ventilation

Certain patients who present with progressively worsening respiratory failure or hemodynamic instability may require invasive mechanical

Table 5.1 Evidence base for non-invasive ventilation in COVID-19

Intervention	Trial	Findings
NIV versus HFNC	HENIVOT (Grieco et al. 2021; Nair et al. 2021)	Reduction in endotracheal intubation for NIV group, but no reduction in hospital mortality or respiratory support-free days within 28 days
	Nair et al. (Grieco et al. 2021)	
	COVIDICUS (Bouadma et al. 2022)	No difference in intubation, mortality, or length of stay
HFNC versus standard oxygen therapy	HENIVOT (Grieco et al. 2021)	Reduction in endotracheal intubation
	Crimi et al. (2022)	No difference in escalation of respiratory support
	Recovery-RS (Perkins et al. 2022)	
	COVIDICUS (Bouadma et al. 2022)	No difference in intubation, mortality or length of stay
CPAP versus standard oxygen therapy	Recovery-RS (Perkins et al. 2022)	Reduction in intubation and mortality
	COVIDICUS (Bouadma et al. 2022)	No difference in intubation, mortality or length of stay
HFNC versus standard oxygen therapy versus CPAP versus BiPAP	Zeraatkar et al. (2022)	HFNC, and CPAP reduce mortality HFNC and mask BiPAP reduce intubation

ventilation (IMV). In those who have been intubated, the initial ventilation settings should follow ARDSNet lung-protective settings (Brower et al. 2000). Some experts have cited a lower threshold for higher levels of PEEP given that COVID-19 primarily causes a hypoxaemic respiratory failure, although the evidence regarding this remains inconclusive (Protti et al. 2022).

It is important to note that IMV using a ventilator was first used in polio nearly 80 years ago—its main indication was respiratory muscle paralysis and difficulty with ventilation. Resources were scarce and there were insufficient negative pressure ventilation devices to support a large number of patients, and positive pressure ventilation was used as a last resort measure. Yet, the most common indication for IMV currently is not an issue of ventilation, but of gas exchange and oxygenation. Because of this, it is important to view modalities such as NIV and HFNC not as options only in the context of resource constraints, but as possibly viable alternatives to IMV in well-selected patients. It is also pertinent to consider the potential complications and harms from IMV including sedation and closer monitoring which requires more resources (Shekar and Ling 2022). More importantly, patients receiving IMV cannot ambulate and are prone to complications including ventilator-associated pneumonia (which is associated with a mortality rate of 30–70%) (Chouhdari et al. 2018), and ventilator-induced lung injury.

Yet, it is equally fundamental to consider the harms of NIV including patient self-inflicted lung injury, patient discomfort and claustrophobia, and facial skin breakdown and ulcers when using face mask NIV (Zeraatkar et al. 2022). Intubation and lung-protective ventilation or escalating respiratory support with ECMO is also a competing event with NIV, and physicians should consider who requires increasing respiratory support more urgently. While the timing of intubation has not been shown to affect mortality (Papoutsi et al. 2021), studies have shown that in patients who receive ECMO, the duration of invasive mechanical ventilation and symptom duration before ECMO (which includes the duration of non-invasive ventilation) are prognostic factors for mortality (Tran et al. 2022). This will likely apply to only a subset of patients who do require rapidly increasing respiratory support, but for these patients, the decision of NIV or escalating respiratory support can make a difference. The risk–benefit ratio of escalating respiratory support varies with demographic factors (age, comorbidities, baseline functional status), centre factors (experience with IMV, NIV, resource constraints), and disease factors (severity of hypoxaemia, respiratory drive). As such, the consideration of which oxygenation strategy to use should be based on all these factors, as well as the patient's preferences where appropriate.

5.7 Adjunctive Therapies in Patients Receiving Mechanical Ventilation

Traditionally, respiratory support is delivered in the supine position. More recently, RCTs and meta-analyses have shown the benefits of prone positioning in acute respiratory failure across a wide spectrum of patients who require supplemental oxygen, have been ventilated, and even while receiving ECMO. There are several postulated mechanisms for this, including equalization of pleural pressure and avoiding alveolar overinflation, homogenization of the lung parenchyma, and improved ventilation and perfusion matching (Shekar and Ling 2022; Poon et al. 2021). Outside of COVID-19, the PROSEVA trial in 2013 showed that prone positioning reduced mortality in patients with moderate-to-severe ARDS receiving invasive mechanical ventilation (Guérin et al. 2013).

Other adjuvant therapies have been investigated before COVID-19, and more studies investigating these therapies in COVID-19 have been reported as well. Inhaled nitric oxide was investigated in non-COVID-19 ARDS and was postulated to improve hypoxic pulmonary vasoconstriction by providing selective vasodilation in well-ventilated areas of the lung, thereby correcting the ventilation-perfusion mismatch in ARDS. However, a Cochrane review in 2016 did not find significant mortality benefits of nitric oxide (Gebistorf et al. 2016). While there was an improvement in oxygenation, this was accompanied by an increase in acute kidney injury. Within COVID-19, a multicentre cohort study found that inhaled nitric oxide did not reduce mortality and was associated with fewer ventilator-free days, longer hospital stays, and acute kidney injury (Al Sulaiman et al. 2022). Another intervention is prostanoids—a meta-analysis in 2015 based on largely observational data showed that inhaled prostaglandins were associated with improvements in oxygenation and reduction in pulmonary artery pressures (Fuller et al. 2015). One RCT investigating prostacyclin infusion in ARDS found a trend towards reduction in mortality and ventilator-free days (Fuller et al. 2015). High frequency oscillatory ventilation (HFOV) was postulated to recruit collapsed alveoli and minimize tidal stress and strain by delivering very small tidal volumes at higher airway pressures. The ATS/ESICM/SCCM guidelines found that there was no mortality benefit in patients receiving HFOV (Fan et al. 2017), and an individual participant data meta-analysis of more than 1500 patients found that HFOV might only be helpful in patients with a PaO_2/FiO_2 ratio of < 100, limiting the role of HFOV in COVID-19 ARDS (Meade et al. 2017). In addition, HFOV requires specialized ventilators and teams well-equipped to run the ventilator. Given the use of VV-ECMO as a rescue therapy, it is likely that HFOV will not have a large role to play in ARDS.

Apart from this, several therapies can be used in COVID-19 related respiratory failure, which were not investigated deeply before. These include the use of steroids. A meta-analysis investigating corticosteroids in both non-COVID-19 and COVID-19 ARDS found a reduction in mortality in non-COVID-19 ARDS (Chaudhuri et al. 2021), with no significant interaction effect in the aetiology of ARDS. In patients with respiratory failure without ARDS, corticosteroids reduced mortality and escalating respiratory support (The W.H.O. Rapid Evidence Appraisal for Covid-Therapies Working Group 2020; Pulakurthi et al. 2021). Another interesting therapy introduced was that of cytokine blockers–specifically interleukin-6 blockers (tocilizumab, sarilumab) and tumour necrosis factor-alpha (TNF-a, baricitinib), all of which have been shown to reduce mortality and increase ventilator-free days in patients with critical COVID-19 requiring mechanical ventilation. The World Health Organisation has released a living guideline for medications to treat COVID-19, this is based on a living network meta-analysis that is periodically updated. As the evidence for and against certain treatments evolves, these recommendations will be updated accordingly.

5.8 Extracorporeal Membrane Oxygenation

Extracorporeal membrane oxygenation (ECMO) is an extracorporeal device that provides cardiorespiratory support in a patient with severe respiratory or cardiac failure that is refractory to conventional therapies. In the context of COVID-19, venovenous (VV) ECMO is most commonly used. This involves oxygenating and removing carbon dioxide from venous blood, and returning it back to the venous circulation (Broman et al. 2019), while resting the diseased lungs. In situations with concomitant cardiac failure, blood can be returned to the arterial circulation in venoarterial (VA) ECMO (Broman et al. 2019). Early in the pandemic, there was little to no information on whether ECMO would be beneficial in COVID-19. Based on previous respiratory virus outbreaks, ECMO had a role to play in patients with critical illness (Cho et al. 2020). Yet, providing a resource-intensive therapy like ECMO required proper planning and centralization of ECMO in higher volume centres. Initial guidelines recommended the consideration of ECMO provided the centre had sufficient experience and resources to provide ECMO, and judicious patient selection (Alhazzani et al. 2020; Shekar et al. 2020).

The initial reports of ECMO for COVID-19 showed dismal outcomes (Henry and Lippi 2020), yet data from experienced centres found better mortality rates similar to non-COVID-19 ARDS (Schmidt et al. 2020). Within the first year of the pandemic, approximately 40% of patients receiving ECMO for COVID-19 died (Barbaro et al. 2020; Ramanathan et al. 2021). Yet with time, mortality rates for ECMO in COVID-19 had gradually increased as the pandemic progressed (Diaz et al. 2021; Karagiannidis et al. 2021; Schmidt et al. 2021; Barbaro et al. 2021; Ling et al. 2022). The exact reason for this is not known—some potential factors investigated include the use of immunomodulators (Ling et al. 2022), the increase and prolonged use of NIV, and evolving viral variants (Tran et al. 2022). Given this worrying signal, can ECMO reduce mortality in COVID-19? While there exist logistical and ethical difficulties in conducting an RCT in patients who require ECMO, observational data can be used to emulate pragmatic RCTs (Groenwold 2021; Matthews et al. 2022). Thus far three target trial emulations (Hajage et al. 2022; Shaefi et al. 2021; Urner et al. 2022) have shown mortality benefits in experienced centres, and in patients with PaO_2/FiO_2 ratios lower than 80–100. This once again echoes both initial and current literature that ECMO can have a role to play, but only if centres have the experience and resources to provide it and can judiciously select for the right patient.

What then, are some insights that we can glean from the use of ECMO in COVID-19 and for future pandemics? First, ECMO has a role to play in patients with refractory ARDS and can provide mortality benefits. However, this is conditional on sufficient resources, manpower and planning, and judicious patient selection (Ramanathan et al. 2020). Second, patient outcomes will evolve alongside an ever-evolving pandemic—viral variants, resource constraints, and management changes will impact mortality outcomes. It is imperative that we re-assess patient outcomes regularly, and adjust the way we manage patients with critical illness accordingly (MacLaren et al. 2022).

5.9 Prone Positioning in ECMO

Prone positioning in ECMO, before COVID-19, was limited to a few observational case series (Guervilly et al. 2014; Kipping et al. 2013). During the pandemic, a surge in the number of patients requiring ECMO helped fill the knowledge gap about prone positioning in COVID-19. Approximately 60% of patients in the ELSO registry received prone positioning before or during ECMO (Barbaro et al. 2020, 2021). Several meta-analyses have been conducted on prone positioning during ECMO for ARDS. The first meta-analysis found that approximately 40% of patients died, and there were no major complications (Poon et al. 2021). The second, an individual participant data meta-analysis and with propensity-score matching, while underpowered, showed that prone positioning may reduce (Giani et al. 2022). The third meta-analysis published in March 2022 found a significant reduction in in-hospital, 60-day, and 90-day mortality (Papazian et al. 2022).

Through advanced imaging including electrical impedance tomography, two subgroups of patients have been defined— patients with a compliance gain of at least 3 ml/cmH_2O (or "responders"), and those without (or "non-responders") (Franchineau et al. 2020). These subgroups showed differences in demographics (BMI), disease characteristics (aetiology of pneumonia), and responses to treatment (including response to prone positioning, duration of ECMO, and ICU length of stay). However, there were no differences in mortality between both groups, and therefore prone positioning can be used in all patients receiving ECMO. Several randomized controlled trials are underway (NCT04139733, NCT04607551, NCT05198986, NCT04995289), and the results of whether prone positioning is effective and safe in patients receiving ECMO are eagerly awaited.

5.10 Lessons Learned for Respiratory Supports from COVID-19

What are some lessons that we can learn from the past three years in managing patients with acute respiratory failure? First, we should depart from the current paradigm of ARDS and respiratory failure as being one broad diagnosis. Instead, a syndromic definition based on validated clinical criteria, and a recognition that ARDS or acute respiratory failure is merely a constellation of signs and symptoms secondary to a wide and extensive list of differential diagnoses is needed to further progress in managing these patients (Gattinoni and Marini 2021; Shekar and Ling 2022). Equally, it is not enough that we decide whether an intervention works, but *when* and *in who* the intervention is effective. Progress is being made to increasingly personalize therapy in acute respiratory failure, but there is a long road ahead. Failing to recognize the importance of personalization can only be detrimental to patients and derail the years of progress the specialty of critical care medicine has made before the pandemic.

Second, collaboration is key to better understanding the disease at hand, and the interventions available to treat the disease. During the pandemic, international trial groups such as RECOVERY and REMAP-CAP have highlighted the immense power of collaborating across nations and centres in determining the effectiveness of interventions for COVID-19 across a wide spectrum of severity (Recovery Group 2022; REMAP-CAP. 2022). Such collaboration must continue in future if we intend to tackle future pandemics or diseases. In addition, registry data, such as that of the Extracorporeal Life Support Organisation, the COVID-19 Critical Care Consortium, the International Severe Acute Respiratory and Merging Infection Consortium, and the Intensive Care National Audit & Research Centre have shown monumental promise in informing future practice in the management of COVID-19 (Barbaro et al. 2020, 2021; Urner et al. 2022; Intensive Care National Audit Research Centre 2022; Abbas et al. 2022). In the context of intensive care medicine, well-collected, longitudinal, retrospective registry data can be very powerful—using statistical tools such as propensity-score matching, target trial emulation, and cluster analysis will prove very useful for determining the benefit of interventions for acute respiratory failure, and who these interventions will benefit.

Finally, it is important for us to re-evaluate our understanding of COVID-19 and other diseases causing acute respiratory failure longitudinally (MacLaren et al. 2022). Studies have shown that phenotypes of COVID-19 in patients requiring mechanical ventilation only appear after longitudinal follow-up (Bos et al. 2021). Similarly, patients who respond to prone positioning only are identified after examining their respiratory mechanics after prone positioning was initiated for some time (Franchineau et al. 2020). For the bedside clinician, it is important to re-assess the patient and re-consider differential diagnoses and therapeutic options for each patient. From an evidence-based medicine point of view, the evidence base for corticosteroids and tocilizumab was not confirmed until the release of the results of the RECOVERY trial, and subsequent meta-analyses with newer evidence (The W.H.O. Rapid Evidence Appraisal for Covid-Therapies Working Group 2020, 2021). Similarly, the evidence base for awake prone positioning, while previously strong, may appear more uncertain in the face of the newly published trials suggesting a lack of benefit. For an evidence-based scientist who bases their clinical practice on current literature, it is important to re-assess the literature and re-consider the utility of such interventions for the patient.

5.11 Concluding Remarks

The COVID-19 pandemic has certainly led to an increasingly rapid understanding of acute respiratory failure and ARDS. Previously obscure interventions have now risen in prominence due to the pandemic, and the field of intensive care medicine has been thrown into the spotlight in the last few years. The volume of publications in respiratory and critical care medicine has expanded immensely, and journals reporting on these topics have seen a huge increase in their reach and impact. While it is still unclear to whom and in what form respiratory support should be given to patients with acute respiratory failure, it is evident that robust patient selection criteria for each intervention are needed. Level-headed and appropriate supportive clinical care that is rooted in physiology and evidence-based medicine is essential. Even as the field of intensive care medicine continues to go from strength to strength, it seems fitting that the old adage "assume nothing, trust no one, give oxygen" still rings true several decades after the specialty was born.

References

Abbas A, Abdukahil SA, Abdulkadir NN et al (2022) ISARIC-COVID-19 dataset: a prospective, standardized, global dataset of patients hospitalized with COVID-19. Sci Data 9(1):454

Al Sulaiman K, Korayem GB, Altebainawi AF et al (2022) Evaluation of inhaled nitric oxide (iNO) treatment for moderate-to-severe ARDS in critically ill patients with COVID-19: a multicenter cohort study. Crit Care 26(1):304

Alhazzani W, Møller MH, Arabi YM et al (2020) Surviving sepsis campaign: guidelines on the management of critically ill adults with coronavirus disease 2019 (COVID-19). Crit Care Med 48(6):e440–e469

Alhazzani W, Parhar KKS, Weatherald J et al (2022) Effect of awake prone positioning on endotracheal intubation in patients With COVID-19 and acute respiratory failure: a randomized clinical trial. JAMA 327(21):2104–2113

Ball L, Silva PL, Giacobbe DR et al (2022) Understanding the pathophysiology of typical acute respiratory distress syndrome and severe COVID-19. Expert Rev Respir Med 16(4):437–446

Barbaro RP, MacLaren G, Boonstra PS et al (2020) Extracorporeal membrane oxygenation support in COVID-19: an international cohort study of the extracorporeal life support organization registry. Lancet 396(10257):1071–1078

Barbaro RP, MacLaren G, Boonstra PS et al (2021) Extracorporeal membrane oxygenation for COVID-19: evolving outcomes from the international extracorporeal life support organization registry. Lancet 398(10307):1230–1238

Bellani G, Laffey JG, Pham T et al (2016) Epidemiology, patterns of care, and mortality for patients with acute respiratory distress syndrome in intensive care units in 50 countries. JAMA 315(8):788–800

Bos LDJ, Sjoding M, Sinha P et al (2021) Longitudinal respiratory subphenotypes in patients with COVID-19-related acute respiratory distress syndrome: results from three observational cohorts. Lancet Respir Med 9(12):1377–1386

Bouadma L, Mekontso-Dessap A, Burdet C et al (2022) High-dose dexamethasone and oxygen support strategies in intensive care unit patients with severe COVID-19 acute hypoxemic respiratory failure: the COVIDICUS randomized clinical trial. JAMA Intern Med 182(9):906–916

Broman LM, Taccone FS, Lorusso R et al (2019) The ELSO maastricht treaty for ECLS nomenclature: abbreviations for cannulation configuration in extracorporeal life support—a position paper of the extracorporeal life support organization. Crit Care 23(1):36

Brower RG, Matthay MA, Morris A, Schoenfeld D, Thompson BT, Wheeler A (2000) Ventilation with lower tidal volumes as compared with traditional tidal volumes for acute lung injury and the acute respiratory distress syndrome. N Engl J Med 342(18):1301–1308

Calfee CS, Delucchi K, Parsons PE, Thompson BT, Ware LB, Matthay MA (2014) Subphenotypes in acute respiratory distress syndrome: latent class analysis of data from two randomised controlled trials. Lancet Respir Med 2(8):611–620

Chaudhuri D, Sasaki K, Karkar A et al (2021) Corticosteroids in COVID-19 and non-COVID-19 ARDS: a systematic review and meta-analysis. Intensive Care Med 47(5):521–537

Cherian R (2021) Clinical risk stratification in COVID-19: the need for a revised approach? Pulm Circ 11(1):2045894020988630

Cho HJ, Heinsar S, Jeong IS et al (2020) ECMO use in COVID-19: lessons from past respiratory virus outbreaks—a narrative review. Crit Care 24(1):301

Chouhdari A, Shokouhi S, Bashar FR et al (2018) Is a low incidence rate of ventilation associated pneumonia associated with lower mortality? A descriptive longitudinal study in Iran. Tanaffos 17(2):110–116

Crimi C, Noto A, Madotto F et al (2022) High-flow nasal oxygen versus conventional oxygen therapy in patients with COVID-19 pneumonia and mild hypoxaemia: a randomised controlled trial. Thorax

Diaz RA, Graf J, Zambrano JM et al (2021) Extracorporeal membrane oxygenation for COVID-19-associated severe acute respiratory distress syndrome in Chile: a nationwide incidence and cohort study. Am J Respir Crit Care Med 204(1):34–43

Duggal A, Kast R, Van Ark E et al (2022) Identification of acute respiratory distress syndrome subphenotypes de novo using routine clinical data: a retrospective analysis of ARDS clinical trials. BMJ Open 12(1):e053297

Ehrmann S, Li J, Ibarra-Estrada M et al (2021) Awake prone positioning for COVID-19 acute hypoxaemic respiratory failure: a randomised, controlled, multinational, open-label meta-trial. Lancet Respir Med 9(12):1387–1395

Famous KR, Delucchi K, Ware LB et al (2017) Acute respiratory distress syndrome subphenotypes respond differently to randomized fluid management strategy. Am J Respir Crit Care Med 195(3):331–338

Fan E, Del Sorbo L, Goligher EC et al (2017) An official American thoracic society/European society of intensive care medicine/society of critical care medicine clinical practice guideline: mechanical ventilation in adult patients with acute respiratory distress syndrome. Am J Respir Crit Care Med 195(9):1253–1263

Ferreyro BL, Angriman F, Munshi L et al (2020) Association of noninvasive oxygenation strategies with all-cause mortality in adults with acute hypoxemic respiratory failure: a systematic review and meta-analysis. JAMA 324(1):57–67

Fish M, Rynne J, Jennings A et al (2022) Coronavirus disease 2019 subphenotypes and differential treatment response to convalescent plasma in critically ill adults: secondary analyses of a randomized clinical trial. Intensive Care Med 1–14

Fralick M, Colacci M, Munshi L et al (2022) Prone positioning of patients with moderate hypoxaemia due to covid-19: multicentre pragmatic randomised trial (COVID-PRONE). BMJ 376:e068585

Franchineau G, Bréchot N, Hekimian G et al (2020) Prone positioning monitored by electrical impedance tomography in patients with severe acute respiratory distress syndrome on veno-venous ECMO. Ann Intensive Care 10(1):12

Fuller BM, Mohr NM, Skrupky L, Fowler S, Kollef MH, Carpenter CR (2015) The use of inhaled prostaglandins in patients with ARDS: a systematic review and meta-analysis. Chest 147(6):1510–1522

Gattinoni L, Marini JJ (2021) Isn't it time to abandon ARDS? The COVID-19 lesson. Crit Care 25(1):326

Gattinoni L, Coppola S, Cressoni M, Busana M, Rossi S, Chiumello D (2020a) COVID-19 does not lead to a "typical" acute respiratory distress syndrome. Am J Respir Crit Care Med 201(10):1299–1300

Gattinoni L, Chiumello D, Caironi P et al (2020b) COVID-19 pneumonia: different respiratory treatments for different phenotypes? Intensive Care Med 46(6):1099–1102

Gebistorf F, Karam O, Wetterslev J, Afshari A (2016) Inhaled nitric oxide for acute respiratory distress syndrome (ARDS) in children and adults. Cochrane Database Syst Rev 6

Giani M, Rezoagli E, Guervilly C et al (2022) Prone positioning during venovenous extracorporeal membrane oxygenation for acute respiratory distress syndrome: a pooled individual patient data analysis. Crit Care 26(1):8

Grieco DL, Menga LS, Cesarano M et al (2021) Effect of helmet noninvasive ventilation vs high-flow nasal oxygen on days free of respiratory support in patients with COVID-19 and moderate to severe hypoxemic respiratory failure: the HENIVOT randomized clinical trial. JAMA 325(17):1731–1743

Groenwold RHH (2021) Trial emulation and real-world evidence. JAMA Netw Open 4(3): e213845-e

Guérin C, Reignier J, Richard JC et al (2013) Prone positioning in severe acute respiratory distress syndrome. N Engl J Med 368(23):2159–2168

Guervilly C, Hraiech S, Gariboldi V et al (2014) Prone positioning during veno-venous extracorporeal membrane oxygenation for severe acute respiratory distress syndrome in adults. Minerva Anestesiol 80(3):307–313

Hajage D, Combes A, Guervilly C et al (2022) Extracorporeal membrane oxygenation for severe acute respiratory distress syndrome associated with COVID-19: an emulated target trial analysis. Am J Respir Crit Care Med 206(3):281–294

Hamilton GS (2021) Aerosol-generating procedures in the COVID era. Respirology 26(5):416–418

Henry BM, Lippi G (2020) Poor survival with extracorporeal membrane oxygenation in acute respiratory distress syndrome (ARDS) due to coronavirus disease 2019 (COVID-19): pooled analysis of early reports. J Crit Care 58:27–28

Intensive Care National Audit & Research Centre (2022) COVID-19 report. 16 Oct 2022. https://www.icnarc.org/Our-Audit/Audits/Cmp/Reports. Accessed 16 Oct 2022

Karagiannidis C, Slutsky AS, Bein T, Windisch W, Weber-Carstens S, Brodie D (2021) Complete countrywide mortality in COVID patients receiving ECMO in Germany throughout the first three waves of the pandemic. Crit Care 25(1):413

Kipping V, Weber-Carstens S, Lojewski C et al (2013) Prone position during ECMO is safe and improves oxygenation. Int J Artif Organs 36(11):821–832

Leisman DE, Ronner L, Pinotti R et al (2020) Cytokine elevation in severe and critical COVID-19: a rapid systematic review, meta-analysis, and comparison with other inflammatory syndromes. Lancet Respir Med 8(12):1233–1244

Li J, Luo J, Pavlov I et al (2022) Awake prone positioning for non-intubated patients with COVID-19-related acute hypoxaemic respiratory failure: a systematic review and meta-analysis. Lancet Respir Med 10(6):573–583

Lim ZJ, Subramaniam A, Ponnapa Reddy M et al (2021) Case fatality rates for patients with COVID-19 requiring invasive mechanical ventilation. A meta-analysis. Am J Respir Crit Care Med 203(1): 54–66

Ling RR, Ramanathan K, Sim JJL et al (2022) Evolving outcomes of extracorporeal membrane oxygenation during the first 2 years of the COVID-19 pandemic: a systematic review and meta-analysis. Crit Care 26(1):147

MacLaren G, Fisher D, Brodie D (2022) Treating the most critically ill patients with COVID-19: the evolving role of extracorporeal membrane oxygenation. JAMA 327(1):31–32

Matthews AA, Danaei G, Islam N, Kurth T (2022) Target trial emulation: applying principles of randomised trials to observational studies. BMJ 378:e071108

Meade MO, Young D, Hanna S et al (2017) Severity of hypoxemia and effect of high-frequency oscillatory ventilation in acute respiratory distress syndrome. Am J Respir Crit Care Med 196(6):727–733

Mitra S, Ling RR, Yang IX et al (2021) Severe COVID-19 and coagulopathy: a systematic review and meta-analysis. Ann Acad Med Singap 50(4):325–335

Nair PR, Haritha D, Behera S et al (2021) Comparison of high-flow nasal cannula and noninvasive ventilation in acute hypoxemic respiratory failure due to severe COVID-19 pneumonia. Respir Care 66(12):1824–1830

National Institutes of Health (2022) Clinical spectrum of SARS-CoV-2 infection. 26 Sept 2022. https://www.covid19treatmentguidelines.nih.gov/overview/clinical-spectrum/. Accessed 15 Oct 2022

Papazian L, Schmidt M, Hajage D et al (2022) Effect of prone positioning on survival in adult patients receiving venovenous extracorporeal membrane oxygenation for acute respiratory distress syndrome: a systematic review and meta-analysis. Intensive Care Med 48(3):270–280

Papoutsi E, Giannakoulis VG, Xourgia E, Routsi C, Kotanidou A, Siempos II (2021) Effect of timing of intubation on clinical outcomes of critically ill patients with COVID-19: a systematic review and meta-analysis of non-randomized cohort studies. Crit Care 25(1):121

Perkins GD, Ji C, Connolly BA et al (2022) Effect of noninvasive respiratory strategies on intubation or mortality among patients with acute hypoxemic respiratory failure and COVID-19: the Recovery-RS randomized clinical trial. JAMA 327(6):546–558

Ponnapa Reddy M, Subramaniam A, Afroz A et al (2021) Prone positioning of nonintubated patients with coronavirus disease 2019-a systematic review and meta-analysis. Crit Care Med 49(10):e1001–e1014

Poon WH, Ramanathan K, Ling RR et al (2021) Prone positioning during venovenous extracorporeal membrane oxygenation for acute respiratory distress syndrome: a systematic review and meta-analysis. Crit Care 25(1):292

Protti A, Santini A, Pennati F et al (2022) Lung response to a higher positive end-expiratory pressure in mechanically ventilated patients with COVID-19. Chest 161(4):979–988

Pulakurthi YS, Pederson JM, Saravu K et al (2021) Corticosteroid therapy for COVID-19: a systematic review and meta-analysis of randomized controlled trials. Medicine 100(20)

Qian ET, Gatto CL, Amusina O et al (2022) Assessment of awake prone positioning in hospitalized adults with COVID-19: a nonrandomized controlled trial. JAMA Intern Med 182(6):612–621

Ramanathan K, Antognini D, Combes A et al (2020) Planning and provision of ECMO services for severe ARDS during the COVID-19 pandemic and other outbreaks of emerging infectious diseases. Lancet Respir Med 8(5):518–526

Ramanathan K, Shekar K, Ling RR et al (2021) Extracorporeal membrane oxygenation for COVID-19: a systematic review and meta-analysis. Crit Care 25(1):211

Recovery Group (2022) Randomised evaluation of COVID-19 therapy. https://www.recoverytrial.net/. Accessed 16 Oct 2022

Reddy MP, Subramaniam A, Chua C et al (2022) Respiratory system mechanics, gas exchange, and outcomes in mechanically ventilated patients with COVID-19-related acute respiratory distress syndrome: a systematic review and meta-analysis. Lancet Respir Med

REMAP-CAP (2022) A randomised, embedded, multi-factorial, adaptive platform trial for community-acquired pneumonia. https://www.remapcap.org/. Accessed 16 Oct 2022

Schmidt M, Hajage D, Lebreton G et al (2020) Extracorporeal membrane oxygenation for severe acute respiratory distress syndrome associated with COVID-19: a retrospective cohort study. Lancet Respir Med 8(11):1121–1131

Schmidt M, Langouet E, Hajage D et al (2021) Evolving outcomes of extracorporeal membrane oxygenation support for severe COVID-19 ARDS in Sorbonne hospitals, Paris. Crit Care 25(1):355

Shaefi S, Brenner SK, Gupta S et al (2021) Extracorporeal membrane oxygenation in patients with severe respiratory failure from COVID-19. Intensive Care Med 47(2):208–221

Shekar K, Ling RR (2022) The pandemic and the great awakening in the management of acute hypoxaemic respiratory failure. Lancet Respir Med 10(6):527–529

Shekar K, Badulak J, Peek G et al (2020) Extracorporeal life support organization coronavirus disease 2019 interim guidelines: a consensus document from an international group of interdisciplinary extracorporeal membrane oxygenation providers. ASAIO J 66(7):707–721

Sinha P, Delucchi KL, Thompson BT, McAuley DF, Matthay MA, Calfee CS (2018) Latent class analysis of ARDS subphenotypes: a secondary analysis of the statins for acutely injured lungs from sepsis (SAILS) study. Intensive Care Med 44(11):1859–1869

Sinha P, Furfaro D, Cummings MJ et al (2021) Latent class analysis reveals COVID-19-related acute respiratory distress syndrome subgroups with differential responses to corticosteroids. Am J Respir Crit Care Med 204(11):1274–1285

Sofizan N, Rahman A, Soon LP, Ly CK, Abdullah NZB (2022) Autopsy findings in COVID-19 infection-related death: a systematic review. Egypt J Forensic Sci 12(1):22

Stanski NL, Wong HR (2020) Prognostic and predictive enrichment in sepsis. Nat Rev Nephrol 16(1):20–31

The ARDS Definition Task Force (2012) Acute respiratory distress syndrome: the Berlin definition. JAMA 307(23):2526–2533

The W.H.O. Rapid Evidence Appraisal for Covid-Therapies Working Group (2020) Association between administration of systemic corticosteroids and mortality among critically ill patients with COVID-19: a meta-analysis. JAMA 324(13):1330–1341

The W.H.O. Rapid Evidence Appraisal for Covid-Therapies Working Group (2021) Association between administration of IL-6 antagonists and mortality among patients hospitalized for COVID-19: a meta-analysis. JAMA 326(6):499–518

Tran A, Fernando SM, Rochwerg B et al (2022) Prognostic factors associated with mortality among patients receiving venovenous extracorporeal membrane oxygenation for COVID-19: a systematic review and meta-analysis. Lancet Respir Med

Urner M, Barnett AG, Bassi GL et al (2022) Venovenous extracorporeal membrane oxygenation in patients with acute covid-19 associated respiratory failure: comparative effectiveness study. BMJ 377:e068723

World Health Organisation (2022) Clinical management of COVID-19: living guideline, 15 Sept 2022. https://www.who.int/publications/i/item/WHO-2019-nCoV-Clinical-2022.2. Accessed 15 Oct 2022

Zeraatkar D, Pitre T, Leung G et al (2022) Non-invasive oxygenation strategies in adult patients with acute hypoxemic respiratory failure: a systematic review and network meta-analysis. SSRN 4204888

Zhang J, Whebell SF, Sanderson B et al (2021) Phenotypes of severe COVID-19 ARDS receiving extracorporeal membrane oxygenation. Br J Anaesth 126(3):e130–e132

Pathophysiology of COVID-19-Related Extrapyramidal Disorders

6

Francesco Cavallieri, Marialuisa Zedde, Valentina Fioravanti, Sara Grisanti, Manuela Napoli, Claudio Moratti, Rosario Pascarella, Elena Moro, and Franco Valzania

Abstract

Coronavirus disease 2019 (COVID-19) has been associated with a variety of neurological manifestations (i.e., anosmia, ageusia, myalgia, headache) and neurological syndromes (i.e., encephalopathy, ischemic and hemorrhagic stroke, myelitis, encephalitis) underlying different pathogenetic mechanisms. COVID-19 has also been associated with various movement disorders including acute or subacute parkinsonism. However, to date, only few cases of parkinsonism linked to COVID-19 have been reported, nevertheless raising the possibility of a post-viral parkinsonian syndrome. Furthermore, various studies in vitro and in animal models have highlighted a close relationship between SARS-CoV-2 virus and α-synuclein, leading to the hypothesis that COVID-19 could represent a factor favoring the long-term development of α-synucleopathies. In this chapter, we will discuss the pathophysiological mechanisms related to movement disorders' manifestations

F. Cavallieri · M. Zedde (✉) · V. Fioravanti · F. Valzania
Neurology Unit, Azienda Unità Sanitaria Locale-IRCCS di Reggio Emilia, Via Amendola 2, 42122 Reggio Emilia, Italy
e-mail: marialuisa.zedde@gmail.com; zedde.marialuisa@ausl.re.it

F. Cavallieri
e-mail: cavallieri.francesco@ausl.re.it

V. Fioravanti
e-mail: fioravanti.valentina@ausl.re.it

F. Valzania
e-mail: valzania.franco@ausl.re.it

M. Zedde
Stroke Unit, Azienda Unità Sanitaria Locale-IRCCS di Reggio Emilia, Via Amendola 2, 42122 Reggio Emilia, Italy

S. Grisanti
Clinical and Experimental Medicine PhD Program, University of Modena and Reggio Emilia, Modena, Italy

M. Napoli · C. Moratti · R. Pascarella
Neuroradiology Unit, Azienda Unità Sanitaria Locale-IRCCS di Reggio Emilia, Via Amendola 2, 42122 Reggio Emilia, Italy
e-mail: napoli.manuela@ausl.re.it

C. Moratti
e-mail: moratti.claudio@ausl.re.it

R. Pascarella
e-mail: pascarella.rosario@ausl.re.it

E. Moro
Division of Neurology, Grenoble Alpes University, Centre Hospitalier Universitaire of Grenoble, Grenoble Institute of Neuroscience, Grenoble, France
e-mail: emoro@chu-grenoble.fr

N. Rezaei (ed.), *The COVID-19 Aftermath*, Advances in Experimental Medicine and Biology 1457,
https://doi.org/10.1007/978-3-031-61939-7_6

of COVID-19 focusing on the possible overlap between pathogenetic mechanisms of Parkinson's Disease and COVID-19.

Keywords

Atypical parkinsonism · COVID-19 · Extrapyramidal · Parkinson's disease · Pathophysiology · Parkinsonism · SARS-CoV-2

6.1 Neuroinflammation

One of the main features of COVID-19 pathogenesis is the systemic inflammatory response which has been subsequently defined as "cytokine storm." Cytokine storm is characterized by elevated levels of several cytokines and chemokines including among the others: IL-1β, tumor necrosis factor-alpha (TNF-α), interleukin-6 (IL-6), chemokine (C–C-motif) ligand 2 (CCL2), and granulocyte–macrophage colony-stimulating factor (GM-CSF) (Kempuraj et al. 2020). However, the systemic inflammatory response may also spread within the CNS into the brain parenchyma (Cavallieri et al. 2022a, c; Soung et al. 2022). Indeed, as showed by preclinical studies (i.e., in vitro cell cultures and in vivo mouse models), coronaviruses may be responsible of a severe inflammatory response within the CNS (Bohmwald et al. 2018). The induction of the several pro-inflammatory cytokines (IL-1b IL-15, IL-6 TNFa, p40 and IL-12) which characterized the "cytokine storm" from different CNS cell types such as astrocytes and microglia plays a key role in this setting (Bohmwald et al. 2018; Bodro et al. 2021; Iadecola et al. 2020). In addition, circulating viral proteins and molecular complexes from damaged cells may also play a role. Indeed, even in the absence of SARS-CoV-2 brain invasion, they could enter the brain through a compromised blood–brain barrier (BBB) leading to the development of an innate immune response within the CNS (Albornoz et al. 2022). SARS-CoV-2 with its spike protein may also lead to the activation of the microglial NLRP3 inflammasome, which is a major driver of neurodegeneration (Albornoz et al. 2022). In addition, the presence of α-synuclein fibrils may sustain this process (Albornoz et al. 2022). The interplay between the virus, microglia activation, and α-synuclein fibrils may represent a possible mechanism which could explain the increased vulnerability to developing neurological symptoms like PD (Albornoz et al. 2022; Beckman et al. 2022).

Based on these premises, many authors have postulated an increased risk of PD after COVID-19 (Beauchamp et al. 2020; Sulzer et al. 2020; Chaná-Cuevas et al. 2020). Obviously, the long-term monitoring for neurological sequelae and data from larger autoptic brain examinations and registries will allow to better understand the role of neuroinflammation and the possible impact of the virus in the risk of PD (Cavallieri et al. 2022a; Beghi et al. 2021, 2023; Cevik et al. 2020; Chen et al. 2021; Chou et al. 2021; Cohen et al. 2020; Cosentino et al. 2021; Desai et al. 2021; Emmi et al. 2023; Fearon et al. 2021; Hamming et al. 2004; Hoffmann et al. 2020; Jackson et al. 2022; Lyoo et al. 2022; Makhoul and Jankovic 2021; Méndez-Guerrero et al. 2020; Merello et al. 2021; Moro et al. 2020; Palacios Cruz et al. 2021; Pavli et al. 2021; Pierce et al. 2022; Raveendran et al. 2021; Reza-Zaldívar et al. 2021; Romero-Sánchez et al. 2020; Sodagar et al. 2022; Song et al. 2021; Tiwari et al. 2021; Yang et al. 2020; Zubair et al. 2020) (Table 6.1).

6.2 Link Between COVID-19 Infection and PD Pathophysiology

Several pathophysiological mechanisms are involved in PD including above the others: conformational changes with malfunction and accumulation of key proteins due to abnormalities in their clearance systems (ubiquitin–proteasome system; lysosome- and chaperone-mediated autophagy); dysregulation of mitochondrial function and oxidative stress; loss of trophic factors; alterations of intracellular Ca^{2+} homeostasis; and finally, neuroinflammation (Cavallieri et al. 2023; Yahya et al. 2023; Bouali-Benazzouz

Table 6.1 Summary of pathogenetic mechanisms of COVID-19-related extrapyramidal disorders

Pathogenetic mechanisms	Parkinson's disease	COVID-19
Neuroinflammation	Systemic inflammation has been related to an increased risk of developing PD, and it may exacerbate ongoing neurodegeneration in PD patients and animal models	The systemic inflammatory response (cytokine storm), characterized by elevated levels of IL-1β, TNF-α, IL-6, CCL2, and GM-CSF, may involve also the CNS leading to neuroinflammation
α-synuclein	The pathological aggregation and spread of α-synuclein is considered the key event in the pathogenesis of PD	Preclinical studies have shown that SARS-CoV-2 may lead to α-synuclein aggregation and accumulation. The neuroinflammation induced by COVID-19 may also play a role, inducing it can trigger α-synuclein misfolding, aggregation, and propagation through the CNS
Mitochondrial dysfunction	Mitochondrial dysfunction represents one of the major pathophysiological mechanisms involved in both monogenic and idiopathic PD. This is supported by the observation that mitochondrial toxins cause PD and that proteins encoded by the autosomal recessively inherited PD genes (i.e., PRKN, PTEN-induced kinase 1 (PINK1), and DJ-1) are involved in mitochondrial pathways	COVID-19 may lead to mitochondrial dysfunction with mitochondrial membrane depolarization, mitochondrial permeability transition pore opening and increased ROS release
Oxidative stress	The role of oxidative stress as a co-factor in the degeneration of dopaminergic neurons in PD is also well known. A number of sources and mechanisms for the generation of reactive oxygen species (ROS) are recognized including the metabolism of dopamine itself, mitochondrial dysfunction, iron, neuroinflammatory cells, calcium, and aging	Oxidative stress is a pivotal point in the pathophysiology of COVID-19 and presumably also in post-COVID syndrome. Inflammation and oxidative stress are mutually reinforcing each other, thus contributing to the systemic hyperinflammatory state and coagulopathy which are cardinal pathological mechanisms of severe stages
Ubiquitin–proteasome pathway	Derangements of the UPS, which normally functions as a type of protein degradation machinery, lead to alterations in α-synuclein homeostasis that could promote the toxic accumulation of this protein detrimental to neuronal survival. Several cellular and animal models of PD that are based on direct disruption of UPS function reproduce the most prominent features of PD	SARS-CoV-2 infection might modulate host immune responses through the ubiquitination of important proteins. Furthermore, the ubiquitin system may promote viral replication
Autophagy-lysosomal system	The pathogenesis of PD has been related to many dysfunctional cellular mechanisms. However, most of its monogenic forms are caused by pathogenic variants in genes involved in endolysosomal function (LRRK2, VPS35, VPS13C, and ATP13A2) and synaptic vesicle trafficking (SNCA, RAB39B, SYNJ1, and DNAJC6)	Recent evidence shows that SARS-CoV-2 virions preferentially exit infected cells using the lysosomal trafficking pathway. Several physiologic lysosomal functions, such as lysosomal acidification, lysosomal enzyme activity, and antigen presentation, were disrupted by viral activities

(continued)

Table 6.1 (continued)

Pathogenetic mechanisms	Parkinson's disease	COVID-19
Chaperones	Dysfunctions in CMA have been associated to the pathogenesis of several neurodegenerative diseases characterized by accumulation and aggregation of proteins identified as CMA substrates including alpha-synuclein and PD. Several mutations associated with familial forms of PD (SNCA, LRRK2, UCHL1 and DJ-1) have been demonstrated to negatively influence the activity of CMA and reduced levels of CMA markers have been observed in postmortem brain samples from PD patients	Few data are available about the effects of Sars-Cov-2 on the CMA. However, it has been recently reported that the stress-inducible ER chaperone GRP78/BiP, which has been identified as a host auxiliary factor for SARS-CoV-2 entry, is upregulated during SARS-CoV-2 infection and acts as a pro-viral protein

CCL2 Chemokine (C–C-motif) ligand 2, *GM-CSF* granulocyte–macrophage colony-stimulating factor, *IL-6* interleukin-6, *TNF-α* tumor necrosis factor-alpha

and Benazzouz 2021). In the following paragraphs, the possible links between these different pathophysiological mechanisms and SARS-CoV-2 will be discussed.

6.2.1 α-synuclein Misfolding, Aggregation, and Toxicity

A-synuclein is widely distributed in healthy brain, and it is involved in several cell functions including neurotransmitters release and synaptic vesicles function (Sulzer and Edwards 2019).

The pathological aggregation and spread of α-synuclein is considered one of the key events in the pathogenesis of PD (Polymeropoulos et al. 1997; Wong and Krainc 2017; Sulzer and Edwards 2019). Starting from a genetic background which predisposes to PD, environmental factors (i.e., viruses, bacteria, toxins, etc.) may act as a second hit starting α-synuclein's misfolding and accumulation (Cavallieri et al. 2023). This follows the path of the Braak dual hit hypothesis, in which the pathogen infection which induces the accumulation of misfolded proteins in peripheral neurons represents the first hit (Hawkes et al. 2007), while the second hit is represented by a multifactorial predisposition (genetic predisposition/metabolic condition/age etc.) necessary to facilitate the accumulation of α-synuclein (Hawkes et al. 2007; Sinha et al. 2021; Cavallieri et al. 2022a).

However, test-tube experiments have shown that the different components of SARS-CoV-2 may act differently and interact with α-synuclein. Indeed, the spike protein (S-protein) did not significantly influence α-synuclein aggregation, while, on the contrary, SARS-CoV-2 nucleocapsid protein (N-protein) facilitated the aggregation process (Semerdzhiev et al. 2022) with a faster formation of fibril (Semerdzhiev et al. 2022). These results corroborate a direct role of the SARS-CoV-2 N-protein in the pathogenesis of parkinsonism (Semerdzhiev et al. 2022). However, it should be taken into account that another study which applied molecular modeling methods found that both monomeric and amyloid forms of alpha-synuclein may interact with the receptor-binding domain of the coronavirus S-protein domain (RBD) of the S-protein (Stroylova et al. 2023). However, even if, according to the molecular modeling data, this interaction made negatively influences the possible further formation of amyloid fibrils (Stroylova et al. 2023). This is due to the less pronounced amyloid properties and cytotoxicity of the newly formed amyloid fibrils (Stroylova et al. 2023). Another study has shown that the SARS-CoV-2 S1 RBD binds to a number of aggregation-prone proteins including TDP-43, prion protein, Aβ, α-synuclein, and tau protein (Idrees and Kumar 2021). In this setting, the heparin-binding site of the S1 protein might assist the interaction between amyloid proteins and the viral surface which may represent the first step of protein aggregation of these proteins (Idrees and Kumar 2021).

Serum and CSF data of COVID-19 patients with neurological symptoms did not reveal significant differences compared with COVID-19 patients without neurological manifestations or healthy controls (Blanco-Palmero et al. 2021). Bioinformatics analysis has confirmed favorable binding affinity between α-Syn and SARS-CoV-2 spike (S) protein and nucleocapsid (N) protein, and direct interactions were further verified in HEK293 cells (Wu et al., 2022) (Wu et al. 2022). SARS-CoV-2 may also cause the development of Lewy-like pathology in the contemporary presence of α-Syn overexpression (Wu et al. 2022), confirming the direct interaction between SARS-CoV-2 and α-Syn (Wu et al. 2022).

Studies on animal models (male and female Syrian golden hamsters) have shown that intranasal SARS-CoV-2 infection may lead to persistent brain pathology (Käufer et al. 2022) leading to significant microglia activation in the olfactory bulb beyond viral clearance (Käufer et al. 2022), while hyperphosphorylated tau and alpha-synuclein accumulated in cortical neurons, without a clear neuroinflammation or neurodegeneration. Another key point was the different involvement of the brain regions, underlying the concept of "selective vulnerability" to the virus effects within the CNS (Käufer et al. 2022). Thus, despite the absence of virus in brain, neurons develop signatures of proteinopathies that may contribute to progressive neuronal dysfunction (Käufer et al. 2022). Another study investigated the brains of rhesus and cynomolgus macaques after COVID-19 pulmonary disease and without overt clinical symptoms (Philippens et al. 2022). Postmortem analysis demonstrated the infiltration of T-cells and activated microglia in the parenchyma of infected animals together with intracellular α-synuclein aggregates (Philippens et al. 2022).

A novel therapeutic approach in PD is based on the increase of degradation of α-synuclein aggregates thanks to the enhancement of autophagy and lysosomal function (Cavallieri et al. 2023). The inhibition of mammalian target of rapamycin (mTOR) signaling and cellular homolog of ABL1 (c-Abl) represent two possible targets of this approach (Gouda et al. 2022). This is a consequence of the inhibition of autophagic properties due to α-synuclein overexpression through the increase of mTOR activity. In this setting, different substances or drugs such as curcumin, piperine rapamycin, or lithium may modulate or inhibit mTOR activity. It should also be mentioned that the inhibition of mTOR activity has also been proposed in different neurodegenerative conditions including ALS and multiple system atrophy (Mandrioli et al. 2018; Ndayisaba and Wenning 2020). Focusing on SARS-CoV-2, dysregulated mTOR activity has been found in COVID-19 patients with hyperactivation of the immune system response, highlighting the possible therapeutic role of mTOR pathways inhibition in COVID-19 patients (Terrazzano et al. 2020). mTOR inhibition and p53 activation have been suggested as effective methods of controlling hyperinflammation in COVID-19 patients because of their roles in promoting the proliferation of pro-inflammatory cytokines, thanks to their abilities to promote the proliferation of pro-inflammatory cytokines (Ramaiah 2020). The mTORNLRP3-IL–1b pathway is suggested to be associated with the cytokine storm through IL-6, a cytokine, which is associated with a poorer prognosis in COVID-19-infected patients (Mehta et al. 2020; Zabetakis et al. 2020; Korakas et al. 2020; Swain et al. 2021). In the context of COVID-19 infection, the dysregulation of mTORC1 may promote both neuroinflammation and downregulation of autophagy, two aspects which are known to be involved in PD pathophysiology (Mehta et al. 2020; Zabetakis et al. 2020; Korakas et al. 2020; Swain et al. 2021) together with COVID-19 induced neuroinflammation. Indeed, it is well known that inflammation can trigger α-synuclein misfolding, aggregation, and propagation through the CNS which can, in turn, induce a pro-inflammatory response with subsequent cell damage (Chaná-Cuevas et al. 2020). In COVID-19 patients, several inflammatory pathways are activated; in particular, cytokine release and microglial activation have been described (Awogbindin et al. 2021).

6.2.2 Mitochondrial Dysfunction

Mitochondria are essential for neuronal function being involved in several cell functions including: production of adenosine triphosphate (ATP) for energy demands; regulation of protein synthesis and synaptic plasticity; regulation of homeostasis of ion levels; regulation of reactive oxygen species that may lead to oxidative cellular stress (Trinh et al. 2021). These roles are crucial for neurons due to their high energy demands and significant susceptibility to oxidative stress (Trinh et al. 2021). Mitochondrial dysfunction represents one of the major pathophysiological mechanisms involved in idiopathic and monogenic PD (Borsche et al. 2021). This statement is supported by several findings which came from the observation that mitochondrial toxins cause PD and that proteins encoded by the autosomal recessively inherited PD genes (i.e., PRKN, PTEN-induced kinase 1 (PINK1), and DJ-1) are involved in mitochondrial pathways (Borsche et al. 2021).

The PRKN gene encodes for the E3 ubiquitin ligase parkin (Seirafi et al. 2015), which acts together with the PINK1 protein in a pathway of the neuroprotective mitochondrial quality control (MQC) system (Seirafi et al. 2015). The MQC has several functions, including the government of mitochondrial morphology, regulation of ATP levels, regulation of the dynamic networks of mitochondria through fusion and fission; control of the mitochondria's turnover (Cavallieri et al. 2023). In this setting, restoring normal mitophagy through targeting the PINK1/Parkin pathway seems to be a promising therapeutic approach in recessive PD (Narendra and Youle 2011; McLelland et al. 2014).

On the other hand, COVID-19 infection may lead to mitochondrial dysfunction. Indeed, mitochondrial dysfunction with mitochondrial membrane depolarization, mitochondrial permeability transition pore opening, and increased ROS release have been found in a fluorescent microscopy-based study (Shang et al. 2021). As a consequence of mitochondrial damage, Pink1 and Parkin proteins accumulated within the mitochondria leading to mitophagy for MQC and virus clearance (Shang et al. 2021) suggesting two important assumptions. Firstly, mitochondria alterations are remarkably involved in SARS-CoV-2 replication, while, secondly, the mitochondrial homeostasis is disrupted by the virus (Shang et al. 2021). SARS-CoV-2 may alter mitochondrial integrity but in turn, mitochondria modulate cellular response against the infection (Bhowal et al. 2022). Mitochondrial integrity is essential for an adequate innate immune system response against the virus. Mitochondrial health thus may determine differential vulnerabilities to SARS-CoV-2 infection rendering markers of mitochondrial functions promising potential biomarkers for SARS-CoV-2 infection risk and severity of outcome (Burtscher et al. 2020).

6.2.3 Oxidative Stress

Oxidative stress is defined by an imbalance between increased levels of reactive oxygen species (ROS) and a low concentration/activity of antioxidants resulting in cellular damages (Preiser 2012). The role of oxidative stress as a co-factor in the degeneration of dopaminergic neurons in PD is well known (Dias et al. 2013). A number of sources and mechanisms are implied in the formation of reactive oxygen species (ROS) including among others, dopamine metabolism, mitochondrial dysfunction, iron, neuroinflammatory cells, calcium, and aging (Dias et al. 2013).

Environmental factors may also play a role in this setting. In particular, neurotoxins, pesticides, and insecticides could lead, together with genetic mutations in PD-associated genes, to mitochondrial dysfunction and subsequent ROS formation (Blesa et al. 2015).

On the other hand, oxidative stress has also a key role in the pathophysiology of COVID-19 and Long-COVID syndrome (Vollbracht and Kraft 2022). Indeed, inflammation and oxidative stress are mutually reinforcing each other, thus contributing to the systemic hyperinflammatory state and coagulopathy characteristic of severe COVID-19 infection (Vollbracht and Kraft 2022).

Epidemiological studies have found that the COVID-19 severity and morbidity are linked to the contemporary presence of specific comorbidities (i.e., diabetes, obesity, cardiovascular diseases, immunosuppressive states) which are associated with elevated ROS levels and oxidative stress (Alam and Czajkowsky 2022). Moreover, accumulating evidence suggests that oxidative stress coupled with the cytokine storm contributes to COVID-19 pathogenesis and immunopathogenesis by causing endotheliitis and endothelial cell dysfunction and by activating the blood clotting cascade that results in blood coagulation and microvascular thrombosis (Alam and Czajkowsky 2022; Gain et al. 2023). The imbalance between reduced antioxidant host responses and increased ROS production triggered by SARS-CoV-2 has the direct consequence of increased redox stress (Alam and Czajkowsky 2022; Gain et al. 2023). Subsequently, the increased redox stress may reduce the antiviral host responses and at the same time increase the virus-induced inflammation and apoptosis with cell and tissue damage and end organ disease (Gain et al. 2023).

6.2.4 Proteostasis

Loss of protein homeostasis (proteostasis) is characterized by the appearance of nonnative protein aggregates in various tissues. Protein aggregation is regulated and suppressed by the proteostasis networks (PNs), which is a collection of several macromolecules that operate in order to maintain proteome integrity across subcellular compartments (Labbadia and Morimoto 2015), preventing protein misfolding, aggregation, and accumulation in the cell compartments. The PNs are composed of several different players which act in concert and include: protein synthesis, molecular chaperones, endoplasmic reticulum (ER), unfolded protein response (UPR), stress response pathways (SRPs), secretions, ubiquitin–proteasome system (UPS), and autophagy-lysosomal pathways (ALPs) (Dong and Cui 2018). In particular, UPS and ALP are the two major pathways, which are crucial for the cellular proteostasis (Li et al. 2019) ensuring the proper synthesis, folding, confirmation, and degradation of trivial proteins such as α-synuclein. The proper functioning of PNs is quite resilient to sustain under the influence of stressors (Kulkarni et al. 2023).

6.2.4.1 Ubiquitin–Proteasome Pathway

The UPS and the autophagy-lysosomal systems are the two main protein degradation systems in the cell. Polyubiquitin chains tag the target proteins which become substrates for the UPS, which feeds unfolded proteins through the barrel of the 26S proteasome and generates small digested peptides (Nedelsky et al. 2008). As early as 30 years ago, anatomopathological studies have found the presence of ubiquitin-immunopositive Lewy bodies in PD patients' brains (Tofaris et al. 2003), followed by the observation of a contemporary impaired proteasomal function in the SN (Bentea et al. 2017). Parkin (E3 ligase), PINK1, and DJ-1 bind to each other to form a complex that promotes degradation of unfolded/misfolded proteins through the ubiquitin–proteasome system (Xiong et al. 2009). In particular, parkin is an E3 ligase mediating multiple forms of ubiquitination and can form an E3 ligase complex with PINK1 and DJ-1 to promote unfolded protein degradation (Xiong et al. 2009). However, if parkin is mutated, it cannot correctly interact with α-synuclein leading to the accumulation of the protein with the subsequent formation of Lewy bodies (Xiong et al. 2009). Derangements of the UPS may lead to the dysfunction of protein homeostasis which in turn promotes the toxic accumulation of proteins detrimental to neuronal survival. Not surprisingly, several cellular and animal PD models based on the disruption of UPS reproduce the most prominent features of PD (Lim and Tan 2007). In this setting, SARS-CoV-2 infection might modulate host immune responses through the ubiquitination of several proteins, including USP5, IQGAP1, TRIM28, and Hsp90. This has been reported in a study which has applied quantitative proteomic technology to assess ubiquitination changes in Vero E6 cells infected by SARS-CoV-2 (Zhang et al. 2021). SARS-Cov-2 may hijack the UPS facilitating assembly and budding of the viral proteins (Seyoum Tola 2023). Ubiquitination of the enzyme keratin-38 (E-K38) residue gave the virion the ability to engage with specific cellular receptors, enhancing virus entry and reproduction (Seyoum Tola 2023). A fraction of infectious viral particles produced during replication has been ubiquitinated (Seyoum Tola 2023). Furthermore, in order to replicate their viral genome after entering the host cell, viruses combine the resources of the host cell with recently generated viral proteins (Seyoum Tola 2023). Considering its important role in multiple steps of the CoV infection cycle, UPS has been identified as a potential drug target to modulate the impact of CoV infection(Raaben et al. 2010).

6.2.4.2 Chaperones

Molecular chaperones represent an important mediator in the PNs, contributing to the proper folding and assembly of other macromolecules. This objective is reached through maturating of unfolded or partially folded macromolecules, refolding of stress-denatured proteins, and modifying oligomeric assembly, otherwise leading to their proteolytic degradation (Khomari et al. 2021). Chaperone-mediated autophagy (CMA) is a selective form of autophagy which plays a key role in the degradation of specific soluble proteins that contain in a pentapeptide motif that is recognized by a cytosolic chaperone which in turn delivers the proteins to the lysosomes for their degradation (Sala et al. 2016). Physiologically, CMA contributes to maintaining crucial cellular functions including energetic balance and protein quality control (Sala et al. 2016). Dysfunctions in CMA have been associated to the pathogenesis of several neurodegenerative diseases characterized by the accumulation and aggregation of proteins identified as CMA substrates including alpha-synuclein and PD (Sala et al. 2016). Several mutations associated with familial forms of PD (SNCA, LRRK2, UCHL1, and DJ-1) have been demonstrated to negatively influence the activity of CMA and reduced levels of CMA markers (Sala et al. 2016). Viruses that use the hosts' gene expression tools and protein synthesis apparatus to survive and replicate are obviously protected by such a host chaperone system (Khomari et al. 2021). This means that many viruses use members of the hosts' chaperoning system to infect the target cells, replicate, and spread (Khomari et al. 2021). During viral infection, the high expression of viral proteins enhances the level of heat shock proteins (HSPs), increases the endoplasmic reticulum (ER) stress, and induces cell apoptosis or necrosis (Khomari et al. 2021). Few data are available about the effects of SARS-CoV-2 on the CMA. However, it has been recently reported that the stress-inducible ER chaperone GRP78/BiP, which has been identified as a host auxiliary factor for SARS-CoV-2 entry, is upregulated during SARS-CoV-2 infection and acts as a pro-viral protein (Shin et al. 2022). More data however are needed to deepen the effects of COVID-19 on CMA system and its possible effects on neurodegeneration.

6.2.4.3 Lysosomes

The autophagy-lysosomal system is comprised of both macroautophagy and microautophagy (Nedelsky et al. 2008). The majority of PD monogenic forms are caused by pathogenic variants in genes involved in endolysosomal function (LRRK2, VPS35, VPS13C, and ATP13A2) and synaptic vesicle trafficking (SNCA, RAB39B, SYNJ1, and DNAJC6) (Yahya et al. 2023). Moreover, an extensive search for PD risk variants revealed strong risk variants in several lysosomal genes (e.g., GBA1, SMPD1, TMEM175, and SCARB2) highlighting the key role of lysosomal dysfunction in PD pathogenesis (Yahya et al. 2023).

β-coronaviruses use lysosomal trafficking for the egression from the host cells. Such non-lytic release of β-coronaviruses results in lysosome deacidification, inactivation of lysosomal degradation enzymes, and disruption of antigen presentation pathways (Ghosh et al. 2020; Calculli et al. 2023; Cavallieri et al. 2022b).

Indeed, recent evidence shows that SARS-CoV-2 virions preferentially exit from the infected cells through the lysosomal trafficking pathway by disrupting lysosomal acidification, lysosomal enzyme activity, and antigen presentation (Ghosh et al. 2020). Normally, the acidic environment within lysosomes helps destroy viruses and other pathogens before they leave cells. This role implies that SARS-CoV-2 might hijack the lysosomal acidification process promoting lysosomal deacidification and finally disrupting both the lysosomal degradative enzymes capabilities and the lysosome-dependent antigen cross-presentation. The deacidification of SARS-CoV-2-hijacked lysosomes may also alter immune system functions to account at least in part for observed COVID-19-associated immune system abnormalities (Weissmann and Dukor 1970; Ghosh et al. 2020; Lan et al. 2022). The viral protein ORF3a strongly inhibited autophagic flux by blocking the fusion of autophagosomes with lysosomes. ORF3a colocalized with lysosomes and interacted with VPS39, a component of the homotypic fusion and protein sorting (HOPS) complex. The ORF3a–VPS39 interaction prohibited the binding of HOPS with RAB7, which prevented the assembly of fusion machinery, leading to the accumulation of unfused autophagosomes (Zhang et al. 2021).

6.3 Conclusions

In conclusion, an overview of the possible pathophysiological mechanisms of COVID-19-related extrapyramidal disorders has been given in this chapter. In particular, as has been shown in the previous paragraphs, it is known that SARS-CoV-2 can influence multiple cellular mechanisms involved in the neurodegeneration and accumulation of alpha-synuclein characteristic of Parkinson's disease. This, combined with various cases of parkinsonism associated with COVID-19, leads us to hypothesize a possible role of the virus in favoring the long-term development of PD. However, to date there is not enough data to allow us to draw conclusions in this sense; therefore, further preclinical and clinical studies will be needed to evaluate the long-term impact of COVID-19 in the development of PD. In this perspective, cohort studies on patients affected by COVID-19 appear vital to evaluate the incidence of PD in the long term.

References

Alam MS, Czajkowsky DM (2022) SARS-CoV-2 infection and oxidative stress: pathophysiological insight into thrombosis and therapeutic opportunities. Cytokine Growth Factor Rev 63:44–57. https://doi.org/10.1016/j.cytogfr.2021.11.001

Albornoz EA, Amarilla AA, Modhiran N et al (2022) SARS-CoV-2 drives NLRP3 inflammasome activation in human microglia through spike protein. Mol Psychiatry. https://doi.org/10.1038/s41380-022-01831-0

Awogbindin IO, Ben-Azu B, Olusola BA et al (2021) Microglial Implications in SARS-CoV-2 Infection and COVID-19: lessons from viral RNA neurotropism and possible relevance to Parkinson's disease. Front Cell Neurosci 15:670298. https://doi.org/10.3389/fncel.2021.670298

Beauchamp LC, Finkelstein DI, Bush AI et al (2020) Parkinsonism as a third wave of the COVID-19 pandemic? J Parkinsons Dis 10:1343–1353. https://doi.org/10.3233/JPD-202211

Beckman D, Bonillas A, Diniz GB et al (2022) SARS-CoV-2 infects neurons and induces neuroinflammation in a non-human primate model of COVID-19. Cell Rep 41:111573. https://doi.org/10.1016/j.celrep.2022.111573

Beghi E, Helbok R, Crean M et al (2021) The European Academy of Neurology COVID-19 registry (ENERGY): an international instrument for surveillance of neurological complications in patients with COVID-19. Eur J Neurol 28:3303–3323. https://doi.org/10.1111/ene.14652

Beghi E, Moro E, Davidescu EI et al (2023) Comparative features and outcomes of major neurological complications of COVID-19. Euro J Neurol 30:413–433. https://doi.org/10.1111/ene.15617

Bentea E, Verbruggen L, Massie A (2017) The Proteasome inhibition model of Parkinson's disease. J Parkinsons Dis 7:31–63. https://doi.org/10.3233/JPD-160921

Bhowal C, Ghosh S, Ghatak D, De R (2022) Pathophysiological involvement of host mitochondria in SARS-CoV-2 infection that causes COVID-19: a comprehensive evidential insight. Mol Cell Biochem. https://doi.org/10.1007/s11010-022-04593-z

Blanco-Palmero VA, Azcárate-Díaz FJ, Ruiz-Ortiz M et al (2021) Serum and CSF alpha-synuclein levels do not change in COVID-19 patients with neurological

symptoms. J Neurol 268:3116–3124. https://doi.org/10.1007/s00415-021-10444-6
Blesa J, Trigo-Damas I, Quiroga-Varela A, Jackson-Lewis VR (2015) Oxidative stress and Parkinson's disease. Front Neuroanat 9. https://doi.org/10.3389/fnana.2015.00091
Bodro M, Compta Y, Sánchez-Valle R (2021) Presentations and mechanisms of CNS disorders related to COVID-19. Neurol Neuroimmunol Neuroinflamm 8:e923. https://doi.org/10.1212/NXI.0000000000000923
Bohmwald K, Gálvez NMS, Ríos M, Kalergis AM (2018) Neurologic alterations due to respiratory virus infections. Front Cell Neurosci 12:386. https://doi.org/10.3389/fncel.2018.00386
Borsche M, Pereira SL, Klein C, Grünewald A (2021) Mitochondria and Parkinson's disease: clinical, molecular, and translational aspects. J Parkinsons Dis 11:45–60. https://doi.org/10.3233/JPD-201981
Bouali-Benazzouz R, Benazzouz A (2021) Covid-19 infection and Parkinsonism: is there a link? Mov Disord 36:1737–1743. https://doi.org/10.1002/mds.28680
Burtscher J, Cappellano G, Omori A et al (2020) Mitochondria: in the cross fire of SARS-CoV-2 and immunity. iScience 23:101631. https://doi.org/10.1016/j.isci.2020.101631
Calculli A, Bocci T, Porcino M et al (2023) Parkinson disease following COVID -19: report of six cases. Euro J Neurol 15732. https://doi.org/10.1111/ene.15732
Cavallieri F, Cury RG, Guimarães T et al (2023) Recent advances in the treatment of genetic forms of Parkinson's disease: hype or hope? Cells 12:764. https://doi.org/10.3390/cells12050764
Cavallieri F, Fioravanti V, Bove F et al (2022a) COVID-19 and parkinsonism: a critical appraisal. Biomolecules 12:970. https://doi.org/10.3390/biom12070970
Cavallieri F, Fioravanti V, Toschi G et al (2022b) COVID-19 and Parkinson's disease: a casual association or a possible second hit in neurodegeneration? J Neurol 269:59–61. https://doi.org/10.1007/s00415-021-10694-4
Cavallieri F, Sellner J, Zedde M, Moro E (2022c) Neurologic complications of coronavirus and other respiratory viral infections. In: Handbook of clinical neurology. Elsevier, Amsterdam, pp 331–358
Cevik M, Kuppalli K, Kindrachuk J, Peiris M (2020) Virology, transmission, and pathogenesis of SARS-CoV-2. BMJ m3862. https://doi.org/10.1136/bmj.m3862
Chaná-Cuevas P, Salles-Gándara P, Rojas-Fernandez A et al (2020) The potential role of SARS-COV-2 in the pathogenesis of Parkinson's disease. Front Neurol 11:1044. https://doi.org/10.3389/fneur.2020.01044
Chen R, Wang K, Yu J et al (2021) The spatial and cell-type distribution of SARS-CoV-2 receptor ACE2 in the human and mouse brains. Front Neurol 11:573095. https://doi.org/10.3389/fneur.2020.573095
Chou SH-Y, Beghi E, Helbok R et al (2021) Global incidence of neurological manifestations among patients hospitalized with COVID-19—a report for the GCS-NeuroCOVID consortium and the energy consortium. JAMA Netw Open 4:e2112131. https://doi.org/10.1001/jamanetworkopen.2021.12131
Cohen ME, Eichel R, Steiner-Birmanns B et al (2020) A case of probable Parkinson's disease after SARS-CoV-2 infection. Lancet Neurol 19:804–805. https://doi.org/10.1016/S1474-4422(20)30305-7
Cosentino G, Todisco M, Hota N et al (2021) Neuropathological findings from COVID-19 patients with neurological symptoms argue against a direct brain invasion of SARS-CoV-2: a critical systematic review. Euro J Neurol 28:3856–3865. https://doi.org/10.1111/ene.15045
Desai I, Manchanda R, Kumar N et al (2021) Neurological manifestations of coronavirus disease 2019: exploring past to understand present. Neurol Sci 42:773–785. https://doi.org/10.1007/s10072-020-04964-8
Dias V, Junn E, Mouradian MM (2013) The Role of Oxidative stress in Parkinson's disease. J Parkinson's Dis 3:461–491. https://doi.org/10.3233/JPD-130230
Dong Z, Cui H (2018) The autophagy-lysosomal pathways and their emerging roles in modulating proteostasis in tumors. Cells 8:4. https://doi.org/10.3390/cells8010004
Emmi A, Rizzo S, Barzon L et al (2023) Detection of SARS-CoV-2 viral proteins and genomic sequences in human brainstem nuclei. NPJ Parkinsons Dis 9:25. https://doi.org/10.1038/s41531-023-00467-3
Fearon C, Mikulis DJ, Lang AE (2021) Parkinsonism as a sequela of SARS-CoV-2 infection: pure hypoxic injury or additional COVID-19-related response? Mov Disord 36:1483–1484. https://doi.org/10.1002/mds.28656
Gain C, Song S, Angtuaco T et al (2023) The role of oxidative stress in the pathogenesis of infections with coronaviruses. Front Microbiol 13:1111930. https://doi.org/10.3389/fmicb.2022.1111930
Ghosh S, Dellibovi-Ragheb TA, Kerviel A et al (2020) β-Coronaviruses use lysosomes for egress instead of the biosynthetic secretory pathway. Cell 183:1520-1535.e14. https://doi.org/10.1016/j.cell.2020.10.039
Gouda NA, Elkamhawy A, Cho J (2022) Emerging therapeutic strategies for Parkinson's disease and future prospects: a 2021 Update. Biomedicines 10:371. https://doi.org/10.3390/biomedicines10020371
Hamming I, Timens W, Bulthuis M et al (2004) Tissue distribution of ACE2 protein, the functional receptor for SARS coronavirus. A first step in understanding SARS pathogenesis. J Pathol 203:631–637. https://doi.org/10.1002/path.1570
Hawkes CH, Del Tredici K, Braak H (2007) Parkinson's disease: a dual-hit hypothesis. Neuropathol

Appl Neurobiol 33:599–614. https://doi.org/10.1111/j.1365-2990.2007.00874.x
Hoffmann M, Kleine-Weber H, Schroeder S et al (2020) SARS-CoV-2 cell entry depends on ACE2 and TMPRSS2 and is blocked by a clinically proven protease inhibitor. Cell 181:271-280.e8. https://doi.org/10.1016/j.cell.2020.02.052
Iadecola C, Anrather J, Kamel H (2020) Effects of COVID-19 on the nervous system. Cell 183:16-27.e1. https://doi.org/10.1016/j.cell.2020.08.028
Idrees D, Kumar V (2021) SARS-CoV-2 spike protein interactions with amyloidogenic proteins: potential clues to neurodegeneration. Biochem Biophys Res Commun 554:94–98. https://doi.org/10.1016/j.bbrc.2021.03.100
Jackson CB, Farzan M, Chen B, Choe H (2022) Mechanisms of SARS-CoV-2 entry into cells. Nat Rev Mol Cell Biol 23:3–20. https://doi.org/10.1038/s41580-021-00418-x
Käufer C, Schreiber CS, Hartke A-S et al (2022) Microgliosis and neuronal proteinopathy in brain persist beyond viral clearance in SARS-CoV-2 hamster model. eBioMedicine 79:103999. https://doi.org/10.1016/j.ebiom.2022.103999
Kempuraj D, Selvakumar GP, Ahmed ME et al (2020) COVID-19, Mast cells, cytokine storm, psychological stress, and neuroinflammation. Neuroscientist 26:402–414. https://doi.org/10.1177/1073858420941476
Khomari F, Nabi-Afjadi M, Yarahmadi S et al (2021) Effects of cell proteostasis network on the survival of SARS-CoV-2. Biol Proced Online 23:8. https://doi.org/10.1186/s12575-021-00145-9
Korakas E, Ikonomidis I, Kousathana F et al (2020) Obesity and COVID-19: immune and metabolic derangement as a possible link to adverse clinical outcomes. Am J Physiol Endocrinol Metabol 319:E105–E109. https://doi.org/10.1152/ajpendo.00198.2020
Kulkarni A, Preeti K, Tryphena KP et al (2023) Proteostasis in Parkinson's disease: recent development and possible implication in diagnosis and therapeutics. Ageing Res Rev 84:101816. https://doi.org/10.1016/j.arr.2022.101816
Labbadia J, Morimoto RI (2015) The biology of proteostasis in aging and disease. Annu Rev Biochem 84:435–464. https://doi.org/10.1146/annurev-biochem-060614-033955
Lan Y, He W, Wang G et al (2022) Potential Antiviral Strategy exploiting dependence of SARS-CoV-2 replication on lysosome-based pathway. Int J Mol Sci 23:6188. https://doi.org/10.3390/ijms23116188
Li C, Wang X, Li X et al (2019) Proteasome inhibition activates autophagy-lysosome pathway associated with TFEB dephosphorylation and nuclear translocation. Front Cell Dev Biol 7:170. https://doi.org/10.3389/fcell.2019.00170
Lim K-L, Tan JMM (2007) Role of the ubiquitin proteasome system in Parkinson's disease. BMC Biochem 8(Suppl1):S13. https://doi.org/10.1186/1471-2091-8-S1-S13
Lyoo K-S, Kim HM, Lee B et al (2022) Direct neuronal infection of SARS-CoV-2 reveals cellular and molecular pathology of chemosensory impairment of COVID-19 patients. Emerg Microbes Infect 11:407–412. https://doi.org/10.1080/22221751.2021.2024095
Makhoul K, Jankovic J (2021) Parkinson's disease after COVID-19. J Neurol Sci 422:117331. https://doi.org/10.1016/j.jns.2021.117331
Mandrioli J, D'Amico R, Zucchi E et al (2018) Rapamycin treatment for amyotrophic lateral sclerosis: protocol for a phase II randomized, double-blind, placebo-controlled, multicenter, clinical trial (RAP-ALS trial). Medicine 97:e11119. https://doi.org/10.1097/MD.0000000000011119
McLelland G-L, Soubannier V, Chen CX et al (2014) Parkin and PINK1 function in a vesicular trafficking pathway regulating mitochondrial quality control. EMBO J 33:282–295. https://doi.org/10.1002/embj.201385902
Mehta P, McAuley DF, Brown M et al (2020) COVID-19: consider cytokine storm syndromes and immunosuppression. Lancet 395:1033–1034. https://doi.org/10.1016/S0140-6736(20)30628-0
Méndez-Guerrero A, Laespada-García MI, Gómez-Grande A et al (2020) Acute hypokinetic-rigid syndrome following SARS-CoV-2 infection. Neurology 95:e2109–e2118. https://doi.org/10.1212/WNL.0000000000010282
Merello M, Bhatia KP, Obeso JA (2021) SARS-CoV-2 and the risk of Parkinson's disease: facts and fantasy. Lancet Neurol 20:94–95. https://doi.org/10.1016/S1474-4422(20)30442-7
Moro E, Priori A, Beghi E et al (2020) The international European Academy of Neurology survey on neurological symptoms in patients with COVID-19 infection. Eur J Neurol 27:1727–1737. https://doi.org/10.1111/ene.14407
Narendra DP, Youle RJ (2011) Targeting mitochondrial dysfunction: role for PINK1 and Parkin in mitochondrial quality control. Antioxid Redox Signal 14:1929–1938. https://doi.org/10.1089/ars.2010.3799
Ndayisaba A, Wenning GK (2020) Inhibition of the mammalian target or rapamycin (mTOR): a potential therapeutic strategy for multiple system atrophy. Clin Auton Res 30:7–8. https://doi.org/10.1007/s10286-019-00662-4
Nedelsky NB, Todd PK, Taylor JP (2008) Autophagy and the ubiquitin-proteasome system: collaborators in neuroprotection. Biochim Biophys Acta 1782:691–699. https://doi.org/10.1016/j.bbadis.2008.10.002
Palacios Cruz M, Santos E, Velázquez Cervantes MA, León Juárez M (2021) COVID-19, a worldwide public health emergency. Revista Clínica Española (English Edition) 221:55–61. https://doi.org/10.1016/j.rceng.2020.03.001
Pavli A, Theodoridou M, Maltezou HC (2021) Post-COVID syndrome: incidence, clinical spectrum,

and challenges for primary healthcare professionals. Arch Med Res 52:575–581. https://doi.org/10.1016/j.arcmed.2021.03.010
Philippens IHCHM, Böszörményi KP, Wubben JAM et al (2022) Brain inflammation and intracellular α-Synuclein aggregates in macaques after SARS-CoV-2 infection. Viruses 14:776. https://doi.org/10.3390/v14040776
Pierce JD, Shen Q, Cintron SA, Hiebert JB (2022) Post-COVID-19 syndrome. Nurs Res 71:164–174. https://doi.org/10.1097/NNR.0000000000000565
Polymeropoulos MH, Lavedan C, Leroy E et al (1997) Mutation in the alpha-synuclein gene identified in families with Parkinson's disease. Science 276:2045–2047. https://doi.org/10.1126/science.276.5321.2045
Preiser J-C (2012) Oxidative stress. JPEN J Parenter Enteral Nutr 36:147–154. https://doi.org/10.1177/0148607111434963
Raaben M, Posthuma CC, Verheije MH et al (2010) The ubiquitin-proteasome system plays an important role during various stages of the coronavirus infection cycle. J Virol 84:7869–7879. https://doi.org/10.1128/JVI.00485-10
Ramaiah MJ (2020) MTOR inhibition and p53 activation, microRNAs: the possible therapy against pandemic COVID-19. Gene Rep 20:100765. https://doi.org/10.1016/j.genrep.2020.100765
Raveendran AV, Jayadevan R, Sashidharan S (2021) Long COVID: an overview. Diabetes Metab Syndr 15:869–875. https://doi.org/10.1016/j.dsx.2021.04.007
Reza-Zaldívar EE, Hernández-Sapiéns MA, Minjarez B et al (2021) Infection mechanism of SARS-COV-2 and its implication on the nervous system. Front Immunol 11:621735. https://doi.org/10.3389/fimmu.2020.621735
Romero-Sánchez CM, Díaz-Maroto I, Fernández-Díaz E et al (2020) Neurologic manifestations in hospitalized patients with COVID-19: the ALBACOVID registry. Neurology 95:e1060–e1070. https://doi.org/10.1212/WNL.0000000000009937
Sala G, Marinig D, Arosio A, Ferrarese C (2016) Role of chaperone-mediated autophagy dysfunctions in the pathogenesis of Parkinson's disease. Front Mol Neurosci 9:157. https://doi.org/10.3389/fnmol.2016.00157
Seirafi M, Kozlov G, Gehring K (2015) Parkin structure and function. FEBS J 282:2076–2088. https://doi.org/10.1111/febs.13249
Semerdzhiev SA, Fakhree MAA, Segers-Nolten I et al (2022) Interactions between SARS-CoV-2 N-Protein and α-Synuclein accelerate amyloid formation. ACS Chem Neurosci 13:143–150. https://doi.org/10.1021/acschemneuro.1c00666
Seyoum Tola F (2023) The role of ubiquitin-proteasome system in the pathogenesis of severe acute respiratory syndrome coronavirus-2 disease. Int J Inflam 2023:6698069. https://doi.org/10.1155/2023/6698069
Shang C, Liu Z, Zhu Y et al (2021) SARS-CoV-2 causes mitochondrial dysfunction and mitophagy impairment. Front Microbiol 12:780768. https://doi.org/10.3389/fmicb.2021.780768
Shin W-J, Ha DP, Machida K, Lee AS (2022) The stress-inducible ER chaperone GRP78/BiP is upregulated during SARS-CoV-2 infection and acts as a pro-viral protein. Nat Commun 13:6551. https://doi.org/10.1038/s41467-022-34065-3
Sinha S, Mittal S, Roy R (2021) Parkinson's disease and the COVID-19 pandemic: a review article on the association between SARS-CoV-2 and α-Synucleinopathy. JMD 14:184–192. https://doi.org/10.14802/jmd.21046
Sodagar A, Javed R, Tahir H et al (2022) Pathological features and neuroinflammatory mechanisms of SARS-CoV-2 in the brain and potential therapeutic approaches. Biomolecules 12:971. https://doi.org/10.3390/biom12070971
Song E, Zhang C, Israelow B et al (2021) Neuroinvasion of SARS-CoV-2 in human and mouse brain. J Exp Med 218:e20202135. https://doi.org/10.1084/jem.20202135
Soung AL, Vanderheiden A, Nordvig AS et al (2022) COVID-19 induces CNS cytokine expression and loss of hippocampal neurogenesis. Brain 145:4193–4201. https://doi.org/10.1093/brain/awac270
Stroylova Y, Konstantinova A, Stroylov V et al (2023) Does the SARS-CoV-2 spike receptor-binding domain hamper the amyloid transformation of Alpha-Synuclein after all? Biomedicines 11:498. https://doi.org/10.3390/biomedicines11020498
Sulzer D, Antonini A, Leta V et al (2020) COVID-19 and possible links with Parkinson's disease and parkinsonism: from bench to bedside. NPJ Parkinsons Dis 6:18. https://doi.org/10.1038/s41531-020-00123-0
Sulzer D, Edwards RH (2019) The physiological role of α-synuclein and its relationship to Parkinson's Disease. J Neurochem 150:475–486. https://doi.org/10.1111/jnc.14810
Swain O, Romano SK, Miryala R et al (2021) SARS-CoV-2 neuronal invasion and complications: potential mechanisms and therapeutic approaches. J Neurosci 41:5338–5349. https://doi.org/10.1523/JNEUROSCI.3188-20.2021
Terrazzano G, Rubino V, Palatucci AT et al (2020) An open question: is it rational to inhibit the mTor-dependent pathway as COVID-19 therapy? Front Pharmacol 11:856. https://doi.org/10.3389/fphar.2020.00856
Tiwari SK, Wang S, Smith D et al (2021) Revealing tissue-specific SARS-CoV-2 infection and host responses using human stem cell-derived lung and cerebral organoids. Stem Cell Rep 16:437–445. https://doi.org/10.1016/j.stemcr.2021.02.005
Tofaris GK, Razzaq A, Ghetti B et al (2003) Ubiquitination of alpha-synuclein in Lewy bodies is a pathological event not associated with impairment of proteasome function. J Biol Chem 278:44405–44411. https://doi.org/10.1074/jbc.M308041200

Trinh D, Israwi AR, Arathoon LR et al (2021) The multi-faceted role of mitochondria in the pathology of Parkinson's disease. J Neurochem 156:715–752. https://doi.org/10.1111/jnc.15154

Vollbracht C, Kraft K (2022) Oxidative stress and hyper-inflammation as major drivers of severe COVID-19 and long COVID: implications for the benefit of high-dose intravenous Vitamin C. Front Pharmacol 13:899198. https://doi.org/10.3389/fphar.2022.899198

Weissmann G, Dukor P (1970) The role of lysosomes in immune responses. In: Advances in immunology. Elsevier, Amsterdam, pp 283–331

Wong YC, Krainc D (2017) α-synuclein toxicity in neurodegeneration: mechanism and therapeutic strategies. Nat Med 23:1–13. https://doi.org/10.1038/nm.4269

Wu Z, Zhang X, Huang Z, Ma K (2022) SARS-CoV-2 Proteins interact with Alpha Synuclein and induce Lewy body-like pathology in vitro. Int J Mol Sci 23:3394. https://doi.org/10.3390/ijms23063394

Xiong H, Wang D, Chen L et al (2009) Parkin, PINK1, and DJ-1 form a ubiquitin E3 ligase complex promoting unfolded protein degradation. J Clin Invest 119:650–660. https://doi.org/10.1172/JCI37617

Yahya V, Di Fonzo A, Monfrini E (2023) Genetic evidence for endolysosomal dysfunction in Parkinson's disease: a critical overview. IJMS 24:6338. https://doi.org/10.3390/ijms24076338

Yang L, Han Y, Nilsson-Payant BE et al (2020) A human pluripotent stem cell-based platform to study SARS-CoV-2 tropism and model virus infection in human cells and organoids. Cell Stem Cell 27:125-136.e7. https://doi.org/10.1016/j.stem.2020.06.015

Zabetakis I, Lordan R, Norton C, Tsoupras A (2020) COVID-19: The inflammation link and the role of nutrition in potential mitigation. Nutrients 12:1466. https://doi.org/10.3390/nu12051466

Zhang Y, Sun H, Pei R et al (2021) The SARS-CoV-2 protein ORF3a inhibits fusion of autophagosomes with lysosomes. Cell Discov 7:31. https://doi.org/10.1038/s41421-021-00268-z

Zubair AS, McAlpine LS, Gardin T et al (2020) Neuropathogenesis and neurologic manifestations of the coronaviruses in the age of coronavirus disease 2019: a Review. JAMA Neurol 77:1018. https://doi.org/10.1001/jamaneurol.2020.2065

Acute Encephalopathies in SARS-CoV-2 Infection: A Clinical and Neuroimaging Perspective

7

Marialuisa Zedde, Maria Sessa, Francesca Romana Pezzella, Manuela Napoli, Claudio Moratti, Simonetta Gerevini, Franco Valzania, and Rosario Pascarella

Abstract

Neurological manifestations are frequent in patients with SARS-CoV-2 infection and can be correlated with different pathogenic mechanisms which can be divided into two categories: direct invasion of the central nervous system by the virus and indirect effects deriving from the severity of the systemic infection and by the inflammatory response correlated with cytokine storm. Among the neurological manifestations, acute encephalopathy is very frequent and its nomenclature has recently been updated. The occurrence of a condition of altered mental status, reduced consciousness, delirium up to coma represents an element associated with a greater severity of the infection and mortality both in an Intensive Care Unit setting and in an Emergency Department setting. The tissue damage mechanisms found in COVID-19 patients' encephalopathy and neuroimaging patterns, as well as histopathology, are similar to those described in sepsis-associated encephalopathy, further confirming the role of indirect mechanisms, with no CNS invasion by the virus. The available data have some limitations, notably the underuse of diagnostic neuroimaging techniques in severely affected patients, particularly in the first wave of the pandemic.

Keywords

COVID-19 · Encephalopathy · Neuroimaging MRI · Vasculitis · SARS-CoV-2 · Sepsis · Microbleeds · Leukoencephalopathy

M. Zedde (✉) · F. Valzania
Neurology Unit, Stroke Unit, Azienda Unità Sanitaria Locale-IRCCS di Reggio Emilia, Via Amendola 2, 42122 Reggio Emilia, Italy
e-mail: zedde.marialuisa@ausl.re.it

F. Valzania
e-mail: valzania.franco@ausl.re.it

M. Sessa
UOC Neurologia, ASST Papa Giovanni XXIII, Bergamo, Italy
e-mail: msessa@asst-pg23.it

F. R. Pezzella
Stroke Unit, San Camillo-Forlanini Hospital, Rome, Italy

M. Napoli · C. Moratti · R. Pascarella
Neuroradiology Unit, Azienda Unità Sanitaria Locale-IRCCS di Reggio Emilia, Via Amendola 2, 42122 Reggio Emilia, Italy
e-mail: napoli.manuela@ausl.re.it

C. Moratti
e-mail: moratti.claudio@ausl.re.it

R. Pascarella
e-mail: pascarella.rosario@ausl.re.it

S. Gerevini
UOC Neuroradiologia, ASST Papa Giovanni XXIII, Bergamo, Italy
e-mail: sgerevini@asst-pg23.it

N. Rezaei (ed.), *The COVID-19 Aftermath*, Advances in Experimental Medicine and Biology 1457,
https://doi.org/10.1007/978-3-031-61939-7_7

7.1 Introduction

One of the main limitations for interpreting the data provided by the literature on central nervous system (CNS) effects of SARS-CoV2 infection is the lack of a uniform definition for general conditions such as "encephalopathy." In fact, although a precise boundary and a clear nosography for this term have been recently suggested, the proposal of a definition and diagnostic criteria did not mean their immediate application, making the approach to the innumerable published papers difficult. Another issue is that elderly patients with chronic medical conditions are at an increased risk of altered mental status in the setting of acute infections, and not all the reported clinical manifestations are related to the biological agent causing COVID-19. The potential neurotropism of SARS-CoV-2 caused a plethora of laboratory, clinical, and imaging findings whose direct association with the virus is not always demonstrated. Unfortunately, in particular during the first wave of the pandemic, only a minority of patients with a presumed COVID-19 encephalopathy underwent a complete diagnostic pathway including CSF analysis and brain MRI and few of the deceased patients underwent autopsy with brain examination. In the next sections, a general picture of the available data will be provided starting from the concept of encephalopathy and comparing COVID-19 manifestations with sepsis-related encephalopathy.

7.2 Operative Definition of Encephalopathy and Clinical Classification

The issue of the lack of a shared and uniform framework (definitions and set of diagnostic criteria) for conditions as acute encephalopathy and delirium has been recently raised by several scientific societies, and the final output was a statement about the updated nomenclature of delirium and acute encephalopathy (Slooter et al. 2020). The actual definition of "encephalopathy" includes the several terms previously used to describe a global cognitive change developed by patients during an acute illness (delirium, acute confusional state, acute brain dysfunction, acute brain failure, and altered mental status) (Morandi et al. 2008). The main aim of the statement was to propose a unique term to collect and substitute the previous names in order to propose a single clinical entity instead of many (Behrouz et al. 2015). Most data about the prevalence of acute cognitive changes come from patients hospitalized in intensive care unit (ICU). In this setting, many predisposing and triggering factors have been identified (Marcantonio 2017). The relevance of acute encephalopathy is not only limited to the symptomatic phase but it has been independently associated with adverse short- and long-term outcomes, including higher mortality and chronic cognitive impairment (Witlox et al. 2010; Pandharipande et al. 2013). A frequent misdistinction is among encephalopathy and delirium, mainly because delirium already had a definition and a set of diagnostic criteria in the fifth version of the Diagnostic and Statistical Manual (DSM-5) of the American Psychiatric Association (2013) and in the eleventh edition of the International Statistical Classification of Diseases and Related Health Problems (ICD-11) (World Health Organization 2019).

The limitation of the term "encephalopathy" is its genericity in describing a global disturbance in brain function, but the several other terms commonly used with the same meaning do not have uniform definitions and diagnostic criteria within formal diagnostic systems. The panelists agreed about the following nomenclature:

- the term acute encephalopathy refers to a rapidly developing ($\leq$ 4 weeks) pathobiological brain process whose clinical features are either subsyndromal delirium, delirium or coma
- encephalopathy may have additional features, such as seizures or extrapyramidal signs
- the term subsyndromal delirium refers to acute cognitive changes that are compatible

with delirium, but do not fulfill all DSM-5 delirium criteria (American Psychiatric Association 2013)

- the term coma refers to a state of severely depressed responsiveness defined using diagnostic systems such as the Glasgow Coma Score (GCS) or the Full Outline of UnResponsiveness (FOUR) score (Teasdale and Jennett 1974; Wijdicks et al. 2005).

The use of different terms in clinical practice or research was discouraged, including acute confusional state, acute brain dysfunction, acute brain failure, or altered mental status. The proposed terminology is congruent with the recent recommendations for the nomenclature of cognitive change associated with anesthesia and surgery (Evered et al. 2018). Delayed neurocognitive recovery after anesthesia and surgery can be regarded as consequence of prolonged postoperative acute encephalopathy.

In the literature about neurological complications of COVID-19, this new nomenclature has almost never been applied outside individual data meta-analysis (Singh et al. 2022), where a definition of encephalopathy similar to the updated nomenclature was used, mediated from a previous published framework of case definitions (Ellul et al. 2020). A minority of patients were qualified as having an acute "encephalopathy," defined accordingly with the updated nomenclature above signaled (Slooter et al. 2020) as a pathobiological process in the brain that usually develops over hours to days and can manifest as changed personality, behavior, cognition, or consciousness (including clinical presentations of delirium or coma).

In an attempt to standardize the definitions of the neurological manifestations to collect data about patients with COVID-19 in the first and second wave, a probabilistic grading in the definitions of the cases was proposed by Ellul et al. (2020), including as separate entities encephalitis (Ellul et al. 2020), encephalopathy, delirium, and coma. Encephalopathy was defined and described using a lower level of diagnostic evidence in comparison with "encephalitis" (grade 1 and 2 of the proposed classification), and the probability levels are the following ones:

- encephalopathy (level 3): acute or subacute (< 4 weeks) alteration in consciousness, cognition (including delirium or coma), personality or behavior persisting for more than 24 h and the absence of an alternative diagnosis for symptoms
- suspected encephalopathy (level 4): an alteration in consciousness, cognition, personality, or behavior with no other diagnosis apparent, but does not fulfill level 3 criteria.

In this context, delirium (Tsai et al. 2004) as a clinical manifestation of encephalopathy should fulfill the following features:

(A) Disturbance in attention and awareness;
(B) Develops over a short period of time and tends to fluctuate in severity during the course of the day;
(C) An additional disturbance in cognition;
(D) Criteria A and C are not explained by another preexisting, established, or evolving neurocognitive disorder and do not occur in the context of a severely reduced level of arousal, like coma;
(E) There is evidence from the history, physical examination, or laboratory findings that the disturbance is a direct physiologic consequence of another medical condition, substance intoxication or withdrawal, or exposure to a toxin, or is because of multiple etiologies.

Where a patient has acute cognitive changes that are compatible with delirium but not all of these features are present, this is defined clinically as subsyndromal delirium.

Most of the patients described, especially in the first wave of the pandemic, had not completed a diagnostic pathway capable of applying the proposed levels of diagnostic evidence, and many of these were described as generically encephalopathic, but without distinctive elements.

7.3 Sepsis-Related Encephalopathy as Model

One of the most defined and studied examples of encephalopathy secondary to CNS involvement during severe systemic infections is sepsis-associated encephalopathy (SAE) representing a model from which to start to understand the mechanisms of acute encephalopathy in patients with COVID-19 (Mazeraud et al. 2016, 2020). Sepsis is characterized by a life-threatening organ dysfunction caused by a dysregulated host response to infection (Singer et al. 2016), and it is one of the leading causes for ICU admission. SAE is present in about the half of patients with sepsis with a variability conditioned from one side by the severity of sepsis and the presence of patient-specific factors (Ely et al. 2004) and from the other side by the absence of consensual definition of SAE. In fact, a disturbance in mental status has been identified as one of the three screening items for detecting sepsis (Seymour et al. 2016). The main risk factors are age, history of previous cognitive impairment, kidney and liver failure, and sepsis severity. Bacteremia is accompanied by changes in neurological status, ranging from lethargy to coma, in about 70% of cases (Eidelman et al. 1996).

SAE is defined by the presence of global cerebral dysfunction in different degrees, ranging from mild delirium to coma, accompanying sepsis without direct CNS infection, structural brain abnormality, or other types of metabolic encephalopathy (for example, hepatic, or renal encephalopathy), as detected and diagnosed through clinical or standard laboratory tests. The mandatory requisites are the presence of sepsis or systemic inflammatory response syndrome (SIRS) (Singer et al. 2016). Sometimes, SAE is a diagnosis of exclusion but, despite lack of direct CNS infection, instrumental and/or laboratory evidence of CNS dysfunction is common in SAE and may manifest as abnormalities in EEG (only 19% of the patients admitted to the ICU showed a normal EEG with alpha rhythm and 80% present alterations of the background rhythm or epileptic anomalies on EEG, with non-convulsive status epilepticus in up to 20% of cases) (Hosokawa et al. 2014; Young et al. 1992; Azabou et al. 2015), somatosensory evoked potentials, levels of biomarkers of CNS injury (such as neuron-specific enolase [NSE] and protein S100b), and in neuroradiological investigations (Iacobone et al. 2009). None of these manifestations is specific for SAE, and this issue may represent the first common point with encephalopathy in COVID-19 patients. In fact, the identification of focal lesions in neuroimaging studies is uncommon, but the subgroup of more severe patients with septic shock show white matter hyperintensities (WMHs) and ischemic stroke in 30% of cases (Hosokawa et al. 2014; Cunningham and Maclullich 2013; Oddo et al. 2009). SAE, as well as COVID-19 encephalopathy, is associated with increased mortality with an unknown or incompletely understood mechanism (Eidelman et al. 1996) and long-term effects, including cognitive impairment (Iwashyna et al. 2010; Jackson et al. 2003), depression, anxiety, and posttraumatic stress disorder (PTSD) (Hatch et al. 2018; Righy et al. 2019). The cognitive dysfunction may be due to an early sepsis-related insults of the hippocampus and frontal lobe (Semmler et al. 2013) and the psychological disorders from those involving the limbic system (Mazeraud et al. 2016).

Systemic inflammation provokes harmful effects on the brain parenchyma resulting in SAE through several pathophysiological mechanisms, including neuroinflammation, vascular changes, and metabolic failure, as demonstrated in animal models (Mazeraud et al. 2016). There is a preferential distribution in the brain involvement which sometimes leads to region-specific lesions, particularly in centers for autonomic controls, arousal, awareness, and behavior. It may take account of the main clinical features of SAE (Cunningham and Maclullich 2013).

Three main processes are responsible for cerebral damage in SAE: neuroinflammation, ischemia, and cellular metabolic stress (Adam et al. 2013). They are activated by specific pathways. Neuroinflammation is associated to blood–brain barrier (BBB) dysfunction, which may be sustained by complement system activation (Flierl et al. 2009). The damage at the level of the neurovascular unit (NVU) (endothelial cells, astrocytes, and BBB) is the leading cause of microvascular ischemia and neuroinflammation through microglia activation; an associated macrovascular component may be due to the prothrombotic issues related to the systemic infection. These mechanisms associated with systemic metabolic derangement may cause metabolic and oxidative stress. As previously written, specific brain regions are particularly sensitive to these processes, e.g., those involved in autonomic control, arousal, defining behavioral response to stress, and more complex cognitive functions such as memory and attention (Hennessy et al. 2017). The hippocampus, the amygdala, the nucleus tractus solitarii, and the locus coeruleus show strong neuronal activation in animal models (Skelly et al. 2013) of SAE. In the same locations, pathological studies found neuronal apoptosis with microglial activation in the brain of patients who died from sepsis (Sharshar et al. 2003). These findings provide the anatomic substrate of acute and long-term consequences of SAE.

In a broader view, the contribution of several pathways induces or triggers several final events through the activation of the three mechanisms above describes: brain signaling, neuroinflammation and microglial activation, astrocytes and BBB dysfunction, ischemic processes, mitochondrial dysfunction, neurotransmitter dysfunction. Systemic inflammation may have effect to the brain through signaling by vagal afferents to the medullary autonomic nuclei and the vagus nerve, but also by the trigeminal nerve and trigeminal nucleus. The medullary autonomic nuclei can modulate more widely the response to the sepsis by their connections to other autonomic neuroendocrine and behavioral centers (Sharshar et al. 2003), in addition, the areas physiologically without BBB (the circumventricular organs and the area postrema), which directly receives inflammatory mediators (Shechter et al. 2013).

Moreover, neuroimaging and neuropathologic data in septic shock confirm the BBB impairment as responsible for vasogenic edema in the white matter (Chrétien et al. 2013; Davies 2002). In addition, the occurrence of posterior reversible encephalopathy syndromes (PRES) in septic patients corroborates these findings (Racchiusa et al. 2019).

The neurotoxic consequences of microglial activation are usually considered secondary to proinflammatory phenotypes (Wolf et al. 2017). Microglial activation might partially explain the neuronal activation and dysfunction. Thus, the neuroinflammatory process contributes to long-term consequences of SAE. In animal models of SAE but not in humans, astrogliosis is a well-reproducible finding (Moraes et al. 2015).

An ischemic mechanism in SAE involves both macrocirculation (hypotension, decreased cerebral flow, and impaired autoregulation) and microcirculatory impairment (direct NVU dysfunction with neurovascular uncoupling impairment, BBB disruption, and coagulation cascade activation). The most dramatic example is septic shock-induced ischemic damage mediated by disseminated intravascular coagulopathy (DIC) and often associated with microhemorrhages both in neuroimaging studies and in neuropathology (Sharshar et al. 2004).

Early mitochondrial dysfunction is one of the proapoptotic mechanisms in SAE, and it has been demonstrated in animal models (Azevedo 2010). Another source of neuronal damage is excitotoxicity (Berg et al. 2011). Excitotoxicity through glutamate receptors (N-methyl D aspartate receptor) induces negative consequences, ranging from functional impairment to cell death, which could account for the clinical and electrophysiological features of SAE and in particular delirium.

Despite the evidence on the mechanisms of damage involved in SAE, there are no homogeneous neuroimaging data on large cohorts of patients with severe or septic shock. There is only one prospective MRI study, performed on 72 septic shock patients who developed an acute brain dysfunction defined by focal neurological sign (18%), seizure (10%), coma (46%), and delirium (49%). The MRI was normal in half of cases in this selected cohort. New WMHs (suggestive of vasogenic edema) and ischemic stroke were observed in half of the remaining cases (roughly 25% of patients) (Polito et al. 2013). As previously considered, sepsis can be complicated by PRES (Racchiusa et al. 2019). Because of the risk of transporting septic patients, brain imaging should be indicated upon relevant clinical deterioration, including focal neurological deficit, seizure, and unexplained impaired consciousness. This issue is even more relevant for COVID-19 patients, in particular during the first wave of the pandemic.

One of the neuroimaging hallmarks of SAE is rapidly progressing atrophy (Nakae et al. 2021), expression of neuronal loss (Orhun et al. 2019). A dedicated MRI study of SAE patients shows normal findings in 29%, brain lesions in 54.9%, and brain atrophy in 16.1% of patients. Voxel-based morphometry (VBM) analysis showed neuronal loss in the insula, cingulate cortex, frontal lobe, precuneus, and thalamus. It is interesting that patients with abnormal MRI findings had worse APACHE II, SOFA, GOSE scores, increased prevalence of delirium, and mortality. In particular, brain atrophy was associated with increased p-tau levels and brain lesions to complement breakdown products (Nakae et al. 2021). Moreover, many ICU patients with severe sepsis who developed prolonged mental status changes and neurological sequelae showed signs of brain atrophy (Nakae et al. 2021). Patients with rapidly progressive brain atrophy were more likely to have required mechanical ventilation.

7.4 Pathophysiology of CNS Involvement in SARS-CoV-2 Infection

The mechanisms of SARS-CoV-2 invasion of the CNS have been extensively studied and discussed, but the matching of the exact mechanism to the final damage is not well defined. Some studies reported the presence of SARS-CoV-2 in the CSF and postmortem brain tissue of COVID-19 patients with encephalitis (Paniz-Mondolfi et al. 2020; Moriguchi et al. 2020; Huang et al. 2020a; Serrano et al. 2021; Freij et al. 2020; Remmelink et al. 2020; Song et al. 2021; Matschke et al. 2020; Barth et al. 2020). However, the neurological complications may be provoked by severe systemic inflammation itself and not the direct invasion of the brain (Toscano et al. 2020; Espindola et al. 2020; Lee et al. 2021; Jensen et al. 2021; Keller et al. 2020; Bihlmaier et al. 2020; Bernard-Valnet et al. 2020; Ye et al. 2020), as detailed in SAE.

The main routes for neuroinvasion of SARS-CoV-2 are similar to the ones used by other coronaviridae, i.e., hematogenous route and retrograde or antegrade transport mechanisms from peripheral nerves to the CNS (Cavallieri et al. 2022). For the purposes of this chapter, the CNS damage mechanisms of greatest interest are those that do not involve direct neuroinjury, but, exactly as for SAE, the effects on the CNS of infection and systemic inflammatory response. In fact, however, COVID-19 patients could also present with acute meningoencephalitis without detectable SARS-CoV-2 or any other virus in the CSF (Ye et al. 2020; Bernard-Valnet et al. 2020; Benameur et al. 2020), and other mechanisms such as severe inflammation could be involved in the development of meningoencephalitis in COVID-19. Another way to provoke CNS damage without direct neuroinvasion is acute disseminated encephalomyelitis (ADEM). It is an acute demyelinating disease of the CNS following viral infections in children and young adults (Yu et al. 2020). It has been reported the autopsy finding of a 71-year-old COVID-19 patient with CNS axonal damage, confirming the MRI diagnosis of ADEM (Reichard et al. 2020).

Acute respiratory distress syndrome (ARDS), which occurs as a consequence of severe SARS-CoV-2 infection, could lead to hypoxia that can have harmful effects on the brain, including edema, congestion, and neuronal degeneration (Huang et al. 2020b; Xu et al. 2020). The extreme consequence of the hypoxia-induced brain damage is called hypoxic encephalopathy which may be complicated and accompanied by ischemic stroke. The opposite pathway of damage, i.e., direct damage of the brain by SARS-CoV-2 leading to respiratory failure and hypoxia, is also true (Li et al. 2020). ARDS is associated to severe inflammation due to the release of an excessive amount of proinflammatory cytokines. It could be responsible for tissue damage in the lungs and other organs, including the brain, through mechanisms similar to the ones responsible for the damage induced by SARS-CoV-2 invasion of the brain, due to the production of proinflammatory cytokines by astrocytes and microglia (Li et al. 2004). Severe inflammation in COVID-19 patients, the infection of endothelial cells, and the activation of coagulation cascade could lead to hypercoagulability and DIC (Pons et al. 2020; Ranucci et al. 2020; Klok et al. 2020; Varga et al. 2020). Cytokines and chemokines play an important role in the antiviral response, but their dysregulation in severe infections may lead to detrimental effects, such as uncontrolled fever, delirium, liver, brain, and other tissue inflammation and damage (Dunai et al. 2022). However, the same mechanisms can be evoked in cytokine release syndrome observed with CAR T cell neurotoxicity (Morris et al. 2022). The main guilty are proinflammatory cytokines: IL-6, TNF, IFNs as well as chemokines that act on neutrophils (IL-8) and monocytes (CCL2 and CCL5) and cytokines that stimulate granulocyte and monocyte production (GM-CSF) (Lucas et al. 2020; Thwaites et al. 2021). Measurements of CSF levels of proteins offer important insight into the CNS. A key early study found elevations of brain injury markers (neurofilament light chain, T-tau, glial fibrillary acidic protein, and IL-8) in the CSF of SARS-CoV-2-associated encephalitis cases (Pilotto et al. 2021). The relevance of these measurements is increased by the awareness that the minority of cases of CNS manifestations (excluding encephalitis and meningitis) had viral presence and virus-specific antibodies in the CSF, but the majority had elevated brain injury markers and proinflammatory cytokines, such as IL-1b, IL-2, IL-6, IL-8, and TNF (Domingues et al. 2022). The presence of inflammation in the absence of virus and virus-specific antibodies implicates indirect effects of the virus on neuropathological findings (Dunai et al. 2022). Moreover, the hyperactivation of the immune system triggers adverse vascular events (microthrombosis and damage to vessel walls with endotheliitis). The neurological complications due to these mechanisms are ischemic and hemorrhagic stroke, cerebral venous sinus thrombosis, and intracerebral endotheliitis. In addition, a BBB damage can start a vicious cycle resulting in autoantigen release in the context of inflammatory mediators and autoantibody production. Cytokine release syndrome can further exacerbate damage to the BBB and cause neurotoxicity as well as vascular complications.

7.5 Clinical Issues in COVID-19 Encephalopathy

Starting from the previously cited clinical framework of cases definitions from Ellul et al. (2020), the COVID-19 patients described as having an "other" acute encephalopathy (i.e., when a formal diagnosis of encephalitis, meningitis or ADEM or CNS vasculitis cannot be proposed) were relatively few. In this group, the authors explicated that brain inflammation has not been proven and there is a wide range of causes other than COVID-19 specific role to consider (hypoxia, drugs, toxins, metabolic derangements, etc.) (Solomon et al. 2012).

With these substantial limitations, encephalopathy has been described in 50% of hospitalized COVID-19 patients (Chen et al. 2020a). In a postmortem study (Chen et al. 2020a), a significant number experienced hypoxic encephalopathy. The limitations of this approach are the retrospective design of the study, the case mix including only severely affected patients who died in hospital because of COVID-19 and its complications, and the limitation to the Chinese population in a single hospital. Deceased patients were older (median age 68 years vs. 51 years), mainly male (73%) and more frequently affected by chronic diseases (chronic hypertension and other cardiovascular comorbidities) than recovered patients. Disorder of consciousness was more common in deceased patients (22% vs. 1%), and the median time from disease onset to death in deceased patients was 16 (interquartile range 12.0–20.0) days. Blood markers of severity of infection and systemic inflammation were more relevantly altered in deceased than in recovered patients. Common complications observed more frequently in deceased patients included, among others, acute respiratory distress syndrome (100%), sepsis (100%), acute cardiac injury (77%), heart failure (49%), acute kidney injury (25%), and hypoxic encephalopathy (20%). The 100% of deceased patients with sepsis is a relevant and consistent information, noteworthy in order to sustain the similarities in mechanisms and findings between SAE and COVID-19 encephalopathy. In fact, encephalopathy was more common in COVID-19 patients with coexisting or previous systemic and/or neurological complications (Mao et al. 2020). The main clinical presentation was with altered mental state and confusion following COVID-19 and without any evidence of CNS infection. Other cohorts considered also patients less severely affected and shown that patients could present with confusion and delirium as early signs of COVID-19 without any of the respiratory symptoms (Butt et al. 2020). A description of the first 2019 cases in Wuhan led to the estimation of a prevalence of 9% for altered mental status in COVID-19 patients (Chen et al. 2020b). The limitations, due to the diagnostic restrictions in the first wave of the pandemic, do not allow recovering further information about the mechanisms for altered mental status, in particular in distinguishing direct invasion of the brain from damage resulting from high levels of inflammatory mediators due to the immune response to SARS-CoV-2 infection. In another country (France), 118 (84.3%) of 140 COVID-19 patients, who were treated in two ICUs, had mental changes including delirium, agitation, and corticospinal tract signs (Helms et al. 2020).

In a prospective, multicenter, observational study on COVID-19 hospitalized patients in the New York City metropolitan area (Frontera et al. 2021a), 606/4491 (13.5%) patients developed a new neurological disorder as diagnosed by a neurologist in a median of two days from COVID-19 symptom onset. The most common diagnoses were toxic/metabolic encephalopathy (6.8%), seizure (1.6%), stroke (1.9%), and hypoxic/ischemic injury (1.4%). In none patients was documented the direct presence of SARS-CoV-2 in the CSF. The patients with neurological disorders were more often older, male, white, hypertensive, diabetic, intubated, and had higher sequential organ failure assessment (SOFA) scores (all $p<0.05$) and an increased risk of in-hospital mortality (hazard ratio [HR] 1.38, 95% confidence interval [CI] 1.17–1.62, $p<0.001$). The authors concluded that many observed neurological disorders may be sequelae of severe systemic illness. A similar approach in the same county with a case–control design (Thakur et al. 2022) identified encephalopathy in 478/532 cases (89.8%), and, in general, cases were more likely than controls to be male (58.6% vs. 52.8%, $p=0.05$), to have baseline neurological comorbidities (36.3% vs. 13.0%, $p<0.0001$), and to be admitted in ICU (62.0% vs. 9.6%, $p<0.0001$). The diagnosis of neurological diseases/manifestations was made by experienced neurologists, but the definition of encephalopathy was not the updated nomenclature but an old similar one (Ball et al. 2002).

Toxic-metabolic encephalopathy (TME) was also present in COVID-19 patients and specifically addressed by Frontera et al. (2021b) in a retrospective, multicenter, observational cohort study excluding patients with structural brain disease, seizures, or primary neurological diagnoses. 559/4491 (12%) patients with COVID-19 were diagnosed with TME, of whom 435/559 (78%) developed encephalopathy immediately prior to hospital admission. It is interesting to note that the most common etiologies were septic encephalopathy in 247/559 (62%) patients, hypoxic-ischemic encephalopathy (HIE) in 331/559 (59%) patients, and uremia in 156/559 (28%) patients, with clear overlapping among categories. In fact, multiple etiologies were present in 435/559 (78%) patients. TME were older (76 vs. 62 years), had dementia (27% vs. 3%) or psychiatric history (20% vs. 10%), were more often intubated (37% vs. 20%), had a longer hospital length of stay (7.9 vs. 6.0 days), and were less often discharged home (25% vs. 66% [all $p<0.001$]). After adjustment for confounders, TME remained associated with increased risk of in-hospital death: 128/425 (30%) patients with TME versus 600/3799 (16%) patients without TME (adjusted hazard ratio [aHR] 1.24, 95% confidence interval [CI] 1.02–1.52, $p=0.031$) died. The highest risk of death was associated to TME due to hypoxemia: 97/233 (42%) patients with HIE died, compared with 631/3991 (16%) patients without HIE (aHR 1.56, 95% CI 1.21–2.00, $p=0.001$).

Delirium as a sign of encephalopathy is well represented in COVID-19 patients, as described by several studies. In particular, in a multicenter cohort of critically ill COVID-19 patients (Pun et al. 2021), the prevalence of delirium and coma and the factors associated to delirium have been investigated including a final cohort of 2088 patients with COVID-19 admitted in 69 adult ICUs across 14 countries. Patients with preexisting mental illness, neurodegenerative disorders, congenital or acquired brain damage, hepatic coma, drug overdose, suicide attempt, or those who were blind or deaf were excluded. The median age was 64 years (IQR 54–71); 1397 (66.9%) were invasively mechanically ventilated on the day of ICU admission, and 1827 (87.5%) were invasively mechanical ventilated at some point during hospitalization. Patients undergoing mechanical ventilation were often treated with intravenous sedatives: 1337 (64.0%) were given benzodiazepines for a median of seven days (4–12), and 1481 (70.9%) were given propofol for a median of seven days (4–11). 1704 (81.6%) patients were comatose for a median of 10 days (6–15), and 1147 (54.9%) were delirious for a median of three days (20–60). The factors significantly associated to the development of delirium the next day were mechanical ventilation, use of restraints, and benzodiazepine, opioid, and vasopressor infusions, and antipsychotics (all $p \leq 0.04$), whereas family visitation (in person or virtual) was associated with a lower risk of delirium ($p<0.0001$). During the 21-day study period, patients were alive without delirium or coma for a median of five days (0–14). At baseline, older age, higher SAPS II scores, male sex, smoking or alcohol abuse, use of vasopressors on day 1, and invasive mechanical ventilation on day 1 were independently associated with fewer days alive and free of delirium and coma (all $p<0.01$). 601 (28.8%) patients died within 28 days of admission, with most of those deaths occurring in the ICU. The authors concluded that acute brain dysfunction was highly prevalent and prolonged in critically ill patients with COVID-19, identifying as modifiable risk factors for delirium benzodiazepine use and lack of family visitation. Delirium was also common in a different setting (ED vs. ICU) (Kennedy et al. 2020) considering a cohort of 817 older patients with COVID-19 with mean (SD) age 77.7 (8.2) years. 226 (28%) of included patients had delirium at presentation, sometimes without typical COVID-19 symptoms or signs (84, 37%), and delirium was the sixth most common of all presenting symptoms and signs. The factors significantly associated with delirium were age older than 75 years (adjusted relative risk [aRR], 1.51; 95% CI, 1.17–1.95), living in a nursing home or assisted living (aRR, 1.23; 95% CI, 0.98–1.55), prior use of psychoactive medication (aRR, 1.42; 95% CI, 1.11–1.81), vision impairment (aRR, 1.98; 95% CI, 1.54–2.54), hearing impairment (aRR, 1.10; 95% CI 0.78–1.55), stroke (aRR, 1.47; 95% CI, 1.15–1.88), and Parkinson disease (aRR, 1.88; 95% CI, 1.30–2.58). Delirium was associated with intensive care unit stay (aRR, 1.67; 95% CI, 1.30–2.15) and death (aRR, 1.24; 95% CI, 1.00–1.55).

A particular type of encephalopathy is acute necrotizing encephalopathy (ANE), which presents as neurological symptoms following viral infection, toxemia, and hypoxia (Wu et al. 2020; Sharifian-Dorche et al. 2020), and it has been reported after SARS-CoV-2 infection too, in particular in highly comorbid patients.

The recommendations for treating and managing encephalopathy are potentially applicable to COVID-19 patients, and there are not different evident-based statements. Because encephalopathy has been identified as a frequent finding among older COVID-19 patients (Kennedy et al. 2020) and is associated with poorer outcomes (Pun et al. 2021), actually patients presenting with encephalopathy, including delirium, should be tested for COVID-19.

7.6 Neuroimaging and Neuropathology Issues in COVID-19 Encephalopathies

As for SAE, there is not a single neuroimaging hallmark specific pathognomonic of COVID-19-related encephalopathy. It is hard to separate neuroimaging (mainly MRI) descriptions of patients with encephalopathy (accordingly to the updated nomenclature) from patients with encephalitis, vasculitis, and stroke, which are outside the topic of this chapter.

Unfortunately, most of critically ill patients were not studied with MRI, so a complete information is missed. In the study by Frontera et al. (2021b), all brain neuroimaging was performed in 397/559 (71%) of TME patients and in 507/3932 (13%) of no TME patients ($p<0.001$) with head CT scan acquired, respectively, in 396/559 (71%) and 492/3932 (13%) patients ($p<0.001$) and MRI acquired, respectively, in 51/559 (9%) and 79/3932 (2%) patients ($p<0.001$). Actually, the recommendations (Kremer et al. 2022) from the Subspecialty Committee on Diagnostic Neuroradiology of the European Society of Neuroradiology (ESNR) are to perform neuroimaging in patients with COVID-19 and neurological manifestations with a basic MRI protocol including T2-weighted, fluid-attenuated inversion recovery (FLAIR) (preferably 3D), and diffusion-weighted images, as well as hemorrhage-sensitive sequence (preferably susceptibility weighted imaging or SWI), and at least for the initial investigation pre- and post-contrast T1-weighted images. 3D FLAIR should be acquired after gadolinium administration in order to optimize the detection of leptomeningeal contrast enhancement.

From a neuroradiological point of view, the most addressed form of encephalopathy is hypoxic-ischemic encephalopathy (Fitsiori et al. 2020; Fanou et al. 2017; Riech et al. 2015; Agarwal et al. 2020; Radmanesh et al. 2020) with supportive MRI imaging of diffusion restriction in primarily gray matter structures (basal ganglia, thalami, cerebellum, hippocampus, cerebral cortex), cerebral microbleeds, and leukoencephalopathy and delayed posthypoxic leukoencephalopathy. Hypoxic-ischemic brain injury is not always evident on CT imaging and may require additional neuroimaging; for this reason, it is probably underestimated in COVID-19 patients. Typical posthypoxic MRI brain findings of leukoencephalopathy, white matter cytotoxic edema, and cerebral microbleeds have been described in COVID-19 patients with hypoxic-ischemic encephalopathy (Fitsiori et al. 2020; Fanou et al. 2017; Riech et al. 2015; Agarwal et al. 2020; Radmanesh et al. 2020). In addition, delayed posthypoxic leukoencephalopathy (DPHL) has been observed in COVID-19 patients with a profound neurological deterioration occurring 1–4 weeks after the initial hypoxic insult (Radmanesh et al. 2020). The glutamate excitotoxicity usually drives acute hypoxic-ischemic brain injury, whereas DPHL pathophysiology is related to necrosis of oligodendrocytes in arterial border zones leading to widespread, axonal-sparing demyelination (Radmanesh et al. 2020; Siegler et al. 2020). Addressing more deeply the MRI phenotype of hypoxic-ischemic encephalopathy (Agarwal et al. 2020), 35/115 (30.4%) patients, who had an MRI of the brain acquisition, presented leukoencephalopathy and/or cerebral microbleeds and they required longer ventilator support (34.6 vs. 9.1 days; $p<0.001$) and were more likely to have moderate and severe acute respiratory distress syndrome score (88.6% vs. 23.8%, $p<0.001$) with poorer prognosis (mortality 20% vs. 9%; $p=0.144$). Among the 35 patients with leukoencephalopathy and/or cerebral microbleeds, 30 (85.5%) had leukoencephalopathy and 16 (53.3%) of these 30 patients showed restricted diffusion on apparent diffusion coefficient maps. 25 (71.5%) of the 35 patients had microbleeds and 20 (57%) patients had both leukoencephalopathy and microbleeds. A characteristic perirolandic leukoencephalopathy was seen in 24 (80%) of the patients with leukoencephalopathy and eight (33%) of these 24 also showed restricted diffusion on apparent diffusion coefficient maps. Microbleeds had a lobar distribution in 15 (60%) of 25 patients and 10 (40%) had diffuse involvement. Cerebral microbleeds were seen in all the segments of corpus callosum in 15 (60%) patients whereas four (16%) patients had microbleeds involving only the splenium. No patient with leukoencephalopathy and/or cerebral microbleeds demonstrated superficial siderosis. This neuroimaging pattern was strongly associated with a critical illness and not directly provoked by CNS invasion by SARS-CoV-2. In fact, the majority of cerebral microbleeds were located in subcortical white matter and corpus callosum as in patients with high-altitude cerebral edema and ARDS (Fanou et al. 2017; Riech et al. 2015). The main common mechanism for these diseases is hypoxemia, and hypoxia-induced hydrostatic or chemical effects on the blood–brain barrier could potentially account for extravasation of erythrocytes and lead to cerebral microbleeds, although alternative hypotheses have been raised such as a specific relation with neuroinflammation (Napolitano et al. 2022). Leukoencephalopathy and white matter cytotoxic edema can be seen in critically ill patients in a variety of conditions including patients with diffuse hypoxic-ischemic injury, PRES, septic shock, ADEM, and various toxic-metabolic causes (Polito et al. 2013; Okumura et al. 2009). Several factors may contribute to leukoencephalopathy from the disruption of the blood–brain barrier by endothelial cell activation (Varga et al. 2020) to cerebral hypoxia and ischemic injury. Neuroinflammatory mechanism was also underlying the finding of significantly increased ADC values in the white matter and in several gray matter regions in COVID-19 patients with neurological manifestations versus controls ($p<0.001$). Moreover, hospitalized patients showed significantly higher ADC alteration than non-hospitalized patients in all brain tissues, and COVID-19 patients with neurological disorders showed significantly higher ADC values than those with olfactory loss only (Caroli et al. 2023). An example of brain MRI in COVID-19 related encephalopathy is provided in Figs. 7.1, 7.2 and 7.3.

At histopathology, the consequences of prolonged hypoxemia include oligodendrocyte cell injury, BBB disruption, and microhemorrhage formation (Fitsiori et al. 2020). Microscopic evidence of acute hypoxic injury was found in 18 COVID-19 patients in which clinical course was complicated by persistent confusional state or decreased arousal (Solomon et al. 2020). Histopathologic findings of neuronal loss were in the typical CNS locations most affected by hypoxia including the cerebral cortex, hippocampus, and cerebellar Purkinje cell layers (Matschke et al. 2020; Solomon et al. 2020; Deigendesch et al. 2020). In particular, in the series from Solomon et al. (2020) including

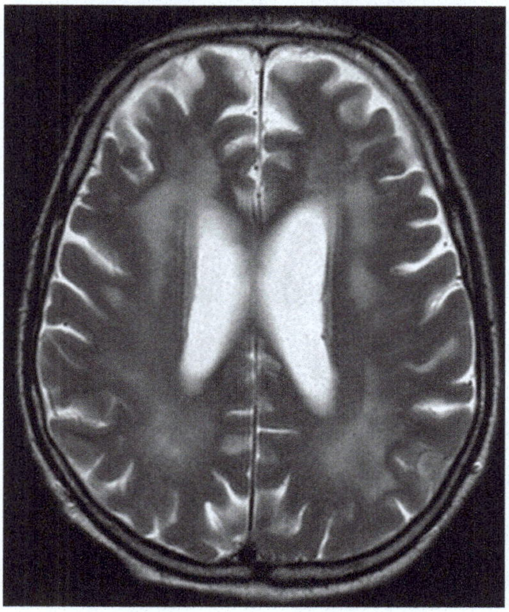

Fig. 7.1 The patient is a 59 years old man with an acute encephalopathy during COVID-19 illness. Axial T2W-MRI at the level of the centrum semiovale showing extensive confluent hyperintense signal changes in the subcortical white matter close to the cortical-subcortical junction

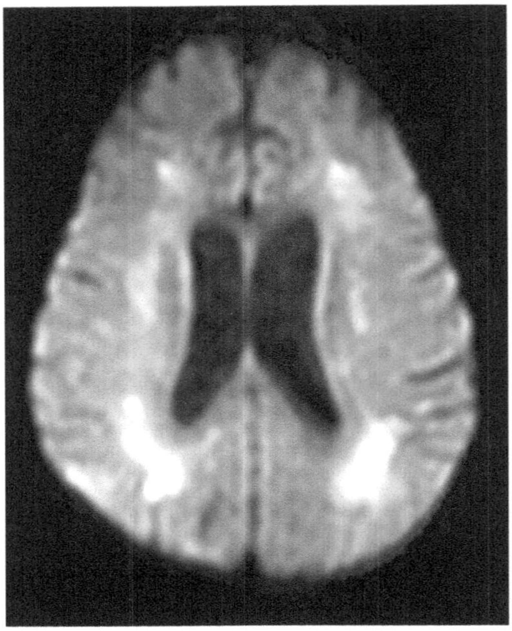

Fig. 7.2 The patient is the same of in Fig. 7.1. Axial DWI-MRI at the level of the centrum semiovale showing that the hypentese T2 lesions are hyperintense on DWI too, the finding is suggestive of an inflammatory process

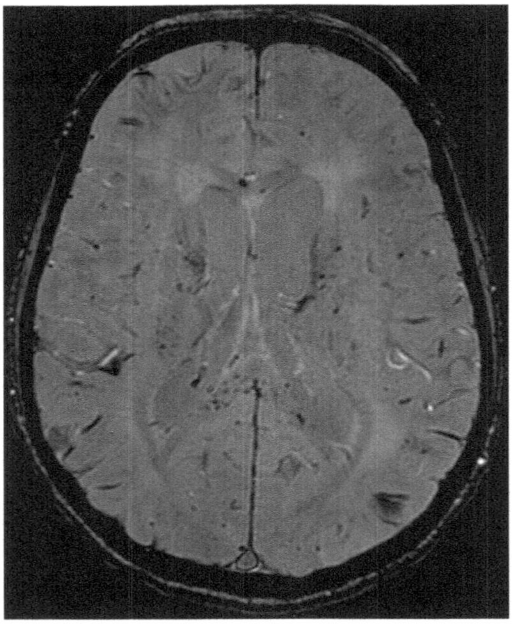

Fig. 7.3 The patient is the same of Figs. 7.1 and 7.2. Axial SWI (susceptibility weighted imaging)-MRI at the level of the centrum semiovale showing several hypointense signal changes both dotted and linear, corresponding to microbleeds and cortical superficial siderosis, respectively

consecutive COVID-19 patients deceased during hospitalization and having had a confusional state or decreased arousal from sedation for ventilation, according to a retrospective chart review by neurologists, it provides relevant histopathological information but without supporting MRI study. Acute ischemic-hypoxic change was present in all specimens in both the cerebrum and the cerebellum, with loss of neurons in the cerebral cortex, hippocampus, and cerebellar Purkinje cell layer, but no thrombi or vasculitis. Neuroinflammatory signs were reported in three patients (two with rare foci of perivascular lymphocytes and one with focal leptomeningeal inflammation). In a different case series (Matschke et al. 2020), neuropathological changes in patients with COVID-19 were reported as mild and 37 (86%) patients had astrogliosis in all assessed regions. Significant neuroinflammatory changes (activation of microglia and infiltration by cytotoxic T lymphocytes) were localized in the brainstem and cerebellum, and meningeal cytotoxic T lymphocyte infiltration was seen in 34 (79%) patients, most likely indicative of non-specific meningeal reaction rather than viral meningitis. In this series, as a main difference in comparison with the previous one (Solomon et al. 2020), SARS-CoV-2 could be detected in the brains of 21 (53%) of 40 examined patients, with SARS-CoV-2 viral proteins found in cranial nerves originating from the lower brainstem and in isolated cells of the brainstem. The presence of SARS-CoV-2 in the CNS was not associated with the severity of neuropathological changes. The majority of patients deceased by pneumonia and/or pulmonary embolism. In a more balanced view (Deigendesch et al. 2020), specifically aimed to assess the brain changes associated to severe inflammatory response rather than direct virus invasion, activated microglia was searched because of the hypothesis of their contribution to the neurological phenotype of COVID-19 (Dantzer 2004). The authors found that microglia activation in the brainstem of COVID-19 patients was significantly more pronounced than in non-septic controls, but the global amount of microglial activation in all specimens did not show differences in COVID-19 and septic patients. The microglia activation observed in COVID-19 patients represents a histopathological correlate of a critical illness-related encephalopathy and is not a disease-specific finding.

In the study by Helms et al. (2020), in an ICU setting, the brain MRI performed in 28/140 patients with altered mental status and delirium shown the following findings: leptomeningeal post-GBCA enhancement in 17/28 patients (60.7%), intraparenchymal, predominantly white matter abnormalities in 8/28 patients, white matter microhemorrhages in 7/28, FLAIR hyperintensities with small foci of contrast enhancement in two cases, and diffusion hyperintensities in two cases. Seventeen patients (60.7%) presented with FLAIR and T1 contrast enhancement into subarachnoid spaces, with specific hyperintensities that were not present on FLAIR images. Twenty-six patients underwent perfusion MRI (arterial spin labeling-ASL). Cerebral blood flow (CBF) maps demonstrated abnormal CBF pattern in 17/26 patients (65.4%) and perfusion abnormalities in 17/26 patients (65.4%). Cerebrospinal fluid examination revealed inflammatory disturbances in 18/28 patients, including oligoclonal bands with mirror pattern and elevated IL-6. The CSF RT-PCR SARS-CoV-2 was positive in one patient.

In case of ANE, brain MRI of patients showed bilateral hemorrhagic rim-enhancing lesions in the thalamic temporal lobes and subinsular regions (Radmanesh et al. 2020; Virhammar et al. 2020; Poyiadji et al. 2020). The cytokine storm that is associated with SARS-CoV-2 infection is believed to damage the BBB and cause brain necrosis in patients with severe COVID-19.

7.7 Conclusions

In patients with COVID-19, encephalopathy is frequent and probably underestimated, being conditioned by the different definitions used and by the use of diagnostic techniques including brain MRI. The underlying mechanisms are mostly secondary to the severity of the infection and the systemic inflammatory response and not to direct CNS invasion by SARS-CoV-2. The clinical and neuroimaging patterns are non-specific but closely resemble the model of sepsis-associated encephalopathy, including toxic-metabolic and hypoxic-ischemic aspects. The occurrence of encephalopathic symptoms is associated with greater severity and with a worse prognosis in patients with COVID-19.

References

Adam N, Kandelman S, Mantz J, Chrétien F, Sharshar T (2013) Sepsis-induced brain dysfunction. Expert Rev Anti Infect Ther 11:211–221

Agarwal S, Jain R, Dogra S et al (2020) Cerebral microbleeds and leukoencephalopathy in critically ill patients with COVID-19. Stroke 51:2649–2655

American Psychiatric Association (2013) Diagnostic and statistical manual of mental disorders, 5th edn. American Psychiatric Association, Arlington

Azabou E, Magalhaes E, Braconnier A et al (2015) Early standard electroencephalogram abnormalities predict mortality in septic intensive care unit patients. PLoS ONE 10:e0139969

Azevedo LCP (2010) Mitochondrial dysfunction during sepsis. Endocr Metab Immune Disord Drug Targets 10:214–223

Ball RHN, Braun MM, Moulton LH et al (2002) Development of case definitions for acute encephalopathy, encephalitis, and multiple sclerosis reports to the vaccine: adverse event reporting system. J Clin Epidemiol 55(8):819–824

Barth RF, Buja LM, Parwani AV (2020) The spectrum of pathological findings in coronavirus disease (COVID-19) and the pathogenesis of SARS-CoV-2. Diagn Pathol 15(1):85. https://doi.org/10.1186/s13000-020-00999-9

Behrouz R, Godoy DA, Azarpazhooh MR, Di Napoli M (2015) Altered mental status in the neurocritical care unit. J Crit Care 30:1272–1277

Benameur K, Agarwal A, Auld SC, Butters MP, Webster AS, Ozturk T, Howell JC, Bassit LC, Velasquez A, Schinazi RF et al (2020) Encephalopathy and encephalitis associated with cerebrospinal fluid cytokine alterations and coronavirus disease, Atlanta, Georgia, USA, 2020. Emerg Infect Dis 26:2016–2021

Berg RMG, Møller K, Bailey DM (2011) Neuro-oxidative-nitrosative stress in sepsis. J Cereb Blood Flow Metab 31:1532–1544

Bernard-Valnet R, Pizzarotti B, Anichini A, Demars Y, Russo E, Schmidhauser M, Cerutti-Sola J, Rossetti AO, Du Pasquier R (2020) Two patients with acute meningoencephalitis concomitant with SARS-CoV-2 infection. Eur J Neurol 27:e43–e44

Bihlmaier K, Coras R, Willam C, Grampp S, Jabari S, Eichhorn P, Haller F, Kuramatsu J, Schwab S, Castellanos I et al (2020) Disseminated multifocal intracerebral bleeding events in three coronavirus disease 2019 patients on extracorporeal membrane oxygenation as rescue therapy. Crit Care Explor 2:e0218

Butt I, Sawlani V, Geberhiwot T (2020) Prolonged confusional state as first manifestation of COVID-19. Ann Clin Transl Neurol 7:1450–1452

Caroli A, Capelli S, Napolitano A, Cabrini G, Arrigoni A, Pezzetti G, Previtali M, Longhi LG, Zangari R, Lorini FL, Sessa M, Remuzzi A, Gerevini S (2023) Brain diffusion alterations in patients with COVID-19 pathology and neurological manifestations. Neuroimage Clin 28(37):103338. https://doi.org/10.1016/j.nicl.2023.103338

Cavallieri F, Sellner J, Zedde M, Moro E (2022) Neurologic complications of coronavirus and other respiratory viral infections. Handb Clin Neurol 189:331–358. https://doi.org/10.1016/B978-0-323-91532-8.00004-5

Chen T, Wu D, Chen H, Yan W, Yang D, Chen G, Ma K, Xu D, Yu H, Wang H et al (2020a) Clinical characteristics of 113 deceased patients with coronavirus disease 2019: retrospective study. BMJ 368:m1091

Chen N, Zhou M, Dong X, Qu J, Gong F, Han Y, Qiu Y, Wang J, Liu Y, Wei Y et al (2020b) Epidemiological and clinical characteristics of 99 cases of 2019 novel coronavirus pneumonia in Wuhan, China: a descriptive study. Lancet 395:507–513

Chrétien F, Stevens RD, Carlier R, Sharshar T (2013) Pattern of brain injury in the acute setting of human septic shock. Crit Care Lond Engl 17:R204

Cunningham C, Maclullich AMJ (2013) At the extreme end of the psychoneuroimmunological spectrum: delirium as a maladaptive sickness behaviour response. Brain Behav Immun 28:1–13

Dantzer R (2004) Cytokine-induced sickness behaviour: a neuroimmune response to activation of innate immunity. Eur J Pharmacol 500:399–411. https://doi.org/10.1016/j.ejphar.2004.07.040

Davies D (2002) Blood–brain barrier breakdown in septic encephalopathy and brain tumours. J Anat 200:639–646

Deigendesch N, Sironi L, Kutza M et al (2020) Correlates of critical illness-related encephalopathy predominate postmortem COVID-19 neuropathology. Acta Neuropathol 140:583–586

Domingues RB, de Moura Leite FBV, Senne C (2022) Cerebrospinal fluid analysis in patients with COVID-19-associated central nervous system manifestations: a systematic review. Arq Neuropsiquiatr 2002:S0004-282X2022005004205. https://doi.org/10.1590/0004-282X-ANP-2021-0117

Dunai C, Collie C, Michael BD (2022) Immune-mediated mechanisms of COVID-19 neuropathology. Front Neurol 13:882905

Eidelman LA, Putterman D, Putterman C, Sprung CL (1996) The spectrum of septic encephalopathy. Definitions, etiologies, and mortalities. JAMA 275:470–473

Ellul MA, Benjamin L, Singh B, Lant S, Michael BD, Easton A et al (2020) Neurological associations of COVID-19. Lancet Neurol 19(9):767–783. PMID: 32622375

Ely EW, Shintani A, Truman B, Speroff T, Gordon SM, Harrell FE, Inouye SK, Bernard GR, Dittus RS (2004) Delirium as a predictor of mortality in mechanically ventilated patients in the intensive care unit. JAMA 291:1753–1762

Espindola OM, Siqueira M, Soares CN, Lima M, Leite A, Araujo AQC, Brandao CO, Silva MTT (2020) Patients with COVID-19 and neurological manifestations show undetectable SARS-CoV-2 RNA levels in the cerebrospinal fluid. Int J Infect Dis 96:567–569

Evered L, Silbert B, Knopman DS et al (2018) Recommendations for the nomenclature of cognitive change associated with anaesthesia and surgery. Br J Anaesth 121:1005–1012

Fanou EM, Coutinho JM, Shannon P et al (2017) Critical illness-associated cerebral microbleeds. Stroke 48:1085–1087

Fitsiori A, Pugin D, Thieffry C et al (2020) COVID-19 is associated with an unusual pattern of brain microbleeds in critically ill patients. J Neuroimaging 30:593–597

Flierl MA, Stahel PF, Rittirsch D et al (2009) Inhibition of complement C5a prevents breakdown of the blood-brain barrier and pituitary dysfunction in experimental sepsis. Crit Care Lond Engl 13:R12

Freij BJ, Gebara BM, Tariq R, Wang AM, Gibson J, El-Wiher N, Krasan G, Patek PM, Levasseur KA, Amin M et al (2020) Fatal central nervous system co-infection with SARS-CoV-2 and tuberculosis in a healthy child. BMC Pediatr 20:429

Frontera JA, Sabadia S, Lalchan R, Fang T, Flusty B, Millar-Vernetti P, Snyder T, Berger S, Yang D, Granger A, Morgan N, Patel P, Gutman J, Melmed K, Agarwal S, Bokhari M, Andino A, Valdes E, Omari M, Kvernland A, Lillemoe K, Chou SH, McNett M, Helbok R, Mainali S, Fink EL, Robertson C, Schober M, Suarez JI, Ziai W, Menon D, Friedman D, Friedman D, Holmes M, Huang J, Thawani S, Howard J, Abou-Fayssal N, Krieger P, Lewis A, Lord AS, Zhou T, Kahn DE, Czeisler BM, Torres J, Yaghi S, Ishida K, Scher E, de Havenon A, Placantonakis D, Liu M, Wisniewski T, Troxel AB, Balcer L, Galetta S (2021a) A prospective study of neurologic disorders in hospitalized patients with COVID-19 in New York city. Neurology 96(4):e575–e586. https://doi.org/10.1212/WNL.0000000000010979

Frontera JA, Melmed K, Fang T, Granger A, Lin J, Yaghi S, Zhou T, Lewis A, Kurz S, Kahn DE, de Havenon A, Huang J, Czeisler BM, Lord A, Meropol SB, Troxel AB, Wisniewski T, Balcer L, Galetta S (2021b) Toxic metabolic encephalopathy in hospitalized patients with COVID-19. Neurocrit Care 35(3):693–706. https://doi.org/10.1007/s12028-021-01220-5

Hatch R, Young D, Barber V, Griffiths J, Harrison DA, Watkinson P (2018) Anxiety, depression and post traumatic stress disorder after critical illness: a UK-wide prospective cohort study. Crit Care Lond Engl 22:310

Helms J, Kremer S, Merdji H, Schenck M, Severac F, Clere-Jehl R, Studer A, Radosavljevic M, Kummerlen C, Monnier A et al (2020) Delirium and encephalopathy in severe COVID-19: a cohort analysis of ICU patients. Crit Care 24:491

Hennessy E, Gormley S, Lopez-Rodriguez AB, Murray C, Murray C, Cunningham C (2017) Systemic TNF-α produces acute cognitive dysfunction and exaggerated sickness behavior when superimposed upon progressive neurodegeneration. Brain Behav Immun 59:233–244

Hosokawa K, Gaspard N, Su F, Oddo M, Vincent J-L, Taccone FS (2014) Clinical neurophysiological assessment of sepsis associated brain dysfunction: a systematic review. Crit Care. https://doi.org/10.1186/s13054-014-0674-y

Huang YH, Jiang D, Huang JT (2020a) SARS-CoV-2 detected in cerebrospinal fluid by PCR in a case of COVID-19 encephalitis. Brain Behav Immun 87:149

Huang C, Wang Y, Li X, Ren L, Zhao J, Hu Y, Zhang L, Fan G, Xu J, Gu X et al (2020b) Clinical features of patients infected with 2019 novel coronavirus in Wuhan, China. Lancet 395:497–506

Iacobone E et al (2009) Sepsis-associated encephalopathy and its differential diagnosis. Crit Care Med 37:S331–S336

Iwashyna TJ, Ely EW, Smith DM, Langa KM (2010) Long-term cognitive impairment and functional disability among survivors of severe sepsis. JAMA 304:1787–1794

Jackson JC, Hart RP, Gordon SM, Shintani A, Truman B, May L, Ely EW (2003) Six-month neuropsychological outcome of medical intensive care unit patients. Crit Care Med 31:1226–1234

Jensen MP, Le Quesne J, Officer-Jones L, Teodosio A, Thaventhiran J, Ficken C, Goddard M, Smith C, Menon D, Allinson KSJ (2021) Neuropathological findings in two patients with fatal COVID-19. Neuropathol Appl Neurobiol 47:17–25

Keller E, Brandi G, Winklhofer S, Imbach LL, Kirschenbaum D, Frontzek K, Steiger P, Dietler S, Haeberlin M, Willms J et al (2020) Large and small cerebral vessel involvement in severe COVID-19: detailed clinical workup of a case series. Stroke 51:3719–3722

Kennedy M, Helfand BKI, Gou RY, Gartaganis SL, Webb M, Moccia JM, Bruursema SN, Dokic B, McCulloch B, Ring H et al (2020) Delirium in older patients with COVID-19 presenting to the emergency department. JAMA Netw Open 3:e2029540

Klok FA, Kruip M, van der Meer NJM, Arbous MS, Gommers D, Kant KM, Kaptein FHJ, van Paassen J, Stals MAM, Huisman MV et al (2020) Incidence of thrombotic complications in critically ill ICU patients with COVID-19. Thromb Res 191:145–147

Kremer S, Gerevini S, Ramos A, Lersy F, Yousry T, Vernooij MW, Anzalone N, Jäger HR (2022) Neuroimaging in patients with COVID-19: a neuroradiology expert group consensus. Eur Radiol 32(6):3716–3725. https://doi.org/10.1007/s00330-021-08499-0

Lee MH, Perl DP, Nair G, Li W, Maric D, Murray H, Dodd SJ, Koretsky AP, Watts JA, Cheung V et al (2021) Microvascular injury in the brains of patients with COVID-19. N Engl J Med 384:481–483

Li Y, Fu L, Gonzales DM, Lavi E (2004) Coronavirus neurovirulence correlates with the ability of the virus to induce proinflammatory cytokine signals from astrocytes and microglia. J Virol 78:3398–34061

Li YC, Bai WZ, Hashikawa T (2020) The neuroinvasive potential of SARS-CoV2 may play a role in the respiratory failure of COVID-19 patients. J Med Virol 92:552–555

Lucas C, Wong P, Klein J, Castro TBR, Silva J, Sundaram M et al (2020) Longitudinal analyses reveal immunological misfiring in severe COVID-19. Nature 584:463–469. https://doi.org/10.1038/s41586-020-2588-y

Mao L, Jin H, Wang M, Hu Y, Chen S, He Q, Chang J, Hong C, Zhou Y, Wang D et al (2020) Neurologic manifestations of hospitalized patients with coronavirus disease 2019 in Wuhan, China. JAMA Neurol 77:683–690

Marcantonio ER (2017) Delirium in hospitalized older adults. N Engl J Med 377:1456–1466

Matschke J, Lutgehetmann M, Hagel C, Sperhake JP, Schroder AS, Edler C, Mushumba H, Fitzek A, Allweiss L, Dandri M et al (2020) Neuropathology of patients with COVID-19 in Germany: a post-mortem case series. Lancet Neurol 19:919–929

Mazeraud A, Pascal Q, Verdonk F, Heming N, Chrétien F, Sharshar T (2016) Neuroanatomy and physiology of brain dysfunction in sepsis. Clin Chest Med 37:333–345

Mazeraud A, Righy C, Bouchereau E et al (2020) Septic-associated encephalopathy: a comprehensive review. Neurotherapeutics 17:392–403. https://doi.org/10.1007/s13311-020-00862-1

Moraes CA, Santos G, de Sampaio e Spohr TCL, D'Avila JC, Lima FRS, Benjamim CF, Bozza FA, Gomes FCA (2015) Activated microglia-induced deficits in excitatory synapses through IL-1β: implications for cognitive impairment in sepsis. Mol Neurobiol 52:653–663

Morandi A, Pandharipande P, Trabucchi M et al (2008) Understanding international differences in terminology for delirium and other types of acute brain dysfunction in critically ill patients. Intensive Care Med 34:1907–1915

Moriguchi T, Harii N, Goto J, Harada D, Sugawara H, Takamino J, Ueno M, Sakata H, Kondo K, Myose N et al (2020) A first case of meningitis/encephalitis associated with SARS-Coronavirus-2. Int J Infect Dis 94:55–58

Morris EC, Neelapu SS, Giavridis T, Sadelain M (2022) Cytokine release syndrome and associated neurotoxicity in cancer immunotherapy. Nat Rev Immunol 22:85–96. https://doi.org/10.1038/s41577-021-00547-6

Nakae R, Sekine T, Tagami T, Murai Y, Kodani E, Warnock G, Sato H, Morita A, Yokota H, Yokobori S (2021) Rapidly progressive brain atrophy in septic ICU patients: a retrospective descriptive study using semiautomatic CT volumetry. Crit Care 25(1):411. https://doi.org/10.1186/s13054-021-03828-7

Napolitano A, Arrigoni A, Caroli A, Cava M, Remuzzi A, Longhi LG, Barletta A, Zangari R, Lorini FL, Sessa M, Gerevini S (2022) Cerebral microbleeds assessment and quantification in COVID-19 patients with neurological manifestations. Front Neurol 23(13):884449. https://doi.org/10.3389/fneur.2022.884449

Oddo M, Carrera E, Claassen J, Mayer SA, Hirsch LJ (2009) Continuous electroencephalography in the medical intensive care unit. Crit Care Med 37:2051–2056

Okumura A, Kidokoro H, Tsuji T, Suzuki M, Kubota T, Kato T, Komatsu M, Shono T, Hayakawa F, Shimizu T et al (2009) Differences of clinical manifestations according to the patterns of brain lesions in acute encephalopathy with reduced diffusion in the bilateral hemispheres. AJNR Am J Neuroradiol 30:825–830. https://doi.org/10.3174/ajnr.A1431

Orhun G, Esen F, Özcan PE, Sencer S, Bilgiç B, Ulusoy C, Noyan H, Küçükerden M, Ali A, Barburoğlu M, Tüzün E (2019) Neuroimaging findings in sepsis-induced brain dysfunction: association with clinical and laboratory findings. Neurocrit Care 30(1):106–117. https://doi.org/10.1007/s12028-018-0581-1

Pandharipande PP, Girard TD, Jackson JC et al (2013) Long-term cognitive impairment after critical illness. N Engl J Med 369:1306–1316

Paniz-Mondolfi A, Bryce C, Grimes Z, Gordon RE, Reidy J, Lednicky J, Sordillo EM, Fowkes M (2020) Central nervous system involvement by severe acute respiratory syndrome coronavirus-2 (SARS-CoV-2). J Med Virol 92:699–702

Pilotto A, Masciocchi S, Volonghi I, De Giuli V, Caprioli F, Mariotto S et al (2021) Severe acute respiratory syndrome coronavirus 2 (SARS-CoV-2) encephalitis is a cytokine release syndrome: evidences from cerebrospinal fluid analyses. Clin Infect Dis Off Publ Infect Dis Soc Am 73:e3019–e3026. https://doi.org/10.1093/cid/ciaa1933

Polito A, Eischwald F, Maho A-L, Polito A, Azabou E, Annane D, Chrétien F, Stevens RD, Carlier R, Sharshar T (2013) Pattern of brain injury in the acute setting of human septic shock. Crit Care Lond Engl 17:R204

Pons S, Fodil S, Azoulay E, Zafrani L (2020) The vascular endothelium: the cornerstone of organ dysfunction in severe SARS-CoV-2 infection. Crit Care 24:353

Poyiadji N, Shahin G, Noujaim D, Stone M, Patel S, Griffith B (2020) COVID-19-associated acute hemorrhagic necrotizing encephalopathy: imaging features. Radiology 296:E119–E120

Pun BT, Badenes R, Heras La Calle G, Orun OM, Chen W, Raman R, Simpson BK, Wilson-Linville S, Hinojal Olmedillo B, Vallejo de la Cueva A et al (2021) Prevalence and risk factors for delirium in critically ill patients with COVID-19 (COVID-D): a multicentre cohort study. Lancet Respir Med 9:239–250

Racchiusa S, Mormina E, Ax A, Musumeci O, Longo M, Granata F (2019) Posterior reversible encephalopathy syndrome (PRES) and infection: a systematic review of the literature. Neurol Sci Off J Ital Neurol Soc Ital Soc Clin Neurophysiol 40:915–922

Radmanesh A, Derman A, Lui YW et al (2020) COVID-19-associated diffuse leukoencephalopathy and microhemorrhages. Radiology 297:E223–E227

Ranucci M, Ballotta A, Di Dedda U, Bayshnikova E, Dei Poli M, Resta M, Falco M, Albano G, Menicanti L (2020) The procoagulant pattern of patients with COVID-19 acute respiratory distress syndrome. J Thromb Haemost 18:1747–1751

Reichard RR, Kashani KB, Boire NA, Constantopoulos E, Guo Y, Lucchinetti CF (2020) Neuropathology of COVID-19: a spectrum of vascular and acute disseminated encephalomyelitis (ADEM)-like pathology. Acta Neuropathol 140:1–6

Remmelink M, De Mendonca R, D'Haene N, De Clercq S, Verocq C, Lebrun L, Lavis P, Racu ML, Trepant AL, Maris C et al (2020) Unspecific post-mortem findings despite multiorgan viral spread in COVID-19 patients. Crit Care 24:495

Riech S, Kallenberg K, Moerer O et al (2015) The pattern of brain microhemorrhages after severe failure resembles the one seen in high-altitude cerebral edema. Crit Care Med 43:e386–e389

Righy C, Rosa RG, da Silva RTA, Kochhann R, Migliavaca CB, Robinson CC, Teche SP, Teixeira C, Bozza FA, Falavigna M (2019) Prevalence of post-traumatic stress disorder symptoms in adult critical care survivors: a systematic review and meta-analysis. Crit Care Lond Engl 23:213

Semmler A, Widmann CN, Okulla T et al (2013) Persistent cognitive impairment, hippocampal atrophy and EEG changes in sepsis survivors. J Neurol Neurosurg Psychiatry 84:62–69

Serrano GE, Walker JE, Arce R, Glass MJ, Vargas D, Sue LI, Intorcia AJ, Nelson CM, Oliver J, Papa J et al (2021) Mapping of SARS-CoV-2 brain invasion and histopathology in COVID-19 disease. medRxiv

Seymour CW, Liu VX, Iwashyna TJ et al (2016) Assessment of clinical criteria for sepsis: for the third international consensus definitions for sepsis and septic shock (sepsis-3). JAMA 315:762–774

Sharifian-Dorche M, Huot P, Osherov M, Wen D, Saveriano A, Giacomini PS, Antel JP, Mowla A (2020) Neurological complications of coronavirus infection; a comparative review and lessons learned during the COVID-19 pandemic. J Neurol Sci 417:117085

Sharshar T, Gray F, Lorin de la Grandmaison G, Hopkinson NS, Ross E, Dorandeu A, Orlikowski D, Raphael J-C, Gajdos P, Annane D (2003) Apoptosis of neurons in cardiovascular autonomic centres triggered by inducible nitric oxide synthase after death from septic shock. Lancet Lond Engl 362:1799–1805

Sharshar T, Annane D, de la Gradmaison GL, Brouland JP, Hopkinson NS, Gray F (2004) The neuropathology of septic shock. Brain Pathol 14:21–33

Shechter R, Miller O, Yovel G et al (2013) Recruitment of beneficial M2 macrophages to injured spinal cord is orchestrated by remote brain choroid plexus. Immunity 38:555–569

Siegler JE, Cardona P, Arenillas JF et al (2020) Cerebrovascular events and outcomes in hospitalized patients with COVID-19: the SVIN COVID-19 Multinational Registry. Int J Stroke 1747493020959216. https://doi.org/10.1177/1747493020959216

Singer M, Deutschman CS, Seymour CW et al (2016) The third international consensus definitions for sepsis and septic shock (sepsis-3). JAMA 315:801–810

Singh B, Lant S, Cividini S, Cattrall JWS, Goodwin LC, Benjamin L et al (2022) Prognostic indicators and outcomes of hospitalised COVID-19 patients with neurological disease: an individual patient data meta-analysis. PLoS ONE 17(6):e0263595. https://doi.org/10.1371/journal.pone.0263595

Skelly DT, Hennessy E, Dansereau M-A, Cunningham C (2013) A systematic analysis of the peripheral and CNS effects of systemic LPS, IL-1β, [corrected] TNF-α and IL-6 challenges in C57BL/6 mice. PLoS ONE 8:e69123

Slooter AJC, Otte WM, Devlin JW, Arora RC, Bleck TP, Claassen J, Duprey MS, Ely EW, Kaplan PW, Latronico N, Morandi A, Neufeld KJ, Sharshar T, MacLullich AMJ, Stevens RD (2020) Updated nomenclature of delirium and acute encephalopathy: statement of ten societies. Intensive Care Med 46(5):1020–1022. https://doi.org/10.1007/s00134-019-05907-4

Solomon T, Michael BD, Smith PE et al (2012) Management of suspected viral encephalitis in adults—Association of British Neurologists and British Infection Association National Guidelines. J Infect 64:347–373

Solomon IH, Normandin E, Bhattacharyya S et al (2020) Neuropathologic features of COVID-19. N Engl J Med 383:989–992

Song E, Zhang C, Israelow B, Lu-Culligan A, Prado AV, Skriabine S, Lu P, Weizman OE, Liu F, Dai Y et al (2021) Neuroinvasion of SARS-CoV-2 in human and mouse brain. J Exp Med 218

Teasdale G, Jennett B (1974) Assessment of coma and impaired consciousness. A practical scale. Lancet 7872:81–84

Thakur KT, Chu VT, Hughes C, Kim CY, Fleck-Derderian S, Barrett CE, Matthews E, Balbi A, Bilski A, Chomba M, Lieberman O, Jacobson SD, Agarwal S, Roh D, Park S, Ssonko V, Silver WG, Vargas WD, Geneslaw A, Bell M, Waters B, Rao A, Claassen J, Boehme A, Willey JZ, Elkind MSV, Sobieszczyk ME, Zucker J, McCollum A, Sejvar J (2022) Risk factors for new neurologic diagnoses in hospitalized patients with COVID-19: a case-control study in New York city. Neurol Clin Pract 12(4):e66–e74. https://doi.org/10.1212/CPJ.0000000000200006

Thwaites RS, Sanchez Sevilla Uruchurtu A, Siggins MK, Liew F, Russell CD, Moore SC et al (2021) Inflammatory profiles across the spectrum of disease reveal a distinct role for GM-CSF in severe COVID-19. Sci Immunol 6:eabg9873. https://doi.org/10.1126/sciimmunol.abg9873

Toscano G, Palmerini F, Ravaglia S, Ruiz L, Invernizzi P, Cuzzoni MG, Franciotta D, Baldanti F, Daturi R, Postorino P et al (2020) Guillain–Barré syndrome associated with SARS-CoV-2. N Engl J Med 382:2574–2576

Tsai LK, Hsieh ST, Chao CC et al (2004) Neuromuscular disorders in severe acute respiratory syndrome. Arch Neurol 61:1669–1673

Varga Z, Flammer AJ, Steiger P, Haberecker M, Andermatt R, Zinkernagel AS, Mehra MR, Schuepbach RA, Ruschitzka F, Moch H (2020) Endothelial cell infection and endotheliitis in COVID-19. Lancet 395:1417–1418. https://doi.org/10.1016/S0140-6736(20)30937-5

Virhammar J, Kumlien E, Fallmar D, Frithiof R, Jackmann S, Skold MK, Kadir M, Frick J, Lindeberg J, Olivero-Reinius H et al (2020) Acute necrotizing encephalopathy with SARS-CoV-2 RNA confirmed in cerebrospinal fluid. Neurology 95:445–449

Wijdicks EF, Bamlet WR, Maramattom BV et al (2005) Validation of a new coma scale: the FOUR score. Ann Neurol 58:585–593

Witlox J, Eurelings LS, de Jonghe JF et al (2010) Delirium in elderly patients and the risk of postdischarge mortality, institutionalization, and dementia: a meta-analysis. JAMA 304:443–451

Wolf SA, Boddeke HWGM, Kettenmann H (2017) Microglia in physiology and disease. Annu Rev Physiol 79:619–643

World Health Organization (2019) International statistical classification of diseases and related health problems, 11th revision. World Health Organization

Wu Y, Xu X, Chen Z, Duan J, Hashimoto K, Yang L, Liu C, Yang C (2020) Nervous system involvement after infection with COVID-19 and other coronaviruses. Brain Behav Immun 87:18–22

Xu Z, Shi L, Wang Y, Zhang J, Huang L, Zhang C, Liu S, Zhao P, Liu H, Zhu L et al (2020) Pathological findings of COVID-19 associated with acute respiratory distress syndrome. Lancet Respir Med 8:420–422

Ye M, Ren Y, Lv T (2020) Encephalitis as a clinical manifestation of COVID-19. Brain Behav Immun 88:945–946

Young GB, Bolton CF, Archibald YM, Austin TW, Wells GA (1992) The electroencephalogram in sepsis-associated encephalopathy. J Clin Neurophysiol Off Publ Am Electroencephalogr Soc 9:145–152

Yu H, Sun T, Feng J (2020) Complications and pathophysiology of COVID-19 in the nervous system. Front Neurol 11:573421

8 Addressing Long COVID Sequelae and Neurocovid: Neuropsychological Scenarios and Neuroimaging Findings

Alessia Celeghin, Mario Stanziano, and Sara Palermo

Abstract

In the face of increasing reports of CNS involvement in COVID-19 cases, it is likely that the current epidemic may be accompanied by a significant increase in the prevalence of neurological sequelae, cognitive dysfunction, and long-term behavioural alterations affecting quality of life and autonomy in daily life. This is consequential to the neuroinvasion and multi-organ dysfunction, but also to the psychological distress and socio-economic changes that occur. Long COVID and neurocovid are now an established concept worldwide. However, the clinical features of these two entities are still debated. The chapter provides information about the nosographic framing, associated pathophysiological mechanisms, alterations in the central and peripheral nervous systems, and the associated neurocognitive profile, indications about predictor and clinical evaluation according to a patient-centred multidimensional immuno-behavioural approach.

Keywords

Post-COVID conditions · Long COVID · Neurocovid · Coronavirus syndrome · Neuroimaging · Neuropathophysiology · Neurology · Neuropsychology · Bio-psycho-social approach

A. Celeghin · S. Palermo (✉)
Department of Psychology, University of Turin, Turin, Italy
e-mail: sara.palermo@unito.it

A. Celeghin
e-mail: alessia.celeghin@unito.it

M. Stanziano
Neurosciences Department "Rita Levi Montalcini", University of Turin, Turin, Italy
e-mail: mario.stanziano@istituto-besta.it

M. Stanziano · S. Palermo
Neuroradiology Unit, Diagnostic and Technology Department, Fondazione Istituto di Ricovero e Cura a Carattere Scientifico (IRCCS) Istituto Neurologico Carlo Besta, Milan, Italy

8.1 Introduction

Similar to the post-acute viral syndromes described in survivors of other virulent coronavirus outbreaks, there are increasing reports of persistent and prolonged effects after acute COVID-19. Survivors still tend to have worse health and quality of life than the general population. The evidence is especially true for patients suffering from the infection distant sequelae—known as "*Post-acute COVID-19 Syndrome*" or "*long COVID*" or Nalbandian et al. (2021)—who typically still have at least

N. Rezaei (ed.), *The COVID-19 Aftermath*, Advances in Experimental Medicine and Biology 1457,
https://doi.org/10.1007/978-3-031-61939-7_8

one symptom of the disease including muscle weakness, fatigue, reduced pulmonary output, difficulty sleeping, anxiety, and depression even two years after becoming ill and can take a long time to recover (Huang et al. 2021; Nalbandian et al. 2021).

The phenomenon is so pervasive that it has prompted the World Health Organization [WHO] to assign a special emergency code to classify cases of "*Post-COVID conditions, unspecified*" within the International Classification of Diseases, Tenth Revision (ICD-10) (WHO 2022a). Based on a Delphi consensus, WHO stated that "*Post COVID-19 condition occurs in individuals with a history of probable or confirmed SARS-CoV-2 infection, usually 3 months from the onset of COVID-19 with symptoms that last for at least 2 months and cannot be explained by an alternative diagnosis. Common symptoms include fatigue, shortness of breath, and cognitive dysfunction but also others which generally have an impact on everyday functioning. Symptoms may be new onset, following initial recovery from an acute COVID-19 episode, or persist from the initial illness. Symptoms may also fluctuate or relapse over time*" (WHO 2022b, p. 1, lines 21–28).

For several reasons, supporters of long COVID believed that "post-" or "chronic" COVID was inappropriate (Callard and Perego 2021). In addition to preventing diagnosis during the acute phase, it also suggests that the disease occurs after acute COVID and persists, whereas patients' experiences suggest a continuation of acute (or suspected acute) disease and persistent symptoms (Callard and Perego 2021).

Long COVID is often defined by the persistence of symptoms beyond 1 month–24 weeks after acute infection (Stefanou et al. 2022). Long COVID is often defined by the temporal criterion, given by the persistence of symptoms beyond 1 month–24 weeks after an acute infection. Long COVID has no solidified diagnosis criteria to date (Guedes 2022; Stefanou et al. 2022). Yet, long COVID is conceptualised as a multi-organ disorder with a wide spectrum of clinical symptoms that may be indicative of underlying lung, cardiovascular, endocrine, hematologic, renal, gastrointestinal, dermatological, immunological, psychiatric, or neurological conditions (Stefanou et al. 2022; Xu et al. 2022) (Table 8.1). More than 200 symptoms have been associated with the disorder, the severity of which ranges from mild to debilitating.

Table 8.1 Sample list of possible persistent symptoms ascribed to post-COVID syndrome that demonstrate its multisystem disease picture

Clinical domain	Symptoms/syndromes
Respiratory	Shortness of breath, wheezing
Cardiovascular	Chest pain, palpitations, chest constriction, hypertension, arrhythmia
Neurological	Headache, difficulty concentrating, light headedness, insomnia, dizziness, paraesthesia and altered sensation, delirium
Psychological/psychiatric	Anxiety and depression, post-traumatic stress disorders, brain fog, other forms of psychological distress
Gastroenterological	Abdominal pain, nausea, vomiting, diarrhoea, anorexia, and poor appetite
Otolaryngologic	Tinnitus, otalgia, anosmia, and ageusia
Metabolic	Glycaemic imbalance, sarcopenia
Vascular	Thrombo-embolic phenomena
Dermatologic	Skin rashes (maculo-papular, pomphi, or chilblains of the extremities, COVID toe)
General symptoms	Fatigue, asthenia, widespread muscle pain, muscle weakness

Theories about long COVID origins vary and include harmful immune responses, small blood clots, and persistent viral reservoirs in the body. Together these factors contribute to the overall burden of the disease. The need to provide ongoing support for an increasingly significant proportion of people who have had COVID-19 is accompanied by equally ongoing surveillance to understand how vaccines, emerging treatments, and variants affect the clinical course of long COVID (Palermo 2020, 2021a, b).

The long-term health impacts of COVID-19 have remained largely unknown because the two-year longitudinal studies have just ended. In addition, the lack of baseline values of pre-COVID-19 health status and comparisons with the general population in most studies have also made it difficult to determine how well COVID-19 patients have recovered. What is certain is that the COVID-19 pandemic has shown multiple neurological complications, both in the acute stages of the disease and in the weeks and months following. There is now also growing concern about the possible cognitive consequences of COVID-19, with reports of symptoms persisting and cases revealing neurological problems in severely affected patients, although there is still little information regarding the nature and broader prevalence of cognitive problems that persist even in the post-infection phase (Hampshire et al. 2021).

Indeed, the hazard ratio of any neurological sequelae was 1.42 and burden 70.69 per 1000 persons at 12 months COVID-19 pandemic (Xu et al. 2022). With millions of people affected by the virus, nervous system complications pose additional public health challenges that will linger for years and involve rehabilitation and recovery of functional capacity (Palermo 2021a). Here, we provide a synopsis of current knowledge on post-acute COVID-19 neurocognitive sequelae and their pathophysiology in the central nervous system. Finally, we discuss relevant considerations for an integrated multidisciplinary approach to long neurocognitive sequelae and propose a framework for identifying high-risk individuals.

8.2 Remarks on Neurocovid

Recovery from COVID-19 is commonly considered primarily in terms of remission of respiratory symptoms. Nevertheless, COVID-19 outpatients, including those who no longer reported symptoms, showed significant neurological and neuropsychological complaints compared with controls (Hampshire et al. 2021). Hence, the definition *neurocovid* is derived, which alludes to the acute and chronic range of neurological, neuropsychological, or neuropsychiatric abnormalities that can occur in people as a result of the viral attack on the nervous system.

Some form of neurocovid appears to occur in up to 30% of infected patients (Baig et al. 2020; Finsterer and Scorza 2021; Misra et al. 2021). At four months after discharge, more than 40% of patients may experience persistent symptoms and activity/participation limitations, with a profound impact on daily life activities (Wahlgren et al. 2023). The proportion of patients with at least one post-COVID-19 symptom 24 months after acute infection was 59.7% for hospitalised patients and 67.5% for those not requiring hospitalisation (Fernández-de-Las-Peñas et al. 2022). Cognitive, sensorimotor, and fatigue symptoms were the most common persisting symptoms (Wahlgren et al. 2023). Survivors with long COVID symptoms had a lower health-related quality of life, worse exercise capacity, more mental health abnormalities, and greater use of health care after discharge than survivors without prolonged symptoms. They also appeared to suffer from more pain or discomfort, as well as anxiety or depression, compared with controls at 24 months (Huang et al. 2022a).

Moreover, pre-existing neurological diseases may become clinically evident or worsen with SARS-CoV-2 infection (Finsterer and Scorza 2021; Palermo et al. 2023). In particular, it appears to exacerbate chronic neurological disorders such as Parkinson's disease (Chambergo-Michilot et al. 2021; Fearon and Fasano 2021; Palermo et al. 2023). Studies of previous viral outbreaks have shown similar symptomatologic pictures (Ritchie et al. 2020; Palermo 2021a, b; Palermo et al. 2023). Indeed, both clinical and animal studies have shown that coronaviruses can spread to the central and peripheral nervous systems [CNS and PNS] altering structures and functions (Baig et al. 2020; Finsterer and Scorza 2021), or they mediated spinal cord demyelinating diseases (Mondal et al. 2021; Garg et al. 2021). At the same time, the presence of physical-cognitive frailty or neurological disease is recognised as a precipitating and predisposing factor for SARS-CoV-2 infection or COVID-19 disease severity (Palermo et al. 2023). There is growing concern that people already suffering from chronic neurological disorders and those undergoing immunosuppressive or immunomodulating therapies for their underlying diseases are more susceptible to this infectious disease (Roy et al. 2021). Moreover, some authors suggested possible interactions between an executive-metacognitive dysfunction and the ability to protect oneself from infection (Bartoli et al. 2020), associated with the detection of conflict events and inhibition of incorrect responses (Palermo et al. 2018).

8.2.1 The Polyform Entity of Clinical Manifestations

The effects of COVID-19 are becoming better defined over time. In the first pandemic wave, it was thought that neurological and neuropsychological complications occurred mainly in those who developed the infection in a severe form, requiring hospitalisation and intubation, or in frail elderly individuals. Now, it is evident that those disorders also affected young adults who had previously never experienced symptoms, developed even very mild forms of COVID, or did not manifest respiratory problems.

A large case history of neurological manifestations associated with SARS-CoV-2 infection has been reported in the literature from the beginning (Mao et al. 2020; Varatharaj et al. 2020). Initial reports included single and/or regional case series and cohorts with variable data definitions, which limited the ability to accurately estimate the incidence of neurological manifestations of COVID-19 and precluded data sharing and generalisation across populations (Chou et al. 2021).

In accordance with what has been reported by some authors, neurological sequelae can be classified as *self-reported symptoms* (for example, ageusia and anosmia, headache, syncope) or *neurological signs and/or syndromes* assessed by clinical evaluation (for example, abnormal tone and brainstem reflexes, sensory abnormalities, dysautonomia, meningitis and/or encephalitis, myelopathy, stroke, coma, seizure, and so on) (Chou et al. 2021). It would seem that following this classification, 80% of patients experienced at least one new neurological symptom, sign, or syndrome, while 55% had at least one neurological sign or syndrome detected by clinical evaluation (Chou et al. 2021).

Depending on which part of the neuroaxis is involved, COVID-19 para-infectious neurological features can be classified as major CNS or PNS involvement, including skeletal muscle injury (Gupta et al. 2020; Roy et al. 2021) (Table 8.2). Consistently, CNS direct attack can result in meningitis/encephalitis (Gallacher and Seaton 2020; Chou et al. 2021; Finsterer and Scorza 2021; Schweitzer et al. 2021), with symptoms of headaches, seizures, confusion, ataxia, pyramidal signs, or altered consciousness (Palermo 2021a). PNS direct damage normally entails hyposmia or hypogeusia (Gupta et al. 2020; Finsterer and Scorza 2021). However, some non-specific symptoms found in the population (such as "brain fog", fatigue, and sleep disturbances) may be simultaneously ascribed to CNS, PNS, autoimmune, cardiovascular, endocrine, psychiatric, or respiratory diseases (Gupta et al. 2020; Crook et al. 2021; Roy et al. 2021; Guedes 2022) (Fig. 8.1).

The immune reaction to COVID-19 can contribute to the occurrence of overt neurological disorders including acute disseminated encephalomyelitis, acute haemorrhagic myoclonus, and necrotising encephalopathy (Delamarre et al. 2020; Ghosh et al. 2020), as well as cytokine release syndromes and mononeuritis (Perrin et al. 2021), myositis (Zhang et al. 2020),

Table 8.2 Long COVID syndrome, based on nervous system location (central or peripheral nervous system)

Nervous system localization	Clinical manifestation
Central nervous system 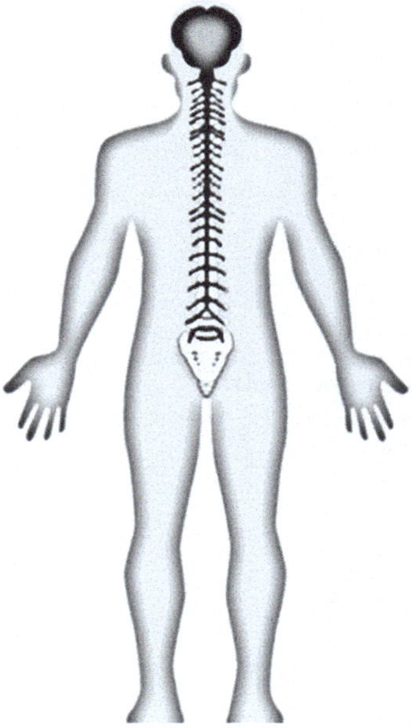	• Brain fog • Cerebellar ataxia • Cerebrovascular events/stroke • Cognitive dysfunction and cognitive complaints • Dizziness • Dysautonomia • Emotional/mood disorders • Fatigue • Headache/cephalalgia • Sleep disorders • Meningoencephalitis and encephalopathy • Myoclonus • Post-infectious myelitis • Seizures
Peripheral nervous system 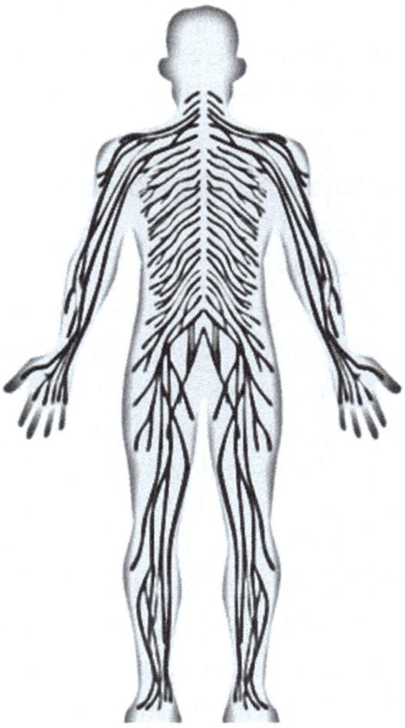	• Ageusia and hypogeusia • Anosmia and hypoanosmia • Hearing loss/tinnitus • Peripheral neuropathy: Guillain–Barré syndrome; Miller–Fisher syndrome; axonal polyneuropathy; and focal neuropathies (single or multiple mononeuropathies, plexopathies) • Muscle weakness • Myalgias • Myopathy • Ophthalmoparesis • Polyradiculopathy • Sensorimotor deficits: hypoesthesia, dysesthesia, tremor

cerebral vasculitis, delirium, psychosis, transverse myelitis, cranial nerve palsy, and Guillain–Barré syndromes (Stefanou et al. 2022). The presence of neurological signs or syndromes concomitant with COVID-19 has been shown to be a negative prognostic factor, with a

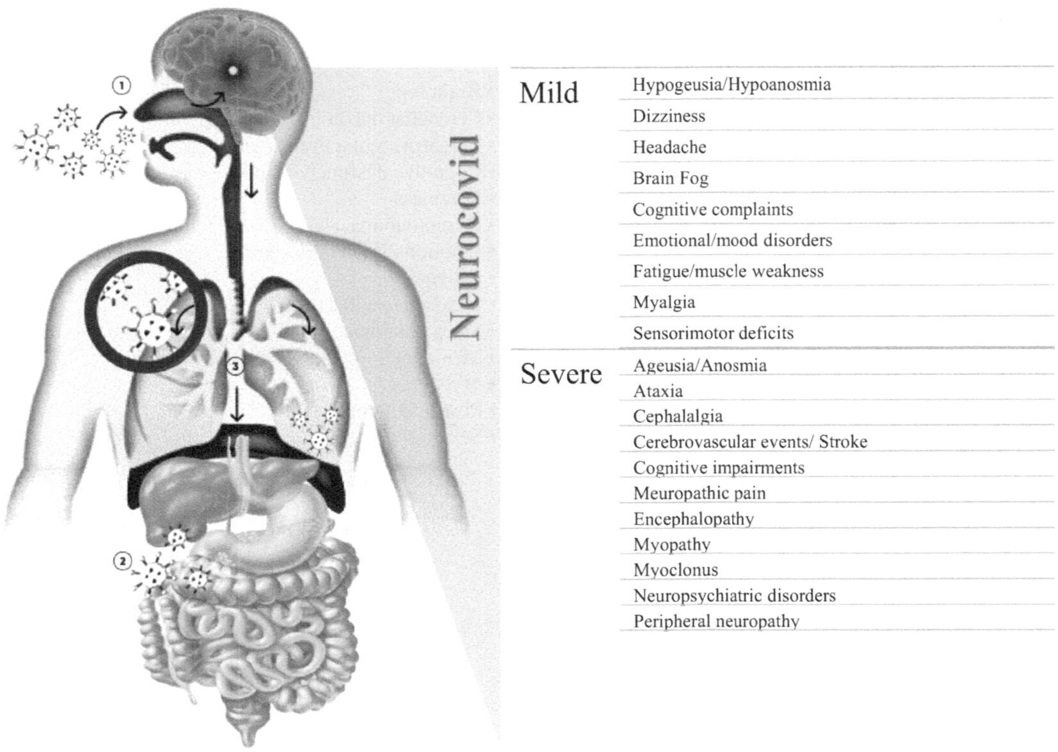

Fig. 8.1 By neuroinvasion, SARS-CoV-2 viruses can directly infect the brain, causing neurological symptoms. COVID-19 may arrive in the brain through several pathways, including (1) the olfactory route through the trigeminal nerve, (2) the vagus nerve in relation to the lungs, and (3) the microbiota-gut-brain axis; or indirectly, as a result of hypoxia or hyperthermia caused by systemic infection. Leading neurocovid manifestations are shown

significantly increased risk of death during the hospitalisation of acute patients (Chou et al. 2021). In addition, pre-existing neurological disorders at infection have been associated with an increased risk of developing new neurological signs and/or syndromes (Chou et al. 2021) or they become clinically evident or worsen with COVID-19 (Finsterer and Scorza 2021; Monje and Iwasaki 2022).

During the pandemic waves, there was also a proliferation of reports of inherent depression, anxiety disorders, suicidal ideation, fear of survival, stigmatisation, reduced quality of life, psychological distress, and post-traumatic stress symptoms. Furthermore, organic hallucinations, persecution delusions, temporal-spatial disorientation, and hypomania have been reported. In some cases, these manifestations were secondary to steroid therapy (Palermo 2021a).

8.3 The Neuropathophysiology of Long COVID and Neurocovid

It has been observed from previous outbreaks that respiratory coronaviruses can enter the brain and cerebrospinal fluid. A series of autopsies of SARS-CoV victims following the 2003 outbreak, for example, revealed SARS-CoV genome sequences throughout the cortex and hypothalamus (Gu et al. 2005). Widespread lesions have been identified in the white matter and subcortical areas of the frontal, temporal, and parietal lobes in patients infected with Middle East respiratory syndrome (MERS-CoV) (Arabi et al. 2015). Two main mechanisms of neuroinvasion have been proposed. The first hypothesis considers passage through the blood–brain barrier compromising the tight junctions between cerebral microvascular endothelial cells (Koyuncu et al. 2013; Miner and Diamond 2016). Alternatively, a second channel of access would be periphery neurons or olfactory sensory neurons, using axonal transport to access the CNS (Dahm et al. 2016).

Although cause-and-effect relationships between SARS-CoV-2 infection and neurological-neuropsychological symptomatology were initially unclear, there is now increasingly corroborated scientific evidence on the matter (Fig. 8.1). They are likely due to interacting causes, particularly direct virus damage to cortical and subcortical structures, as well as indirect effects due to non-CNS systemic impairment (i.e., the respiratory system) (Ritchie et al. 2020). Seventy percent of critically ill patients admitted to the ICU with COVID-19 for respiratory complications required mechanical ventilation (Arentz et al. 2020; Chang et al. 2021), developing acute respiratory distress syndrome within three days. Analysing neuropsychological studies of long-term outcomes in adults who underwent forced ventilation ascertained alteration in attention, memory, verbal fluency, and processing speed, and in executive functioning (Hopkins et al. 2017; Pfoh et al. 2015), in 78% of patients one year after discharge and about half of the patients up to two years (Hopkins et al. 1999, 2005; Mikkelsen et al. 2012).

Long COVID and neurocovid conditions can actually result from viral attacks on neurons, glial cells, blood vessels, or blood–brain barriers, as well as the immune response to the virus, or the side effects of treatment applied during acute infection, or caused by lung, heart, or kidney disease (Ritchie et al. 2020; Crook et al. 2021; Monje and Iwasaki 2022; Stefanou et al. 2022). The possibility of virus invasion into the central nervous system, where viral proteins have been shown, has emerged, but most of the damage appears to be related to vascular or inflammatory mechanisms, with activation of microglial cells, a mechanism known as *neuroinflammation* (Stefanou et al. 2022).

As of now, there are six possible mechanisms of action that are emerging (Fig. 8.2) (Monje and Iwasaki 2022): (1) despite being rare, SARS-CoV-2 can directly infect the nervous system; (2) SARS-CoV-2 triggers neuroinflammation in the respiratory system by increasing cytokines, chemokines, and immune cell trafficking in the brain, and causing reactive states in microglia and other immune cells resident in the brain; (3) cerebrovascular and thrombotic diseases can disrupt blood flow, interfere with blood–brain barrier functions, and lead to further neuroinflammation and/or neuroischemia; (4) SARS-CoV-2 may elicit an autoimmune response to the nervous system; (5) reactivation of latent herpesviruses, such as Epstein–Barr, can lead to neuropathology; (6) severe COVID-19 can lead to pulmonary and multi-organ dysfunction, which adversely affects neural cells thorough hypoxia, hypotension, and metabolic disturbances.

CNS and PNS damage can occur in a variety of ways, and these six mechanisms cannot be considered mutually exclusive. The frequency and timing of these mechanisms may vary in COVID-19 patients. For instance, a more frequent mechanistic principle associated with mild COVID-19, which occurs even in its acute phase, is neuroinflammation triggered by an immune response to respiratory infection and the subsequent dysregulation of homeostasis and neural plasticity. Conversely, severe COVID-19 is likely to be associated with a rare mechanism of direct brain infection.

8.3.1 Possible Mechanisms Contributing to COVID-19-Related Neuropsychological Dysfunctions

No matter how severe a patient's condition may be, it is important to consider CNS post-COVID-19-related changes when dealing with their reports of neuropsychological impairment. Indeed, neural-immune interactions and cognitive function are at the basis of the observed clinical syndromes (Monje and Iwasaki 2022).

New evidence suggests that peripheral organs influence immune responses in the CNS, and

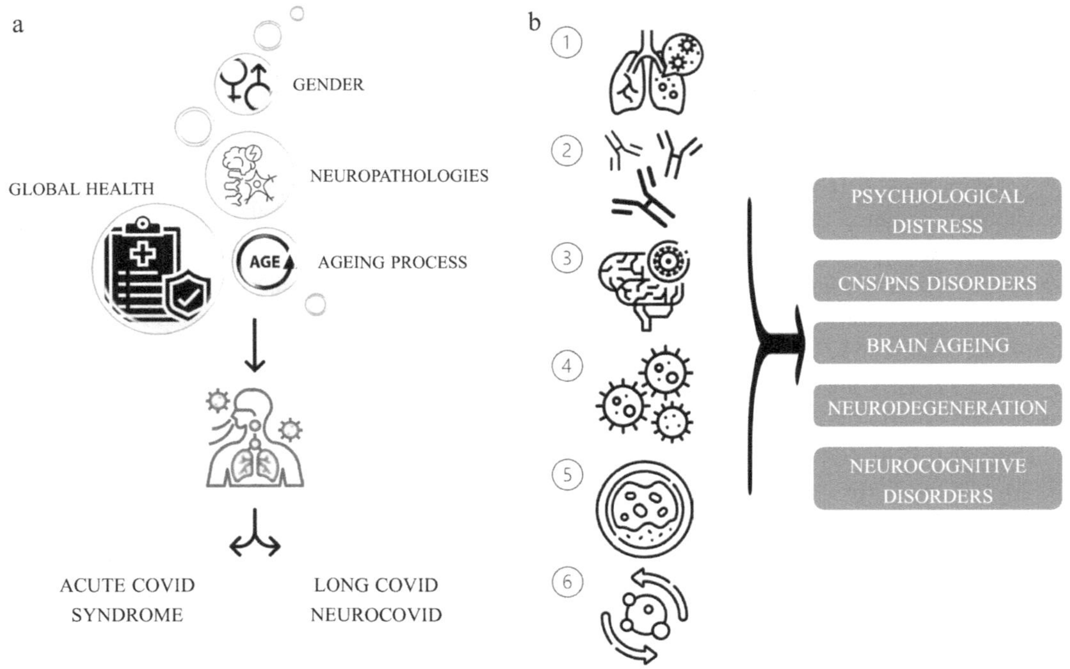

Fig. 8.2 Mechanisms contributing to neurocovid. **a** There are infection susceptibility factors that are also expressed as facilitating factors for the onset of post-COVID conditions, which in turn are expressed as long COVID and neurocovid. **b** The infectious process is able to attack the CNS and PNS through six main mechanisms of action: 1 respiratory system inflammation; 2 autoimmunity; 3 direct neuroinvasive infection; 4 reactivation of latent herpesviruses; 5 thrombolytic and microvascular disease destroying the blood–brain barrier; 6 hypoxia and metabolic disturbances

a close link between the lung microbiota and brain autoimmune responses has recently been discovered (Hosang et al. 2022). A neuropathological cascade leading to cognitive dysfunction is thought to be triggered by inflammation (Gorelick 2010; Han and Mallampalli 2015). The respiratory SARS-CoV-2 infection can cause immune injury in the CNS as a result of excessive peripheral inflammation (Li et al. 2022). It is associated with impaired hippocampal neurogenesis, persistent oligodendrocyte loss, and myelination axon loss when CSF CCL11 and white matter microglial reactivity are elevated (Li et al. 2022). This could be the link between mild respiratory COVID-19 and cognitive impairment.

In the acute phase of COVID-19, the virus enters the CNS facilitated by haematogenous or directly trans-synaptic pathways through ACE2 engagement (Stefanou et al. 2022). ACE2 is typically present on the surface of several cell types, including endothelial cells and smooth muscle cells of cerebral blood vessels. The cytokine storm induced by SARS-CoV-2 can disrupt tight junctions in the endothelial lining of the blood–brain barrier. This leads to increased blood–brain barrier permeability and allows virus-infected leukocyte transmigration. In addition, cytokine release precipitates platelet activation and adhesion, causing further endothelial impairment leading to increased thrombotic risk observed in acute patients. Once cytokines and leukocytes cross the blood–brain barrier, they activate microglial cells, which in turn trigger apoptotic cascades and the demyelination of neurons (Stefanou et al. 2022). In "long COVID", chronic inflammatory and secondary degenerative processes are believed to prevail, which cause "long-term" neurological sequelae (Stefanou et al. 2022). In the

"ACE2-rich" brain areas, including those in the somatosensory cortex, rectus/orbital gyrus, temporal lobe, hypothalamus/thalamus, brainstem, and cerebellum, significant hypometabolism is found (Stefanou et al. 2022). The reduced glucose metabolism observed in these areas may further promote oxidative stress, mitochondrial dysfunction, or impaired brain autoregulation. These conditions are typically secondary to SARS-CoV-2 infection (Stefanou et al. 2022).

Using logistic regression analysis, impaired functioning profiles were predicted by acute COVID-19 symptoms, current depression score, number of medical comorbidities, and subjective cognitive impairment for memory, language, and executive functions (Ferrando et al. 2022). COVID symptoms, medical comorbidities, fatigue, and inversely executive function were all associated with interleukin-6. Symptoms of COVID and depression score were positively associated with C-reactive proteins, while quality of life was negatively associated (Ferrando et al. 2022).

Post-acute sequelae of COVID-19 have been reported in most patients with viral antigen persistence. Indeed, SARS-CoV-2 antigens persist in the intestinal mucosa for months after acute COVID-19 in most patients with inflammatory bowel disease, regardless of immunosuppressive therapy or intestinal inflammation (Zollner et al. 2022). Neurological impairments associated with alterations of the microbiota-gut-brain axis are recognised (Dinan and Cryan 2017; Kowalski and Mulak 2019; Escobar et al. 2022), and increasingly hypothesised following infection (Palermo et al. 2023).

In summary, acute respiratory distress syndrome can also involve inflammatory responses (Han and Mallampalli 2015), as well as anaemia and ischemia, resulting in cardiovascular and hepatic failure (Matthay and Zemans 2011). This cascade of neurological and physiological events can further exacerbate neurological damage in the acute stages and promote chronic cognitive dysfunction. Cognitive impairment following mild respiratory COVID-19 is associated with persistent neuroinflammation, which has allowed us to identify pathophysiological similarities between "COVID fog" and "chemo fog" syndromes (Fernández-Castañeda et al. 2022). As a consequence, antiinflammatory therapies targeting microglia or CCL11 signalling can be useful in preventing and correcting these multicellular dysregulations (Li et al. 2022).

8.4 Human Neuroimaging Findings

SARS-CoV-2 infection is well known to cause brain changes (Mankad et al. 2020; de Erausquin et al. 2021), even in subjects with only anosmia as a symptom (Politi et al. 2020). Indeed, alterations in olfactory areas, along with nearby brain regions, including prefrontal and limbic regions, have been associated with SARS-CoV-2 contraction (Najt et al. 2021).

Postmortem neuroradiological evaluation detected asymmetrical olfactory bulb structures, abnormalities of the brain parenchyma, micro/macro haemorrhages in subcortical regions, cortico-subcortical oedema, and non-specific deep white matter changes (Coolen et al. 2020). Homologous findings were shown in hospitalised patients who survived (Jain et al. 2020; Md Noh 2020; Radmanesh et al. 2020; Katal et al. 2021). Indeed, patients with COVID-19 recovery can be effectively tracked using [18F] fluorodeoxyglucose positron emission tomography (18F-FDG-PET), structural (MRI) and functional magnetic resonance imaging (fMRI) (Huang et al. 2022b).

The most common *in vivo* neuroimaging findings include abnormalities in cortical signal on fluid-attenuated inversion recovery images (de Erausquin et al. 2021), which may reflect the presence of encephalitis, or hypoglycemia, hypoxia, or seizures (Kandemirli et al. 2020). Acute demyelinating lesions have also been observed (Abdi et al. 2020; Brun et al. 2020; Novi et al. 2020; Zanin et al. 2020). Decreased brain metabolism was mostly found in the frontal lobe region using 18F-FDG-PET (Rudroff et al. 2021). A hypometabolic long COVID pattern involving frontal-orbital and olfactory regions, other limbic/paralimbic regions, the brainstem, and the cerebellum has been reported (Guedj et al. 2021). Decreased metabolism was also found in the right parahippocampal gyrus and thalamus in long COVID patients compared with control subjects without long COVID (Sollini et al. 2021). Associations have been suggested between decreased metabolism in the parahippocampal and orbitofrontal areas and persistent anosmia, and associations between decreased metabolism in the parahippocampal gyrus, brainstem, and thalamus and persistent fatigue (Sollini et al. 2021). Divergent results were obtained in an 18F-FDG-PET study in which comparisons were made by normalising brain metabolism on the white matter: in this case, no decrease in metabolism was observed in patients with subjective disorders, either at the group or individual level (Dressing et al. 2021).

Tissue damage in regions that are functionally connected to the primary olfactory cortex, a greater reduction in grey matter thickness, and tissue contrast in the orbitofrontal cortex and parahippocampal gyrus have been now demonstrated (Douaud et al. 2022). Evidence suggests that SARS-CoV-2 infection can reduce grey matter as much as 10 years of ageing in people aged 51–81 years who are infected but not hospitalised. Indeed, in the four-month follow-up acquisition, the infected usually show between 0.2 and 2% more grey matter loss than healthy controls (Douaud et al. 2022). Importantly, patients who go through these changes in brain volume also incur cognitive decline (Douaud et al. 2022). It is unclear what the implications of grey matter volume changes really are, and no cause-and-effect correlations with permanent damage or dysfunction in thinking, memory, or other brain functions have been demonstrated by this study.

Grey matter volume changes in limbic and secondary olfactory areas are also present in neuropsychiatric long COVID patients and could be dynamic over time (Besteher et al. 2022). Increased volume in the frontotemporal areas, insula, hippocampus, amygdala, basal ganglia, and thalamus in both hemispheres has been detected through voxel-based morphometry techniques (Besteher et al. 2022).

Additional MRI abnormalities were found two months after acute COVID-19 infection in non-hospitalised subjects (Silva et al. 2021). The structural changes in COVID-19 included increased diffusivity in the corpus callosum, corticospinal, dorsal cingulate, parahippocampal cingulate, uncinate fascicles, and inferior longitudinal fascicles (Silva et al. 2021). Such evidence appears to be robust and reduced grey matter volume in cortical, limbic, and cerebellar areas, and alterations in axial and mean white matter diffusivity have been confirmed over time (Díez-Cirarda et al. 2022). Loss of grey matter volume showed significant associations with cognitive dysfunction. These cognitive and brain alterations were more pronounced in hospitalised patients than in non-hospitalised patients (Díez-Cirarda et al. 2022). Moreover, patients with neurocovid may show alterations in functional connectivity, characterised by high connectivity between the right and left parahippocampal areas and between bilateral orbitofrontal and cerebellar areas compared with controls (Díez-Cirarda et al. 2022).

Up to 71% of patients showed multiple subcortical white matter lesions, located in the cerebral hemispheres near the junction between grey and white matter, particularly in the frontal and parietal lobes (Hellgren et al. 2021). White matter lesions are often found with increasing age and may be a marker of cerebrovascular disease. This suggests that there appears to be an age-associated vulnerability for cerebrovascular disease as a complication of COVID-19 infection (Hellgren et al. 2021). In addition, all patients who underwent MRI during the acute phase of COVID-19 acquired new multiple white matter lesions when examined at follow-up. COVID-19 infection can therefore attack the brain even after acute and subacute viral infection (Hellgren et al. 2021). Compared with matched controls, individuals with SARS-CoV-2 seem to be also characterised by elevated global measures of mean diffusivity and extracellular free water also in the white matter (Petersen et al. 2022). It has been proposed that the increase in free water and mean diffusivity in patients with long COVID may be an indirect sign of a prolonged neuroinflammatory response. Indeed, activated local immune responses involving microglia and astrocytes are involved in the production of cytokines, which promote water osmosis from the blood to the extracellular space (Petersen et al. 2022). Importantly, white matter hyperintensities appear potentially correlated with verbal memory deficits in surviving subjects (Cecchetti et al. 2022).

It was also found that about a quarter of mild COVID-19 individuals had specific visuoconstructive deficits, accompanied by changes in molecular and structural brain imaging (Hellgren et al. 2021), as well as upregulation of peripheral immune markers (de Paula et al. 2023). In addition, functional connectivity analysis detected atypical connections altered connections in the visuospatial network, with 29 atypical connections involving the superior parietal lobe/supramarginal gyrus/postcentral gyrus/angular gyrus (Silva et al. 2021). In an already sizable and growing percentage of infected persons, cognitive impairment seems to be caused by the presence of a neuroinflammatory burden (de Paula et al. 2023).

Advanced neuroimaging techniques may be increasingly useful in answering outstanding questions about the effects of the SARS-CoV-2 infection on the brain: whether these effects can be partially reversed; whether they will persist until they become chronic; lastly, whether there is neurocognitive damage due to structural brain alterations (Palermo, 2022).

8.4.1 Staging Neurocovid: A Possible MRI Classification System

Based on the analysis of potential pathophysiological mechanisms involved in the neurocognitive manifestations of SARS-CoV-2 and detectable by advanced bioimaging, a three-stage classification system has been proposed that can serve as a basis for future discussions and investigations (Fotuhi et al. 2020) (Table 8.3). The classification system starts from the observation that many COVID-19-positive patients will not initially present with neurocovid symptoms, while others will manifest them before the more commonly expected COVID-19 symptoms (fever, cough, and shortness of breath) (Fotuhi et al. 2020).

Stage I of neurocovid affects primarily the nose and mouth epithelium. Blood clots in the brain and autoantibodies damaging peripheral nerves and muscles are common in neurocovid stage II. The cytokine storm in neurocovid stage III can damage the blood–brain barrier, resulting in seizures, coma, or encephalopathy.

8.5 Neurocovid: Neuropsychological Findings

There is growing evidence that COVID-19 can cause lasting cognitive and mental health problems, with recovered patients reporting various symptoms (Hampshire et al. 2021; Arica-Polat et al. 2022; Delgado-Alonso et al. 2022; Guo et al. 2022a; May 2022). These symptoms are collectively referred to as "*brain fog*". Together with the increased incidence of anxiety, depression, sleep disturbances, and fatigue, this cognitive impairment syndrome contributes substantially to the morbidity of post-COVID conditions (Ferrando et al. 2022; Venkataramani and Winkler 2022).

Fatigue was reported in up to 74% of patients (Hellgren et al. 2021). Fatigue/Mixed symptoms during the initial COVID-19 infection and ongoing neurological symptoms were predictive of reduced memory performance and slower reaction time on executive function performance (Guo et al. 2022b). Consistently, cognitive abnormalities are one of the most commonly reported symptoms by those with long COVID (Guo et al. 2022a).

Actually, one in four patients had neurocognitive test results corresponding to severe impairment compared with pre-infection baseline levels. These results imply that, in some individuals, a COVID-19 infection could have a negative impact on cognition lasting at least several months after hospital discharge (Hellgren

Table 8.3 Neurological manifestations of SARS-Cov2 can be grouped into three stages (neurocovid stages I, II, and III), each characterised by neuroimaging and clinical evidence

Neurocovid	Neuroimaging evidence	Clinical manifestation
Stage I	The binding of SARS-Cov-2 to angiotensin-converting enzyme 2 receptors is restricted to nasal and gustatory epithelial cells The virus-activated cytokine storm remains low and controlled	Patients may present only with olfactory or taste disturbances and often recover without any intervention
Stage II	A robust immune response is activated by SARS-CoV-2, resulting in increased ferritin, C-reactive protein, and D-dimer levels. Consequently, blood clots occur as a result of hypercoagulability The higher immune response leads also to vasculitis in the muscles and nerves, and immune-mediated “molecular mimicry” that damages neurological tissues (cranial and peripheral nerves, and/or muscle)	Patients may go into a stroke, due to arterial occlusion or venous thrombosis Cranial nerve disorders can affect smell, taste, vision, facial sensitivity, facial expressions, hearing, balance, speech, swallowing, and neck muscles Peripheral neuropathies (sensory, motor, vegetative) may occur
Stage III	SARS-CoV-2 cytokine storm damages the blood–brain barrier, infiltrating the brain with inflammatory factors and other blood contents (including viral particles) Angiotensin II levels increase as a result of high viral titers occupying angiotensin-converting enzyme 2 in the blood	Oedema and brain injury can lead to delirium, encephalopathy, and/or seizures An increase in peripheral vascular resistance and hypertension increases intracranial haemorrhage risk

et al. 2021). Cognitive deficits after severe COVID-19 relate most strongly to acute illness severity, persist long into the chronic phase, and recover slowly if at all, with a characteristic profile highlighting higher cognitive functions and processing speed (Hampshire et al. 2021). Partly different evidence is reached by another study whereby disease severity, premorbid conditions, pulmonary function tests, and hypoxia did not contribute to cognitive performance (Hadad et al. 2022). Cognitive decline can affect otherwise healthy patients after COVID, regardless of disease severity, with abnormalities in executive function, attention, and phonemic fluency (Hadad et al. 2022). Nevertheless, a gradient of cognitive impairment indicating different neuropsychological phenotypes was sometimes observed (Voruz et al. 2022). Subjects with mild infection were more stressed, anxious, and depressed. Subjects with moderate infection had poorer emotion recognition, which correlated positively with persistent olfactory dysfunction. Those with severe infection performed poorer on long-term episodic memory tests and showed greater anosognosia than the other two groups (Voruz et al. 2022).

Immediate memory and delayed memory were indices in which most patients scored below the cut-off (Hellgren et al. 2021), a result in agreement with previous research showing altered results for memory in relation to COVID-19 infection (Almeria et al. 2020). Indeed, most cognitive alterations were found in attention and working memory (up to 44.2%), and long-term memory (up to 40.7%) (Díez-Cirarda et al. 2022). Deficits were also found in executive functions (up to 39.5%), followed by visuospatial ability (up to 36%) and language (up to 18.6%) (Díez-Cirarda et al. 2022). This evidence is confirmed by a meta-analysis through which it was corroborated how the most affected cognitive domains were executive functions, attention, and episodic memory (Tavares-Júnior et al. 2022). A recurrent pattern of cognitive impairment sees implication of information processing speed, inattention, and executive dysfunction assessed through working memory (Delgado-Alonso et al. 2022;

Tavares-Júnior et al. 2022). Patients exhibit alterations in processing speed, divided attention, selective attention, visual vigilance, intrinsic vigilance, working memory and inhibition, episodic memory, and visuospatial processing (Delgado-Alonso et al. 2022; Ferrando et al. 2022), often correlated with hyposmia/anosmia, poor sleep quality, and anxious states (Delgado-Alonso et al. 2022).

There are indications of trends in cognitive recovery (May 2022) from studies that examined patients at different cross-sectional time points (e.g. Poletti et al. 2022) or that followed patients for a short period after infection (e.g. Alemanno et al. 2021).

8.5.1 Brain Fog

"Brain fog" is a descriptive term that is usually used by patients to explain what they experience with a range of problems in intellectual function (Nouraeinejad 2023). Therefore, "brain fog" is an informal term for a common intellectual function disorder among patients with post-acute COVID-19 (Krishnan et al. 2022). It is an umbrella term used to explain the constellation of cognitive abnormalities subjectively encountered, such as dizziness, declines in concentration, mental fatigue, confusion, short-term memory loss, and slowing of cognitive function (Krishnan et al. 2022; Nouraeinejad 2023). Subjects who suffer from brain fog constantly experience a feeling of mental confusion making it difficult to perform daily activities especially the most demanding ones: driving, doing important economic transactions, and working. To understand the feeling these people experience, one must think about how one feels as soon as they wake up. This physiological sensation is called *sleep inertia*.

Brain fog is difficult to diagnose and separate from other causes of symptoms in a single patient because longitudinal neurocognitive data from patients are rarely available (May 2022; Venkataramani and Winkler 2022). Those subjects usually presented cognitive impairment in screening tests and comprehensive neurocognitive batteries, although the pattern of these abnormalities was not aetiology-specific (May 2022). Moreover, cognitive deficits should be interpreted with caution because of potential confounding/contributing factors (weaknesses may partly reflect psychiatric sequelae or be overinterpreted because of lack of data about premorbid functioning) (May 2022).

8.5.2 Neuropsychiatric and Behavioural Abnormalities

A study of SARS patients at the time confirmed high levels of psychological distress with 59% diagnosed with psychiatric disorders and persistence of symptoms in 33% at 30-month follow-up (Mak et al. 2009). The severity of psychological symptoms was found to correlate with disease severity and functional impairment (Mak et al. 2009; Wing and Leung 2012). The same phenomenon has been detected worldwide during the current pandemic (Gouraud et al. 2021; Mattioli et al. 2021).

"Coronavirus syndrome" is a mental disorder caused by the global problem of COVID-19 (Soloveva et al. 2021). It affects approximately 10% of the population and is characterised by acute stress reactions caused by infections spreading and changes to ordinary lifestyles. Most of the psychological effects are secondary to pandemic-associated regulatory, socioeconomic and daily life changes (Roy et al. 2021). It is likely that the most severe response will occur later on: here, the phenomenon is completely superimposed on post-traumatic stress disorder (Soloveva et al. 2021).

A multitude of factors can be held responsible for these behavioural and mood alterations found. Social isolation, fear of infection, loss of near and dear ones, enforcement of strict public health measures that restrict personal freedom, increasing insecurity from loss of income and jobs, and financial losses are the main stressors responsible for this psychological upheaval (Roy et al. 2021). A sense of social isolation and loneliness can contribute to deterioration in cognitive performance, faster cognitive decline, impaired executive functioning, and greater sensitivity to social threats (Morese and Palermo 2022). Pandemic conditions have exacerbated these effects because of the social distancing, fiduciary isolation, and quarantine imposed to reduce the spread of infection. The consequences for the elderly population, especially those vulnerable and less resilient to SARS-CoV-2 infection, are of particular interest (Amanzio et al. 2021; Palermo 2020, 2021a, b).

While some psychological distress may be reactive to exposure to traumatic events, there may also be people whose cognitive and psychological disorders are directly related to brain changes due to human coronaviruses (Ferrando et al. 2022). In this case, it would be interesting to see whether these people would respond to standard drug treatment, e.g. with antidepressants, anxiolytics, and cognitive therapies.

8.6 Clinical Evaluation of the Patient with Long COVID and Neurocovid

From a research perspective, there are no defined standards for neuropsychological function assessment in COVID-19 patients (May 2022). *NeuroCOVID International Neuropsychology Taskforce* was formed in 2020 to help harmonise neuropsychological methods and procedures (Cysique et al. 2022). As part of harmonisation guidelines, it is important to identify methods for assessing the consequences of COVID-19, including measuring the breadth and severity of potential neuropsychological weaknesses, separating neuropsychological disorders from psychological distress, assessing neuropsychological presentations during the acute and subacute phases of COVID-19, and assessing premorbidity, comorbidity, performance validity, and other relevant factors (Cysique et al. 2022). The guidelines for harmonisation include adapting methods to situations such as pandemic lockdowns, quarantines, or hospitalisations or alertness (for telehealth, computerised, remote, pen-and-pencil assessments), as well as the use of screenings in lieu of comprehensive evaluations (Cysique et al. 2022). Finally, methods and procedures that are appropriate for international purposes must be identified, including tests with cross-cultural validity or that are widely available (Cysique et al. 2022). For adults with COVID-19, the proposed recommendations provide rationale and methodological guidance.

8.6.1 Factors Predicting Physical, Neurological, and Cognitive Symptoms in Long COVID

The risk of cognitive impairment appears evident in all COVID-19 patients analysed, regardless of age, sex, obesity, hypertension, hyperlipidemia, smoking history, or area deprivation index (Xu et al. 2022).

There are a number of mechanisms by which COVID-19 infection can lead to neurological symptoms and structural and functional changes in the brain, and it is reasonable to expect that many of these will result in cognitive problems. Therefore, it becomes critical to intercept factors that may act as predisposing, precipitating, or aggravating factors in the neurocognitive profile of those who encounter such infection. The "COVID and Cognition Study" (COVCOG) is a part cross-sectional, part longitudinal study that aimed to explore which factors may be predictive of ongoing neurological and neuropsychological symptoms and their severity in patients with long COVID (Guo et al. 2022a).

The main finding of COVCOG1 is that severity of initial illness is a significant predictor of the presence and severity of ongoing symptoms and that some symptoms during initial illness, particularly limb weakness, may be more common in individuals with more severe ongoing symptoms (Guo et al. 2022a). Among patients experiencing symptoms at the time of the analyses, 77.8% reported difficulty concentrating, 69% reported brain fog, 67.5% reported forgetfulness, 59.5% reported problems finding words with the tip of the tongue, and 43.7% reported semantic dysfluency (Guo et al. 2022a). The possible disease profiles can be defined in terms of six components, which reflect the variety of this highly heterogeneous condition experienced by the individual: neurological, gastrointestinal/autoimmune, cardiopulmonary/affective, dermatological/fever, loss of appetite, and mood. Notably, neurological/psychiatric and fatigue/mixed symptoms during the initial illness and neurological, gastrointestinal and cardiopulmonary symptoms/fatigue during the ongoing illness predict the occurrence of neuropsychological symptomatology (Guo et al. 2022a). This suggests different types of neurological involvement, potentially reflecting neuroinflammation (fatigue/mixed factor) and encephalitis (neurological/psychiatric factor), respectively (Guo et al. 2022a).

In COVCOG2, the authors then found a consistent pattern of memory deficits in those who had experienced COVID-19 infection, with objective neuropsychological impairments on tests increasing with the severity of self-reported ongoing infection symptoms (Guo et al. 2022b). The particular difficulties with language and verbal memory were consistent with the frequency of self-reported deficits in these cognitive domains in other studies on long COVID (e.g. Davis et al. 2021; Ziauddeen et al. 2022) and with the finding of grey matter loss in the left hemisphere (Douaud et al. 2022).

Demographic and social determinants of cognitive dysfunction after hospitalisation for COVID-19 were also suggested. Valdes and collaborators (2022) found that the best univariate predictors of alterations on screening tests for global cognition included older age, education ≤ 12 years, pre-COVID unemployment, black race, and a history of pre-COVID cognitive impairment (all $p<0.05$). In their multivariable analyses, education ≤ 12 years and the interaction between functional status at baseline and pre-COVID unemployment were significantly associated with abnormal profiles after adjustment for age, history of dementia, language, neurological complications, income, and disposition at discharge (Valdes et al. 2022). Contrary to expectations, the authors found that cognitive impairment was independent of any hospitalisation metrics or recorded measures of disease severity, including ventilator status, sequential organ failure assessment (SOFA) scores, medications administered during hospitalisation, inflammatory markers, medical or neurological complications, and discharge disposition (Valdes et al. 2022). The finding is controversial. In fact, post-exposure viral shedding and symptom severity have been found to be associated with a new measure of pre-exposure cognitive performance variability, a measure of the person's cognitive dissonance over time along any dimension and defined before viral exposure occurs (Zhai et al. 2022).

8.6.2 Multidimensional Evaluation

The complexity of COVID-19 pathology and the impact on the brain require an appropriate multidimensional clinical approach to treat and prevent related sequelae. Proper stratification of COVID-19 patients based on an inflammatory panel and the direct effects of SARS-Cov-2 on the CNS and immune systems, together with a focus on psychosocial factors could improve clinical outcome. The use of cognitive performance markers together with measures of stress and fatigue can predict the likelihood of severe symptoms and viral shedding, with clinical and epidemiological implications (Zhai et al. 2022).

The pattern of dysfunction presented by long COVID and neurocovid patients suggests an executive pattern (Becker et al. 2021). This pattern is consistent with early reports describing a dysexecutive syndrome after COVID-19 and has significant implications for occupational, psychological, and functional outcomes (Helms et al. 2020).

Follow-up of people who have recovered from COVID-19 should be as comprehensive as possible and check functional, nutritional, cognitive, and affective status. The selection of patients for follow-up can make use of specific questionnaires. A synthetic questionnaire should be administered, as the first screening tool, to all individuals recovered from the acute phase of COVID-19, after about 12 weeks or earlier if the patient is still symptomatic. If the synthetic questionnaire is found to be negative, the same can be repeated after an additional three months. Otherwise, it would be appropriate to put the subject through an advanced protocol. It should be based on detailed medical history, physical examination, medical and pharmacological history, lifestyle factors, vaccination status, information on the acute phase of COVID-19 disease (signs and symptoms suffered, both during hospital stay and at home), assessment of the impact of COVID-19 and long COVID on nutritional aspects, changes in body weight and loss of interest in eating and drinking, assessment of the presence of cognitive/neuropsychiatric symptoms, and alterations in mood. Cognitive screening test such as the NIH Toolbox and MOCA are useful for the detection of cognitive dysfunction in COVID-19 (Aiello et al. 2022).

8.6.3 Neuroimaging Examinations

It will be recommended that patients be monitored for life with 18F-FDG-PET, which will not only reveal the pathophysiological mechanism of post-COVID-19 at the molecular level, but will also help detect metabolic abnormalities in multiple organs (Huang et al. 2022b). Individual patient characteristics and confounding factors must be considered when interpreting cortical metabolic abnormalities (Verger et al. 2022).

According to the protocol of the Karolinska neurocovid study (Nelson et al. 2022), in patients with COVID-19 and neurological manifestations, neuroimaging should be performed at least once during the acute phase to identify the underlying causal pathology. If the initial computer tomography (CT) scan does not provide an explanation for the neurological symptoms, an MRI can be performed before concluding that the neuroimaging is negative (Kremer et al. 2022). The basic recommended MRI protocol includes T2-weighted, FLAIR (preferably 3D) and diffusion-weighted images, as well as haemorrhage-sensitive sequences (preferably SWI) and, at least for the initial investigation, T1-weighted pre- and post-contrast images. The 3D FLAIR should be acquired after gadolinium administration to optimise detection of leptomeningeal contrast enhancement (Kremer et al. 2022). It is also recommended that MRI of the spinal cord be considered in patients with major clinical presentation of the spinal cord, but also in patients with demyelinating brain lesions (Kremer et al. 2022).

Perfusion-weighted imaging, especially arterial spin labelling, diffusion tensor imaging, spectroscopy, and functional MRI, especially resting state MRI, are an integral part of research investigations in patients with COVID-19 (Kremer et al. 2022; Palermo, 2022). Yet, their value in contributing to routine clinical examination has not yet been demonstrated (Kremer et al. 2022; Nelson et al. 2022).

Advanced MRI evaluation of the brain and lungs at 3- and 12-month follow-ups, supplementing routine chest radiography and CT, is recommended to document any residual lesions and resolution or progression of initial lesions (Kremer et al. 2022; Nelson et al. 2022)

8.7 Conclusion

Long COVID and neurocovid are now an established concept worldwide. However, the clinical features of the two entities are still debated. The associated pathophysiological mechanisms are not yet fully understood, although the evidence primarily involves immune dysfunction.

Clinicians have to plan and manage post-COVID care to avoid the risk of repeating what happened with SARS: with patients who had chronic neurocognitive disorders that arose consequently to the epidemic. Long COVID and neurocovid have the power to affect as much (if not more) than acute COVID. Continuous follow-up of COVID-19 survivors, particularly those with long COVID, is essential to understand the long-term course of the disease, as well as to better analyse the benefits of outpatients' treatment and rehabilitation programs. The need to provide ongoing support for an increasingly significant proportion of people who have had COVID-19 is coupled with equally ongoing surveillance to enable researchers to understand how vaccines, emerging treatments, and variants affect the clinical course of the disease. The long-term health impacts of COVID-19 are only now beginning to emerge clearly, as longer follow-up studies are now exceeding two years (Fernández-de-Las-Peñas et al. 2022; Huang et al. 2022a; Wahlgren et al. 2023).

Difficulties in understanding the impact of long COVID and neurocovid also depend on the lack of baseline values of pre-COVID-19 health status and comparisons with the general population. All of this makes it difficult to determine how well COVID-19 patients have recovered, whether long COVID is a remissive disorder or chronicisation of the primary disease.

8.7.1 To Identify Targets for Treatment

Since the beginning of the pandemic, a neuroscience-based clinical approach has been advocated to reduce the mental sequelae related to SARS-CoV-2 (Pallanti et al. 2020). Even though we can use knowledge of disease mechanisms and therapies acquired before the pandemic to manage patients with COVID-19 neurological manifestations, we need to gain a better understanding of the disease-specific mechanisms of COVID-19 and develop therapies for its neurological morbidity (Graham et al. 2022).

Screening and ruling out alternative causes of cognitive dysfunction are recommended, including B12 or folate deficiency, HIV infection, syphilis, thyroid dysfunction, and obstructive sleep apnoea (Graham et al. 2022).

A new immuno-behavioural approach to the treatment of behavioural disorders is also needed. If the psychological disorder is related to inflammation and not just event-related stress, patients-though clinically similar will be distinguished based on different inflammatory involvement, with blood tests of the inflammation activation profile. Pharmacological treatments have to be designed based on the effect on traditional targets (serotonin, dopamine) and based on the inflammatory and immune response.

8.7.2 Final Considerations

After the third coronavirus outbreak in the past decade (SARS, MERS, and COVID-19), attention will have to focus on the long-term consequences of COVID-19 infection and their rehabilitation or cure. New pandemics loom in the future. Being able to contain and manage the acute phase of the disease is no longer enough. In the face of increasing reports of CNS involvement in COVID-19 cases, it is likely that the current epidemic may be accompanied by a significant increase in the prevalence of neurological sequelae, cognitive dysfunction, and long-term behavioural alterations that affect the quality of life and autonomy in daily living. This is sequential to the consequences of neuroinvasion and multi-organ dysfunction, but also to psychological distress and intervening socioeconomic changes. At present, the predisposing-precipitating factors nor the protective factors operating in cases that do not experience long COVID and neurocovid are unclear.

Neuropsychologists will benefit from approaching the assessment and rehabilitation of individuals considering cognition, emotional functioning, behaviour, and the potential socioeconomic impact of the pandemic as interacting variables affecting functional independence, quality of life, and emotional well-being (Cysique et al. 2022). A bio-psycho-social approach is the first step to be taken to ensure the best recovery of the greatest number of sensory-motor functions and cognitive-behavioural domains, the impairment of which could over the years greatly affect the health status and level of autonomy of those recovered from infection. Moving now, anticipating patients', and frailty vulnerable people's needs with good screening and targeted personalised intervention programs could make all the difference.

References

Abdi S, Ghorbani A, Fatehi F (2020) The association of SARS-CoV-2 infection and acute disseminated encephalomyelitis without prominent clinical pulmonary symptoms. J Neurol Sci 416:117001. https://doi.org/10.1016/j.jns.2020.117001

Aiello EN, Fiabane E, Manera MR et al (2022) Screening for cognitive sequelae of SARS-CoV-2 infection: a comparison between the Mini-Mental State Examination (MMSE) and the Montreal Cognitive Assessment (MoCA). Neurol Sci 43(1):81–84. https://doi.org/10.1007/s10072-021-05630-3

Alemanno F, Houdayer E, Parma A et al (2021) COVID-19 cognitive deficits after respiratory assistance in the subacute phase: a COVID-rehabilitation unit experience. PLoS ONE 16(2):e0246590. https://doi.org/10.1371/journal.pone.0246590

Almeria M, Cejudo JC, Sotoca J et al (2020) Cognitive profile following COVID-19 infection: clinical predictors leading to neuropsychological impairment. Brain Behav Immun Health 9:100163. https://doi.org/10.1016/j.bbih.2020.100163

Amanzio M, Canessa N, Bartoli M et al (2021) Lockdown effects on healthy cognitive aging during the COVID-19 pandemic: a longitudinal study. Front Psychol 12:685180. https://doi.org/10.3389/fpsyg.2021.685180

Arabi YM, Harthi A, Hussein J et al (2015) Severe neurologic syndrome associated with Middle East respiratory syndrome corona virus (MERS-CoV). Infection 43(4):495–501. https://doi.org/10.1007/s15010-015-0720-y

Arentz M, Yim E, Klaff L et al (2020) Characteristics and outcomes of 21 critically ill patients with COVID-19 in Washington state. JAMA 323(16):1612–1614. https://doi.org/10.1001/jama.2020.4326

Arica-Polat BS, Gündoğdu AA, Cinar N et al (2022) Evaluation of cognitive deficits in patients infected with COVID-19. Eur Rev Med Pharmacol Sci 26(2):678–685. https://doi.org/10.26355/eurrev_202201_27894

Baig AM, Khaleeq A, Ali U et al (2020) Evidence of the COVID-19 virus targeting the CNS: tissue distribution, host-virus interaction, and proposed neurotropic mechanisms. ACS Chem Neurosci 11(7):995–998. https://doi.org/10.1021/acschemneuro.0c00122

Bartoli M, Palermo S, Stanziano M, Cipriani GE, Leotta D, Valentini MC et al (2020) Reduced self-awareness following a combined polar and paramedian bilateral thalamic infarction. A possible relationship with SARS-CoV-2 risk of contagion? Front. Psychol. 11:570160. https://doi.org/10.3389/fpsyg.2020.570160

Becker JH, Lin JJ, Doernberg M et al (2021) Assessment of cognitive function in patients after COVID-19 infection. JAMA Netw Open 4(10):e2130645. https://doi.org/10.1001/jamanetworkopen.2021.30645

Besteher B, Machnik M, Troll M et al (2022) Larger gray matter volumes in neuropsychiatric long-COVID syndrome. Psychiatry Res 317:114836. https://doi.org/10.1016/j.psychres.2022.114836

Brun G, Hak JF, Coze S et al (2020) COVID-19-white matter and globus pallidum lesions: demyelination or small-vessel vasculitis? Neurol Neuroimmunol Neuroinflamm 7(4):e777. https://doi.org/10.1212/NXI.0000000000000777

Callard F, Perego E (2021) How and why patients made long COVID. Soc Sci Med 268:113426. https://doi.org/10.1016/j.socscimed.2020.113426

Cecchetti G, Agosta F, Canu E et al (2022) Cognitive, EEG, and MRI features of COVID-19 survivors: a 10-month study. J Neurol 269(7):3400–3412. https://doi.org/10.1007/s00415-022-11047-5

Chambergo-Michilot D, Barros-Sevillano S, Rivera-Torrejón O, De la Cruz-Ku GA, Custodio N (2021) Factors associated with COVID-19 in people with Parkinson's disease: a systematic review and meta-analysis. Eur J Neurol 28(10):3467–3477. https://doi.org/10.1111/ene.14912

Chang R, Elhusseiny KM, Yeh YC et al (2021) COVID-19 ICU and mechanical ventilation patient characteristics and outcomes—a systematic review and meta-analysis. PLoS ONE 16(2):e0246318. https://doi.org/10.1371/journal.pone.0246318

Chou SH, Beghi E, Helbok R et al (2021) Global incidence of neurological manifestations among patients hospitalized with COVID-19—a report for the GCS-NeuroCOVID Consortium and the ENERGY Consortium. JAMA Netw Open 4(5):e2112131. https://doi.org/10.1001/jamanetworkopen.2021.12131

Coolen T, Lolli V, Sadeghi N et al (2020) Early postmortem brain MRI findings in COVID-19 non-survivors. Neurology 95(14):e2016–e2027. https://doi.org/10.1001/10.1212/WNL.0000000000010116

Crook H, Raza S, Nowel J et al (2021) Long COVID-mechanisms, risk factors, and management. BMJ 374:n1648. https://doi.org/10.1136/bmj.n1648

Cysique LA, Łojek E, Cheung TC et al (2022) Assessment of neurocognitive functions, olfaction, taste, mental, and psychosocial health in COVID-19 in adults: recommendations for harmonization of research and implications for clinical practice. J Int Neuropsychol Soc 28(6):642–660. https://doi.org/10.1017/S1355617721000862

Dahm T, Rudolph H, Schwerk C et al (2016) Neuroinvasion and inflammation in viral central nervous system infections. Mediat Inflamm 2016:8562805. https://doi.org/10.1155/2016/8562805

Davis HE, Assaf GS, McCorkell L et al (2021) Characterizing long COVID in an international cohort: 7 months of symptoms and their impact. Eclin Med 38:3820561. https://doi.org/10.1016/j.eclinm.2021.101019

de Erausquin GA, Snyder H, Carrillo M et al (2021) The chronic neuropsychiatric sequelae of COVID-19: the need for a prospective study of viral impact on brain functioning. Alzheimers Dement 17(6):1056–1065. https://doi.org/10.1002/alz.12255

de Paula JJ, Paiva RERP, Souza-Silva NG et al (2023) Selective visuoconstructional impairment following mild COVID-19 with inflammatory and neuroimaging correlation findings. Mol Psychiatry 28(2):553–563. https://doi.org/10.1038/s41380-022-01632-5

Delamarre L, Gollion C, Grouteau G et al (2020) COVID-19-associated acute necrotising encephalopathy successfully treated with steroids and polyvalent immunoglobulin with unusual IgG targeting the cerebral fibre network. J Neurol Neurosurg Psychiatry 91(9):1004–1006. https://doi.org/10.1136/jnnp-2020-323678

Delgado-Alonso C, Valles-Salgado M, Delgado-Álvarez A et al (2022) Cognitive dysfunction associated with COVID-19: a comprehensive neuropsychological study. J Psychiatr Res 2150:40–46. https://doi.org/10.1016/j.jpsychires.2022.03.033

Díez-Cirarda M, Yus M, Gómez-Ruiz N et al (2022) Multimodal neuroimaging in post-COVID syndrome and correlation with cognition. Brain awac384. https://doi.org/10.1093/brain/awac384

Dinan TG, Cryan JF (2017) The microbiome-gut-brain axis in health and disease. Gastroenterol Clin North Am 46(1):77–89. https://doi.org/10.1016/j.gtc.2016.09.007

Douaud G, Lee S, Alfaro-Almagro F et al (2022) SARS-CoV-2 is associated with changes in brain structure in UK Biobank. Nature 604(7907):697–707. https://doi.org/10.1038/s41586-022-04569-5

Dressing A, Bormann T, Blazhenets G et al (2021) Neuropsychologic profiles and cerebral glucose metabolism in neurocognitive long COVID syndrome. J Nucl Med 63(7):1058–1063. https://doi.org/10.2967/jnumed.121.262677

Escobar YH, O'Piela D, Wold LE et al (2022) Influence of the microbiota-gut-brain axis on cognition in Alzheimer's disease. J Alzheimer's Dis 87(1):17–31. https://doi.org/10.3233/JAD-215290

Fearon C, Fasano A (2021) Parkinson's disease and the COVID-19 pandemic. J Parkinson's Dis 11(2):431–444. https://doi.org/10.3233/JPD-202320

Fernández-de-Las-Peñas C, Rodríguez-Jiménez J, Cancela-Cilleruelo I et al (2022) Post-COVID-19 symptoms 2 years after SARS-CoV-2 infection among hospitalized vs nonhospitalized patients. JAMA Netw Open 5(11):e2242106. https://doi.org/10.1001/jamanetworkopen.2022.42106

Fernández-Castañeda A, Lu P, Geraghty AC, Song E, Lee M-H, Wood J et al (2022) Mild respiratoryCOVID can cause multi-lineage neural cell and myelin dysregulation ll Mild respiratory COVID can cause multilineageneural cell and myelin dysregulation. Cell. 185, e008. https://doi.org/10.1016/j.cell.2022.06.008

Ferrando SJ, Dornbush R, Lynch S et al (2022) Neuropsychological, medical, and psychiatric findings after recovery from acute COVID-19: a cross-sectional study. J Acad Consult Liaison Psychiatry 63(5):474–484. https://doi.org/10.1016/j.jaclp.2022.01.003

Finsterer J, Scorza FA (2021) Clinical and pathophysiologic spectrum of neuro-COVID. Mol Neurobiol 58(8):3787–3791. https://doi.org/10.1007/s12035-021-02383-0

Fotuhi M, Mian A, Meysami S, Raji CA (2020) Neurobiology of COVID-19. J Alzheimer's Dis 76(1):3–19. https://doi.org/10.3233/JAD-200581

Gallacher SD, Seaton A (2020) Meningococcal meningitis and COVID-19 co-infection. BMJ Case Rep 13(8):e237366. https://doi.org/10.1136/bcr-2020-237366

Garg RK, Paliwal VK, Gupta A (2021) Spinal cord involvement in COVID-19: a review. J Spinal Cord Med 1–15. https://doi.org/10.1080/10790268.2021.1888022

Ghosh R, Dubey S, Finsterer J et al (2020) SARS-CoV-2-associated acute hemorrhagic, necrotizing encephalitis (AHNE) presenting with cognitive impairment in a 44-year-old woman without comorbidities: a case report. Am J Case Rep 21:e925641. https://doi.org/10.12659/AJCR.925641

Gorelick PB (2010) Role of inflammation in cognitive impairment: results of observational epidemiological studies and clinical trials. Ann N Y Acad Sci 1207:155–162. https://doi.org/10.1111/j.1749-6632.2010.05726.x

Gouraud C, Bottemanne H, Lahlou-Laforêt K et al (2021) Association between psychological distress, cognitive complaints, and neuropsychological status after a severe COVID-19 episode: a cross-sectional study. Front Psychiatry 12:725861. https://doi.org/10.3389/fpsyt.2021.725861

Graham EL, Koralnik IJ, Liotta EM (2022) Therapeutic approaches to the neurologic manifestations of COVID-19. Neurotherapeutics 19(5):1435–1466. https://doi.org/10.1007/s13311-022-01267-y

Gu J, Gong E, Zhang B et al (2005) Multiple organ infection and the pathogenesis of SARS. J Exp Med 202(3):415–424. https://doi.org/10.1084/jem.20050828

Guedes BF (2022) NeuroCOVID-19: a critical review. Arq Neuropsiquiatr 80(5 Suppl 1):281–289. https://doi.org/10.1590/0004-282X-ANP-2022-S136

Guedj E, Campion JY, Dudouet P et al (2021) 18F-FDG brain PET hypometabolism in patients with long COVID. Eur J Nucl Med Mol Imaging 48(9):2823–2833. https://doi.org/10.1007/s00259-021-05215-4

Guo P, Benito Ballesteros A, Yeung SP et al (2022a) COVCOG 1: factors predicting physical, neurological and cognitive symptoms in long COVID in a community sample. A first publication from the COVID and cognition study. Front Aging Neurosci 14:804922. https://doi.org/10.3389/fnagi.2022.804922

Guo P, Benito Ballesteros A, Yeung SP et al (2022b) COVCOG 2: cognitive and memory deficits in long COVID: a second publication from the COVID and cognition study. Front Aging Neurosci 14:804937. https://doi.org/10.3389/fnagi.2022.804937

Gupta A, Madhavan MV, Sehgal K et al (2020) Extrapulmonary manifestations of COVID-19. Nat Med 26:1017–1032. https://doi.org/10.1038/s41591-020-0968-3

Hadad R, Khoury J, Stanger C et al (2022) Cognitive dysfunction following COVID-19 infection. J Neurovirol 28(3):430–437. https://doi.org/10.1007/s13365-022-01079-y

Hampshire A, Trender W, Chamberlain SR et al (2021) Cognitive deficits in people who have recovered from COVID-19. EClinicalMedicine 39:101044. https://doi.org/10.1016/j.eclinm.2021.101044

Han S, Mallampalli RK (2015) The acute respiratory distress syndrome: from mechanism to translation [published correction appears in J Immunol. 2015 Jun 1;194(11):5569]. J Immunol 194(3):855–860. https://doi.org/10.4049/jimmunol.1402513

Hellgren L, Birberg Thornberg U, Samuelsson K et al (2021) Brain MRI and neuropsychological findings at long-term follow-up after COVID-19 hospitalisation: an observational cohort study. BMJ Open 11(10):e055164. https://doi.org/10.1136/bmjopen-2021-055164

Helms J, Kremer S, Merdji H et al (2020) Neurologic features in severe SARS-CoV-2 infection. N Engl J Med 382(23):2268–2270. https://doi.org/10.1056/NEJMc2008597

Hopkins RO, Weaver LK, Pope D et al (1999) Neuropsychological sequelae and impaired health status in survivors of severe acute respiratory distress syndrome. Am J Respir Crit Care Med 160(1):50–56. https://doi.org/10.1164/ajrccm.160.1.9708059

Hopkins RO, Weaver LK, Collingridge D et al (2005) Two-year cognitive, emotional, and quality-of-life outcomes in acute respiratory distress syndrome. Am J Respir Crit Care Med 171(4):340–347. https://doi.org/10.1164/rccm.200406-763OC

Hopkins RO, Wade D, Jackson JC (2017) What's new in cognitive function in ICU survivors. Intensive Care Med 43(2):223–225. https://doi.org/10.1007/s00134-016-4550-x

Hosang L, Canals RC, van der Flier FJ et al (2022) The lung microbiome regulates brain autoimmunity. Nature 603(7899):138–144. https://doi.org/10.1038/s41586-022-04427-4

Huang C, Huang L, Wang Y et al (2021) 6-month consequences of COVID-19 in patients discharged from hospital: a cohort study. Lancet (Lond Engl) 397(10270):220–232. https://doi.org/10.1016/S0140-6736(20)32656-8

Huang L, Li X, Gu X et al (2022a) Health outcomes in people 2 years after surviving hospitalisation with COVID-19: a longitudinal cohort study. Lancet Respir Med 10(9):863–876. https://doi.org/10.1016/S2213-2600(22)00126-6

Huang Y, Ling Q, Manyande et al (2022b) Brain imaging changes in patients recovered from COVID-19: a narrative review. Front Neurosci 16:855868. https://doi.org/10.3389/fnins.2022.855868

Jain R, Young M, Dogra S et al (2020) COVID-19 related neuroimaging findings: a signal of thromboembolic complications and a strong prognostic marker of poor patient outcome. J Neurol Sci 414:116923. https://doi.org/10.1016/j.jns.2020.116923

Kandemirli SG, Dogan L, Sarikaya ZT et al (2020) Brain MRI findings in patients in the intensive care unit with COVID-19 infection. Radiology 297(1):E232–E235. https://doi.org/10.1148/radiol.2020201697

Katal S, Balakrishnan S, Gholamrezanezhad A (2021) Neuroimaging and neurologic findings in COVID-19 and other coronavirus infections: a systematic review in 116 patients. J Neuroradiol 48(1):43–50. https://doi.org/10.1016/j.neurad.2020.06.007

Kowalski K, Mulak A (2019) Brain-gut-microbiota axis in Alzheimer's disease. J Neurogastroenterol Motil 25(1):48–60. https://doi.org/10.5056/jnm18087

Koyuncu OO, Hogue IB, Enquist LW (2013) Virus infections in the nervous system. Cell Host Microbe 13(4):379–393. https://doi.org/10.1016/j.chom.2013.03.010

Kremer S, Gerevini S, Ramos A et al (2022) Neuroimaging in patients with COVID-19: a neuroradiology expert group consensus. Eur Radiol 32(6):3716–3725. https://doi.org/10.1007/s00330-021-08499-0

Krishnan K, Lin Y, Prewitt KM et al (2022) Multidisciplinary approach to brain fog and related persisting symptoms post COVID-19. J Health Serv Psychol 48(1):31–38. https://doi.org/10.1007/s42843-022-00056-7

Li Q, Dang C, Wang LH (2022) Neuroinflammation in mild respiratory COVID-19: insights into cognitive impairment in milder cases. Mil Med Res 9(1):72. https://doi.org/10.1186/s40779-022-00431-x

Mak IW, Chu CM, Pan PC et al (2009) Long-term psychiatric morbidities among SARS survivors. Gen Hosp Psychiatry 31(4):318–326. https://doi.org/10.1016/j.genhosppsych.2009.03.001

Mankad K, Perry MD, Mirsky DM et al (2020) COVID-19: a primer for neuroradiologists. Neuroradiology 62(6):647–648. https://doi.org/10.1007/s00234-020-02437-5

Mao L, Jin H, Wang M et al (2020) Neurologic manifestations of hospitalized patients with coronavirus disease 2019 in Wuhan, China. JAMA Neurol 77(6):683–690. https://doi.org/10.1001/jamaneurol.2020.1127

Matthay MA, Zemans RL (2011) The acute respiratory distress syndrome: pathogenesis and treatment. Annu Rev Pathol 6:147–163. https://doi.org/10.1146/annurev-pathol-011110-130158

Mattioli F, Stampatori C, Righetti F et al (2021) Neurological and cognitive sequelae of COVID-19: a four month follow-up. J Neurol 268(12):4422–4428. https://doi.org/10.1007/s00415-021-10579-6

May PE (2022) Neuropsychological outcomes in adult patients and survivors of COVID-19. Pathogens 11(4):465. https://doi.org/10.3390/pathogens11040465

Md Noh MSF (2020) Brain imaging findings in COVID-19: what do we know so far? J Neuroradiol 47(5):329–330. https://doi.org/10.1016/j.neurad.2020.05.004

Mikkelsen ME, Christie JD, Lanken PN et al (2012) The adult respiratory distress syndrome cognitive outcomes study: long-term neuropsychological function in survivors of acute lung injury. Am J Respir Crit Care Med 185(12):1307–1315. https://doi.org/10.1164/rccm.201111-2025OC

Miner JJ, Diamond MS (2016) Mechanisms of restriction of viral neuroinvasion at the blood–brain barrier. Curr Opin Immunol 38:18–23. https://doi.org/10.1016/j.coi.2015.10.008

Misra S, Kolappa K, Prasad M et al (2021) Frequency of neurologic manifestations in COVID-19: a systematic review and meta-analysis. Neurology 97(23): e2269–e2281. https://doi.org/10.1212/WNL.0000000000012930

Mondal R, Deb S, Shome G et al (2021) COVID-19 and emerging spinal cord complications: a systematic review. Mult Scler Relat Disord 51:102917. https://doi.org/10.1016/j.msard.2021.102917

Monje M, Iwasaki A (2022) The neurobiology of long COVID. Neuron 110(21):3484–3496. https://doi.org/10.1016/j.neuron.2022.10.006

Morese R, Palermo S (2022) Feelings of loneliness and isolation: social brain and social cognition in the elderly and Alzheimer's disease. Front Aging Neurosci 14:896218. https://doi.org/10.3389/fnagi.2022.896218

Najt P, Richards HL, Fortune DG (2021) Brain imaging in patients with COVID-19: a systematic review. Brain Behav Immun Health 16:100290. https://doi.org/10.1016/j.bbih.2021.100290

Nalbandian A, Sehgal K, Gupta A et al (2021) Post-acute COVID-19 syndrome. Nat Med 27(4):601–615. https://doi.org/10.1038/s41591-021-01283-z

Nelson DW, Granberg T, Andersen P et al (2022) The Karolinska NeuroCOVID study protocol: neurocognitive impairment, biomarkers and advanced imaging in critical care survivors. Acta Anaesthesiol Scand 66(6):759–766. https://doi.org/10.1111/aas.14062

Nouraeinejad A (2023) Brain fog as a long-term sequela of COVID-19. SN Compr Clin Med 5(1):9. https://doi.org/10.1007/s42399-022-01352-5

Novi G, Rossi T, Pedemonte E et al (2020) Acute disseminated encephalomyelitis after SARS-CoV-2 infection [published correction appears in Neurol Neuroimmunol Neuroinflamm. 2020 Dec 15;8(1)]. Neurol Neuroimmunol Neuroinflamm 7(5):e797. https://doi.org/10.1212/NXI.0000000000000797

Palermo S (2020) COVID-19 pandemic: maximizing future vaccination treatments considering aging and frailty. Front Med (Lausanne) 7:558835. https://doi.org/10.3389/fmed.2020.558835

Palermo S (2021a) "NeuroCOVID": an analysis of the impact of COVID-19 on the older adults. Evolving psychological and neuropsychological understanding, chap 14. In: Baddour MM (ed) Fighting the COVID-19 pandemic. IntechOpen Limited, London. https://doi.org/10.5772/intechopen.99414

Palermo S (2021b) Frailty, vulnerability, and plasticity: towards a new medicine of complexity, Chap 1. In: Palermo S (ed) Frailty in the elderly—understanding and managing complexity. IntechOpen Limited, London. https://doi.org/10.5772/intechopen.96244

Palermo S (2022) Neurocognition and placebo analgesia: linking in functional magnetic resonance imaging, Chap 28. In: Rajendram R, Patel VB, Preedy VR, Martin CR (eds) The neurobiology, physiology, and psychology of pain. Academic Press, pp 307–317. ISBN 9780128205891. https://doi.org/10.1016/B978-0-12-820589-1.00028-2

Palermo S, Stanziano M, Morese R (2018) Commentary: anterior cingulate cortex and response conflict: effects of frequency, inhibition and errors. Front Behav Neurosci 12:171. https://doi.org/10.3389/fnbeh.2018.00171

Palermo S, Stanziano M, Nigri A et al (2023) Parkinson's disease, SARS-CoV-2, and frailty: is there a vicious cycle related to hypovitaminosis D? Brain Sci 13:528. https://doi.org/10.3390/brainsci13040528

Perrin P, Collongues N, Baloglu S et al (2021) Cytokine release syndrome-associated encephalopathy in patients with COVID-19. Eur J Neurol 28(1):248–258. https://doi.org/10.1111/ene.14491

Petersen M, Nägele FL, Mayer C et al (2022) Brain imaging and neuropsychological assessment of individuals recovered from mild to moderate SARS-CoV-2 infection. medRxiv. https://doi.org/10.1101/2022.07.08.22277420

Pfoh ER, Chan KS, Dinglas VD et al (2015) Cognitive screening among acute respiratory failure survivors: a cross-sectional evaluation of the Mini-Mental State Examination. Crit Care 19(1):220. https://doi.org/10.1186/s13054-015-0934-5

Poletti S, Palladini M, Mazza MG et al (2022) Long-term consequences of COVID-19 on cognitive functioning up to 6 months after discharge: role of depression and impact on quality of life. Eur Arch Psychiatry Clin Neurosci 272(5):773–782. https://doi.org/10.1007/s00406-021-01346-9

Politi LS, Salsano E, Grimaldi M (2020) Magnetic resonance imaging alteration of the brain in a patient with coronavirus disease 2019 (COVID-19) and anosmia. JAMA Neurol 77(8):1028–1029. https://doi.org/10.1001/jamaneurol.2020.2125

Radmanesh A, Raz E, Zan E et al (2020) Brain imaging use and findings in COVID-19: a single academic center experience in the epicenter of disease in the United States. AJNR Am J Neuroradiol 41(7):1179–1183. https://doi.org/10.3174/ajnr.A6610

Ritchie K, Chan D, Watermeyer T (2020) The cognitive consequences of the COVID-19 epidemic: collateral damage? Brain Commun 2(2):fcaa069. https://doi.org/10.1093/braincomms/fcaa069

Roy D, Ghosh R, Dubey S et al (2021) Neurological and neuropsychiatric impacts of COVID-19 pandemic. Can J Neurol Sci 48(1):9–24. https://doi.org/10.1017/cjn.2020.173

Rudroff T, Workman CD, Ponto LLB (2021) 18F-FDG-PET imaging for post-COVID-19 brain and skeletal muscle alterations. Viruses 13(11):2283. https://doi.org/10.3390/v13112283

Schweitzer F, Kleineberg NN, Göreci Y et al (2021) Neuro-COVID-19 is more than anosmia: clinical presentation, neurodiagnostics, therapies, and prognosis. Curr Opin Neurol 34(3):423–431. https://doi.org/10.1097/WCO.0000000000000930

Silva LS, Joao RB, Nogueira MH et al (2021) Functional and microstructural brain abnormalities, fatigue, and cognitive dysfunction after mild COVID-19. medRxiv. https://doi.org/10.1101/2021.03.20.21253414

Sollini M, Morbelli S, Ciccarelli M et al (2021) Long COVID hallmarks on [18F]FDG-PET/CT: a case-control study. Eur J Nucl Med Mol Imaging 48(10):3187–3197. https://doi.org/10.1007/s00259-021-05294-3

Soloveva NV, Makarova EV, Kichuk IV (2021) Coronavirus syndrome: COVID-19 psychotrauma. Eur J Transl Myol 30(4):9302. https://doi.org/10.4081/ejtm.2020.9302

Stefanou MI, Palaiodimou L, Bakola E et al (2022) Neurological manifestations of long-COVID syndrome: a narrative review. Ther Adv Chronic Dis 13:20406223221076890. https://doi.org/10.1177/20406223221076890

Tavares-Júnior JWL, de Souza ACC, Borges JWP et al (2022) COVID-19 associated cognitive impairment: a systematic review. Cortex 152:77–97. https://doi.org/10.1016/j.cortex.2022.04.006

Valdes E, Fuchs B, Morrison C et al (2022) Demographic and social determinants of cognitive dysfunction following hospitalization for COVID-19. J Neurol Sci 438:120146. https://doi.org/10.1016/j.jns.2022.120146

Varatharaj A, Thomas N, Ellul MA et al (2020) Neurological and neuropsychiatric complications of COVID-19 in 153 patients: a UK-wide surveillance study. Lancet Psychiatry 7(10):875–882. https://doi.org/10.1016/S2215-0366(20)30287-X

Venkataramani V, Winkler F (2022) Cognitive deficits in long COVID-19. N Engl J Med 387(19):1813–1815. https://doi.org/10.1056/NEJMcibr2210069

Verger A, Barthel H, Tolboom N et al (2022) 2-[18F]-FDG PET for imaging brain involvement in patients with long COVID: perspective of the EANM Neuroimaging Committee. Eur J Nucl Med Mol Imaging 49(11):3599–3606. https://doi.org/10.1007/s00259-022-05913-7

Voruz P, Allali G, Benzakour L et al (2022) Long COVID neuropsychological deficits after severe, moderate, or mild infection. Clin Transl Neurosci 26(2):9. https://doi.org/10.3390/ctn6020009

Wahlgren C, Forsberg G, Divanoglou A et al (2023) Two-year follow-up of patients with post-COVID-19 condition in Sweden: a prospective cohort study [published online ahead of print, 2023 Feb 24]. Lancet Reg Health Eur 100595. https://doi.org/10.1016/j.lanepe.2023.100595

Wing YK, Leung CM (2012) Mental health impact of severe acute respiratory syndrome: a prospective study. Hong Kong Med J 18(Suppl 3):24–27

World Health Organization (2022a) Emergency use ICD codes for COVID-19 disease outbreak 2020. Available from: https://www.who.int/standards/classifications/classification-of-diseases/emergency-use-icd-codes-for-covid-19-disease-outbreak

World Health Organization (2022b) A clinical case definition of post COVID-19 condition by a Delphi consensus 2021. Internet. Available from: https://www.who.int/publications-detail-redirect/WHO-2019-nCoV-Post_COVID-19_condition-Clinical_case_definition-2021.1

Xu E, Xie Y, Al-Aly Z (2022) Long-term neurologic outcomes of COVID-19. Nat Med 28(11):2406–2415. https://doi.org/10.1038/s41591-022-02001-z

Zanin L, Saraceno G, Panciani PP et al (2020) SARS-CoV-2 can induce brain and spine demyelinating lesions. Acta Neurochir (Wien) 162(7):1491–1494. https://doi.org/10.1007/s00701-020-04374-x

Zhai Y, Doraiswamy PM, Woods CW et al (2022) Pre-exposure cognitive performance variability is associated with severity of respiratory infection. Sci Rep 12(1):22589. https://doi.org/10.1038/s41598-022-26081-6

Zhang H, Charmchi Z, Seidman RJ et al (2020) COVID-19-associated myositis with severe proximal and bulbar weakness. Muscle Nerve 62(3):E57–E60. https://doi.org/10.1002/mus.27003

Ziauddeen N, Gurdasani D, O'Hara ME et al (2022) Characteristics and impact of long COVID: findings from an online survey. PLoS ONE 17(3):e0264331. https://doi.org/10.1371/journal.pone.0264331

Zollner A, Koch R, Jukic A et al (2022) Postacute COVID-19 is characterized by gut viral antigen persistence in inflammatory bowel diseases. Gastroenterology 163(2):495-506.e8. https://doi.org/10.1053/j.gastro.2022.04.037

Vaccination and Therapeutics

9

Walid Shaban Abdella, Ahmed Nabil Selim, Sheeren Diab, Esraa Qobeia, and Akram Fekry Elgazzar

Abstract

In the wake of the novel Coronavirus, it has become imperative to develop vaccines that can alert our immune system to the virus, hence reducing the severity of disease if exposed to it. Different types of vaccines have been studied in this respect, such as nucleic acid vaccines (mRNA and DNA vaccines), vector-based vaccines, whole-virus vaccines (inactivated virus and live-attenuated vaccine), as well as protein subunits vaccines. The results have demonstrated that these vaccines are efficient against both the original strain and emerging variants. Furthermore, they also proved their safety with no grave adverse events. Despite this, hesitancy toward taking these vaccines is still present among certain groups in society due to various factors. Special emphasis has been placed on studies concerning pregnant women, children, elderly people, and immunocompromised individuals where efficacy and safety were proven.

Keywords

COVID-19 · Coronavirus · Vaccines · DNA · Adenovirus · EpiVacCorona · Glycoprotein · NVX-CoV2373 · Ad26.COV2.S · mRNA · Novavax

Authors Walid Shaban Abdella and Ahmed Nabil Selim contributed similarly.

W. S. Abdella (✉) · A. F. Elgazzar
Department of Ophthalmology, Faculty of Medicine, Al-Azhar University, Dimette, Egypt
e-mail: Walid.Abdelaa.1997@gmail.com

A. N. Selim
Faculty of Medicine, Al-Azhar University, Assiut, Egypt

S. Diab
Faculty of Medicine, Ain Shams University, Cairo, Egypt

E. Qobeia
Faculty of Medicine, Al-Azhar University for Girls, Cairo, Egypt

9.1 Introduction

Globally, in April 2023, over 700 million people had been infected with SARS-CoV, resulting in 6.8 million reported deaths (WHO 2023). This virus is highly contagious due to its potential to spread asymptomatically and the difficulty of identifying infected individuals (Han et al. 2021). This led to a global call for vaccines. Within a year from the beginning of the infection spread, various vaccines came onto the market and began to be administered (Tregoning et al. 2021). By April 2023, more than 13 billion vaccine doses had been dispensed (WHO 2023).

N. Rezaei (ed.), *The COVID-19 Aftermath*, Advances in Experimental Medicine and Biology 1457,
https://doi.org/10.1007/978-3-031-61939-7_9

9.1.1 Immunity Development Against COVID-19 Vaccines

To know how variable COVID-19 vaccines can establish immunity, we should shed light on how our bodies form immunity against the COVID-19 virus.

Immunity against COVID-19 begins when the virus spike protein attaches to the ACE-2 receptor that is expressed on cellular surfaces and enables viral entry. After entering, viral RNA is encapsulated and encodes several genes that are required for its replication (Mortaz et al. 2020) (Fig. 9.1).

Pattern recognition receptors on antigen-presenting cells identify the infected cells. APCs present a portion of the viral peptide on its surface to naive T cells to be differentiated into helper T cells and cytotoxic T cells. Helper T cells which in turn activate B cells to plasma cells to have the ability to produce antibodies against the infected virus (Farzi et al. 2022). These antibodies effectively block the virus's entry into host cells and consequently diminish the chance of viral relapse (Röltgen and Boyd 2021), while cytotoxic T cells prevent virus progression by killing virally infected cells. Understanding immunity development against COVID-19 will get help to understand how different vaccines will help to establish subsequent immunity. Because after immunity developed, either as a sequel of vaccination or previous infection memory B and T cells will be ready to strongly respond. In this way, vaccines will help to establish an immune response against different COVID-19 proteins to enable the immune system to defend the body when infected with the virus later (Mortaz et al. 2020; Röltgen and Boyd 2021; Farzi et al. 2022) (Fig. 9.2).

COVID-19 vaccines are divided into four major types as illustrated in

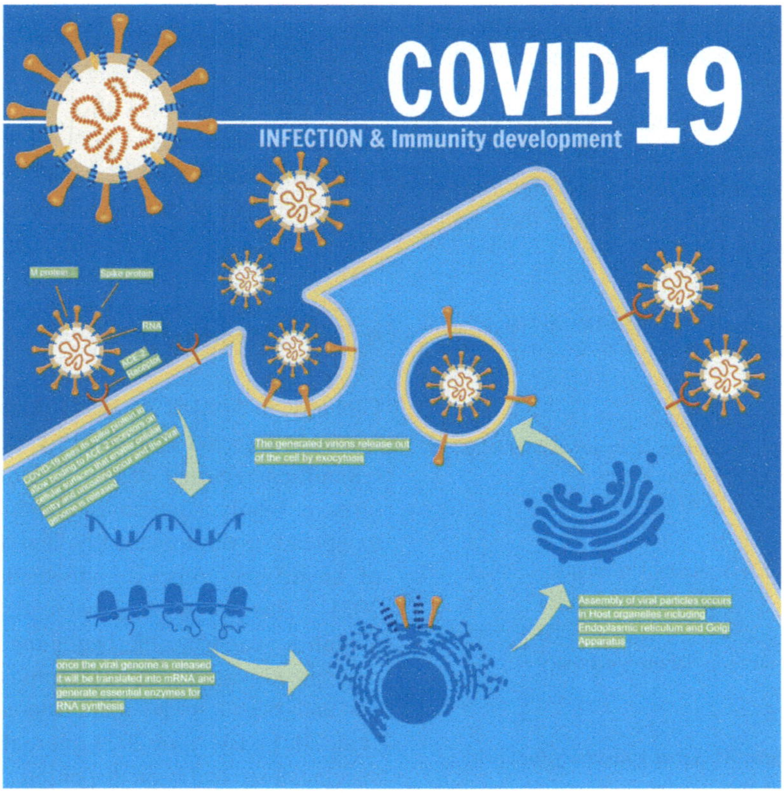

Fig. 9.1 COVID-19 virus replication cycle

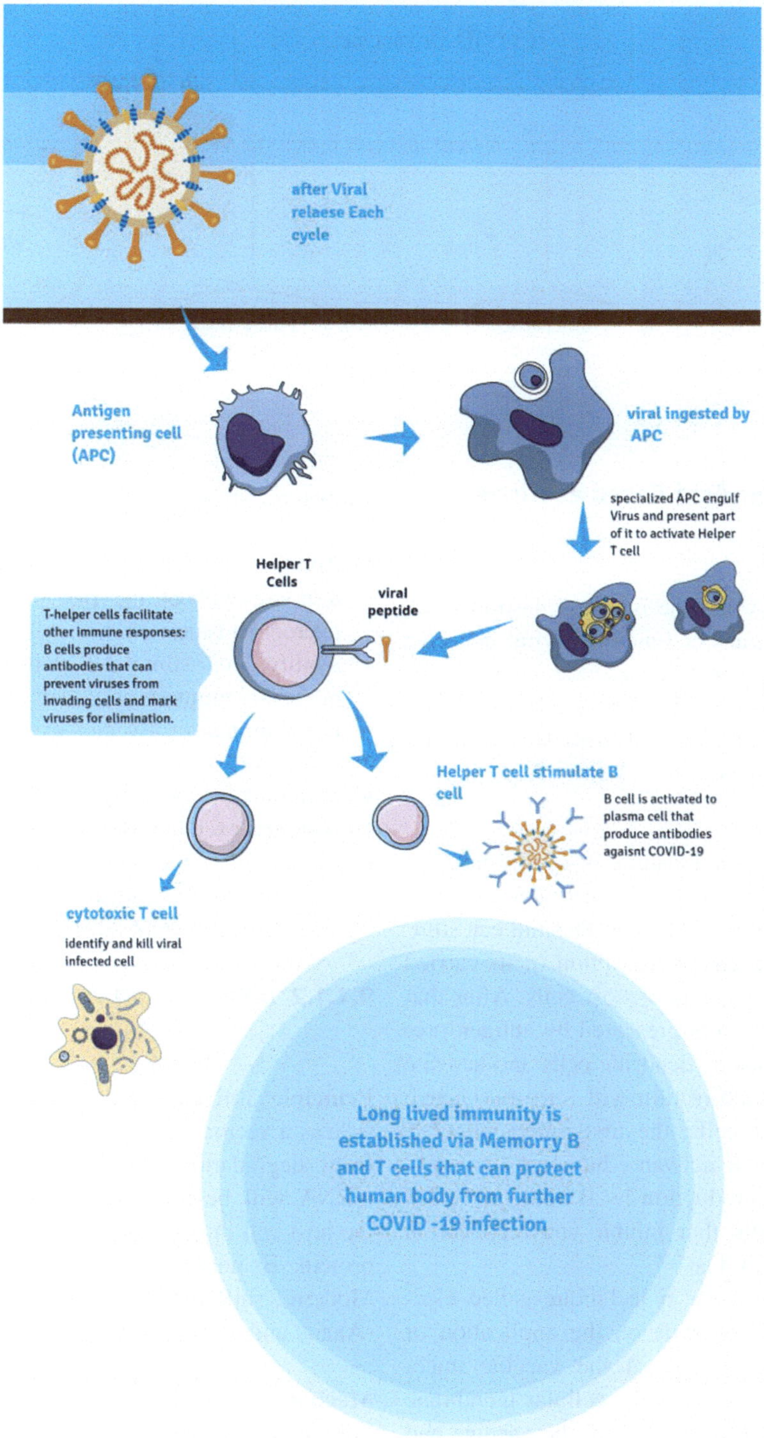

Fig. 9.2 Immunity development against COVID-19

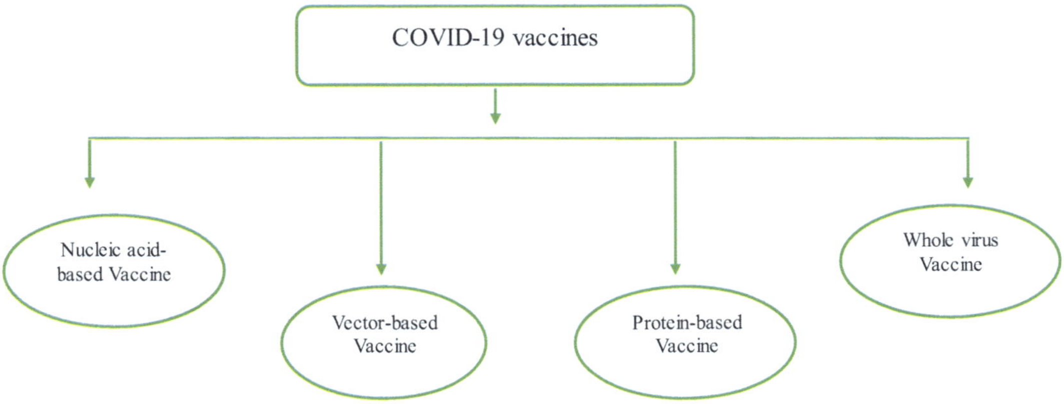

9.1.1.1 Nucleic Acid-Based Vaccines

DNA-Based Vaccines

Principle: DNA vaccine is using plasmid DNA as a vector to transfer functional viral genes or portions of them that can encode immunological antigens, to host cells. This method will trigger both humoral and cell-mediated immunity (Silveira et al. 2021).

Mechanism: DNA plasmid is synthesized by bacteria, supplied by human promotor, and able to encode antigens of interest. Plasmid can stimulate translation and transcription of its carried gene using the nucleus of host cells. After that synthesized protein is presented by antigen-presenting cells such as dendritic cells and activate naive T cell which in turn will form two active population of T cells, the first is the helper T cell (CD4) which activates humoral immunity and antibodies production by B cells, and CD8 cytotoxic T cells that inhibit viral replication (Silveira et al. 2021).

DNA vaccines use a technique called electroporation that depends on the application of electrical pulse activity at the vaccine injection site which increases the cellular membrane permeability of the host cell to the vaccine and subsequent improved immune response (Silveira et al. 2021) (Fig. 9.3).

Characteristics

- Ability to trigger a broad immune response without the risk of reactivation of microorganisms.
- Ability to stimulate cell-mediated and humoral immunity.
- Development of a vector expressing multiple antigens in a single vaccine.
- Manufacture on a big scale at a reasonable cost; and excellent storage stability.
- Extremely stable and require less refrigeration, making them ideal for usage in endemic areas (Silveira et al. 2021).

9.1.1.2 mRNA-Based Vaccine

Principle: mRNA vaccine using lipid nanoparticle as a vector to transfer RNA, protect mRNA from degradation, and facilitate endocytosis. mRNA will be translated in the cytoplasm of the host cell into a protein of interest, i.e., spike protein. Examples of this type of vaccine are Moderna (mRNA-1273) and Pfizer (BNT162) (Anand and Stahel 2021).

Mechanism: mRNA vaccine is protected by lipid nanoparticles, a positively charged molecule that ensures the transfer of RNA via negative charge cell membrane and facilitates

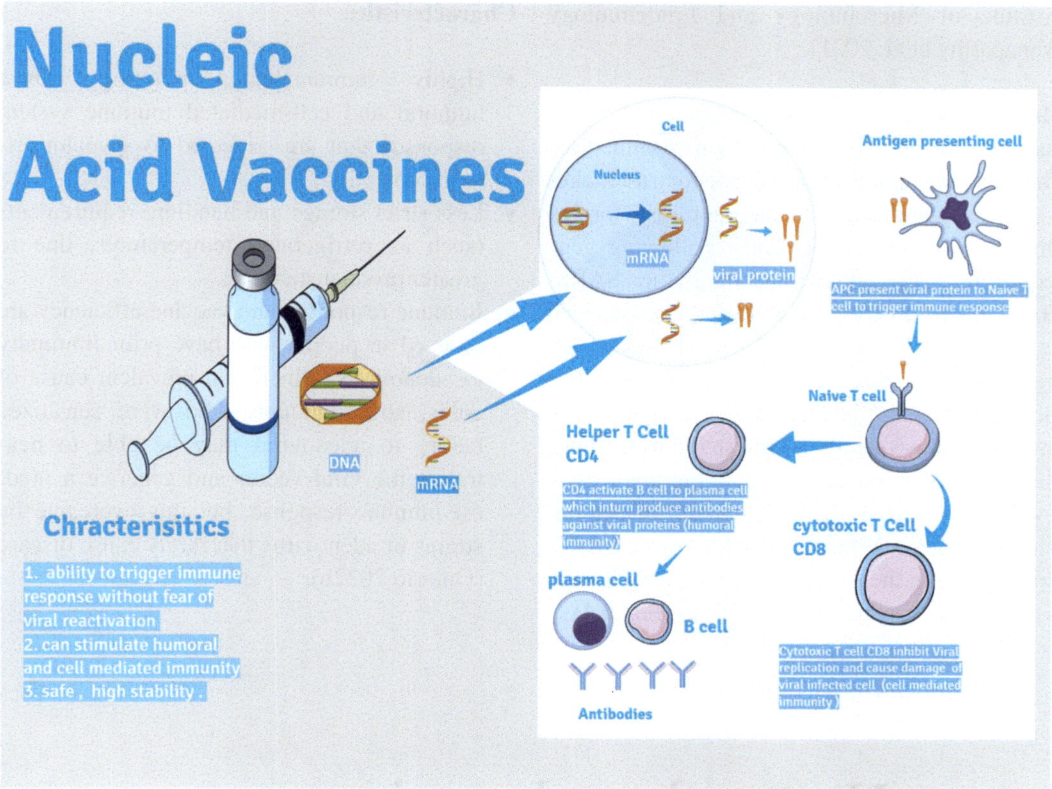

Fig. 9.3 Preview of nucleic acid vaccines

endocytosis. After release from the lipid nanoparticle, mRNA will be translated into spike protein. The antigen-presenting cell will identify the synthesized viral protein and trigger both humoral and cell-mediated immune responses like the DNA vaccines (Anand and Stahel 2021) (Fig. 9.3).

Characteristics

- Ability to trigger an immune response without the risk of reactivation of microorganisms.
- Ability to stimulate cell-mediated and humoral immunity.
- Fast in development and design.
- Not reach the nucleus and translated to a specific protein, unlike DNA which has a risk to reach the nucleus and is integrated into the genome of the host cell (Anand and Stahel 2021; Park et al. 2021).

9.1.1.3 Vector-Based Vaccine

Principle: This approach employs a vector, such as adenovirus that has been genetically engineered to be incapable of causing disease. These modified viruses convey antigen-specific genetic material to host cells to develop immunity. There are two types of vector-based vaccines: replicating and non-replicating. Replicating virus infects the host cell and will cause the production of a new vaccine antigen (antigen of interest) and new viral particles while the non-replicating virus after host cell infection only produces a new vaccine antigen with a lacking ability to produce new viral particles. Examples of non-replicating adenovirus vector-based vaccines are AZD1222 by Oxford-AstraZeneca, JNJ-78435735 by Johnson and Johnson, Sputnik V, and Sputnik Light by Gamaleya Research

Institute of Microbiology and Epidemiology (Vanaparthy et al. 2021).

Mechanism: Deletion of elongation 1 gene that results in viral replication termination. The genomic compaction of adenovirus makes it a good candidate for incorporating foreign DNA (Deoxyribonucleic acid). Following gene sequencing, the genes encoding the SARS-CoV-2 spike protein are extracted and inserted into the adenovirus vector. These viruses proliferate into specialized cells and are purified further before being developed into vaccines. The vaccine is subsequently administered to the host, eliciting a humoral and cellular immunological response as well as the development of neutralizing antibodies. When exposed to a true SARS-CoV-2 virus in the future, the SARS-CoV-2 virus is destroyed (Vanaparthy et al. 2021) (Fig. 9.4).

Characteristics

- Highly immunogenic, eliciting robust humoral and cell-mediated immune system responses that are expected to give longer-term protection.
- Less strict storage and handling requirements (such as refrigerated temperatures, due to greater product stability).
- Immune responses and vaccine efficiency are reduced in people who have prior immunity to adenovirus, which is a prevalent cause of colds, so individuals who were sensitized before to adenoviral may be able to neutralize the viral vector and generate a modest immune response, but this overcame by strains of adenovirus that rarely cause disease (Ontario 2022b).

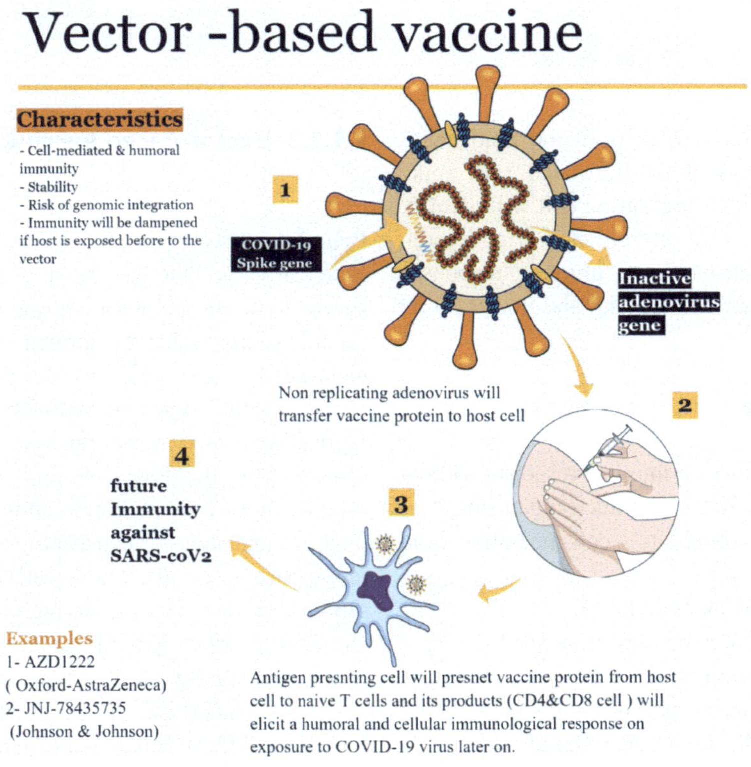

Fig. 9.4 Preview of vector-based vaccines

9.1.1.4 Protein-Based Vaccine

Principle: Instead of the entire viral particle, a protein component containing a specific virus product is used due to its ability to trigger an immune response with no tendency to cause infection. This type of vaccine has adjuvants which are additional agents added to vaccine antigens aiming to increase the efficacy and durability of the immune response; example of a vaccine based on this type of vaccination is EpiVacCorona (Ontario 2022c).

Mechanism: A specific piece of the COVID-19 virus's genetic material (DNA) that contains instructions for making the viral protein, the SARS-CoV-2 spike (S) glycoprotein, is synthesized using recombinant DNA technology, and the synthesized antigens that resemble the native structure of the SARS-CoV-2 S glycoprotein are then assembled as nanoparticles. Then, these highly immunogenic nanoparticles will be used to activate an immunological response in our bodies. The SARS-CoV-2 recombinant spike S glycoprotein is combined with an adjuvant to increase immunity's response to the protein antigen. This adjuvant causes an influx of antigen-presenting cells (APCs), which in turn help process and present the protein antigen to the rest of the immune system. APCs will stimulate increasing antibodies and induce long-lived B memory cells in addition to T cell recruitment which resulted in improving immunological response to the vaccine. On future infection by the virus, memory B cells will be ready to produce antibodies that fight it and prevent entry to host cells (Ontario 2022c) (Fig. 9.5).

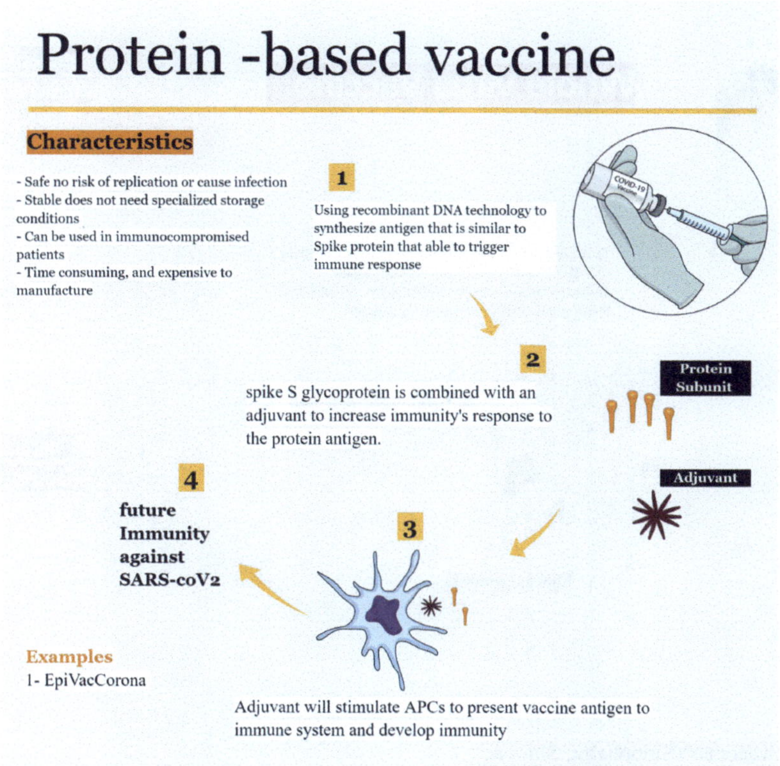

Fig. 9.5 Preview of protein-based vaccines

Characteristics

- Safe, as this type of vaccine has no genetic materials so no risk of replicating within the host and subsequently no risk of infection occurrence.
- Stable, does not need specialized storage requirements to maintain their potency such as very low temperature.
- Can be used in immunocompromised patients due to lacking ability to replicate or cause infection.
- More complex in production, time-consuming, and expensive to manufacture (Efuntoye et al. 2021).

9.1.1.5 Whole-Virus Vaccine

Principles: From inactivated or live-attenuated viral particles, whole-virus vaccines are made. Live-attenuated vaccines are created by exposing disease-causing pathogens to cell culture in a foreign host to remove their pathogenic properties. This causes the pathogen to cause a mild infection that the recipient's immune system can quickly clear up and which then triggers the development of a potent immunogenic response. Inactivated SARS-CoV-2 virus is grown in Vero cell culture to produce COVID-19 vaccines, which are then chemically inactivated to render the virus inactive but still capable of inducing an immune response in the recipient (Efuntoye et al. 2021).

Mechanism: Since the pathogens in inactivated vaccines retain part of their integrity, the immune system can identify and track them, which triggers an adaptive immune response, and the vaccine cannot be infectious when it is manufactured properly. But the killed vaccine cannot replicate, so adjuvants and periodic

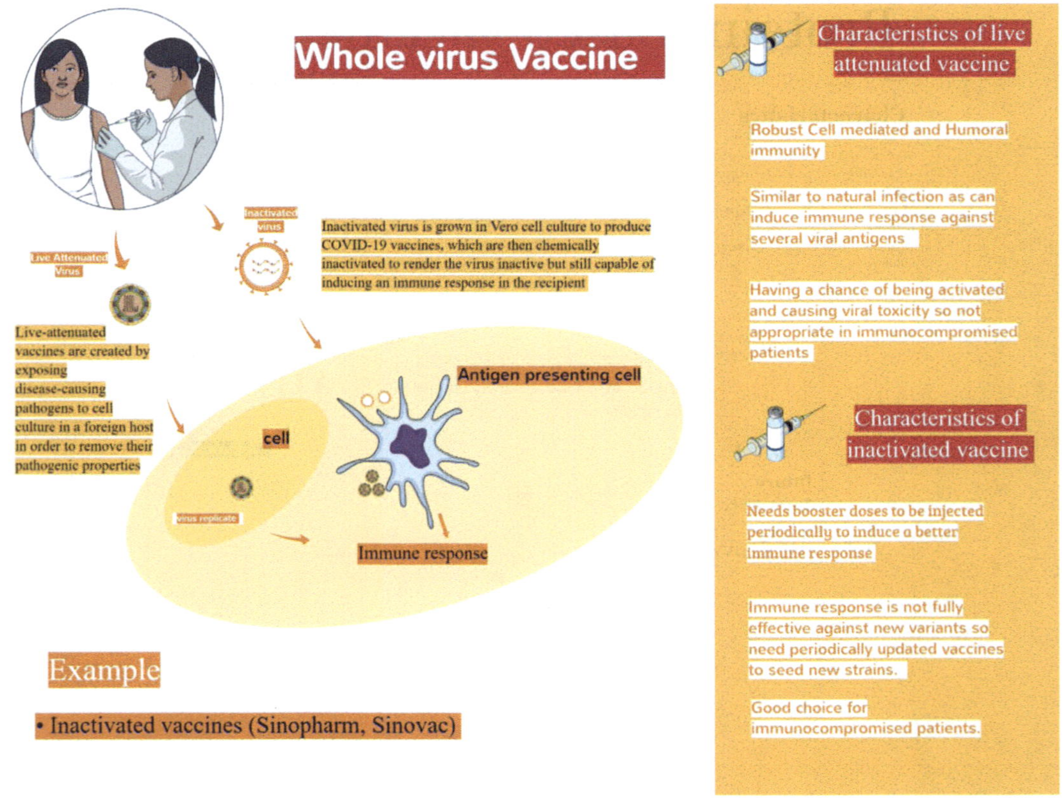

Fig. 9.6 Preview of whole-virus vaccines

booster doses are required to reinforce its immune response. As the recipient's immune system comes into touch with the SARS-COV-2 virus as a whole, the immune response to these vaccines targets the virus's envelope, spike, matrix, and nuclear proteins (Okamura and Ebina 2021) (Fig. 9.6).

Characteristics of Inactivated (Killed) Vaccine

- Inactivated (killed) vaccine does not replicate giving it a short-lived immune response and needs booster doses to be injected periodically to induce a better immune response.
- The immune response is not fully effective against new variants, so need periodically updated vaccines to seed new strains.
- No risk of causing an infection which makes it a good choice for immunocompromised patients.
- Not appropriate for large-scale vaccination due to the complexity of manufacturing (Efuntoye et al. 2021; Okamura and Ebina 2021).

Characteristics of Live-Attenuated Vaccines

- Regardless of its slower replication, it is effective enough to induce a robust humoral and cell-mediated immune response.
- Can induce immunity against several antigens, unlike other vaccines that are directed to elicit immunity against viral spike protein.
- Its disadvantages concluded in having a chance of being virus regaining its toxicity due to mutations after vaccination which make it contraindicated to be used in immunocompromised patients (Okamura and Ebina 2021).

9.2 Vaccines Efficacy

Vaccines to be approved by WHO must achieve an efficacy rate of 50% and more in clinical trials; this means that half and more of the vaccinated participants who are usually diseased and received the vaccine should develop outcomes of interest compared to participants who received a placebo and developed the same outcomes (WHO 2021).

After discussion about how vaccines can develop immunity, the types, mechanism of their actions, and the characteristics of each one, we will move to spotlight about efficacy and safety of commonly available vaccines in facing different COVID-19 variants.

9.2.1 NVX-CoV2373

9.2.1.1 B.1.1.7 Variant

Regarding the clinical trial conducted by the Biotech firm, Novavax, it was found that NVX-CoV2373 was very effective in eliciting immune-mediated antibody release against the B variant (Callaway and Mallapaty 2021).

9.2.1.2 B.1.351 Variant

A randomized clinical trial that was conducted on 4387 participants who received two doses of NVX-CoV2373 found that the vaccine developed an efficacy of about 49.4%, 95% (confidence interval), CI (6.1–72.8) (Shinde et al. 2021).

9.2.1.3 B.1.1.529 Variant and D164G Mutation

Regarding the omicrons variant, it was found that participants who received two doses of Novavax were not able to neutralize this strain sub-lineages BA.1 and BA.4/BA.5; however, after three doses, the finding changed and there were high titers of neutralizing antibodies with an immune response like was elicited post-three doses of mRNA vaccine (Bhiman et al. 2023). Also, it was found that a higher antibodies titer after two doses of vaccine was detected against the D164G mutation variant compared to beta and omicron sub-lineages, and these titers were boosted after the third dose (Bhiman et al. 2023).

9.2.2 Ad26.COV2.S

9.2.2.1 Original Wuhan-Hu-1 and D164G Mutation

Sadoff et al. conducted a randomized, double-blind, placebo-controlled phase-three trial which showed that Ad26.COV2.S vaccine had 66.9% effectiveness (95% CI, 59.0–73.4) 14 days after dose administration in a study population of 19,630 participants and remained at 66.1% (95% CI, 55.0–74.8) 28 days after the dose was administered. In addition, effectiveness against severe COVID-19 increased to 76.7% (95% CI, 54.6–89.1) for $\geq$ 14 days and 85.4% (95% CI, 54.2–96.9) for $\geq$ 28 days (Sadoff et al. 2021).

9.2.2.2 B.1.351 Variant

In a clinical trial of 19,630 participants in South Africa, who received the vaccine, the B.1.351 sequence constituted 94.5%. These vaccines displayed a 52% efficacy for moderate and severe conditions of COVID-19, and 73.1% efficacy for severe critical cases 14 days after vaccination. At 28 days post-vaccination, the efficacy had increased to 64% and 81.7%, respectively.

In Brazil, 69% of the samples demonstrated the P.2 lineage, carrying an E484K mutation. Despite now being exposed to varied variants, these vaccines still maintained their high effectiveness as they were formulated based on the Wuhan-Hu-1 strain—signifying that they offer cross-protective immunity against newer variants in South Africa and Brazil (Sadoff et al. 2021).

9.2.2.3 B.1.1.529 Variant

A study of 162,637 PCR tests conducted during the omicron surge in South Africa revealed that of the 93,854 participants who received two doses of either the BNT162b2 vaccine (42 days apart) or the Ad26.COV2.S vaccine (4–6 months apart), 34% tested positive—with 1.6% requiring hospitalization and 0.5% critical with ICU admission. Results showed that after receiving two doses of Ad26.COV2.S vaccine, recipients experienced immunity against hospitalization for the disease at 55% (95% CI, 22–74) within 13 days, 74% (95% CI, 57–84) between 14 and 27 days, and 72% (95% CI, 59–81) within 1–2 months; while protection against ICU admission was 69% (95%, 26–87) at 14–27 days and 82% (95%, 57–93) within 1–2 months post-administration of the second dose (Gray et al. 2022).

9.2.3 AZD 1222

9.2.3.1 B.1.1.7 Variant

The vaccine showed reduced titer levels for neutralization with the B.1.1.7 lineage, yet adequate clinical efficacy against features of the disease was achieved, indicating that even low levels of neutralizing antibodies can provide protection (Emary et al. 2021).

9.2.3.2 B.1.351 Variant

A randomized control trial with 2026 participants conducted in Africa concluded that two doses of the vaccine did not affect mild to moderate disease caused by B.1.351, but fortunately, there were no reports of severe disease either. The reduced efficacy should be taken into account, bearing in mind that the first dose of the vaccine proved to have 75% efficacy (95% CI 8.7–95.5) in guarding against mild to moderate cases of COVID-19 before the emergence of this variant (Madhi et al. 2021).

9.2.3.3 B.1.1.529 Variant

In a retrospective analysis, 886,774 individuals infected with omicron were identified and their effectiveness was studied; it was noted that no effect was seen after they received two doses of the AZD-1222 vaccine. The efficacy of the AZD-1222 primary course increased up to 70.1% (95% CI, 69.5–70.7) after 2–4 weeks when an mRNA-1273 booster was administered, but reduced to 60.9% (95% CI, 59.7–62.1) after 5–9 weeks (Andrews et al. 2022).

9.2.4 mRNA-1273

9.2.4.1 B.1.1.7 and B.1.351 Variants

In a study in Qatar, 181,304 recipients were administered the full two-dose regimen of the vaccine. The efficacy of the vaccine against

B.1.1.7 was limited in the first two weeks after the first dose, but increased significantly to 81.6% (95% CI, 73.1–87.8%) in the third week and 94.4% (95% CI, 89.1–97.5%) in the fourth week, reaching 99.2% (95% CI, 95.3–100%) by the second week after receiving the second dose and 100% (95% CI, 91.8–100%) 14 days post-second dose. Also B.1.351 variants were collected from PCR-positive samples, and the efficacy was limited for two weeks after the first dose before increasing to 47.9% (95% CI, 39.5–55.2%) at three weeks; this increased further to 73.7% (95% CI, 67.6–78.8%) in week four before receiving a second dose and achieved 96.4% (95% CI, 94.3–97.9%) efficacy by two weeks post-second dose (Chemaitelly et al. 2021).

9.2.4.2 B.1.1.529 Variant

Booster doses of mRNA vaccines RNA-1273 and BNT162b2 may not be as effective in protecting against omicron infection, but they do provide strong protection against COVID-19-related deaths and hospitalization (Abu-Raddad et al. 2022).

9.2.5 BNT162b2

9.2.5.1 B.1.1.7 and B.1.351 Variants

A study conducted in Qatar found that two-dose regimen efficacy against the B.1.1.7 variant was 89.5% (95% CI, 85.9–92.3) < 14 days post-second dose, whereas efficacy against B.1.351 variant was 75% (95% CI, 70/5–78/9) (Abu-Raddad et al. 2021).

9.2.5.2 B.1.617.1 and 2 Variants

In the study conducted by Liu et al. to compare the reduction in the neutralization of variants B.1.617.1 and 2, they found that compared to the original Wuhan-related strains, neutralization of both B.1.351 and P.1 against B.1.617.2 has decreased significantly, though they have not seen any widespread antibody escape as with B.1.351. This suggests that prior infection with these variants may leave people more vulnerable to reinfection from B.1.617.2 (Meo et al. 2021).

9.2.5.3 B.1.1.529 Variant

Andrew et al. conducted a case–control study to estimate vaccine efficacy against the omicron variant. They found that mRNA boosters were shown to be highly effective against infection with the delta variant yet were slightly less successful in combating the omicron variant. Nonetheless, these boosters provided strong protection against COVID-19-associated hospitalization and mortality for both variants (Abu-Raddad et al. 2022).

9.3 Safety of COVID-19 Vaccines

9.3.1 Novavax

A study by Health et al. looked into how effective and safe the NVX-Co2373 vaccine was on 7500 participants who received the vaccine. They experienced numerous unfavorable effects, going from gentle yet common symptoms like exhaustion, migraine, and muscle soreness to more severe systemic unfavorable events like a fever ranging from 38 to 40 °C in 2.4% of participants after the first dose and 5.4% after the second dose, with two people even getting a fever >40 °C following one and two doses, respectively. Also noteworthy is a certain severe unwanted event—myocarditis—that was reported in one participant three days after their second dose and is believed to be an immune-mediated reaction. Older participants had fewer serious adverse events, and they occurred less frequently. Due to the number of participants, it is impossible to rule out the possibility of rare adverse reactions occurring (Abu-Raddad et al. 2021).

9.3.2 mRNA Vaccines (Pfizer and Moderna)

Et al. conducted a study comparing the pharmacology, indications, contraindications, and adverse effects of both mRNA vaccines the Pfizer/BioNTech and Moderna; it was determined that they are advantageous in providing

COVID-19 immunity. They are both recommended for individuals over the age of sixteen and provide immunity for around 119 days. The Pfizer/BioNTech vaccine is 95% effective, and Moderna is 94.t% effective in preventing COVID-19. Potential adverse effects include redness, swelling at the injection site, muscle or joint pain, fever, fatigue, headache, nausea, vomiting, itching, and chills; rarely an anaphylactic shock has been reported. The occurrence of adverse effects was generally lower for the Pfizer/BioNTech vaccine; since the Moderna vaccine is less temperature sensitive, it is easier to store and transport (Meo et al. 2021). A rare but serious adverse event was noted from the use of either vaccine in USA—1226 reports of myocarditis between December 2020 and June 2021; although no casualties were reported, there was a review by the CDC but ultimately the benefits outweighed any risks, so its use continued (Gargano et al. 2021).

9.3.3 AZD1222

It has been observed that the AstraZeneca (vector vaccine) has been known to cause a rare form of a blood clot, called vaccine-induced immune thrombotic thrombocytopenia (VITT), between 4 and 28 days after vaccination. Globally, it is estimated that VITT appears in 1 out of every 26,000–100,000 people who have received their first dose of either AstraZeneca. Health authorities both locally and abroad are closely monitoring the situation. Furthermore, cases of capillary leak syndrome—a condition that causes fluid leakage from the capillaries—have also been reported following vaccination with these two vaccines; for this reason, individuals with a history of capillary leak syndrome should not get these vaccines (WHO 2022a).

9.3.4 Johnson & Johnson (J&J)/Janssen

It was found that a very rare side effect occurred in a small group of women who had been administered J&J/Janssen's vaccine. Thrombosis with thrombocytopenia syndrome was reported in roughly 7 cases per 1,000,000 vaccinated women between the ages of 18 and 49, and even rarer in those over 50. An analysis ultimately concluded that the benefits of the vaccine outweighed its potential risks. Although uncommon and treatable, this syndrome can be quite severe. Additional symptoms linked to the vaccine include persistent headaches or blurred vision, breathing difficulties, chest pain, leg swelling, prolonged abdominal pain, and bruising or pinpoint spots near the injection site (Sharma et al. 2022).

9.4 Medical Exemption of COVID-19 Vaccines

The following are the conditions that may be qualified to be accepted as medical exemption of vaccination and vaccine should not be administrated due to the hazards of vaccination outweighing its potential therapeutic benefits.

I. Pre-existing history of myocarditis before mRNA vaccines and, history of severe allergic reaction or anaphylaxis to any COVID-19 components (Ontario 2022a).

II. Presence of one contraindication condition to initiate AstraZeneca or Janssen vaccines including history of capillary leak syndrome, history of cerebral venous sinus thrombosis (CVST) with thrombocytopenia, history of heparin-induced thrombocytopenia (HIT), and history of major venous and/or arterial thrombosis with thrombocytopenia. These conditions if present mRNA vaccines should be used instead (Ontario 2022a).

III. If severe adverse effects were developed post-first dose vaccinations, thrombosis with thrombocytopenia syndrome (TTS)/VITT3 following the AstraZeneca or Janssen COVID-19 vaccine and myocarditis, pericarditis following an mRNA COVID-19 vaccine, severe allergic reaction (including anaphylaxis) following a COVID-19 vaccine, and serious adverse event following COVID-19 immunization (e.g., results in hospitalization, persistent or significant disability/incapacity) (Ontario 2022a).

IV. Actively receiving monoclonal antibody therapy or convalescent plasma therapy for the treatment or prevention of COVID-19, this is considered a temporary precaution; once the intervention is discontinued, vaccine can be administered (Ontario 2022a).

V. Actively receiving or recently completed immunosuppressing therapy is anticipated to significantly blunt vaccine response, and the vaccine may be administered after these therapies are discontinued. Timing of administration differs according to the mechanism of each drug as the following is recommended to postpone vaccination to six months after Anti CD20, to three months after Chimeric antigen response (CAR) t-cell therapy and hematopoietic stem cell transplant, and to one month after anti-thymocyte globulin and Prednisone $\geq$ 1 mg/kg (Ontario 2022a).

9.5 Booster Dose

The potential of recurrent mutations in the virus and the appearance of new strains have called into question whether a booster dose is necessary. Currently, the vaccination plan consists of two shots given at least two weeks apart, so that adaptive immunity can form; the second dose is given to make sure it does not dwindle, and that the immune system can provide long-term protection. So, ***are we in need to receive booster doses*?**

Multiple studies explored the significance of booster doses in different vaccines. In the clinical trial conducted by Chenchula et al. to investigate the immunogenicity of seven different vaccines as booster doses after two doses of AZD1222 or BNT162b2, they found high immunogenicity in all vaccines which underwent the investigation (Chenchula et al. 2022). Also, in studying the efficacy of the third dose of the BNT162b2 vaccine by Moreira et al., they found that the third dose elevated immunity by 95.5% compared to the two-dose regimen (Moreira et al. 2022). These findings are similar to what was reported by Yorsaeng et al. who found that there was a higher immunogenicity in participants who received booster doses compared to ones who received only two doses (Yorsaeng et al. 2022). One of the causes of the need to boost our immune system is the outbreak of new variants such as alpha (B.1.1.7), beta (B.1.351), gamma (formerly known as B.1.1.28.1), and delta (B.1.617.2) of the virus; these variants are more transmissible and cause a greater severity of illness than the original SARS-CoV-2 strain (Harvey et al. 2021). Fortunately, current vaccinations appear to be effective against these variants, as suggested by Pegu et al.'s study of antibody neutralization from vaccinated individuals after mRNA-127 vaccination: all had binding antibodies to all variants, with most keeping an active level of protection six months after the second dose of vaccination was completed (Pegu et al. 2021).

9.5.1 Booster Doses Candidates

Immunocompromised individuals are of particular interest, as they may not have sufficient immune responses following two doses of the vaccine. Kamar et al.'s research on immunocompromise in solid-organ transplant patients found antibodies to SARS-CoV-2 in only 40% (forty out of ninety-nine) four weeks after the second dose. However, that figure rose to 68% (sixty-seven out of ninety-nine) 4 weeks after the third dose (Kamar et al. 2021). On August 12, 2021, the U.S. Food and Drug Administration modified emergency use approvals for both the Pfizer-BioNTech COVID-19 vaccine and the Moderna COVID-19 vaccine to permit one extra dose for certain immunocompromised individuals post-solid organ transplantation or for those with conditions that have a similar level of immunosuppression (FDA 2021). According to the CDC, three weeks should pass between the second and third doses for the Pfizer vaccine, and four weeks for Moderna's vaccine. As of October 7, 2021, adults aged 65 and above, residents in long-term care facilities, those with underlying medical issues, or employees/residents in riskier settings such as first responders, educators, food/agriculture staff, corrections workers, or public transit workers are all qualified to receive a third dose of the COVID-19 vaccine (Oliver 2021).

9.6 Hesitancy of Vaccination

Many circumstances can influence people's decisions to accept, decline, or delay immunizations. Vaccination hesitancy is multifactorial complex condition that can be predisposed by the following circumstances:

- **Gender**: Studies revealed that women were more likely than men to be reluctant to get the COVID-19 vaccine (Badr et al. 2021; Chen et al. 2021; Du et al. 2021; Griva et al. 2021; Guillon and Kergall 2021; Lee and You 2022; Rehati et al. 2022; Walsh et al. 2022).
- **Age**: Age was observed to influence COVID-19 vaccine reluctance; however, the results were inconclusive. For example, younger persons, particularly those under 30 years old, were more vaccine-reluctant (Toth-Manikowski et al. 2022; Walsh et al. 2022). This is in line with what was observed by Du et al. (2021) who noted that those under the age of 45 were more likely to be hesitant to get a COVID-19 vaccine. In contrast Lee and You (2022) showed that people aged 50 and over were more vaccine-hesitant.
- **Education**: Education and vaccine reluctance were substantially associated; however, the findings are inconsistent (Du et al. 2021; Hossain et al. 2021; Lee and You 2022; Wang et al. 2022). It is interesting to note that Lee and You (2022) observed that college students were more likely to be vaccination reluctant, but other studies revealed that people with only a high school degree or less were more reluctant to get vaccinated against COVID-19 (Du et al. 2021; Hossain et al. 2021; Wang et al. 2022).
- **Personal Belief**: Participants who had personal anti-vaccine beliefs had lower acceptance (Troiano and Nardi 2021).
- **Occupation**: According to studies, occupation influences COVID-19 vaccine reluctance (Chen et al. 2021; Toth-Manikowski et al. 2022). Nonmedical workers, for instance, were more likely than medical personnel to be vaccine-hesitant (Chen et al. 2021). Regarding healthcare workers, compared to non-physician staff including nurses, office employees, and healthcare technicians, physicians were less likely to be vaccine-reluctant (Toth-Manikowski et al. 2022).
- **The Individual's Income**: It was found that vaccine reluctance was related to the participant's income. Contrary to Chen et al., who observed that participants were more likely to be vaccine-hesitant if they had a greater income (Chen et al. 2021), Badr et al. reported that people with higher incomes were less likely to be vaccine-hesitant (Badr et al. 2021).
- **Working Status**: Unemployed individuals are more prone to be vaccine-hesitant (Troiano and Nardi 2021).
- **Race**: Black or African Americans were more likely to be reluctant to receive the COVID-19 vaccination (Toth-Manikowski et al. 2022).
- **Knowledge of COVID-19**: Individuals who have low knowledge about COVID-19 are likely to be vaccine-hesitant (Badr et al. 2021; Du et al. 2021; Hossain et al. 2021; Huynh et al. 2022).
- **Concern About COVID-19**: Those who were really frightened about becoming infected were less likely to decline the vaccination (Troiano and Nardi 2021).
- **Prior Vaccination for Flu**: Those who had already received immunizations (particularly influenza) had higher acceptance (Troiano and Nardi 2021; Lee and You 2022; Toth-Manikowski et al. 2022).
- **History of Asymptomatic COVID-19 Condition**: People who did not self-report symptoms were shown to be more prone to vaccine reluctance than those who did report symptoms, and they were even related to a COVID-19 diagnosis (Oliveira et al. 2021).

9.7 Vaccination in Specific Population

It is worth noting that the COVID-19 pandemic has posed a threat to many, especially a certain segment that includes pregnant women, children, the Elderly, those who suffer from immunodeficiency, and organ transplant patients.

9.7.1 Pregnant

Pregnancy and the immediate postnatal period are associated with an increased risk of complications and mortality due to infectious diseases in women and their babies that can be prevented by vaccines, including COVID-19 (Sebghati and Khalil 2021). Furthermore, the immune system of pregnant women changes during these crucial periods to accommodate fetal cells from foreign origins, making them more prone to infection (Jamieson et al. 2006). A cohort study conducted in the UK evaluated the outcomes of hospitalized pregnant women with asymptomatic or symptomatic COVID-19 in comparison with hospitalized women without COVID-19. It was discovered that hospitalization due to symptomatic COVID-19 infection increases the odds of admission into an intensive care unit (Vousden et al. 2021). Moreover, according to the Centers for Disease Control and Prevention (CDC), pregnant women with COVID-19 infection experience an elevated risk ratio for death when compared to non-pregnant individuals within a comparable reproductive age range. Following a record number of 22 deaths among pregnant persons in the USA during August 2021, CDC issued an urgent public health advisory on September 29th urging people who are or may become pregnant to obtain vaccinations (Zambrano et al. 2020; CDC 2021).

The vaccine appears just as effective during pregnancy as it is for non-pregnant individuals—this is excellent news! After administration of the second dose, antibodies induced by vaccination are notably higher than those instigated by infection (Gray et al. 2021). Additionally, vaccine efficacy during pregnancy is comparable with that of the general population with a 96% rate documented for any infection, 97% for symptomatic infection, and 89% for hospitalization involving alpha and ancestral variants (Dagan et al. 2021). Concerning safety, no elevation or change has been reported regarding adverse events post-vaccination in pregnancy which had already been recorded prior to the outbreak of the pandemic; these findings are supported by a study encompassing over 500 pregnant persons who have been vaccinated throughout pregnancy indicating no increase in obstetric complications (Bookstein Peretz et al. 2021; Shimabukuro et al. 2021). The importance of vaccination is further reinforced given its ability to provide immunological protection against SARS-CoV-2 through antibody transmission from the mother (via umbilical cord blood or breast milk) following administration of mRNA vaccines during pregnancy; this increased the need for urging vaccinations toward improved maternal and neonatal health outcomes (Ai-ris et al. 2021; Gray et al. 2021).

9.7.2 Children

Children are a population group with special needs and features, and the attitude of parents or guardians toward the COVID-19 vaccine is an essential factor affecting children's vaccination. In addition to the mild presentation of their symptoms, this makes their vaccination decision disputed. There are currently three pediatric COVID-19 vaccines available: Pfizer-BioNTech, Moderna, and Novavax. The first two are mRNA vaccines that can be given to kids from 6 months to 17 years old, while Novavax is a protein subunit vaccine for kids aged 12–17 (CDC 2022). In October 2022, the CDC's Advisory Committee on Immunization Practices recommended that these vaccines be included in the routine childhood vaccination schedule for 2023 (Kim et al. 2017). The good news is that recent research shows that mRNA vaccines are effective at protecting children from SARS-CoV-2 infection, severe symptoms, and hospitalization—plus they appear safe too with only 1.8 out of every million vaccines having mild myocarditis. Not only do the vaccines defend against major health issues like MIS-C during both delta and omicron waves, but the possibility of myocarditis is very small too (Watanabe et al. 2023). The available efficacy information of the mRNA vaccines in July 2022 was 68–74% effective in preventing hospitalizations up to two months after vaccination for fully vaccinated children and adolescents, and up to 92% effective in adolescents up to 44 weeks after vaccination. In infants younger than 6 months old whose mothers were vaccinated with the mRNA vaccine, it was 70% effective against intensive care unit admission and 47% effective against non-intensive care unit hospitalization—but almost twice as effective if it was given after 20 weeks of pregnancy (DeZure 2023). As far as safety goes, all three COVID-19 vaccines have passed FDA's rigorous standards, but side effects vary depending on age group and can range from pain at the injection site to swollen lymph nodes; tiredness; headache; muscle/joint pain; chills; allergic reactions (though rare); thrombosis with thrombocytopenia syndrome; myocarditis; pericarditis; and Guillain–Barré syndrome (more common in adults) (DeZure 2023). So far there have been no causally associated deaths associated with COVID-19 vaccinations in children (DeZure 2023).

9.7.3 Elderly

Due to their vulnerability and a heightened need for the coronavirus disease 2019 (COVID-19) vaccine, older adults are more prone to incurring adverse events from its administration (Farge et al. 2021). Contributing factors such as multimorbidity, frailty, disability, and immunosenescence (an age-related decline in the immune system) all pose a challenge for successful COVID-19 vaccinations in elderly individuals (Osterholm et al. 2012). In comparison with younger adults, older people generally demonstrate decreased immunogenicity and effectiveness of vaccines due to factors such as immunosenescence, inflammation, and shifted immune responses to infections and vaccines (Ciabattini et al. 2018). Enhancing vaccine immunogenicity and efficacy in older adults can be accomplished through increased vaccine dose or incorporation of suitable adjuvants (Andrew et al. 2019; Cunningham et al. 2021). Although limited evidence exists confirming the safety and efficacy of COVID-19 vaccinations for frail elderly people in long-term care settings, infection control practices should still be observed post-vaccination among care staff in these settings (Liang et al. 2022). To mitigate the presence of SARS-CoV-2 infections and achieve immunity against the virus, it is recommended that a primary series of COVID-19 vaccines followed by a booster dose after a period of 4–6 months is completed (Liang et al. 2022). Finally, geriatricians can play an influential role in ensuring successful adult vaccination strategies against COVID-19 by actively taking part in vaccine development and related policy formation (Liang et al. 2022).

9.7.4 Immunocompromised Patients

Immunocompromised individuals are particularly prone to deficient vaccine responses. Solid organ transplant recipients, those with hematological malignancies, and those taking corticosteroids, immunosuppressants, or anti-CD20 agents are of particular concern (Galmiche et al. 2022). Compared to non-immunocompromised individuals, a primary series of two doses does not generate an adequate magnitude of immune response in immunocompromised patients (Boyarsky et al. 2021). The World Health Organization (WHO) places these patients in the highest priority category and recommends two booster doses for all immunocompromised people regardless of age (WHO 2022b). Relying on safety and efficacy assessments, all cancer patients should receive a vaccine except those on anti-B cell therapies such as Anti CD20 or Anti CD38 monoclonal antibodies; for these individuals, a four to a six-month interval between medication and injection is advised (Negahdaripour et al. 2021). It is crucial to complete the vaccination schedule. All types of vaccines are safe and effective for this population except for live-attenuated vaccines (Negahdaripour et al. 2021). Prior to transplantation—if possible—solid organ transplant recipients should be vaccinated in order to strengthen their immune response against COVID-19; however, acute cellular rejection episodes must be resolved, and high-dose corticosteroid use discontinued before vaccination can begin (Negahdaripour et al. 2021). In order to improve its effectiveness, it is advised that six weeks pass after surgery before administering the second dose of the vaccine (Negahdaripour et al. 2021).

9.8 Other Therapeutics Modalities for COVID-19 Management

9.8.1 Plasma

Filippatos et al. (2023) presented a meta-analysis of 34 randomized clinical trials which reported that patients with COVID-19 who received convalescent plasma in comparison with those treated with the standard of care had a 26% lower risk of requiring hospital care [RR = 0.74, 95% CI (0.56, 0.99)] (1), in addition to an 8% less risk of ICU-related disease progression in comparison with those treated with the standard of care ICU-related disease progression (Seth et al. 2020). However, there was no association with decreased 28 days mortality or improved hospital discharge (Seth et al. 2020). This suggests potential advantages when used early to stop the progression of a severe illness. In trials conducted in Europe, convalescent plasma was also significantly linked to better ICU-related outcomes. In the post-pandemic era, well-conducted prospective studies could clarify its potential advantages for subpopulations.

9.8.2 Interferons

Interferons are the most conservative and evolved system for fighting viruses. IFN-a, IFN-b, IFN-w, IFN-k, and IFN-t are examples of type I IFNs (de Veer et al. 2001). These IFNs engage the JAK/STAT pathway, which activates a variety of genes known as interferon-stimulated genes (ISGs) (de Veer et al. 2001). SARS-CoV-2 replication, viral antigen expression, viral load decrease, and plaque reduction are all reduced by type I IFNs (de Veer et al. 2001). In vitro, recombinant human interferon-a1b inhibits SARS-CoV-2 more effectively than IFN-a2b (de Veer et al. 2001). To induce synergistic consolidations, IFN-a can be coupled with medicines that influence viral RNA transcription, protein synthesis, and processing (de Veer et al. 2001).

FN-a treatment in patients with moderate-to-severe COVID-19 increased the number of patients discharged from the hospital (OR 26.6; 95% CI, 2.7–254.3; I 2 = 95%) (Mantlo et al. 2020). However, there is no significant change in the mortality rate (Mantlo et al. 2020).

9.8.3 Hydroxychloroquine (HCQ), Azithromycin

Because of encouraging findings from small trials, hydroxychloroquine (HCQ) has gained widespread attention as a possible therapy for COVID-19 (Devaux et al. 2020; Meo et al. 2020). Since then, HCQ has been regarded as a viable treatment drug for individuals with COVID-19, whether coupled or not with azithromycin (Meo et al. 2020).

Moraes et al. (2022) presented a systematic review and meta-analysis of nine randomized trials of COVID-19 patients treated with hydroxychloroquine (Moraes et al. 2022). Reported that there was no significant reduction in the mortality rate of COVID-19 in comparison with the standard of care (Moraes et al. 2022), while the rate of serious adverse effects was comparable in both HCQ and the standard of care groups (3.7% vs. 2.9%; RR 1.22; 95% CI: 0.76–1.96) (Moraes et al. 2022).

Fiolet et al. (2021) presented a systematic review and meta-analysis of 29 articles that compared the efficacy of chloroquine or hydroxychloroquine with or without azithromycin (Fiolet et al. 2021). The study reported an increased risk of mortality in the hydroxychloroquine with azithromycin group (RR = 1.27; 95% CI 1.04–1.54, $n = 7$ studies), compared to hydroxychloroquine group (RR = 0.83; 95% CI 0.65–1.06, $n = 17$ studies) (Fiolet et al. 2021).

9.8.4 Non-steroids

There is no evidence of a link between the use of NSAIDs, particularly ibuprofen, and the SARS-CoV-2 infection or its sequelae. The reasons for the recommendation against using ibuprofen to relieve pain (especially headache) and fever during SARS-CoV-2 infection included a suspicion that ibuprofen might increase the risk of acquiring SARS-CoV-2 infection through ACE-2 upregulation (Qiao et al. 2015) and a risk of greater severity of the disease, including an increased risk of death, theoretical arguments that were extended to all NSAIDs (Micallef et al. 2020a, b; Moore et al. 2020).

9.9 Conclusion

Various vaccinations have demonstrated their efficacy and safety in battling the COVID-19 pandemic, along with the necessary booster doses that help to confront new variants emerging throughout the world. To further reduce hesitation among those most likely to experience symptoms, responsible health institutions must effectively educate individuals about the benefits of taking these vaccines for strong immunity for all members of society.

References

Abu-Raddad LJ et al (2021) Effectiveness of the BNT162b2 COVID-19 vaccine against the B.1.1.7 and B.1.351 variants. N Engl J Med 385(2):187–189

Abu-Raddad LJ et al (2022) Effect of mRNA vaccine boosters against SARS-CoV-2 Omicron infection in Qatar. N Engl J Med 386(19):1804–1816

Ai-ris YC et al (2021) Immunogenicity of COVID-19 mRNA vaccines in pregnant and lactating women. JAMA 325(23):2370–2380

Anand P, Stahel VP (2021) The safety of COVID-19 mRNA vaccines: a review. Patient Saf Surg 15(1):1–9

Andrew MK et al (2019) Influenza vaccination in older adults: recent innovations and practical applications. Drugs Aging 36:29–37

Andrews N et al (2022) COVID-19 vaccine effectiveness against the Omicron (B.1.1.529) variant. N Engl J Med 386(16):1532–1546

Badr H et al (2021) Overcoming COVID-19 vaccine hesitancy: insights from an online population-based survey in the United States. Vaccines 9(10):1100

Bhiman JN et al (2023) Novavax NVX-COV2373 triggers neutralization of Omicron sub-lineages. Sci Rep 13(1):1222

Bookstein Peretz S et al (2021) Short-term outcome of pregnant women vaccinated with BNT162b2 mRNA COVID-19 vaccine. Ultrasound Obstet Gynecol 58(3):450–456

Boyarsky BJ et al (2021) Antibody response to 2-dose SARS-CoV-2 mRNA vaccine series in solid organ transplant recipients. JAMA 325(21):2204–2206

Callaway E, Mallapaty S (2021) Novavax COVID vaccine protects people against variants. Nature 590(17):10–38

CDC (2021) COVID-19 vaccination for pregnant people to prevent serious illness, deaths, and adverse pregnancy outcomes from COVID-19. Accessed 18 Apr 2023. https://emergency.cdc.gov/han/2021/han00453.asp

CDC (2022) Overview of COVID-19 vaccines. Accessed 19 Apr 2023. https://www.cdc.gov/coronavirus/2019-ncov/vaccines/different-vaccines/overview-COVID-19-vaccines.html

Chemaitelly H et al (2021) MRNA-1273 COVID-19 vaccine effectiveness against the B.1.1.7 and B.1.351 variants and severe COVID-19 disease in Qatar. Nat Med 27(9):1614–1621

Chen H et al (2021) Health belief model perspective on the control of COVID-19 vaccine hesitancy and the promotion of vaccination in China: web-based cross-sectional study. J Med Internet Res 23(9):e29329

Chenchula S et al (2022) Current evidence on efficacy of COVID-19 booster dose vaccination against the Omicron variant: a systematic review. J Med Virol 94(7):2969–2976

Ciabattini A et al (2018) Vaccination in the elderly: the challenge of immune changes with aging. Sem Immunol

Cunningham AL et al (2021) Vaccines for older adults. BMJ 372

Dagan N et al (2021) Effectiveness of the BNT162b2 mRNA COVID-19 vaccine in pregnancy. Nat Med 27(10):1693–1695

de Veer MJ et al (2001) Functional classification of interferon-stimulated genes identified using microarrays. J Leukoc Biol 69(6):912–920

Devaux CA et al (2020) New insights on the antiviral effects of chloroquine against coronavirus: what to expect for COVID-19? Int J Antimicrob Agents 55(5):105938

DeZure C (2023) COVID-19 vaccines in children. Pediatr Ann 52(1):e4–e7
Du M et al (2021) The association between risk perception and COVID-19 vaccine hesitancy for children among reproductive women in China: an online survey. Front Med 8:741298
Efuntoye O et al (2021) Review of COVID-19 vaccine. Niger J Med 34(4):356–361
Emary KR et al (2021) Efficacy of ChAdOx1 nCoV-19 (AZD1222) vaccine against SARS-CoV-2 variant of concern 202012/01 (B.1.1.7): an exploratory analysis of a randomised controlled trial. Lancet 397(10282):1351–1362
Farge G et al (2021) The health professional's view on the inclusion of age in the recommendations for pneumococcal vaccination: results of a cross-sectional survey in France. Geriatrics 7(1):4
Farzi R et al (2022) The role of antigen-presenting cells in the pathogenesis of COVID-19. Pathol Res Pract 233:153848
FDA (2021) Coronavirus (COVID-19) update: FDA authorizes additional vaccine dose for certain immunocompromised individuals. Accessed 17 Apr 2023. https://www.fda.gov/news-events/press-announcements/coronavirus-covid-19-update-fda-authorizes-additional-vaccine-dose-certain-immunocompromised
Fiolet T et al (2021) Effect of hydroxychloroquine with or without azithromycin on the mortality of coronavirus disease 2019 (COVID-19) patients: a systematic review and meta-analysis. Clin Microbiol Infect 27(1):19–27
Galmiche S et al (2022) Immunological and clinical efficacy of COVID-19 vaccines in immunocompromised populations: a systematic review. Clin Microbiol Infect 28(2):163–177
Gargano JW et al (2021) Use of mRNA COVID-19 vaccine after reports of myocarditis among vaccine recipients: update from the Advisory Committee on Immunization Practices—United States, June 2021. MMWR Morb Mortal Wkly Rep 70(27):977
Gray KJ et al (2021) Coronavirus disease 2019 vaccine response in pregnant and lactating women: a cohort study. Am J Obstetr Gynecol 225(3):303.e1–303.e17
Gray G et al (2022) Effectiveness of Ad26.COV2.S and BNT162b2 vaccines against Omicron variant in South Africa. N Engl J Med 386(23):2243–2245
Griva K et al (2021) Evaluating rates and determinants of COVID-19 vaccine hesitancy for adults and children in the Singapore population: strengthening our community's resilience against threats from emerging infections (SOCRATEs) cohort. Vaccines 9(12):1415
Guillon M, Kergall P (2021) Factors associated with COVID-19 vaccination intentions and attitudes in France. Public Health 198:200–207
Han X et al (2021) Analysis of COVID-19 vaccines: types, thoughts, and application. J Clin Lab Anal 35(9):e23937
Harvey WT et al (2021) SARS-CoV-2 variants, spike mutations and immune escape. Nat Rev Microbiol 19(7):409–424
Hossain MB et al (2021) COVID-19 vaccine hesitancy among the adult population in Bangladesh: a nationwide cross-sectional survey. PLoS ONE 16(12):e0260821
Huynh G et al (2022) Determinants of COVID-19 vaccine hesitancy among parents in Ho Chi Minh City, Vietnam. Postgrad Med 134(3):303–308
Jamieson DJ et al (2006) Emerging infections and pregnancy. Emerg Infect Dis 12(11):1638
Kamar N et al (2021) Three doses of an mRNA COVID-19 vaccine in solid-organ transplant recipients. N Engl J Med 385(7):661–662
Kim DK et al (2017) Advisory Committee on Immunization Practices recommended immunization schedule for adults aged 19 years or older—United States, 2017. MMWR Morb Mortal Wkly Rep 66(5):136
Lee M, You M (2022) Direct and indirect associations of media use with COVID-19 vaccine hesitancy in South Korea: cross-sectional web-based survey. J Med Internet Res 24(1):e32329
Liang C-K et al (2022) COVID-19 vaccines in older adults: challenges in vaccine development and policy making. Clin Geriatr Med 38(3):605–620
Madhi SA et al (2021) Efficacy of the ChAdOx1 nCoV-19 COVID-19 vaccine against the B.1.351 variant. N Engl J Med 384(20):1885–1898
Mantlo E et al (2020) Antiviral activities of type I interferons to SARS-CoV-2 infection. Antiviral Res 179:104811
Meo S et al (2020) Efficacy of chloroquine and hydroxychloroquine in the treatment of COVID-19. Eur Rev Med Pharmacol Sci 4539–4547
Meo S et al (2021) COVID-19 vaccines: comparison of biological, pharmacological characteristics and adverse effects of Pfizer/BioNTech and Moderna vaccines. Eur Rev Med Pharmacol Sci 1663–1669
Micallef J et al (2020a) COVID-19 and NSAIDs: primum non nocere. Therapies 75(5):514–515
Micallef J et al (2020b) Non-steroidal anti-inflammatory drugs, pharmacology, and COVID-19 infection. Therapies 75(4):355–362
Moore N et al (2020) Does ibuprofen worsen COVID-19? Drug Saf 43:611–614
Moraes VYD et al (2022) Hydroxychloroquine for treatment of COVID-19 patients: a systematic review and meta-analysis of randomized controlled trials. einstein (São Paulo) 20
Moreira ED Jr et al (2022) Safety and efficacy of a third dose of BNT162b2 COVID-19 vaccine. N Engl J Med 386(20):1910–1921
Mortaz E et al (2020) The immune response and immunopathology of COVID-19. Front Immunol 11:2037
Negahdaripour M et al (2021) Administration of COVID-19 vaccines in immunocompromised patients. Int Immunopharmacol 99:108021
Okamura S, Ebina H (2021) Could live attenuated vaccines better control COVID-19? Vaccine 39(39):5719–5726

Oliveira BLCA et al (2021) Prevalence and factors associated with COVID-19 vaccine hesitancy in Maranhão, Brazil. Rev Saúde Públ 55:12
Oliver S (2021) Evidence to recommendation framework: Pfizer-BioNTech COVID-19 booster dose. https://www.cdc.gov/vaccines/acip/meetings/downloads/slides-2021-9-23/03-COVID-Oliver.pdf
Ontario Ministry of Health (2022a) Medical exemptions to COVID-19 vaccination. Accessed 17 Apr 2023. https://health.gov.on.ca/en/pro/programs/publichealth/coronavirus/docs/vaccine/medical_exemptions_to_vaccination.pdf
Ontario Public Health (2022b) Focus on: COVID-19 vaccines: viral vector-based vaccines. Accessed 16 Apr 2023. https://www.publichealthontario.ca/-/media/documents/ncov/vaccines/2021/03/covid-19-focus-on-viral-vector-based-vaccines.pdf?la=en
Ontario Public Health (2022c) Focus on COVID-19 vaccines: protein subunit vaccines. Accessed 16 Apr 2023. https://www.publichealthontario.ca/-/media/Documents/nCoV/Vaccines/2022/05/focus-on-protein-subunit-vaccine.pdf?rev=6c02ca1793794b6aa257e605abc1645c&sc_lang=en
Osterholm MT et al (2012) Efficacy and effectiveness of influenza vaccines: a systematic review and meta-analysis. Lancet Infect Dis 12(1):36–44
Park JW et al (2021) MRNA vaccines for COVID-19: what, why and how. Int J Biol Sci 17(6):1446
Pegu A et al (2021) Durability of mRNA-1273-induced antibodies against SARS-CoV-2 variants (preprint)
Qiao W et al (2015) Ibuprofen attenuates cardiac fibrosis in streptozotocin-induced diabetic rats. Cardiology 131(2):97–106
Rehati P et al (2022) COVID-19 vaccine hesitancy among adolescents: cross-sectional school survey in four Chinese cities prior to vaccine availability. Vaccines 10(3):452
Röltgen K, Boyd SD (2021) Antibody and B cell responses to SARS-CoV-2 infection and vaccination. Cell Host Microbe 29(7):1063–1075
Sadoff J et al (2021) Safety and efficacy of single-dose Ad26.COV2.S vaccine against COVID-19. N Engl J Med 384(23):2187–2201
Sebghati M, Khalil A (2021) Uptake of vaccination in pregnancy. Best Pract Res Clin Obstet Gynaecol 76:53–65
Seth T et al (2020) Convalescent plasma therapy for COVID-19: a systematic review. Natl Med J India 33(4)
Sharma E et al (2022) Efficacy and safety of COVID-19 vaccines—an update. Diseases 10(4):112
Shimabukuro TT et al (2021) Preliminary findings of mRNA COVID-19 vaccine safety in pregnant persons. N Engl J Med
Shinde V et al (2021) Efficacy of NVX-CoV2373 COVID-19 vaccine against the B.1.351 variant. N Engl J Med 384(20):1899–1909
Silveira MM et al (2021) DNA vaccines against COVID-19: perspectives and challenges. Life Sci 267:118919
Toth-Manikowski SM et al (2022) COVID-19 vaccination hesitancy among health care workers, communication, and policy-making. Am J Infect Control 50(1):20–25
Tregoning JS et al (2021) Progress of the COVID-19 vaccine effort: viruses, vaccines and variants versus efficacy, effectiveness and escape. Nat Rev Immunol 21(10):626–636
Troiano G, Nardi A (2021) Vaccine hesitancy in the era of COVID-19. Public Health 194:245–251
Vanaparthy R et al (2021) Review of COVID-19 viral vector-based vaccines and COVID-19 variants. Infez Med 29(3):328
Vousden N et al (2021) The incidence, characteristics and outcomes of pregnant women hospitalized with symptomatic and asymptomatic SARS-CoV-2 infection in the UK from March to September 2020: a national cohort study using the UK Obstetric Surveillance System (UKOSS). PLoS ONE 16(5):e0251123
Walsh JC et al (2022) The psychological and behavioural correlates of COVID-19 vaccine hesitancy and resistance in Ireland and the UK. Acta Physiol (Oxf) 225:103550
Wang Y et al (2022) COVID-19 vaccine hesitancy and associated factors among diabetes patients: a cross-sectional survey in Changzhi, Shanxi, China. Vaccines 10(1):129
Watanabe A et al (2023) Assessment of efficacy and safety of mRNA COVID-19 vaccines in children aged 5 to 11 years: a systematic review and meta-analysis. JAMA Pediatr
WHO (2021) Vaccine efficacy, effectiveness and protection. Accessed 16 Apr 2023. https://www.who.int/news-room/feature-stories/detail/vaccine-efficacy-effectiveness-and-protection
WHO (2022a) The Oxford/AstraZeneca (ChAdOx1-S [recombinant] vaccine) COVID-19 vaccine: what you need to know. Accessed 16 Apr 2023. https://www.who.int/news-room/feature-stories/detail/the-oxford-astrazeneca-covid-19-vaccine-what-you-need-to-know
WHO (2022b) Good practice statement on the use of second booster doses for COVID 19 vaccines, 18 Aug 2022. World Health Organization
WHO (2023) WHO COVID-19 dashboard. Accessed 14 Apr 2023. https://covid19.who.int/
Yorsaeng R et al (2022) Immunogenicity of a third dose viral-vectored COVID-19 vaccine after receiving two-dose inactivated vaccines in healthy adults. Vaccine 40(3):524–530
Zambrano LD et al (2020) Update: characteristics of symptomatic women of reproductive age with laboratory-confirmed SARS-CoV-2 infection by pregnancy status—United States, January 22–October 3, 2020. MMWR Morb Mortal Wkly Rep 69(44):1641

10 Controversies in COVID-19: Diagnostic Evaluation and Management in Neuro-COVID

Jordan Yakoby and Juda Zurndorfer

Abstract

SARS-CoV-2 is a virus in the coronavirus family originating out of Wuhan, China clinically known as COVID-19. While traditionally thought of as a respiratory virus, COVID-19 can have a multi-organ impact due to its invasion and widespread distribution throughout the body and via the angiotensin converting enzyme. Neurologic events due to COVID-19 are common, especially in the critically ill, and are called Neuro-COVID. Among these events are peripheral and central nervous system diseases such as Guillain–Barré, ischemic stroke, and various types of encephalitis. The impact of Neuro-COVID is devastating and is often far more severe than its non-COVID-19 form. Immunosuppressive or immunomodulatory therapy is often a mainstay of treatment, such as for encephalitis and Guillain–Barré, respectively, while management may fall in line with conventional processes in most cases, such as ischemic stroke. Much remains to be studied about the evaluation and management of Neuro-COVID.

Keywords

Neuro-COVID · COVID neurologic complications · COVID Guillain–Barré · COVID encephalitis · COVID stroke · COVID diagnosis

10.1 Introduction

The worldwide coronavirus disease 2019 (COVID-19) outbreak unleashed devastation in its wake not seen to this degree in living memory. Neurological symptoms following infection have been extensively documented and range from minor symptoms such as anosmia and ageusia, due to debilitating illnesses such as stroke, encephalitis, and demyelinating disease. Together, the various neurologic syndromes resulting from COVID-19 have become known as Neuro-COVID. This chapter will address the phenomena of Guillain–Barré syndrome, ischemic stroke, and meningoencephalitis due to SARS-CoV-2 infection.

J. Yakoby (✉)
Department of Nursing, New York School of Career and Applied Studies, Touro University, 3 Times Square, 3rd Floor, Room 322, New York, NY 10044, USA
e-mail: jyakoby@touro.edu

J. Zurndorfer
Department of Nursing, New York School of Career and Applied Studies, Touro University, 3 Times Square, 3rd Floor, Room 320, New York, NY 10044, USA
e-mail: jzurndor@touro.edu

N. Rezaei (ed.), *The COVID-19 Aftermath*, Advances in Experimental Medicine and Biology 1457,
https://doi.org/10.1007/978-3-031-61939-7_10

10.2 Guillain–Barré Syndrome

10.2.1 Background

Guillain–Barré syndrome (GBS) is one of various clinical entities falling under the umbrella term Neuro-COVID. GBS is an autoimmune disorder characterized by polyneuropathy and progressive, usually symmetric ascending muscle weakness and absent or depressed tendon reflexes. However, this symptomatology is reflective of the most common variant, acute inflammatory demyelinating polyneuropathy. Other signs include dysautonomia, and variable degrees of weakness of bulbar, facial, and respiratory paresis. Patients often present up to 10 days after infection, though it is well-documented that presentation may occur as late as up to 24 days after infection (Yakoby et al. 2021; Chan et al. 2020). It is important to recognize that the term GBS is an umbrella term which encompasses variants of the disease, such as acute inflammatory demyelinating polyneuropathy (AIDP), acute motor axonal neuropathy (AMAN), acute motor and sensory axonal neuropathy (AMSAN), Miller-Fisher syndrome, pharyngeal-cervical-brachial variant, and Bickerstaff brainstem encephalitis (Sriwastava et al. 2021). The pathophysiology is believed to be cross-reactivity to the axons of peripheral nerves and demyelination after immunologic response to the viral infection. The acute phase of GBS may last up to 2 weeks (Chan et al. 2020).

The most common cause of acute flaccid paralysis across the globe is GBS. Most cases are seen in patients who first developed a viral upper respiratory illness with onset of subsequent GBS symptoms as mentioned above, though diarrheal illness is well-documented as well. The incidence of GBS is about 0.16–4 per 100,000 individuals (Ghazanfar et al. 2020). Especially in AIDP, *Campylobacter jejuni* is the most common offending pathogen, though Epstein-Barr virus, Zika virus, and cytomegalovirus have well-defined relationships with GBS as well (Yuki and Hartung 2012).

GBS Subtypes. AMAN is a pure motor disorder that also results in progressive symmetric weakness. Respiratory failure is common, and pediatric patients are primarily affected. Like AIDP, *C. jejuni* is a common offending pathogen. Additionally, many patients demonstrate evidence of anti-ganglioside autoantibodies, specifically against GD1a, GD1b, and GM1. AMSAN differs from AMAN in that sensory nerve roots are also involved, and those primarily affected are adults rather than children with poor recovery as a common feature. Miller-Fisher syndrome accounts for about 5% of all GBS causes and classically presents with ophthalmoplegia, areflexia, and ataxia. Other signs of bulbar dysfunction are common. This syndrome is characterized usually by the presence of anti-GQ1b ganglioside antibodies with recovery lasting weeks to months.

COVID-19 has now taken its place as one of various microbiological insults known to cause GBS. Because GBS usually presents after a recent illness, clinicians should suspect COVID-19 as the offending pathogen in any patient presenting with possible GBS who reports a recent COVID-19 infection. Since the beginning of the pandemic, there have been many studies, case reports, case series, and other evidence pointing to the association of peripheral nervous system involvement and COVID-19. COVID-19 has since been linked to GBS and various neuropathies and myopathies (Taga and Lauria 2022). While the literature indicates that classic signs of GBS infection include areflexia and albuminocytologic dissociation on cerebrospinal fluid (CSF) analysis, this is only present in about half of cases (Yuki and Hartung 2012).

Other coronavirus outbreaks have caused GBS, also with onset extending up to three weeks after viral symptom onset. Angiotensin converting enzyme (ACE2) is widely expressed throughout the body, including in the ventrolateral medulla, and respiratory and gastrointestinal tracts. COVID-19 associated Neuro-COVID may occur due to direct neuroinvasion through ACE2 or perhaps through an inflammatory mechanism involving inflammatory cytokines. ACE2 is also involved in the viral cycle of SARS-CoV-2 since the virus has a spike protein surface unit that binds to this protein. In previous coronavirus outbreaks, such as Middle East respiratory-related coronavirus (MERS) in the first decade of this century, GBS was likewise seen as a neurologic complication (van den Bergh and Pieret 2004). Trigeminal, olfactory, or gustative nerve endings may be a portal of entry to the nervous system, given the prominent symptoms of anosmia and ageusia.

10.2.2 Diagnostic Approach

Diagnosis of GBS has classically included CSF analysis and neurodiagnostic studies such as electromyography. When COVID-19 is suspected as an offending pathogen, a PCR test for SARS-CoV-2 should be obtained promptly. Otherwise, COVID-19 as an etiology of GBS does not alter the diagnostic pathway. GBS is primarily a clinical diagnosis, with diagnostic testing playing an assistive role, rather than any one test being able to definitively confirm or refute the diagnosis (Table 10.1). The early phase of the disease may present with normal or inconclusive CSF, serologic, and electrodiagnostic results (Rahimi 2020; Saeed et al. 2019; van den Bergh and Pieret 2004). Nevertheless, these studies have remained an important part of the diagnosis of GBS.

Instrumental to the diagnosis of GBS is the Brighton criteria (Table 10.2). The Brighton criteria is the gold standard to evaluate the degree of certainty for a diagnosis of GBS, with level 1 certainty indicating the highest degree of certainty. The Brighton criteria require the treating clinician to evaluate the combination of neurologic exam findings, electrodiagnostic results, and CSF analysis. It should be noted that there are various levels of certainty, which do not necessarily negate the diagnosis. GBS is a clinical diagnosis and there are numerous situations in which a recommended neurodiagnostic test may not be possible or practical at the time of presentation due to one or more patient or logistical factors.

Table 10.1 Suggested diagnostic work up in COVID-19 associated GBS

Category	Diagnostic test	Rationale or comments
Cerebrospinal fluid	Cell count Glucose Red blood cells Protein Culture	Evaluate for albuminocytologic dissociation (elevated CSF protein)
Complete blood count	Lymphopenia Thrombocytopenia	Elevations or derangements may be present if CNS involvement
Metabolic panel	Azotemia	
Serum inflammatory markers	Erythrocyte sedimentation rate (ESR) C-reactive protein (CRP) D-dimer Ferritin	
Thyroid studies	Thyroid stimulating hormone	Exclude hypothyroidism as a cause of neuropathic symptoms
Vitamin studies	Folate Vitamin B12	Hypovitaminosis B12 may cause demyelination, producing peripheral neuropathy (Leishear et al. 2013). Folate and B12 metabolism are intricately linked
Heavy metals	Mercury Cadmium Lead Arsenic	Consider if in developing country or in those with increased likelihood of occupational exposure to these neurotoxicants (Valentine 2021)
Other serologic tests	Glycosylated hemoglobin Anti-nuclear antibody HIV 1 and 2 by PCR	

Table 10.2 Brighton criteria

Diagnostic criteria	Level of diagnostic certainty			
Bilateral and flaccid weakness of limbs	1	2	3	4
Decreased or absent deep tendon reflexes in weak limbs	+	+	+	±
Monophasic course and time between onset-nadir 12 h to 28 days	+	+	+	±
CSF cell count < 50/μl	+	+[a]	−	±
CSF protein concentration > normal value	+	±[a]	−	±
NCS findings consistent with one of the subtypes of GBS	+	±	−	±
Absence of alternative diagnosis for weakness	+	+	+	+

[a] If CSF is not collected or results not available, nerve electrophysiology results must be consistent with the diagnosis Guillain–Barré syndrome (Fokke et al. 2013)
+ Present; − absent; ± present or absent

Cerebrospinal Fluid Analysis. A key investigation in the diagnosis of GBS is CSF analysis. Classically, albuminocytologic dissociation is found in GBS, though this may be absent in some cases, particularly early in the disease course (Saeed et al. 2019; Rahimi 2020). Specifically, as it relates to COVID-19, attention has been given to CSF testing for SARS-CoV-2 who tested positive by nasopharyngeal swab. One study found that none of the affected patients who tested positive by nasopharyngeal polymerase chain reaction (PCR) testing were found to have positive SARS-CoV-2 CSF results (Toscano et al. 2020). Likewise, a multi-center study seeking to characterize COVID-19 patients' CSF profiles with various levels of neurologic involvement failed to demonstrate a single positive CSF SARS-CoV-2 result out of 76 samples. However, exceptions to this have been noted and there have been positive CSF SARS-CoV-2 results in some patients. Notwithstanding, neuropathologic investigation has found direct evidence of brain invasion with SARS-CoV-2 with negative CSF studies Therefore, the utility of routine SARS-CoV-2 testing in the CSF is unclear (Taga and Lauria 2022; Lou et al. 2021).

These findings are not surprising given that there is also a frequent lack of serologic evidence of a systemic inflammatory process in such peripheral neuropathies compared to other COVID-19 presentations, in which a markedly elevation in systemic inflammatory markers is noted (Mao et al. 2020). The investigation of other CSF cytokines in the diagnostic evaluation of GBS is still being investigated. The routine analysis of CSF SARS-CoV-2 assays cannot be used to refute or confirm the diagnosis of COVID-19 associated GBS. Until a body of literature on these issues is more fully developed, the use of these tests may confound the clinical picture more than any potential benefit. Likewise, with the increasing number of patients on anticoagulant medications or with other bleeding diatheses, prompt CSF analysis may not be possible on a significant number of affected patients.

Electrodiagnostic Studies. Nerve conduction studies (NCS) feature various electrodes placed along different locations of a peripheral nerve distribution, with the nerve stimulated at one site and recorded at the other. Along with NCS testing is the electromyogram (EMG), which provides valuable diagnostic information. The analysis includes F wave, H reflex, sensory nerve conduction, and motor nerve conduction which can be interpreted according to latency, velocity, and amplitude to assist in reaching a diagnosis of neuromuscular disease. Typically, electrodiagnostic findings include depressed or absent F or H responses, conduction block, and a reduction in velocity on NCS consistent with a demyelinating pattern (Nagappa et al. 2016).

Evidence of demyelination in these investigations might also include delayed distal latency and temporal dispersion of waveforms. It is essential to note that findings may be normal initially, with a delay of a few weeks before fibrillation develops and thus a normal study does not exclude the diagnosis of GBS (Nagappa et al. 2016).

Anti-ganglioside Antibodies. While CSF analysis is a primary study of interest, there has been some interest in analysis of anti-ganglioside antibody levels to assist in diagnosis. This has been controversial. A prospective study evaluating whether analysis of anti-ganglioside antibodies in patients suspected to have GBS found that serum anti-ganglioside antibodies were negative in all patients, leading to the conclusion that anti-ganglioside antibodies cannot be used in the diagnostic evaluation of GBS (Saeed et al. 2019). Among the threats to the study was low statistical power given the small sample ($n = 15$), as well as possible selection threat given that only patients with equivocal CSF or nerve conduction studies (NCS) results were analyzed. Therefore, from this study alone, we cannot generalize whether ganglioside autoantibody measurement can be helpful in the evaluation of GBS.

Nevertheless, in this author's experience, anti-ganglioside antibody measurement was likewise unhelpful in Neuro-COVID of peripheral nervous system etiology, such as GBS (Yakoby et al. 2021). However, while positive serum anti-ganglioside antibodies were found in a sample of GBS associated with Zika virus, it was likewise found in Zika virus patients without GBS manifestations, though patients who subsequently went on to develop GBS had higher levels of anti-ganglioside antibodies (Rivera-Correa et al. 2019). Given the relative rareness of the disease, it is unlikely that a randomized control trial will be forthcoming to elucidate this clinical problem. The current lack of strong levels of evidence and the conflicting existing evidence makes it difficult to recommend the routine use of anti-ganglioside antibodies to confirm or refute the diagnosis of COVID-19 Associated GBS.

Serum Laboratory Studies. Lymphopenia, thrombocytopenia, azotemia, and elevated inflammatory markers, such as ferritin, d-dimer, CRP, and ESR have been recommended as part of the work up for COVID-19 and trending throughout the course of the disease. However, these findings are more commonly found in patients with central nervous system involvement rather than those with peripheral nervous involvement, such as GBS (Mao et al. 2020; Yakoby 2021). Therefore, the routine evaluation of inflammatory markers in COVID-19 Associated GBS is unlikely to be helpful.

However, there are a number of other etiologies of polyradiculopathy and myelopathy that should be considered in the patients with such symptoms. Vitamin B12 and folate are intimately linked, and hypovitaminosis B12 may result in peripheral nerve demyelination resulting in symptoms that parallel those of GBS (Leishear et al. 2013). Therefore, vitamin B12 and folate should be routinely checked to exclude these vitamin derangements as potential confounders.

10.2.3 Treatment

IVIg. The fundamental principle of the management of GBS does not vary according to whether it is COVID-19 associated. Two options are available: either intravenous immunoglobulin therapy (IVIg) or plasmapheresis. IVIg is the preferred initial treatment regimen for many clinicians since it requires fewer preparatory arrangements to institute. The dosage is most commonly 2 g/kg in five divided daily doses. In patients who initially appear to demonstrate inadequate response, some advocate for a second round of IVIg, though this is not a universally accepted practice. There is an elevated risk of thrombosis and patients should be tested for immunoglobulin A deficiency prior to initiation of therapy. Renal function should be monitored closely for the duration of therapy and thereafter. Close monitoring by nursing staff is especially recommended during initiation of therapy, with special attention after the first several minutes of infusion initiation.

Plasmapheresis. Plasma exchange is the other possible definitive management strategies. To date, there is no strong scientific literature indicating whether it is more advisable to initiate plasmapheresis or IVIg first. Most such decisions come down to clinician preference, though it is theorized that the initiation of plasmapheresis after apparent failure of IVIg may remove the IVIg from the circulation, interfering with any possible additive effect of the two interventions together. Still others argue that the benefits of IVIg may occur in a delayed fashion, with too often a hasty declaration of failure made in clinical practice.

Nevertheless, plasmapheresis may result in the removal of autoantibodies responsible for axonal demyelination, removal of circulating cytotoxic elements as well as autoimmune complexes. Plasmapheresis may be implemented over a ten day to two-week course using either 5% albumin or fresh frozen plasma. Plasmapheresis is equally as effective as IVIg, though some studies have indicated a speedier recovery of neurologic function. Plasmapheresis requires insertion of a central venous catheter, which can be associated with iatrogenic risks of infection, pneumothorax, pain, and vascular injury. Furthermore, fibrinogen levels should be frequently monitored when albumin is used as the replacement product. There is also a risk of transfusion reaction, and the patient should be monitored for evidence of hypocalcemia.

Corticosteroids. Corticosteroids have no proven benefit in the management of GBS. However, corticosteroids may be indicated to manage other concomitant COVID-19 associated issues, including any pulmonary involvement resulting thereof. The use of corticosteroids in those with severe COVID-19 infection has been found to promote longer ventilator-free days and lower mortality rather than those not treated. Additionally, clinicians may choose to start Remdesivir, a viral inhibitor for several viral families. Therapy may begin at an initial intravenous dose of 200 mg followed by 100 mg daily for five to ten days and has been demonstrated to improve time to clinical improvement and perhaps reduce the incidence of further adverse events (Ribeiro da Silva et al. 2022).

Hemodynamic Complications and Ventilator Management. Nearly a third of GBS patients require mechanical ventilation. Respiratory failure in GBS occurs due to demyelination of the phrenic nerve leading to diaphragmatic weakness and respiratory failure (Khanna et al. 2017). Bulbar dysfunction leading to dysphagia and weakness of the upper airway structures markedly increases the risk of aspiration, decreased pulmonary compliance, and poor secretion management, further contributing to respiratory failure. Clinicians should be aware of physical examination changes. Early manifestations of impending respiratory failure often do not include classic signs such as accessory muscle recruitment and hypoxemia.

Rather, early manifestations may include an inability to speak in full sentences, tachypnea, and tachycardia. Predictors of the need for mechanical ventilation include rapid symptom progression, bulbar weakness, and evidence of progressive respiratory failure, as measured by negative inspiratory force (NIF) and vital capacity measurements (VC). These pulmonary function tests should be performed at least every four to six hours. Normal NIF values in the adult patient range from − 60 to 100 cm H_2O and normal VC values are 60–70 mL/kg.

Special attention should be given to the patient with VC < 20 mL/kg, NIF < 30 cm H_2O, or maximum expiratory pressure < 40 cm H_2O, or the so-called 20–30–40 rule (Lawn et al. 2001). This rule proposed that patients who meet any of those criteria should be electively intubated, though the rule has since been heavily criticized on the basis of research methodological issues in the paper. Clinicians should use their clinical judgment, using the combination of physical assessment findings along with pulmonary function testing to determine the appropriateness of mechanical ventilation in GBS patients. Unlike certain pulmonary disease states, in a pure GBS presentation, there is no special ventilatory strategy required. Patients requiring intubation and mechanical ventilation should be assessed for extubation potential as soon as practical, though it should be noted that many such patients will require tracheostomy due to ongoing diaphragmatic or pharyngeal and laryngeal neuromuscular weakness.

Tachyarrhythmia and bradyarrhythmia can occur due to autonomic involvement. Other signs of autonomic derangement include wide variability in blood pressure, orthostasis, thermoregulatory dysfunction, urinary retention, and ileus. Supportive care should be instituted to manage these complications. These issues are often self-limited to the acute phase of the illness.

10.3 Acute Stroke in Neuro-COVID

10.3.1 Background

Ischemic stroke is a common cerebrovascular disease in the United States. Ischemic stroke occurs when a cerebral artery experiences sudden deprivation of blood flow. This results in blood and oxygen deprivation to the affected area of the brain resulting in sudden loss of neurologic function. Among the top causes of disability, it is the fifth leading cause of death and afflicts approximately 800,000 individuals annually, with about 140,000 of these cases resulting in death. Among these patients, African American patients have a higher incidence of stroke and stroke-related death (Yang et al. 2017). Risk factors include hypertension, hyperlipidemia, diabetes, and smoking, especially for stroke due to small vessel disease. In addition to these risk factors, cardiogenic stroke, as in the case of atrial fibrillation, is a major cause of large vessel occlusion (Grotta et al. 2021). Risk factors are classified as modifiable and non-modifiable risk factors. Modifiable risk factors include those which are manageable, such as hypertension, hyperlipidemia, diabetes, smoking, obesity, and arrhythmia. Non-modifiable risk factors predisposing to stroke include older age, African American race, male gender, history of prior stroke, or strong family history of cerebrovascular disease (Yang et al. 2017).

However, COVID-19 associated stroke is a special subset of this disease and has a different pathophysiology. The incidence of stroke in COVID-19 patients is approximately 1.5%, though the literature indicates that the incidence of stroke in these patients may vary according to acuity, drawing a distinction between critically ill patients and those in less acute settings (Sagris et al. 2021). In fact, those with severe COVID-19 are five times more likely to suffer from a stroke (Nannoni et al. 2021), with a predilection for patients who are younger and more male compared to the general stroke population. Furthermore, such patients often have poorer outcomes and more extensive neuroradiographic findings. This is potentially because of a prothrombotic state due to COVID-19 infection. Immunologically mediated hypercoagulopathy is a commonly understood etiology of stroke, though inflammatory mechanisms, cardioembolism, and COVID-19 cardiac injury, renin–angiotensin–aldosterone system activation are alternative pathophysiologies (Sagris et al. 2021).

As stated previously, the spike protein allows SARS-CoV-2 to bind to ACE2, allowing for viral entry into host cells, including the alveoli and various other bodily structures. The release of the virus into the systemic circulation after intracellular replication triggers the cytokine release syndrome, involving interleukins 1β and 6 as well as tumor necrosis factor. It has been found that critically ill patients have higher levels of inflammatory markers, and its binding to ACE2 includes the endothelium, likely explaining the circulatory issues associated with COVID-19. This endothelial injury also results in increased von Willebrand factor and thrombomodulin and platelet activation. The extrinsic coagulation cascade results in increased thrombin formation. The result is an immunologically mediated coagulation cascade in overdrive resulting in ischemic complications, including stroke (Sagris et al. 2021; Nannoni et al. 2021).

The renin–angiotensin–aldosterone system is a potential culprit of COVID-19 associated stroke. Renin is a protein which is responsible for the conversion of the progenitor angiotensinogen to angiotensin I, which is itself converted into angiotensin II by ACE2, which is the receptor of choice for SARS-CoV-2 as indicated earlier. ACE2 is downregulated by SARS-CoV-2 which is of great importance since ACE2 has a key role in decreasing platelet adhesion and aggregation. Therefore, in addition to the intense inflammatory effect of cytokine release, which results in endothelial injury and inflammation, the downregulation of ACE2 results in increased thrombotic potential which greatly contributes to stroke risk and other thromboembolic events. For these same reasons, cardiac injury can result in coronary ischemia; as related to stroke inflammatory effects on the heart can result in intracardiac thrombi with cerebrovascular embolization (Zhang et al. 2021).

In the general population, though, ischemic stroke is instead frequently a result of atherosclerosis or embolism and is the more common type of stroke as compared to the pathophysiology described above or intracerebral hemorrhage. This is further categorized into large vessel, small vessel (or so-called lacunar), and cardioembolic infarcts (Adams et al. 1999). The Circle of Willis is an anastomotic connection of the various large cerebral arteries, providing alternate pathway for blood flow in case of incompetency or blockage (Fig. 10.1). The signs and symptoms of ischemic stroke vary according to the vascular territory affected, though affected patients may present with signs including sudden alteration in consciousness, aphasia, dysarthria, facial droop, focal weakness, and visual field defects (Benjamin et al. 2017).

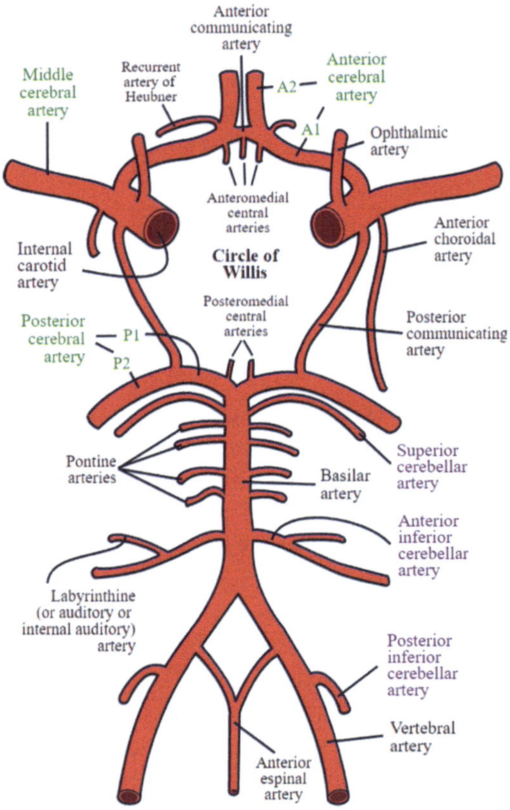

Fig. 10.1 Circle of Willis. From https://www.ncbi.nlm.nih.gov/books/NBK534861

10.3.2 Diagnostic Approach

Initially, patients presenting with neurologic deficit should receive a non-contrast computed tomography (CT) scan. Vascular neuroimaging such as a CT angiogram and CT perfusion scan should also be obtained to determine whether a large vessel occlusion is present. Facilities without these capabilities should transfer patients to the nearest comprehensive stroke center for more advanced evaluation and management. Magnetic resonance imaging (MRI) can be obtained during the hospitalization to further characterize the stroke.

10.3.3 Treatment

In the hyperacute phase of ischemic stroke, clinicians should proceed with emergent revascularization therapies as with any stroke patient. These therapies include endovascular thrombectomy or intravenous fibrinolysis with alteplase or tenecteplase, though it should be noted that the latter does not yet have US Food and Drug Administration approval for stroke (Hailu et al. 2022). Criteria for intravenous thrombolysis continues to include, but is not limited to onset of less than 4.5 h, absence of subarachnoid hemorrhage symptomatology, no recent myocardial infarction, novel oral anticoagulant use, INR > 1.7, history of intracranial hemorrhage, and thrombocytopenia (Adams et al. 2007). Endovascular thrombectomy up to 24 h post-symptom onset with favorable tissue-based imaging profiles may also be offered to patients with large vessel occlusions amenable to treatment (Nogueira et al. 2018). Permissive hypertension is generally recommended in ischemic stroke, with systolic blood pressure less the 220 mmHg. There is no recommendation to establish a lower blood pressure parameter in cases of large hemispheric stroke. Corticosteroid therapy is best avoided in ischemic stroke, with guidelines indicating a strong recommendation for this point (Torbey et al. 2015).

Literature during the early phase of the pandemic suggested prophylactic therapeutic anticoagulation in patients afflicted with severe COVID-19 and evidence of coagulopathy. Results indicated better prognosis and lower mortality in patients treated with therapeutic anticoagulation (Tang et al. 2020; Thachil 2020). However, correlation does not denote causality, and despite the predilection for younger patients, stroke in this setting is often confounded by other comorbid conditions which could plausibly serve as the etiology of a patient's stroke. Nevertheless, given the prothrombotic state, anticoagulation has been recommended (Wang et al. 2020). At this time, there is no concrete evidence to routinely recommend systemic anticoagulation on every patient suffering ischemic stroke in the presence of COVID-19. It is imperative for the clinician to make an individual determination as to the likelihood of hypercoagulable or cardioembolic etiologies and the appropriateness of therapeutic anticoagulation on a case-by-case basis.

Clinicians must also judge whether the detection of SARS-CoV-2 by nasopharyngeal swab is a finding incidental to routine testing or whether the patient truly has severe COVID-19. The implications of these possibilities may drive the management of the patient. Anti-platelet choices include aspirin, clopidogrel, and cilostazol. The timing of anticoagulation following large ischemic stroke, if appropriate, is the subject of ongoing debate, especially for large hemispheric stroke (Torbey et al. 2015). In cases where anticoagulation is appropriate, the decision on when to 'step down' to anti-platelet therapy can vary due is unclear or may be driven by concomitant thromboses elsewhere, such as venous thromboembolism. In such cases, clinicians may choose to de-escalate from anticoagulation according to established guidelines surrounding the management of provoked venous thromboembolism.

10.4 Meningoencephalitis in Neuro-COVID

10.4.1 Background

Encephalitis has likewise been demonstrated as a complication of COVID-19 infection. While overall incidence was only approximately 0.215%, this number spiked to 6.7% among critically ill COVID-19 patients. The typical onset was about two weeks after initial onset of COVID-19 symptoms (Siow et al. 2021). As is typical for most encephalitis presentations, symptoms included seizures, depressed consciousness, focal weakness, and headache, though other deficits were reported in varying degrees of frequency. Furthermore, COVID-19 has been associated with various types of encephalitis.

Limbic encephalitis, anti-N-methyl-D-aspartate (NMDA) receptor encephalitis, acute disseminated encephalomyelitis (ADEM), necrotizing encephalitis, and unspecified types have all been reported in associated with COVID-19 (Nabizadeh et al. 2022). Cytokine storm associated with severe COVID-19 infection potentially may disrupt an otherwise intact blood–brain barrier allowing for direct viral invasion into the CNS. Alternatively, the severe inflammatory cascade unleashed by cytokine release syndrome could cause blood–brain barrier breakdown and direct neuroinflammation in the absence of viral infection into CNS tissues, which could explain case reports of negative CSF results in some encephalitis patients. Another possibility as mentioned previously is olfactory transmission resulting in direct neuroinvasion is a plausible mechanism as well. It should also be noted that anti-NMDA receptor antibodies have been provoked by SARS-CoV-2 resulting in encephalitis. In addition to the usual elevations in serum inflammatory markers and cytokine levels well-documented in the literature, CSF findings are those classic of an encephalitic picture with elevated protein, preserved glucose levels, but also elevated red blood cell and white blood cell levels, though this finding was not neutrophilic. Furthermore, oligoclonal bands were reported in a few studies (Siow et al. 2021).

10.4.2 Diagnostic Approach

Consideration should be given to the measurement and trending of serum inflammatory markers such as ferritin, d-dimer, lactate dehydrogenase, and CRP. Neuroimaging is a key investigation in the patient with suspected encephalitis. Findings may include high signal on FLAIR and T2 along with hemosiderin deposition on gradient echo sequences (Kihira et al. 2020). Likewise, CSF should be sampled, provided there is no contraindication, such as clinical or radiographic evidence of elevated intracranial pressure. Studies should include protein, glucose, cell count, and culture. Specialized tests to exclude herpes simplex virus and other viral encephalitides should be strongly considered.

Suggested CSF analyses for investigation also include oligoclonal bands, anti-NMDA receptor antibodies, cytomegalovirus, varicella, and Epstein-Barr. Some tertiary or quaternary centers may have specialized meningitis/encephalitis panels available for analysis. Measurement of CSF cytokines, such as interleukin 1, interleukin 6 among others are possible, though the likely need to send to a reference laboratory likely mitigate against the clinical utility of these tests. Furthermore, there is no evidence as of yet that this has any real diagnostic utility or yields any differences in outcomes (Guasp et al. 2022). Finally, strong consideration should be given to continuous electroencephalography (EEG) given the propensity toward seizures, including subclinical epileptiform activity, seizures, or status epilepticus.

10.4.3 Treatment

Various regimens for the management of patients with encephalitis have been proposed. Immunosuppression is generally recommended, and one regimen has included pulse dose corticosteroid therapy with 1 g methylprednisolone daily for up to 10 days. However, IVIg and plasma exchange have both been utilized with positive patient outcomes (Siow et al. 2021). The regimen for these two management strategies is that which was delineated in the discussion on GBS. Finally, the successful use of the chemotherapeutic agent rituximab has been documented in a case report (Haider et al. 2020). It should be noted that the recommended treatment pathways above have not yet been supported with randomized control trails and still rely on lower levels of scientific evidence. Nevertheless, this type of scientific investigation is not likely forthcoming imminently and the benefits of this therapy are likely beneficial and outweigh the associated risks of immunosuppression given the critical brain injury and high morbidity and mortality associated with the disease process, particularly left untreated.

Patients found to be in status epilepticus should be treated emergently. Attention should first be turned to airway, breathing, and circulation. First line therapy for status epilepticus is a benzodiazepine such as lorazepam 0.1 mg/kg up to 4 mg maximum. Alternatively, diazepam 0.15 mg/kg up to 10 mg can be given intravenously or intramuscularly. Midazolam is a preferred intramuscular agent with a dose of 0.2 mg/kg. Clinicians should be aware that hypotension and respiratory insufficiency can occur with the administration of these medications and should again be prepared to support airway and breathing as part of the management of the patient. Second-line agents should be prepared concurrently and can also be used in the management of status epilepticus. Preferred agents and dosages include the intravenous administration of levetiracetam 60 mg/kg, fosphenytoin 20 mg PE/kg, valproate sodium 40 mg/kg. Other agents recommended in the literature include intravenous lacosamide and phenobarbital, though there should be a low threshold at this juncture to proceed to endotracheal intubation with general anesthesia for patients with refractory status epilepticus (Brophy et al. 2012). Ongoing management should be tailored under continuous EEG guidance.

10.5 Conclusion

Neuro-COVID is an umbrella term for a variety of severe neurologic disease resulting from COVID-19 infection. These include ischemic stroke, Guillain–Barré syndrome, and encephalitis. The incidence of Neuro-COVID varies according to the specific underlying pathology, though is more common in patients afflicted with severe disease. There is great controversy surrounding diagnostic strategies, particularly as they relate to the detection of SARS-CoV-2 in CSF. To date, there is no strong literature supporting the routine use of SARS-CoV-2 testing in any other fashion aside from nasopharyngeal testing, as the clinical picture is the key factor in making the determination that a neurologic illness is secondary to COVID-19. While specialized testing with cytokines has been investigated, there is no strong support to date to support its routine use at the bedside. Treatment of Neuro-COVID is largely supportive, though there is mounting evidence that treatment modalities such as plasmapheresis, IVIg are beneficial in certain disease states, while more conventional management is more appropriate in cases such as ischemic stroke. Nevertheless, ongoing rigorous research designs are needed to study the diagnostic approach and management of Neuro-COVID further. It is imperative that clinicians remain vigilant to the development of neurologic symptoms, as these most often follow onset of symptoms by many days to a couple weeks. Vulnerable, critically ill patients in whom multiple confounding factors are present may be diagnosed late. Neuro-COVID is especially prevalent among the critically ill, and monitoring for neurologic involvement can lead to quicker detection and treatment, potentially improving patient outcomes.

Abbreviations

ACE2	Angiotensin converting enzyme 2
AIDP	Acute inflammatory demyelinating polyneuropathy
AMAN	Acute motor axonal neuropathy
AMSAN	Acute motor and sensory axonal neuropathy
CNS	Central nervous system
CRP	C-reactive protein
CSF	Cerebrospinal fluid
CT	Computed tomography
EEG	Electroencephalogram
EMG	Electromyogram
ESR	Erythrocyte sedimentation rate
GBS	Guillain–Barré syndrome
MERS	Middle East respiratory-related coronavirus syndrome
NCS	Nerve conduction study
NIF	Negative inspiratory force
NMDA	N-methyl-D-aspartate
PCR	Polymerase chain reaction
VC	Vital capacity

References

Adams H, Davis P, Leira E, Chang K, Bendixen B, Clarke W et al (1999) Baseline NIH stroke scale score strongly predicts outcome after stroke: a report of the trial of ORG 10172 in acute stroke treatment (TOAST). Neurology 53(1):126–131

Adams H, del Zoppo G, Alberts MJ, Bhatt DL, Brass L, Furlan A et al (2007) Guidelines for the early management of adults with ischemic stroke: a guideline from the American Heart Association/American Stroke Association Stroke Council, Clinical Cardiology Council, Cardiovascular Radiology and Intervention Council, and the Atherosclerotic Peripheral Vascular Disease and Quality of Care Outcomes in Research Interdisciplinary Working Groups. Stroke 38(5):1655–1711

Benjamin E, Blaha M, Chiuve S, Muntner P et al (2017) American Heart Association statistics committee and stroke statistics subcommittee. Heart disease and stroke statistics—2017 update: a report

from the American Heart Association. Circulation 135(10):e146–e603
Brophy G, Bell R, Claassen J et al (2012) Guidelines for the evaluation and management of status epilepticus. Neurocrit Care 17(1):3–23
Chan M, Han SC, Kelly S, Tamimi M, Giglio B, Lewis A (2020) A case series of Guillain–Barré syndrome following COVID-19 infection in New York. Neurol Clin Pract 11(4):e576–e578
Fokke C, van den Berg B, Drenthen J, Walgaard C, van Doorn P, Jacobs B (2013) Diagnosis of Guillain–Barré syndrome and validation of Brighton criteria. Brain 137(1):33–43
Garg M (2017) Respiratory involvement in Guillain–Barré syndrome: the uncharted road to recovery. J Neurosci Rural Pract 8(3):325–326
Ghazanfar H, Qazi R, Ghazanfar A, Iftekhar S (2020) Significance of Brighton criteria in the early diagnosis and management of Guillain–Barré syndrome. Cureus 12(5):e8318
Grotta J, Albers G, Broderick J, Day A, Kasner S, Lo E, Sacco R, Wong L (2021) Stroke: pathophysiology, diagnosis, and management, 7th edn. Elsevier, Philadelphia
Guasp M, Munoz-Sanchez G, Martinez-Hernandez E et al (2022) CSF biomarkers in COVID-19 associated encephalopathy and encephalitis predict long-term outcome. Front Immunol 13:866153
Haider A, Siddiqa A, Ali N, Dhallu M (2020) COVID-19 and the brain: acute encephalitis as a clinical manifestation. Cureus 12(10):e10784
Hailu K, Cannon C, Hayes S (2022) Tenecteplase use in the management of acute ischemic stroke: literature review and clinical considerations. Am J Health Syst Pharm 12(15):944–949
Khanna M, Rawat N, Gupta A, Nagappa M, Taly A, Rukmani MR, Sathyaprabha TN, Haldar P (2017) Pulmonary involvement in patients with Guillain–Barré syndrome in subacute phase. J Neurosci Rural Pract 8(3):412–416
Kihira S, Delman B, Belani P et al (2020) Imaging features of acute encephalopathy in patients with COVID-19: a case series. Am J Neuroradiol 41(10):1804–1808
Lawn ND, Fletcher DD, Henderson RD, Wolter TD, Wijdicks EF (2001) Anticipating mechanical ventilation in Guillain–Barré syndrome. Arch Neurol 58(6):893–898
Leishear K, Boudreau R, Studenski S, Ferrucci L et al (2013) The relationship of vitamin B12 and sensory and motor peripheral nerve function in older adults. J Am Geriatr Soc 60(6):1057–1063
Lou J, Movassaghi M, Gordy D, Olson M, Zhang T, Khurana M, Chen Z, Perez-Rosendahl M, Thammachantha S, Singer E, Magaki S, Vinters H, Yong W (2021) Neuropathology of COVID-19 (neuro-COVID): clinicopathological update. Free Neuropathol 2(2)
Mao L, Jin H, Wang M et al (2020) Neurologic manifestations of hospitalized patients with coronavirus disease 2019 in Wuhan, China. JAMA Neurol 77:683–690
Nabizadeh F, Balabandian M, Sodeifian F, Rezaei N, Rostami M, Moghadasi A (2022) Autoimmune encephalitis associated with COVID-19: a systematic review. Mult Scler Relat Disord 62:103795
Nagappa M, Netto AB, Taly AB, Kulkarni GB, Rao U, Periyavan S, Rao S (2016) Electrophysiological observations in critically ill Guillain–Barré syndrome. Neurol India 64(5)
Nannoni S, de Groot R, Bell S, Markus HS (2021) Stroke in COVID-19: a systematic review and meta-analysis. Int J Stroke 16(2):137–149
Nogueira R, Jadhav A, Haussen D, Bonafe A et al (2018) Thrombectomy 6 to 24 hours after stroke with a mismatch between deficit and infarct. N Engl J Med 378:11–21
Rahimi K (2020) Guillain–Barré syndrome during COVID-19 pandemic: an overview of the reports. Neurol Sci 41:3149–3156
Ribeiro da Silva S, Frutuoso do Nascimento JC, Germano Mendes RP, Miranda Guarines K, Targino Alves da Silva C, Gomes da Silva P, Ferraz de Magalhaes JJ, Vigar JRJ, Silva-Junior A, Kohl A, Pardee K, Pena L (2022) Two years into the COVID-19 pandemic: lessons learned. ACS Infect Dis 8(9):1758–1814
Rivera-Correa J, de Siqueira IC, Mota S, do Rosario M, de Jesus P, Alcantara L, Ernst J, Rodriguez A (2019) Anti-ganglioside antibodies in patients with Zika virus infection-associated Guillain–Barré syndrome in Brazil. PLoS Negl Trop Dis 13(9):e0007695
Saeed ML, Baloch BZ, Mahmud SN, Khan M, Qureshi M, Shad Z, Hussain S, Munawar K, Qadeer A, Abdullah A (2019) Role of anti-ganglioside antibodies in the diagnosis of Guillain–Barré syndrome as an alternate investigation. Cureus 11(5):e4625
Sagris D, Papanikolaou A, Kvernland A, Korompoki E, Frontera J, Troxel A, Gavriatopoulou M, Milionis H, Lip G, Michel P, Yaghi S, Ntaios G (2021) COVID-19 and ischemic stroke. Eur J Neurol 28(11):3826–3836
Siow I, Lee K, Zhang J, Saffari S, Ng A (2021) Encephalitis as a neurological complication of COVID-19: a systematic review and meta-analysis of incidence, outcomes, and predictors. Eur J Neurol 28(10):3491–3502
Sriwastava S, Kataria S, Tandon M et al (2021) Guillain–Barré syndrome and its variants as a manifestation of COVID-19: a systematic review of case reports and case series. J Neurol Sci 420:117263
Taga A, Lauria G (2022) COVID-19 and the peripheral nervous system. A 2-year review from the pandemic to the vaccine era. J Peripher Nerv Syst 27(1):4–30
Tang N, Bai H, Chen X, Gong J, Li D, Sun Z (2020) Anticoagulant treatment is associated with decreased

mortality in severe coronavirus disease 2019 patients with coagulopathy. J Thromb Haemost 18(5):1094–1099

Thachil J (2020) The versatile heparin in COVID-19. J Thromb Haemost 18(5):1020–1022

Torbey M, Bosel J, Rhoney D, Rincon F, Staykov D, Amar A, Varelas P et al (2015) Evidence-based guidelines for the management of large hemispheric infarction. Neurocrit Care 22:146–164

Toscano G, Palmerini F, Ravaglia S et al (2020) Guillain–Barré syndrome associated with SARS-CoV-2. N Engl J Med 382:2574–2576

Valentine W (2021) Toxic peripheral neuropathies: agents and mechanisms. Toxicol Pathol 48(1):152–173

van den Bergh YK, Pieret F (2004) Electrodiagnostic criterion for acute and chronic inflammatory demyelinating polyradiculoneuropathy. Muscle Nerve 29:565–574

Wang Z, Yang Y, Lian X, Gao B, Liu M, Li W, Chen Z, Wang Z (2020) COVID-19 associated ischemic stroke and hemorrhagic stroke: incidence, potential pathological mechanism, and management. Front Neurol 11:51996

Yakoby J (2021) SARS-COV-2-associated Guillain–Barré syndrome requires appropriate exclusion of possible differentials. J Emerg Med 61(6):794–796

Yakoby J, Litvak I, Yu E (2021) Guillain–Barré syndrome after novel coronavirus disease 2019. J Emerg Med 61(4):67–70

Yang Q, Tong X, Schieb L et al (2017) Vital signs: recent trends in stroke death rates—United States, 2000–2015. MMWR Morb Mortal Wkly Rep 66(35):933–939

Yuki N, Hartung H-P (2012) Guillain–Barré syndrome. N Engl J Med 366:2294–2304

Zhang S, Zhang J, Wang C, Chen X, Zhao X, Jing H, Liu H, Li Z, Wang L, Shi J (2021) COVID-19 and ischemic stroke: mechanisms of hypercoagulability (review). Int J Mol Med 47(3):21

Computational Drug Design Strategies for Fighting the COVID-19 Pandemic

11

Mohammed Hakmi, El Mehdi Bouricha, Abdellatif Soussi, Ilias Abdeslam Bzioui, Lahcen Belyamani, and Azeddine Ibrahimi

Abstract

The advent of COVID-19 has brought the use of computer tools to the fore in health research. In recent years, computational methods have proven to be highly effective in a variety of areas, including genomic surveillance, host range prediction, drug target identification, and vaccine development. They were also instrumental in identifying new antiviral compounds and repurposing existing therapeutics to treat COVID-19. Using computational approaches, researchers have made significant advances in understanding the molecular mechanisms of COVID-19 and have developed several promising drug candidates and vaccines. This chapter highlights the critical importance of computational drug design strategies in elucidating various aspects of COVID-19 and their contribution to advancing global drug design efforts during the pandemic. Ultimately, the use of computing tools will continue to play an essential role in health research, enabling researchers to develop innovative solutions to combat new and emerging diseases.

M. Hakmi (✉) · E. M. Bouricha · A. Ibrahimi
Medical Biotechnology Laboratory (MedBiotech), Faculty of Medicine and Pharmacy, Bioinova Research Center, Mohammed Vth University, Rabat, Morocco
e-mail: gml.hakmi@gmail.com

M. Hakmi · E. M. Bouricha · L. Belyamani · A. Ibrahimi
Mohammed VI Center for Research and Innovation (CM6), Rabat, Morocco

A. Soussi
Department of Informatics, Bioengineering, Robotics and Systems Engineering, University of Genoa, 16145 Genova, Italy

I. A. Bzioui
Department of Gynecology and Obstetrics, Faculty of Medicine, Abdelmalek Essaâdi University Hospital, Tangier, Morocco

L. Belyamani · A. Ibrahimi
Mohammed VI University of Health Sciences (UM6SS), Casablanca, Morocco

L. Belyamani
Emergency Department, Medical and Pharmacy School, Military Hospital Mohammed V, Mohammed V University, Rabat, Morocco

Keywords

COVID-19 · Pandemic · SARS-CoV-2 · CADD · QSAR · Vaccine · Drug

11.1 Introduction

The *Coronaviridae* family comprises of enveloped, positive-strand RNA viruses that infect amphibians, birds, and mammals. Depending on the host species, coronavirus infections can lead to varying symptoms and tissue tropism. These

N. Rezaei (ed.), *The COVID-19 Aftermath*, Advances in Experimental Medicine and Biology 1457,
https://doi.org/10.1007/978-3-031-61939-7_11

viruses are the largest RNA viruses known, with genomes ranging from 25 to 32 kilobases in length. The family is divided into two subfamilies, *Letovirinae* and *Orthocoronavirinae*, which contain four genera: *Alphacoronavirus*, *Betacoronavirus*, *Gammacoronavirus*, and *Deltacoronavirus* (Payne 2017). Human coronaviruses, including HCoV-229E and HCoV-OC43, have been known to circulate in the population for a long time and are responsible for seasonal respiratory tract infections that are typically mild and associated with symptoms of the common cold. However, in contrast, three highly pathogenic coronaviruses that have emerged in the last two decades—severe acute respiratory syndrome coronavirus (SARS-CoV), Middle East respiratory Syndrome coronavirus (MERS-CoV), and SARS-CoV-2—can cause severe and life-threatening respiratory pathologies and lung injuries in humans. These viruses infect cells in the bronchial epithelium, pneumocytes, and upper respiratory tract, and no approved prophylactic or therapeutic treatments for these diseases currently exist. The initial stages of coronavirus infection are marked by the specific binding of the coronavirus spike (S) protein to cellular entry receptors, which have been identified for several coronaviruses. These receptors include human aminopeptidase N(APN) for HCoV-229E, angiotensin-converting enzyme 2 (ACE2) for HCoV-NL63, SARS-CoV and SARS-CoV-2, and dipeptidyl peptidase 4 (DPP4) for MERS-CoV. The expression and tissue distribution of these entry receptors play a crucial role in determining viral tropism and pathogenicity (Chen et al. 2020; V'kovski et al. 2021).

SARS-CoV-2 is responsible for the COVID-19 pandemic that has had a significant impact on global health and society. The COVID-19 outbreak was initially detected in Wuhan, China, in December 2019, and rapidly spread to become a global pandemic. On January 30, 2020, the World Health Organization (WHO) declared the outbreak a Public Health Emergency of International Concern, and on March 11, 2020, it was declared a pandemic (Ciotti et al. 2020). The virus mainly spreads through respiratory droplets when an infected person talks, coughs, or sneezes, but can also be transmitted by touching a contaminated surface and then touching one's face. Symptoms of COVID-19 can range from mild to severe and include fever, cough, shortness of breath, and loss of taste or smell. In severe cases, the virus can lead to pneumonia, acute respiratory distress syndrome, and even death. The literature suggests that young adults of reproductive age are more susceptible to contracting COVID-19 than older individuals, with males having slightly higher rates of infection than females. Mortality rates tend to be higher in those of advanced age and in those with underlying comorbidities, including hypertension, diabetes, and obesity. Pregnant women in particular are at risk of developing similar comorbidities during pregnancy, including gestational diabetes, excessive weight gain, and hypertensive disorders such as gestational hypertension, preeclampsia, eclampsia, and HELLP syndrome (hemolysis, elevated liver enzymes, and low platelet count). Despite this knowledge, the impact of COVID-19 on the health of pregnant women remains poorly understood (Cavalcante et al. 2021; Dashraath et al. 2020; Smith et al. 2020).

As of April 2023, the pandemic has resulted in over 764 million confirmed cases and over 6.9 million deaths worldwide (WHO 2023), with the United States, India, and Brazil reporting the highest number of confirmed cases and deaths. The pandemic has also caused significant disruptions to economies, education, travel, and daily life (Esakandari et al. 2020; Rosa Mesquita et al. 2021). Numerous countries have taken measures such as social distancing, quarantine, and travel restrictions to counteract the pandemic. Furthermore, the development of vaccines and their authorization for emergency use has played a crucial role in reducing the spread of the virus. Despite this progress, there has been unequal distribution of the vaccines, leading to difficulties for many low-income countries in accessing them. The pandemic has emphasized the importance of global collaboration and preparedness in the face of infectious diseases, while also revealing disparities in healthcare systems and the need for increased

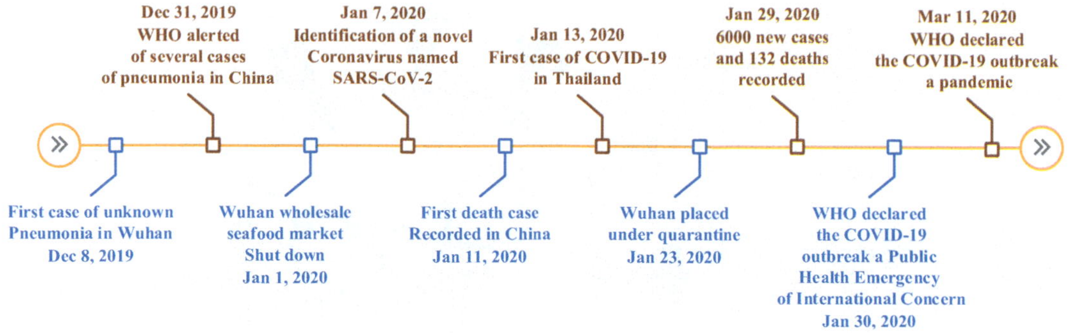

Fig. 11.1 Key events in the COVID-19 timeline

investment in public health infrastructure. As the pandemic persists, ongoing research and surveillance efforts are vital for monitoring the virus and devising novel treatments and vaccines to combat it (Ayouni et al. 2021; Girum et al. 2021) (Fig. 11.1).

11.2 Understanding the SARS-CoV-2 Virus

11.2.1 Structure and Life Cycle of the SARS-CoV-2 Virus

SARS-CoV-2 is an enveloped positive-sense single-stranded RNA virus that has a diameter ranging from 60 to 140 nm and possesses 9- to 12-nm-long spikes giving it a crown-like appearance, which is the origin of the name "coronavirus." The virus particle consists of several structural proteins, including the nucleocapsid (N), envelope (E), membrane (M), and spike (S) proteins (Fig. 11.2). The N protein is the only structural protein found inside the virion, while other structural proteins are located within the lipidic viral envelope. The N protein is associated with the viral genomic single-stranded RNA (ssRNA) and modulates its unwinding after the virus enters the cell. The E protein is involved in viral assembly, while the M protein plays a critical role in incorporating essential viral components into new virions during morphogenesis. Finally, the S protein binds, by its receptor-binding domain (RBD), to angiotensin-converting enzyme 2 (ACE2) that is expressed by host cells and facilitates the fusion of viral and cellular membranes (Pizzato 2022).

The S glycoprotein of the virus plays a pivotal role in its life cycle by binding to ACE2 (Laporte et al. 2021). This glycoprotein is composed of two subunits, S1 and S2, which have distinct functions in the viral entry process. Specifically, the S1 subunit binds the virion to the host cell receptor at the N terminus, while the S2 subunit mediates the fusion of the viral and cellular membranes at the C terminus. This allows the virus to fuse with the host cell membrane and release its RNA genome into the host cell. The S glycoprotein also has a polybasic furin cleavage site, known as S2′, between the S1 and S2 subunits, which is cleaved by cellular proteases. This cleavage results in the formation of a viral fusion peptide that is activated during virus endocytosis. Notably, the presence of such a furin cleavage site distinguishes SARS-CoV-2 from other coronaviruses, such as SAR-CoV-1, which has a monobasic cleavage site between the S1/S2 subunits. As a result, the entry of SARS-CoV-2 into susceptible cells is a complex process that involves receptor binding and proteolytic processing of S proteins (Haque et al. 2020; Almehdi et al. 2021).

Once the virus successfully enters the host cell, it relies on the host's cellular machinery to replicate its RNA genome and synthesize viral proteins. The viral RNA genome is translated into two large polyproteins, pp1a and pp1ab, which are subsequently cleaved by viral proteases to

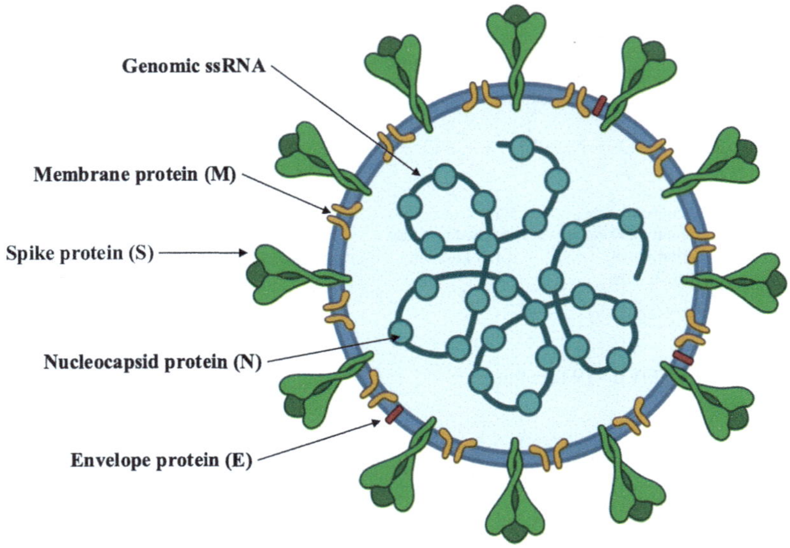

Fig. 11.2 SARS-CoV-2 structure. Image created with BioRender

form individual non-structural proteins (nsps). Polyproteins pp1a and pp1ab contain the nsps 1–11 and 1–16, respectively. These nsps perform crucial functions such as regulating transcriptional activity and suppressing the host cell's immune response. The viral polymerase initiates the transcription of a series of subgenomic messenger RNAs through a process known as discontinuous transcription, which are then translated into the viral structural proteins. The N protein forms a complex with the viral genomic RNA, while the S, E, and M proteins are integrated into the viral envelope at the intermediate compartments of the endoplasmic reticulum and Golgi. The newly assembled viral particles are released from the infected cells via the process of exocytosis (Pizzato 2022; Wang et al. 2020b; Brant et al. 2021). The manner in which virions are transported from the Golgi is uncertain, as it remains unclear whether the virus utilizes the same pathway as large cargo transport or employs a distinct and novel route (Perlman and Netland 2009; Fehr and Perlman 2015).

Coronaviruses encode two key proteases which facilitate the cleavage of the replicase polyproteins. These include the papain-like proteases (PLpro) that are encoded within nsp3, and a serine-type protease called the main protease (Mpro) that is encoded by nsp5. While most coronaviruses encode two PLpros within nsp3, the γ-coronaviruses, SARS-CoV, and MERS-CoV only express one PLpro. The PLpros are responsible for cleaving the nsp1/2, nsp2/3, and nsp3/4 boundaries, while the Mpro is involved in the remaining 11 cleavage events. As such, Mpro has great potential as a target for antiviral drugs (Barretto et al. 2005; Ho et al. 2022).

In addition to its core structural and non-structural proteins, SARS-CoV-2 encodes multiple accessory proteins. Among these, ORF3b, ORF6, ORF7a, and ORF8 have been identified as interferon-I (IFN-I) antagonists, thus inhibiting the host's immune response. ORF3a is implicated in apoptosis, while ORF9b and ORF9c interact with cellular organelles, resulting in the suppression of the antiviral response in infected cells. Nevertheless, the precise functions of ORF7b and ORF10 remain incompletely understood (Redondo et al. 2021).

11.2.2 Potential Drug Targets of SARS-CoV-2

Understanding the unique life cycle and key viral proteins of SARS-CoV-2 is indispensable for the development of effective treatments and vaccines. A thorough comprehension of the viral structure and function is therefore critical to facilitate drug discovery efforts. This includes identifying potential targets that can interrupt the viral life cycle and impede viral replication. One promising strategy for designing drugs against SARS-CoV-2 involves targeting viral proteins that play crucial roles in viral replication and pathogenesis. For example, the spike protein is an attractive target for drug design, as several inhibitors have been discovered that can bind to it and block its interaction with the host cell receptor. Another vital viral protein that can be targeted by inhibitors is the main protease (Mpro), which cleaves the majority of viral polyproteins into individual functional proteins. Numerous M^{pro} inhibitors have been identified through drug discovery efforts, some of which have demonstrated promising results in preclinical studies (Gil et al. 2020; Ayele et al. 2021) (Fig. 11.3).

Moreover, comprehending the viral structure and function can guide the development of vaccines against SARS-CoV-2. Vaccines can be designed to target specific viral proteins or portions of viral proteins that are critical for viral life cycle. For instance, vaccines based on the spike protein have been developed and have exhibited encouraging outcomes in clinical trials (Soraci et al. 2022).

Furthermore, besides targeting the viral proteins directly, an understanding of the viral structure and function can help design drugs that target host cell pathways or proteins necessary for viral replication. For instance, drugs that impede host cell proteases or modulate the host immune response can be effective in inhibiting viral replication and reducing tissue damage (Kaur et al. 2021).

11.3 Overview of Computational Drug Design Approaches

The use of computational techniques in drug design has significantly transformed the modern drug discovery process, making it possible to identify promising drug candidates with enhanced efficacy and specificity. The conventional approach to drug discovery, which involves extensive experimental screening and testing of large chemical libraries, is known to be time-consuming, expensive, and often results in compounds with suboptimal properties. Computational methods, on the other hand, provide a cost-effective and efficient alternative, allowing researchers to swiftly explore massive chemical spaces and pinpoint potential drug candidates with favorable properties (Hung and Chen 2014).

One of the key advantages of computational methods in drug design is their ability to

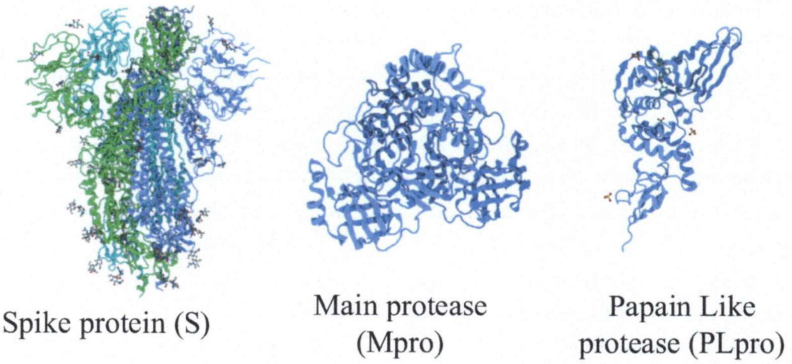

Fig. 11.3 Main molecular targets for COVID-19 drug discovery

predict the activity and specificity of candidate compounds with high accuracy. By leveraging advanced computational algorithms and techniques, researchers can simulate the interactions between potential drug candidates and their target proteins or enzyme active sites. These simulations can reveal valuable insights into the binding affinity, pharmacokinetics, and mechanism of action of candidate compounds, allowing researchers to prioritize and optimize the most promising candidates for further development (Yu and MacKerell 2017).

Moreover, computational methods have demonstrated remarkable efficacy in designing drugs for intricate diseases like cancer and neurological disorders, where conventional drug discovery approaches have frequently fallen short (De et al. 2019; Baig et al. 2018; Dorahy et al. 2023). By taking advantage of the vast amounts of data generated by high-throughput experiments and omics technologies, computational methods can identify new drug targets and design molecules that interact with them in a highly specific and effective manner. Such precision in drug design offers significant advantages over traditional approaches, which often rely on trial-and-error experimentation. Essentially, computational methods have opened new avenues for drug discovery by enabling a data-driven, hypothesis-testing approach that has revolutionized the field (Tayara et al. 2021; Wang et al. 2013).

11.3.1 Structure-Based Approaches

One of the most employed structure-based methods in drug design is molecular docking. Molecular docking involves the prediction of the binding orientation and affinity of a small molecule with a target biomolecule. This approach leverages knowledge of the protein structure and the three-dimensional conformation of the ligand to produce a model of the protein–ligand complex. The model is subsequently assessed using diverse scoring functions, such as empirical scoring methods, to ascertain the binding affinity and select the most promising candidate compounds for further evaluation. The molecular docking approach enables researchers to perform virtual screening of many candidate molecules against a target protein, facilitating rapid identification of lead compounds that can be further optimized through medicinal chemistry approaches (Torres et al. 2019; Saikia and Bordoloi 2019).

Another important method in structure-based drug design is molecular dynamics (MD) simulations. MD simulations allow the simulation of the motion of atoms over time, providing insight into the dynamics of protein–ligand interactions and the stability of protein–ligand complexes. Molecular dynamics simulations are often employed to refine the results obtained through molecular docking and to identify the crucial binding interactions between the protein and ligand, consequently optimizing the binding affinity and selectivity of candidate compounds. This powerful combination of molecular dynamics simulations and molecular docking enables researchers to design drug molecules that interact with their target proteins with precision, specificity, and high affinity (Liu et al. 2018; Bera and Payghan 2019).

11.3.2 Ligand-Based Approaches

In addition to structure-based methods, ligand-based methods are also used in rational drug design. Ligand-based methods utilize the knowledge of the chemical properties and structure of known ligands to predict the activity and specificity of new compounds. Quantitative structure–activity relationship (QSAR) and pharmacophore modeling are two commonly used ligand-based methods. QSAR involves the development of quantitative relationships between the chemical structures of a set of compounds and their corresponding biological activities. This method is used to predict the biological activity of novel compounds based on their chemical structure. On the other hand, pharmacophore modeling involves the identification of common chemical features, or pharmacophores, that are present in active compounds but absent in inactive compounds. These pharmacophores are used to design new compounds with similar structural features and, hence, similar biological activity (Achary 2020; Favia 2011).

11.3.3 Artificial Intelligence (AI) and Machine Learning (ML) Techniques

Artificial intelligence (AI) has emerged as a powerful tool in drug design and discovery. The use of machine learning (ML) algorithms and other computational techniques has facilitated the identification of novel drug candidates. One significant application of AI is drug screening, which utilizes ML algorithms to predict the binding affinity of large chemical libraries to a specific target, consequently reducing the duration and expenses associated with traditional high-throughput screening methods (Gentile et al. 2022; Carpenter et al. 2018). De novo drug design is another area where AI methods have been applied, allowing the generation of new compounds with desired properties that can interact with a target protein (Bhisetti and Fang 2022). Additionally, AI and ML techniques have been used in lead optimization to predict the physicochemical properties of lead compounds and assess their potential for further development. Finally, AI and ML can be applied to toxicity prediction to identify potential safety risks of drug candidates and design safer drugs (McNair 2023; Zhang et al. 2017). The utilization of AI methods in drug design has the potential to accelerate the drug discovery process and enhance the efficacy of developing new drugs to treat a wide range of diseases (Paul et al. 2021).

In rational drug design, a combination of diverse computational methods is frequently employed to optimize the selection and design of promising drug candidates. These methods offer researchers the ability to efficiently search vast chemical libraries and predict the activity and specificity of candidate compounds, with the goal of developing new therapeutics for treating a wide range of diseases, including COVID-19.

11.4 Recent Advances in Computational Drug Design for COVID-19

In the quest to identify suitable compounds that can specifically target the causative agent of COVID-19, the SARS-CoV-2 virus, computer-aided drug design (CADD) approaches have gained wide acceptance. These studies can be classified into three broad categories: repurposing studies, de novo design studies, and synergistic drug combination studies. Repurposing studies entail evaluating the efficacy of existing drugs in treating COVID-19 by analyzing their binding affinity to the target viral proteins. De novo design studies employ molecular modeling techniques to develop new drugs from scratch and optimize their binding affinity and selectivity to the target protein. On the other hand, synergistic drug combination studies entail screening and evaluating the effectiveness of combinations of existing drugs against COVID-19 by analyzing their potential interactions and enhancing their antiviral activity. These computational approaches have immense potential for the swift identification and development of effective drugs to mitigate the COVID-19 pandemic.

11.4.1 Computational Drug Repurposing for COVID-19

The lack of a specific and approved therapy for treating the disease has led to the urgent need to identify effective treatments using existing drugs, which has become a well-established strategy. Since the regulatory process for newly approved drugs is time-consuming, the drug repurposing (DR) strategy heavily relies on identifying alternative treatment methods by utilizing drugs that have already been approved (Pushpakom et al. 2019). Repurposing drugs offers several advantages over developing new drugs from scratch, including an abundance of safety data, faster approval timelines, and established manufacturing systems (Parvathaneni et al. 2019). However, identifying the most suitable candidates from a vast pool of thousands of drugs is a challenging task. This challenge is further complicated when investigating drug combinations, which significantly expands the search space and introduces the potential for drug interactions.

Computational methods have been extensively used to repurpose several approved drugs (Wang 2020; Hakmi et al. 2020; Galindez et al. 2021). These drugs have been selected to target various stages of the SARS-CoV-2 life cycle, with cepharanthine being an example of an anti-inflammatory compound that has shown efficacy in mitigating the cytopathic effects of SARS-CoV-2 (Rogosnitzky et al. 2020). Another drug, corilagin, which has potential as an antifibrosis agent, has demonstrated effectiveness against SARS-CoV-2 infection by targeting RBD-ACE2 binding (Yang et al. 2021). Similarly, Luo et al. employed a combination of molecular docking, free energy perturbation techniques, and bioassay validation to identify potential therapeutic agents capable of inhibiting the viral M^{pro}. Specifically, the researchers investigated a dataset comprising 2500 FDA-approved drugs to identify promising candidates. Out of the 25 agents identified, 15 displayed robust in vitro inhibitory activity against Mpro. Notably, dipyridamole emerged as the most potent agent with a Ki value of 0.04 μM (Li et al. 2020).

In a separate investigation, a virtual screening of the DrugBank database was conducted using quantitative structure–activity relationship (QSAR) techniques to identify potential inhibitors against SARS-CoV-2. This screening procedure produced 42 compounds, out of which 11 were subjected to further analysis. Of these 11 compounds, three demonstrated noteworthy inhibitory activity against the virus in vitro: cenicriviroc (with an AC50 of 8.9 μM), proglumetacin (with AC50 values of 8.9 μM and 12.5 μM in two independent tests), and sufugolix (with an AC50 of 12.6 μM) (Alves et al. 2021).

Other research groups adopted an integrative methodology that utilized machine learning to investigate drug repositioning against SARS-CoV-2. This method involved the integration and analysis of large-scale knowledge graphs, literature, and transcriptome data. Specifically, their study focused on the PARP1 inhibitor, CVL218. Their in vitro experiments showed that CVL218 effectively inhibited SARS-CoV-2 replication without inducing significant cytopathic effects. Moreover, their findings suggested that CVL218 bound effectively to the nucleocapsid (N) protein of SARS-CoV-2 and was able to suppress the production of several inflammatory cytokines that are highly relevant to preventing immunopathology induced by SARS-CoV-2 infection, following lipopolysaccharide (LPS)-induced stimulation (Ge et al. 2021).

Natural products were also of particular interest for COVID-19 drug repurposing due to their novelty and diverse molecular properties. Among them, ivermectin was found to bind tightly to the RdRp binding pocket (Eweas et al. 2021; Piplani et al. 2021). An in vitro infection study on Vero/hSLAM cells infected with the SARS-CoV-2 showed a 93% reduction in viral RNA at 24 h and a 99.98% reduction at 48 h upon administration of 5 μM ivermectin (Heidary and Gharebaghi 2020). One natural product that has also been predicted to possess strong activity against RdRp is digoxin (Qayed et al. 2022). Previous in vitro investigations have confirmed the potent antiviral effects of digoxin against SARS-CoV-2 in Vero cells, as evidenced by IC50 values of 0.043 and 0.19 μM reported by two separate studies (Cho et al. 2020; Jeon et al. 2020).

11.4.2 De Novo Design of Anti-SARS-CoV-2 Compounds

Upon the emergence of COVID-19, researchers employed the structure-based virtual screening (SBVS) technique to explore the potential of repurposing existing drugs or natural substances as therapeutic agents against SARS-CoV-2. While the characterization and optimization of natural compounds pose challenges, they can be used as a valuable reference for the development of new treatments. The virtual screening process primarily utilized publicly available databases that contained drugs in late-phase clinical trials, as well as antiviral, anticancer, and antifungal compounds. However, empirical validation of the in silico-generated hits has demonstrated that most conventional drugs, and natural substances lack the required potency for targeting SARS-CoV-2 (Ma and Wang 2022). Instead, the search for high-potency hits against SARS-CoV-2 should focus on screening novel chemical entities. This approach was demonstrated by Yuan et al. who identified a potential inhibitor from a collection of 50,080 small-molecule compounds with drug-like properties (Yuan et al. 2022). Furthermore, two small-molecule inhibitors were discovered by means of a small-scale SBVS approach followed by biological confirmation. One out of two hits showed a structure consistent with that reported previously by Shen et al. through a rigorous medicinal chemistry approach (Zhao et al. 2022; Tian 2022).

In a separate investigation, Ma and colleagues employed high-throughput screening (HTS) to screen a large library of lead-like compounds and identified two highly effective hit compounds (Ma et al. 2021). These hits shared similar properties with GRL-0617, emphasizing the importance of screening even larger libraries to discover novel hits. However, conventional HTS of ultra-large libraries containing millions of compounds is often impractical (Szymański et al. 2011). Consequently, conducting large-scale SBVS offers a more cost-effective and efficient option. This approach was exemplified by Elseginy and colleagues, who virtually screened substantial commercial libraries consisting of 500,000 molecules and identified two potential lead compounds (compounds 7 and 13) which demonstrated efficient suppression of SARS-CoV-2-infection in Vero E6 cells; with IC50 values of 0.77 and 0.11 μM, respectively (Elseginy et al. 2021; Elseginy and Anwar 2021). Nevertheless, these compounds may have a poor gut permeability which limits their clinical value. Hence, additional optimization is necessary to enhance their clinical utility (Ton et al. 2022).

In a previous investigation, we explored the potential of small-molecule alpha helix mimetics as inhibitors of the interaction between the viral spike protein and the ACE2 receptor. Given that the N-terminal alpha helix of ACE2 is primarily responsible for this interaction, we focused on this class of molecules. However, the relatively flat RBD–ACE2 interface presents a challenge for small-molecule inhibitors, which tend to bind weakly and diffuse outside the binding site (Unni et al. 2020; Chikhale et al. 2021; Deganutti et al. 2021). To overcome this limitation, we applied MMGBSA decomposition to divide the RBD into two hot spot regions and then targeted each region with alpha helix mimetic small molecules. Our results were promising, with the small molecules forming stable complexes with the RBD domain of the spike protein, as confirmed by molecular dynamics simulations. Therefore, these candidate molecules may serve as potential scaffolds for the development of novel compounds that could disrupt viral entry into target cells (Hakmi et al. 2022).

The modern computer-aided drug design (CADD) pipeline mentioned above effectively detected new small-molecule therapeutics that could be suitable for further hit-to-lead optimization, whereas experimental methods encountered challenges. These preliminary investigations underscored the increasing role of CADD in the drug discovery of SARS-CoV-2. It is worth noting that in the drug discovery of SARS-CoV-1, most of the relevant compounds were initially identified through expensive experimental high-throughput screening (HTS) campaigns, and computationally driven techniques did not play a significant role in the discovery stage (Ton et al. 2022). In the SARS and MERS pandemics, computational methods such as molecular modeling were utilized to gain insight into binding interactions and facilitate the optimization of compounds, rather than to discover new hits. However, in the context of COVID-19, there was an urgent requirement to identify novel antiviral therapeutics, and there was a shift in focus from hit optimization to the rapid discovery of small molecular hits. In response to SARS-CoV-2, CADD projects could generate effective and new chemotypes, thus highlighting the utility of computational pipelines in the fight against the virus (Gentile et al. 2021). There has been a notable change in methodology, where the traditional approach of experimental screening of compounds on a small scale followed by computational validation of binding hypotheses has been replaced by a more extensive computational screening of chemical libraries, followed by experimental validation of on-target effects on a larger scale. This shift has been emphasized, where a higher number of compounds are being screened computationally, and the validation of the findings is done experimentally (Ton et al. 2022; Lyu et al. 2019). Moreover, the development of deep learning (DL) generative models offers a promising avenue for the design of novel anti-COVID-19 drugs. These models utilize reported hits to automatically generate analogs that are tailored to the target of interest and optimized for enhanced binding features, making them highly practical for this purpose.

11.4.3 Discovery of Synergistic Drug Combinations (Drug Cocktails) for COVID-19

Combination therapies have demonstrated greater efficacy than single drugs in treating various diseases. These combinations can result in a synergistic effect, enhancing therapeutic potency and efficacy, leading to more robust therapeutic effects, or reducing the required dosage, thereby minimizing side effects. In the context of the COVID-19 pandemic and future pandemics, identifying effective combinations of approved drugs has an additional advantage over developing a novel single-agent therapy by reducing time to clinical implementation. Approved drugs are readily available in large quantities, have well-established toxicity profiles, and may be used off-label in critical situations. Taken together, these considerations emphasize the importance of discovering novel synergistic drug combinations for treating COVID-19. One illustrative instance involves the work of Bobrowski and colleagues, who employed knowledge-based techniques to produce a list of potential drug combinations and experimentally verified their efficacy in generating antiviral synergy against SARS-CoV-2 (Bobrowski et al. 2021).

Another study utilized a DL model called ComboNet for in silico repurposing of existing drugs and subsequently tested 30 drug combinations through experimentation. This empirical testing revealed two drug combinations (remdesivir and reserpine; remdesivir and IQ-1S) that displayed potent synergy in vitro (Jin, et al. 2021). Reserpine is an FDA-approved drug primarily used as a peripheral antihypertensive. Its potency against SARS-CoV-2 has been reported to be moderate, with an IC50 of 11.2 μM in Vero E6 cells (Chen 2021) and 6.4 μM in HeLa-ACE2 cells (Janes et al. 2018). Meanwhile, IQ-1S is a JNK inhibitor that has demonstrated an IC50 of 6.3 μM against SARS-CoV-2 in a Vero E6 cell cytopathic effect assay.

By using computational methods, researchers have discovered and tested 73 combinations of 32 drugs for their potential activity against SARS-CoV-2. Out of these, 16 combinations showed synergistic effects while eight showed antagonistic effects. Notably, combinations of nitazoxanide, an FDA-approved drug, with remdesivir, amodiaquine, or umifenovir showed significant synergy against SARS-CoV-2 in cell models. Surprisingly, the combination of remdesivir and lysosomotropic drugs exhibited strong antagonism, with hydroxychloroquine completely extinguishing the antiviral activity of remdesivir at a concentration of 10 μM in vitro (Bobrowski et al. 2021).

11.4.4 Vaccine and Epitope Prediction

Immunization is a vital measure in the prevention of numerous infections and has been instrumental in the elimination of certain illnesses, resulting in the preservation of millions of human lives (Greenwood 2014). In order to design a vaccine, researchers utilize in silico studies to predict epitopes that can stimulate both T-cell and B-cell responses, leading to both cellular and humoral immune responses. Immunoinformatics is a subfield of bioinformatics that investigates the relationship between immune responses and predicted epitopes. The field has made significant progress, resulting in the development of numerous tools for predicting epitopes recognized by B-cells or T-cells and their corresponding antigenic responses (Yurina and Adianingsih 2022).

An in silico approach combining immunoinformatics and deep neural network (DeepVacPred) was proposed to predict and design a multiepitope vaccine for COVID-19. DeepVacPred identified 26 potential vaccine subunits that were further evaluated and constructed. These subunits were selected based on their high likelihood of containing B-cell epitopes and multiple T-cell epitopes, as well as their high antigenicity and low allergenicity. This artificial intelligence (AI)-based vaccine discovery method accelerated the vaccine design process and resulted in a 694 amino acid multiepitope vaccine that contained 16 B-cell epitopes, 82 cytotoxic T lymphocytes (CTL) epitopes, and 89 helper T lymphocyte (HTL) epitopes. The 3-D structure of the designed vaccine was subsequently predicted, refined, and validated using other in silico tools (Yang et al. 2021).

A separate investigation employed proteomic data from human cell lines infected with SARS-CoV-2 and samples from COVID-19 patients to identify peptides that could be useful for vaccine development and diagnostics. The researchers analyzed the host responses to the virus in both colon carcinoma-2 and H1299 cell lines, as well as in naturally infected COVID-19 patient samples. A total of 323 and 143 human peptides were identified in the cell lines and patients, respectively. To ensure that only suitable peptides were chosen for diagnostic purposes, the authors verified the expression of the selected peptides using RNA-Seq data analysis of the infected cell lines. Four antigenic peptides were then selected and docked to recognized viral T-cell receptors (TCR), class I and II peptide major histocompatibility complexes (pMHC), and paratopes sequences that had been identified. The resulting antigenic peptides show great potential for generating antibodies that are specific to SARS-CoV-2, and the paratope peptides can be directly used for developing a COVID-19 diagnostic assay (Kesarwani et al. 2021).

A different class of vaccines developed using in silico methods are nucleic acid-based vaccines. One such vaccine was constructed using the codon-optimized consensus sequence of the spike glycoprotein as a DNA-based vaccine. The vaccine's immunogenicity was evaluated in two mouse models, and both models showed a significant induction of spike-specific IgG antibodies and neutralizing antibodies four weeks after receiving 100 μg of the in silico designed vaccine via intramuscular injections (Alamri 2021).

In silico tools are being utilized to investigate COVID-19 candidate vaccines in various ways. One of these ways is through modeling and simulation, which can actively improve the design of vaccine prototypes, support decision-making processes, reduce experimental costs, and time, and ultimately enhance the success rates of vaccine trials. A recent study employed the Universal Immune System Simulator (UISS) platform to predict the effectiveness of a potential cross-neutralizing antibody against COVID-19, which was initially proposed by Wang et al. (2020a). The results of the study showed that the computational framework accurately reproduced the in vitro findings reported by the original authors and successfully predicted the efficacy and mechanism of action of the antibody (Russo et al. 2020).

11.4.5 Genomic Surveillance and Computational Host Range Prediction

COVID-19 genomic surveillance is a process of systematically sequencing and analyzing the genomes of SARS-CoV-2 virus to track its evolution and spread. It has played a critical role in identifying new variants of the virus, including the Delta and Omicron variants, which have raised concerns due to their increased transmissibility and potential to reduce vaccine efficacy (Tegally et al. 1979). To control the spread of the disease and develop a universal vaccine, it is essential to understand and monitor the genetic evolution, geographical characteristics, and stability. Comparative genomics is a key tool that helps elucidate biodiversity and evolution at the genetic level through the analysis of multiple genome sequences. By using this approach, researchers can investigate the whole-genome mutation profiles of SARS-CoV-2 and the frequency of each mutation in the population under study (Laamarti et al. 2020; Alouane et al. 2020). These findings will enable better preparedness and prompt implementation of effective measures to prevent and control the spread of emerging viruses like SARS-CoV-2 variants.

The genetic diversity of coronaviruses results in different infectivity profiles, a wide range of clinical manifestations, and different modes of spread and transmission patterns within a given host (Tang et al. 2006; Phan 2020). Therefore, understanding the host range of the SARS-CoV-2 virus is crucial to determine its potential for zoonotic spillover and to design effective countermeasures. The binding of SARS-CoV-2 spike protein to ACE2 is the critical determinant of viral host range and cross-species infection (Jackson et al. 2022). In a recent study, computational approaches have been used to predict the potential animal range with high susceptibility to SARS-CoV-2 by studying the spike–ACE2 interaction of 22 domestic and wild animals. A specific substitution of tyrosine to histidine at position 41 in ACE2 was identified as potentially reducing the affinity to SARS-CoV-2 in some animal species including horses and greater horseshoe bats (Bouricha et al. 2020). Notably, another study by surface plasmon resonance (SPR) spectroscopy showed that equine ACE2-H41 accounts for the lower binding affinities compared with human ACE2 (Lan et al. 2022), validating the results of the in silico study. These findings may provide significant insights into SARS-CoV-2 host range, aiding in controlling its spread and identifying animal models for screening antiviral drugs or vaccine candidates. However, zoonotic risk prediction is a complex task that requires a thorough understanding of the biology of the virus and potential host species. Further research is needed to improve the accuracy of host range predictions and provide insights into the mechanisms underlying viral transmission between species. The development of accurate host range prediction tools can aid in the identification of potential zoonotic hosts, improve our understanding of virus evolution and pathogenesis, and inform strategies to prevent and combat future pandemics (Carlson et al. 2021; Pandit et al. 2022).

11.5 Conclusion

The COVID-19 pandemic has had a profound impact on human life and presented unprecedented challenges to the scientific community. Nevertheless, computer-aided methods have proven to be a valuable tool in the search for effective therapies and countermeasures in this time-critical situation. Using various computational methods, researchers have identified numerous promising drug candidates that have shown potential in vitro and in vivo. In addition, computational techniques coupled with virology, molecular biology, and biophysics have enhanced our understanding of the evolution and molecular mechanisms involved in the life cycle of the SARS-CoV-2 virus. These approaches have proven their effectiveness in COVID-19 research, and further advances in this field may help manage future pandemics.

References

Achary PGR (2020) Applications of quantitative structure-activity relationships (QSAR) based virtual screening in drug design: a review. Mini-Rev Med Chem 20(14):1375–1388. https://doi.org/10.2174/1389557520666200429102334

Alamri SS et al (2021) Synthetic SARS-CoV-2 spike-based DNA vaccine elicits robust and long-lasting Th1 humoral and cellular immunity in mice. Front Microbiol 12. https://doi.org/10.3389/fmicb.2021.727455

Almehdi AM, Khoder G, Alchakee AS, Alsayyid AT, Sarg NH, Soliman SSM (2021) SARS-CoV-2 spike protein: pathogenesis, vaccines, and potential therapies. Infection 49(5):855–876. https://doi.org/10.1007/s15010-021-01677-8

Alouane T et al (2020) Genomic diversity and hotspot mutations in 30,983 SARS-CoV-2 genomes: moving toward a universal vaccine for the 'confined virus'? Pathogens 9(10):829. https://doi.org/10.3390/pathogens9100829

Alves VM et al (2021) QSAR modeling of SARS-CoV Mpro inhibitors identifies sufugolix, cenicriviroc, proglumetacin, and other drugs as candidates for repurposing against SARS-CoV-2. Mol Inform 40(1):2000113. https://doi.org/10.1002/minf.202000113

Ayele AG, Enyew EF, Kifle ZD (2021) Roles of existing drug and drug targets for COVID-19 management. Metabol Open 11:100103. https://doi.org/10.1016/j.metop.2021.100103

Ayouni I et al (2021) Effective public health measures to mitigate the spread of COVID-19: a systematic review. BMC Public Health 21(1):1015. https://doi.org/10.1186/s12889-021-11111-1

Baig MH, Ahmad K, Rabbani G, Danishuddin M, Choi I (2018) Computer aided drug design and its application to the development of potential drugs for neurodegenerative disorders. Curr Neuropharmacol 16(6):740–748. https://doi.org/10.2174/1570159X15666171016163510

Barretto N, Jukneliene D, Ratia K, Chen Z, Mesecar AD, Baker SC (2005) The papain-like protease of severe acute respiratory syndrome coronavirus has deubiquitinating activity. J Virol 79(24):15189–15198. https://doi.org/10.1128/JVI.79.24.15189-15198.2005

Bera I, Payghan PV (2019) Use of molecular dynamics simulations in structure-based drug discovery. Curr Pharm Des 25(31):3339–3349. https://doi.org/10.2174/1381612825666190903153043

Bhisetti G, Fang C (2022) Artificial intelligence–enabled De Novo design of novel compounds that are synthesizable, pp 409–419. https://doi.org/10.1007/978-1-0716-1787-8_17

Bobrowski T et al (2021) Synergistic and antagonistic drug combinations against SARS-CoV-2. Mol Ther 29(2):873–885. https://doi.org/10.1016/j.ymthe.2020.12.016

Bouricha EM, Hakmi M, Akachar J, Belyamani L, Ibrahimi A (2020) In silico analysis of ACE2 orthologues to predict animal host range with high susceptibility to SARS-CoV-2. 3 Biotech 10(11):483. https://doi.org/10.1007/s13205-020-02471-3

Brant AC, Tian W, Majerciak V, Yang W, Zheng Z-M (2021) SARS-CoV-2: from its discovery to genome structure, transcription, and replication. Cell Biosci 11(1):136. https://doi.org/10.1186/s13578-021-00643-z

Carlson CJ et al (2021) The future of zoonotic risk prediction. Philos Trans R Soc B: Biol Sci 376(1837):20200358. https://doi.org/10.1098/rstb.2020.0358

Carpenter KA, Cohen DS, Jarrell JT, Huang X (2018) Deep learning and virtual drug screening. Future Med Chem 10(21):2557–2567. https://doi.org/10.4155/fmc-2018-0314

Cavalcante MB, de Melo Bezerra Cavalcante CT, Sarno M, Barini R, Kwak-Kim J (2021) Maternal immune responses and obstetrical outcomes of pregnant women with COVID-19 and possible health risks of offspring. J Reprod Immunol 143:103250. https://doi.org/10.1016/j.jri.2020.103250.

Chen B et al (2020) Overview of lethal human coronaviruses. Signal Transduct Target Ther 5(1):89. https://doi.org/10.1038/s41392-020-0190-2

Chen CZ et al (2021) Drug repurposing screen for compounds inhibiting the cytopathic effect of SARS-CoV-2. Front Pharmacol 11. https://doi.org/10.3389/fphar.2020.592737

Chikhale RV et al (2021) Sars-cov-2 host entry and replication inhibitors from Indian ginseng: an in-silico approach. J Biomol Struct Dyn 39(12):4510–4521. https://doi.org/10.1080/07391102.2020.1778539

Cho J et al (2020) Antiviral activity of digoxin and ouabain against SARS-CoV-2 infection and its implication for COVID-19. Sci Rep 10(1):16200. https://doi.org/10.1038/s41598-020-72879-7

Ciotti M, Ciccozzi M, Terrinoni A, Jiang W-C, Wang C-B, Bernardini S (2020) The COVID-19 pandemic. Crit Rev Clin Lab Sci 57(6):365–388. https://doi.org/10.1080/10408363.2020.1783198

da Rosa Mesquita R et al (2021) Clinical manifestations of COVID-19 in the general population: systematic review,. Wien Klin Wochenschr 133(7–8):377–382. https://doi.org/10.1007/s00508-020-01760-4

Dashraath P et al (2020) Coronavirus disease 2019 (COVID-19) pandemic and pregnancy. Am J Obstet Gynecol 222(6):521–531. https://doi.org/10.1016/j.ajog.2020.03.021

De B, Bhandari K, Mendonça FJB, Scotti MT, Scotti L (2019) Computational studies in drug design against cancer. Anticancer Agents Med Chem 19(5):587–591. https://doi.org/10.2174/1871520618666180911125700

Deganutti G, Prischi F, Reynolds CA (2021) Supervised molecular dynamics for exploring the druggability of the SARS-CoV-2 spike protein. J Comput Aided Mol Des 35(2):195–207. https://doi.org/10.1007/s10822-020-00356-4

Dorahy G, Chen JZ, Balle T (2023) Computer-aided drug design towards new psychotropic and neurological drugs. Molecules 28(3):1324. https://doi.org/10.3390/molecules28031324

Elseginy SA, Anwar MM (2021) In silico analysis of SARS-CoV-2 papain-like protease potential inhibitors. RSC Adv 11(61):38616–38631. https://doi.org/10.1039/D1RA07845C

Elseginy SA et al (2021) Promising anti-SARS-CoV-2 drugs by effective dual targeting against the viral and host proteases. Bioorg Med Chem Lett 43:128099. https://doi.org/10.1016/j.bmcl.2021.128099

Esakandari H, Nabi-Afjadi M, Fakkari-Afjadi J, Farahmandian N, Miresmaeili S-M, Bahreini E (2020) A comprehensive review of COVID-19 characteristics. Biol Proced Online 22(1):19. https://doi.org/10.1186/s12575-020-00128-2

Eweas AF, Alhossary AA, Abdel-Moneim AS (2021) Molecular docking reveals ivermectin and remdesivir as potential repurposed drugs against SARS-CoV-2. Front Microbiol 11. https://doi.org/10.3389/fmicb.2020.592908

Favia AD (2011) Theoretical and computational approaches to ligand-based drug discovery. Front Biosci 16(1):1276. https://doi.org/10.2741/3788

Fehr AR, Perlman S (2015) Coronaviruses: an overview of their replication and pathogenesis, pp 1–23. https://doi.org/10.1007/978-1-4939-2438-7_1

Galindez G et al (2021) Lessons from the COVID-19 pandemic for advancing computational drug repurposing strategies. Nat Comput Sci 1(1):33–41. https://doi.org/10.1038/s43588-020-00007-6

Ge Y et al (2021) An integrative drug repositioning framework discovered a potential therapeutic agent targeting COVID-19. Signal Transduct Target Ther 6(1):165. https://doi.org/10.1038/s41392-021-00568-6

Gentile F et al (2021) Automated discovery of noncovalent inhibitors of SARS-CoV-2 main protease by consensus Deep Docking of 40 billion small molecules. Chem Sci 12(48):15960–15974. https://doi.org/10.1039/D1SC05579H

Gentile F et al (2022) Artificial intelligence–enabled virtual screening of ultra-large chemical libraries with deep docking. Nat Protoc 17(3):672–697. https://doi.org/10.1038/s41596-021-00659-2

Gil C et al (2020) COVID-19: drug targets and potential treatments. J Med Chem 63(21):12359–12386. https://doi.org/10.1021/acs.jmedchem.0c00606

Girum T, Lentiro K, Geremew M, Migora B, Shewamare S, Shimbre MS (2021) Optimal strategies for COVID-19 prevention from global evidence achieved through social distancing, stay at home, travel restriction and lockdown: a systematic review. Arch Public Health 79(1):150. https://doi.org/10.1186/s13690-021-00663-8

Greenwood B (2014) The contribution of vaccination to global health: past, present and future. Philosop Trans R Soc B Biol Sci 369(1645):20130433. https://doi.org/10.1098/rstb.2013.0433

Hakmi M et al (2020) Repurposing of known anti-virals as potential inhibitors for SARS-CoV-2 main protease using molecular docking analysis. Bioinformation 16(4):301. https://doi.org/10.6026/97320630016301

Hakmi M et al (2022) In silico exploration of small-molecule α-helix mimetics as inhibitors of SARS-COV-2 attachment to ACE2. J Biomol Struct Dyn 40(4):1546–1557. https://doi.org/10.1080/07391102.2020.1830175

Haque SM, Ashwaq O, Sarief A, Azad John Mohamed AK (2020) A comprehensive review about SARS-CoV-2. Future Virol 15(9):625–648. https://doi.org/10.2217/fvl-2020-0124

Heidary F, Gharebaghi R (2020) Ivermectin: a systematic review from antiviral effects to COVID-19 complementary regimen. J Antibiot (Tokyo) 73(9):593–602. https://doi.org/10.1038/s41429-020-0336-z

Ho C-Y et al (2022) A structural comparison of SARS-CoV-2 main protease and animal coronaviral main protease reveals species-specific ligand binding and dimerization mechanism. Int J Mol Sci 23(10):5669. https://doi.org/10.3390/ijms23105669

Hung C-L, Chen C-C (2014) Computational approaches for drug discovery. Drug Dev Res 75(6):412–418. https://doi.org/10.1002/ddr.21222

Jackson CB, Farzan M, Chen B, Choe H (2022) Mechanisms of SARS-CoV-2 entry into cells. Nat Rev Mol Cell Biol 23(1):3–20. https://doi.org/10.1038/s41580-021-00418-x

Janes J et al (2018) The ReFRAME library as a comprehensive drug repurposing library and its application to the treatment of cryptosporidiosis. Proc Natl Acad Sci 115(42):10750–10755. https://doi.org/10.1073/pnas.1810137115

Jeon S et al (2020) Identification of antiviral drug candidates against SARS-CoV-2 from FDA-approved drugs. Antimicrob Agents Chemother 64(7). https://doi.org/10.1128/AAC.00819-20

Jin W et al (2021) Deep learning identifies synergistic drug combinations for treating COVID-19. Proc Nat Acad Sci 118(39). https://doi.org/10.1073/pnas.2105070118

Kaur U et al (2021) Targeting host cell proteases to prevent SARS-CoV-2 invasion. Curr Drug Targets 22(2):192–201. https://doi.org/10.2174/1389450121666200924113243

Kesarwani V, Gupta R, Vetukuri RR, Kushwaha SK, Gandhi S (2021) Identification of unique peptides for SARS-CoV-2 diagnostics and vaccine development by an in silico proteomics approach. Front Immunol 12. https://doi.org/10.3389/fimmu.2021.725240

Laamarti M et al (2020) Large scale genomic analysis of 3067 SARS-CoV-2 genomes reveals a clonal geo-distribution and a rich genetic variations of hotspots mutations. PLoS ONE 15(11):e0240345. https://doi.org/10.1371/journal.pone.0240345

Lan J et al (2022) Structural insights into the binding of SARS-CoV-2, SARS-CoV, and hCoV-NL63 spike receptor-binding domain to horse ACE2. Structure 30(10):1432-1442.e4. https://doi.org/10.1016/j.str.2022.07.005

Laporte M et al (2021) The SARS-CoV-2 and other human coronavirus spike proteins are fine-tuned towards temperature and proteases of the human airways. PLoS Pathog 17(4):e1009500. https://doi.org/10.1371/journal.ppat.1009500

Li Z et al (2020) Identify potent SARS-CoV-2 main protease inhibitors via accelerated free energy perturbation-based virtual screening of existing drugs. Proc Natl Acad Sci 117(44):27381–27387. https://doi.org/10.1073/pnas.2010470117

Liu X, Shi D, Zhou S, Liu H, Liu H, Yao X (2018) Molecular dynamics simulations and novel drug discovery. Expert Opin Drug Discov 13(1):23–37. https://doi.org/10.1080/17460441.2018.1403419

Lyu J et al (2019) Ultra-large library docking for discovering new chemotypes. Nature 566(7743):224–229. https://doi.org/10.1038/s41586-019-0917-9

Ma C, Wang J (2022) Validation and invalidation of SARS-CoV-2 papain-like protease inhibitors. ACS Pharmacol Transl Sci 5(2):102–109. https://doi.org/10.1021/acsptsci.1c00240

Ma C et al (2021) Discovery of SARS-CoV-2 papain-like protease inhibitors through a combination of high-throughput screening and a FlipGFP-based reporter assay. ACS Cent Sci 7(7):1245–1260. https://doi.org/10.1021/acscentsci.1c00519

McNair D (2023) Artificial intelligence and machine learning for lead-to-candidate decision-making and beyond. Annu Rev Pharmacol Toxicol 63(1):77–97. https://doi.org/10.1146/annurev-pharmtox-051921-023255

Pandit PS et al (2022) Predicting the potential for zoonotic transmission and host associations for novel viruses. Commun Biol 5(1):844. https://doi.org/10.1038/s42003-022-03797-9

Parvathaneni V, Kulkarni NS, Muth A, Gupta V (2019) Drug repurposing: a promising tool to accelerate the drug discovery process. Drug Discov Today 24(10):2076–2085. https://doi.org/10.1016/j.drudis.2019.06.014

Paul D, Sanap G, Shenoy S, Kalyane D, Kalia K, Tekade RK (2021) Artificial intelligence in drug discovery and development. Drug Discov Today 26(1):80–93. https://doi.org/10.1016/j.drudis.2020.10.010

Payne S (2017) Family Coronaviridae. In: Viruses. Elsevier, Amsterdam, pp 149–158. https://doi.org/10.1016/B978-0-12-803109-4.00017-9

Perlman S, Netland J (2009) Coronaviruses post-SARS: update on replication and pathogenesis. Nat Rev Microbiol 7(6):439–450. https://doi.org/10.1038/nrmicro2147

Phan T (2020) Genetic diversity and evolution of SARS-CoV-2. Infect Genet Evol 81:104260. https://doi.org/10.1016/j.meegid.2020.104260

Piplani S, Singh PK, Winkler DA, Petrovsky N (2021) Computationally repurposed drugs and natural products against RNA dependent RNA polymerase as potential COVID-19 therapies. Molecular Biomedicine 2(1):28. https://doi.org/10.1186/s43556-021-00050-3

Pizzato M et al (2022) SARS-CoV-2 and the host cell: a tale of interactions. Front Virol 1. https://doi.org/10.3389/fviro.2021.815388

Pushpakom S et al (2019) Drug repurposing: progress, challenges and recommendations. Nat Rev Drug Discov 18(1):41–58. https://doi.org/10.1038/nrd.2018.168

Qayed WS, Ferreira RS, Silva JRA (2022) In silico study towards repositioning of FDA-approved drug candidates for anticoronaviral therapy: molecular docking, molecular dynamics and binding free energy calculations. Molecules 27(18):5988. https://doi.org/10.3390/molecules27185988

Redondo N, Zaldívar-López S, Garrido JJ, Montoya M (2021) SARS-CoV-2 accessory proteins in viral pathogenesis: knowns and unknowns. Front Immunol 12. https://doi.org/10.3389/fimmu.2021.708264

Rogosnitzky M, Okediji P, Koman I (2020) Cepharanthine: a review of the antiviral potential of a Japanese-approved alopecia drug in COVID-19. Pharmacol Rep 72(6):1509–1516. https://doi.org/10.1007/s43440-020-00132-z

Russo G et al (2020) In silico trial to test COVID-19 candidate vaccines: a case study with UISS platform. BMC Bioinform 21(S17):527. https://doi.org/10.1186/s12859-020-03872-0

Saikia S, Bordoloi M (2019) Molecular docking: challenges, advances and its use in drug discovery perspective. Curr Drug Targets 20(5):501–521. https://doi.org/10.2174/1389450119666181022153016

Smith V et al (2020) Maternal and neonatal outcomes associated with COVID-19 infection: a systematic review. PLoS ONE 15(6):e0234187. https://doi.org/10.1371/journal.pone.0234187

Soraci L et al (2022) COVID-19 vaccines: current and future perspectives. Vaccines (basel) 10(4):608. https://doi.org/10.3390/vaccines10040608

Szymański P, Markowicz M, Mikiciuk-Olasik E (2011) Adaptation of high-throughput screening in drug discovery—toxicological screening tests. Int J Mol Sci 13(1):427–452. https://doi.org/10.3390/ijms13010427

Tang XC et al (2006) Prevalence and genetic diversity of coronaviruses in bats from China. J Virol 80(15):7481–7490. https://doi.org/10.1128/JVI.00697-06

Tayara H, Abdelbaky I, To Chong K (2021) Recent omics-based computational methods for COVID-19 drug discovery and repurposing. Brief Bioinform 22(6). https://doi.org/10.1093/bib/bbab339

Tegally H et al (1979) The evolving SARS-CoV-2 epidemic in Africa: insights from rapidly expanding genomic surveillance. Science 378(6615). https://doi.org/10.1126/science.abq5358

Tian X et al (2022) Discovery of novel and highly potent inhibitors of SARS CoV-2 papain-like protease through structure-based pharmacophore modeling, virtual screening, molecular docking, molecular dynamics simulations, and biological evaluation. Front Pharmacol 13. https://doi.org/10.3389/fphar.2022.817715.

Ton A-T, Pandey M, Smith JR, Ban F, Fernandez M, Cherkasov A (2022) Targeting SARS-CoV-2 papain-like protease in the postvaccine era. Trends

Pharmacol Sci 43(11):906–919. https://doi.org/10.1016/j.tips.2022.08.008
Torres PHM, Sodero ACR, Jofily P, Silva-Jr FP (2019) Key topics in molecular docking for drug design. Int J Mol Sci 20(18):4574. https://doi.org/10.3390/ijms20184574
Unni S, Aouti S, Thiyagarajan S, Padmanabhan B (2020) Identification of a repurposed drug as an inhibitor of Spike protein of human coronavirus SARS-CoV-2 by computational methods. J Biosci 45(1):130. https://doi.org/10.1007/s12038-020-00102-w
V'kovski P, Kratzel A, Steiner S, Stalder H, Thiel V (2021) Coronavirus biology and replication: implications for SARS-CoV-2. Nat Rev Microbiol 19(3):155–170. https://doi.org/10.1038/s41579-020-00468-6
Wang J (2020) Fast identification of possible drug treatment of coronavirus disease-19 (COVID-19) through computational drug repurposing study. J Chem Inf Model 60(6):3277–3286. https://doi.org/10.1021/acs.jcim.0c00179
Wang YC, Deng N, Chen S, Wang Y (2013) Computational study of drugs by integrating omics data with kernel methods. Mol Inform 32(11–12):930–941. https://doi.org/10.1002/minf.201300090
Wang C et al (2020a) A human monoclonal antibody blocking SARS-CoV-2 infection. Nat Commun 11(1):2251. https://doi.org/10.1038/s41467-020-16256-y
Wang Y, Grunewald M, Perlman S (2020b) Coronaviruses: an updated overview of their replication and pathogenesis, pp 1–29. https://doi.org/10.1007/978-1-0716-0900-2_1
WHO (2023) WHO coronavirus (COVID-19) dashboard. WHO Coronavirus (COVID-19) Dashboard With Vaccination Data. https://covid19.who.int/. Accessed 30 Apr 2023
Yang LJ et al (2021) Corilagin prevents SARS-CoV-2 infection by targeting RBD-ACE2 binding. Phytomedicine 87:153591. https://doi.org/10.1016/j.phymed.2021.153591
Yang Z, Bogdan P, Nazarian S (2021) An in silico deep learning approach to multi-epitope vaccine design: a SARS-CoV-2 case study. Sci Rep 11(1):3238. https://doi.org/10.1038/s41598-021-81749-9
Yu W, MacKerell AD (2017) Computer-aided drug design methods, pp 85–106. https://doi.org/10.1007/978-1-4939-6634-9_5
Yuan S et al (2022) Targeting papain-like protease for broad-spectrum coronavirus inhibition. Protein Cell 13(12):940–953. https://doi.org/10.1007/s13238-022-00909-3
Yurina V, Adianingsih OR (2022) Predicting epitopes for vaccine development using bioinformatics tools. Ther Adv Vaccines Immunother 10:251513552211002. https://doi.org/10.1177/25151355221100218
Zhang H et al (2017) Development of novel prediction model for drug-induced mitochondrial toxicity by using naïve Bayes classifier method. Food Chem Toxicol 110:122–129. https://doi.org/10.1016/j.fct.2017.10.021
Zhao G et al (2022) Hydrogen bonding penalty used for virtual screening to discover potent inhibitors for Papain-Like cysteine proteases of SARS-CoV-2. Chem Biol Drug Des 100(4):502–514. https://doi.org/10.1111/cbdd.14115

The Potential of Natural Products in the Management of COVID-19

12

Mona M. Saber, Maha M. Salama, and Osama A. Badary

Abstract

Severe acute respiratory syndrome coronavirus 2 (SARS-CoV-2) was identified as the novel coronavirus that caused a life-threatening viral illness (COVID-19) at the end of 2019. Within a short period of time, this virus spread leading to tremendous loss of life and economic damage. Medications to treat this virus are not yet established, and the process of implementing new strategies for medications is time-consuming. Recent clinical studies revealed the abandonment of the most promising candidates, who later became potential leads. Only through comprehensive study for safety and efficacy the medications, which have already received approval, be repurposed for use in different therapeutic purposes. Natural sources are being used arbitrarily as antiviral drugs and immunity boosters because there are no clear therapies on the horizon. It has long been known that most natural compounds have strong antiviral properties including SARS-CoV-2. Natural remedies have been demonstrated to have inhibitory effects on MERS-CoV and SARS-CoV infections. The non-structural proteins of the virus, such as PLPRO, MPRO, and RdRp, as well as structural proteins like the spike (S) protein, have been demonstrated to have a substantial binding affinity and an inhibitory effect by a variety of natural products, according to in silico research. The virus also demonstrates to be a legitimate target for therapeutic development since it makes use of the host cell's transmembrane ACE2 receptor. In this chapter, we highlight on the potential of alkaloids, phenolic and polyphenolic compounds, flavonoids, terpenoids, cardiac glycosides, and natural products from marine sources against the human coronavirus via different mode of actions. Most of the studied metabolites act either by inhibiting virus replication or by blocking the active site of the protein of the virus

M. M. Saber
Faculty of Pharmacy, Department of Pharmacology and Toxicology, Cairo University, Kasr El-Aini St, Cairo 11562, Egypt

M. M. Salama
Faculty of Pharmacy, Department of Pharmacognosy, Cairo University, Kasr El-Aini St, Cairo 11562, Egypt

M. M. Salama
Faculty of Pharmacy, Department of Pharmacognosy, The British University in Egypt, El Sherouk City, Suez Desert Road, Cairo 11837, Egypt

O. A. Badary (✉)
Faculty of Pharmacy, Department of Clinical Pharmacy, Misr University for Science and Technology, Cairo, Egypt
e-mail: osama.badary@bue.ed

O. A. Badary
Faculty of Pharmacy, Department of Clinical Pharmacy, Ain Shams University, Cairo, Egypt

N. Rezaei (ed.), *The COVID-19 Aftermath*, Advances in Experimental Medicine and Biology 1457,
https://doi.org/10.1007/978-3-031-61939-7_12

either in silico or ex vivo. This review serves as a topic for further study and to discover other secondary metabolites for COVID-19 management.

Keywords

COVID-19 · Management · Natural products · Phytotherapy · Marine sources

12.1 COVID-19 Background

The severe acute respiratory syndrome coronavirus 2 (SARS-CoV-2) was first detected in Wuhan, China, in December 2019 (Boni et al. 2020). The World Health Organization (WHO) stated COVID-19 to be a global pandemic in March 2020 (Cucinotta and Vanelli 2020). On April 9, 2023, over 762 million cases of COVID-19 were reported worldwide, with over 6.8 million deaths. Because of the virus's rapid spread through these events and the lack of effective pharmaceutical therapies for the sickness, substantial control measures have been undertaken around the world (Benke et al. 2020; Badary 2021; Hamza et al. 2021; Hamza et al. 2022). Comorbid chronic diseases and acute organ injuries are significantly associated with disease severity and mortality in COVID-19 patients (Wang et al. 2020). COVID-19 has a wide range of clinical characteristics, from asymptomatic to acute respiratory distress syndrome (ARDS) and multiple organ failure. COVID-19 symptoms include fever, cough, sore throat, headache, weariness, myalgia, and shortness of breath (Wu et al. 2020).

Currently present treatments are divided into many therapeutic categories (Murakami et al. 2023). Some target the host's response to infection (inflammation, thrombosis, ARDS, and RAAS activation), while others target the virus directly. SARS-CoV-2 infection can cause hyperinflammation, which is characterized by high circulating levels of proinflammatory cytokines such as IL-6. Immunosuppressive medications like glucocorticoids and anti-IL-6 receptor antibodies are examples of anti-inflammatory therapies. Numerous randomized clinical trials (RCTs) have been conducted to evaluate the safety and efficacy of these agents, which include both novel and repurposed medications. It is worth noting that several COVID-19 vaccinations are presently available (Forni et al. 2021). One of the most difficult aspects of pandemic containment is making the remedy available to everyone as quickly as possible. One of the challenges in global COVID-19 control is the high expense of vaccinations (Yarnoff et al. 2021). In addition to extensive vaccination efforts, the development of extremely effective COVID-19 treatments has most likely led to lower rates of hospitalization, critical illness, and death observed now compared to the early stages of the pandemic (Lluch et al. 2022; Jeong et al. 2023).

12.2 Natural Therapeutic Approach

Natural products existed in history from antiquity. There are documents of its utilization dating several thousand years ago, when Egyptian papyri (Ancient Egyptian Medicine—PubMed) described that herbs played a significant role in Egyptian medicine. The plant remedies described in the Ebers Papyrus include opium, cannabis, myrrh, frankincense, fennel, cassia, senna, thyme, henna, juniper, aloe, linseed, and castor oil. Garlic cloves have been discovered in Egyptian burial sites, including Tutankhamen's tomb and the famous underground bull temple at Saqqara. Furthermore, Chinese Encyclopedias covered the medical benefits of various plants, fungi, and even animal exudations (Zhang et al. 2020). Natural products' potential is also visible today; for example, salicin, a glycoside found in the white willow tree (*Salix alba*), is metabolized into salicylic acid (Moors et al. 2014). The WHO acknowledges the availability and utility of phytotherapy, especially in low-income countries (WHO Traditional Medicine Strategy 2002–2005, n.d.).

In this chapter, we shed light on the positive benefits of natural compounds in SARS-CoV-2. Many natural compounds are currently being evaluated for possible anti-COVID-19 medicines using in silico assays (Antonio et al. 2021;Antonio et al. 020;Sayed et al. 2020).The in silico tests are important as they are not time-consuming as the laboratorial procedures. They can control illness symptoms or forecast which metabolites will be capable of affecting the various targets associated with the SARS-CoV-2 life cycle (Kalhori et al. 2021). PLPRO, MPRO, and RdRp are examples of the non-structural proteins of the virus, while spike (S) protein is an example of a structural protein. In silico research has demonstrated that a variety of natural products inhibit these proteins because of their substantial binding affinity. The virus also demonstrates its validity as a target for therapeutic development because it makes use of the host cell's transmembrane ACE2 receptor. Many natural compounds have been tested for their effectiveness versus the main COVID-19 therapy targets, which include NSP25, Nsp15, 3CL pro/Mpro, TMPRSS2, Nsp2, PLPro, RdRp, Cathepsin L, ACE2 receptor, Spike (s) protein, and nucleocapsid (N) protein. The binding energy or docking score of a few natural compounds, such as steroids, quinones, flavanones, flavonols, terpenes, fatty acids, and alkaloids, was analogous to that of repurposed drugs with proteins implicated in COVID-19 replication (Antonio et al. 2020). To counteract COVID-19, a variety of natural products are effective against different targets. A small number of substances, including gingerol, asparagoside-C and asparagoside-D, diosgenin, allicin, curcumin, quinadoline B, procyanidin, luteolin, gallocatechin gallate, kaempferol, and quinadoline B, have the ability to act against COVID-19 in many ways. Numerous investigations have contributed to the understanding of the mechanism of action of these natural products as well as prospective targets such as viral entry, copying, construction, release, and host-specific interactions(Cheng et al. 2006; Lin et al. 2014). Important SARS-CoV-2 viral proteins that are possible targets of a natural product are shown in Table 12.1.

There have been some instances of traditional herbal medication helping to combat COVID-19. Human immune cells are modulated by ginseng (*Panax ginseng*) (Kang and Min 2012). According to Q. Cheng et al. (2019), ginger (*Zingiber officinale*) possesses antioxidant, anti-apoptotic, anti-tumor, anti-inflammatory, anti-hyperglycemic, and analgesic qualities. *Allium sativum*, or garlic, boosts immunity (Daliri et al. 2019). Antimicrobial and antioxidant properties are present in echinacea extract (*Echinacea purpurea* (L.) *Moench*) (Sharifi-Rad et al. 2018). According to Cheng et al. (2006), extracts from *Artemisia annua*, *Lycoris radiate*, *Pyrrosia lingua*, and *Lindera aggregate*, as well as triterpene glycosides derived from *Heteromorpha arborescens*, have been shown to be useful in lowering infectious conditions. Natural suppressors of the SARS-CoV enzymes were found, including the nsP13 helicase, 3CL protease, and phenolic compounds from *Isatis indigotica* and *Torreya nucifera*, as well as myricetin and scutellarein (Yu et al. 2012). Furthermore, it has been reported by Cinatl et al. (2003) that glycyrrhizin exhibited a noteworthy antiviral effect on COVID-19. There has been recent research suggesting that *Nigella sativa*, or black seed, may be beneficial in alleviating symptoms and treating COVID-19 patients (Badary et al. 2021; Abdelrahim et al. 2022; Kulyar et al. 2021; Taysi et al. 2022).When the action of Nigella sativa seeds' ethanolic extract and other extracts (*Anthemis hyalina* and *Citrus sinensis* peels) was evaluated for their effects on the expression of TRP genes during the infection and on the replication of the virus, the results revealed a significant shift in both gene expression and a decrease in the viral load after treatment (Ulasli et al. 2014). A randomized, open-label clinical study assessed Nigella sativa oil's potential as a herbal supplement for COVID-19 treatment. For ten days, the patients were given 500 mg of the oil twice a day. When compared to the traditional treatment, the groups receiving NSO showed superior recovery, according to the findings (Koshak et al. 2021).

Table 12.1 SARS-CoV-2 viral proteins that may act as targets for potential natural products

Name of natural product	Class of the compound	Potential target(s)	Possible mechanism of action of the compounds	Refs.
Curcumin Hesperetin Hesperidin Tangeretin	Flavonoid Polyphenol	MERS-CoV S protein	Host cell viral entry inhibition	Yamamoto et al. (2016)
Quercetin	Flavonoid	SARS-CoV S protein SARS-CoV MPRO	Host cell viral entry and viral life cycle inhibition	Nguyen et al. (2012), Ghosh et al. (2021)
Heptafuhalol A	Marine phorethol	SARS-CoV-2 MPRO	Intracellular viral lifecycle inhibition	Gentile et al. (2020)
Naringenin	Flavanone	SARS-CoV-2 S protein	Host cell viral entry inhibition	Tutunchi et al. (2020)
Phlorizin, Rutin	Flavonoid Glycoside	SARS-CoV-2 MPRO, host TMPRSS2 SARS-CoV-2 RdRp	Host cell viral entry and viral life cycle inhibition	Singh et al. (2020)
Berberine	Isoquinoline alkaloid	SARS-CoV Nucleic acid intercalation	Viral replication inhibition	(Schmeller et al. 1997)
Thymoquinone	Monoterpene	SARS-CoV MPRO	Intracellular viral lifecycle inhibition	Zhang et al. (2021)
Salvinorin A	Terpenoid	SARS-CoV MPRO	Intracellular viral lifecycle inhibition	Zhang et al. (2021)
Ginkgolide A	Diterpene	SARS-CoV MPRO	Intracellular viral lifecycle inhibition	Zhang et al. (2021)
Phycocyanobilin	Marine tetrapyrrole chromophore	SARS-CoV-2 MPRO, SARS-CoV-2 RdRp	Intracellular viral lifecycle inhibition	Naz et al. (2023)
Puupehedione	Marine natural product	SARS-CoV-2 MPRO	Intracellular viral lifecycle inhibition	Khan et al. (2020)

MPRO(3CLPRO) Main protease (3C-like protease), *TMPRSS2* transmembrane protease serine 2,*RdRp* RNA-dependent RNA polymerase

12.3 Secondary Metabolites in COVID-19 Management

Significant natural products such as alkaloids, polyphenols, flavonoids, terpenoids, cardiac glycosides, and marine products are discussed in this review that may be suggested for the treatment of COVID-19. Plant secondary metabolites have diverse chemical structures with varying biological activity that place them superior to treat serious diseases. Several studies reported secondary metabolites as promising alternative medicines and lead compounds for drug repurposing and discovery. This diversity might suggest novel antibiotics multi-drug-resistant microbes.

12.3.1 Alkaloids (Fig. 12.1)

Alkaloids are basic plant secondary metabolites (PSM) that include at least one nitrogen atom in their structure. Higher terrestrial plants (Nussbaum 2003), fungi like psilocybin from Psilocybe, mammals like bufotenin from toad skin, and marine species all generate alkaloids (Fattorusso and Taglialatela-Scafati 2007). Other small compounds with broad-spectrum antiviral

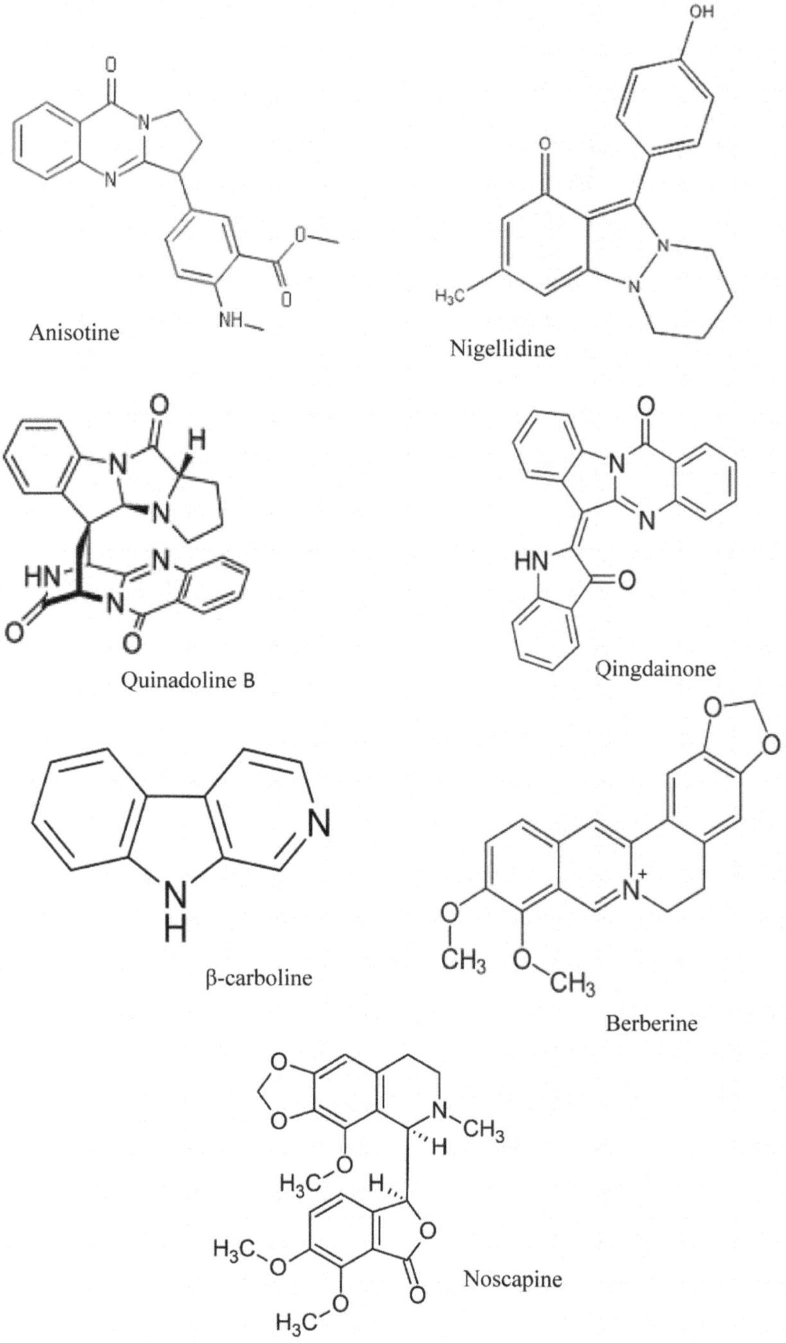

Fig. 12.1 Selected alkaloids structures withanti-COVID-19 activity

activity being studied as possible COVID-19 treatments include plant-derived alkaloids including emetine, homoharringtonine, and cepharanthine (Wang and Yang 2020). Cepharanthine inhibits the attachment of the virus to the angiotensin-converting enzyme 2 (ACE2) receptor, hence preventing it from infecting the host cell (Fan et al. 2020).

Transmembrane protease serine 2 (TMPRSS2) is a host cell receptor that could control SARS-CoV-2 entrance. Qingdainone's affinity for TMPRSS2 was investigated by Vivek-Ananth and colleagues and was found to have effectiveness against the virus (Vivek-Ananth et al. 2020). Qingdainone may be obtained from *Yarrowialipolytica* (Jahng 2013) and *Isatis tinctoria* L. (Wu et al. 1997). Oxoturkiyenine, from *Hypecoum pendulum* L. (Mete and GöZlerr 1988), has a decreased affinity for cathepsin L, which is a necessary host cell receptor for the virus entrance. It was discovered that the *Cinchona calisaya* Wedd. derived 3,17-cinchophylline inhibited the virus via the cathepsin L receptor (Vivek-Ananth et al. 2020).

RNA-dependent RNA polymerase (RdRp), papain-like protease (PLpro), chymotrypsin-like protease (3CLpro), and spike (S) protein are some of the important SARS-CoV-2 proteins that infect the host cell.

The alkaloids isoquinoline, quinolone, and β-carboline have powerful antiviral activity and intercalating properties (Schmeller et al. 1997; Wink 2007, 2015). *Aspergillus giganteus* has been used to extract quinadoline B, which has also been obtained from *Ulva lactuca*, a marine alga (Chen et al. 2020) and *Aspergillus* sp. FKI-1746 (Koyama et al. 2008). Because of its strong propensity for binding to viral proteins, it has been demonstrated to exhibit efficacy against SARS-CoV2. The viral proliferation was also suppressed by isoquinoline alkaloids like dicentrine, berberine, berbamine, berberrubine, palmatine, coptisine, tetrandrine, fangchinoline, jatrorrhizine, and cepharanthine. Hydrastis canadensis L. (Berberidaceae) roots, rhizomes, stems, and bark can all be used to extract berberine. When the probable suppressor of 3CLpro, the primary protease in SARS-CoV-2, was docked and screened for viruses, it revealed a significantly inferior binding affinity to 3CLpro than other compounds obtained from *Tinospora cordifolia*. Furthermore, the docking simulation indicates that the berberine: 3CLpro complex is more stable than other inhibitors and that it effectively inhibits COVID-19 by blocking 3CLpro's activity (Chowdhury 2021). *Berberis vulgaris* L. (Freile et al. 2003), *Berberis aquifolium* Pursh (Cernakova and Koššálová 2002), *Berberis vulgaris* L. (Imanshahidi & Hosseinzadeh 2008), and *Corydalis chaerophylla* DC. (Ameer Basha et al. 2002) are other documented sources of berberine.

Intercalators are believed to be efficient antiviral medicines for COVID-19 since the FDA-approved treatment of choice for the existing epidemic, chloroquine, is a derivative of the intercalating alkaloid quinine (Wink 2007). Thorough in vitro and in vivo research should be done before the intercalating alkaloids are identified as additional therapy possibilities. Schizanthine Z, which is a tropane alkaloid derived from *Schizanthus porrigens*, can block the viral protease according to a molecular docking research study (Alfaro et al. 2020). An additional in silico study demonstrated a substantial binding affinity for the viral Mpro by the alkaloids 10-hydroxyusambarensine (Frédérich et al. 1999) found in the roots of *Strychnos usambarensis* and by cryptoquindoline and cryptospirolepine (Cimanga et al. 1998) found in *Cryptolepis sanguinolenta*. Mpro is a critical coronavirus enzyme that mediates transcription and replication of the virus. According to the same study, 10-hydroxyusambarensine exhibited the highest associations with the virus's 3CLpro, whereas cryptospirolepine had the best binding affinity and selectivity for the viral 3CLpro (Gyebi et al. 2021).

Speciophylline has a stronger affinity for 3CLpro than Nucleocapsid (N)3, which is the known inhibitor of 3CLpro. *Uncaria tomentosa* is the source of cadambine, which has a strong affinity for 3CLpro (Yepes-Pérez et al. 2020).

The alkaloids from *Cryptolepis sanguinolenta*, biscryptolepine, cryptomisrine, cryptoquindoline, and, cryptospirolepine revealed high attachment to the RdRp in another in silico investigation, indicating their potential as RdRp inhibitors (Borquaye et al. 2020). Alkaloids from *Justicia adhatoda* leaves anisotine, adhatodine, vasicoline, and vasicine also inhibited the viral Mpro by molecular dynamics investigation, making them a potential protease inhibitors (Kar et al. 2022).

Nigellidine, obtained from *Nigella sativa* L. seeds, was previously noted for its antibacterial, anti-hypertensive, anti-inflammatory, and immunomodulatory properties. According to Maiti and colleagues' docking investigation, nigellidine could prevent the viral N protein from performing its role. By binding to Cys240, nigellidine may possibly prevent the viral Nsp2, which may be concerned with the mitochondria's integrity and resistance to various stressors in the host cell. Comparing noscapine to chloroquine, ribavirin, and favipiravir reveals that noscapine binds to viral proteins with a greater affinity and a significantly lower affinity. However, according to Kumar et al. (2020), it was a promising natural treatment for COVID-19. Apart from its widely isolated natural source, *Papaver somniferum* L., there is sufficient data on noscapine and the production of its derivatives (Dang and Facchini 2012; Devine et al. 2018; Ni et al. 2011).

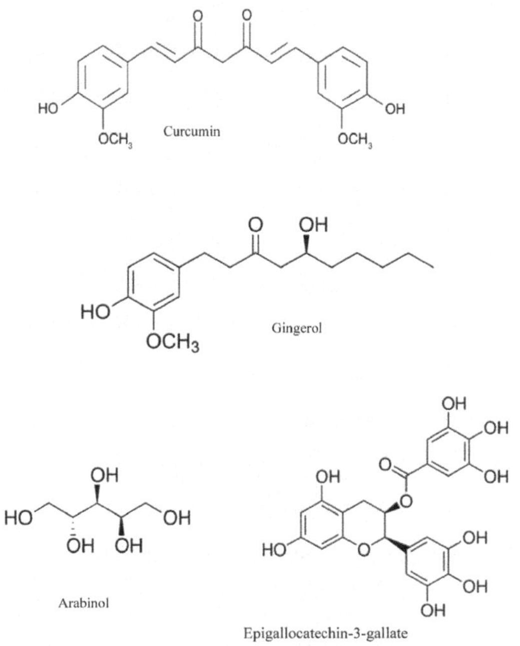

Fig. 12.2 Structures of polyphenols withanti-COVID-19 activity

12.3.2 Phenolic and Polyphenolic Compounds (Fig. 12.2)

Curcumin, the most significant phytonutrient in turmeric (*Curcuma longa* L.) evidenced, through previous research, that it opposes both HIV integrase and protease enzymes, therefore may have anti-AIDS activity and working in concert with antiviral medications (Gupta et al. 2020; Prasad and Tyagi 2015)(Gupta et al. 2020; Prasad and Tyagi 2015). On the other hand, in case of influenza A virus (Ciavarella et al. 2020; Gupta et al. 2020), it was discovered that H1N1 lowers the expression of nucleoproteins, preventing the influenza virus from infecting cells (Gupta et al. 2020; Lai et al. 2020; Richart et al. 2018). Collective studies effectively recommended that curcumin has powerful antiviral activity. This prompted in silico investigation to determine affinities toward COVID-19-associated proteases (cathepsin K, COVID-19 main protease, and SARS-CoV 3CLpro) by curcumin. According to the study findings, curcumin exhibited high affinity of binding to the protein target, with the SARS-CoV 3CLpro being the best (Oso et al. 2022).

Khan and colleagues discovered that Syn-16, a coumarin derivative, had the intriguing potential that it might stop the reproduction and growth of SARS-CoV-2 after conducting computational methods involving structural analysis, dynamic simulations, and evaluation of binding energetics (Khan et al. 2021a, b). It was also previously reported that the influenza A virus and norovirus was inactivated by gallocatechin gallate derived from the *Saxifraga spinulosa*. Takeda and colleagues researched the efficacy of a pyrogallol-enriched fraction against SARS-CoV-2. The results revealed that this fraction exhibited a rapid inhibition of cDNA reverse transcription compared to other fractions, reduced the S2 subunit of the S protein, and deactivated 99.53% of SARS-CoV-2 (Takeda et al. 2020).

In the docking investigation, arabinol found in *Senna occidentalis* showed that the strongest affinity for cathepsin L. Ararobinol is thought to have antiviral characteristics, according to earlier investigations. With cathepsin L residues, ararobinol can form hydrogen bonds (Vivek-Ananth et al. 2020). A study was conducted on gingerol, another natural bioactive compound which is capable of binding and inhibiting COVID-19-associated proteases. Gingerol is major phytoconstituent of *Zingiber officinale* Roscoe, and it is responsible for the pungent taste and medicinal properties of ginger. According to the study findings, gingerol has a strong affinity of binding to the enzyme target, particularly cathepsin K. They also found that hydrogen bonds link between catK amino acids and the SARS-CoV 3CLpro in their subsequent tests (Oso et al. 2022).

12.3.2.1 Flavonoids and Flavonoid Conjugates (Fig. 12.3)

Vegetables, seeds, and fruits are rich sources of flavonoids. More than 6000 structures of flavonoids were identified. Flavonoids are hydroxylated phenolic molecules with potent antioxidant activity due to the functional –OH groups. The unique structure of flavonoids is responsible for different pharmacological properties such as antiviral and anti-microbial, antioxidant, as well as anti-inflammatory combined with their capability to modulate key cellular enzyme function. The decrease in cardiovascular rate of mortality and prevention of coronary heart disease due to flavonoids drew the attention of researchers to their importance (Panche et al. 2016). Moreover, flavonoids have been studied extensively against DNA and RNA viruses (Lalani and Poh 2020). After the pandemic outbreak of COVID-19, the concerning finding alternative potential therapeutics from natural sources to treat-SARS-CoV-2 virus was raised.

Cytokine storm is one of the harshest cases of COVID-19 consequences, which is considered a critical medical condition accompanied by the acute respiratory distress syndrome which could lead to high percent of hospitalized patients as well as high mortality rate (Cron et al. 2021). Flavonoids are reputed for their significant immunomodulatory and anti-inflammatory effects.

Quercetin (a flavonol)reported to have several health management benefits including oxidative stress relief, regulation of the immune system. It also suppresses the proinflammatory and mediators responsible for inflammation, as well as histamine propagation inhibition (Mlcek et al. 2016). Inflammasomes are a group of multiple proteins found in the cytosol that come together as a reaction to cytosolic PAMP and DAMPs. Creating active forms of cytokines is their main target. NLRP3(excitation /prevention) inflammasome is influenced by TXNIP, SIRT1, and NRF2 regulators (Zhao and Zhao 2020). Quercetin hindered fructose-induced NOD-like receptor protein 3 activation via delaying the transportation of TXNIP in macrophage cell lines and accordingly prevented the aggravation of inflammation (Choe and Kim 2017). Hence, it was suggested that quercetin compound could be an important remedy for the severe inflammation accompanying COVID-19 that could be the main life threating for the patients (Saeedi-Boroujeni and Mahmoudian-Sani 2021). A clinical trial, ID code (IRCT2020041 9047128N2), was also conducted on high-risk COVID-19 patients treated with two tablets (500 mg quercetin) controlled by follow-up examination and measuring inflammatory cytokines; however, the results are not yet published (Saeedi-Boroujeni & Mahmoudian-Sani 2021).

Seven flavonoids; flavonols: fisetin, myricetin, kaempferol, and quercetin, flavones: luteolin and apigenin, and isoflavone: genistein was screened for its power to avoid the replication of SARS-CoV-2 via cell-based assays by pursuing all the proteins that played an essential role in SARS-CoV-2 pathogenicity. Findings revealed that the concentration values (EC) of flavonols were 50 times lower than those of flavones and isoflavones, where fisetin and myricetin reported the highest EC values, suggesting that these two compounds are potential targets for preventing SARS CoV-2 replication. Moreover, (CC_{50}) was 60-fold higher than the corresponding EC_{50} values for most of the flavonoids tested. It showed a selectivity index (SI = 787) aligned with the therapeutic effects. The cell-based assays

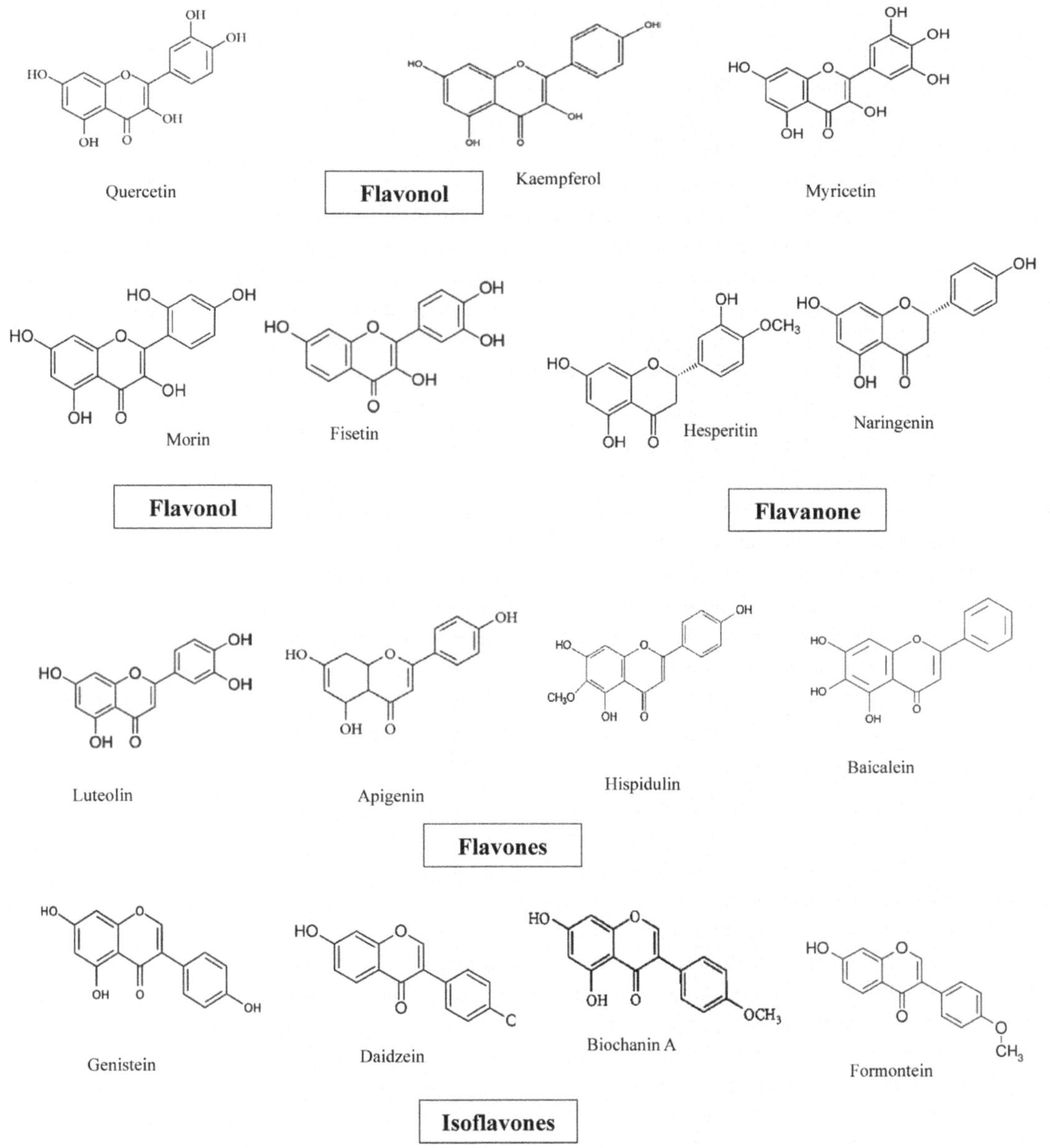

Fig. 12.3 Structures of flavonoids withanti-COVID-19 activity

showed that only genistein reduced (IL-6) (TNF-α) levels, while other flavonoids affected only TNF-α (Chaves et al. 2022).

Bioinformatics were implemented to specify the anti-virusputative genes of kaempferol (flavonol) in LPS-activated macrophages model. Significant results indicated a decrease in the levels of AKT1, JNK, and JUN transcription in lipopolysaccharide (LPS)-activated macrophages following treatment with kaempferol. Moreover, kaempferol attenuated the release of inflammatory mediators by lipopolysaccharide-induced macrophage, modulated the MAPK/NF-κB signaling pathway, and influenced macrophage opposition toward the M2 phenotype. Additionally, it mitigated the cytokine storm induced by endotoxins and enhanced survival in murine models. Notably, molecular docking analysis revealed the potential of kaempferol to interact with the COVID-19 protein 5R84. These findings suggest that kaempferol may hold promise in combating COVID-19 by ameliorating COVID-19-induced physiological dysregulation, modulating molecule–protein interactions, and identifying crucial target genes that could serve as molecular markers for COVID-19 detection (Sun et al. 2023).

Hesperitin and naringenin are flavanones which are present in citrus fruits and peels. A study was conducted on mouse adipocytes (in vitro) (Yoshida et al. 2010) both hesperitin and naringenin reported a downregulation in TNF-α inflammatory mediator release and accordingly ERK and NF-κB pathways were blocked, eventually hindering IL-6 transcription.

An in silico study was conducted to obstruct the main protease (Mpro) (critical for the replication of SARS-CoV) and spike (S) glycoprotein by hesperidin and other bioactive compounds with the aim that these compounds could be used as active oral drugs. The findings indicated that hesperidin exhibited superior inhibitory activity against the spike glycoprotein compared to nelfinavir, chloroquine, and hydroxychloroquine. Additionally, hesperidin demonstrated comparable efficacy to nelfinavir but outperformed chloroquine and hydroxychloroquine sulfate as inhibitors of the main protease (Mpro). The authors proposed that hesperidin could be a potential candidate for managing COVID-19 due to its strong binding affinity relative to nelfinavir, chloroquine, and hydroxychloroquine sulfate (Tallei et al. 2020).

Elfiky (2021) reported by molecular computation and protein–ligand docking the affinity power for isoflavones (phytoestrogens): formontein, genistein, daidzein, and biochanin A, against HSPA5 substrate-binding domain b (SBDb), which is regarded as ligand binding position for the SARS-CoV-2. Results revealed that the phytoestrogens showed the best binding HSPA5; therefore, these compounds may interfere with SARS-CoV-2 coherence to the cells under stress. Accordingly, it was suggested that phytoestrogens could be used as anti-COVID-19 for patients liable to stress. These isoflavones can be effective as anti-COVID-19 agents for people prone to cell stress such as cancer patients and old age.

In a recent study, in the year 2023, hispidulin (flavone with potent antiepileptic activity), morin (flavonol isolated from the family Moraceae) alongside with baicalein (trihydroxyflavone) (100 μg/mL) were tested in silico to identify their strong attraction toward SARS-CoV-2 protease 3CLpro enzymatic activity which is an important key target for drug modification for COVID-19 remedy. The toxicity-tested metabolites were also tested in vitro and, for their antiviral activity against SARS-CoV-2, in variable cell lines. The results revealed that baicalein evidenced potent 3CLpro inhibition, while hispidulin and morin demonstrated limited effect. Furthermore, baicalein prevented more than 50% of SARS-CoV-2 replication, thus confirming baicalein anti-SARS-CoV-2(Mori et al. 2023).

SARS-CoV-2 can cause substantial inflammation in the gastrointestinal (GI) tract of infected individuals, thus leading to many clinical symptoms including diarrhea. A study was achieved to test the anti-diarrheal effect of flavonoids. In this study, the flavonoids, especially epigallocatechin-3-gallate, potentiate the Nrf2 transcription factor that leads to the downregulation ACE2 expression from the cell surfaces. As regards COVID-19-induced diarrhea, flavonoids can relieve COVID-19 induced diarrhea. Alternatively, quercetin delays the interface of the SARS-CoV-2 spike protein with ACE2 receptor and restrains the inflammatory markers. However, preclinical and clinical studies are recommended (Paul et al. 2021).

12.3.3 Terpenoids (Fig. 12.4)

There are a considerable and diverse class of bioactive compounds generated from herbal sources. Terpenoids have the potential to be therapeutic agents, evidenced from several bioassays. Terpenoids have long been regarded as the primary ingredient in essential oils. Certain terpenoids are derived from marine species, while

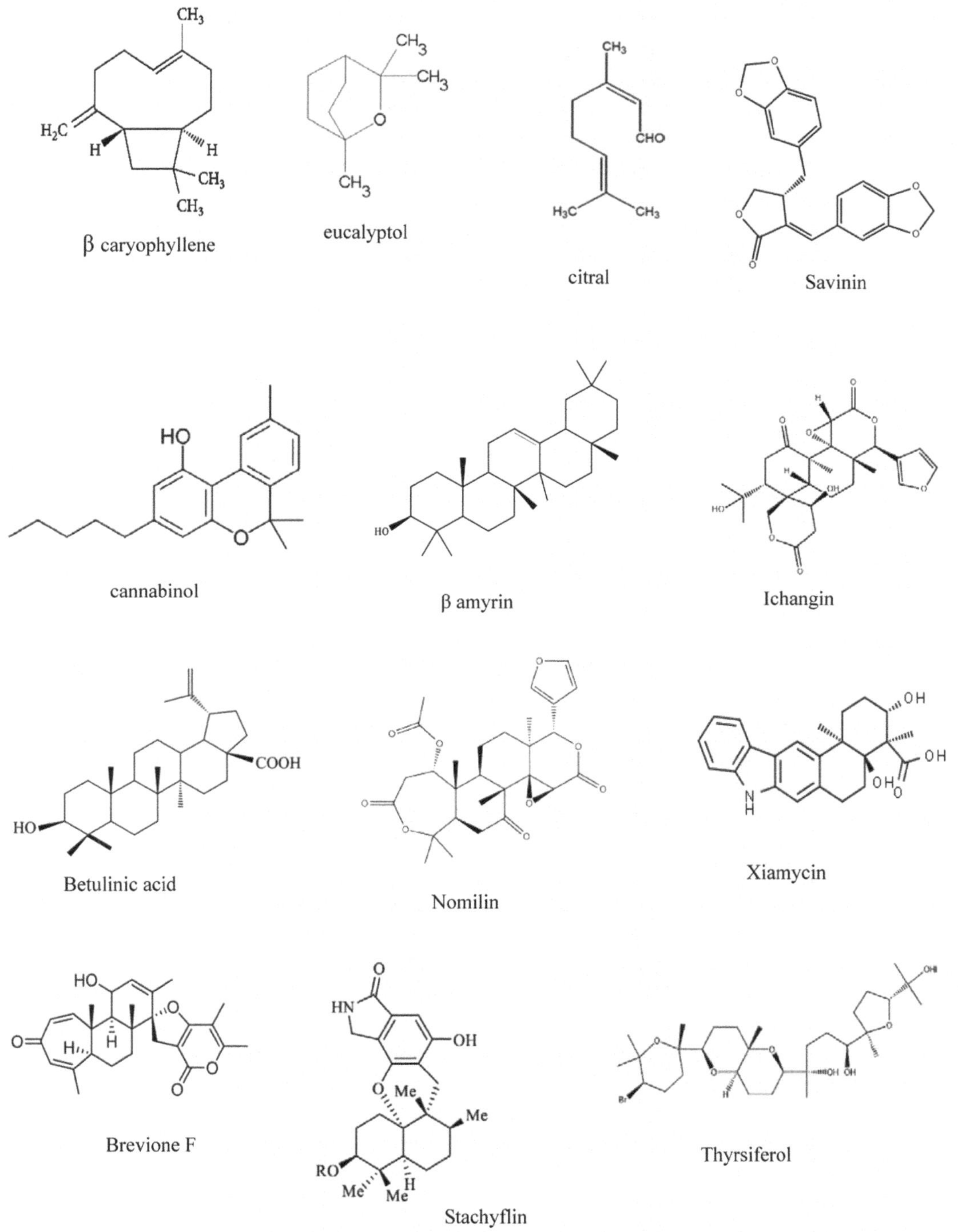

Fig. 12.4 Structures of terpenoids with anti-COVID-19 activity

most terpenoids are derived from plants through the mevalonate pathway. They are classified according to the number of carbons: isoprene, monoterpenes, sesquiterpenes, diterpenes, triterpenes, (C5), (C10), (C15), (C20), (C 30), respectively. Wen et al. (2007) investigated the impact on respiratory distress (acute case) accompanying coronavirus (SARS-CoV) using more than 200 isolated terpenes from several plant, and this was achieved by assessing SARS-CoV-induced cytopathogenic effect on Vero E6 cells. Fourteen terpenoids: 10 diterpenoids, 2 sesquiterpenoids, 2 triterpenoids; 5 lignoids; curcumin in addition to niclosamide and valinomycin as reference drugs were selected and revised for the down-regulation of the protease enzyme 3CL action. Results revealed that the most active compounds were from diterpene class, one triterpene (betulinic acid) and one lignoid (savinin). Moreover, betulinic acid and savinin inhibited the tested enzyme. The authors recommended abietane-type diterpenoids and lignoids as potent anti-SARS-CoV (Wen et al. 2007). In another research, a new terpene-based formula (β caryophyllene, eucalyptol, and citral) was tested alongside with human strain 229E (MRC-5 cells) in combination with cannabinol compared to pyrazofurin and glycyrrhizin (positive control). Results evidenced that the terpene-based formula + cannabinol were more potent antiviral as regards the positive controls (Chatow et al. 2021).

Recently, a computational study was conducted on limonoids and terpenoids to review their potency to prevent the most important therapeutic target proteins of SARS-CoV-2. The limonoids (de-acetylnomilin, ichangin and nomilin) and terpenoids (β-amyrin) evidenced superior binding affinity with SARS-CoV-2 3CL hydrolase (Mpro) advocating their role in preventing virus replication and growth. Further in vitro and in vivo investigations were recommended to get a good insight of the inhibitory potential (Giofrè et al. 2021).

In view of finding alternative solutions for COVID-19 treatment, which is usually accompanied by severe acute respiratory syndrome, terpenoids are the major constituents of marines with antiviral activity. A study was conducted on 68 terpenoids having broad-spectrum antiviral effect, applying computational docking against two tested enzymes (Mpro or 3CLpro) and RNA-dependent RNA polymerase (RdRp). Results revealed that out of the sixty-eight marine terpenoids, both brevione F and stachyflin showed almost the same activity as lopinavir and darunavir (reference antiviral drugs) against the target SARS−CoV−Mpro. Likewise, xiamycin, thyrsiferol, liouvilloside A & B, and stachyflin evidenced relatively more docking counts than the antiviral drugs remdesivir and favipiravir against the target SARS-CoV-2 – RdRp. This study spotted stachyflin as a lead compound against SARS-CoV-2(Sahoo et al. 2021).

12.3.4 Cardiac Glycosides (Fig. 12.5)

Cardiac glycosides are secondary metabolites with unique structure that are made up of three parts: a steroid ring, a lactone ring with 5–6 carbons, and a sugar moiety. The plants rich in cardiac glycosides are Digitalis leaves, Strophanthus seeds, and the squill bulb plant. Digoxin is a cardenolide isolated from *Digitalis lanata* leaves. Likewise, ouabain isolated from strophanthus seeds are used to treat cardiac insufficiency via Na^+/K^+-ATPase (NKA) modulation (Diederich et al. 2017). Cardiac glycosides' antiviral activity, fordigoxin and digitoxin, was correlated to the inhibition pf NKA, and this was suggested as treatment for SARS-CoV-2 infection((Burkard et al. 2015; Souza e Souza et al. 2021).

Cardiac glycosides play a key role in hindering the propagation process of the virus by blocking the incorporation of transmembrane of the virus as well as the viral RNA and pre-RNA bond; this is vital step for the creation of viral proteins formation (Amarelle et al. 2019). Since nuclear factor kappa B (NFκB) is crucial transcriptive factors for SARS-CoV-2 infection, accordingly by inhibiting NFκB is necessary route for SARS-CoV-2 treatment, cardiac glycosides interfere completely with SARS-CoV-2 by inhibiting NFκB (Souza e Souza et al. 2021;Yang et al. 2005). Further, digitoxin eliminates the production of proinflammatory cytokines: (TNFα), (IL-1β), (MIP2-α), e (MCP1), (TGFβ), and (IFNγ) that are expressed during lung epithelial cell infection (Yang et al. 2005).

Fig. 12.5 Structures of cardiac glycosides with anti-COVID-19 activity

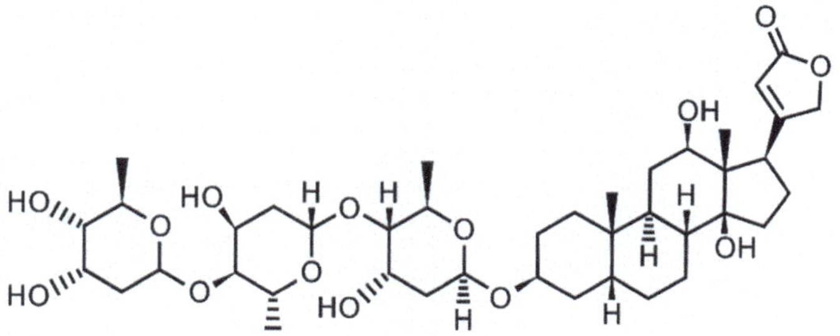

Digoxin

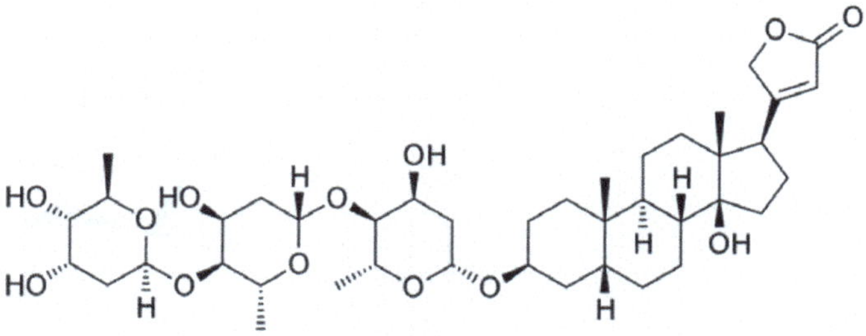

Digitoxin

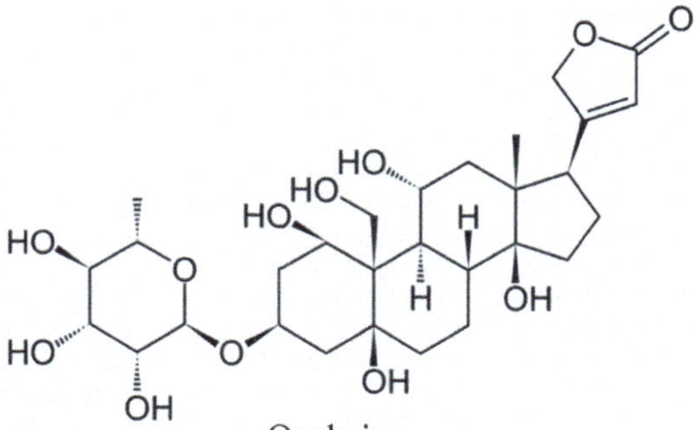

Ouabain

Digoxin also decreases the expression of proinflammatory cytokines accompanying arthritis (Saeed et al. 2020). Ouabain prevents the expression of Integrin beta chain-2 (CD18) (in vivo)in monocytes; this results in hindering the monocyte from reaching inflamed injury (Cavalcante-Silva et al. 2020) Additionally, the eruption of fluid, leukocyte permeation, as well as the cytokines IL-1β and TNFα measures are decreased by Ouabain, in mice with zymosan-induced peritoneal inflammation (Leite et al. 2015).The antiviral activity of cardiac glycosides, as previously discussed, to ameliorate COVID-19 is correlated to cell signaling activation and anti-inflammatory activity this is manifested by abolishing viral replication and modulating inflammatory expression. Accordingly, adjusting the dose is a crucial step to activate cell signaling without affecting Na^+/K^+-ATPase modulation activity (Souza e Souza et al. 2021).

12.3.5 Natural Products from Marine Sources (Fig. 12.6)

Marines are natural ingredients that have are considered as prospective medicinal means. These bioactive compounds evidenced potent antibacterial, anticancer, anti-inflammatory, and

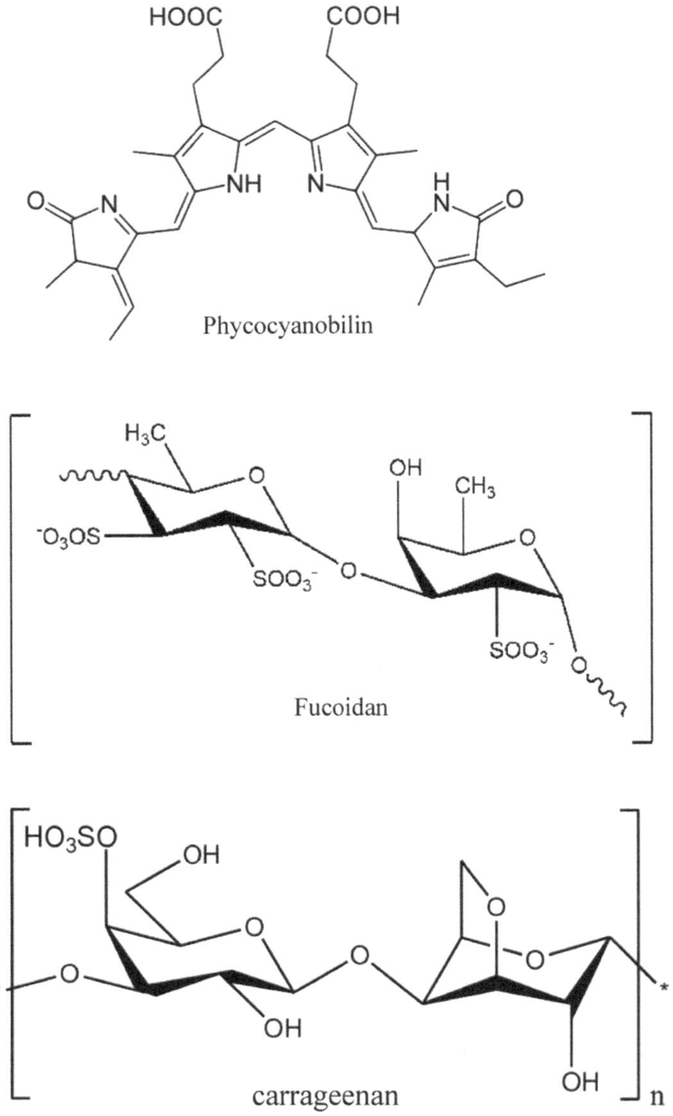

Fig. 12.6 Selected structures of marine metabolites as SARS-CoV-2 inhibitors

other pharmacological actions, and the most active reported compounds are phycocyanin, lutein, polysaccharides, vitamins, and phenolics, in addition to other bioactive substances present in microalgae of Rhodophyta and Phaeophyta (Sathasivam et al. 2019). In 2010, Hirata et al. investigated the antiviral and antioxidant properties of phycocyanobilins, which is a tetrapyrrole chromophores in cyanobacteria (Hirata et al. 2000). On the other hand, computational docking investigations revealed that phycocyanobilins exhibit a strong attraction for the MPRO and RdRpofSARS-CoV-2 (Pendyala et al. 2020). Alternatively, the antiviral action against hepatitis C and HIV-1 was revealed by griffithsin (a lectin produced from red algae)(Lusvarghi and Bewley 2016; Meuleman et al. 2011) and hence was suggested for possible applications. Moreover, griffinsin inhibited MERS-CoV, based on an in vitro study by Millet et al. (2016). The brown microalgae *Sargassum henslowianum*is rich in (sulfated polysaccharides); this compound revealed strong in vitro inhibitory effect on HSV (Sun et al. 2020). Fucoidans were also extracted from *Saccharina japonica* (macroalgae) and proven considerable antiviral activity against SARS-CoV-2 where one the fucoidans (out of 2 implied in the investigation) showed more antiviral efficacy than remdesivir. Accordingly, it was assumed that fucoidans can be provided alongside other antiviral agents to effectively inhibit SARS-CoV-2 (Kwon et al. 2020). Further, the compound ethyl esculetate derived from *Axinella corrugata* (marine sponge) exhibited robust interactions with the SARS-CoV-2 protease, suggesting its potential utility as a remedy against the virus (Vijayaraj et al. 2020). A group of marine sulfated polysaccharides known as carrageenan is thought to inhibit several viruses. The study was conducted by hindering the virus's attachment to host cells and subsequent incorporation. To avoid COVID-19 infection, Nagle et al. conjectured that these compounds might be applied as coating materials on hygienic goods (Nagle et al. 2020).In silico surveys recently assisted to find potential lead chemicals for the creation of drugs to combat the COVID-19 pandemic. In another report, five highly effective compounds of marine nature were studied as—SARS-CoV-2 MPRO—inhibitors, by analyzing their molecular conformational changes, that could be applied to impede the viral life cycle in the host (Khan et al. 2021a, b). The scarcity, logistical challenges associated with collection, and relative insufficiency of marine continue to pose significant obstacles in the discovery of potent bioactive compounds from this vast and underexplored reservoir. However, the past few decades have witnessed a stream in the identification of marine natural products exhibiting a remarkable diversity of biological activities. This momentum provides a compelling rationale for further exploration of marine biodiversity as a potential treasure trove for novel therapeutic interventions.

12.4 Conclusion and Future Prospective

The SARS-CoV-2 pandemic results in about 762 million human's death, where the major causes of death are acute respiratory distress and organ injury especially lung damage due to severe inflammation and cytokine storm. The principal recommended conventional drugs for the management of this disease are remdesivir, chloroquine, and dexamethasone. Vaccines targeting COVID-19 are available; however, their high cost and side effects limit their use. The search for safe and cost-effective alternative remedies for COVID-19 is a crucial step, especially in low-income countries. Repurposing of natural products has been advocated by researchers and approved by some countries' COVID-19 management. Phytotherapy is now recognized by WHO as many natural herbs and their secondary metabolites that provided evidence as lead molecules and complementary medicines in drug discovery.

This review highlights the importance of metabolites of natural resource such as alkaloids, polyphenols, flavonoids, terpenoids, cardiac glycosides, and marine metabolites in COVID-19 management. These metabolites act either by inhibiting proinflammatory markers that are responsible for the severe inflammation or by blocking the active sites that bind with the SARS-CoV-2 and hence hinder the virus propagation and replication as well as reducing the viral load. Their mechanisms are proved either in silico or in vitro; therefore, further preclinical and quantifiable trials are requested to justify their application in the marketplace.

This review paves the way for implementing new therapeutic models from natural sources such as plants, bacteria, and fungi with lower side effects and affordable costs. Additionally, combinatorial therapy between conventional and natural compounds could be implemented to lessen the side effects of the former and increase the efficacy of the latter.

References

Abdelrahim M, Esmail A, Al Saadi N, Zsigmond E, Al Najjar E, Bugazia D, Al-Rawi H, Alsaadi A, Kaseb AO (2022) Thymoquinone's antiviral effects: it is time to be proven in the Covid-19 pandemic era and its omicron variant surge. Front Pharmacol 13. https://doi.org/10.3389/fphar.2022.848676

Alfaro M, Alfaro I, Angel C (2020) Identification of potential inhibitors of SARS-CoV-2 papain-like protease from tropane alkaloids from Schizanthus porrigens: a molecular docking study. Chem Phys Lett 761. https://doi.org/10.1016/J.CPLETT.2020.138068

Amarelle L, Katzen J, Shigemura M, Welch LC, Cajigas H, Peteranderl C, Celli D, Herold S, Lecuona E, Sznajder JI (2019) Cardiac glycosides decrease influenza virus replication by inhibiting cell protein translational machinery. Am J Physiol Lung Cell Mol Physiol 316:1094–1106. https://doi.org/10.1152/ajplung.00173.2018.-Cardiac

Ameer Basha S, Mishra RK, Jha RN, Pandey VB, Singh UP (2002) Effect of berberine and (±)-bicuculline Isolated from *Corydalis chaerophylla* on spore germination of some fungi. Folia Microbiol 47(2):161–165. https://doi.org/10.1007/BF02817675/METRICS

Ancient Egyptian medicine—PubMed (n.d.) Retrieved 25 April 2023, from https://pubmed.ncbi.nlm.nih.gov/10326089/

Antonio ADS, Wiedemann LSM, Veiga-Junior VF (2020) Natural products' role against COVID-19. RSC Adv 10(39):23379–23393. https://doi.org/10.1039/d0ra03774e

Antonio AS, Wiedemann LSM, Galante EBF, Guimarães AC, Matharu AS, Veiga-Junior VF (2021) Efficacy and sustainability of natural products in COVID-19 treatment development: opportunities and challenges in using agro-industrial waste from citrus and apple. Heliyon 7(8). https://doi.org/10.1016/j.heliyon.2021.e07816

Badary OA, Hamza MS, Tikamdas R (2021) Thymoquinone: a promising natural compound with potential benefits for COVID-19 prevention and cure. In: Drug design, development and therapy, vol 15. Dove Medical Press Ltd., pp. 1819–1833. https://doi.org/10.2147/DDDT.S308863

Badary OA (2021) Pharmacogenomics and COVID-19: clinical implications of human genome interactions with repurposed drugs. Pharmacogenom J 21(3):275–284. https://doi.org/10.1038/s41397-021-00209-9

Benke C, Autenrieth LK, Asselmann E, Pané-Farré CA (2020) Lockdown, quarantine measures, and social distancing: associations with depression, anxiety and distress at the beginning of the COVID-19 pandemic among adults from Germany. Psychiatry Res 293. https://doi.org/10.1016/j.psychres.2020.113462

Boni MF, Lemey P, Jiang X, Lam TTY, Perry BW, Castoe TA, Rambaut A, Robertson DL (2020) Evolutionary origins of the SARS-CoV-2 sarbecovirus lineage responsible for the COVID-19 pandemic. Nat Microbiol 5(11):1408–1417. https://doi.org/10.1038/s41564-020-0771-4

Borquaye LS, Gasu EN, Ampomah GB, Kyei LK, Amarh MA, Mensah CN, Nartey D, Commodore M, Adomako AK, Acheampong P, Mensah JO, Mormor DB, Aboagye CI (2020) Alkaloids from *Cryptolepis sanguinolenta* as potential inhibitors of SARS-CoV-2 viral proteins: an in silico study. BioMed Res Int. https://doi.org/10.1155/2020/5324560

Burkard C, Verheije MH, Haagmans BL, van Kuppeveld FJ, Rottier PJM, Bosch B-J, de Haan CAM (2015) ATP1A1-mediated Src signaling inhibits coronavirus entry into host cells. J Virol 89(8):4434–4448. https://doi.org/10.1128/jvi.03274-14

Cavalcante-Silva LHA, de Lima ÉA, Carvalho DCM, Rodrigues-Mascarenhas S (2020) Ouabain reduces the expression of the adhesion molecule CD18 in neutrophils. Inflammopharmacology 28(3):787–793. https://doi.org/10.1007/s10787-019-00602-8

Cernakova M, Košťálová D (2002) Antimicrobial activity of berberine—a constituent of *Mahonia aquifolium*. Folia Microbiol 47(4):375–378. https://doi.org/10.1007/BF02818693

Chatow L, Nudel A, Nesher I, Hemo DH, Rozenberg P, Voropaev H, Winkler I, Levy R, Kerem Z, Yaniv Z, Eyal N (2021) In vitro evaluation of the activity of terpenes and cannabidiol against human coronavirus E229. Life 11(4). https://doi.org/10.3390/life11040290

Chaves OA, Fintelman-Rodrigues N, Wang X, Sacramento CQ, Temerozo JR, Ferreira AC, Mattos M, Pereira-Dutra F, Bozza PT, Castro-Faria-neto HC, Russo JJ, Ju J, Souza TML (2022) Commercially available flavonols are better SARS-CoV-2 inhibitors than isoflavone and flavones. Viruses 14(7). https://doi.org/10.3390/v14071458

Chen JJ, Wang SW, Chiang YR, Pang KL, Kuo YH, Shih TY, Lee TH (2020) Highly oxygenated constituents from a marine alga-derived fungus *Aspergillus giganteus* NTU967. Mar Drugs 18(6). https://doi.org/10.3390/MD18060303

Cheng PW, Ng LT, Chiang LC, Lin CC (2006) Antiviral effects of saikosaponins on human coronavirus 229E in vitro. Clin Exp Pharmacol Physiol 33(7):612–616. https://doi.org/10.1111/j.1440-1681.2006.04415.x

Cheng Q, Shahrajabian MH, Sun W (2019) The power of natural chinese medicine, ginger and ginseng root in an organic life. Middle-East J Sci Res 27(1):64–71. https://doi.org/10.5829/idosi.mejsr.2019.64.71

Choe JY, Kim SK (2017) Quercetin and ascorbic acid suppress fructose-induced NLRP3 inflammasome activation by blocking intracellular shuttling of TXNIP in human macrophage cell lines. Inflammation 40(3):980–994. https://doi.org/10.1007/s10753-017-0542-4

Chowdhury P (2021) In silico investigation of phytoconstituents from Indian medicinal herb '*Tinospora cordifolia* (giloy)' against SARS-CoV-2 (COVID-19) by molecular dynamics approach. J Biomol Struct Dyn 39(17):6792–6809. https://doi.org/10.1080/07391102.2020.1803968

Ciavarella C, Motta I, Valente S, Pasquinelli G (2020) Pharmacological (or synthetic) and nutritional agonists of PPAR-γ as candidates for cytokine storm modulation in COVID-19 disease. Molecules (Basel, Switzerland) 25(9). https://doi.org/10.3390/MOLECULES25092076

Cimanga K, De Bruyne T, Pieters L, Totte J, Tona L, Kambu K, Vanden Berghe D, Vlietinck AJ (1998) Antibacterial and antifungal activities of neocryptolepine, biscryptolepine and cryptoquindoline, alkaloids isolated from *Cryptolepis sanguinolenta*. Phytomed Int J Phytotherapy Phytopharmacol 5(3):209–214. https://doi.org/10.1016/S0944-7113(98)80030-5

Cinatl J, Morgenstern B, Bauer G, Chandra P, Rabenau H, Doerr HW (2003) Glycyrrhizin, an active component of liquorice roots, and replication of SARS-associated coronavirus. Lancet 361:2045–2048. https://doi.org/10.1016/s0140-6736(03)13615-x

Cron RQ, Caricchio R, Chatham WW (2021) Calming the cytokine storm in COVID-19. Nat Med 27(10):1674–1675. https://doi.org/10.1038/s41591-021-01500-9

Cucinotta D, Vanelli M (2020) WHO declares COVID-19 a pandemic. Acta Biomedica 91(1):157–160. https://doi.org/10.23750/abm.v91i1.9397

Daliri EBM, Kim SH, Park BJ, Kim HS, Kim JM, Kim HS, Oh DH (2019) Effects of different processing methods on the antioxidant and immune stimulating abilities of garlic. Food Sci Nutr 7(4):1222–1229. https://doi.org/10.1002/fsn3.942

Dang TTT, Facchini PJ (2012) Characterization of three O-methyltransferases involved in noscapine biosynthesis in opium poppy. Plant Physiol 159(2):618–631. https://doi.org/10.1104/PP.112.194886

Devine SM, Yong C, Amenuvegbe D, Aurelio L, Muthiah D, Pouton C, Callaghan R, Capuano B, Scammells PJ (2018) Synthesis and pharmacological evaluation of noscapine-inspired 5-substituted tetrahydroisoquinolines as cytotoxic agents. J Med Chem 61(18):8444–8456. https://doi.org/10.1021/ACS.JMEDCHEM.8B00986/SUPPL_FILE/JM8B00986_SI_002.CSV

Diederich M, Muller F, Cerella C (2017) Cardiac glycosides: from molecular targets to immunogenic cell death. Biochem Pharmacol 125:1–11. https://doi.org/10.1016/j.bcp.2016.08.017

Elfiky AA (2021) Natural products may interfere with SARS-CoV-2 attachment to the host cell. J Biomol Struct Dyn 39(9):3194–3203. https://doi.org/10.1080/07391102.2020.1761881

Fan HH, Wang LQ, Liu WL, An XP, Liu ZD, He XQ, Song LH, Tong YG (2020) Repurposing of clinically approved drugs for treatment of coronavirus disease 2019 in a 2019-novel coronavirus-related coronavirus model. Chin Med J 133(9):1051. https://doi.org/10.1097/CM9.0000000000000797

Fattorusso E, Taglialatela-Scafati O (2007) Modern alkaloids: structure, isolation, synthesis and biology, pp 1–665. https://doi.org/10.1002/9783527621071

Forni G., Mantovani A, Moretta L, Rappuoli R, Rezza G, Bagnasco A, Barsacchi G, Bussolati G, Cacciari M, Cappuccinelli P, Cheli E, Guarini R, Bacci ML, Mancini M, Marcuzzo C, Morrone MC, Parisi G et al (2021) COVID-19 vaccines: where we stand and challenges ahead. Cell Death Different 28(2):626–639. https://doi.org/10.1038/s41418-020-00720-9

Frédérich M, Tits M, Hayette MP, Brandt V, Penelle J, DeMol P, Llabrès G, Angenot L (1999) 10'-Hydroxyusambarensine, a new antimalarial bisindole alkaloid from the roots of *Strychnos usambarensis*. J Nat Prod 62(4):619–621. https://doi.org/10.1021/NP980375M

Freile ML, Giannini F, Pucci G, Sturniolo A, Rodero L, Pucci O, Balzareti V, Enriz RD (2003) Antimicrobial activity of aqueous extracts and of berberine isolated from *Berberis heterophylla*. Fitoterapia 74(7–8):702–705. https://doi.org/10.1016/S0367-326X(03)00156-4

Gentile D, Patamia V, Scala A, Sciortino MT, Piperno A, Rescifina A (2020) Putative inhibitors of SARS-COV-2 main protease from a library of marine natural products: a virtual screening and molecular modeling study. Mar Drugs 18(4). https://doi.org/10.3390/md18040225

Ghosh R, Chakraborty A, Biswas A, Chowdhuri S (2021) Evaluation of green tea polyphenols as novel corona virus (SARS CoV-2) main protease (Mpro) inhibitors–an in silico docking and molecular dynamics simulation study. J Biomol Struct Dyn 39(12):4362–4374. https://doi.org/10.1080/07391102.2020.1779818

Giofrè SV, Napoli E, Iraci N, Speciale A, Cimino F, Muscarà C, Molonia MS, Ruberto G, Saija A (2021) Interaction of selected terpenoids with two SARS-CoV-2 key therapeutic targets: an in silico study through molecular docking and dynamics simulations. Comput Biol Med 134. https://doi.org/10.1016/j.compbiomed.2021.104538

Gupta H, Gupta M, Bhargava S (2020) Potential use of turmeric in COVID-19. Clin Exp Dermatol 45(7):902–903. https://doi.org/10.1111/CED.14357

Gyebi GA, Ogunro OB, Adegunloye AP, Ogunyemi OM, Afolabi SO (2021) Potential inhibitors of coronavirus 3-chymotrypsin-like protease (3CLpro): an in silico screening of alkaloids and terpenoids from African medicinal plants. J Biomol Struct Dyn 39(9):3396–3408. https://doi.org/10.1080/07391102.2020.1764868

Hamza MS, Badary OA, Elmazar MM (2021) Cross-sectional study on awareness and knowledge of COVID-19 among senior pharmacy students. J Community Health 46(1):139–146. https://doi.org/10.1007/s10900-020-00859-z

Hamza MS, Tikamdas R, Baghdady NSE, Sayed M, Elbarazi AS, Badary OA, Elmazar MM (2022) Safety and effectiveness of COVID-19 vaccines: results from a cross-sectional survey among staff, workers and students at an Egyptian University. Vaccines 10(6). https://doi.org/10.3390/vaccines10060846

Hirata T, Tanaka M, Ooike M, Tsunomura T, Sakaguchi M (2000) Antioxidant activities of phycocyanobilin prepared from *Spirulina platensis*. J Appl Phycol 12(3–5):435–439. https://doi.org/10.1023/A:1008175217194/METRICS

Imanshahidi M, Hosseinzadeh H (2008) Pharmacological and therapeutic effects of *Berberis vulgaris* and its active constituent, berberine. Phytotherapy Res (PTR) 22(8):999–1012. https://doi.org/10.1002/PTR.2399

Jahng Y (2013) Progress in the studies on tryptanthrin, an alkaloid of history. Arch Pharmacal Res 36(5):517–535. https://doi.org/10.1007/S12272-013-0091-9

Jeong YJ, Wi YM, Park H, Lee JE, Kim SH, Lee KS (2023) Current and emerging knowledge in COVID-19. Radiology 306(2):e222462. https://doi.org/10.1148/radiol.222462

Kalhori MR, Saadatpour F, Arefian E, Soleimani M, Farzaei MH, Aneva IY, Echeverría J (2021) The potential therapeutic effect of RNA interference and natural products on COVID-19: a review of the coronaviruses infection. Front Pharmacol 12. https://doi.org/10.3389/FPHAR.2021.616993

Kang S, Min H (2012) Ginseng, the "immunity boost": the effects of *Panax ginseng* on immune system. J Ginseng Res 36(4):354–368. https://doi.org/10.5142/jgr.2012.36.4.354

Kar P, Kumar V, Vellingiri B, Sen A, Jaishee N, Anandraj A, Malhotra H, Bhattacharyya S, Mukhopadhyay S, Kinoshita M, Govindasamy V, Roy A, Naidoo D, Subramaniam MD (2022) Anisotine and amarogentin as promising inhibitory candidates against SARS-CoV-2 proteins: a computational investigation. J Biomol Struct Dyn 40(10):4532–4542. https://doi.org/10.1080/07391102.2020.1860133

Khan MT, Ali A, Wang Q, Irfan M, Khan A, Zeb MT, Zhang YJ, Chinnasamy S, Wei DQ (2021a) Marine natural compounds as potents inhibitors against the main protease of SARS-CoV-2-a molecular dynamic study. J Biomol Struct Dyn 39(10):1–11. https://doi.org/10.1080/07391102.2020.1769733

Khan SA, Zia K, Ashraf S, Uddin R, Ul-Haq Z (2021b) Identification of chymotrypsin-like protease inhibitors of SARS-CoV-2 via integrated computational approach. J Biomol Struct Dyn 39(7):2607–2616. https://doi.org/10.1080/07391102.2020.1751298

Khan MT, Ali A, Wang Q, Irfan M, Khan A, Zeb MT, Zhang YJ, Chinnasamy S, Wei DQ (2020) Marine natural compounds as potents inhibitors against the main protease of SARS-CoV-2—a molecular dynamic study. J Biomol Struct Dynam 1–11. https://doi.org/10.1080/07391102.2020.1769733

Koshak AE, Koshak EA, Mobeireek AF, Badawi MA, Wali SO, Malibary HM, Atwah A, Alhamdan MM, Almalki RA, Madani TA (2021) *Nigella sativa* for the treatment of COVID-19: an open-label randomized controlled clinical trial. Complement Therap Med 61. https://doi.org/10.1016/j.ctim.2021.102769

Koyama N, Inoue Y, Sekine M, Hayakawa Y, Homma H, Omura S, Tomoda H (2008) Relative and absolute stereochemistry of quinadoline B, an inhibitor of lipid droplet synthesis in macrophages. Org Lett 10(22):5273–5276. https://doi.org/10.1021/OL802089P

Kulyar MFA, Li R, Mehmood K, Waqas M, Li K, Li J (2021) Potential influence of *Nagella sativa* (Black cumin) in reinforcing immune system: a hope to decelerate the COVID-19 pandemic. Phytomedicine 85:153277. https://doi.org/10.1016/J.PHYMED.2020.153277

Kumar N, Sood D, van der Spek PJ, Sharma HS, Chandra R (2020) Molecular binding mechanism and pharmacology comparative analysis of noscapine for repurposing against SARS-CoV-2 protease. J Proteome Res 19(11):4678–4689. https://doi.org/10.1021/ACS.JPROTEOME.0C00367

Kwon PS, Oh H, Kwon SJ, Jin W, Zhang F, Fraser K, Hong JJ, Linhardt RJ, Dordick JS (2020) Sulfated polysaccharides effectively inhibit SARS-CoV-2 in vitro. Cell Discov 6(1):1–4. https://doi.org/10.1038/s41421-020-00192-8

Lai Y, Yan Y, Liao S, Li Y, Ye Y, Liu N, Zhao F, Xu P (2020) 3D-quantitative structure-activity relationship and antiviral effects of curcumin derivatives as potent inhibitors of influenza H1N1 neuraminidase. Arch Pharmacal Res 43(5):489–502. https://doi.org/10.1007/S12272-020-01230-5

Lalani S, Poh CL (2020) Flavonoids as antiviral agents for enterovirus A71 (EV-A71). Viruses 12(2). https://doi.org/10.3390/v12020184

Leite JA, De Abreu Alves AK, Galvão JGM, Teixeira MP, Cavalcante-Silva LHA, Scavone C, Morrot A, Rumjanek VM, Rodrigues-Mascarenhas S (2015) Ouabain modulates zymosan-induced peritonitis in mice. Mediators Inflamm 2015. https://doi.org/10.1155/2015/265798

Li SY, Chen C, Zhang HQ, Guo HY, Wang H, Wang L, Zhang X, Hua SN, Yu J, Xiao PG, Li RS, Tan X (2005) Identification of natural compounds with antiviral activities against SARS-associated coronavirus. Antiviral Res 67(1):18–23. https://doi.org/10.1016/J.ANTIVIRAL.2005.02.007

Lin LT, Hsu WC, Lin CC (2014) Antiviral natural products and herbal medicines. J Tradit Complement Med 4(1):24. https://doi.org/10.4103/2225-4110.124335

Lluch MTG, Vega BP, Becerra MC, Rodríguez Á, Zalvide AE, Sánchez SÁ (2022) Retrospective observational study of the persistence of SARS-CoV-2 infection in patients previously treated with rituximab. Revista Espanola de Quimioterapia 35(3):260–264. https://doi.org/10.37201/req/122.2022

Lusvarghi S, Bewley CA (2016) Griffithsin: an antiviral lectin with outstanding therapeutic potential. Viruses 8(10). https://doi.org/10.3390/V8100296

Mete IE, GöZlerr T (1988) (+)-Oxoturkiyenine: an isoquinoline-derived alkaloid from *Hypecoum pendulum*. J Nat Prod 51(2):272–274. https://doi.org/10.1021/NP50056A013/ASSET/NP50056A013.FP.PNG_V03

Meuleman P, Albecka A, Belouzard S, Vercauteren K, Verhoye L, Wychowski C, Leroux-Roels G, Palmer KE, Dubuisson J (2011) Griffithsin has antiviral activity against hepatitis C virus. Antimicrob Agents Chemother 55(11):5159. https://doi.org/10.1128/AAC.00633-11

Millet JK, Séron K, Labitt RN, Danneels A, Palmer KE, Whittaker GR, Dubuisson J, Belouzard S (2016) Middle East respiratory syndrome coronavirus infection is inhibited by griffithsin. Antiviral Res 133:1–8. https://doi.org/10.1016/J.ANTIVIRAL.2016.07.011

Mlcek J, Jurikova T, Skrovankova S, Sochor J (2016) Quercetin and its anti-allergic immune response. Molecules 21(5). https://doi.org/10.3390/molecules21050623

Moors EHM, Cohen AF, Schellekens H (2014) Towards a sustainable system of drug development. Drug Discovery Today 19(11):1711–1720. https://doi.org/10.1016/j.drudis.2014.03.004

Mori M, Quaglio D, Calcaterra A, Ghirga F, Sorrentino L, Cammarone S, Fracella M, D'Auria A, Frasca F, Criscuolo E, Clementi N, Mancini N, Botta B, Antonelli G, Pierangeli A, Scagnolari C (2023) Natural flavonoid derivatives have pan-coronavirus antiviral activity. Microorganisms 11(2). https://doi.org/10.3390/microorganisms11020314

Nagle V, Gaikwad M, Pawar Y, Dasgupta S (2020) Marine red alga *Porphyridium* sp. as a source of sulfated polysaccharides (SPs) for combating against COVID-19. Preprints.

Naz F, Khan I, Baammi S, Islam A (2023) Investigation of the interactions of HSA and SARS-CoV-2 papain-like protease against eugenol for novel COVID-19 drug discovery: spectroscopic and insilico study. J Biomol Struct Dynam. 1–10. https://doi.org/10.1080/07391102.2022.2164062

Nguyen TTH, Woo HJ, Kang HK, Nguyen VD, Kim YM, Kim DW, Ahn SA, Xia Y, Kim D (2012) Flavonoid-mediated inhibition of SARS coronavirus 3C-like protease expressed in Pichia pastoris. Biotech Lett 34(5):831–838. https://doi.org/10.1007/s10529-011-0845-8

Ni J, Xiao H, Weng L, Wei X, Xu Y (2011) Blocking group-directed diastereoselective total synthesis of (±)-α-noscapine. Tetrahedron 67(29):5162–5167. https://doi.org/10.1016/J.TET.2011.05.060

von Nussbaum F (2003) Alkaloids. Nature's curse or blessing? By Manfred Hesse. AngewandteChemie Int Ed 42(40):4852–4854. https://doi.org/10.1002/ANIE.200385928

Oso BJ, Adeoye AO, Olaoye IF (2022) Pharmacoinformatics and hypothetical studies on allicin, curcumin, and gingerol as potential candidates against COVID-19-associated proteases. J Biomol Struct Dyn 40(1):389–400. https://doi.org/10.1080/07391102.2020.1813630

Panche AN, Diwan AD, Chandra SR (2016) Flavonoids: an overview. J Nutrition Sci 5. https://doi.org/10.1017/jns.2016.41

Paul AK, Jahan R, Bondhon TA, Jannat K, Hasan A, Rahmatullah M, Nissapatorn V, Pereira ML, Wiart C (2021) Potential role of flavonoids against sars-cov-2 induced diarrhea. Trop Biomed 38(3):360–365. https://doi.org/10.47665/TB.38.3.079 (Malaysian Society for Parasitology)

Pendyala B, Patras A, Professor A (2020) In silico screening of food bioactive compounds to predict potential inhibitors of COVID-19 main protease (Mpro) and RNA-dependent RNA polymerase (RdRp). https://doi.org/10.26434/CHEMRXIV.12051927.V2

Prasad S, Tyagi AK (2015) Curcumin and its analogues: a potential natural compound against HIV infection and AIDS. Food Funct 6(11):3412–3419. https://doi.org/10.1039/C5FO00485C

Richart SM, Li YL, Mizushina Y, Chang YY, Chung TY, Chen GH, Tzen JTC, Shia KS, Hsu WL (2018) Synergic effect of curcumin and its structural analogue (*Monoacetyl curcumin*) on anti-influenza virus infection. J Food Drug Anal 26(3):1015–1023. https://doi.org/10.1016/j.jfda.2017.12.006

Saeed H, Mateen S, Moin S, Khan AQ, Owais M (2020) Cardiac glycoside digoxin ameliorates pro-inflammatory cytokines in PBMCs of rheumatoid arthritis patients in vitro. Int Immunopharmacol 82. https://doi.org/10.1016/j.intimp.2020.106331

Saeedi-Boroujeni A, Mahmoudian-Sani MR (2021) Anti-inflammatory potential of Quercetin in COVID-19 treatment. J Inflamm (United Kingdom) 18(1). https://doi.org/10.1186/s12950-021-00268-6

Sahoo A, Fuloria S, Swain SS, Panda SK, Sekar M, Subramaniyan V, Panda M, Jena AK, Sathasivam KV, Fuloria NK (2021) Potential of marine terpenoids against sars-cov-2: an in silico drug development approach. Biomedicines 9(11). https://doi.org/10.3390/biomedicines9111505

Sathasivam R, Radhakrishnan R, Hashem A, Abd Allah EF (2019) Microalgae metabolites: a rich source for food and medicine. Saudi J Biol Sci 26(4): 709–722. https://doi.org/10.1016/J.SJBS.2017.11.003

Sayed AM, Khattab AR, AboulMagd AM, Hassan HM, Rateb ME, Zaid H, Abdelmohsen UR (2020) Nature as a treasure trove of potential anti-SARS-CoV drug leads: a structural/mechanistic rationale. RSC Adv 10(34):19790–19802. https://doi.org/10.1039/D0RA04199H

Schmeller T, Latz-Brüning B, Wink M (1997) Biochemical activities of berberine, palmatine and sanguinarine mediating chemical defence against microorganisms and herbivores. Phytochemistry 44(2):257–266. https://doi.org/10.1016/S0031-9422(96)00545-6

Sharifi-Rad M, Mnayer D, Morais-Braga MFB, Carneiro JNP, Bezerra CF, Coutinho HDM, Salehi B, Martorell M, del Mar Contreras M, Soltani-Nejad A, Uribe YAH, Yousaf Z, Iriti M, Sharifi-Rad J (2018) Echinacea plants as antioxidant and antibacterial agents: from traditional medicine to biotechnological applications. Phytother Res 32(9):1653–1663. https://doi.org/10.1002/ptr.6101

Singh R, Gautam A, Chandel S, Ghosh A, Dey D Roy, S., Ravichandiran, V., & Ghosh, D. (2020). Protease inhibitory effect of natural polyphenolic compounds on SARS-CoV-2: an in silico study. Molecules 25(20). https://doi.org/10.3390/molecules25204604

Souza e Souza KFC, Moraes BPT, de Paixão ICN, Burth P, Silva AR, Gonçalves-de-Albuquerque CF (2021) Na+/K+-ATPase as a target of cardiac glycosides for the treatment of SARS-CoV-2 infection. Front Pharmacol 12. https://doi.org/10.3389/fphar.2021.624704

Sun QL, Li Y, Ni LQ, Li YX, Cui YS, Jiang SL, Xie EY, Du J, Deng F, Dong CX (2020) Structural characterization and antiviral activity of two fucoidans from the brown algae *Sargassum henslowianum*. Carbohydrate Polymers 229. https://doi.org/10.1016/J.CARBPOL.2019.115487

Sun, Y., Tao, Q., Cao, Y., Yang, T., Zhang, L., Luo, Y., & Wang, L. (2023). Kaempferol has potential anti-coronavirus disease 2019 (COVID-19) targets based on bioinformatics analyses and pharmacological effects on endotoxin-induced cytokine storm. *Phytotherapy Research*, 1–10.

Takeda Y, Murata T, Jamsransuren D, Suganuma K, Kazami Y, Batkhuu J, Badral D, Ogawa H (2020) *Saxifraga spinulosa*-derived components rapidly inactivate multiple viruses including SARS-CoV-2. Viruses 12(7). https://doi.org/10.3390/V12070699

Tallei TE, Tumilaar SG, Niode NJ, Fatimawali, Kepel, BJ, Idroes R, Effendi Y, Sakib SA, Emran TB (2020). Potential of Plant Bioactive Compounds as SARS-CoV-2 main protease (Mpro) and spike (S) glycoprotein inhibitors: a molecular docking study. Scientifica. https://doi.org/10.1155/2020/6307457

Taysi S, Algburi FS, Mohammed ZR, Ali OA, Taysi ME (2022) Thymoquinone: a review on its pharmacological importance, and its association with oxidative stress, COVID-19, and radiotherapy. Mini Rev Med Chem 22(14):1847–1875. https://doi.org/10.2174/1389557522666220104151225

Tutunchi H, Naeini F, Ostadrahimi A, Hosseinzadeh-Attar MJ (2020) Naringenin, a flavanone with antiviral and anti-inflammatory effects: a promising treatment strategy against COVID-19. Phytotherapy Res 34(12):3137–3147. https://doi.org/10.1002/ptr.6781

Ulasli M, Gurses SA, Bayraktar R, Yumrutas O, Oztuzcu S, Igci M, Igci YZ, Cakmak EA, Arslan A (2014) The effects of *Nigella sativa* (Ns), *Anthemis hyalina* (Ah) and *Citrus sinensis* (Cs) extracts on the replication of coronavirus and the expression of TRP genes family. Mol Biol Rep 41(3):1703–1711. https://doi.org/10.1007/s11033-014-3019-7

Vijayaraj R, Altaff K, Rosita AS, Ramadevi S, Revathy J (2020). Bioactive compounds from marine resources against novel corona virus (2019-nCoV): in silico study for corona viral drug. 1–5. https://doi.org/10.1080/14786419.2020.1791115

Vivek-Ananth RP, Rana A, Rajan N, Biswal HS, Samal A (2020). In silico identification of potential natural product inhibitors of human proteases key to SARS-CoV-2 infection. Molecules 25(17):3822. https://doi.org/10.3390/MOLECULES25173822

Wang X, Fang X, Cai Z, Wu X, Gao X, Min J, Wang F (2020) Comorbid chronic diseases and acute organ injuries are strongly correlated with disease severity and mortality among COVID-19 patients: a systemic review and meta-analysis. Research. https://doi.org/10.34133/2020/2402961

Wang Z, Yang L (2020) Turning the tide: natural products and natural-product-inspired chemicals as potential counters to SARS-CoV-2 infection. Front Pharmacol 11:1013. https://doi.org/10.3389/FPHAR.2020.01013/BIBTEX

Wen CC, Kuo YH, Jan JT, Liang PH, Wang SY, Liu HG, Lee CK, Chang ST, Kuo CJ, Lee SS, Hou CC, Hsiao PW, Chien SC, Shyur LF, Yang NS (2007) Specific plant terpenoids and lignoids possess potent antiviral activities against severe acute respiratory syndrome coronavirus. J Med Chem 50(17):4087–4095. https://doi.org/10.1021/jm070295s

WHO Traditional Medicine Strategy 2002–2005 (n.d.) Retrieved 25Apri 2023, from https://apps.who.int/iris/handle/10665/67163

Wink M (2015) Modes of action of herbal medicines and plant secondary metabolites. Medicines 2(3):251. https://doi.org/10.3390/MEDICINES2030251

Wink M (2007) Molecular modes of action of cytotoxic alkaloids: from DNA intercalation, spindle poisoning, topoisomerase inhibition to apoptosis and multiple drug resistance. The Alkaloids Chem Biol 64(SUPPL.: COMPLETE):1–47. https://doi.org/10.1016/S1099-4831(07)64001-2

Wu X, Liu Y, Sheng W, Sun J, Qin G (1997) Chemical constituents of *Isatis indigotica*. Planta Med 63(1):55–57. https://doi.org/10.1055/S-2006-957604

Wu D, Wu T, Liu Q, Yang Z (2020) The SARS-CoV-2 outbreak: what we know. Int J Infect Diseases 94:44–48. https://doi.org/10.1016/j.ijid.2020.03.004

Yamamoto M, Matsuyama S, Li X, Takeda M, Kawaguchi Y, Inoue JI, Matsuda Z (2016) Identification of nafamostat as a potent inhibitor of middle east respiratory syndrome coronavirus s protein-mediated membrane fusion using the split-protein-based cell-cell fusion assay. Antimicrob Agents Chemother 60(11):6532–6539. https://doi.org/10.1128/AAC.01043-16

Yang Q, Huang W, Jozwik C, Lin Y, Glasman M, Caohuy H, Srivastava M, Esposito D, Gillette W, Hartley J, Pollard HB (2005) Cardiac glycosides inhibit TNF-NF-B signaling by blocking recruitment of TNF receptor-associated death domain to the TNF receptor. 10.1073pnas.0504097102. www.pnas.org

Yarnoff B, Bodhaine S, Cohen E, Buck PO (2021) Time and cost of administering COVID-19 mRNA vaccines in the United States. Hum Vaccin Immunother 17(11):3871–3875. https://doi.org/10.1080/21645515.2021.1974289

Yepes-Pérez AF, Herrera-Calderon O, Sánchez-Aparicio JE, Tiessler-Sala L, Maréchal JD, Cardona GW (2020) Investigating potential inhibitory effect of *Uncaria tomentosa* (Cat's Claw) against the main protease 3CLpro of SARS-CoV-2 by molecular modeling. In: Evidence-based complementary and alternative medicine (ECAM, 2020). https://doi.org/10.1155/2020/4932572

Yoshida H, Takamura N, Shuto T, Ogata K, Tokunaga J, Kawai K, Kai H (2010) The citrus flavonoids hesperetin and naringenin block the lipolytic actions of TNF-α in mouse adipocytes. Biochem Biophys Res Commun 394(3):728–732. https://doi.org/10.1016/j.bbrc.2010.03.060

Yu MS, Lee J, Lee JM, Kim Y, Chin YW, Jee JG, Keum YS, Jeong YJ (2012) Identification of myricetin and scutellarein as novel chemical inhibitors of the SARS coronavirus helicase, nsP13. Bioorg Med Chem Lett 22(12):4049–4054. https://doi.org/10.1016/J.BMCL.2012.04.081

Zhang D, Wu K, Zhang X, Deng S, Peng B (2020) In silico screening of Chinese herbal medicines with the potential to directly inhibit 2019 novel coronavirus. J Integr Med 18(2):152–158. https://doi.org/10.1016/J.JOIM.2020.02.005

Zhang D, Hamdoun S, Chen R, Yang L, Ip CK, Qu Y, Li R, Jiang H, Yang Z, Chung SK, Liu L, Wong VKW (2021) Identification of natural compounds as SARS-CoV-2 entry inhibitors by molecular docking-based virtual screening with bio-layer interferometry. Pharmacol Res 172. https://doi.org/10.1016/j.phrs.2021.105820

Zhao C, Zhao W (2020) NLRP3 Inflammasome—a key player in antiviral responses. Front Immunology 11. https://doi.org/10.3389/fimmu.2020.00211

Role of Imaging in Diagnosis and Management of COVID-19: Evidence-Based Approaches

13

Dante L. Pezzutti and Mina S. Makary

Abstract

Imaging has been demonstrated to play a crucial role in both the diagnosis and management of COVID-19. Depending on resources, pre-test probability, and risk factors for severe disease progression, real-time polymerase chain reaction (RT-PCR) testing may be followed by chest radiography (CXR) or chest computed tomography (CT) to further aid in diagnosis or excluding COVID-19 disease. SARS-CoV-2 has been shown not only to pathologically impact the pulmonary system, but also the cardiovascular, gastrointestinal, and neurological systems to name a few. Imaging has again proven useful in further investigating and managing extrapulmonary disease, with the use of echocardiogram, CT angiography of the cardiovascular and cerebrovascular structures, MRI of the brain, as well as ultrasound of the abdomen and CT of the abdomen and pelvis proving particularly useful. Research in artificial intelligence and its application in the diagnosis of COVID-19 and disease severity prediction is underway, and point-of-care ultrasound is an emerging bedside technique that may allow for more efficient and timely diagnosis of COVID-19.

Keywords

Radiology · Imaging · COVID-19 · SARS-CoV-2 · Chest radiography · CXR · Ultrasound · Computed tomography · CT · Magnetic resonance imaging · MRI · Computed tomography angiography · CTA · Echocardiogram · Pneumonia · Pandemic · Artificial intelligence · Point of care ultrasound (POCUS)

13.1 Introduction

The first reports of respiratory-related infections with a novel coronavirus appeared in December of 2019 (Meirelles 2020). Severe acute respiratory syndrome coronavirus 2 (SARS-CoV-2) and the disease it causes, coronavirus disease-2019 (COVID-19), then quickly spread around the world. Presently, COVID-19 is the third leading cause of death within the United States behind only cancer and cardiovascular disease (Zhu et al. 2020).

Initially, COVID-19 was thought to be primarily a respiratory tract disease. However, further investigation into its pathophysiology revealed SARS-CoV-2 causes direct viral damage

D. L. Pezzutti · M. S. Makary (✉)
Department of Radiology, The Ohio State University Wexner Medical Center, 395 W. 12th Ave, 4th Floor, Columbus, OH 43210, USA
e-mail: mina.makary@osumc.edu

N. Rezaei (ed.), *The COVID-19 Aftermath*, Advances in Experimental Medicine and Biology 1457,
https://doi.org/10.1007/978-3-031-61939-7_13

to extrapulmonary tissues through binding the largely distributed ACE2 receptors, leading to not only pulmonary manifestations of disease but also myocardial, renal, liver, thromboembolic, and neurological manifestations (Gupta et al. 2020).

From the outset of the pandemic, imaging has played a vital role in both the diagnosis and management of COVID-19. Although initially, radiological imaging was employed largely for evaluation of the pulmonary system, imaging modalities proved to also be useful in the evaluation of the neurological, cardiovascular, and gastrointestinal complications of COVID-19 as well (Manna et al. 2020).

13.2 Role of Imaging in Diagnosis of COVID-19

13.2.1 Deciding When to Image

The initial test of choice in the diagnosis of COVID-19 is a real-time reverse transcriptase PCR (RT-PCR) test from a nasopharyngeal swab. The specificity of RT-PCR is high, but the sensitivity is low, reported somewhere between 80 and 90% (Wong et al. 2020). A high number of false-negatives are obtained as a result of these testing characteristics, and in combination with nonspecific biomarkers for COVID-19 including lactate dehydrogenase, C-reactive Protein (CRP), and erythrocyte sedimentation rate (ESR), imaging has become a critical tool in the detection of early stages of COVID-19, as well as risk stratification of the disease (Rubin et al. 2020).

More specific details regarding the indication for imaging following RT-PCR testing was provided by the Fleischner Society in July of 2020 (Rubin et al. 2020). Based on the pretest probability, availability of resources in the surrounding area, and the suspected severity of disease, guidelines were recommended to either obtain or not obtain chest imaging to assist further in the diagnostic workup of a suspected COVID-19 patient. Imaging is recommended in the following scenarios: (1) patients with a positive RT-PCR test or in a patient with high pre-test probability without a RT-PCR test and with high risk factors for progression of the disease (i.e., immunocompromised, comorbidities such as diabetes mellitus, chronic lung disease, cardiovascular disease, hypertension, and those > 65 years of age); (2) patients with moderate to severe symptoms of COVID-19, irrespective of RT-PCR testing; (3) patients with moderate to severe symptoms of COVID-19 in an area of high prevalence of the disease, regardless of RT-PCR results (Rubin et al. 2020; Kwee and Kwee 2020; Roshkovan et al. 2020; Floridi et al. 2020).

13.2.2 Chest Radiography (CXR)

With these general guidelines in mind, it is now prudent to discuss the advantages and disadvantages of the various chest imaging modalities in the initial workup and diagnosis of COVID-19. Although in the early stages of disease chest radiography yields a high number of false negative results, it is a widely utilized technique for screening and disease severity assessment because of its portability, relatively quick turnaround time, and the low dose radiation associated with its use (Wong et al. 2020; Zu et al. 2020). Findings on chest radiograph most often include consolidation(s) or hazy opacifications, usually seen in the peripheral regions of the lung and are often in a basilar distribution (Ng et al. 2020; Gravell et al. 2020). As the disease progresses, the opacities can grow in size and density and can become multi-lobar in nature, sometimes leading to acute respiratory distress syndrome (ARDS), which is characterized by diffuse bilateral lung opacities and thickened interlobular septa (Kwee and Kwee 2020; Gandhi et al. 2020; Wong et al. 2020).

13.2.3 Ultrasound

Ultrasound is another imaging modality that is both easily accessible and widely available, like chest radiography, and is commonly used to assist in the diagnosis of patients with suspected

COVID-19. Lung ultrasound is known to have a sensitivity of 94% and specificity of 85% in diagnosing viral pneumonia, making it a valuable diagnostic tool for clinicians with the skillset to utilize bedside ultrasound (Soldati et al. 2020; Kulkarni et al. 2020; Vetrugno et al. 2020). Common lung ultrasound findings for COVID-19 include pleural thickening, B-lines, non-lobar and trans-lobar consolidations with air bronchograms, subpleural consolidations, and pleural effusions that are smaller and localized. More specifically, in the early stages of the disease, pleural thickening with B-lines are observed, while in later stages, air bronchograms and subpleural consolidations are more predominant (Xing et al. 2020).

13.2.4 Computed Tomography (CT)

Chest computed tomography is the most sensitive imaging modality relative to ultrasound and chest radiography in detecting viral pneumonia, with a sensitivity of 98% (Varadarajan et al. 2021; Soldati et al. 2020; Altmayer et al. 2020). The most widely identified findings on chest CT of patients with COVID-19 are in a peripheral, bilateral, and posterior distribution and include ground-glass opacity (GGO), reticular and crazy-paving pattern, and consolidation (Bao et al. 2020; Sun et al. 2020; Yang et al. 2020; Ye et al. 2020). The course of the disease on chest CT is as follows: (1) GGO form, lack of broncho-vascular margins with a subpleural involvement in the lower lobes emerge, and hazy regions of increased density form; (2) multifocal patchy or segmental areas of consolidations appear and are an indicator of disease progression; (3) GGOs with a crazy-paving pattern emerge along with thickened interlobular septa and lines from alveolar edema and/or interstitial inflammation; (4) resolution of findings in the recovery stage with persistence of GGO (Pan et al. 2019; Wang et al. 2020; Fan et al. 2020). CT can be utilized to in the evaluation of complications from COVID-19 that are not differentiated on CXR, such as pulmonary emboli, ARDS, acute lung edema, myocarditis, and abscesses (Salehi et al. 2020; Guan et al. 2019; Fields et al. 2020). Although CT has superior sensitivity and resolution as compared to ultrasound and chest radiography, it involves more members of the staff within the hospital and therefore increases the risk of spread of the virus (Hani et al. 2020; Pontone et al. 2020; Nagpal et al. 2020). Additionally, it can delay the diagnosis of critical findings and increases the radiation exposure to the patient. Presently, the American College of Radiology recommends against CT as an initial diagnostic test for suspected COVID-19 patients and instead suggests it be used for hospitalized patients with a specific clinical indication. However, there are circumstances in facilities in which lack of testing necessitates the use of chest CT in the diagnosis of a patient with COVID-19 (Song et al. 2020; Shi et al. 2020).

13.2.5 MRI

Pulmonary MRI is less commonly employed for the initial diagnosis of COVID-19 given strict precautionary measures taken to limit the spread of infection. Additionally, it is not as widely available, costs more, and has a long image attainment time. Even still, there have been case reports of MRIs being obtained for indications outside of the pulmonary system that have captured incidental findings of COVID-19. The "cloudy sky sign" can be visualized, which represents focal areas of lung tissue with abnormal intensities on either T1 or T2. These areas can be correlated to the GGO identified on CT (Vasilev et al. 2021). Furthermore, in patients such as pregnant patients or pediatric patients, where ionizing radiation should be minimized, MRI may be a viable option for further diagnostic workup in a suspected COVID-19 patient. In general, MRI is used for the extrapulmonary complications associated with COVID-19, such as myocarditis or acute encephalopathy or thromboembolic stroke (Ates et al. 2020; Yang et al. 2020; Torkian et al. 2021).

13.2.6 Nuclear Medicine

Fluorine-18-Fluorodeoxyglucose Positron Emission Tomography/CT (18 F-FDG-PET/CT) is a less commonly employed imaging technique for the diagnosis of COVID-19 and one that cannot be used in the emergency setting. However, in instances where the diagnosis is unclear and symptoms are nonspecific, it can be a useful additional diagnostic tool (Polverari et al. 2019; Qin et al. 2020). This imaging method works by highlighting areas of higher metabolic demand, which in turn aids in localizing sites of inflammation and/or infectious processes (Fontana et al. 2020). For example, a number of studies have demonstrated increased uptake of (Vetrugno et al. 2020) F-FDG in COVID-19 patients within the areas of lung consolidation and GGO (Zou and Zhu 2020). Even still, the extended image acquisition time, cost, excess radiation exposure, and involvement of a large amount of medical staff limit the use of this imaging modality for diagnostic purposes in COVID-19 patients (Pezzutti et al. 2021; Guedj et al. 2020a, b).

13.3 Role of Imaging in Management of COVID-19 Patients

As mentioned previously, although COVID-19 is primarily a disease of the lower respiratory tract, several other pathologies have been found to be associated with the disease. These complications involve the gastrointestinal, vascular, dermatological, neurologic, and cardiac systems and are thought to be related to the pathophysiology of the virus involving coagulation system dysfunction, extreme hypoxia, and an elevated inflammatory response (Revzin et al. 2020; Behzad et al. 2020). SARS-CoV-2 method of entry to cells through the widely distributed angiotensin-cell converting enzyme II receptor may also explain its multisystem pathology (Zhang et al. 2020). As such, clinicians are often called upon to exclude extrapulmonary complications in patients with COVID-19, and alongside appropriate laboratory testing, imaging becomes a pivotal cornerstone in dictating the management of these patients.

13.3.1 Cardiovascular Imaging (Echocardiogram)

Cardiovascular complications from COVID-19 include a systemic inflammatory response leading to a hyperdynamic cardiovascular response (Mahmoud-Elsayed et al. 2020). In general, echocardiography is the first line imaging modality to better assess cardiac function as it provides an objective evaluation of both global and regional cardiac performance (Wang et al. 2020). A commonly observed sign on echocardiogram in complicated COVID-19 patients includes dilation of the right ventricle, which could be secondary to a pulmonary embolism, left ventricular systolic dysfunction, pericardial effusion, or hypoxic respiratory failure (Lan et al. 2021). Importantly, global left ventricular dysfunction and wall motion abnormalities have been documented in all severe cases of COVID-19 (Krishnamoorthy et al. 2021). Formation of a pulmonary embolus because of local inflammation of the lung from SARS-CoV-2, a phenomenon denotated "pulmonary thrombosis in situ" has been studied as a potential complication of COVID-19 disease (Peng et al. 2020). While echocardiography is not sensitive in the detection of a pulmonary embolus, it is useful in identifying downstream signs of a PE, which most commonly involves right ventricular dysfunction (Bonow et al. 2020; Inciardi et al. 2020; Madjid et al. 2020). Finally, there have also been several studies that have reported incidences of Takotsubo cardiomyopathy detected on echocardiography in COVID-19 patients, a stress-induced cardiomyopathy that leads to ballooning of the left ventricle, akinesis of the apex, and hyperkinesis of the basal segments (Ghasemi et al. 2023).

13.3.2 Cardiovascular Imaging (MRI)

Cardiac magnetic resonance imaging (CMR) is the most suitable modality for examining and characterizing myocardial tissue, but it is not a feasible method of image acquisition in the acute setting (Han et al. 2020). Even still, it is important to understand the different features of disease seen on cardiac MRI upon initial presentation in the case of an undifferentiated patient. These findings include thickening of the ventricular walls, hypokinesis of the apical segments, marked LV dysfunction, and pericardial effusion formation (Torkian et al. 2021). Furthermore, myocarditis is a well-documented complication of COVID-19 disease, and myocardial inflammation can be detected on CMR by identifying an increased signal intensity in the LV myocardium in both T2-weighted images or native T1 images (Varadarajan et al. 2021; Lan et al. 2021).

13.3.3 Cardiovascular Imaging (CT-Angiogram)

Patients with COVID-19 often have elevated cardiac biomarkers, which leads clinicians to further investigate whether there is an atheroembolic cause of this elevation in cardiac enzymes or if it is related to increased cardiac demand secondary to infection with SARS-CoV-2 (Cardiology TFftmoC-otESo 2022). Computed tomography angiography (CTA) has emerged as a fast, non-invasive, and remarkably accurate method to image patients with suspected coronary disease (Siripanthong et al. 2020). Indeed, CTA has been taking the place of coronary angiography due to its minimally invasive nature, lower cost, and lower radiation dose exposure. CTA can also visualize filling defects in the pulmonary vascular system and assist in the diagnosis of a PE, and CT pulmonary angiography should be obtained in patients with COVID-19 disease that develop hypoxia (Ricks et al. 2021). Furthermore, late contrast-enhanced cardiac CT is an alternative modality to CMR for diagnosis of myocarditis, especially in those patients that are already scheduled to receive a lung CT. Characteristics of myocarditis on cardiac CT include subepicardial late iodine enhancement, which represents myocardial edema, along with a sub-endocardial perfusion abnormality (Varadarajan et al. 2021).

13.3.4 Neurological Imaging (CT)

Neurological presentations can often accompany infection with COVID-19, with symptoms varying from anosmia, dysgeusia, and dizziness during the early parts of the disease to the development of strokes, cranial neuropathies, vasculitides, or encephalomyelitis later in the disease course in severely advanced cases (Fotuhi et al. 2020). There is a higher incidence of strokes in those patients with pre-existing risk factors such as hyperlipidemia or hypertension, and the first line imaging modality to work up neurological complications, especially stroke, in COVID-19 patients is CT (Baracchini et al. 2020; Conklin et al. 2020; Kremer et al. 2020; Qureshi et al. 2021). Furthermore, marked subarachnoid and intraventricular hemorrhages along with cerebral swelling have been reported at later stages in the disease, along with hypoattenuation in the bilateral medial thalami leading to cerebral venous thrombi and acute necrotizing encephalopathy (Egbert et al. 2020; Wu et al. 2020).

13.3.5 Neurological Imaging (MRI)

MRI fluid-attenuated inversion recovery (FLAIR) demonstrates bilateral olfactory bulb hyperintensity bulbs in the early stages of disease in patients with anosmia (Politi et al. 2020). Further progression of disease then leads to strokes due to the inflammatory and hypercoagulable pathophysiology of the disease, which on brain MRI manifest as infarcts in the cerebral vascular territories and stenosis of cerebral arteries on magnetic resonance angiogram (MRA) (Avula et al. 2020). Finally, the most advanced disease leads to extensive white matter microhemorrhages, as well as acute necrotizing encephalopathy, which is characterized by rim enhancement on a T1 post-contrast image (Radmanesh et al. 2020).

13.3.6 Gastrointestinal Imaging

Gastrointestinal symptoms are often not the sole presenting symptoms of COVID-19 patients, but nearly half of COVID-19 positive patients have both pulmonary and gastrointestinal symptoms (Gu et al. 2020). Furthermore, those patients with digestive symptoms often have an increased severity of the disease that warrants further investigation with imaging of the abdomen and pelvis (Dane et al. 2020; Bhayana et al. 2020). It is thought that SARS-CoV-2 leads to disruption of the permeability of both the large and small intestine, which leads to gastrointestinal distress.

13.3.7 Gastrointestinal Imaging (Ultrasound and CT)

Findings on abdominal ultrasound can demonstrate gallbladder wall thickening and distension, consistent with acute cholestasis (Carvalho et al. 2020). There can also be increased echogenicity of the liver parenchyma as well as mild to severe pancreatitis with ductal dilation and focal necrosis of the pancreas (Hamilton et al. 2020). CT findings in the gastrointestinal tract in patients with COVID-19 include bowel wall thickening, intestinal lumen filled with fluid, and enhancement of the mucosa. More severe cases can lead to segmental bowel ischemia and pneumatosis intestinalis (Varadarajan et al. 2021). Abdominal CT can also assist in the diagnosis of splenic infarcts, as well as detecting renal infarct due to thromboembolism through demonstration of wedge-shaped low attenuating regions in the kidneys (Dane et al. 2020). CT angiography in the abdomen and pelvis also proves useful when searching for arterial or venous thrombi in the mesentery, a complication not uncommonly seen in COVID-19.

13.3.8 Gastrointestinal Imaging (MRI)

The liver is the second most common organ involved in COVID-19 disease, behind only the lungs (Pezzutti et al. 2021). It is hypothesized that severe inflammation and stress lead to hepatitis. MRI can be used to visualize injury to the liver, and one method for quantifying the severity of hepatitis in COVID-19 patients is with hepatic T1 mapping MRI, where liver fat proportions of up to 10% have been shown to be associated with infection with SARS-CoV-2 (Xu et al. 2020).

13.4 Future Directions

13.4.1 Artificial Intelligence

Ongoing efforts are being applied toward artificial intelligence (AI) with the goal of utilizing AI to help diagnose COVID-19 and predict disease severity (Long and Ehrenfeld 2020; Li et al. 2020). One example of the use of AI in COVID-19 is with CXR and CT, where several studies have demonstrated that certain AI models have successfully differentiated COVID-19 from community acquired pneumonia (Shaikh et al. 2020; Apostolopoulos and Mpesiana 2020). Furthermore, as more standardized reporting methods emerge, AI models will become more precise at predicting clinical outcomes based on relevant imaging findings such as vascular changes and the number of opacities present (Shi et al. 2020). Despite these exciting lines of research, there remains more room for growth and consistency in the datasets; these AI models are utilizing, as well as more investigation into the use of AI for the extrapulmonary manifestations of COVID-19 (Wang et al. 2020).

13.4.2 Point-of-Care Diagnostics

Apart from RT-PCR, point-of-care ultrasound (POCUS) has been demonstrated to be useful in the diagnosis and management of COVID-19 patients (Bitar et al. 2019; Bikdeli et al. 2020; Copetti et al. 2008). As mentioned previously, POCUS has a low image acquisition time, offers no ionizing radiation, and is portable and can be done with the patient lying in bed (Kumar et al. 2020; Tusman et al. 2016). It answers many of the same questions the chest CT and chest radiography answer and can be utilized to monitor the progression of disease with low effort and cost, including evaluating for ARDS, pericardial and pleural effusion, pneumothorax, pulmonary edema, and deep vein thrombosis (DVT) formation (Duggan et al. 2020). The use of POCUS is not without limitations, however. For example, POCUS is highly operator-dependent, and thus, detection rate will vary greatly from clinician to clinician depending upon their skillset (Moore and Gardiner 2020; Jackson et al. 2021). Furthermore, many of the studies conducted using POCUS were in areas of high prevalence of disease, thereby possibly overestimating its diagnostic accuracy (Vetrugno et al. 2020; Piscaglia et al. 2020). Despite these limitations, POCUS is a useful point-of-care diagnostic for which further prospective research needs to be conducted to determine its true clinical utility in the diagnosis and management COVID-19.

References

Altmayer S, Zanon M, Pacini GS et al (2020) Comparison of the computed tomography findings in COVID-19 and other viral pneumonia in immunocompetent adults: a systematic review and meta-analysis. Eur Radiol 30(12):6485–6496. https://doi.org/10.1007/s00330-020-07018-x

Apostolopoulos ID, Mpesiana TA (2020) Covid-19: automatic detection from X-ray images utilizing transfer learning with convolutional neural networks. Phys Eng Sci Med 43(2):635–640. https://doi.org/10.1007/s13246-020-00865-4

Ates OF, Taydas O, Dheir H (2020) Thorax magnetic resonance imaging findings in patients with coronavirus disease (COVID-19). Acad Radiol 27(10):1373–1378. https://doi.org/10.1016/j.acra.2020.08.009

Avula A, Nalleballe K, Narula N et al (2020) COVID-19 presenting as stroke. Brain Behav Immun 87:115–119. https://doi.org/10.1016/j.bbi.2020.04.077

Bao C, Liu X, Zhang H, Li Y, Liu J (2020) Coronavirus Disease 2019 (COVID-19) CT findings: a systematic review and meta-analysis. J Am Coll Radiol 17(6):701–709. https://doi.org/10.1016/j.jacr.2020.03.006

Baracchini C, Pieroni A, Viaro F et al (2020) Acute stroke management pathway during Coronavirus-19 pandemic. Neurol Sci 41(5):1003–1005. https://doi.org/10.1007/s10072-020-04375-9

Behzad S, Aghaghazvini L, Radmard AR, Gholamrezanezhad A (2020) Extrapulmonary manifestations of COVID-19: radiologic and clinical overview. Clin Imaging 66:35–41. https://doi.org/10.1016/j.clinimag.2020.05.013

Bhayana R, Som A, Li MD et al (2020) Abdominal imaging findings in COVID-19: preliminary observations. Radiology 297(1):E207–E215. https://doi.org/10.1148/radiol.2020201908

Bikdeli B, Madhavan MV, Jimenez D et al (2020) COVID-19 and thrombotic or thromboembolic disease: implications for prevention, antithrombotic therapy, and follow-up: JACC state-of-the-art review. J Am Coll Cardiol 75(23):2950–2973. https://doi.org/10.1016/j.jacc.2020.04.031

Bitar ZI, Maadarani OS, El-Shably AM, Al-Ajmi MJ (2019) Diagnostic accuracy of chest ultrasound in patients with pneumonia in the intensive care unit: a single-hospital study. Health Sci Rep 2(1):e102. https://doi.org/10.1002/hsr2.102

Bonow RO, Fonarow GC, O'Gara PT, Yancy CW (2020) Association of coronavirus disease 2019 (COVID-19) with myocardial injury and mortality. JAMA Cardiol 5(7):751–753. https://doi.org/10.1001/jamacardio.2020.1105

Cardiology TFftmoC-otESo (2022) ESC guidance for the diagnosis and management of cardiovascular disease during the COVID-19 pandemic: part 2-care pathways, treatment, and follow-up. Cardiovasc Res 118(7):1618–1666. https://doi.org/10.1093/cvr/cvab343

Carvalho A, Alqusairi R, Adams A et al (2020) SARS-CoV-2 gastrointestinal infection causing hemorrhagic colitis: implications for detection and transmission of COVID-19 disease. Am J Gastroenterol 115(6):942–946. https://doi.org/10.14309/ajg.0000000000000667

Conklin J, Frosch MP, Mukerji S et al (2020) Cerebral microvascular injury in severe COVID-19. medRxiv. https://doi.org/10.1101/2020.07.21.20159376

Copetti R, Soldati G, Copetti P (2008) Chest sonography: a useful tool to differentiate acute cardiogenic pulmonary edema from acute respiratory distress syndrome. Cardiovasc Ultrasound 6:16. https://doi.org/10.1186/1476-7120-6-16

Dane B, Brusca-Augello G, Kim D, Katz DS (2020) Unexpected findings of coronavirus disease (COVID-19) at the lung bases on abdominopelvic CT. AJR Am J Roentgenol 215(3):603–606. https://doi.org/10.2214/AJR.20.23240

Duggan NM, Shokoohi H, Liteplo AS, Huang C, Goldsmith AJ (2020) Best practice recommendations for point-of-care lung ultrasound in patients with suspected COVID-19. J Emerg Med 59(4):515–520. https://doi.org/10.1016/j.jemermed.2020.06.033

Egbert AR, Cankurtaran S, Karpiak S (2020) Brain abnormalities in COVID-19 acute/subacute phase: a rapid systematic review. Brain Behav Immun

10(89):543–554. https://doi.org/10.1016/j.bbi.2020.07.014
Fan L, Li D, Xue H et al (2020) Progress and prospect on imaging diagnosis of COVID-19. Chin J Acad Radiol 1–10. https://doi.org/10.1007/s42058-020-00031-5
Fields BKK, Demirjian NL, Dadgar H, Gholamrezanezhad A (2020) Imaging of COVID-19: CT, MRI, and PET. Semin Nucl Med. https://doi.org/10.1053/j.semnuclmed.2020.11.003
Floridi C, Fogante M, Agostini A et al (2020) Radiological diagnosis of coronavirus disease 2019 (COVID-19): a practical guide. Acta Biomed 91(8-S):51–59. https://doi.org/10.23750/abm.v91i8-S.9973
Fontana IC, Bongarzone S, Gee A, Souza DO, Zimmer ER (2020) PET imaging as a tool for assessing COVID-19 brain changes. Trends Neurosci 43(12):935–938. https://doi.org/10.1016/j.tins.2020.10.010
Fotuhi M, Mian A, Meysami S, Raji CA (2020) Neurobiology of COVID-19. J Alzheimers Dis 76(1):3–19. https://doi.org/10.3233/JAD-200581
Gandhi D, Ahuja K, Grover H et al (2020) Review of X-ray and computed tomography scan findings with a promising role of point of care ultrasound in COVID-19 pandemic. World J Radiol 12(9):195–203. https://doi.org/10.4329/wjr.v12.i9.195
Ghasemi H, Kazemian S, Nejadghaderi SA, Shafie M (2023) Takotsubo syndrome and COVID-19: a systematic review. Health Sci Rep 6(1):e972. https://doi.org/10.1002/hsr2.972
Gravell RJ, Theodoreson MD, Buonsenso D, Curtis J (2020) Radiological manifestations of COVID-19: key points for the physician. Br J Hosp Med (Lond) 81(6):1–11. https://doi.org/10.12968/hmed.2020.0231
Gu J, Han B, Wang J (2020) COVID-19: gastrointestinal manifestations and potential fecal-oral transmission. Gastroenterology 158(6):1518–1519. https://doi.org/10.1053/j.gastro.2020.02.054
Guan WJ, Ni ZY, Hu Y et al (2019) Clinical characteristics of coronavirus disease 2019 in China. N Engl J Med 382(18):1708–1720. https://doi.org/10.1056/NEJMoa2002032
Guedj E, Million M, Dudouet P et al (2020) F-FDG brain PET hypometabolism in post-SARS-CoV-2 infection: substrate for persistent/delayed disorders? Eur J Nucl Med Mol Imaging. https://doi.org/10.1007/s00259-020-04973-x
Guedj E, Verger A, Cammilleri S (2020b) PET imaging of COVID-19: the target and the number. Eur J Nucl Med Mol Imaging 47(7):1636–1637. https://doi.org/10.1007/s00259-020-04820-z
Gupta A, Madhavan MV, Sehgal K et al (2020) Extrapulmonary manifestations of COVID-19. Nat Med 26(7):1017–1032. https://doi.org/10.1038/s41591-020-0968-3
Hamilton NE, Adam GH, Ifan DL et al (2020) Diagnostic utility of additional whole-chest CT as part of an acute abdominal pain CT imaging pathway during the COVID-19 pandemic. Clin Radiol 75(8):592–598. https://doi.org/10.1016/j.crad.2020.06.002
Han Y, Chen T, Bryant J et al (2020) Society for cardiovascular magnetic resonance (SCMR) guidance for the practice of cardiovascular magnetic resonance during the COVID-19 pandemic. J Cardiovasc Magn Reson 22(1):26. https://doi.org/10.1186/s12968-020-00628-w
Hani C, Trieu NH, Saab I et al (2020) COVID-19 pneumonia: A review of typical CT findings and differential diagnosis. Diagn Interv Imaging 101(5):263–268. https://doi.org/10.1016/j.diii.2020.03.014
Inciardi RM, Lupi L, Zaccone G et al (2020) Cardiac Involvement in a patient with coronavirus disease 2019 (COVID-19). JAMA Cardiol 5(7):819–824. https://doi.org/10.1001/jamacardio.2020.1096
Jackson K, Butler R, Aujayeb A (2021) Lung ultrasound in the COVID-19 pandemic. Postgrad Med J 97(1143):34–39. https://doi.org/10.1136/postgradmedj-2020-138137
Kremer S, Lersy F, de Sèze J et al (2020) Brain MRI findings in severe COVID-19: a retrospective observational study. Radiology 297(2):E242–E251. https://doi.org/10.1148/radiol.2020202222
Krishnamoorthy P, Croft LB, Ro R et al (2021) Biventricular strain by speckle tracking echocardiography in COVID-19: findings and possible prognostic implications. Future Cardiol 17(4):663–667. https://doi.org/10.2217/fca-2020-0100
Kulkarni S, Down B, Jha S (2020) Point-of-care lung ultrasound in intensive care during the COVID-19 pandemic. Clin Radiol 75(9):710.e1–710.e4. https://doi.org/10.1016/j.crad.2020.05.001
Kumar S, Kumar A, Goel P, Vyas S, Baitha U, Wig N (2020) Use of ultrasonography in COVID-19: probing for success. J Family Med Prim Care 9(8):3863–3866. https://doi.org/10.4103/jfmpc.jfmpc_764_20
Kwee TC, Kwee RM (2020) Chest CT in COVID-19: what the radiologist needs to know. Radiographics 40(7):1848–1865. https://doi.org/10.1148/rg.2020200159
Lan Y, Liu W, Zhou Y (2021) Right ventricular damage in COVID-19: association between myocardial injury and COVID-19. Front Cardiovasc Med 8:606318. https://doi.org/10.3389/fcvm.2021.606318
Li L, Qin L, Xu Z et al (2020) Using artificial intelligence to detect COVID-19 and community-acquired pneumonia based on pulmonary CT: evaluation of the diagnostic accuracy. Radiology 296(2):E65–E71. https://doi.org/10.1148/radiol.2020200905
Long JB, Ehrenfeld JM (2020) The role of augmented intelligence (AI) in detecting and preventing the spread of novel coronavirus. J Med Syst 44(3):59. https://doi.org/10.1007/s10916-020-1536-6
Madjid M, Safavi-Naeini P, Solomon SD, Vardeny O (2020) Potential effects of coronaviruses on the cardiovascular system: a review. JAMA Cardiol 5(7):831–840. https://doi.org/10.1001/jamacardio.2020.1286
Mahmoud-Elsayed HM, Moody WE, Bradlow WM et al (2020) Echocardiographic findings in patients with COVID-19 pneumonia. Can J

Cardiol 36(8):1203–1207. https://doi.org/10.1016/j.cjca.2020.05.030
Manna S, Wruble J, Maron SZ et al (2020) COVID-19: a multimodality review of radiologic techniques, clinical utility, and imaging features. Radiol Cardiothorac Imaging 2(3):e200210. https://doi.org/10.1148/ryct.2020200210
Meirelles GSP (2020) COVID-19: a brief update for radiologists. Radiol Bras 53(5):320–328. https://doi.org/10.1590/0100-3984.2020.0074
Moore S, Gardiner E (2020) Point of care and intensive care lung ultrasound: a reference guide for practitioners during COVID-19. Radiography (Lond) 26(4):e297–e302. https://doi.org/10.1016/j.radi.2020.04.005
Nagpal P, Narayanasamy S, Vidholia A et al (2020) Imaging of COVID-19 pneumonia: patterns, pathogenesis, and advances. Br J Radiol 93(1113):20200538. https://doi.org/10.1259/bjr.20200538
Ng MY, Lee EYP, Yang J et al (2020) Imaging profile of the covid-19 infection: radiologic findings and literature review. Radiol Cardiothorac Imaging 2(1):e200034. https://doi.org/10.1148/ryct.2020200034
Pan F, Ye T, Sun P et al (2019) Time course of lung changes at chest CT during recovery from coronavirus disease 2019 (COVID-19). Radiology 295(3):715–721. https://doi.org/10.1148/radiol.2020200370
Peng QY, Wang XT, Zhang LN, CCCUSG (2020) Using echocardiography to guide the treatment of novel coronavirus pneumonia. Crit Care 24(1):143. https://doi.org/10.1186/s13054-020-02856-z
Pezzutti DL, Wadhwa V, Makary MS (2021) COVID-19 imaging: diagnostic approaches, challenges, and evolving advances. World J Radiol. 13(6):171–191. https://doi.org/10.4329/wjr.v13.i6.171
Piscaglia F, Stefanini F, Cantisani V et al (2020) Benefits, open questions and challenges of the use of ultrasound in the COVID-19 pandemic era. The views of a panel of worldwide international experts. Ultraschall Med 41(3):228–236. https://doi.org/10.1055/a-1149-9872
Politi LS, Salsano E, Grimaldi M (2020) Magnetic resonance imaging alteration of the brain in a patient with coronavirus disease 2019 (COVID-19) and anosmia. JAMA Neurol 77(8):1028–1029. https://doi.org/10.1001/jamaneurol.2020.2125
Polverari G, Arena V, Ceci F et al (2019) F-fluorodeoxyglucose uptake in patient with asymptomatic severe acute respiratory syndrome coronavirus 2 (coronavirus disease 2019) referred to positron emission tomography/computed tomography for NSCLC restaging. J Thorac Oncol 15(6):1078–1080. https://doi.org/10.1016/j.jtho.2020.03.022
Pontone G, Scafuri S, Mancini ME et al (2020) Role of computed tomography in COVID-19. J Cardiovasc Comput Tomogr. https://doi.org/10.1016/j.jcct.2020.08.013
Qin C, Liu F, Yen TC, Lan X (2020) F-FDG PET/CT findings of COVID-19: a series of four highly suspected cases. Eur J Nucl Med Mol Imaging 47(5):1281–1286. https://doi.org/10.1007/s00259-020-04734-w
Qureshi AI, Baskett WI, Huang W et al (2021) Acute ischemic stroke and COVID-19: an analysis of 27 676 patients. Stroke 52(3):905–912. https://doi.org/10.1161/STROKEAHA.120.031786
Radmanesh A, Derman A, Lui YW et al (2020) COVID-19-associated diffuse leukoencephalopathy and microhemorrhages. Radiology 297(1):E223–E227. https://doi.org/10.1148/radiol.2020202040
Revzin MV, Raza S, Srivastava NC et al (2020) Multisystem imaging manifestations of COVID-19, part 2: from cardiac complications to pediatric manifestations. Radiographics 40(7):1866–1892. https://doi.org/10.1148/rg.2020200195
Ricks E, Wahed A, Dasgupta A, Buja LM (2021) COVID-19 cardiac injury: an important cause of COVID-19 related morbidityand mortality. Ann Clin Lab Sci 51(2):156–162
Roshkovan L, Chatterjee N, Galperin-Aizenberg M et al (2020) The Role of imaging in the management of suspected or known COVID-19 pneumonia. A multidisciplinary perspective. Ann Am Thorac Soc 17(11):1358–1365. https://doi.org/10.1513/AnnalsATS.202006-600FR
Rubin GD, Ryerson CJ, Haramati LB et al (2020) The role of chest imaging in patient management during the COVID-19 pandemic: a multinational consensus statement from the fleischner society. Chest 158(1):106–116. https://doi.org/10.1016/j.chest.2020.04.003
Salehi S, Abedi A, Balakrishnan S, Gholamrezanezhad A (2020) Coronavirus disease 2019 (COVID-19): a systematic review of imaging findings in 919 patients. AJR Am J Roentgenol 215(1):87–93. https://doi.org/10.2214/AJR.20.23034
Shaikh F, Andersen MB, Sohail MR et al (2020) Current landscape of imaging and the potential role for artificial intelligence in the management of COVID-19. Curr Probl Diagn Radiol. https://doi.org/10.1067/j.cpradiol.2020.06.009
Shi H, Han X, Jiang N et al (2020) Radiological findings from 81 patients with COVID-19 pneumonia in Wuhan, China: a descriptive study. Lancet Infect Dis 20(4):425–434. https://doi.org/10.1016/S1473-3099(20)30086-4
Shi F, Wang J, Shi J et al (2020) Review of artificial intelligence techniques in imaging data acquisition, segmentation and diagnosis for COVID-19. IEEE Rev Biomed Eng. https://doi.org/10.1109/RBME.2020.2987975
Siripanthong B, Nazarian S, Muser D et al (2020) Recognizing COVID-19-related myocarditis: the possible pathophysiology and proposed guideline for diagnosis and management. Heart Rhythm 17(9):1463–1471. https://doi.org/10.1016/j.hrthm.2020.05.001
Soldati G, Smargiassi A, Inchingolo R et al (2020) Is there a role for lung ultrasound during the COVID-19 pandemic? J Ultrasound Med 39(7):1459–1462. https://doi.org/10.1002/jum.15284

Soldati G, Smargiassi A, Inchingolo R et al (2020) Proposal for international standardization of the use of lung ultrasound for patients with COVID-19: a simple, quantitative, reproducible method. J Ultrasound Med 39(7):1413–1419. https://doi.org/10.1002/jum.15285

Song F, Shi N, Shan F et al (2020) Emerging 2019 novel coronavirus (2019-nCoV) pneumonia. Radiology 297(3):E346. https://doi.org/10.1148/radiol.2020209021

Sun Z, Zhang N, Li Y, Xu X (2020) A systematic review of chest imaging findings in COVID-19. Quant Imaging Med Surg 10(5):1058–1079. https://doi.org/10.21037/qims-20-564

Torkian P, Rajebi H, Zamani T, Ramezani N, Kiani P, Akhlaghpoor S (2021) Magnetic resonance imaging features of coronavirus disease 2019 (COVID-19) pneumonia: the first preliminary case series. Clin Imaging 69:261–265. https://doi.org/10.1016/j.clinimag.2020.09.002

Tusman G, Acosta CM, Costantini M (2016) Ultrasonography for the assessment of lung recruitment maneuvers. Crit Ultrasound J 8(1):8. https://doi.org/10.1186/s13089-016-0045-9

Varadarajan V, Shabani M, Ambale Venkatesh B, Lima JAC (2021). Role of imaging in diagnosis and management of COVID-19: a multiorgan multimodality imaging review. Front Med (Lausanne) 8:765975. https://doi.org/10.3389/fmed.2021.765975

Vasilev YA, Sergunova KA, Bazhin AV et al (2021) Chest MRI of patients with COVID-19. Magn Reson Imaging 79:13–19. https://doi.org/10.1016/j.mri.2021.03.005

Vetrugno L, Bove T, Orso D et al (2020) Our Italian experience using lung ultrasound for identification, grading and serial follow-up of severity of lung involvement for management of patients with COVID-19. Echocardiography 37(4):625–627. https://doi.org/10.1111/echo.14664

Wang TKM, Tang WHW, Flamm SD et al (2020) The role of cardiac imaging in hospitalized COVID-19-positive patients. Cleve Clin J Med. https://doi.org/10.3949/ccjm.87a.ccc020

Wang Y, Dong C, Hu Y et al (2020) Temporal changes of CT findings in 90 patients with COVID-19 pneumonia: a longitudinal study. Radiology 296(2):E55–E64. https://doi.org/10.1148/radiol.2020200843

Wang L, Lin ZQ, Wong A (2020) COVID-Net: a tailored deep convolutional neural network design for detection of COVID-19 cases from chest X-ray images. Sci Rep 10(1):19549. https://doi.org/10.1038/s41598-020-76550-z

Wong HYF, Lam HYS, Fong AH et al (2020) Frequency and distribution of chest radiographic findings in patients positive for covid-19. Radiology 296(2):E72–E78. https://doi.org/10.1148/radiol.2020201160

Wong MD, Thai T, Li Y, Liu H (2020) The role of chest computed tomography in the management of COVID-19: a review of results and recommendations. Exp Biol Med (Maywood) 245(13):1096–1103. https://doi.org/10.1177/1535370220938315

Wu Y, Xu X, Chen Z et al (2020) Nervous system involvement after infection with COVID-19 and other coronaviruses. Brain Behav Immun 07(87):18–22. https://doi.org/10.1016/j.bbi.2020.03.031

Xing C, Li Q, Du H, Kang W, Lian J, Yuan L (2020) Lung ultrasound findings in patients with COVID-19 pneumonia. Crit Care 24(1):174. https://doi.org/10.1186/s13054-020-02876-9

Xu L, Liu J, Lu M, Yang D, Zheng X (2020) Liver injury during highly pathogenic human coronavirus infections. Liver Int 40(5):998–1004. https://doi.org/10.1111/liv.14435

Yang W, Sirajuddin A, Zhang X et al (2020) The role of imaging in 2019 novel coronavirus pneumonia (COVID-19). Eur Radiol 30(9):4874–4882. https://doi.org/10.1007/s00330-020-06827-4

Yang S, Zhang Y, Shen J et al (2020) Clinical potential of UTE-MRI for assessing COVID-19: patient- and lesion-based comparative analysis. J Magn Reson Imaging 52(2):397–406. https://doi.org/10.1002/jmri.27208

Ye Z, Zhang Y, Wang Y, Huang Z, Song B (2020) Chest CT manifestations of new coronavirus disease 2019 (COVID-19): a pictorial review. Eur Radiol 30(8):4381–4389. https://doi.org/10.1007/s00330-020-06801-0

Zhang L, Feng X, Zhang D et al (2020) Deep vein thrombosis in hospitalized patients with COVID-19 in Wuhan, China: prevalence, risk factors, and outcome. Circulation 142(2):114–128. https://doi.org/10.1161/CIRCULATIONAHA.120.046702

Zhu N, Zhang D, Wang W et al (2020) A novel coronavirus from patients with pneumonia in China. N Engl J Med 382(8):727–733. https://doi.org/10.1056/NEJMoa2001017

Zou S, Zhu X (2020) FDG PET/CT of COVID-19. Radiology 296(2):E118. https://doi.org/10.1148/radiol.2020200770

Zu ZY, Jiang MD, Xu PP et al (2020) Coronavirus disease 2019 (COVID-19): a perspective from China. Radiology 296(2):E15–E25. https://doi.org/10.1148/radiol.2020200490

Does the COVID-19 Vaccination Reduce the Risk to Transmit SARS-CoV-2 to Others?

14

Günter Kampf

Abstract

It has been assumed that the COVID-19 vaccination reduces the risk of transmission to others. Results during the delta predominance show that the viral load in the vaccinated population is not consistently lower compared to the unvaccinated, and during the omicron predominance, the viral load was even somewhat higher. Levels of infectious SARS-CoV-2 were partly lower in the vaccinated population. Viral loads were mostly lower in re-infections compared to breakthrough infections. Viral clearance including the detection of infectious virus has mostly been described to be faster in the vaccinated population suggesting a shorter duration as a possible source for transmission. The epidemiological relevance of this finding remains uncertain. Approximately half of the transmission studies found lower secondary attack rates from the fully vaccinated population, but the results are probably best explained by the vaccination status of the contact population. Public health data from the UK show that the number of COVID-19 cases is higher among the fully vaccinated and boosted population who might be possible sources, in contrast to lower case numbers within the first three months among the vaccinated obtained in phase 3 trials on symptomatic cases. Overall, there is no convincing evidence that the COVID-19 vaccination significantly reduces the risk to transmit SARS-CoV-2 to others.

Keywords

COVID-19 · Vaccination · SARS-CoV-2 · Transmission · Viral load

G. Kampf (✉)
University Medicine Greifswald,
Ferdinand-Sauerbruch-Strasse, 17475 Greifswald,
Germany
e-mail: guenter.kampf@uni-greifswald.de

14.1 Introduction

COVID-19 vaccines have been developed and approved by medicinal product authorities with the aim to reduce the individual risk for symptomatic disease as described in the phase III trials (WHO 2022). This effect has been well established. The vaccination reduces the risk for symptomatic disease (Korang et al. 2022), severe disease (Ghazy et al. 2022) and fatal disease (Ghazy et al. 2022). The protective effect against SARS-CoV-2 infection, however, is waning from 83% after one month to only 22% after five months (Ssentongo et al. 2022) and with the omicron variant even faster (Accorsi et al. 2022). The protection of others has been

N. Rezaei (ed.), *The COVID-19 Aftermath*, Advances in Experimental Medicine and Biology 1457,
https://doi.org/10.1007/978-3-031-61939-7_14

advocated by some politicians in a simplified way by saying that the vaccinated part of the population does not pose a danger to others (Der Spiegel 2021). This public message may have affected the personal motivation for COVID-19 vaccination. A relevant part of priority occupational groups wanted to receive the COVID-19 vaccination mainly to protect their family members (76.2%) or colleagues (72.3%) and less frequently to protect themselves (49.1%) (Štěpánek et al. 2021). Public official campaigns in some countries repeatedly alleged that the COVID-19 vaccination protects others (Bundesministerium für Gesundheit 2022; Felix 2021).

Various outbreaks provide evidence that vaccinated healthcare workers, for example, can still spread SARS-CoV-2 to colleagues, patients or residents in nursing homes (Zürcher et al. 2022; Park et al. 2022; Shitrit et al. 2021). The WHO indicated in 2021 that vaccines provide some protection from transmission (World Health Organization 2021). That is why this chapter looks at the evidence if the vaccinated population adds significant protection from infection to their contacts.

14.2 Levels of Viral Load

One hypothesis for a protection of others was that the fully vaccinated presumably have a significantly lower viral load during a breakthrough infection resulting in a lower probability to transmit SARS-CoV-2 to others. Numerous studies addressed comparative viral loads in unvaccinated and fully vaccinated subjects.

14.2.1 RNA Levels (Ct Values)

14.2.1.1 Beta and Alpha Variant of SARS-CoV-2

During the beta wave, two data sets with different vaccination protocols described a significantly higher viral load (lower Ct values) for the unvaccinated population (Fig. 14.1). During the alpha wave, four of five studies demonstrated a higher viral load among the unvaccinated subjects with a significant difference in two of the studies (Muhsen et al. 2022; Regev-Yochay et al. 2021). In one study, no significant difference was found (Luo et al. 2022), and in another

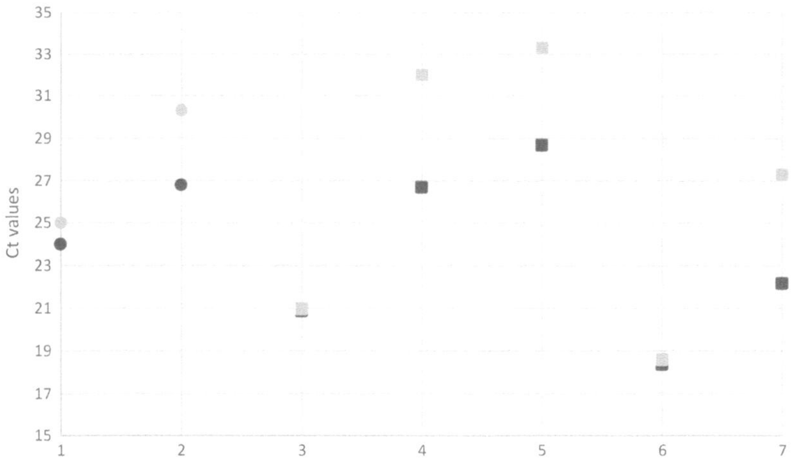

Fig. 14.1 Viral load in unvaccinated (dark grey) and fully vaccinated (light grey) COVID-19 cases described as mean or median Ct values; ○ describes mainly beta variant; □ describes mainly alpha variant; seven data sets obtained from six studies; data sets 1 and 2 (Abu-Raddad et al. 2022); data set 3 (Luo et al. 2022); data set 4 (Muhsen et al. 2022); data set 5 (Pouwels et al. 2021); data set 6 (Ioannou et al. 2021); data set 7 (Regev-Yochay et al. 2021); significant differences in data sets 1, 2, 5 and 7; no significant difference in data sets 3 and 6; no statistical evaluation in data set 4

study, no comparative statistical evaluation was done (Pouwels et al. 2021). Finally, one study found the same level of viral load in both groups (Ioannou et al. 2021). Two studies looked at the RNA level after one dose of vaccine. The viral load was higher in the unvaccinated in one study (median: 28.7 versus 31.6; p-value not described) (Pouwels et al. 2021), and in the other study, the viral load was lower among the unvaccinated (median: 26.6 vs. 25.9; $p = 0.158$) (Shrotri et al. 2021).

Additional studies provide further findings on viral load. The evaluation of 85 unvaccinated and 165 vaccinated cases during an outbreak in a nursing home with the alpha variant of SARS-CoV-2 revealed a lower, albeit not significantly, mean Ct value among the vaccinated residents (3.04 cycles lower; no description of mean or median Ct values per group), but the difference was not significant ($p = 0.38$) (Meyer et al. 2022). Among nursing home residents, a significantly lower viral load was found among the vaccinated individuals but only when the vaccination was completed at least 28 days ago (Shrotri et al. 2021).

14.2.1.2 Delta Variant of SARS-CoV-2

Data obtained during the delta wave revealed an inconsistent picture. Overall, 15 data sets obtained in 12 studies described a lower viral load among the vaccinated compared to the unvaccinated. The difference was significant in five data sets from four studies (Bramante et al. 2022; Bollinger et al. 2022; Levine-Tiefenbrun et al. 2021; Elliott et al. 2021), not significant in seven data sets from seven studies (Brown et al. 2021; Park et al. 2022; Caserta et al. 2022; Plante et al. 2022; Chia et al. 2022; Levine-Tiefenbrun et al. 2021; Acharya et al. 2022) and without reporting results of a statistical evaluation in three data sets of two studies (Griffin et al. 2021; Pouwels et al. 2021). On the other side, 16 data sets in 12 studies described a higher viral load among the vaccinated compared to the unvaccinated. The difference was significant in three data sets from three studies (Riemersma et al. 2022; Jung et al. 2022; Levine-Tiefenbrun et al. 2021), not significant in 11 data sets from eight studies (Altawalah et al. 2021; Bollinger et al. 2022; Li et al. 2022; Koh et al. 2022; Jung et al. 2022; Fall et al. 2022; Luo et al. 2022; Acharya et al. 2022) and without reporting results of a statistical evaluation in two studies (Pouwels et al. 2021; Sriraman et al. 2022).

Recipients of a third vaccine dose had in three data sets of one study a significantly lower viral load compared to unvaccinated individuals (Levine-Tiefenbrun et al. 2021). Recipients of only one vaccine dose had a significantly lower viral load during the early delta period but a slightly higher viral load during the late delta period (Pouwels et al. 2021). Similar results were reported by another research group (Griffin et al. 2021).

Among the boosted population (having received one dose of booster after the completion of the original course of vaccination), the effect of a lower viral load is insignificant within 61–120 days (Levine-Tiefenbrun et al. 2022) (Fig. 14.2).

14.2.1.3 Omicron Variants of SARS-CoV-2

In COVID-19 cases caused by the omicron variant the viral load was higher in the fully vaccinated population in all four data sets from two studies, the differences were, however, not significant (Fig. 14.3).

14.2.1.4 Multiple Variants of SARS-CoV-2

The viral load was higher among the fully vaccinated in four of five data sets obtained from COVID-19 cases caused by various variants of SARS-CoV-2 (Fig. 14.3). The difference was significant in one data set and not significant in the remaining data sets. In one study the viral load was marginally higher among the fully vaccinated subjects.

14.2.2 RNA Levels (Viral Copies)

During the delta predominance, the viral load was determined in five studies as the $\log_{10}$

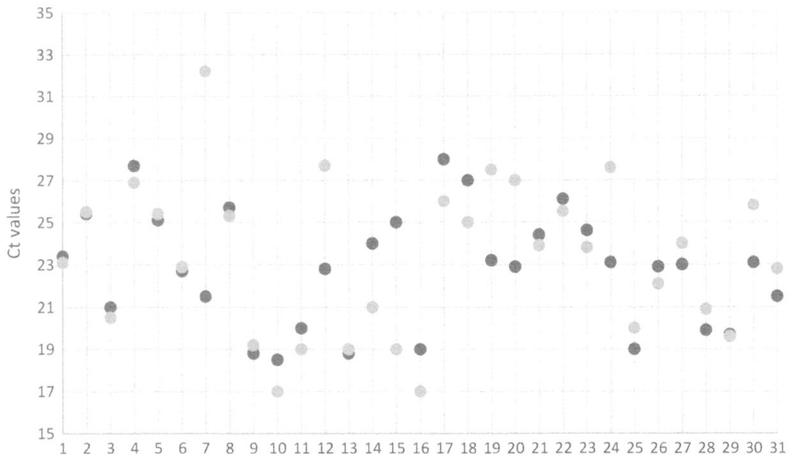

Fig. 14.2 Viral load in unvaccinated (dark grey) and fully vaccinated (light grey) COVID-19 cases described as mean or median Ct values during the delta predominance; 31 data sets obtained from 20 studies; data sets 1 and 2 (Acharya et al. 2022); data set 3 (Luo et al. 2022); data sets 4, 5 and 6 (Levine-Tiefenbrun et al. 2021); data set 7 and 8 (Pouwels et al. 2021); data set 9 (Chia et al. 2022); data set 10 (Fall et al. 2022); data sets 11 and 15 (Jung et al. 2022); data sets 12 and 13 (Griffin et al. 2021); data set 14 (Sriraman et al. 2022); data set 16 (Koh et al. 2022); data sets 17 and 18 (Li et al. 2022); data sets 19–23 (Bollinger et al. 2022); data set 24 (Elliott et al. 2021); data set 25 (Plante et al. 2022); data set 26 (Riemersma et al. 2022); data set 27 (Caserta et al. 2022); data set 28 (Park et al. 2022); data set 29 (Altawalah et al. 2021); data set 30 (Bramante et al. 2022); data set 31 (Brown et al. 2021); significant differences in data sets 4, 5, 15, 19, 20, 24, 26 and 30; no significant differences in data sets 1, 2, 3, 6, 9, 10, 11, 16, 17, 18, 21, 22, 23, 25, 27, 28, 29 and 31; no statistical evaluation in data sets 7, 8, 12, 13 and 14

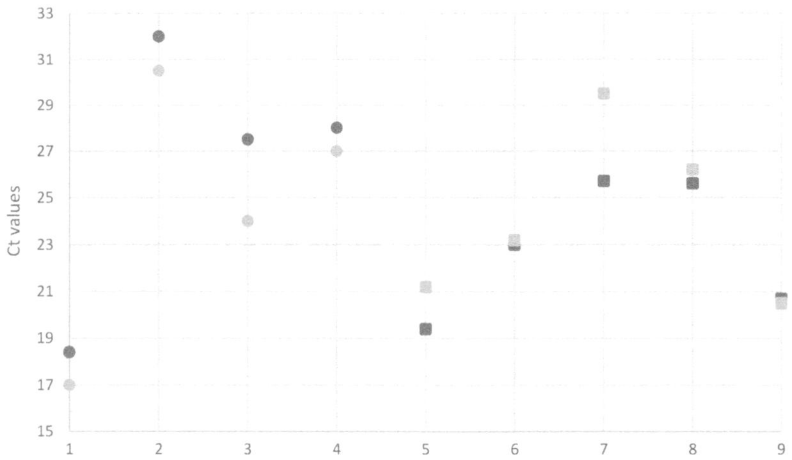

Fig. 14.3 Viral load in unvaccinated (dark grey) and fully vaccinated (light grey) COVID-19 cases described as mean or median Ct values; ○ describes mainly omicron variant; □ describes various variants; nine data sets obtained from six studies; data set 1 (Fall et al. 2022); data sets 2, 3 and 4 (Laura et al. 2022); data set 5 (Colavita et al. 2022); data set 6 (Servellita et al. 2022); data sets 7 and 8 (Hsu et al. 2021); data set 9 (Kissler et al. 2021); significant difference in data set 7; no significant differences in data sets 1, 2, 3, 4, 5, 6, 8 and 9

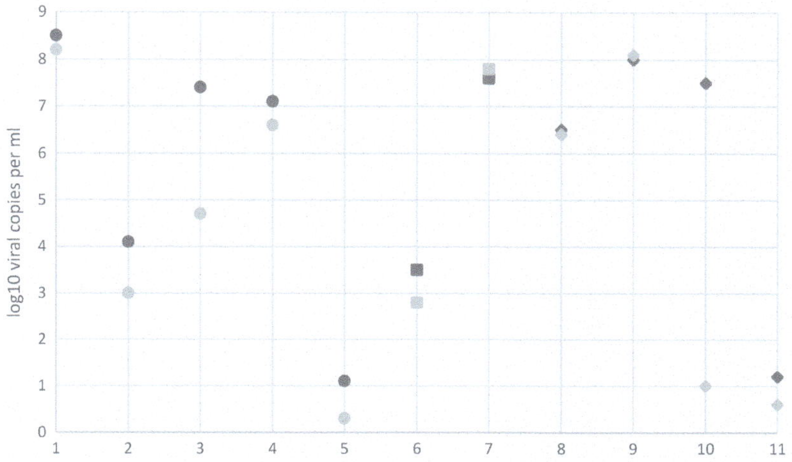

Fig. 14.4 Viral load in unvaccinated (dark grey) and vaccinated (light grey) COVID-19 cases described as mean or median $\log_{10}$ viral copies per ml; ○ describes mainly delta variant; □ describes mainly omicron variant; ◊ describes various variants; eleven data sets from eight studies; data sets 1 and 7 (Puhach et al. 2022); data sets 2 and 6 (Thompson et al. 2022); data set 3 (Rife Magalis et al. 2022); data set 4 (Migueres et al. 2022); data set 5 (Bramante et al. 2022); data set 8 (Servellita et al. 2022); data set 9 (Kissler et al. 2021); data sets 10 and 11 (Garcia-Knight et al. 2022); significant differences in data sets 1, 2, 3, 4, 5 and 6; no significant differences in data sets 7, 8, 9, 10 and 11

number of viral copies per ml. It was significantly higher in the unvaccinated population in all data sets. During the omicron predominance, one study described a significantly higher viral load among the unvaccinated, and another study found a lower viral load among the unvaccinated without a significant difference. When multiple variants were investigated, the viral load was more or less the same for unvaccinated and fully vaccinated subjects in two data sets, whereas higher viral load was described in two other studies but without a significant difference (Fig. 14.4).

14.2.3 Levels of Infectious SARS-CoV-2

The titer of infectious SARS-CoV-2 was determined in five studies. During the delta period, the viral load was mostly higher in the unvaccinated population with a significant difference in two data sets. During the omicron predominance, the difference between the fully vaccinated and unvaccinated was marginal. When multiple variants were included, the viral load was higher among the unvaccinated, but the difference was not significant (Fig. 14.5).

14.2.4 Subjects with a Previous COVID-19 Infection

Table 14.1 shows that patients with a COVID-19 reinfection mostly had a lower viral load as indicated by higher Ct values compared to unvaccinated cases with a primary infection (Maier et al. 2022; Abu-Raddad et al. 2022; Pouwels et al. 2021). Another study, however, demonstrated a higher viral load for the same comparison (Pouwels et al. 2021). The comparison of viral load in reinfection and fully vaccinated but previously uninfected revealed a lower viral load in reinfection in two data sets (Abu-Raddad et al. 2022) and a higher viral load in three data sets (Pouwels et al. 2021).

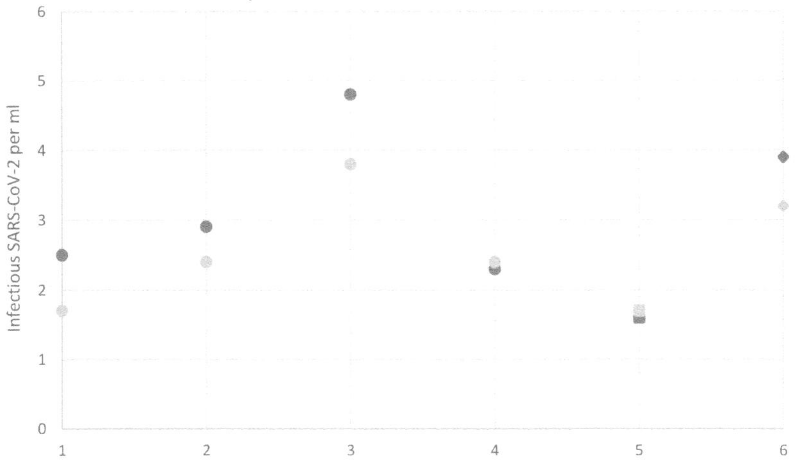

Fig. 14.5 Viral load in unvaccinated (dark grey) and vaccinated (light grey) COVID-19 cases described as infectious SARS-CoV-2 per ml; ○ describes mainly delta variant; □ describes mainly omicron variant; ◊ describes various variants; six data sets from five studies; data sets 1 and 5 (Puhach et al. 2022); data set 2 (Plante et al. 2022); data set 3 (Thompson et al. 2022); data set 4 (Riemersma et al. 2022); data set 6 (Garcia-Knight et al. 2022); significant differences in data sets 1 and 2; no significant differences in data sets 3, 4, 5 and 6

Table 14.1 Ct values from respiratory tract samples of unvaccinated COVID-19 cases having a reinfection compared unvaccinated COVID-19 cases (primary infection) and to fully vaccinated COVID-19 cases with a breakthrough infection

SARS-CoV-2 variant	Type of control	Ct values (reinfection in unvaccinated)	Ct values (control group)	References
Mainly beta	Primary infection in unvaccinated	29.9* **	23.6* **	Abu-Raddad et al. (2022)
Mainly beta	BNT162b2 breakthrough infections	29.5* **	26.4* **	Abu-Raddad et al. (2022)
Mainly beta	mRNA-1273 breakthrough infections	33.0* **	31.6* **	Abu-Raddad et al. (2022)
Alpha B.1.1.7	Primary infection in unvaccinated	32.8*	28.7*	Pouwels et al. (2021)
Alpha B.1.1.7	BNT162b2 and ChAdOx1 breakthrough infections	32.8*	33.3*	Pouwels et al. (2021)
Delta B.1.617.2 (early period)	Primary infection in unvaccinated	30.8*	21.5*	Pouwels et al. (2021)
Delta B.1.617.2 (early period)	BNT162b2 and ChAdOx1 breakthrough infections	30.8*	32.2*	Pouwels et al. (2021)
Delta B.1.617.2 (late period)	Primary infection in unvaccinated	22.3*	25.7*	Pouwels et al. (2021)
Delta B.1.617.2 (late period)	BNT162b2 and ChAdOx1 breakthrough infections	22.3*	25.3*	Pouwels et al. (2021)
Multiple variants	Primary infection in mostly unvaccinated	29.8** ***	28.0** ***	Maier et al. (2022)

* median; ** significant difference; *** mean

14.2.5 Relevance of Viral Load for Transmission

Caution has been advised when interpreting Ct values as a proxy for infectivity (Hawken et al. 2022). It is likely that additional factors contribute to transmission. An analysis of 602 PCR test samples was performed from COVID-19 cases among students. Contact tracing information was available for 195 index cases with 101 of them being non-spreaders and 94 of them being spreaders. Mean Ct values of the spreaders and the non-spreaders were nearly identical suggesting that Ct values do not predict transmissibility of SARS-CoV-2 (Tian et al. 2021).

14.3 Duration of Viral Shedding

Another hypothesis for a protection of others was that the fully vaccinated presumably have a shorter duration of viral shedding resulting in a lower probability to transmit SARS-CoV-2 to others. Some studies addressed the comparative duration of viral shedding in unvaccinated and fully vaccinated subjects.

14.3.1 Viral RNA

Two studies with delta variants show a shorter duration of viral RNA shedding for those who have received one or two doses of COVID-19 vaccine. The difference was significant in one study (Koh et al. 2022). When multiple variants were investigated in four studies, the duration of viral RNA detection was one, two or 15 days shorter for the fully vaccinated with three of them showing a significant difference (Garcia-Knight et al. 2022; Tian et al. 2022; Maier et al. 2022). A single dose also reduced the duration of RNA shedding by six days which was significant (Thompson et al. 2021) (Fig. 14.6).

14.3.2 Infectious SARS-CoV-2

The duration of shedding of infectious SARS-CoV-2 according to the vaccination status was described in four studies. It is much shorter compared to the duration of shedding of viral RNA. During the delta predominance, the unvaccinated subjects were longer viral carriers than those who had received one or two doses of COVID-19 vaccine with a significant difference in one of the two data sets. When multiple variants were evaluated, viral shedding was longer from the unvaccinated compared to the fully vaccinated with a significant difference in one of the two data sets (Fig. 14.7).

14.3.3 Relevance of Duration of Viral Shedding for Transmission

Viral clearance including the detection of infectious virus has been described to be faster in the vaccinated population so that their expectable duration as a possible source for transmission is mostly shorter. Viral shedding, however, is continuously decreasing with time over the course of infection (Boucau et al. 2022). That is why the overall relevance of a somewhat faster clearance of viral RNA or infectious SARS-CoV-2 on the transmissibility of SARS-CoV-2 is not clear.

14.4 Transmission Studies

Another hypothesis for a protection of others was that the fully vaccinated presumably have fewer secondary cases in their surroundings. Some studies addressed the secondary attack rates in defined settings among unvaccinated and fully vaccinated subjects.

Without any vaccination, the secondary attack rate in households was as high as 64.3% (Kolodziej et al. 2022). During the alpha variant predominance, secondary attack rates were significantly lower when the index case was vaccinated compared to unvaccinated index cases (Layan et al. 2022; Meyer et al. 2022). In households with children, the vaccination status of the parents was also associated with a lower COVID-19 case number (mainly alpha variant) among the unvaccinated children (risk reduction when one parent is vaccinated: − 26.0%;

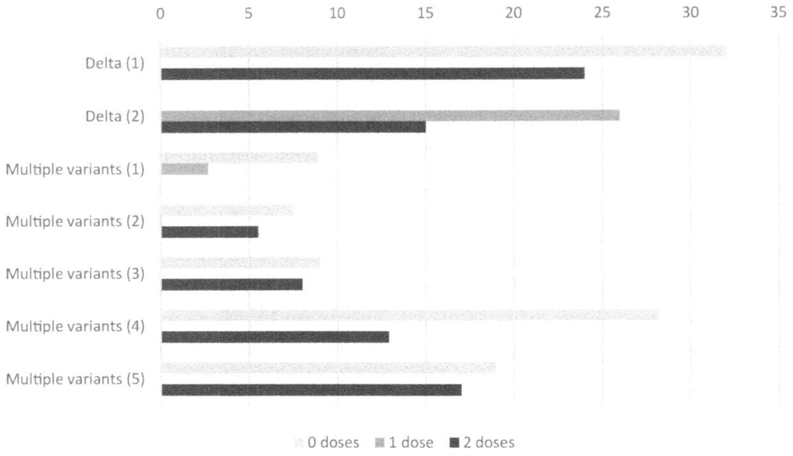

Fig. 14.6 Duration of detection of SARS-CoV-2 RNA from COVID-19 cases in days according to their vaccination status; delta (1) from (Koh et al. 2022); delta (2) from (Kang et al. 2022); multiple variants (1) from (Thompson et al. 2021); multiple variants (2) from (Kissler et al. 2021); multiple variants (3) from (Garcia-Knight et al. 2022); multiple variants (4) from (Maier et al. 2022); multiple variants (5) from (Tian et al. 2022); significant differences in data sets delta (1), multiple variants (1), (2), (4) and (5); no significant differences in data sets delta (2) and multiple variants (3)

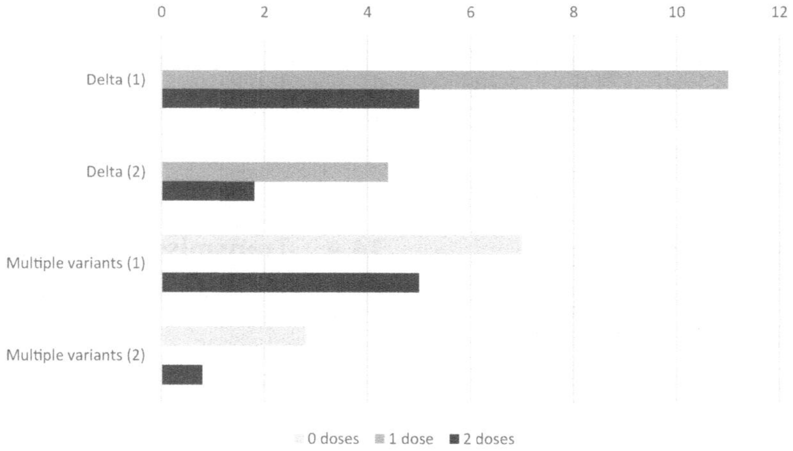

Fig. 14.7 Duration of detection of infectious SARS-CoV-2 from COVID-19 cases in days according to their vaccination status; delta (1) from (Kang et al. 2022); delta (2) from (Jung et al. 2022); multiple variants (1) from (Garcia-Knight et al. 2022); multiple variants (2) from (Ke et al. 2022); significant differences in data sets delta (1) and multiple variants (1); no significant differences in data sets delta (2) and multiple variants (2)

risk reduction when both parents are vaccinated: − 71.7%) (Hayek et al. 2022). In addition, in households of healthcare workers less secondary cases were found among unvaccinated household members when the index case was vaccinated (Salo et al. 2022). Significantly lower secondary attack rates were also found during the period when both alpha and delta variant were present for both the fully and partly vaccinated index cases (Eyre et al. 2022).

During the delta predominance, secondary attack rates were lower from fully vaccinated index cases in all studies described in Table 14.2 with three of them being statistically significant (Kang et al. 2022; Jung et al. 2022; Baker et al. 2022b) and four of them without a statistical evaluation (de Gier et al. 2021a, b; Lyngse et al. 2022a, b). In one study, the secondary attack rate was higher from fully vaccinated index cases, but the difference was not significant (Singanayagam et al. 2022).

An additional transmission analysis was done during the delta wave between index cases and contacts in households. The proportion of COVID-19 cases among contacts was not significantly different between vaccinated and unvaccinated index cases (Ng et al. 2021). In households with children, the vaccination status of the parents was associated with a lower COVID-19 case number (mainly delta variant) among the unvaccinated children (risk reduction when one parent is vaccinated: – 20.8%; risk reduction when both parents are vaccinated: – 58.1%) (Hayek et al. 2022).

When the omicron variant became predominant in the second half of 2021, a lower secondary attack rate was found from fully vaccinated individuals compared to the unvaccinated (Lyngse et al. 2022b; Baker et al. 2022a) and in one study a higher secondary attack rate (Grüne et al. 2022). Boosted index cases revealed a lower secondary attack rate compared to fully vaccinated index cases in two studies (Baker et al. 2022a; Grüne et al. 2022)) and a higher secondary attack rate in one study (Lyngse et al. 2022b). Three more studies with multiple SARS-CoV-2 variants described a lower secondary attack rate from vaccinated index cases compared to the unvaccinated in two studies (Kelly et al. 2022; Hsu et al. 2021)) and a higher rate in the other study (Muadchimkaew et al. 2022) with one of them being significant (Hsu et al. 2021).

In four studies, the majority of contact persons of fully vaccinated index cases were also fully vaccinated, whereas the vaccination rate in households from unvaccinated index cases was much lower (Kelly et al. 2022; de Gier et al. 2021a; Baker et al. 2022b; Muadchimkaew et al. 2022). In one study, the proportion of vaccinated households contacts was similar for vaccinated and unvaccinated index cases (Singanayagam et al. 2022).

The lower secondary attack rates described for the vaccinated subjects probably reflect the vaccination status of the contact persons. In five studies, the vaccination coverage of the contact persons was described according to the vaccination status of the index cases. Four of the studies indicate that contact persons were more often vaccinated when the index case was also vaccinated (Kelly et al. 2022; de Gier et al. 2021a; Baker et al. 2022b; Muadchimkaew et al. 2022). The differences in vaccination rates varied between 11.6% (Muadchimkaew et al. 2022) and 65.3% (Kelly et al. 2022). A finding from South Korea supports the relevance of the contact persons vaccination status. A population-based study during the omicron predominance indicated that fully vaccinated and boosted contacts have a lower incidence ratio compared to unvaccinated contacts (Lim et al. 2022). It appears therefore plausible to assume that a lower secondary attack rate in the households of a vaccinated index case is better explained by the protection of the vaccinated household members rather than the infectiousness of the vaccinated index case. It is known that those who encouraged their families and friends to get vaccinated are more likely to accept a COVID-19 vaccination (Cai et al. 2021). The presence of symptoms may also be relevant for transmission as symptomatic infections have been described to increase transmissibility from the index cases (Wright et al. 2022). Finally, the behaviour may be different between households of vaccinated and unvaccinated index cases. It has been described that those who are willing to get vaccinated have a higher adherence to physical distancing guidelines compared to the unvaccinated (Freeman et al. 2022). In addition, mitigation behaviours such as masking, physical distancing and hand hygiene is significantly more performed by the vaccinated (Hall et al. 2022). Keeping more physical distance may also have a relevant impact on the incidence of secondary cases (Chu et al. 2020). It is therefore highly questionable if the partly observed lower secondary attack rates in contact populations of vaccinated index cases is indeed explained by

Table 14.2 Secondary attack rates from COVID-19 cases with different vaccination status infected with different SARS-CoV-2 variants; #significant difference to unvaccinated subjects

SARS-CoV-2 variant	Vaccination status of index cases (n)	Type of contact	Secondary attack rate per group	Proportion of fully vaccinated contacts	Reference
Mainly alpha	Unvaccinated (200) Fully vaccinated (15)	Household contacts of healthcare workers	42.0% 19.0%#	–	Layan et al. (2022)
Mainly alpha	Unvaccinated (11) Fully vaccinated (5)	Household contacts	66.7% 22.2%#	0%* 0%*	Meyer et al. (2022)
Alpha and delta	Unvaccinated (218,946) Partially vaccinated (91,394) Fully vaccinated (63,775)	"Close contacts" of infected adult patients	46.4% 28.0% 26.8%	27.8%*	Eyre et al. (2022)
Delta	Unvaccinated (63) Partially vaccinated (25) Fully vaccinated (50)	Mainly household contacts	23.0% 37.1% 24.6%	63.0%** 54.3%*** 62.3%****	Singanayagam et al. (2022)
Delta	Unvaccinated (110,872) Partially vaccinated (2088) Fully vaccinated (622)	Household contacts	30.8% 28.9% 11.2%	2.1%*	de Gier et al. (2021a)
Delta	Unvaccinated (16,431) Fully vaccinated (8262)	Household contacts	22.3% 20.1%	–	Lyngse et al. (2022a)
Delta	Unvaccinated (61) Fully and partially vaccinated (22)	Household contacts	57.8% 16.7%#	35.7%** 75.0%****	Baker et al. (2022b)
Delta	Unvaccinated (8611) Fully vaccinated, no previous infection (8765) Boosted (528) Unvaccinated, previous infection (145) Fully vaccinated, previous infection (58)	Household contacts	21.5% 20.2% 18.0% 7.5% 7.5%	48.7%*	Lyngse et al. (2022b)
Delta	Unvaccinated (100) Partially vaccinated (10) Fully vaccinated (43)	Healthcare setting	27.0% 20.0% 7.0%#	–	Jung et al. (2022)
Delta	Unvaccinated (2641) Partially vaccinated (540) Fully vaccinated (1740)	Household contacts	17.7% 7.9% 12.4%	35.3%** 67.3%*** 79.6%****	de Gier et al. (2021b)
Delta	Partially vaccinated or unvaccinated (30) Fully vaccinated (19)	"Close contacts"	14.1% 3.7%#	–	Kang et al. (2022)

(continued)

Table 14.2 (continued)

SARS-CoV-2 variant	Vaccination status of index cases (n)	Type of contact	Secondary attack rate per group	Proportion of fully vaccinated contacts	Reference
Omicron	Unvaccinated (1166) Fully vaccinated, no previous infection (6934) Boosted (468) Unvaccinated, previous infection (128) Fully vaccinated, previous infection (414)	Household contacts	30.8% 29.5% 31.8% 15.0% 16.0%	48.7%*	Lyngse et al. (2022b)
Omicron	Unvaccinated (36) Fully vaccinated (12) Boosted (57)	Household contacts	63.9% 43.6% 42.7%	35.1%*	Baker et al. (2022a)
Omicron	Unvaccinated (202) Fully vaccinated (202) Boosted (204)	"Close contacts"	47.9% 49.5% 34.8%	–	Grüne et al. (2022)
Multiple variants	Unvaccinated (21) Fully vaccinated (21)	Household contacts	45.0% 41.4%	20.0%** 85.3%****	Kelly et al. (2022)
Multiple variants	Unvaccinated (177) Fully vaccinated (231)	Household contacts	46.8% 50.8%	3.1%** 14.7%****	Muadchimkaew et al. (2022)
Multiple variants	Unvaccinated (357) Fully vaccinated (357)	"Close contacts"	37.8% 10.1%#	–	Hsu et al. (2021)

* all contacts; ** contacts of unvaccinated index cases; *** contacts of partially vaccinated index cases; **** contacts of fully vaccinated index cases

their vaccination status because other factors explain the observed differences more plausible.

Another aspect with relevance for transmission may be a previous COVID-19 infection of the index case resulting in natural immunity. A previous COVID-19 infection reduced the duration of viral RNA detection from 26.4 to 13.7 days (Tian et al. 2022), a magnitude of reduction not observed among the fully vaccinated (Fig. 14.6). Household transmission studies described a lower secondary attack rate from unvaccinated index cases with a previous infection compared to unvaccinated index cases without a previous infection, for example, 7.5% versus 21.5% with the delta variant (Lyngse et al. 2022b), 15.0% versus 30.8% with the omicron variant (Lyngse et al. 2022b) and 40.9% versus 63.9% for the omicron variant (Baker et al. 2022a). Importantly, the secondary attack rate from vaccinated index cases without a previous infection was consistently higher compared to those from unvaccinated index cases with a previous infection (delta variant: 20.2% versus 7.5% (Lyngse et al. 2022b); omicron variant: 29.5% versus 15.0% (Lyngse et al. 2022b); omicron variant: 43.6% versus 40.9% (Baker et al. 2022a)). Natural immunity has been described to be at least as effective as vaccination (Sette and Crotty 2022). One of the reasons may be that natural immunity results in local tissue immunity (nose, mouth, saliva, and lungs) as a first line defence against SARS-CoV-2 which is not generated by intramuscular injection (Sette and Crotty 2022).

14.5 Population-Based Case Numbers

A further hypothesis for a protection of others was that the number of COVID-19 cases among the fully vaccinated population is lower compared to unvaccinated individuals resulting in fewer possible sources for transmission. Some studies and some epidemiological data addressed the population-based case numbers among unvaccinated and fully vaccinated subjects.

14.5.1 Phase III Trials of COVID-19 Vaccines

Four trials on the efficacy of COVID-19 vaccines described a significant reduction of symptomatic COVID-19 cases within the first two to three months after completion of the vaccination. In originally seronegative people, the vaccination reduced the number of symptomatic sources (Table 14.3). In one study, a subgroup of asymptomatic cases was analysed in both study arms. Asymptomatic viral carriers were identified within the first three months after vaccination in a similar proportion among both the unvaccinated and fully vaccinated study participants (Table 14.3). These data provide evidence that the number of symptomatic COVID-19 cases was successfully reduced for a few months, whereas the proportion of asymptomatic COVID-19 cases was not reduced by one of the vaccines.

14.5.2 Population-Based Data

Data are available from the United Kingdom from which the numbers of cases per 100,000 people were summarized for each four weeks and broken down according to their vaccination status. All COVID-19 cases are included, regardless of their symptoms. Figure 14.8 summarizes the UK case numbers up in the age group between 40 and 49 years for the period between calendar week 36 in 2021 and 13 in 2022. Case numbers were consistently higher among the fully vaccinated and boosted population compared to the unvaccinated. With the start of the omicron wave at the beginning of 2022, there were suddenly significantly more cases among the fully vaccinated. Even after the start of booster vaccinations, there were consistently more COVID-19 cases in the group of people with a booster vaccination compared to the unvaccinated (Fig. 14.8).

The higher proportion of COVID-19 cases among the fully vaccinated was found in most age groups. In Fig. 14.9, the UK data are summarized for calendar weeks 39–42 in 2021. These population-based data clearly show that in real life the vaccination did not reduce the number COVID-19 cases and thereby did not reduce the number of possible sources for transmission.

14.6 Lessons Learned

14.6.1 Fully Vaccinated but no Consistently Lower Viral Load

One hypothesis for a protection of others was that the fully vaccinated might have a significantly lower viral load during a breakthrough infection resulting in a lower probability to transmit SARS-CoV-2 to others. An evaluation of the levels of RNA and infectious SARS-CoV-2 provides evidence that the viral load was partly lower and also partly higher among the vaccinated. The entire picture is not as clear as hypothesized. The viral load seems also to be dependent on the SARS-CoV-2 variant and the severity of disease (Puhach et al. 2023).

14.6.2 Fully Vaccinated and Partly Shorter Duration of Viral Shedding

Viral clearance including the detection of infectious virus has been described to be faster in the vaccinated population so that their expectable

Table 14.3 Number of symptomatic (subjects with at least one symptom indicative of COVID-19 and a positive PCR test for detection of SARS-CoV-2 RNA) or asymptomatic COVID-19 cases in phase 3 vaccination trials

Vaccine (manufacturer)	Vaccinated subjects	Proportion (%)	Control subjects	Proportion (%)	References
Symptomatic COVID-19 cases					
AZD1222 (AstraZeneca)	74 of 7201	1.0	197 of 7178	2.7	Voysey et al. (2021)
mRNA 1273 (Moderna)	11 of 15,210	0.07	185 of 15,210	1.2	Baden et al. (2021)
BNT162b2 (BioNTech)	8 of 18,198	0.04	162 of 18,325	0.9	Polack et al. (2020)
Sputnik V (Gamaleya)	16 of 14,964	0.1	62 of 4902	1.3	Logunov et al. (2021)
COVID-19 cases without symptoms or with unknown symptoms					
AZD1222 (AstraZeneca)	22 of 2168	1.0	23 of 2223	1.0%	Voysey et al. (2021)

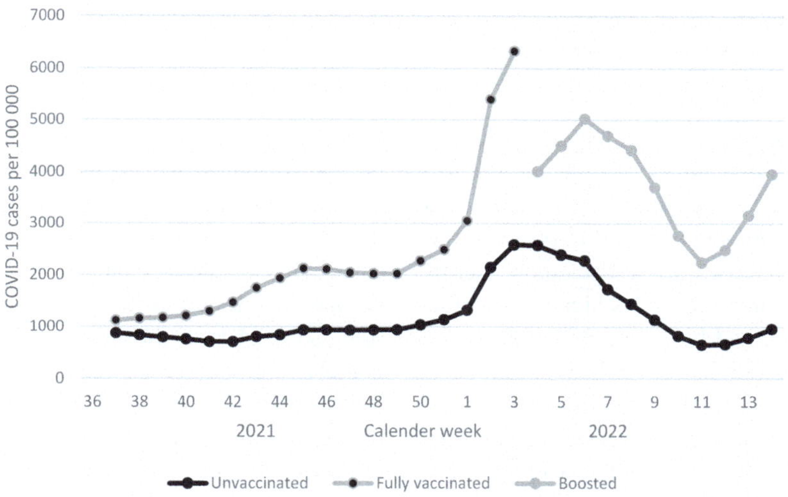

Fig. 14.8 COVID-19 cases per 100,000 unvaccinated, fully vaccinated and boosted individuals in the 40–49 age group in the UK; data from calendar weeks 36 in 2021 to 13 in 2022; sources: weekly reports on COVID 19 vaccination in the UK, published by the UK Health Security Agency

duration as a possible source for transmission is mostly shorter. Viral shedding, however, is continuously decreasing with time over the course of infection (Boucau et al. 2022). That is why the overall relevance of a somewhat faster clearance of viral RNA or infectious SARS-CoV-2 on the transmissibility of SARS-CoV-2 is not clear.

14.6.3 Fully Vaccinated but no Favourable Effect on Case Numbers

Public health reporting should consider to collect, analyse and publish data on symptomatic and asymptomatic cases per population unit (e.g. 100,000). The reports from the UK are a positive

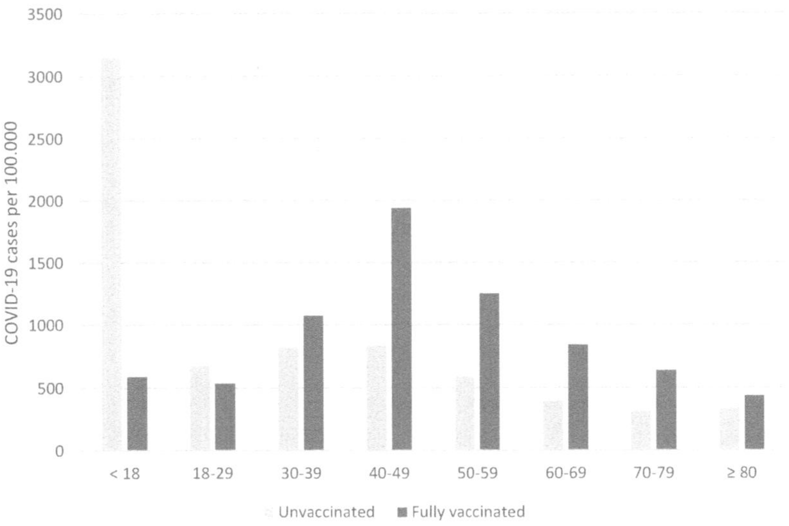

Fig. 14.9 COVID-19 cases per 100,000 unvaccinated or fully vaccinated in the UK by age group; data from calendar weeks 39–42 in 2021 (UK Health Security Agency 2021)

example in that respect. This measure allows to verify all possible sources of transmission irrespective of their vaccination or immunity status. In contrast to data obtained in phase 3 trials on symptomatic cases, these data show that the number of possible sources is higher among the fully vaccinated and boosted population. In that respect, vaccination has not helped to reduce the number of possible sources for transmission. On the contrary, it may even have increased the number of cases because of the waning immunity with new SARS-CoV-2 variants such as omicron.

14.6.4 Many Transmission Studies Narrowed Regarding the Causality of Outcome

Some studies found lower secondary attack rates from the fully vaccinated population. The findings have supported the public view that vaccination might protect close contact persons. The results are, however, probably best explained by the vaccination status of the contact population and the compliance with mitigation behaviours such as masking, physical distancing and hand hygiene. A previous infection of the contact person may also explain the lower risk to acquire COVID-19. But researchers narrowed their focus on the vaccination status of the index cases. This should be changed in future research so that all relevant criteria such as the vaccination status of both the index persons and their contacts are analysed, compliance with mitigation behaviour is evaluated in each setting, and the natural immunity of the contact persons is taken into account. Any analysis without considering these and may be more relevant criteria is likely to be narrowed or even biased regarding the causality of the outcome.

14.7 Conclusions

Most of the collected evidence indicates that the COVID-19 vaccination does not reduce the risk to transmit SARS-CoV-2 to others in a relevant proportion. On the contrary: epidemiological data from the UK indicate an even higher number of COVID-19 cases among the vaccinated and boosted population suggesting a higher risk for transmission from them. Public health messages from organizations and politicians should therefore not exaggerate or whitewash the expectable evidence-based health benefits of a COVID-19 vaccination.

References

Abu-Raddad LJ, Chemaitelly H, Ayoub HH et al (2022) Relative infectiousness of SARS-CoV-2 vaccine breakthrough infections, reinfections, and primary infections. Nat Commun 13:532. https://doi.org/10.1038/S41467-022-28199-7

Accorsi EK, Britton A, Fleming-Dutra KE et al (2022) Association between 3 doses of mRNA COVID-19 vaccine and symptomatic infection caused by the SARS-CoV-2 omicron and delta variants. JAMA 327:639–651. https://doi.org/10.1001/JAMA.2022.0470

Acharya CB, Schrom J, Mitchell AM et al (2022) Viral load among vaccinated and unvaccinated, asymptomatic and symptomatic persons infected with the SARS-CoV-2 delta variant. Open Forum Infect Dis 9:ofac135. https://doi.org/10.1093/OFID/OFAC135

Altawalah H, Alfouzan W, Al-Fadalah T, Ezzikouri S (2021) Diagnostic performance of automated SARS-CoV-2 antigen assay in nasal swab during COVID-19 vaccination campaign. Diagnostics 11:2110. https://doi.org/10.3390/DIAGNOSTICS11112110

Baden LR, El Sahly HM, Essink B et al (2021) Efficacy and safety of the mRNA-1273 SARS-CoV-2 vaccine. N Engl J Med 384:403–416. https://doi.org/10.1056/NEJMoa2035389

Baker JM, Nakayama JY, O'Hegarty M et al (2022a) SARS-CoV-2 B.1.1.529 (omicron) variant transmission within households—Four U.S. Jurisdictions, Nov 2021–Feb 2022. Morb Mortal Wkly Rep 71:341–346. https://doi.org/10.15585/MMWR.MM7109E1

Baker JM, Shah MM, O'Hegarty M et al (2022b) Primary and secondary attack rates by vaccination status after a SARS-CoV-2 B.1.617.2 (delta) variant outbreak at a Youth Summer Camp—Texas, June 2021. J Pediatric Infect Dis Soc piac086. https://doi.org/10.1093/JPIDS/PIAC086

Bollinger M, Saile P, Shapeton AD et al (2022) Sensitivity of severe acute respiratory syndrome coronavirus type 2 rapid antigen point-of-care tests in vaccinated patients. Eur J Emerg Med 29:285–290. https://doi.org/10.1097/MEJ.0000000000000928

Boucau J, Marino C, Regan J et al (2022) Duration of shedding of culturable virus in SARS-CoV-2 omicron (BA.1) infection. N Engl J Med 387:275–277. https://doi.org/10.1056/NEJMC2202092

Bramante CT, Proper JL, Boulware DR et al (2022) Vaccination Against SARS-CoV-2 is associated with a lower viral load and likelihood of systemic symptoms. Open Forum Infect Dis 9:ofac066. https://doi.org/10.1093/OFID/OFAC066

Brown CM, Vostok J, Johnson H et al (2021) Outbreak of SARS-CoV-2 infections, including COVID-19 vaccine breakthrough infections, associated with large public gatherings—Barnstable County, Massachusetts, July 2021. MMWR Morb Mortal Wkly Rep 70:1059–1062. https://doi.org/10.15585/mmwr.mm7031e2

Bundesministerium für Gesundheit (2022) Gemeinschaftsschutz durch Corona-Schutzimpfung|Zusammen gegen Corona. https://www.zusammengegencorona.de/impfen/gemeinschaftsschutz-solidaritaet-in-der-coronavirus-pandemie/. Accessed 30 Nov 2022

Cai H, Bai W, Liu S et al (2021) Attitudes toward COVID-19 vaccines in Chinese adolescents. Front Med 8:691079. https://doi.org/10.3389/FMED.2021.691079

Caserta LC, Martins M, Cronk B et al (2022) Infection and transmission of SARS-CoV-2 B.1.617.2 lineage (delta variant) among fully vaccinated individuals. Microbiol Spectr 10:e0056322. https://doi.org/10.1128/SPECTRUM.00563-22

Chia PY, Ong SWX, Chiew CJ et al (2022) Virological and serological kinetics of SARS-CoV-2 delta variant vaccine breakthrough infections: a multicentre cohort study. Clin Microbiol Infect 28:612.e1-612.e7. https://doi.org/10.1016/J.CMI.2021.11.010

Chu DK, Akl EA, Duda S et al (2020) Physical distancing, face masks, and eye protection to prevent person-to-person transmission of SARS-CoV-2 and COVID-19: a systematic review and meta-analysis. Lancet (London, England) 395:1973–1987. https://doi.org/10.1016/S0140-6736(20)31142-9

Colavita F, Meschi S, Gruber CEM et al (2022) Virological and serological characterisation of SARS-CoV-2 infections diagnosed after mRNA BNT162b2 vaccination between December 2020 and March 2021. Front Med 8:815870. https://doi.org/10.3389/FMED.2021.815870/FULL

de Gier B, Andeweg S, Backer JA, et al (2021a) Vaccine effectiveness against SARS-CoV-2 transmission to household contacts during dominance of Delta variant (B.1.617.2), the Netherlands, August to September 2021. Euro Surveill 26:2100977. https://doi.org/10.2807/1560-7917.ES.2021.26.44.2100977

de Gier B, Andeweg S, Joosten R et al (2021b) Vaccine effectiveness against SARS-CoV-2 transmission and infections among household and other close contacts of confirmed cases, the Netherlands, February to May 2021. Eurosurveillance 26:2100640. https://doi.org/10.2807/1560-7917.ES.2021.26.31.2100640

Der Spiegel (2021) Corona: Markus Söder (CSU) für Unterschiede zwischen Geimpften und Nicht-Geimpften. https://www.spiegel.de/politik/deutschland/corona-markus-soeder-csu-fuer-unterschiede-zwischen-geimpften-und-nicht-geimpften-a-6cbfd0a0-104b-4f6b-b25e-05cb4f33abff. Accessed 29 Dec 2022

Elliott P, Haw D, Wang H et al (2021) Exponential growth, high prevalence of SARS-CoV-2, and vaccine effectiveness associated with the Delta variant. Science 374:eabl9551. https://doi.org/10.1126/SCIENCE.ABL9551

Eyre DW, Taylor D, Purver M et al (2022) Effect of Covid-19 vaccination on transmission of alpha and delta variants. N Engl J Med 386:744–756. https://doi.org/10.1056/NEJMOA2116597

Fall A, Eldesouki RE, Sachithanandham J et al (2022) The displacement of the SARS-CoV-2 variant delta with omicron: an investigation of hospital admissions and upper respiratory viral loads. EBioMedicine 79:104008. https://doi.org/10.1016/J.EBIOM.2022.104008

Felix H (2021) Weitergabe des Virus - Warum man mit einer Corona-Impfung auch andere schützt. https://www.derbund.ch/warum-man-mit-einer-corona-impfung-auch-andere-schuetzt-414879400098. Accessed 30 Nov 2022

Freeman D, Loe BS, Chadwick A et al (2022) COVID-19 vaccine hesitancy in the UK: the Oxford coronavirus explanations, attitudes, and narratives survey (Oceans) II. Psychol Med 52:3127–3141. https://doi.org/10.1017/S0033291720005188

Garcia-Knight M, Anglin K, Tassetto M et al (2022) Infectious viral shedding of SARS-CoV-2 delta following vaccination: a longitudinal cohort study. PLoS Pathog 18:e1010802. https://doi.org/10.1371/JOURNAL.PPAT.1010802

Ghazy RM, Ashmawy R, Hamdy NA et al (2022) Efficacy and effectiveness of SARS-CoV-2 vaccines: a systematic review and meta-analysis. Vaccines 10:350. https://doi.org/10.3390/VACCINES10030350/S1

Griffin JB, Haddix M, Danza P et al (2021) SARS-CoV-2 infections and hospitalizations among persons aged ≥16 years, by vaccination status—Los Angeles County, California, May 1-July 25, 2021. MMWR Morb Mortal Wkly Rep 70:1170–1176. https://doi.org/10.15585/MMWR.MM7034E5

Grüne B, Grüne J, Kossow A, Joisten C (2022) Vaccination and transmission risk during the Outbreak of B.1.1.529 (omicron). Vaccines 10:1003. https://doi.org/10.3390/VACCINES10071003

Hall PA, Meng G, Sakib MN et al (2023) Do the vaccinated perform less distancing, mask wearing and hand hygiene? A test of the risk compensation hypothesis in a representative sample during the COVID-19 pandemic. Vaccine 41:4027-4030. https://doi.org/10.1016/J.VACCINE.2022.10.028

Hawken SE, Sellers SA, Smedberg JR et al (2022) Longitudinal SARS-CoV-2 testing among the unvaccinated is punctuated by intermittent positivity and variable rates of increasing cycle threshold values. Microbiol Spectr 10:e0271521. https://doi.org/10.1128/SPECTRUM.02715-21

Hayek S, Shaham G, Ben-Shlomo Y et al (2022) Indirect protection of children from SARS-CoV-2 infection through parental vaccination. Science (80-) 375:1155–1159. https://doi.org/10.1126/SCIENCE.ABM3087

Hsu L, Grüne B, Buess M et al (2021) Covid-19 breakthrough infections and transmission risk: real-world data analyses from Germany's largest public health department (Cologne). Vaccines 9:1267. https://doi.org/10.3390/VACCINES9111267

Ioannou P, Karakonstantis S, Astrinaki E et al (2021) Transmission of SARS-CoV-2 variant B.1.1.7 among vaccinated health care workers. Infect Dis 53:876–879. https://doi.org/10.1080/23744235.2021.1945139

Jung J, Kim JY, Park H et al (2022) Transmission and infectious SARS-CoV-2 shedding kinetics in vaccinated and unvaccinated individuals. JAMA Netw Open 5:e2213606. https://doi.org/10.1001/JAMANETWORKOPEN.2022.13606

Kang SW, Kim JY, Park H et al (2022) Comparison of outward transmission potential between vaccinated and partially vaccinated or unvaccinated individuals with the SARS-CoV-2 delta variant infection. J Infect 85:e69. https://doi.org/10.1016/J.JINF.2022.06.002

Ke R, Martinez PP, Smith RL et al (2022) Longitudinal analysis of SARS-CoV-2 vaccine breakthrough infections reveals limited infectious virus shedding and restricted tissue distribution. Open Forum Infect Dis 9:ofac192. https://doi.org/10.1093/OFID/OFAC192

Kelly JD, Lu S, Anglin K et al (2022) Magnitude and determinants of SARS-CoV-2 household transmission: a longitudinal cohort study. Clin Infect Dis 75:S193–S204. https://doi.org/10.1093/CID/CIAC545

Kissler SM, Fauver JR, Mack C et al (2021) Viral Dynamics of SARS-CoV-2 variants in vaccinated and unvaccinated persons. N Engl J Med 385:2489–2491. https://doi.org/10.1056/NEJMc2102507

Koh T, Ooi XY, Vasoo S, Yeo SC (2022) Impact of Variant of concern and vaccination status on COVID-19 infection virological dynamics in end stage kidney disease patients receiving haemodialysis. Nephrology 27:804–809. https://doi.org/10.1111/NEP.14084

Kolodziej LM, van Lelyveld SFL, Haverkort ME et al (2022) High severe acute respiratory syndrome coronavirus 2 (SARS-CoV-2) household transmission rates detected by Dense Saliva Sampling. Clin Infect Dis 75:e10–e19. https://doi.org/10.1093/CID/CIAC261

Korang SK, von Rohden E, Veroniki AA et al (2022) Vaccines to prevent COVID-19: a living systematic review with trial sequential analysis and network meta-analysis of randomized clinical trials. PLoS ONE 17:e0260733. https://doi.org/10.1371/JOURNAL.PONE.0260733

Laura L, Dalmatin-Dragišić M, Martinović K et al (2023) Does pre-existing immunity determine the course of SARS-CoV-2 infection in health-care workers? Single-Center Experience. Infection 51:323-330. https://doi.org/10.1007/S15010-022-01859-Y

Layan M, Gilboa M, Gonen T et al (2022) Impact of BNT162b2 vaccination and isolation on SARS-CoV-2 transmission in Israeli households: an observational study. Am J Epidemiol 191:1224–1234. https://doi.org/10.1093/AJE/KWAC042

Levine-Tiefenbrun M, Yelin I, Alapi H et al (2021) Viral loads of delta-variant SARS-CoV-2 breakthrough

infections after vaccination and booster with BNT162b2. Nat Med 27:2108–2110. https://doi.org/10.1038/S41591-021-01575-4
Levine-Tiefenbrun M, Yelin I, Alapi H et al (2022) Waning of SARS-CoV-2 booster viral-load reduction effectiveness. Nat Commun 13:1237. https://doi.org/10.1038/S41467-022-28936-Y
Li J, Zhang Y, Jiang L et al (2022) Similar aerosol emission rates and viral loads in upper respiratory tracts for COVID-19 patients with delta and omicron variant infection. Virol Sin 37:762–764. https://doi.org/10.1016/J.VIRS.2022.07.010
Lim DS, Choe YJ, Kim YM et al (2022) Household secondary attack rates of SARS-CoV-2 omicron variant, South Korea, February 2022. Emerg Infect Dis 28:1731–1734. https://doi.org/10.3201/EID2808.220384
Logunov DY, Dolzhikova IV, Shcheblyakov DV et al (2021) Safety and efficacy of an rAd26 and rAd5 vector-based heterologous prime-boost COVID-19 vaccine: an interim analysis of a randomised controlled phase 3 trial in Russia. Lancet 397:671–681. https://doi.org/10.1016/s0140-6736(21)00234-8
Luo CH, Morris CP, Sachithanandham J et al (2022) Infection with the severe acute respiratory syndrome coronavirus 2 (SARS-CoV-2) delta variant is associated with higher recovery of infectious virus compared to the alpha variant in both unvaccinated and vaccinated individuals. Clin Infect Dis 75:e715–e725. https://doi.org/10.1093/CID/CIAB986
Lyngse FP, Mølbak K, Denwood M et al (2022a) Effect of vaccination on household transmission of SARS-CoV-2 Delta variant of concern. Nat Commun 13:3764. https://doi.org/10.1038/S41467-022-31494-Y
Lyngse FP, Mortensen LH, Denwood MJ et al (2022b) Household transmission of the SARS-CoV-2 omicron variant in Denmark. Nat Commun 13:5573. https://doi.org/10.1038/S41467-022-33328-3
Maier HE, Plazaola M, Lopez R et al (2022) SARS-CoV-2 infection-induced immunity and the duration of viral shedding: results from a Nicaraguan household cohort study. Influenza Other Respi Viruses 17:e13074. https://doi.org/10.1111/IRV.13074
Meyer ED, Sandfort M, Bender J et al (2023) BNT162b2 vaccination reduced infections and transmission in a COVID-19 outbreak in a nursing home in Germany, 2021. Influenza Other Respi Viruses 17:e13051. https://doi.org/10.1111/IRV.13051
Migueres M, Dimeglio C, Trémeaux P et al (2022) Influence of the delta variant and vaccination on the SARS-CoV-2 viral load. Viruses 14:323. https://doi.org/10.3390/V14020323
Muadchimkaew M, Siripongboonsitti T, Wongpatcharawarakul S et al (2022) Effect of inactivated SARS-CoV-2 vaccines and ChAdOx1 nCoV-19 vaccination to prevent COVID-19 in Thai households (VacPrevent trial). Int J Infect Dis 124:190–198. https://doi.org/10.1016/J.IJID.2022.09.032
Muhsen K, Maimon N, Mizrahi A et al (2022) Effectiveness of BNT162b2 mRNA COVID-19 vaccine against acquisitions of SARS-CoV-2 among health care workers in long-term care facilities: a prospective cohort study. Clin Infect Dis 75:e755–e763. https://doi.org/10.1093/CID/CIAB918
Ng OT, Koh V, Chiew CJ et al (2021) Impact of delta variant and vaccination on SARS-CoV-2 secondary attack rate among household close contacts. Lancet Reg Heal West Pacific 17:100299. https://doi.org/10.1016/J.LANWPC.2021.100299
Park SY, Kim TH, Lee E et al (2022) A SARS-CoV-2 outbreak associated with vaccine breakthrough in an acute care hospital. Am J Infect Control 50:1006–1012. https://doi.org/10.1016/J.AJIC.2022.05.010
Plante JA, Machado RRG, Mitchell BM et al (2022) Vaccination decreases the infectious viral load of delta variant SARS-CoV-2 in asymptomatic patients. Viruses 14:2071. https://doi.org/10.3390/V14092071
Polack FP, Thomas SJ, Kitchin N et al (2020) Safety and efficacy of the BNT162b2 mRNA Covid-19 vaccine. N Engl J Med 383:2603–2615. https://doi.org/10.1056/NEJMoa2034577
Pouwels KB, Pritchard E, Matthews PC et al (2021) Effect of delta variant on viral burden and vaccine effectiveness against new SARS-CoV-2 infections in the UK. Nat Med 27:2127–2135. https://doi.org/10.1038/S41591-021-01548-7
Puhach O, Adea K, Hulo N et al (2022) Infectious viral load in unvaccinated and vaccinated individuals infected with ancestral, delta or omicron SARS-CoV-2. Nat Med 28:1491–1500. https://doi.org/10.1038/s41591-022-01816-0
Puhach O, Meyer B, Eckerle I (2023) SARS-CoV-2 viral load and shedding kinetics. Nat Rev Microbiol 21:147–161. https://doi.org/10.1038/S41579-022-00822-W
Regev-Yochay G, Amit S, Bergwerk M et al (2021) Decreased infectivity following BNT162b2 vaccination: a prospective cohort study in Israel. Lancet Reg Heal - Eur 7:100150. https://doi.org/10.1016/J.LANEPE.2021.100150
Riemersma KK, Haddock LA, Wilson NA et al (2022) Shedding of infectious SARS-CoV-2 despite vaccination. PLoS Pathog 18:e1010876. https://doi.org/10.1371/JOURNAL.PPAT.1010876
Rife Magalis B, Rich S, Tagliamonte MS et al (2022) Severe acute respiratory syndrome coronavirus 2 delta vaccine breakthrough transmissibility in Alachua County, Florida. Clin Infect Dis 75:1618–1627. https://doi.org/10.1093/CID/CIAC197
Salo J, Hägg M, Kortelainen M et al (2022) The indirect effect of mRNA-based COVID-19 vaccination on healthcare workers' unvaccinated household members. Nat Commun 13:1162. https://doi.org/10.1038/S41467-022-28825-4

Servellita V, Morris MK, Sotomayor-Gonzalez A et al (2022) Predominance of antibody-resistant SARS-CoV-2 variants in vaccine breakthrough cases from the San Francisco Bay Area, California. Nat Microbiol 7:277–288. https://doi.org/10.1038/S41564-021-01041-4

Sette A, Crotty S (2022) Immunological memory to SARS-CoV-2 infection and COVID-19 vaccines. Immunol Rev 310:27–46. https://doi.org/10.1111/IMR.13089

Shitrit P, Zuckerman NS, Mor O et al (2021) Nosocomial outbreak caused by the SARS-CoV-2 delta variant in a highly vaccinated population, Israel, July 2021. Euro Surveill 26:2100822. https://doi.org/10.2807/1560-7917.es.2021.26.39.2100822

Shrotri M, Krutikov M, Palmer T et al (2021) Vaccine effectiveness of the first dose of ChAdOx1 nCoV-19 and BNT162b2 against SARS-CoV-2 infection in residents of long-term care facilities in England (VIVALDI): a prospective cohort study. Lancet Infect Dis 21:1529–1538. https://doi.org/10.1016/S1473-3099(21)00289-9

Singanayagam A, Hakki S, Dunning J et al (2022) Community transmission and viral load kinetics of the SARS-CoV-2 delta (B.1.617.2) variant in vaccinated and unvaccinated individuals in the UK: a prospective, longitudinal, cohort study. Lancet Infect Dis 22:183–195. https://doi.org/10.1016/S1473-3099(21)00648-4

Sriraman K, Shaikh A, Vaswani S et al (2023) Impact of COVID-19 vaccination on transmission risk of breakthrough infections: lessons from adapted N95 mask sampling for emerging variants and interventions. J Med Virol 95:e28188. https://doi.org/10.1002/JMV.28188

Ssentongo P, Ssentongo AE, Voleti N et al (2022) SARS-CoV-2 vaccine effectiveness against infection, symptomatic and severe COVID-19: a systematic review and meta-analysis. BMC Infect Dis 22:439. https://doi.org/10.1186/s12879-022-07418-y

Štěpánek L, Janošíková M, Nakládalová M et al (2021) Motivation for COVID-19 vaccination in priority occupational groups: a cross-sectional survey. Int J Environ Res Public Health 18:11726. https://doi.org/10.3390/IJERPH182111726/S1

Thompson MG, Burgess JL, Naleway AL et al (2021) Prevention and attenuation of Covid-19 with the BNT162b2 and mRNA-1273 vaccines. N Engl J Med 385:320–329. https://doi.org/10.1056/NEJMOA2107058

Thompson MG, Yoon SK, Naleway AL et al (2022) Association of mRNA vaccination with clinical and virologic features of COVID-19 among US essential and frontline workers. JAMA 328:1523–1533. https://doi.org/10.1001/JAMA.2022.18550

Tian D, Lin Z, Kriner EM et al (2021) Ct values do not predict severe acute respiratory syndrome coronavirus 2 (SARS-CoV-2) transmissibility in college students. J Mol Diagnostics 23:1078–1084. https://doi.org/10.1016/J.JMOLDX.2021.05.012

Tian X, Zhang Y, Wang W et al (2022) The impacts of vaccination status and host factors during early infection on SARS-CoV-2 persistence:a retrospective single-center cohort study. Int Immunopharmacol 114:109534. https://doi.org/10.1016/J.INTIMP.2022.109534

UK Health Security Agency (2021) COVID-19 vaccine surveillance report. Week 43. https://assets.publishing.service.gov.uk/government/uploads/system/uploads/attachment_data/file/1029606/Vaccine-surveillance-report-week-43.pdf. Accessed 23 Jan 2023

Voysey M, Clemens SAC, Madhi SA et al (2021) Safety and efficacy of the ChAdOx1 nCoV-19 vaccine (AZD1222) against SARS-CoV-2: an interim analysis of four randomised controlled trials in Brazil, South Africa, and the UK. Lancet 397:99–111. https://doi.org/10.1016/s0140-6736(20)32661-1

WHO (2022) Vaccines and immunization: What is vaccination? https://www.who.int/news-room/questions-and-answers/item/vaccines-and-immunization-what-is-vaccination. Accessed 1 Aug 2022

World Health Organization (2021) Vaccine efficacy, effectiveness and protection. https://www.who.int/news-room/feature-stories/detail/vaccine-efficacy-effectiveness-and-protection. Accessed 30 Nov 2022

Wright E, Pollard G, Robertson H, Anuradha S (2022) Household transmission of the delta COVID-19 variant in Queensland, Australia: a case series. Epidemiol Infect 150:e173. https://doi.org/10.1017/S0950268822001546

Zürcher K, Abela IA, Stange M et al (2023) Alpha variant coronavirus outbreak in a nursing home despite high vaccination coverage: molecular, epidemiological and immunological studies. Clin Infect Dis 77:537-546. https://doi.org/10.1093/CID/CIAB1005

Sudden Hearing Loss Waves: The Effect of COVID-19 Infection and Vaccination on the Inner Ear

15

Karen Tawk, Anthony Tawk, and Mehdi Abouzari

Abstract

Sudden sensorineural hearing loss (SSNHL) has emerged as a potential complication of COVID-19 infection and vaccination. Various mechanisms by which the SARS-CoV-2 virus can cause hearing loss have been reported, including direct viral invasion, neuroinflammation, blood flow disturbances, and immune-mediated response. However, the temporal relationship between COVID-19 infection and SSNHL remains unclear, with mixed findings and conflicting results reported in different studies. Similarly, while anecdotal reports have linked COVID-19 vaccination to SSNHL, evidence remains scarce. Establishing a correlation between COVID-19 vaccines and SSNHL implies a complex and multifactorial pathogenesis involving interactions between the immune system and the body's stress response. Nevertheless, it is important to consider the overwhelming evidence of the vaccines' safety and efficacy in limiting the spread of the disease and remains the primordial tool in reducing death.

Keywords

Sudden sensorineural hearing loss · SARS-CoV-2 · COVID-19 infection · COVID-19 vaccine · Inner ear · Mechanisms

15.1 Introduction

In December 2019, the first case of coronavirus (COVID-19) was reported in Wuhan, China. However, the World Health Organization did not officially declare COVID-19 disease as a pandemic until March 11, 2020 (WHO 2020). Patients affected may experience a wide range of symptoms going from no symptoms to lung failure and death. Common symptoms included nasal congestion, cough, sore throat, dyspnea, headaches, fatigue, fever, myalgia, arthralgia, and diarrhea (Guan et al. 2020). Additionally, some patients have reported non-respiratory symptoms such as olfactory and gustatory dysfunction (Jacob et al. 2020; Agyeman et al. 2020; Rocke et al. 2020), cochleovestibular symptoms (Maharaj et al. 2020; Jafari et al. 2022; Wang et al. 2022), and neurological disorders (Abboud et al. 2020). The UK National Institute for Health and Care Excellence has issued clinical definitions to identify COVID-19's acute and long-term effects. These definitions were divided into three phases: (1) acute symptoms persisting for up to 4 weeks, (2) ongoing symptoms for 4–12 weeks,

K. Tawk · A. Tawk · M. Abouzari (✉)
Department of Otolaryngology-Head and Neck Surgery, University of California, Irvine, 333 City Blvd. West, Suite 525, Orange, CA 92868, USA
e-mail: mabouzar@hs.uci.edu

N. Rezaei (ed.), *The COVID-19 Aftermath*, Advances in Experimental Medicine and Biology 1457,
https://doi.org/10.1007/978-3-031-61939-7_15

and (3) post-infection syndrome with symptoms persisting for more than 12 weeks (NICE 2020). The most common long-term effects included cochleovestibular symptoms such as dizziness, tinnitus, and otalgia (NICE 2020; Ng et al. 2022). In a systemic review and meta-analysis, the main otologic symptoms reported after SARS-CoV-2 infection were dizziness (12.2%) followed by tinnitus (4.5%) and hearing loss (3.2%) (Jafari et al. 2022). Moreover, individuals have reported experiencing similar symptoms after receiving the COVID-19 vaccine (Wichova et al. 2021; Yanir et al. 2022).

Sudden sensorineural hearing loss (SSNHL) is an acute and significant reduction of at least 30 dB in hearing across three consecutive frequencies on a pure-tone audiogram over a 3-day period (Wilson et al. 1980). Approximately 66,000 new cases are reported each year in the United States, resulting in an annual incidence rate of 27 cases per 100,000 people (Alexander and Harris 2013). However, this number may not accurately reflect the actual incidence since some cases spontaneously resolved without medical intervention. SSNHL can be caused by various factors such as viral infections (e.g., mumps, herpes zoster, and cytomegalovirus), autoimmune and neurologic conditions, metabolic disorders, neoplastic lesions, otologic (e.g., Meniere's disease) and vascular (e.g., stroke) diseases, and ototoxic drugs (e.g., salicylates, aminoglycosides, non-steroid anti-inflammatory drugs, and chemotherapy agents) (Kuhn et al. 2011). However, around 71% of cases are considered idiopathic (Chau et al. 2010). It is an urgent medical situation that requires timely intervention to prevent permanent hearing loss and the resulting negative impact on quality of life (Dallan et al. 2014).

Ever since Almufarrij et al. published the first systematic review reporting audio-vestibular symptoms reportedly associated with COVID-19 infection (Almufarrij et al. 2020), numerous additional reports on SSNHL post COVID-19 infection have subsequently emerged. Research has shown that coronaviruses have neuroinvasive, neurovirulent, and neurotropism properties (Bauer et al. 2022). For instance, by using the angiotensin-converting enzyme 2 (ACE2) receptor, SARS-CoV-2 virus is capable of infecting multiple organs, including the central nervous system, peripheral nervous system, vessels feeding the ear, and cochlea, thus leading to SSNHL through multiple pathways (Jeong et al. 2021; Salamanna et al. 2020). On the other hand, as the rate of vaccination rises, so did the number of adverse events. During the clinical trials of the vaccines, otologic symptoms were not listed as a potential adverse effect. However, after the vaccine became available, healthcare providers observed a rise in patients with new onset SSNHL (Wichova et al. 2021). Although the literature highlights a possible impact of COVID-19 infection and/or vaccination on the cochleovestibular function, there is no strong published evidence, and the pathophysiology remains not well understood or investigated. The present chapter intends to summarize the current knowledge about the epidemiological and/or mechanistic association between SSNHL and COVID-19 infection and vaccination. The chapter also highlights the implications of these findings for clinical practice.

15.2 COVID-19 Infection and Sudden Sensorineural Hearing Loss

SSNHL has emerged as a potential complication of COVID-19 infection. While the findings have been mixed, with some studies reporting a significant association between COVID-19 infection and SSNHL, most of the data is based on case reports, small-sized studies, and studies relying on self-reported hearing loss. The first reported case of SNHL in a COVID-19 patient was an old woman in Thailand. Although the patient recovered from her COVID-19 infection, her hearing did not show any improvement (Sriwijitalai and Wiwanitkit 2020). Since then, similar case reports have emerged regarding COVID-19 patients with SSNHL (Table 15.1). For instance, Swain and Pani found that 24 (5.08%) of their patients with COVID-19 had hearing loss, with 60.71% of those cases being SSNHL (Kumar Swain and Ranjan Pani 2021). Another study conducted in a tertiary hospital in China found that the number of

Table 15.1 Studies reporting SSNHL as an adverse effect of COVID-19 infection

Study	Study nature	Sample size	RT-PCR for SARS-CoV-2	Symptoms	Treatment	Recovery
Sriwijitalai and Wiwanitkit, Thailand Sriwijitalai and Wiwanitkit (2020)	Case report	1	Positive	Sensorineural hearing loss	Respiratory care	No improvement
Kilic et al., Turkey (Kilic et al. 2020)	Cross-sectional	5	Positive in one patient Negative in four patients	SSNHL	Patient positive for COVID-19 received oral hydroxychloroquine The other four patients received oral prednisolone	Patient who had Covid: total recovery at 1 month after treatment Other patients: complete recovery in 3 patients
Degen et al., Germany (Degen et al. 2020)	Case report	1	Positive	Deafness on the right Profound sensorineural hearing loss on the left	Cochlear implant on the right Intratympanic triamcinolone on the left	N/A
Abdel Rhman and Abdel Wahid, Egypt (Abdel Rhman and Abdel Wahid 2020)	Case report	1	Positive	SSNHL with tinnitus	Intratympanic steroid injections	Improvement of hearing levels
Koumpa et al., United Kingdom (Koumpa et al. 2020)	Case report	1	Positive	SSNHL	Oral prednisolone and intratympanic steroid injections	Partial improvement
Lamounier et al., Brazil (Lamounier et al. 2020)	Case report	1	Positive	SSNHL with tinnitus	Oral prednisolone and intratympanic steroid injections	Partial improvement
Karimi-Galougahi et al., Iran (Karimi-Galougahi et al. 2020)	Cross-sectional	6	Positive in three patients Pending in one patient Negative in two patients	SSNHL in all patients Tinnitus in 4 patients Vertigo in two patients	N/A	N/A

(continued)

Table 15.1 (continued)

Study	Study nature	Sample size	RT-PCR for SARS-CoV-2	Symptoms	Treatment	Recovery
Takahashi et al., Japan (Takahashi et al. 2020)	Case report	1	Positive (7 days after the hearing loss)	SSNHL	Prednisone	N/A
Swain and Pani, India (Kumar Swain and Ranjan Pani 2021)	Cross-sectional	28 patients, 17 with SSNHL	Positive	SSNHL	Oral steroids	11 out of the 17 improved to normal
Yaseen et al., Iraq (Yaseen et al. 2021)	Observational retrospective study	29	Positive	SSNHL with tinnitus and vertigo in some patients	Prednisolone, vitamin B complex, and intratympanic dexamethasone injections if no improvement	Partial improvement
Lang et al. Ireland (Lang et al. 2019)	Case report	1	Positive	SSNHL	Oral prednisolone	No improvement
Chirakkal et al., Qatar (Chirakkal et al. 2021)	Case report	1	Positive	SSNHL with tinnitus	N/A	N/A
Chern et al., United States of America (Chern et al. 2021)	Case report	1	Positive	SSNHL with aural fullness and vertigo	Oral prednisone and intratympanic dexamethasone injection	Partial improvement
Perret et al., France (Perret et al. 2021)	Case report	1	Positive	SSNHL	Oral prednisone	Progressive recovery
Akbar et al., United Kingdom (Akbar et al. 2022)	Case report	1	Positive	Profound SSNHL with tinnitus	Intravenous dexamethasone and oral prednisolone	No improvement

SSNHL Sudden sensorineural hearing loss

patients with sudden deafness in its outpatient and emergency departments increased during the pandemic (Jin et al. 2021), while a study in Turkey reported an increased incidence of SSNHL during the pandemic with 60.3% of patients diagnosed with COVID-19 (Fidan et al. 2021). Parrino et al. investigated the frequency of SSNHL in a specialized medical center of audiology in Italy. They found that the yearly incidence of SSNHL and combined cochlear-vestibular considered together was higher and that the pure-tone audiometry was 10 dB worse than in previous times. However, when considering the incidence of SSNHL as a single entity, the increase was not significant when compared to previous years (Parrino et al. 2022). According to a retrospective study which gathered self-reported hearing symptoms among patients with COVID-19 in the Denmark, 59 (10.7%) individuals with chemosensory loss (smell, taste, or both) also experienced hearing loss. However, the study's validity is questionable since the patients did not undergo any objective diagnosis or audiological tests (Thrane et al. 2022). Chern et al. described a case of bilateral SSNHL with bilateral intralabyrinthine hemorrhage in a COVID-19 positive patient (Chern et al. 2021). Although this patient experienced intralabyrinthine hemorrhage, the majority of patients with COVID-19 related SSNHL displayed normal MRI results. Finally, Mustafa compared the audiologic profile of 20 patients with asymptomatic confirmed COVID-19 infection to 20 healthy controls. They found significantly worse high-frequency pure-tone thresholds and transitory evoked otoacoustic emission amplitude (TEOAE) in the COVID-19 positive group. These findings suggest that COVID-19 infection can damage the hair cells, especially the outer hair cells as evidenced by the TEOAE test, even in the absence of major symptoms (Mustafa 2020).

On the other hand, in a recent study based on a global federated research network, Africa et al. reported no statistically significant difference in sudden idiopathic hearing loss between matched cohorts of COVID-19 positive and COVID-19 negative patients. The authors analyzed a large population with real-world data from multiple healthcare institutions (Africa et al. 2023). Similarly, van Rijssen et al. examined 25 patients from a large single institution who reported audiometrically confirmed SSNHL. None of these patients tested positive for active COVID-19 infection, with two patients testing positive three and eight months before the onset of SSNHL (van Rijssen et al. 2022). Doweck et al. included more than 3 million individuals from Clalit Health Services to compare the incidence of SSNHL during the pandemic with the years 2018 and 2019. The authors found that the incidence of SSNHL was significantly lower during the pandemic, with a standardized incidence ratio of 0.8 (95% CI, 0.75–0.86) for 2020 versus 2018 and 0.78 (95% CI, 0.72–0.83) for 2020 versus 2019. They attributed these findings to fewer patients visiting physicians and emergency departments (Doweck et al. 2022). Aslan and Çiçek failed to find a significant difference in the incidence of SSNHL between the period before the pandemic and during the pandemic (49 patients during one-year period before the pandemic vs. 42 patients during one-year period during the pandemic, with $p = 0.746$). They also hypothesized that the decreased incidence of SSNHL during the pandemic could be due to a reduction in patients seeking medical attention, but further suggested that isolation measures may have played a role in reducing SSNHL with a viral etiology (Aslan and Çiçek 2021). Similarly, Chari et al. showcased a significant reduction in the absolute number of SSNHL patients during the COVID-19 outbreak compared to a similar period in the previous year (71 vs. 13 patients), but a relatively constant percentage (1.77% in 2019 vs. 19.1% in 2020) of SSNHL diagnoses between the two time periods (Chari et al. 2021). Finally, Hafrén et al. found no changes in SSNHL incidence during the pandemic (Hafrén et al. 2022).

15.3 Exploring the Link Between COVID-19 Infection and Sudden Sensorineural Hearing Loss

Several mechanisms by which a virus can cause hearing loss have been reported. For example, cytomegaloviruses trigger an immune-mediated response that leads to SSNHL (Schraff et al. 2007). Other viruses, such as rubella and varicella zoster, can directly damage the cochlea or vestibulocochlear nerve (Sweeney and Gilden 2001). Neuroinvasion of the peripheral and central auditory pathways, as seen with the West Nile virus, has also been demonstrated (Parrino et al. 2019). In addition, viruses can cause conductive hearing loss due to middle ear effusion. Although coronaviruses are known to cause otitis media (Sebothoma and Khoza-Shangase 2022), the mechanism by which COVID-19 infection could cause SSNHL is not fully understood and might be heterogeneous. One plausible hypothesis for SSNHL is a direct virus infection of the inner ear and cochleovestibular nerve. Studies have revealed neurotropic and neuroinvasive characteristics of COVID-19 (Beyerstedt et al. 2021). The viral spike protein of SARS-CoV-2 can access the inner ear since ACE2 receptors, which are required for virus entry, are expressed in the organ of Corti, spiral ganglion neurons, and stria vascularis (Jeong et al. 2021). Once bound to its receptor, cytokines and inflammatory mediators such as tumor necrosis factor-α, interleukin 1, and interleukin 6 are released, which may lead to hearing loss by increasing oxidative damage (Saniasiaya 2021). By directly invading the cochlea and the cochlear nerve, SARS-CoV-2 can cause cochleitis and neuritis (Harenberg et al. 2020). Alternatively, hearing loss can occur due to the damage of the stria vascularis and the subsequent disruption of the endocochlear potential and fluid and ions homeostasis (Jeong et al. 2021).

In addition, the ACE2 receptors are found throughout the vestibulocochlear pathway including the temporal lobe and the medulla oblongata (Kaliyappan et al. 2022). Recent studies reported that SARS-CoV-2 can cross the blood–brain barrier via the ACE2 receptors expressed by the vascular endothelial cells, pericytes, and astrocytes (Chen et al. 2021). Consequently, the neuroinflammation of the blood–brain barrier—especially the auditory cortex in the temporal lobe—can be responsible for central hearing loss during COVID-19 infection (Luca et al. 2021; Welcome and Mastorakis 2021). Moreover, SARS-CoV-2 is related to a hypercoagulable state by binding to ACE2 receptors of the vascular smooth muscles (Lee et al. 2020). Therefore, by infecting the vessels that feed the hearing center or the inner ear, COVID-19 causes endotheliitis and subsequent endothelial dysfunction, platelet activation, and widespread thrombosis, which lead to hearing loss (Spence et al. 2020; Bikdeli et al. 2020). In addition, SARS-CoV-2 infects erythrocytes by binding to the band-3 protein on the red blood cell membrane, thus leading to hypoxia by deoxygenating the binding erythrocytes and a decline in hearing thresholds (Khawaja et al. 2021; Russo et al. 2022; Sonbay Yilmaz et al. 2019). On the other hand, SARS-CoV-2 shares 13 out of 24 pentapeptides with lung surfactant protein which can result in cross-reactivity with pulmonary surfactant and subsequent lung disease (Kanduc and Shoenfeld 2020). Based on these findings, it is possible to speculate that a similar mechanism may occur in the inner ear, leading to an immune-mediated response against antigens in the inner ear and causing SSNHL.

In addition to the cytokine storm and state of hyperinflammation, COVID-19 has had a significant impact on mental health worldwide. The pandemic has caused increased panic, anxiety, and stress (Silva Andrade et al. 2021). Both physical and psychological stressors are known triggers for otologic migraine which includes symptoms such as SSNHL (Abouzari et al. 2021), tinnitus (Goshtasbi et al. 2021a), otalgia (Benjamin et al. 2022), aural fullness (Moshtaghi et al. 2018), and hyperacusis (Abouzari et al. 2020). Therefore, SSNHL related to acute or long COVID could be a migraine-related process. Finally, different drugs used to treat COVID-19 infection were proven to be ototoxic and cause SSNHL. These drugs include chloroquine, hydroxychloroquine, azithromycin, Lopinavir, Favipiravir, and Remdesivir (Bortoli and Santiago 2007; Kaliyappan et al. 2022). Chloroquine and hydroxychloroquine accumulation in the melanocytes of the inner ear causes sensory hair cell destruction, loss of neurons and supporting cells, and atrophy of the stria vascularis (Bortoli and

Santiago 2007). Little is known about the mechanism of azithromycin ototoxicity. Studies have suggested that it can cause outer hair cells damage with systemic use and inner hair cells damage with topical application (Rybak et al. 2021).

SSNHL could be a manifestation of COVID-19 infection, and its underlying mechanisms are complex and multifactorial (Fig. 15.1). Direct viral invasion, neuroinflammation, blood flow disturbances, and immune-mediated response all could contribute to the development of SSNHL in COVID-19 patients. Therefore, it is important to be aware of this potential complication, particularly in patients with severe and long-lasting COVID-19 symptoms. Further research is needed to fully understand the causative link between COVID-19 and SSNHL.

15.4 COVID-19 Vaccination and Sudden Sensorineural Hearing Loss

At present, more than 750 million confirmed cases of COVID-19 and more than 6 million death were reported worldwide, while the number of vaccines administered has surpassed 13 billion (WHO). As soon as the genetic code of the COVID-19 virus was discovered (Wu et al. 2020), scientists, researchers, and physicians rushed to design an effective and safe vaccine against COVID-19. For most diseases, it takes more than 10 years to develop a vaccine. However, advancements in genomic sequencing, existing data on coronaviruses, emergency funding, and global collaborations enabled the

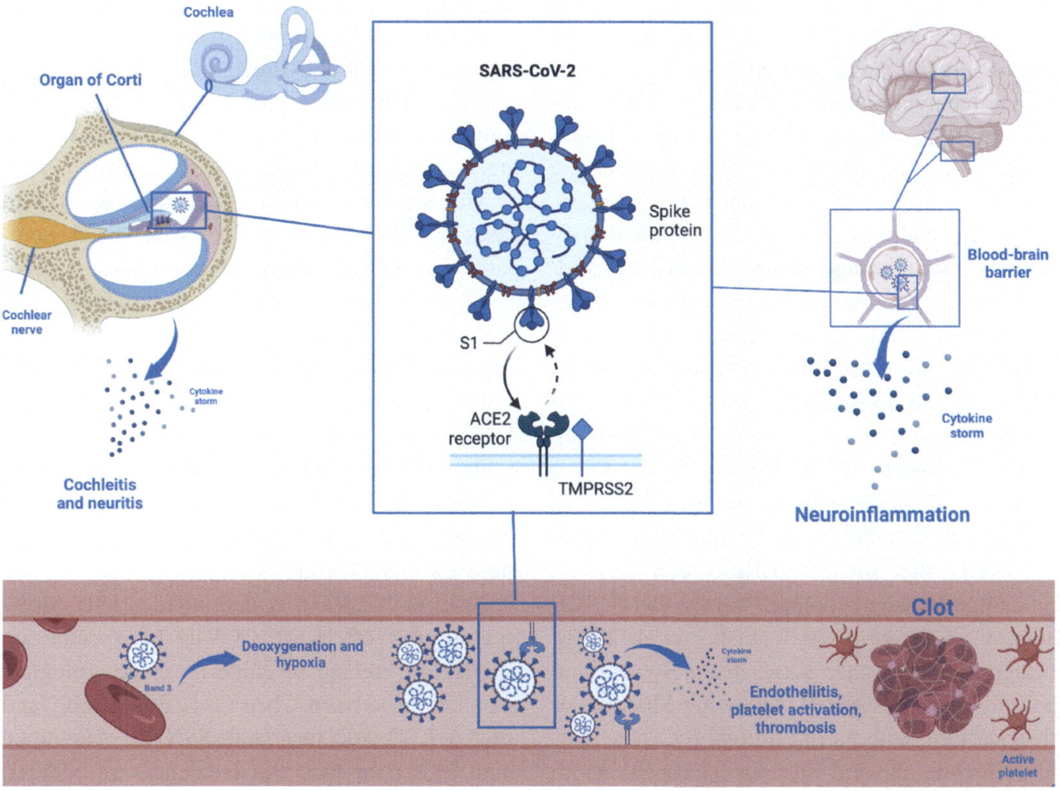

Fig. 15.1 Mechanisms of SARS-CoV-2 invasion of the central nervous system, inner ear, cochlear nerve, and vessels by using its spike protein to bind ACE2 receptors and the red blood cell (RBC) by binding the band 3 protein. The cytokine storm associated with the invasion is believed to cause neuroinflammation, cochleitis, neuritis, and endotheliitis which can cause sudden hearing loss. In addition, binding the RBC results in hypoxia, which is another mechanism of hearing loss. This figure was created by BioRender.com

fast development of COVID-19 vaccines. On December 11 and 18, 2020, the Food and Drug Administration (FDA) issued an emergency use authorization for BNT162b2 mRNA Pfizer-BioNTech (Polack et al. 2020) and mRNA-1273 SARS-CoV-2 Moderna (Baden et al. 2020) COVID-19 vaccines, respectively. As the percentage of the vaccinated population increases, adverse effects that might be due to the vaccination have been recorded. Otological symptoms, such as vertigo, dizziness, tinnitus, and SSNHL, have been reported following COVID-19 vaccination in the Vaccine Adverse Events Reporting System (VAERS). However, evidence remains scarce and anecdotal on the association between COVID-19 vaccination and SSNHL.

As of February 2022, a literature review indicated that there have been at least 1178 reported cases of SNHL following COVID-19 vaccination (Finsterer et al. 2022). In a retrospective chart review, Wichova et al. reported a twofold increase in idiopathic SSNHL in vaccinated patients receiving either Moderna or Pfizer vaccine within 30-day period when compared with the same interval in 2019 (1.60% in 2019 *vs.* 2.44% and 3.85% in 2020 and 2021, respectively). In addition, patients with previously stabilized autoimmune inner ear disease and/or Meniere's disease presented with exacerbation of their symptoms (Wichova et al. 2021). Jeong et al. were the first to report a case of SSNHL due to an intralabyrinthine hemorrhage in a 61-year-old female patient after receiving the third booster vaccination of Pfizer (Jeong and Yoon 2022). Two studies conducted in Israel showed a slight increase in the annual incidence of SSNHL after the Pfizer-BioNTech vaccination campaign when compared to the 2018–2020 database (Yanir et al. 2022; Fisher et al. 2023). Furthermore, two case series from Switzerland and Italy reported SSNHL following Moderna or Pfizer vaccine with partial to total recovery after intratympanic or oral steroids (Ekobena et al. 2022; Zoccali et al. 2022). Another case series described five patients experiencing SSNHL within a few days of receiving the Astra Zeneca vaccine (Canales Medina and Ramirez Gómez 2022). Tsetsos et al. reported another case of SSNHL two days after receiving the second dose of Astra Zeneca vaccine that recovered after two courses of corticosteroids (IV then oral) and acetylosalicid acid (Tsetsos et al. 2021). Whereas Pisani et al. described a case of SSNHL that occurred two days after the Astra Zeneca vaccine and partially recovered after a course of corticosteroids only (IV then oral) (Pisani et al. 2021). Finally, the Sinovac-related SSNHL was only reported by Zhao et al. with two cases of severe unilateral sudden sensorineural hearing loss presenting four days after receiving the vaccine. Unfortunately, the average threshold improvement was less than 10 dB in both patients after receiving intravenous corticosteroids (Zhao et al. 2022). Although the authors of the mentioned studies highlighted an epidemiological association, they could not prove unequivocally a causal link between COVID-19 vaccination and SSNHL (Table 15.2).

A study analyzing data from the Centers for Disease Control and Prevention (CDC) VAERS between December 2020 and March 2021 showed that the incidence of SSNHL following the two mRNA messenger vaccines (Pfizer or Moderna) was similar or even lower (0.6–28 cases per 100,000 people per year) when compared to the unvaccinated population (11–77 cases per 100,000 people per year) (Formeister et al. 2022). However, it is important to note that this study relied on self-reported symptoms without any objective measurements of hearing. Additionally, VAERS reports may be unverifiable, biased, and incomplete. Nonetheless, a study that included audiometry-confirmed SSNHL cases demonstrated no increase in these cases in Pfizer vaccine recipients (Cohen Michael et al. 2023). A nationwide study investigating the effect of COVID-19 vaccine on SSNHL in Finland revealed no epidemiological association between ChAdOx1, BNT162b2, and mRNA.1273 vaccines and SSNHL (Nieminen et al. 2022). In fact, the incidences of SSNHL within the first 54 days following the first dose of ChAdOx1, BNT162b2, and mRNA.1273 were 22.1 (95% CI, 11.4–38.7), 21.2 (95% CI, 17.5–25.6), and 18.5 (95% CI, 10.3–30.5), per 100,000 person-years, respectively, with an

Table 15.2 SSNHL as an adverse effect following COVID-19 vaccines

Study	Study nature	Sample size	Type of COVID-19 vaccine	Symptoms	Treatment	Recovery
Kahn et al. (2021)	Case report	1	Pfizer-BioNTech	SSNHL two days after the first dose	High dose glucocorticoids	No recovery
Yanir et al. (2022)	Retrospective cohort	170	Pfizer-BioNTech	SSNHL 21–60 days after dose 1 (91 patients) and dose 2 (79 patients)	Prednisone	N/A
Fisher et al. (2023)	Retrospective cohort	13	Pfizer-BioNTech	SSNHL 96 h to 30 days after dose 1, 2, or 3	N/A	N/A
Wichova et al. (2021)	Retrospective cohort	25	11 Pfizer-BioNTech14 Moderna	SSNHL, tinnitus, dizziness, and vertigo 10 ± 9 days after dose 1 or dose 2	Oral, intratympanic steroids, and antivirals	N/A
Ekobena et al. (2022)	Case series	2	Moderna	SSNHL and diplacusis 12 h after dose 2	Short course prednisone	Complete recovery, no reoccurrence after Pfizer booster
			Pfizer-BioNTech	SSNHL, tinnitus, and ear fullness 24 h after dose 2	Prednisone, betahistine, and intratympanic steroids	Partial recovery
Zoccali et al. (2022)	Case series	2	Pfizer-BioNTech	SSNHL 5 days after dose 3	Intratympanic dexamethasone 5 mg once a day for three alternate days, repeated for two cycles	Complete recovery
			Moderna	SSNHL 7 days after dose 3	Intratympanic dexamethasone 5 mg once a day for three alternate days, repeated for two cycles	Significant improvement
Jeong and Choi (2021)	Case series	3	1 Oxford-AstraZeneca	SSNHL one day after the first dose of Oxford-AstraZeneca	Prednisolone 60 mg per day for five days, which was tapered for the next five days + salvage treatment with intratympanic steroid injection (dexamethasone 5 mg every third day for three cycles) was performed	Complete recovery

(continued)

Table 15.2 (continued)

Study	Study nature	Sample size	Type of COVID-19 vaccine	Symptoms	Treatment	Recovery
			1 Pfizer-BioNTech	SSNHL on the day of the first dose	Prednisolone 60 mg per day for five days, which was tapered for the next five days + salvage treatment with intratympanic steroid injection (dexamethasone 5 mg every third day for five cycles) was performed	Partial improvement
			1 Pfizer-BioNTech	SSNHL two days after the second dose	Prednisolone 60 mg per day for five days, which was tapered for the next five days). Intratympanic steroid injection (dexamethasone 5 mg every third day for five cycles) was administered simultaneously due to severe hearing loss	Hearing threshold worsened
Canales and Medina (2022)	Case series	4	Oxford-AstraZeneca	SSNHL and tinnitus 10 days after dose 2	Intratympanic dexamethasone	Significant improvement
			Oxford-AstraZeneca	SSNHL and tinnitus 10 days after dose 2	Oral prednisone 1 mg/kg/day	Complete recovery
			Oxford-AstraZeneca	Bilateral SSNHL 18 days after dose 2	Oral prednisone 1 mg/kg/day single dose	Complete recovery
			Oxford-AstraZeneca	SSNHL and tinnitus 11 days after dose 1	Oral prednisone 1 mg/kg/day single dose	Complete recovery
			Oxford-AstraZeneca	SSNHL and tinnitus 14 days after dose 2	Oral prednisone 1 mg/kg/day single dose	No improvement at one week follow-up
Tsetsos et al. (2021)	Case report	1	Oxford-AstraZeneca	SSNHL 2 days after dose 2	Dexamethasone 8 mg three times per day for three days with tapering over a week + oral glucocorticosteroids after discharge for 9 days + 100 mg of acetylsalicylic acid once daily until 9 days after discharge	Almost full recovery

(continued)

Table 15.2 (continued)

Study	Study nature	Sample size	Type of COVID-19 vaccine	Symptoms	Treatment	Recovery
Pisani et al. (2021)	Case report	1	Oxford-AstraZeneca	SSNHL 2 days after dose 1	12-day cycle of therapy with methylprednisolone + B-vitamin complex integration of 400 mcg/day for 2 months and a 500mcg/day integration of folic acid	Almost full recovery
Zhao et al. (2022)	Case series	2	Sinovac	SSNHL, tinnitus, and dizziness 4 days after dose 1	Methylprednisone IV, Ginkgo biloba, and Mecobalamin	No improvement
			Sinovac	SSNHL, tinnitus, and dizziness 4 days after dose 1	Methylprednisone IV, Ginkgo biloba, and Mecobalamin	No improvement

IV Intravenous, *SSNHL* sudden sensorineural hearing loss

adjusted incidence rate ratio (aIRR) less than 1. Likewise, the aIRR following the second and third doses varied between 0.6 and 1.2 with all confidence intervals including 1, thus, showing no increase in the incidence of hearing loss when compared to the pre-pandemic unvaccinated persons (Nieminen et al. 2022) (Table 15.3).

On October 29, 2021, the FDA authorized the administration of BNT162b2 mRNA Pfizer-BioNTech to children between the age of 5 and 11 years old (Fergie et al. 2022; Hause et al. 2021). Approximately, 4249 of 8.7 million cases reported advert events of which 97.6% were not serious (Hause et al. 2021). Aldè et al. were the only ones to date to evaluate the prevalence of audio-vestibular symptoms in children aged 5–11 years following COVID-19 vaccination. In their study, no cases of SSNHL were reported among the vaccine recipients (Aldè et al. 2023).

15.5 Exploring the Mechanisms of COVID-19 Vaccination and Sudden Sensorineural Hearing Loss

The mechanism of SSNHL following COVID-19 vaccination is believed to be caused by an immune-mediated response. In fact, mRNA vaccines (Pfizer and Moderna) elicit the production of neutralizing antibodies against the viral spike protein and antigen-specific CD4+and CD8+cells (Nagy and Alhatlani 2021). Therefore, it is hypothesized that due to the phenomenon of molecular mimicry, these antibodies may also recognize self-proteins in the inner ear; thus leading to an autoimmune reaction that results in SSNHL (Vojdani and Kharrazian 2020). This hypothesis is corroborated by hearing loss occurring 10–14 days after getting the vaccine (Wichova et al. 2021). This period aligns with the commencement of immunoglobulin G (IgG) synthesis following vaccination, which are better neutralizing antibodies than IgM (Clem 2011). In addition, the excessive production of proinflammatory cytokines during this immune-mediated response can also damage the inner ear (Andrade et al. 2022). Another mechanism involves the production of antibodies that triggers an abnormal cellular stress pathway in the inner ear such as the sphingosine-1-phosphate signaling pathway and the NF-kB activation pathway. This mechanism, known as the "stress response hypothesis", is associated with reduced blood flow in the cochlea, thrombogenic profile, and systemic stress response, which results in SSNHL (Masuda et al. 2012; Chen et al. 2019). It is worth noting that some authors have postulated a thrombotic etiology of SSNHL cases. For instance, Tsetsos et al. treated their patient with a combination of corticosteroids and antiplatelets, implying that adding the latter may have contributed to the recovery of hearing loss (Tsetsos et al. 2021). This is particularly significant given the association between thrombotic events and the AstraZeneca vaccine with several countries temporarily halting its administration (Oldenburg et al. 2021; Wise 2021). Furthermore, the viral spike protein, expressed by the recipient's cells after receiving the mRNA vaccines (Pfizer or Moderna), might be capable of activating the alternative pathway of complement, which plays a role in endothelial damage, platelets aggregation, and formation of blood clots (Montano 2022). However, given that patients with SSNHL following vaccine administration responded positively to corticosteroids (oral, IV, or intratympanic), which do not have an anti-thrombotic effect, careful consideration is crucial when interpreting and disseminating information regarding thromboembolism in the pathogenesis of SSNHL.

Another consideration is that the vaccination itself is considered physical stress to the body, and both physical and psychological stress are known triggers for migraine. Therefore, it is plausible that SSNHL could be a manifestation of migraine in the inner ear. This can occur due to the vasoconstriction of the cochlear vasculature following the activation of the trigeminal nerve by the migraine phenomenon. (Abouzari et al. 2021; Goshtasbi et al. 2021b). Migraine prophylactic medication and modifications in diet and

Table 15.3 SSNHL incidence following COVID-19 vaccines

Study	Country	Vaccination period	Source	Type of COVID-19 vaccine	SSNHL cases following COVID-19 vaccination	Incidence per 100,000 person-years (95% CI)
Formeister et al. (2022)	USA	December 2020 to July 2021	Centers for Disease Control and Prevention Vaccine Adverse Events Reporting System	Pfizer-BioNTech, Moderna, Janssen/Johnson & Johnson	555	0.6–28 cases (vs. 11–77 cases in unvaccinated people)
Cohen Michael et al. (2023)	Israel	January 2021 to October 2021	Samson Assuta Ashdod University Hospital	Pfizer-BioNTech	34	0.58 (0.46–1.67) compared with unvaccinated, period 1; $p = 0.98$ 0.61 (0.48–1.74) compared with unvaccinated, period 2; $p = 0.92$
Nieminen et al. (2022)	Finland	Early 2021 to April 2022	Finnish Institute for Health and Welfare	Pfizer-BioNTech, Moderna, Oxford-AstraZeneca	–	14–21 cases

lifestyle have been proven efficacious in treating acute and long-term SSNHL (Abouzari et al. 2021; Goshtasbi et al. 2021b). Finally, SSNHL following vaccination may be attributed to the psychological stress associated with immunization (Taylor and Asmundson 2021). In fact, in a study by Zhang et al., adverse effects following the Sinovac vaccination were also noted in the placebo group. In the phase 1 trial, the incidence of adverse events occurred in 8% and 13% for the days 0 and 14 cohort and the days 0 and 28 cohort, respectively. In the phase 2 trial, the incidence of adverse effects reported was 22% and 18% for the days 0 and 14 cohort and the days 0 and 28 cohort, respectively. In addition, three participants in the placebo group exhibited seroconversion of neutralizing antibodies (Zhang et al. 2021). Overall, establishing a correlation between COVID-19 vaccination and SSNHL implies that the pathogenesis of the condition is complex and multifactorial, involving interactions between our immune system and our stress response. Further research is needed to understand the precise mechanisms involved in order to develop preventive and curative strategies for treatment.

15.6 Treatment of COVID-19-Related Sudden Hearing Loss

Despite the unclear association between SSNHL and COVID-19 infection and vaccination, patients should still be referred to otolaryngologists for treatment. Whether or not COVID-19 infection or vaccination is the cause of SSNHL, the treatment remains the same (Meng et al. 2022; Li et al. 2022). Clinical guidelines indicated that patients should be offered corticosteroids within 2 weeks of symptom onset and intratympanic steroid injections in case of incomplete recovery 2–6 weeks from symptom onset (Chandrasekhar et al. 2019). In one study, patients received oral mesoglycan and hyperbaric oxygen therapy as an adjunctive therapy (Ricciardiello et al. 2021). Patients who do not show improvement in their hearing may be considered for hearing aids or cochlear implants (Meng et al. 2022). On the other hand, Maliah et al. found that prolonged high-dose steroid treatment interfered with CD4 T-cell activation, despite high antibody titers (Maliah et al. 2022). Therefore, some physicians favor intratympanic steroids over systemic steroids after COVID-19 vaccination. However, it is important to note that the mRNA vaccines (Pfizer and Moderna) and viral vector vaccines (Johnson & Johnson and AstraZeneca) contain only the genetic code necessary to produce the spike protein of SARS-CoV-2. Thus, while systemic corticosteroids can impair the immune response and promote viral replication, that is not the case for COVID-19 vaccines as there is no delivery of live and replicable virus. Therefore, vaccination does not preclude of systemic corticosteroids when indicated (Chang et al. 2022).

15.7 Conclusion

Despite some reports of otologic side effects following COVID-19 infection or vaccination, the temporal relationship remains unclear, with conflicting results reported in different studies. Nevertheless, in cases of SSNHL occurring after the infection or vaccination, the treatment remains the same. It is important, however, to consider the overwhelming evidence of the vaccines' safety and efficacy in limiting the spread of the disease. It is worth noting that COVID-19 vaccines have undergone extensive clinical trials and have been proven to be safe and effective in preventing COVID-19. Additionally, real-world data demonstrated significant reductions in COVID-19 cases and deaths following mass vaccination efforts worldwide. Vaccination remains a primordial tool in reducing the spread of the infection and protecting public health. By May, 2021, COVID-19 vaccination may have prevented 139 393 deaths which translate to a value per statistical life of $625 billion to $1.4 trillion (Gupta et al. 2021).

References

Abboud H, Abboud FZ, Kharbouch H et al (2020) COVID-19 and SARS-Cov-2 infection: pathophysiology and clinical effects on the nervous system. World Neurosurg 140:49–53. https://doi.org/10.1016/j.wneu.2020.05.193

Abdel Rhman S, Abdel Wahid A (2020) COVID -19 and sudden sensorineural hearing loss, a case report. Otolaryngol Case Rep 16:100198. https://doi.org/10.1016/j.xocr.2020.100198

Abouzari M, Goshtasbi K, Chua JT et al (2021) Adjuvant migraine medications in the treatment of sudden sensorineural hearing loss. Laryngoscope 131:E283–E288. https://doi.org/10.1002/lary.28618

Abouzari M, Tan D, Sarna B et al (2020) Efficacy of multi-modal migraine prophylaxis therapy on hyperacusis patients. Ann Otol Rhinol Laryngol 129:421–427. https://doi.org/10.1177/0003489419892997

Africa RE, Westenhaver ZK, Zimmerer RE, McKinnon BJ (2023) Evaluation of disturbances in hearing, tinnitus, and dizziness as signs of COVID-19 Infection. Otol Neurotol 44:126–133. https://doi.org/10.1097/MAO.0000000000003787

Agyeman AA, Chin KL, Landersdorfer CB et al (2020) Smell and taste dysfunction in patients with COVID-19: a systematic review and meta-analysis. Mayo Clin Proc 95:1621–1631. https://doi.org/10.1016/j.mayocp.2020.05.030

Akbar S, Gopikrishna D, Leong WJ (2022) Unilateral, permanent hearing loss following severe COVID-19 infection. BMJ Case Rep 15:e251717. https://doi.org/10.1136/bcr-2022-251717

Aldè M, Di Berardino F, Ambrosetti U et al (2023) Audiological and vestibular symptoms following SARS-CoV-2 infection and COVID-19 vaccination in children aged 5–11 years. Am J Otolaryngol 44:103669. https://doi.org/10.1016/j.amjoto.2022.103669

Alexander TH, Harris JP (2013) Incidence of sudden sensorineural hearing loss. Otol Neurotol 34:1586–1589. https://doi.org/10.1097/MAO.0000000000000222

Almufarrij I, Uus K, Munro KJ (2020) Does coronavirus affect the audio-vestibular system? A rapid systematic review. Int J Audiol 59:487–491. https://doi.org/10.1080/14992027.2020.1776406

Andrade J, Sessa L, Ephrat M et al (2022) A case report of sudden sensorineural hearing loss (SSNHL) after administration of the COVID-19 vaccine. J Pharm Pract. https://doi.org/10.1177/08971900221147584

Aslan M, Çiçek MT (2021) Can isolated sudden sensorineural hearing loss (SSNHL) and idiopathic acute facial paralysis (Bell's palsy) be symptoms of COVID-19? Am J Otolaryngol 42:103129. https://doi.org/10.1016/j.amjoto.2021.103129

Baden LR, El Sahly HM, Essink B, et al (2020) Efficacy and safety of the mRNA-1273 SARS-CoV-2 vaccine. N Engl J Med. https://doi.org/10.1056/NEJMoa2035389

Bauer L, Laksono BM, de Vrij FMS et al (2022) The neuroinvasiveness, neurotropism, and neurovirulence of SARS-CoV-2. Trends Neurosci 45:358–368. https://doi.org/10.1016/j.tins.2022.02.006

Benjamin T, Gillard D, Abouzari M et al (2022) Vestibular and auditory manifestations of migraine. Curr Opin Neurol 35:84–89. https://doi.org/10.1097/WCO.0000000000001024

Beyerstedt S, Casaro EB, Rangel ÉB (2021) COVID-19: angiotensin-converting enzyme 2 (ACE2) expression and tissue susceptibility to SARS-CoV-2 infection. Eur J Clin Microbiol Infect Dis 40:905–919. https://doi.org/10.1007/s10096-020-04138-6

Bikdeli B, Madhavan MV, Jimenez D et al (2020) COVID-19 and thrombotic or thromboembolic disease: implications for prevention, antithrombotic therapy, and follow-up. J Am Coll Cardiol 75:2950–2973. https://doi.org/10.1016/j.jacc.2020.04.031

Bortoli R, Santiago M (2007) Chloroquine ototoxicity. Clin Rheumatol 26:1809–1810. https://doi.org/10.1007/s10067-007-0662-6

Canales Medina M, Ramirez Gómez M Tinnitus (2022) Sudden sensorineural hearing loss, and vestibular neuritis as complications of the Astra Zeneca COVID-19 vaccine. Cureus 14:e20906. https://doi.org/10.7759/cureus.20906

Chandrasekhar SS, Tsai Do BS, Schwartz SR et al (2019) Clinical practice guideline: sudden hearing loss (update). Otolaryngol Head Neck Surg 161:S1–S45. https://doi.org/10.1177/0194599819859885

Chang CWD, McCoul ED, Briggs SE et al (2022) Corticosteroid use in otolaryngology: current considerations during the COVID-19 era. Otolaryngol Head Neck Surg 167:803–820. https://doi.org/10.1177/01945998211064275

Chari DA, Parikh A, Kozin ED et al (2021) Impact of COVID-19 on presentation of sudden sensorineural hearing loss at a single institution. Otolaryngol Head Neck Surg 165:163–165. https://doi.org/10.1177/0194599820974685

Chau JK, Lin JRJ, Atashband S et al (2010) Systematic review of the evidence for the etiology of adult sudden sensorineural hearing loss. Laryngoscope 120:1011–1021. https://doi.org/10.1002/lary.20873

Chen R, Wang K, Yu J et al (2021) The spatial and cell-type distribution of sars-cov-2 receptor ace2 in the human and mouse brains. Front Neurol 11:573095. https://doi.org/10.3389/fneur.2020.573095

Chen X, Fu Y, Zhang T (2019) Role of viral infection in sudden hearing loss. J Int Med Res 47:2865–2872. https://doi.org/10.1177/0300060519847860

Chern A, Famuyide AO, Moonis G, Lalwani AK (2021) Bilateral sudden sensorineural hearing loss and intralabyrinthine hemorrhage in a patient with COVID-19. Otol Neurotol 42:e10–e14. https://doi.org/10.1097/MAO.0000000000002860

Chirakkal P, Al Hail AN, Zada N, Vijayakumar DS (2021) COVID-19 and tinnitus. Ear Nose Throat J 100:160S-162S. https://doi.org/10.1177/0145561320974849

Clem AS (2011) Fundamentals of vaccine immunology. J Glob Infect Dis 3:73–78. https://doi.org/10.4103/0974-777X.77299
Cohen Michael O, Tamir SO, O'Rourke N, Marom T (2023) Audiometry-confirmed sudden sensorineural hearing loss incidence among COVID-19 patients and BNT162b2 vaccine recipients. Otol Neurotol 44:e68–e72. https://doi.org/10.1097/MAO.0000000000003777
Dallan I, Fortunato S, Casani AP et al (2014) Long-term follow up of sudden sensorineural hearing loss patients treated with intratympanic steroids: audiological and quality of life evaluation. J Laryngol Otol 128:669–673. https://doi.org/10.1017/S0022215114001595
Degen C, Lenarz T, Willenborg K (2020) Acute profound sensorineural hearing loss after COVID-19 pneumonia. Mayo Clin Proc 95:1801–1803. https://doi.org/10.1016/j.mayocp.2020.05.034
Doweck I, Yanir Y, Najjar-Debbiny R et al (2022) Sudden sensorineural hearing loss during the COVID-19 pandemic. JAMA Otolaryngol Head Neck Surg 148:373–375. https://doi.org/10.1001/jamaoto.2021.4105
Ekobena P, Rothuizen LE, Bedussi F et al (2022) Four cases of audio-vestibular disorders related to immunisation with SARS-CoV-2 mRNA vaccines. Int J Audiol 1–5. https://doi.org/10.1080/14992027.2022.2056721
Fergie J, Moran MM, Cane A et al (2022) COVID-19 epidemiology, immunity, and vaccine development in children: a review. Vaccines (basel) 10:2039. https://doi.org/10.3390/vaccines10122039
Fidan V, Akin O, Koyuncu H (2021) Rised sudden sensorineural hearing loss during COVID-19 widespread. Am J Otolaryngol 42:102996. https://doi.org/10.1016/j.amjoto.2021.102996
Finsterer J, Scorza FA, Fiorini AC (2022) Impaired hearing following SARS-CoV-2 vaccinations. Int J Infect Dis 115:215–216. https://doi.org/10.1016/j.ijid.2021.12.317
Fisher R, Tarnovsky Y, Hirshoren N et al (2023) The association between COVID-19 vaccination and idiopathic sudden sensorineural hearing loss, clinical manifestation and outcomes. Eur Arch Otorhinolaryngol 1–5. https://doi.org/10.1007/s00405-023-07869-2
Formeister EJ, Wu MJ, Chari DA et al (2022) Assessment of sudden sensorineural hearing loss after COVID-19 vaccination. JAMA Otolaryngol Head Neck Surg 148:307–315. https://doi.org/10.1001/jamaoto.2021.4414
Goshtasbi K, Abouzari M, Risbud A et al (2021a) Tinnitus and subjective hearing loss are more common in migraine: a cross-sectional NHANES analysis. Otol Neurotol 42:1329–1333. https://doi.org/10.1097/MAO.0000000000003247
Goshtasbi K, Chua JT, Risbud A et al (2021b) Treatment of long-term sudden sensorineural hearing loss as an otologic migraine phenomenon. Otol Neurotol 42:1001–1007. https://doi.org/10.1097/MAO.0000000000003111
Guan W, Ni Z, Hu Y et al (2020) Clinical characteristics of coronavirus disease 2019 in China. N Engl J Med. https://doi.org/10.1056/NEJMoa2002032
Gupta S, Cantor J, Simon KI et al (2021) Vaccinations against COVID-19 may have averted up to 140,000 deaths in the United States. Health Aff (millwood) 40:1465–1472. https://doi.org/10.1377/hlthaff.2021.00619
Hafrén L, Saarinen R, Lundberg M (2022) Effects of social distancing on the incidence of Bell's palsy and sudden sensorineural hearing loss. Acta Otolaryngol 142:220–223. https://doi.org/10.1080/00016489.2021.2025425
Harenberg J, Jonas JB, Trecca EMC (2020) A liaison between sudden sensorineural hearing loss and SARS-CoV-2 infection. Thromb Haemost 120:1237–1239. https://doi.org/10.1055/s-0040-1714370
Hause AM, Baggs J, Marquez P et al (2021) COVID-19 vaccine safety in children aged 5–11 years—United States, November 3–December 19, 2021. MMWR Morb Mortal Wkly Rep 70:1755–1760. https://doi.org/10.15585/mmwr.mm705152a1
Jacob J, Flannery W, Mostert C (2020) Novel ENT triad of anosmia, ageusia and hearing impairment in COVID-19. Intern Med J 50:1155. https://doi.org/10.1111/imj.14880
Jafari Z, Kolb BE, Mohajerani MH (2022) Hearing loss, tinnitus, and dizziness in COVID-19: a systematic review and meta-analysis. Can J Neurol Sci 49:184–195. https://doi.org/10.1017/cjn.2021.63
Jeong J, Choi HS (2021) Sudden sensorineural hearing loss after COVID-19 vaccination. Int J Infect Dis 113:341–343. https://doi.org/10.1016/j.ijid.2021.10.025
Jeong J, Yoon PH (2022) Sudden sensorineural hearing loss with intralabyrinthine hemorrhage after COVID-19 vaccination. Hum Vaccin Immunother 18:2097462. https://doi.org/10.1080/21645515.2022.2097462
Jeong M, Ocwieja KE, Han D et al (2021) Direct SARS-CoV-2 infection of the human inner ear may underlie COVID-19-associated audiovestibular dysfunction. Commun Med 1:1–14. https://doi.org/10.1038/s43856-021-00044-w
Jin L, Fan K, Tan S et al (2021) Analysis of the characteristics of outpatient and emergency diseases in the department of otolaryngology during the "COVID-19" pandemic. Sci Prog 104:00368504211036319. https://doi.org/10.1177/00368504211036319
Kahn B, Apostolidis SA, Bhatt V et al (2021) Multisystem inflammation and organ dysfunction after BNT162b2 messenger RNA coronavirus disease 2019 Vaccination. Crit Care Explor 3:e0578. https://doi.org/10.1097/CCE.0000000000000578
Kaliyappan K, Chen Y-C, Krishnan Muthaiah VP (2022) Vestibular cochlear manifestations in COVID-19 cases. Front Neurol 13:850337. https://doi.org/10.3389/fneur.2022.850337

Kanduc D, Shoenfeld Y (2020) On the molecular determinants of the SARS-CoV-2 attack. Clin Immunol 215:108426. https://doi.org/10.1016/j.clim.2020.108426

Karimi-Galougahi M, Naeini AS, Raad N et al (2020) Vertigo and hearing loss during the COVID-19 pandemic—is there an association? Acta Otorhinolaryngol Ital 40:463–465. https://doi.org/10.14639/0392-100X-N0820

Khawaja UA, Shamsoddin E, Desideri LF, Tovani-Palone MR Infection of red blood cells by SARS-CoV-2: new evidence. Einstein (Sao Paulo) 19:eCE6285. https://doi.org/10.31744/einstein_journal/2021CE6285

Kilic O, Kalcioglu MT, Cag Y et al (2020) Could sudden sensorineural hearing loss be the sole manifestation of COVID-19? An investigation into SARS-COV-2 in the etiology of sudden sensorineural hearing loss. Int J Infect Dis 97:208–211. https://doi.org/10.1016/j.ijid.2020.06.023

Koumpa FS, Forde CT, Manjaly JG (2020) Sudden irreversible hearing loss post COVID-19. BMJ Case Rep 13:e238419. https://doi.org/10.1136/bcr-2020-238419

Kuhn M, Heman-Ackah SE, Shaikh JA, Roehm PC (2011) Sudden sensorineural hearing loss. Trends Amplif 15:91–105. https://doi.org/10.1177/1084713811408349

Kumar Swain S, Ranjan Pani S (2021) Incidence of hearing loss in COVID-19 patients: a COVID hospital-based Study in the eastern part of India. IJCRR 13:103–107. https://doi.org/10.31782/IJCRR.2021.13329

Lamounier P, Gonçalves VF, Ramos HVL et al (2020) A 67-year-old woman with sudden hearing loss associated with SARS-CoV-2 infection. Am J Case Rep 21:e927519-1–e927519-6. https://doi.org/10.12659/AJCR.927519

Lang B, Hintze J, Conlon B (2019) Coronavirus disease 2019 and sudden sensorineural hearing loss. J Laryngol Otol 1–3. https://doi.org/10.1017/S0022215120002145

Lee SG, Fralick M, Sholzberg M (2020) Coagulopathy associated with COVID-19. CMAJ 192:E583. https://doi.org/10.1503/cmaj.200685

Li W, Micco A, Fang D, Liu H (2022) Intratympanic steroid treatments rescued recurrent hearing loss following COVID-19 vaccination and detection of an intralabyrinthine schwannoma. BMJ Case Rep 15:e249316. https://doi.org/10.1136/bcr-2022-249316

Luca PD, Scarpa A, Ralli M et al (2021) Auditory disturbances and SARS-CoV-2 infection: brain inflammation or cochlear affection? Systematic review and discussion of potential pathogenesis. Front Neurol 12. https://doi.org/10.3389/fneur.2021.707207

Maharaj S, Bello Alvarez M, Mungul S, Hari K (2020) Otologic dysfunction in patients with COVID-19: a systematic review. Laryngoscope Investig Otolaryngol 5:1192–1196. https://doi.org/10.1002/lio2.498

Maliah A, Parikh R, Tayer-Shifman OE et al (2022) Steroid treatment suppresses the CD4+ T-cell response to the third dose of mRNA COVID-19 vaccine in systemic autoimmune rheumatic disease patients. Sci Rep 12:21056. https://doi.org/10.1038/s41598-022-25642-z

Masuda M, Kanzaki S, Minami S et al (2012) Correlations of inflammatory biomarkers with the onset and prognosis of idiopathic sudden sensorineural hearing loss. Otol Neurotol 33:1142–1150. https://doi.org/10.1097/MAO.0b013e3182635417

Meng X, Wang J, Sun J, Zhu K (2022) COVID-19 and sudden sensorineural hearing loss: a systematic review. Front Neurol 13:883749. https://doi.org/10.3389/fneur.2022.883749

Montano D (2022) Frequency and associations of adverse reactions of COVID-19 vaccines reported to pharmacovigilance systems in the European Union and the United States. Front Public Health 9:756633. https://doi.org/10.3389/fpubh.2021.756633

Moshtaghi O, Ghavami Y, Mahboubi H et al (2018) Migraine-related aural fullness: a potential clinical entity. Otolaryngol Head Neck Surg 158:100–102. https://doi.org/10.1177/0194599817739255

Mustafa MWM (2020) Audiological profile of asymptomatic COVID-19 PCR-positive cases. Am J Otolaryngol 41:102483. https://doi.org/10.1016/j.amjoto.2020.102483

Nagy A, Alhatlani B (2021) An overview of current COVID-19 vaccine platforms. Comput Struct Biotechnol J 19:2508–2517. https://doi.org/10.1016/j.csbj.2021.04.061

Ng R, Vargas G, Jashar DT et al (2022) Neurocognitive and psychosocial characteristics of pediatric patients with post-acute/long-COVID: a retrospective clinical case series. Arch Clin Neuropsychol 37:1633–1643. https://doi.org/10.1093/arclin/acac056

Nieminen TA, Kivekäs I, Artama M et al (2022) Sudden hearing loss following vaccination against COVID-19. JAMA Otolaryngol-Head Neck Surg. https://doi.org/10.1001/jamaoto.2022.4154

Oldenburg J, Klamroth R, Langer F et al (2021) Diagnosis and management of vaccine-related thrombosis following AstraZeneca COVID-19 vaccination: guidance statement from the GTH. Hamostaseologie 41:184–189. https://doi.org/10.1055/a-1469-7481

Parrino D, Brescia G, Trimarchi MV et al (2019) Cochlear-vestibular impairment due to West Nile virus infection. Ann Otol Rhinol Laryngol 128:1198–1202. https://doi.org/10.1177/0003489419866219

Parrino D, Frosolini A, Toninato D et al (2022) Sudden hearing loss and vestibular disorders during and before COVID-19 pandemic: an audiology tertiary referral centre experience. Am J Otolaryngol 43:103241. https://doi.org/10.1016/j.amjoto.2021.103241

Perret M, Bernard A, Rahmani A et al (2021) Acute labyrinthitis revealing COVID-19. Diagnostics (basel) 11:482. https://doi.org/10.3390/diagnostics11030482

Pisani D, Leopardi G, Viola P et al (2021) Sudden sensorineural hearing loss after covid-19 vaccine; a possible adverse reaction? Otolaryngol Case Rep 21:100384. https://doi.org/10.1016/j.xocr.2021.100384

Polack FP, Thomas SJ, Kitchin N et al (2020) Safety and efficacy of the BNT162b2 mRNA Covid-19 vaccine. N Engl J Med. https://doi.org/10.1056/NEJMoa2034577

Ricciardiello F, Pisani D, Viola P et al (2021) Sudden sensorineural hearing loss in mild COVID-19: case series and analysis of the literature. Audiol Res 11:313–326. https://doi.org/10.3390/audiolres11030029

Rocke J, Hopkins C, Philpott C, Kumar N (2020) Is loss of sense of smell a diagnostic marker in COVID-19: a systematic review and meta-analysis. Clin Otolaryngol 45:914–922. https://doi.org/10.1111/coa.13620

Russo A, Tellone E, Barreca D et al (2022) Implication of COVID-19 on erythrocytes functionality: red blood cell biochemical implications and morpho-functional aspects. Int J Mol Sci 23:2171. https://doi.org/10.3390/ijms23042171

Rybak LP, Ramkumar V, Mukherjea D (2021) Ototoxicity of non-aminoglycoside antibiotics. Front Neurol 12:652674. https://doi.org/10.3389/fneur.2021.652674

Salamanna F, Maglio M, Landini MP, Fini M (2020) Body localization of ACE-2: on the trail of the key-hole of SARS-CoV-2. Front Med 7

Saniasiaya J (2021) Hearing loss in SARS-CoV-2: what do we know? Ear Nose Throat J 100:152S-154S. https://doi.org/10.1177/0145561320946902

Schraff SA, Schleiss MR, Brown DK et al (2007) Macrophage inflammatory proteins in cytomegalovirus-related inner ear injury. Otolaryngol Head Neck Surg 137:612–618. https://doi.org/10.1016/j.otohns.2007.03.044

Sebothoma B, Khoza-Shangase K (2022) Middle ear status—structure, function and pathology: a scoping review on middle ear status of COVID-19 positive patients. S Afr J Commun Disord 69:910. https://doi.org/10.4102/sajcd.v69i2.910

Silva Andrade B, Siqueira S, de Assis Soares WR et al (2021) Long-COVID and post-COVID health complications: an up-to-date review on clinical conditions and their possible molecular mechanisms. Viruses 13:700. https://doi.org/10.3390/v13040700

Sonbay Yilmaz ND, Saka C, Oktay Arslan B et al (2019) The effect of hypoxia on hearing function. Turk J Med Sci 49:1450–1454. https://doi.org/10.3906/sag-1902-210

Spence JD, De Freitas GR, Pettigrew LC et al (2020) Mechanisms of stroke in COVID-19. Cerebrovasc Dis 1–8. https://doi.org/10.1159/000509581

Sriwijitalai W, Wiwanitkit V (2020) Hearing loss and COVID-19: a note. Am J Otolaryngol 102473. https://doi.org/10.1016/j.amjoto.2020.102473

Sweeney C, Gilden D (2001) Ramsay hunt syndrome. J Neurol Neurosurg Psychiatry 71:149–154. https://doi.org/10.1136/jnnp.71.2.149

Takahashi N, Abe R, Hattori N et al (2020) Clinical course of a critically ill patient with severe acute respiratory syndrome coronavirus 2 (SARS-CoV-2). J Artif Organs 23:397–400. https://doi.org/10.1007/s10047-020-01183-y

Taylor S, Asmundson GJG (2021) Immunization stress-related responses: implications for vaccination hesitancy and vaccination processes during the COVID-19 pandemic. J Anxiety Disord 84:102489. https://doi.org/10.1016/j.janxdis.2021.102489

Thrane JF, Britze A, Fjaeldstad AW (2022) Incidence and duration of self-reported hearing loss and tinnitus in a cohort of COVID-19 patients with sudden chemosensory loss: a strobe observational study. Eur Ann Otorhinolaryngol Head Neck Dis 139:125–128. https://doi.org/10.1016/j.anorl.2021.07.012

Tsetsos N, Poutoglidis A, Vlachtsis K et al (2021) Sudden sensorineural hearing loss following the second dose of COVID-19 vaccine. Cureus 13:e17435. https://doi.org/10.7759/cureus.17435

van Rijssen LB, Derks W, Hoffmans R et al (2022) No COVID-19 in patients with sudden sensorineural hearing loss (SSNHL). Otol Neurotol 43:170–173. https://doi.org/10.1097/MAO.0000000000003438

Vojdani A, Kharrazian D (2020) Potential antigenic cross-reactivity between SARS-CoV-2 and human tissue with a possible link to an increase in autoimmune diseases. Clin Immunol 217:108480. https://doi.org/10.1016/j.clim.2020.108480

Wang J, Liang S, Yu M, Gong Z (2022) COVID-19 from the perspective of otorhinolaryngology: an analysis of bibliometrics. Front Public Health 10:1002686. https://doi.org/10.3389/fpubh.2022.1002686

Welcome MO, Mastorakis NE (2021) Neuropathophysiology of coronavirus disease 2019: neuroinflammation and blood brain barrier disruption are critical pathophysiological processes that contribute to the clinical symptoms of SARS-CoV-2 infection. Inflammopharmacology 29:939–963. https://doi.org/10.1007/s10787-021-00806-x

Wichova H, Miller ME, Derebery MJ (2021) Otologic manifestations after COVID-19 vaccination: the house ear clinic experience. Otol Neurotol 42:e1213–e1218. https://doi.org/10.1097/MAO.0000000000003275

Wilson WR, Byl FM, Laird N (1980) The efficacy of steroids in the treatment of idiopathic sudden hearing loss. A double-blind clinical study. Arch Otolaryngol 106:772–776. https://doi.org/10.1001/archotol.1980.00790360050013

Wise J (2021) Covid-19: European countries suspend use of Oxford-AstraZeneca vaccine after reports of blood clots. BMJ 372:n699. https://doi.org/10.1136/bmj.n699

Wu A, Peng Y, Huang B et al (2020) Genome composition and divergence of the novel coronavirus

(2019-nCoV) originating in China. Cell Host Microbe 27:325–328. https://doi.org/10.1016/j.chom.2020.02.001

Yanir Y, Doweck I, Shibli R et al (2022) Association between the BNT162b2 messenger RNA COVID-19 vaccine and the risk of sudden sensorineural hearing loss. JAMA Otolaryngol Head Neck Surg 148:299–306. https://doi.org/10.1001/jamaoto.2021.4278

Yaseen NK, Al-Ani RM, Ali Rashid R (2021) COVID-19—related sudden sensorineural hearing loss. Qatar Med J 2021:58. https://doi.org/10.5339/qmj.2021.58

Zhang Y, Zeng G, Pan H et al (2021) Safety, tolerability, and immunogenicity of an inactivated SARS-CoV-2 vaccine in healthy adults aged 18–59 years: a randomised, double-blind, placebo-controlled, phase 1/2 clinical trial. Lancet Infect Dis 21:181–192. https://doi.org/10.1016/S1473-3099(20)30843-4

Zhao H, Li Y, Wang Z (2022) Adverse event of sinovac coronavirus vaccine: deafness. Vaccine 40:521–523. https://doi.org/10.1016/j.vaccine.2021.11.091

Zoccali F, Cambria F, Colizza A et al (2022) Sudden sensorineural hearing loss after third dose booster of COVID-19 vaccine administration. Diagnostics (bAsel) 12:2039. https://doi.org/10.3390/diagnostics12092039

The Views of the Citizens Regarding COVID-19 Vaccination

16

Veljko Turanjanin

Abstract

The pandemic caused by the SARS-CoV-2 virus has been ongoing for over three years. After the initial months of confusion and searching for different solutions in the fight against this disease, several different vaccines were developed. Although they were eagerly awaited, social networks were flooded with negative and discouraging comments about vaccines. Citizens' views on this issue varied not only from country to country but also within the same country. At the same time, a certain number of officials in some countries contributed to a negative view of vaccines. Therefore, the first goal of this paper is to review the results of research conducted in different countries and then analyze the results of research conducted in Serbia. The sample consisted of 301 respondents. The results of the research show that the majority of respondents refused to be vaccinated, as they do not believe in the effectiveness or safety of the vaccine against COVID-19, nor do they trust the state authorities in their fight against the pandemic.

V. Turanjanin (✉)
Faculty of Law, University of Kragujevac, Kragujevac, Serbia
e-mail: vturanjanin@jura.kg.ac.rs; turanjaninveljko@gmail.com

Keywords

COVID-19 · Vaccine · Hesitation · Compulsory vaccination · Serbia

16.1 Introduction

The new coronavirus epidemic (Taherizadeh et al. 2020), caused by SARS-CoV-2 (Farhud and Zokaei 2020)—a virus with high transmission potential (Sheikhshahrokh et al. 2020)—has been raging around the world for over three years. The symptoms are diverse from patient to patient, while some patients remain asymptomatic (Chen et al. 2021). Forms of state opposition to a global pandemic vary from region to region. The epidemic affected the physical and mental health of the people as well as the economy of almost every state in the world (Mukhtar et al. 2021). At the same time, technological sector advances have been made (Ishmaev et al. 2021). After a diverse range of measures taken by the states in the initial phase of the appearance of the coronavirus (Turanjanin and Radulović 2020), as well as the initial misunderstandings, it seems that the population has become accustomed to a slightly different way of living and working.

Derogations from certain human rights in times of crisis, like H1N1 in 2009 (Richardson and Devine 2020), are no novelty in modern

N. Rezaei (ed.), *The COVID-19 Aftermath*, Advances in Experimental Medicine and Biology 1457,
https://doi.org/10.1007/978-3-031-61939-7_16

times. Around the world, we have witnessed voluntary and involuntary quarantine (Quintana and Quintana 2021), and legal stay-at-home orders (Stickle and Felson 2020), while some countries have quarantined the entire population completely regardless of their medical status (Turanjanin 2021a). The COVID-19 pandemic indicates that countries must prepare and adopt effective and efficient public health policies in order to combat this and future highly dangerous infectious diseases.

Vaccination against this infectious disease is certainly one of the current topics. One of the primary objectives of vaccination is the protection of people receiving the vaccine as well as those who cannot be vaccinated for various reasons. After the period of announcing the vaccine, several types have appeared to date, but which have not been accepted as expected. Yet, more than 100 vaccines are in clinical trials (Savulescu 2021; Velásquez 2022). On the one hand, China, the country where the pandemic occurred, produces *Sinopharm* and *Sinovac*, while Russia produces *Sputnik V*. On the other hand, *Pfizer-BioNTech*, *Moderna*, *Johnson and Johnson's Janssen*, *AstraZeneca*, and *Novavax* COVID-19 vaccines also appeared.

Another problem is vaccine hesitancy as a barrier to complete inoculation of the population (Dror et al. 2020). Serbia even provides a nearly EUR 30 reward to everyone who gets a coronavirus vaccine. During the pandemic, Serbia faced numerous problems, which were also considered by the Constitutional Court, starting with curfews and ending with injuries in criminal and misdemeanor proceedings (Turanjanin 2021b, 2022).

There have been more and more requests for mandatory vaccination, worldwide. Numerous advantages and disadvantages are also outlined in the theory (Zen 2020; Giubilini 2020). For instance, one of the reasons for compulsory vaccination is to compare vaccination with tax evasion (Giubilini 2020). Additionally, the issue of compulsory vaccination was raised before the European Court of Human Rights.

This paper is divided into two key parts. The part deals with the issue of compulsory vaccination and public health. The first part deals with the research on citizens' attitudes regarding vaccination against COVID-19 in certain countries. In the second part, the results of the research conducted in Serbia shall be presented. The research was conducted in the period December 2022–February 2023.

16.2 The Views of the Citizens Regarding COVID-19 Vaccination Worldwide

Citizens' views on this issue varied not only from country to country but also within the same country. The special problem in almost all countries is vaccine hesitancy. The literature defines vaccine hesitancy as a delay in vaccine uptake or a refusal to receive immunization despite vaccine services being available (Alam et al. 2022; Dubé et al. 2015). The study of Biswas et al. shows that the factors for vaccine hesitancy are complex and varying through time and sociodemographic variables (Biswas et al. 2021; see also Aw et al. 2021). The vaccine hesitancy is a concern and significant threat to public health (Raza et al. 2023). Key factors for hesitance and refusing COVID-19 vaccines are related to conspiracy theories, but also sources of information about this infectious disease.

The surveys show that men over 55 are most likely to have received a vaccine, while the percentage of men is slightly higher than that of women. Furthermore, the largest group that is vaccine-hesitant is a group that contains a population of 18- to 24-year-olds (Neumann-Böhme et al. 2020). According to a survey of adults in Australia, 86% of respondents generally accept vaccination against infectious diseases for both themselves and their children, while a high percentage (78%) of respondents expressed a desire for vaccination against COVID-19 (Enticott et al. 2022). The important fact is that the results of research in Australia show a great willingness by citizens to continue with measures against the spread of the virus, such as wearing masks and maintaining physical distance even after vaccination (Enticott et al. 2022). In Japan, the percentage of the respondents who would like to get COVID-19 vaccine was 62.1% (Machida et al. 2021).

On the one hand, according to research from Italy (Biasio et al. 2021) and Portugal (Fernandes et al. 2021), the majority of the population has a positive attitude toward vaccination against COVID-19. On the other hand, the research in Romania shows that 40% of the respondents rejects the COVID-19 vaccine (Maftei and Holman 2021). One study conducted in India shows a high degree of desire among subjects to be vaccinated (Kishore et al. 2021). However, another study conducted in the same country shows a high degree of reluctance toward vaccination, while the percentage of people who refuse vaccination is low (Danabal et al. 2021).

Singapore has a high level of knowledge about vaccines, as well as high rates of vaccination (Subramaniam et al. 2023). In the United States, a survey shows that 69% of respondents over the age of 18 want to be vaccinated against COVID-19 (Reiter et al. 2020). The Latin America and the Caribbean provides similar situation (Salazar-Fernández et al. 2022).

Social networks played a huge role in the negative campaign against vaccination, by disseminating a series of comments both before and after the availability of vaccines, which deterred a certain percentage of the world's population from vaccination (Wawrzuta et al. 2021). This percentage varies from country to country. According to a study from Bangladesh, 34.74% of students use the Internet as a key source of information about COVID-19, while a slightly smaller percentage (33.41%) use social networks, and an even smaller percentage use electronic media (25.61%) (Rahman et al. 2022). Less than 50% of respondents believe that vaccination can stop the pandemic, although 54.34% believe that COVID-19 vaccines are safe and effective (Rahman et al. 2022). In Italy, the key source of information is the Internet (72%), followed by social networks (47%) and television (49%) (Biasio et al. 2021).

Research conducted in Ethiopia shows that a higher percentage of the population has a positive attitude toward vaccination against COVID-19, but reasons for vaccine refusal include lack of information about vaccines, lack of conviction about vaccine safety, and concerns about the time spent on vaccine development (Asres and Umeta 2022). Another study from the same country conducted among clinical practitioners indicates a high acceptance of vaccines by respondents (84.4%) with an even higher percentage of recommendations to others [86.1% (Abay et al. 2022)]. Research carried out in Turkey shows high rates of COVID-19 vaccine hesitancy, while the percentage in Great Britain is somewhat lower (Salali and Uysal 2020).

In Saudi Arabia, the respondents' perceptions on COVID-19 showed that they do not believe in seriousness of the disease, with lack of trust in efficacy and safety of the vaccine (Alrajeh et al. 2021). The lack of trust and concerns around the vaccine safety is present in the research among population in Kashmir (Khan et al. 2022) and Syria (Mohamad et al. 2021).

The research from 2020 in the United Kingdom emphasized distrust generally in vaccines and mistrust in government, which raised vaccine hesitancy and conspiracy theories (Jennings et al. 2021). Another survey shows a high percentage of the population willing to be vaccinated (71.7% against, 16.6% very unsure, and 11.7% strongly hesitant) (Freeman et al. 2022; see also Freeman et al. 2021).

The national survey in the United States revealed that approximately 30% of the respondents is not sure would they accept vaccine (Fisher et al. 2020). Another research points out that conspiracy theory is the strongest reason for vaccine hesitancy (Farhart et al. 2022). In Italy, the research pointed out the fact that people who believe in conspiracy theory do not trust medicine (Simione et al. 2021). Some studies indicate a connection between digital literacy and vaccine hesitancy due to the creation of conspiracy theories (Gao et al. 2023). And a study conducted in Pakistan, Saudi Arabia, India, Malaysia, Sudan, and Egypt shows belief in a conspiracy and vaccine hesitancy, with the fact that almost 52% of respondents would take the vaccine only if it were free, and only 24% would pay for it (Salman et al. 2022). Another research in Pakistan has shown similar results regarding the connection between conspiracy theories and vaccination hesitancy (Arshad et al. 2021). Research in Jordan and Kuwait has shown that low acceptance of the COVID-19 vaccine is also linked to conspiracy theories (Sallam et al. 2021a, b). The similar situation is in Hungary (Bíró-Nagy and Szászi 2022). In China, Zhang et al. found that conspiracy beliefs have direct and indirect effects on vaccine hesitancy (Zhang et al. 2021). The research in the United Kingdom recognized also the link between conspiracy theories and vaccination hesitancy (Eberhardt and Ling 2021). Another survey in this country shows the fact that reducing conspiracy theories may have positive effect to reducing vaccine hesitancy (Allington et al. 2021).

16.3 Current Study in Serbia

16.3.1 Methods

The research was conducted in 7 cities in Serbia: Kragujevac, Novi Sad, Kraljevo, Užice, Kruševac, Čačak, and Jagodina. Respondents were selected randomly. All respondents who participated in the research gave their consent. The participation was completely voluntary. Written consent was obtained from all participants.

The initial hypotheses are:

1. The respondents' trust in the virus protection measures implemented by Serbia is weak.
2. The respondents have a higher level of trust in vaccines against other infectious diseases than the vaccine against COVID-19.
3. The respondents mostly do not believe the statements of government officials regarding vaccines.
4. The respondents trust the statements and opinions of doctors more than civil servants.
5. The majority of the respondents believe that vaccination against the COVID-19 virus should not be mandatory and refusal should not be punishable.

The questionnaire was written in Serbian language and consisted of 7 parts. The first section gathers sociodemographic characteristics: age, sex, a family member diagnosed with COVID-19, and whether the subject was infected before the availability of the COVID-19 vaccine. The second section deals with the assessment of the effectiveness of the recommendations of state authorities and the Crisis Staff during the duration of the coronavirus pandemic (i.e., wearing masks, maintaining physical distance, and curfews). The third section of the questionnaire deals with attitudes toward vaccination (whether the respondent was vaccinated, and if so, with which vaccine; what side effects occurred, if any; and if the respondent was not vaccinated, what the reason for that was). The questionnaire offered suggested answers, with the ability to add a new answer. All respondents selected one or more of the suggested answers. The fourth section deals with the attitude toward vaccines against infectious diseases in general and the vaccine against COVID-19 in particular. The fifth section deals with the respondents' trust in the statements of medical workers and state representatives regarding vaccines against the COVID-19 virus, and the seventh section deals with the views of the respondents regarding obligation, coercion and the possibility of punishment for unvaccinated people. Statistical analysis was performed using SPSS Statistic 26.0.

16.3.2 Results and Discussion

The study discovers more female survey respondents (59.5%) than male (40.2%). There is no statistically significant difference in terms of vaccination between men and women in the research sample. The study also shows that a large number of respondents are neither married nor in an extramarital union. Percentage-wise, the largest number of respondents have obtained a bachelor's degree (43.9%) or completed high school (43.5%). In the research, there were no respondents without completed school, while there were also respondents with completed master's or doctoral studies. The largest percentage of respondents are under 30 years of age. The largest number does not have any chronic disease. At the same time, 68.4% of respondents were not infected with COVID-19 before the availability of the vaccine in Serbia, while 55.1% did not have a family member who was infected with this virus. Table 16.1 contains more detailed information on respondents.

Section two deals with the assessment of the effectiveness of the recommendations of the state authorities and the Crisis Staff during the duration of the coronavirus pandemic (i.e., wearing masks, maintaining physical distance, and curfews). Respondents evaluated the success of the measure on five-point scale statements with options of "no results"; "the results are weak"; "the results are average"; "the results are good"; and "the results are excellent". Regardless of whether the respondent was vaccinated or not, a huge percentage of respondents believe that the results of recommendations of the state authorities and the Crisis Staff during the duration of the coronavirus pandemic are weak or average (Table 16.2).

The third section of the questionnaire deals with attitudes toward vaccination. In this sample, 34.2% of respondents have been vaccinated, 9.3% have not but are considering doing so, and 56.1% of respondents refuse to be vaccinated. Vaccinated subjects were vaccinated with 4 vaccines: *Pfizer* (15%), *Sinopharm* (15%), *Sputnik V* (3.7%), and *AstraZeneca* (1%). The majority

Table 16.1 Basic characteristics of the respondents

	Frequency	Percent	Valid percent	Cumulative percent	
Age	18–30	250	83.1	83.1	83.1
	31–45	29	9.6	9.6	92.7
	46–59	15	5.0	5.0	97.7
	Over 60 years	7	2.3	2.3	100.0
	Total	301	100.0	100.0	
Sex	Male	121	40.2	40.3	40.3
	Female	179	59.5	59.7	100.0
	Total	300	99.7	100.0	
Marital status	Married	45	15.0	15.0	15.0
	Not married	252	83.7	83.7	98.7
	Divorced	4	1.3	1.3	100.0
Education	Elementary school	3	1.0	1.0	1.0
	High school	131	43.5	44.3	45.3
	Bachelor's degree	132	43.9	44.6	89.9
	Master's degree	25	8.3	8.4	98.3
	Ph.D. degree	5	1.7	1.7	100.0
Chronic disease	Yes	21	7.0	7.0	7.0
	No	277	92.0	93.0	100.0
	Total	298	99.0	100.0	
Family member	Yes	133	44.2	44.5	44.5
	No	166	55.1	55.5	100.0
	Total	299	99.3	100.0	
Infection before vaccine	Yes	94	31.2	31.3	31.3
	No	206	68.4	68.7	100.0
	Total	300	99.7	100.0	

Table 16.2 Effectiveness of the recommendations

	Frequency	Percent	Valid percent	Cumulative percent	
Effectiveness	No results	55	18.3	18.6	18.6
	The results are weak	82	27.2	27.7	46.3
	The results are average	123	40.9	41.6	87.8
	The results are good	27	9.0	9.1	97.0
	The results are excellent	9	3.0	3.0	100.0
	Total	296	98.3	100.0	

of the subjects had no adverse events, while some of them appeared after the first dose (in 11 subjects), after the second dose (in 14 subjects), or after both doses without a difference (in 10 subjects). At the same time, serious adverse events were reported by two respondents, while 3 had medium-intensity adverse events. Adverse events were mild in the rest of the respondents. The most commonly reported adverse reactions were fatigue (in 12 subjects); muscle pain (in 10 subjects); fever (in 6 subjects); headache (in 5 subjects); joint pain (in 5 subjects); pain in extremities (in 5 subjects); extreme sleepiness (in 4 subjects); shivering (in 3 subjects); temperature (in 3 subjects); itching at the injection site (in 2 subjects); redness at the injection site (in 4 subjects), and sleep disorders, dizziness, cold and bad memory (in one subject each). Fatigue, headache, muscle pain, and extreme sleepiness occurred most often in subjects vaccinated with the *Pfizer* vaccine, while other symptoms were almost equally distributed in subjects vaccinated with the *Sinopharm* and *Sputnik V* vaccines.

Respondents who refused to be vaccinated cited several reasons for such thinking. All the reasons were stated in advance in the questionnaire, and no reason appeared outside of the provided ones. Essentially, 33.2% of respondents are not sure of the fact that the vaccine against the coronavirus is safe, 18.6% believe that they do not have sufficient information about vaccines, 15.6% do not trust the effectiveness of the vaccine, 3.7% of respondents believe that COVID-19 is not a serious illness, 6% think COVID-19 is a conspiracy, 2% think they will get sick from COVID-19 if they receive the vaccine, 16.9% are worried about the side effects of the vaccination, 4% think they are naturally immune to COVID-19, 4.3% claim to use other types of protection against COVID-19, 3.3% report fear of needles as the reason they refuse vaccination, and 1.3% believe that vaccines are not stored properly in Serbia. What we can emphasize is that a small percentage of respondents believe that it is a conspiracy or that COVID-19 is not a serious disease.

Afterward, we examined the attitudes of non-vaccinated participants about the circumstances under which they would receive the vaccine against COVID-19. In this regard, 20.3% of respondents would decide to take that step if the state declared it mandatory, 16.3% if it was recommended by the doctor, 6.3% if the respondent's family and friends were vaccinated, 1% if there was another method of vaccination other than injection, 2.3% if it becomes mandatory for the respondent's job, 1% due to travel, and even 36.5% do not want to be vaccinated under any conditions.

A special question was about the main sources of information related to COVID-19. The largest number of respondents mentioned the official websites of the Ministry of Health of Serbia as the main source (29.6%). This is followed by Internet sources (16.6%); social networks (15.3%); family and friends (13.3%); WHO (9%), and scientific journals (6.3%). It is noticeable that 10% of respondents did not answer this question. These data are somewhat different compared to those in the Italy (Biasio et al. 2021).

The fourth section deals with the attitude toward vaccines against infectious diseases in general and the vaccine against COVID-19 in particular. This group of questions are related to the examination of respondents' attitudes regarding vaccination in general and about vaccination against COVID-19. In this context, as many as 73.4% of respondents believe that in general vaccination against infectious diseases is important for their health, while 65.4% of respondents believe that their vaccination against infectious diseases is important for the health of other persons. However, in the field of vaccination against COVID-19, only 29.6% of respondents believe that vaccination is important for their health, and 34.2% believe that vaccination is important for the health of other people. This information says a lot about the fact that the majority of respondents believe that vaccination against COVID-19 is not important for the health of the vaccinated, nor for other people.

The majority of respondents believe that vaccines against infectious diseases are safe (66.8%), but only 35.9% of respondents consider the vaccine against COVID-19 to be safe. However, data cross-checking indicates that out of 103 vaccinated respondents, 22 respondents believe that vaccine against the COVID-19 is not safe, which is not a negligible percentage.

Confidence is even lower when looking at the data on the effectiveness of the vaccine against COVID-19, where only 29.2% of respondents consider the vaccine to be effective. At the same time, an even smaller percentage of respondents (19.3%) believe that vaccination protects them from getting sick from COVID-19. It is interesting to note that out of 103 vaccinated respondents, 34 believe that the vaccine is not effective, and even 52, judging by the statistical data, correctly believe that the vaccine does not protect them from COVID-19 infection.

Similarly, only 21.9% of respondents are of the opinion that vaccination is the best way to protect against COVID-19 infection, but at the same time it is interesting that as many as 60.8% of respondents think that it is not necessary to strengthen citizens' awareness of the importance of COVID-19 vaccination. Finally, a special question concerns the overall attitude toward COVID-19 vaccination, which is predominantly negative (Table 16.3).

A special group of questions are related to citizens' trust in the statements of health workers and authorities (Table 16.4). As expected, a larger percentage of respondents trust the statements and claims of doctors more than authorities, but still, that percentage is not very different. The place where the difference is noticed is that a higher percentage of respondents answered clearly negatively to the question of trust in authorities, while the percentage is approximate for respondents who doubt the truth of the claims.

The last 4 questions are related to respondents' attitudes regarding mandatory and compulsory vaccination, and the potential punishment of people who do not want to be vaccinated (Table 16.5). In the first place, we have to notice that an International Covenant on Economic, Social and Cultural Rights recognizes everyone's right to the enjoyment of the highest attainable standard of physical and mental health and therefore, the UN Committee on Economic, Social and Cultural Rights emphasizes that diseases control relating to the implementation or enhancement of immunization programs and other infectious disease control strategies. The UN Committee on Economic, Social and Cultural Rights has repeatedly emphasized preventive vaccination in its concluding

Table 16.3 General attitude toward vaccination

		Frequency	Percent	Valid percent	Cumulative percent
Valid	Positive	115	38.2	40.1	40.1
	Negative	172	57.1	59.9	100.0
	Total	287	95.3	100.0	
Missing	System	14	4.7		
Total		301	100.0		

Table 16.4 Trust in health workers and state authorities

	Frequency	Percent	Valid percent	Cumulative percent
I trust medical professionals with information regarding COVID-19				
Yes	109	36.2	37.2	37.2
No	63	20.9	21.5	58.7
I doubt the veracity of their information	121	40.2	41.3	100.0
Total	293	97.3	100.0	
I trust the authorities with information regarding COVID-19				
Yes	62	20.6	21.2	21.2
No	112	37.2	38.4	59.6
I doubt the veracity of their information	118	39.2	40.4	100.0
Total	292	97.0	100.0	
Missing system	9	3.0		

Table 16.5 Attitudes toward mandatory and compulsory vaccination

	Frequency	Percent	Valid percent	Cumulative percent
Do you think the vaccine should be mandatory?				
Yes	46	15.3	15.6	15.6
No	249	82.7	84.4	100.0
Total	295	98.0	100.0	
Missing system	6	2.0		
Do you think the vaccine should be compulsory?				
Yes	14	4.7	4.7	4.7
No	281	93.4	95.3	100.0
Total	295	98.0	100.0	
Missing system	6	2.0		
Do you think that refusing to get vaccinated against COVID-19 should be punishable by a fine?				
Yes	25	8.3	8.5	8.5
No	269	89.4	91.5	100.0
Total	294	97.7	100.0	
Missing system	7	2.3		
Do you think that refusing to get vaccinated against COVID-19 should be a criminal offense?				
Yes	12	4.0	4.1	4.1
No	282	93.7	95.9	100.0
Total	294	97.7	100.0	
Missing system	7	2.3		

observations as part of the periodic review of individual states, and the goal is the high possible percentage of the population. Furthermore, in its "Global Vaccine Action Plan" published in 2013 the WHO pointed out that vaccines had saved a large number of lives, primarily thanks to the process of immunization, therefore, consequently, immunization should be one of the components of the right to health (WHO 2013). The European Commission Report on the state of vaccine confidence is very important for these considerations (Larson et al. 2018), which states that high confidence in vaccination programs is crucial in the immunization process, especially for maintaining high coverage rates.

The vaccination is largely related to the autonomy of the patient's will, which appears in various forms at the intersection of medical and criminal law. It is generally considered one of the three or four leading principles of medical ethics, and there do not seem to be many ethical problems in which it does not play an important part (Schermer 2002). We can say that patient autonomy is the main principle of contemporary rights-oriented bioethics (Todorović and Protić 2017), and there are numerous sources of autonomy in legal, medical, moral, and human rights philosophy (Turanjanin 2023; Lohmus 2015; O'Neil 2002; McLean 2010; Foster 2009; Marshall 2009).

It should be emphasized that certain legislations that are already in the constitutions emphasize the principles of autonomy of the will during medical procedures. For instance, Article 68 of The Constitution of the Republic of Serbia stipulates that everyone shall have the right to protect their mental and physical health, while the Code of Medical Ethics of the Medical Chamber of Serbia prescribes four principles, of which the principle of respect for autonomy and the rights of the patient is in the third place. However, we believe that the constitutional provisions are essentially too broad. Although most constitutions do not have a specific provision on vaccination, we believe that nowadays it is necessary and desirable to include provisions in the world constitutions to regulate the issue of vaccination by special laws. In this way, one of the most important health issues rises to the rank of a constitutional provision.

The problem of vaccination, however, is not quite a question of the autonomy of will. As judge Lemmens emphasized in ECHR's judgment Vavřička and Others v. the Czech Republic (2021), the right to health is a fundamental human right, while the obligation to vaccinate is one of the ways in which the authorities can fulfill their positive obligation to protect this right. When it comes to vaccination, some countries have different approaches. Moreover, some countries introduced mandatory vaccination in schools in order to prevent the spread of infectious diseases (Tomsick 2020). This is an issue when parents decide for their children; for example, their refusal to vaccinate cannot only endanger their children's health. There is also a danger to the health of other people who have not been vaccinated due to medical contraindications. It is denying their right to a healthy life.

For example, in France, vaccination of children is compulsory for three diseases, while it is recommended for eight. The prescribed fine for a breach of vaccination duty is a term of imprisonment of up to two years and a fine of up to EUR 30,000. It is important to emphasize that vaccination duty is a necessary condition for enrollment in the educational system. In the Czech Republic, Act no. 258/2000 Coll., requires all permanent residents and all foreigners authorized to reside in the country on a long-term basis to undergo a series of routine vaccinations, while for children under the age of fifteen their legal representatives are responsible for compliance with this duty. Although vaccination is a legal duty in the Czech Republic, compliance with it cannot be directly imposed. Simply, there is no provision that allows for coercive vaccination, while the compulsory nature of vaccination is carried out through the application of sanctions. It is an administrative fine that may only be imposed once. In Germany, compulsory vaccination is not the coercive administration of a vaccine—it is a duty to receive vaccination in clearly defined cases. The prescribed fine is up to EUR 2500, which may be repeated. Additionally, exclusion from educational institutions is prescribed. In Poland, vaccination against eleven diseases is mandatory for anyone residing in this country. Slovakia recognizes vaccination duty as constitutional because every country is considered to have a legal duty and positive obligation to protect public health. Although there is no mechanism for coercive vaccine application, non-vaccination is a misdemeanor punishable by a fine of up to EUR 331. In Serbia and Slovenia, we can find prescribed compulsory vaccination for several diseases. In other countries, vaccination is not compulsory, but the question is whether compulsory vaccination is useful and necessary in modern societies (Vavřička and Others v. the Czech Republic 2021).

In the research sample, a significant percentage of respondents believe that the vaccine should not be mandatory or compulsory. Additionally, the cross-sectional data shows that all respondents who believe that the vaccine should be compulsory have been vaccinated, while out of 46 respondents who believe that the COVID-19 vaccine should be mandatory, only 7 of them have not been vaccinated, but are considering doing so.

This issue is significant for criminal law. The role of criminal law can be twofold. On the one hand, the state can prescribe by law that anyone who refuses a vaccine should be punished. The law can subject someone to a fine or even imprisonment for refusing or neglecting to vaccinate. On the other hand, the state may consider submitting to vaccination as a measure to combat infectious diseases. Therefore, any failure to vaccinate constitutes a criminal offense. At this point, if some countries decide to introduce mandatory adult vaccination, they are likely to opt for the first model, which will be threatened with fines that can be repeated until the person is vaccinated. The second model, although possible, still represents an excessive amount of encroachment on a personality, and may lead to the application of coercive vaccination measures. A significantly smaller percentage of respondents believe that refusing vaccination against COVID-19 should be punishable. As illustrated in the Table 16.5, 8.3% of respondents believe that this refusal should be punishable by a fine, and an even smaller percentage (4%) that refusing vaccination should be a criminal offense. Data cross-checking indicates that all respondents who believe that refusing vaccination should be punishable by a fine and prescribed as a criminal offense have been vaccinated.

16.4 Conclusion

Research conducted in Serbia shows that the majority of citizens do not trust the vaccine against this infectious disease. All initial hypotheses are confirmed. It is not clear who is to blame for this situation. At the time of writing, 3,354,075 people have received the first dose in Serbia, 3,278,198 people have received the second dose, and only 1,902,412 residents have received the third dose. Even now, on the Ministry's official website, it states that the vaccine is a simple, safe, and effective way to protect people from diseases. However, when considering the number of people who got infected despite the vaccination, the accuracy of this statement is open to questioning, and it is clear that in the opinion of the respondents, the vaccine does not protect them from the disease, nor is it effective. In view of the number of deaths in the world caused by this virus, it is reasonable to ask whether the lesson has really been learned; in Serbia, it appears that it has not.

References

Abay ES, Belew MD, Ketsela BS, Mengistu EE, Getachew LS, Teferi YA, Zerihun AB (2022) Assessment of attitude towards COVID-19 vaccine and associated factors among clinical practitioners in Ethiopia: a cross-sectional study. PLoS ONE 17(6):e0269923. https://doi.org/10.1371/journal.pone.0269923

Alam W, Manzoor F, Farnaz N, Aktar B, Rashid SF (2022) Perception and attitudes towards COVID-19 vaccination among urban slum dwellers in Dhaka, Bangladesh. Lancet Glob Health 10:S3. https://doi.org/10.1016/S2214-109X(22)00132-2

Allington D, McAndrew S, Moxham-Hall V, Duffy B (2021) Coronavirus conspiracy suspicions, general vaccine attitudes, trust and coronavirus information source as predictors of vaccine hesitancy among UK residents during the COVID-19 pandemic. Psychol Med 53:1–12. https://doi.org/10.1017/S0033291721001434

Alrajeh AM, Daghash H, Buanz SF, Altharman HA, Belal S (2021) COVID-19 vaccine hesitancy among the adult population in Saudi Arabia. Cureus 13(12):e20197. https://doi.org/10.7759/cureus.20197

Arshad MS, Hussain I, Mahmood T, Hayat K, Majeed A, Imran I, Saeed H, Iqbal MO, Uzair M, Rehman AU, Ashraf W, Usman A, Syed SK, Akbar M, Chaudhry MO, Ramzan B, Islam M, Saleem MU, Shakeel W, Iqbal I, Hashmi F, Rasool MF (2021) A national survey to assess the COVID-19 vaccine-related conspiracy beliefs, acceptability, preference, and willingness to pay among the general population of Pakistan. Vaccines (Basel) 9(7):720. https://doi.org/10.3390/vaccines9070720

Asres F, Umeta B (2022) COVID-19 vaccines: awareness, attitude and acceptance among undergraduate University students. J Pharm Policy Pract 15:32. https://doi.org/10.1186/s40545-021-00397-6

Aw J, Seng JJB, Seah SSY, Low LL (2021) COVID-19 vaccine hesitancy—a scoping review of literature in high-income countries. Vaccines (Basel) 9(8):900. https://doi.org/10.3390/vaccines9080900

Biasio LR, Bonaccorsi G, Lorini C, Pecorelli S (2021) Assessing COVID-19 vaccine literacy: a preliminary online survey. Hum Vaccin Immunother 17:1304–1312

Bíró-Nagy A, Szászi ÁJ (2022) The roots of COVID-19 vaccine hesitancy: evidence from Hungary. J Behav Med 46:1–16. https://doi.org/10.1007/s10865-022-00314-5

Biswas MR, Alzubaidi MS, Shah U, Abd-Alrazaq AA, Shah Z (2021) A scoping review to find out worldwide COVID-19 vaccine hesitancy and its underlying determinants. Vaccines (Basel) 9(11):1243. https://doi.org/10.3390/vaccines9111243

Chen F, Liu Y, Ya B, He J, Leng T, Sun Y, Xin Q (2021) Latest effective measures to combat COVID-19: a review. Iran J Public Health 50(4):640–648

Danabal, KGM, Magesh, SS, Saravanan, S et al (2021) Attitude towards COVID 19 vaccines and vaccine hesitancy in urban and rural communities in Tamil Nadu, India—a community based survey. BMC Health Serv Res 21:994. https://doi.org/10.1186/s12913-021-07037-4

Dror AA, Eisenbach N, Taiber S, Morozov NG, Mizrachi M, Zigron A, Srouji S, Sela E (2020) Vaccine hesitancy: the next challenge in the fight against COVID-19. Eur J Epidemiol 35(8):775–779

Dubé E, Vivion M, MacDonald NE (2015) Vaccine hesitancy, vaccine refusal and the anti-vaccine movement: influence, impact and implications. Expert Rev Vaccines 14(1):99–117. https://doi.org/10.1586/14760584.2015.964212

Eberhardt J, Ling J (2021) Predicting COVID-19 vaccination intention using protection motivation theory and conspiracy beliefs. Vaccine 39(42):6269–6275. https://doi.org/10.1016/j.vaccine.2021.09.010

Enticott J, Singh Gill J, Bacon S, Lavoie KL, Epstein D, Dawadi S, Teede H, Boyle J (2022) Attitudes towards vaccines and intention to vaccinate against COVID-19: a cross-sectional analysis—implications for public health communications in Australia. BMJ Open 12:e057127. https://doi.org/10.1136/bmjopen-2021-057127

Farhart CE, Douglas-Durham E, Lunz Trujillo K, Vitriol JA (2022) Vax attacks: how conspiracy theory belief undermines vaccine support. Prog Mol Biol Transl Sci 188(1):135–169. https://doi.org/10.1016/bs.pmbts.2021.11.001

Farhud D, Zokaei S (2020) Fight against viruses (COVID-19): peace among nations. Iran J Public Health 49(Suppl. 1):1–3

Fernandes N, Costa D, Costa D, Keating J, Arantes J (2021) Predicting COVID-19 vaccination intention: the determinants of vaccine hesitancy. Vaccines (Basel) 9(10):1161. https://doi.org/10.3390/vaccines9101161

Fisher KA, Bloomstone SJ, Walder J, Crawford S, Fouayzi H, Mazor KM (2020) Attitudes toward a potential SARS-CoV-2 vaccine: a survey of U.S. adults. Ann Intern Med 173(12):964–973. https://doi.org/10.7326/M20-3569

Foster C (2009) Choosing life, choosing death: the Tyranny of autonomy in medical ethics and law. Hart Publishing, Oxford

Freeman D, Loe BS, Yu LM, Freeman J, Chadwick A, Vaccari C, Shanyinde M, Harris V, Waite F, Rosebrock L, Petit A, Vanderslott S, Lewandowsky S, Larkin M, Innocenti S, Pollard AJ, McShane H, Lambe S (2021) Effects of different types of written vaccination information on COVID-19 vaccine hesitancy in the UK (OCEANS-III): a single-blind, parallel-group, randomised controlled trial. Lancet Publ Health 6(6):e416–e427. https://doi.org/10.1016/S2468-2667(21)00096-7

Freeman D, Loe BS, Chadwick A, Vaccari C, Waite F, Rosebrock L, Jenner L, Petit A, Lewandowsky S, Vanderslott S, Innocenti S, Larkin M, Giubilini A, Yu LM, McShane H, Pollard AJ, Lambe S (2022) COVID-19 vaccine hesitancy in the UK: the Oxford coronavirus explanations, attitudes, and narratives survey (Oceans) II. Psychol Med 52(14):3127–3141. https://doi.org/10.1017/S0033291720005188

Gao J, Raza SH, Yousaf M, Shah AA, Hussain I, Malik A (2023) How does digital media search for COVID-19 influence vaccine hesitancy? Exploring the trade-off between google trends, infodemics, conspiracy beliefs and religious fatalism. Vaccines (Basel) 11(1):114. https://doi.org/10.3390/vaccines11010114

Giubilini A (2020) An argument for compulsory vaccination: the taxation analogy. J Appl Philos 37(3):446–466

Ishmaev G, Dennis M, Jeroen van den Hoven M (2021) Ethics in the COVID-19 pandemic: myths, false dilemmas, and moral overload. Ethics Inf Technol. https://doi.org/10.1007/s10676-020-09568-6

Jennings W, Stoker G, Bunting H, Valgarðsson VO, Gaskell J, Devine D, McKay L, Mills MC (2021) Lack of trust, conspiracy beliefs, and social media use predict COVID-19 vaccine hesitancy. Vaccines (Basel) 9(6):593. https://doi.org/10.3390/vaccines9060593

Khan Z, Khursheed SQ, Dar SA, Shah NN, Reagu S, Alabdulla M, Haq I, Ud Din Azad AM, Dar KA, Farooq SS, Wani ZA (2022) Vaccine hesitancy and coronavirus disease-19: where do we stand? J Educ Health Promot 11:59. https://doi.org/10.4103/jehp.jehp_642_21

Kishore J, Venkatesh U, Ghai G, Heena, Kumar P (2021) Perception and attitude towards COVID-19 vaccination: a preliminary online survey from India. J Family Med Prim Care 10(8):3116–3121. https://doi.org/10.4103/jfmpc.jfmpc_2530_20

Larson HJ, de Figueiredo A, Karafillakis E, Rawal M (2018) Report of the European commission on the state of vaccine confidence in 2018. Eur J Publ Health 29(Suppl. 4):ckz184. https://doi.org/10.1093/eurpub/ckz184

Lohmus K (2015) Caring autonomy: European human rights law and the challenge of individualism. Cambridge University Press, Cambridge

Machida M, Nakamura I, Kojima T, Saito R, Nakaya T, Hanibuchi T, Takamiya T, Odagiri Y, Fukushima N, Kikuchi H, Amagasa S, Watanabe H, Inoue S (2021) Acceptance of a COVID-19 vaccine in Japan during the COVID-19 pandemic. Vaccines (Basel) 9(3):210. https://doi.org/10.3390/vaccines9030210

Maftei A, Holman AC (2021) SARS-CoV-2 threat perception and willingness to vaccinate: the mediating role of conspiracy beliefs. Front Psychol 19(12):672634. https://doi.org/10.3389/fpsyg.2021.672634

Marshall J (2009) Personal freedom through human rights law?: Autonomy, identity and integrity under the European convention on human rights. Martinus Nijhoff Publishers, Leiden, Boston

McLean SAM (2010) Autonomy, consent and the law. Routledge-Cavendish, Abingdon

Mohamad O, Zamlout A, AlKhoury N, Mazloum AA, Alsalkini M, Shaaban R (2021) Factors associated with the intention of Syrian adult population to accept COVID19 vaccination: a cross-sectional study. BMC Publ Health 21(1):1310. https://doi.org/10.1186/s12889-021-11361-z

Mukhtar M, Hussain M, Tariq N, Mukhtar R (2021) Evaluation and comparison of awareness and practice of protective measures for COVID-19 between preclinical and clinical medical students in developing country: Pakistan. Techn Soc Sci J 17:104–110

Neumann-Böhme S, Elsem Varghese N, Sabat I, Pita Barros P, Brouwer W, van Exel J, Schreyögg J, Stargardt T (2020) Once we have it, will we use it? A European survey on willingness to be vaccinated against COVID-19. Eur J Health Econ 21(7):977–982

O'Neil O (2002) Autonomy and trust in bioethics. Cambridge University Press, Cambridge

Quintana AF, Quintana MR (2021) Surviving the COVID-19 pandemic of 2020: a constitutional and policy review of involuntary medical quarantine. Wake Forest J Law Policy 11(2):327–394

Rahman MM, Chisty MA, Alam MA, Sakib MS, Quader MA, Shobuj IA et al (2022) Knowledge, attitude, and hesitancy towards COVID-19 vaccine among university students of Bangladesh. PLoS ONE 17(6):e0270684. https://doi.org/10.1371/journal.pone.0270684

Raza SH, Yousaf M, Zaman U, Waheed Khan S, Core R, Malik A (2023) Unlocking infodemics and mysteries in COVID-19 vaccine hesitancy: nexus of conspiracy beliefs, digital informational support, psychological well-being, and religious fatalism. Vaccine 41(10):1703–1715. https://doi.org/10.1016/j.vaccine.2023.01.047

Reiter PL, Pennell ML, Katz ML (2020) Acceptability of a COVID-19 vaccine among adults in the United States: how many people would get vaccinated? Vaccine 38(42):6500–6507. https://doi.org/10.1016/j.vaccine.2020.08.043

Richardson E, Devine C (2020) Emergencies end eventually: how to better analyze human rights restrictions sparked by the COVID-19 pandemic under the international covenant on civil and political rights. Mich J Int Law 42(1):105–176

Salali GD, Uysal MS (2020) COVID-19 vaccine hesitancy is associated with beliefs on the origin of the novel coronavirus in the UK and Turkey. Psychol Med pp 1–3. https://doi.org/10.1017/S0033291720004067

Salazar-Fernández C, Baeza-Rivera MJ, Villanueva M, Bautista JAP, Navarro RM, Pino M (2022) Predictors of COVID-19 vaccine intention: evidence from Chile, Mexico, and Colombia. Vaccines (Basel) 10(7):1129. https://doi.org/10.3390/vaccines10071129

Sallam M, Dababseh D, Eid H, Hasan H, Taim D, Al-Mahzoum K, Al-Haidar A, Yaseen A, Ababneh NA, Assaf A, Bakri FG, Matar S, Mahafzah A (2021a) Low COVID-19 vaccine acceptance is correlated with conspiracy beliefs among University students in Jordan. Int J Environ Res Public Health 18(5):2407. https://doi.org/10.3390/ijerph18052407

Sallam M, Dababseh D, Eid H, Al-Mahzoum K, Al-Haidar A, Taim D, Yaseen A, Ababneh NA, Bakri FG, Mahafzah A (2021b) High rates of COVID-19 vaccine hesitancy and its association with conspiracy beliefs: a study in Jordan and Kuwait among other Arab Countries. Vaccines (Basel) 9(1):42. https://doi.org/10.3390/vaccines9010042

Salman M, Mallhi TH, Tanveer N, Shehzadi N, Khan HM, Ul Mustafa Z, Khan TM, Hussain K, Mohamed MS, Maqbool F, Aftab RA, Butt MH, Panda DS,

Alotaibi NH, Khedr AIM, Alanazi AS, Alatawi AD, Alzarea AI, Sulatana K, Khan YH (2022) Evaluation of conspiracy beliefs, vaccine hesitancy, and willingness to pay towards COVID-19 vaccines in six countries from Asian and African regions: a large multinational analysis. Vaccines (Basel) 10(11):1866. https://doi.org/10.3390/vaccines10111866

Savulescu J (2021) Good reasons to vaccinate: mandatory or payment for risk? J Med Ethics 47(2):78–85

Schermer M (2002) The different faces of autonomy: patient autonomy in ethical theory and hospital practice. Springer-Science+Business Media, B.V., Dordrecht

Sheikhshahrokh A, Ranjbar R, Saeidi E, Sa-Farpoor Dehkordi F, Heiat M, Ghasemi-Dehkordi P, Goodarzi H (2020) Frontier therapeutics and vaccine strategies for SARS-CoV-2 (COVID-19): a review. Iran J Public Health 49(Suppl. 1):18–29

Simione L, Vagni M, Gnagnarella C, Bersani G, Pajardi D (2021) Mistrust and beliefs in conspiracy theories differently mediate the effects of psychological factors on propensity for COVID-19 vaccine. Front Psychol 7(12):683684. https://doi.org/10.3389/fpsyg.2021.683684

Stickle B, Felson M (2020) Crime rates in a pandemic: the largest criminological experiment in history. Am J Crim Justice 45:525–536

Subramaniam M, Abdin E, Shafie S, Shahwan S, Zhang Y, Satghare P, Devi F, Lun P, Yuxuan Ni M, Chong SA (2023) Sociodemographic and psychosocial factors associated with vaccine hesitancy - results from a longitudinal study in Singapore. Hum Vaccin Immunother 19(2):2235964. https://doi.org/10.1080/21645515.2023.2235964

Taherizadeh M, Tabibzadeh A, Panahi M, Safarnezhad Tameshkel F, Golahdooz M, Karbalaie Niya MH (2020) An introduction to SARS coronavirus 2; comparative analysis with MERS and SARS coronaviruses: a brief review. Iran J Public Health 49(Suppl 1):30–37

Todorović Z, Protić D (2017) Patient autonomy and informed consent in critically Ill. Hosp Pharm 4(1):463–468

Tomsick E (2020) The public health demand for revoking non-medical exemptions to compulsory vaccination statutes. J Law Health 34(1):130–156

Turanjanin V (2021a) Unforeseeability and abuse of criminal law during the COVID-19 pandemic in Serbia. EU Compar Law Issues Ser 5:223–246

Turanjanin V (2021b) Social implications caused by state reaction on COVID-19 and human rights in Republic of Serbia. Teme 45:1081–1096

Turanjanin V (2022) Migrants and safety in Serbia during and after coronavirus pandemic. In: EU and comparative law issues and series, pp 410–429

Turanjanin V (2023) Patient autonomy and criminal law: a Serbian perspective. In: Patient autonomy and criminal law. Taylor and Francis, pp 285–301

Turanjanin V, Radulović D (2020) Coronavirus (Covid-19) and possibilities for criminal law reaction in Europe: a review. Iran J Public Health 49(Suppl. 1):4–11

Vavřička and Others v. the Czech Republic (nos. 47621/13 and 5 others), 8 Apr 2021

Velásquez G (2022) Vaccines, medicines and COVID-19: how can WHO be given a stringer voice. Springer

Wawrzuta D, Jaworski M, Gotlib J, Panczyk M (2021) What arguments against COVID-19 vaccines run on Facebook in Poland: content analysis of comments. Vaccines 9:481

WHO (2013) Global Vaccine Action Plan 2011–2020. World Health Organisation

Zhang X, Guo Y, Zhou Q, Tan Z, Cao J (2021) The mediating roles of medical mistrust, knowledge, confidence and complacency of vaccines in the pathways from conspiracy beliefs to vaccine hesitancy. Vaccines (Basel) 9(11):1342. https://doi.org/10.3390/vaccines9111342

Zen F (2020) Why the COVID-19 vaccine should be mandatory. The Jakarta Post. Available: https://www.thejakartapost.com/academia/2020/09/25/why-the-covid-19-vaccine-should-be-mandatory.html. Accessed 1 Oct 2020

The Worrying Phenomenon of COVID-19 Vaccine Hesitancy and Its Negative Impact on Pandemic Control Efforts: Common Themes that Emerged in the Middle East and North Africa (MENA) Region

17

Azmi Mahafzah, Malik Sallam, Faris G. Bakri, and Mohammad S. Mubarak

Abstract

Since the declaration of coronavirus disease 2019 (COVID-19) as a pandemic, intensive measures were taken to mitigate its negative health, psychological, social, and economic impact. COVID-19 continues to pose serious threats globally, with vaccination as the central safe strategy to control the pandemic. However, COVID-19 vaccine hesitancy is a major concern, especially in the Middle East and North Africa (MENA). Concerns regarding vaccine safety, efficacy, and misinformation contribute to vaccine hesitancy. Addressing these concerns and providing accurate information is crucial for increasing COVID-19 vaccine acceptance and uptake in this region, where the coverage is low. Variable rates of COVID-19 vaccine hesitancy were found in the numerous studies conducted in the region. Complex factors contributed to vaccination hesitancy in the region including concerns about COVID-19 vaccine safety and efficacy, low trust in healthcare systems, complacency toward the risks of COVID-19, constraints hindering access to COVID-19 vaccination services, as well as the circulation of misinformation and conspiracy beliefs about COVID-19 and its vaccination. Effective approaches to address COVID-19 vaccine hesitancy in the MENA region rely on developing evidence-based communication strategies that are recommended to build trust in vaccination, highlight the disease risks, and counter COVID-19 vaccine-related misinformation. Ensuring COVID-19 vaccine affordability is also necessary besides the cautious consideration of implementing COVID-19 vaccine mandates. Based on the preceding discussion, this chapter aims to identify the common themes of COVID-19 vaccine hesitancy in the MENA region. In addition, the chapter

A. Mahafzah (✉) · M. Sallam
Department of Pathology, Microbiology, and Forensic Medicine, School of Medicine, The University of Jordan, Amman 11942, Jordan
e-mail: mahafzaa@ju.edu.jo

A. Mahafzah · M. Sallam
Department of Clinical Laboratories and Forensic Medicine, Jordan University Hospital, Amman 11942, Jordan

F. G. Bakri
Department of Internal Medicine, School of Medicine, The University of Jordan, Amman 11942, Jordan

F. G. Bakri
Infectious Diseases and Vaccine Center, The University of Jordan, Amman 11942, Jordan

M. S. Mubarak (✉)
Department of Chemistry, School of Science, The University of Jordan, Amman 11942, Jordan
e-mail: mmubarak@ju.edu.jo

N. Rezaei (ed.), *The COVID-19 Aftermath*, Advances in Experimental Medicine and Biology 1457,
https://doi.org/10.1007/978-3-031-61939-7_17

highlights the importance of understanding the root causes of COVID-19 vaccination hesitancy and its associated determinants to develop effective strategies for promoting COVID-19 vaccine acceptance and uptake in the MENA region. To build community trust, promote community education and awareness, and counter misinformation for better COVID-19 vaccine coverage in the region, it is recommended to involve healthcare professionals and policymakers.

Keywords

Prevention · Vaccine resistance · Vaccine rejection · Vaccine hesitation

17.1 Introduction

Since its first reporting as a pandemic in March 2020, coronavirus disease 2019 (COVID-19) has continued to pose serious threats to different communities around the world (Cucinotta and Vanelli 2020; Dhama et al. 2020; Nature Editorial 2023). The prompt development and distribution of effective vaccines following the characterization of severe acute respiratory syndrome coronavirus 2 (SARS-CoV-2) have been regarded as the most promising strategy to control the negative impact of the pandemic (Kumari et al. 2022; Kyriakidis et al. 2021; Pascual-Iglesias et al. 2021). However, the success of any vaccination strategy in general, and COVID-19 vaccination, depends on the widespread community engagement and uptake of vaccines (Burgess et al. 2021; Larson et al. 2022; Brynne et al. 2022). Subsequently, COVID-19 vaccine hesitancy/resistance emerged as a worrying phenomenon during the pandemic (Sallam et al. 2022a; Sallam 2021; Fajar et al. 2022; Abdelmoneim et al. 2022). The challenge of COVID-19 vaccination hesitancy was particularly evident in the MENA region and could undermine the ability to control the pandemic and hinder the efforts to mitigate its negative health, economic, psychological, and social impacts (Troiano and Nardi 2021; Sallam 2021; Sallam et al. 2022a).

Arab countries in the MENA region experienced significant challenges in the administration of COVID-19 vaccines due to several factors (Galal et al. 2022). However, the emergence of COVID-19 vaccine hesitancy/resistance as a significant barrier to vaccine uptake posed a unique and concerning challenge in the region, which is among the least world regions in COVID-19 vaccine coverage rates (Sallam et al. 2022a; The New York Times 2023). In this respect, the Arab countries of the MENA region include Algeria, Bahrain, Djibouti, Egypt, Iraq, Jordan, Kuwait, Lebanon, Libya, Morocco, Oman, Palestine, Qatar, Saudi Arabia, Sudan, Syria, Tunisia, the United Arabic Emirates (UAE), and Yemen (Galal et al. 2022; Sallam et al. 2021a). Due to the shared language, cultural, sociopolitical, and religious perspectives in the region, the study of different phenomena in the countries of the region as a single unit can be beneficial to unravel common challenges faced by its populations (Sallam et al. 2017; Athamneh et al. 2021).

Worldwide, several common themes that contributed to the phenomenon of COVID-19 vaccination hesitancy have emerged (Lin et al. 2020; Norhayati et al. 2021; Ackah et al. 2022; Perrone et al. 2023). These themes include safety concerns, efficacy concerns, and conspiracy theories, besides the influence of political and religious beliefs (Pertwee et al. 2022; Lazarus et al. 2021, 2023); these were also reported in the MENA region. Within this context, safety concerns have been a significant contributor to vaccine hesitancy in the MENA region like other regions in the world (Kacimi et al. 2022; Sallam et al. 2021b). In addition, the rapid development and emergency use of COVID-19 vaccines may have led to concerns about the long-term safety of these vaccines along with the perception that the vaccines have not been sufficiently tested before being authorized (Kashte et al. 2021; Cole et al. 2022). Thus, the expression of safety concerns was manifested clearly regarding the potential side effects of these vaccines, and people from their perspective were hesitant to receive a vaccine that has not undergone extensive testing (Holzmann-Littig et al. 2022; Siu et al. 2022; Fieselmann et al. 2022). Efficacy concerns have also contributed to vaccine hesitancy in the MENA region similar to other regions (Lazarus et al. 2022b). This issue intensified amid the emergence of new variants of the virus, leading to uncertainty about the effectiveness of the vaccines against these newly emerging variants (Singh et al. 2021). Subsequently, this issue may have contributed to questions regarding the benefits of COVID-19 vaccination from the efficacy perspective (Fiske et al. 2022; Mangla et al. 2021).

Conspiracy theories and political and religious beliefs have also contributed to vaccine hesitancy in the MENA region (Sallam et al. 2021b; Qunaibi et al. 2021a, b). In this respect, misinformation or misconceptions about SARS-CoV-2 and its vaccines on social media and other platforms have contributed to concerns about the safety and efficacy of the vaccines (Sallam et al. 2020; Freeman et al. 2022). This issue included the unsubstantiated claims that the vaccines are introduced for surveillance purposes or to cause sterility to reduce the size of the global population (Islam et al. 2021; Sallam et al. 2021b). Additionally, political ideologies and religious beliefs were reported to play a significant role in shaping individuals' perceptions of vaccination with contribution to their attitudes toward vaccination (Alsuwaidi et al. 2023; Banerjee et al. 2022; Debus and Tosun 2021). In some cases, political or religious leaders expressed skepticism or even opposition to vaccines, leading to concerns and hesitancy among their followers or the people they influence, while the leading figures who spread messages supporting vaccination positively influenced their followers causing more support for vaccination (Albrecht 2022; Gaweł et al. 2021). For example, some religious leaders expressed concerns about the compatibility of different vaccines with Islamic beliefs including the "*halal*" status, contributing to vaccine hesitancy among some communities (Padmawati et al. 2019; Mardian et al. 2021). Therefore, it is critical to address the aforementioned concerns, take steps to combat misinformation, and engage with communities to increase vaccine acceptance and uptake which can subsequently help to provide safe population immunity, thus lessening the negative impact of the pandemic (Betsch et al. 2022). Furthermore, the scientific community must work with local governments, health organizations, and community leaders to provide accurate and accessible information about the safety and efficacy of vaccines (Schoch-Spana et al. 2021). In the MENA region, this information must be tailored to address the specific concerns and beliefs taking into account the specific peculiarities of these communities, since this region is considered one of the lowest vaccine uptake and highest vaccination hesitancy/rejection worldwide (Abdelmoneim et al. 2022; Sallam 2021; Sallam et al. 2022a; The New York Times 2023).

Based on the preceding discussion, this book chapter aims to elucidate the common themes that emerged in the MENA region in the context of COVID-19 vaccine hesitancy. Given the concerns about the negative impact that vaccine hesitancy could have on the ability to control the COVID-19 pandemic and protect the health of communities worldwide; this aim appears timely and relevant. In addition, this chapter literature deals with COVID-19 vaccine hesitancy in the MENA region. In this regard, different stakeholders must collaborate efficiently to overcome the challenges of COVID-19 vaccine hesitancy and ensure that the benefits of the vaccine are accessible to different populations. Building trust in vaccination can help to protect the communities from the negative consequences of the pandemic.

17.2 The Definition of Vaccine Hesitancy: Understanding a Complex Phenomenon to Help Improve the Global Health

Vaccines can be viewed as one of the most outstanding achievements of science and it is the safest and most effective tool in the arsenal used to prevent infectious diseases and reduce their associated morbidity and mortality (Andre et al. 2008; Amanna and Slifka 2020). However, vaccine hesitancy has emerged as a complex and multifaceted phenomenon that threatens to undermine the efforts to control and prevent the spread of vaccine-preventable diseases (VPDs), despite the proven efficacy and safety of the authorized vaccines (Galagali et al. 2022; MacDonald 2015). Thus, to tackle the threats associated with this phenomenon, it is crucial to clearly define and understand vaccine hesitancy which would help to develop strategies that promote vaccine acceptance and uptake (Peters 2022).

Vaccine hesitancy can simply be defined as the delay in the acceptance of vaccines despite the availability of vaccination services (MacDonald 2015). However, this simplified definition should be interpreted in light of the complexity that vaccine hesitancy poses, which entails a wide range of beliefs, attitudes, and behaviors that are shared by a large and diverse portion of the population based on the type of vaccine, target population, and health behavior among other aspects (Peretti-Watel et al. 2015). The decision to get vaccinated is also influenced by psychological factors such as complacency toward the disease risks, convenience in the vaccination experience, confidence in health professionals and policymakers, as well as in the pharmaceutical industry, and a range of sociocultural, political, and economic factors (Dubé et al. 2014; Betsch et al. 2018; Cornelia et al. 2020). Vaccine hesitancy is not a new phenomenon encountered during COVID-19; there is evidence that this phenomenon emerged as early as the first days of vaccination inception (Schwartz 2012). However, vaccine hesitancy has gained new relevance and momentum in the era of COVID-19, where vaccination was essential to control the pandemic, which was also described as an infodemic with the spread of misinformation faster than the virus (Wiysonge et al. 2022; Ferreira Caceres et al. 2022; Depoux et al. 2020).

The significance of investigating vaccine hesitancy is related to its profound and far-reaching consequences (Pandey and Galvani 2023; Nguyen et al. 2022). Vaccination hesitancy or complete resistance to vaccination can lead to decreased rates of vaccine coverage, which could result in VPDs outbreaks, increased morbidity and mortality from infectious disease, and a larger strain on healthcare systems (Rawlings et al. 2022; Anderson et al. 2020). Furthermore, vaccine hesitancy is a global issue that affects all countries and regions, regardless of income or development status; nevertheless, it is more pronounced in certain world regions including the MENA (Frenkel 2021; Nuwarda et al. 2022; Sallam 2021; Sallam et al. 2022a). Thus, it is important to promote a deeper understanding of vaccine hesitancy and its determinants, and to identify and disseminate evidence-based messages to address this complex issue through well-designed research with valid results (Singh et al. 2022). This includes addressing the fine granularity of vaccine hesitancy as an initial step and engaging the stakeholders including healthcare providers, policymakers, and community leaders to understand the root causes of vaccine hesitancy and develop effective strategies to promote vaccine acceptance and uptake (Tuckerman et al. 2022; Larson 2022; Karafillakis et al. 2022).

Addressing vaccine hesitancy requires a multifaceted approach that considers the unique sociocultural, economic, and political contexts of different regions and populations (Hunter et al. 2021; Strully et al. 2021). It also requires building trust, promoting education, and addressing misconceptions and misinformation about vaccines in different communities based on accumulating evidence regarding the factors related to vaccine hesitancy/resistance (Shen and Dubey 2019). Furthermore, addressing the root causes of vaccine hesitancy, such as the psychological, social, and economic determinants of health is necessary to create more equitable vaccine distribution and uptake with a subsequent positive impact on public health (Dada et al. 2022; Rawlings et al. 2022). In this regard, the complexity and multifaceted nature of vaccine hesitancy require rigorous concerted efforts from healthcare providers, policymakers, and the scientific community to identify and disseminate evidence-based strategies to promote vaccine acceptance and uptake (Lazarus et al. 2022a). Thus, the study of vaccination hesitancy can play an important role in promoting global health and preventing the spread of infectious diseases.

17.3 The Investigation of Vaccine Hesitancy and Its Associated Determinants: The Necessity for Evidence-Based Strategies to Promote Vaccine Acceptance and Uptake

As stated earlier, vaccine hesitancy is a complex and multifaceted phenomenon that entails decision-making based on several factors (Cerda and García 2021). It is important to recognize the urgent need to investigate and understand the determinants of vaccine hesitancy, to develop well-informed effective strategies that could help to promote vaccine acceptance and uptake (Kumar et al. 2016). Thus, investigation of vaccine hesitancy requires a multidisciplinary approach that considers the unique psychological, sociocultural, economic, and political contexts of different regions and populations (Nuwarda et al. 2022). It requires robustly designed studies to understand the root causes of vaccine hesitancy as an initial step, followed by proper follow-up to identify evidence-based strategies to address the concerns correlated with vaccination hesitancy/resistance (Bussink-Voorend et al. 2022).

A range of determinants have been associated with vaccine hesitancy, including sociodemographic characteristics, psychological factors, perceived susceptibility and efficacy, benefits, and cues to action as framed by the health belief model (Patwary et al. 2021; Schmid et al. 2017). Additionally, vaccine hesitancy is influenced by a range of sociocultural, political, and economic factors, including trust in healthcare providers and institutions, historical experiences of discrimination and mistreatment, and political and ideological beliefs including the embrace of conspiratorial ideas and compliance with health regulations as framed in the 7C model (Geiger et al. 2021). Therefore, a deep understanding of the factors linked to willingness to get vaccinated is a prerequisite to informing future vaccine promotion strategies (Wang et al. 2021). Analyzing vaccination hesitancy and its associated factors can help to design well-informed interventions and educational measures aimed at promoting vaccination, which can lead to achieving safe population immunity and proper control of infectious diseases (Soares et al. 2021). The accurate depiction of the scope of vaccination hesitancy is the first critical step in addressing this widespread and detrimental phenomenon (Kumar et al. 2016).

Accurate characterization of the factors associated with vaccine hesitancy can help to develop cost-effective evidence-based tools tailored to the needs of vaccine-hesitant populations, such as parents and health professionals (Larson 2022). Therefore, the significant role of developing and implementing validated tools to investigate vaccine hesitancy cannot be overstated (Larson et al. 2015). Currently, there are several valid survey instrument tools commonly used to investigate vaccine hesitancy (Oduwole et al. 2022). Some of these tools include: (a) Vaccine Hesitancy Scale (VHS): This is a validated scale developed by the WHO-SAGE Working Group on Vaccines to measure vaccine hesitancy. It assesses factors such as vaccine confidence (Larson et al. 2015; Luyten et al. 2019). (b) Vaccination Attitudes Examination (VAX) Scale: A short and simple survey tool to assess the intentions for vaccination (Martin and Petrie 2017). (c) Parent Attitudes About Childhood Vaccines (PACV): This survey tool is used to measure parental attitudes toward childhood vaccinations. It evaluates factors such as vaccine safety, efficacy, and perceived risks (Opel et al. 2011). (d) The Health Belief Model (HBM): This model is used to understand health-related behaviors, including vaccine hesitancy. It assesses factors such as perceived susceptibility, perceived severity, perceived benefits, and perceived barriers to vaccination (Jones et al. 2015). (e) The Vaccine Conspiracy

Beliefs Scale (VCBS): A tool used to measure the extent to which individuals believe in conspiracy theories related to vaccines developed initially to test the conspiratorial attitude in the context of human papillomavirus vaccination (Shapiro et al. 2016).

These survey instrument tools have been used in various studies and contexts to investigate vaccine hesitancy and to identify the possible factors that could contribute to low vaccine acceptance and uptake (Oduwole et al. 2022). These tools are valuable for understanding the scope of vaccine hesitancy which subsequently can aid in developing evidence-based interventions to address this complex issue with negative health consequences (Bussink-Voorend et al. 2022; MacDonald 2015). During the COVID-19 pandemic, several survey instrument tools including the aforementioned validated tools were adopted to investigate vaccine hesitancy (Oduwole et al. 2022). These tools assessed factors such as the perceived vaccine safety, efficacy, access to vaccination services and constraints challenging this access, and the trust in healthcare providers and institutions (Oduwole et al. 2022; Larson et al. 2022). Consequently, these tools have been instrumental in developing evidence-based interventions to address vaccine hesitancy and to promote COVID-19 vaccine acceptance and uptake (Marzo et al. 2022; Sallam 2021; Poland and Ratishvili 2022; Tuckerman et al. 2022; Omer et al. 2021).

17.4 COVID-19 Vaccine Hesitancy as a Notable Phenomenon: Understanding Its Implications as a Threat to the Global Health

The development of highly effective COVID-19 vaccines has been a remarkable achievement in the global efforts to control the pandemic (Tregoning et al. 2021). However, vaccine hesitancy emerged as a significant obstacle to achieving safe population immunity and reducing the morbidity and mortality associated with the disease (Gerretsen et al. 2021). The impact of COVID-19 vaccine hesitancy on global health underscores the need to fully understand its implications (Gerretsen et al. 2021; Kricorian et al. 2022). COVID-19 vaccine hesitancy has been investigated extensively and results pointed to the complex interplay of several factors in determining the willingness to get COVID-19 vaccination (Piltch-Loeb et al. 2022). A recent review compiled the rates of COVID-19 vaccine acceptance from numerous survey-based studies conducted in 114 countries/territories worldwide (Sallam et al. 2022a). Per region, the highest acceptance rates were reported in Ethiopia (92%) in East and Southern Africa, Niger (93%) in West/Central Africa, Nepal, and Vietnam (97%) in Asia and the Pacific, Montenegro (69%) and Kazakhstan (64%) in Eastern Europe/Central Asia, Mexico (88%) in Latin America and the Caribbean, Tunisia (92%) in the MENA region, and Canada (91%) and Norway (89%) in Western/Central Europe and North America (Urrunaga-Pastor et al. 2021; Sallam et al. 2022a).

On the other hand, the lowest COVID-19 vaccine acceptance rates per region were reported in Zimbabwe in East and Southern Africa (50%), Cameroon in West/Central Africa (15%), Hong Kong in Asia and the Pacific (42%), Russia in Eastern Europe/Central Asia (30%), Haiti in Latin America and the Caribbean (43%), Iraq in the MENA region (13%), and Cyprus and Portugal in Western/Central Europe and North America (35%) (Sallam et al. 2022a; Urrunaga-Pastor et al. 2021). COVID-19 vaccine acceptance rates $\geq 60\%$ were observed in 72/114 countries/territories worldwide, with vaccine hesitancy being more pronounced as a phenomenon in the MENA region, Europe and Central Asia, and Western/Central Africa, highlighting the need for further studies in these regions to understand vaccine acceptance intentions (Sallam et al. 2022a). Additionally, a recent meta-analysis pointed to a global prevalence of COVID-19 vaccine hesitancy/rejection at a rate of 25% highlighting the common occurrence of this worrying phenomenon (Fajar et al. 2022).

COVID-19 vaccine hesitancy has been found to vary based on region, age, and socioeconomic status (Lee and Huang 2022; Wang et al. 2023; Piltch-Loeb et al. 2022). The reasons for COVID-19 vaccine hesitancy and rejection were reported to be complex and multifaceted, and evidence from studies conducted globally has shown that the rapid development and emergency use authorization of COVID-19 vaccines have caused concerns about the safety and efficacy of these vaccines highlighting the role of confidence as a major determinant for COVID-19 vaccine acceptance (Ghare et al. 2023; El-Elimat et al. 2021; Sapienza and Falcone 2022). Moreover, the rapid spread of misinformation and conspiracy theories related to COVID-19 vaccines has also been one of the major factors contributing to COVID-19 vaccine hesitancy (Sallam et al. 2021b; Zimmerman et al. 2023). The proliferation of social media platforms besides the reliance on these sources for vaccine information and the ease by which information can be disseminated through these channels significantly contributed to the widespread circulation of false claims and unconfirmed information about COVID-19 vaccines (Skafle et al. 2022; Ngai et al. 2022). These claims ranged from unsubstantiated allegations questioning the vaccines' efficacy and safety to bizarre conspiracy theories alleging that the vaccines are part of a malicious plot to control the population sizes or to implant microchips into individuals who receive them for surveillance purposes (Sallam et al. 2021b; Wonodi et al. 2022). These claims can appeal to the existing biases and fears among people, and harmful notions can spread quickly through social networks, leading to confusion and mistrust among people toward vaccines (Cascini et al. 2022; Dobson 2021). Moreover, social media algorithms could end up amplifying and reinforcing these messages serving as echo chambers and making it difficult for individuals to distinguish between legitimate and false information (Gisondi et al. 2022; Wang et al. 2022).

The impact of this misinformation on COVID-19 vaccine hesitancy has been well-documented, with studies showing that exposure to false claims and conspiracy theories can end up reducing the COVID-19 vaccine acceptance rates (Lee et al. 2022; Al-Sanafi and Sallam 2021). Moreover, such vaccine misinformation can undermine the public confidence in the healthcare systems and contribute to broader mistrust in health professionals and policymakers issuing guidelines to protect public health (Garett and Young 2021; Marco-Franco et al. 2021). Given the scale and complexity of the challenges posed by vaccine misinformation and conspiracy theories, addressing COVID-19 vaccine hesitancy requires sustained efforts from public health officials, healthcare providers, and the scientific community to delineate the scope of this problem and provide proper interventions to tackle such a harmful issue (Knudsen et al. 2023). These efforts should involve targeted education and outreach campaigns that should focus on building trust and confidence in COVID-19 vaccines and correcting false claims and misinformation (Cerda and García 2021). Vaccine promotion efforts also require coordination to combat the online spread of misinformation and conspiracy theories, including the importance of collaboration with social media platforms and digital agencies to develop strategies for identifying and removing false claims and promoting accurate information (Joseph et al. 2022; Germani et al. 2022). By addressing these challenges posed by misinformation and conspiracy theories, it may be possible to promote COVID-19 vaccine acceptance rates and achieve the goal of safe population immunity against the virus (Ullah et al. 2021).

Equitable and easy access to vaccines has also been a significant factor driving COVID-19 vaccine acceptance, with barriers such as cost, transportation, and availability hindering vaccination efforts (Wouters et al. 2021). Additionally, disparities in vaccine access and uptake based on socioeconomic status, geographic location, or racial background have been observed contributing to lower COVID-19 vaccine uptake (Shearn and Krockow 2023; Na et al. 2023). In light of the aforementioned challenges, concerted efforts are needed to promote COVID-19 vaccine acceptance and uptake. Healthcare providers, policymakers, and the scientific community must work together with community engagement to understand the determinants of vaccine hesitancy and to develop effective strategies. By promoting education, building trust, and addressing concerns and misconceptions, the aim of higher COVID-19 vaccine acceptance and uptake can be achieved, ultimately contributing to the control of the COVID-19 pandemic (Rodrigues et al. 2022; Adhikari et al. 2022).

17.5 The Prevalence of COVID-19 Vaccine Hesitancy in the MENA Region

The MENA region has been significantly affected by the COVID-19 pandemic, with high rates of infections and mortality (Wehbe et al. 2021). Further complicating this issue, COVID-19 vaccine hesitancy emerged as a notable challenge in the MENA region, with potential implications locally and on global health (Sallam 2021; Sallam et al. 2022a). The high prevalence of COVID-19 vaccination hesitancy in the MENA region is reflected in the low uptake of the vaccine in the region as illustrated in Fig. 17.1.

Numerous studies indicated that COVID-19 vaccine hesitancy rates in the MENA region are among the highest globally, with contributing factors such as low trust in healthcare systems, concerns about vaccine safety and efficacy, and conspiracy beliefs (Qunaibi et al. 2021a, b; Sallam 2021; Sallam et al. 2021b, 2022a). Resonating the findings from several world regions, the factors contributing to COVID-19 vaccine hesitancy in the MENA region were complex, including concerns about vaccine safety and efficacy, mistrust in healthcare systems, and sociocultural beliefs (Galal et al. 2022). Addressing the root causes of COVID-19 vaccine hesitancy in the MENA region is essential for developing effective communication strategies to promote vaccine acceptance and uptake (Kacimi et al. 2022).

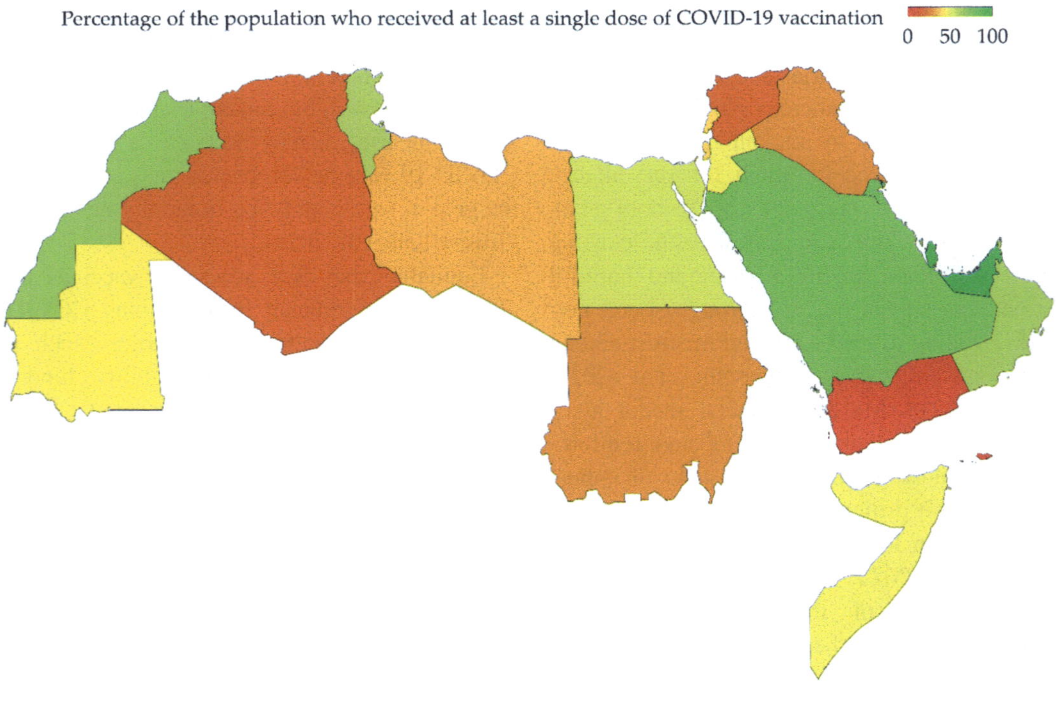

Fig. 17.1 Percentage of the population who received at least a single dose of COVID-19 vaccination in the Middle East and North Africa. The source of data is the Johns Hopkins University & Medicine coronavirus resource center (Johns Hopkins University & Medicine 2023)

Varying rates of COVID-19 vaccine acceptance were reported in the studies conducted in the MENA region. For example, vaccine acceptance rates as low as 20–25% were reported among adults in Lebanon, Egypt, and Jordan (El-Elimat et al. 2021; Sallam et al. 2020; Kasrine Al Halabi et al. 2021; Omar and Hani 2021). On the other hand, higher rates close to 89% were reported in a recent study from Iraq (Al-Qerem et al. 2022b). Moderate rates ranging between 40 and 80% were reported in Saudi Arabia, Oman, the UAE, Qatar, Somalia, and Yemen (Albahri et al. 2021; Bitar et al. 2021; Al-Mansour et al. 2021; Khaled et al. 2021; Al-Marshoudi et al. 2021). It is important to interpret results of the studies conducted in the MENA region in light of the survey instruments used to assess COVID-19 vaccine hesitancy and its associated factors, the timing of the surveys, and the population investigated which can result in large variations in the rates reported for COVID-19 vaccine hesitancy (Sallam et al. 2022a).

Considering the peculiarities of the MENA region in terms of language, culture, economic, and sociocultural perspectives, it is important to consider these factors in addressing COVID-19 vaccine hesitancy in the region (Kacimi et al. 2022; Qunaibi et al. 2021a, b; Sallam et al. 2021b). This can help in developing culturally sensitive and evidence-based communication strategies that address concerns and misconceptions about vaccines that could vary based on different regions (Ackah et al. 2022). These strategies should be tailored to specific communities and must consider the sociocultural factors that could contribute to COVID-19 vaccine hesitancy (Al-Ghuraibi et al. 2022). Additionally, healthcare professionals, researchers, and policymakers should work collaboratively to build trust and promote education and awareness about the benefits of vaccination in the MENA region like other world regions (Kruk et al. 2018).

17.6 Common Themes that Could Have Resulted in COVID-19 Vaccine Hesitancy in the MENA Region

It is crucial to recognize the complex factors that have contributed to COVID-19 vaccine hesitancy in the MENA region since this problem poses a significant challenge that threatens to undermine efforts to control the pandemic and its negative impacts in the region. The common themes that emerged from the investigation of COVID-19 vaccination hesitancy in the MENA region are summarized in Fig. 17.2. Results pointed to the following as the major determinants driving this phenomenon on public health: (1) Lack of confidence in vaccine safety and efficacy besides mistrust in health professionals, institutions, and policymakers; (2) Complacency toward the risks posed by COVID-19; (3) Constraints hindering convenient access to COVID-19 vaccination; and (4) Widespread embrace of misinformation and conspiracy beliefs regarding SARS-CoV-2 and its vaccines.

One of the most significant factors fueling COVID-19 vaccine hesitancy was the high prevalence of concerns about vaccine safety and efficacy significantly contributing to COVID-19 vaccine hesitancy in the MENA region (El-Elimat et al. 2021; Qunaibi et al. 2021a, b). As previously mentioned, the rapid development and emergency use authorization of COVID-19 vaccines could have led to concerns about the long-term safety of these vaccines driving hesitancy to receive such vaccines. Moreover, the spread of misinformation and conspiracy theories about vaccines on social media has also contributed to COVID-19 vaccine hesitancy (Sallam et al. 2020, 2021b), in addition to the mistrust in vaccines and healthcare systems and institutions (Charura et al. 2022). Historical events and systemic issues in the MENA region, such as the political instabilities and ongoing conflicts and civil wars may have contributed

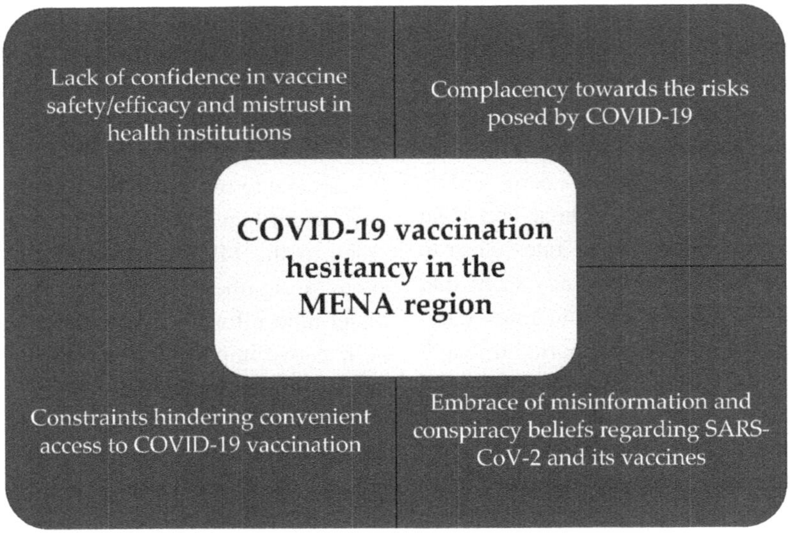

Fig. 17.2 Common factors that resulted in COVID-19 vaccination hesitancy in the Middle East and North Africa. COVID-19: coronavirus disease 2019; SARS-CoV-2: severe acute respiratory syndrome coronavirus 2

to a lack of trust in healthcare providers and institutions, making it challenging to promote COVID-19 vaccine acceptance (Bizri et al. 2021; Tang et al. 2022).

Similarly, sociocultural beliefs have also played a significant role in vaccine hesitancy in the MENA region. Concerns about the compatibility of vaccines with religious or cultural beliefs may have contributed to the high prevalence of vaccination hesitancy (Magadmi and Kamel 2021; Galal et al. 2022; Kibongani Volet et al. 2022). Moreover, reliance on traditional medicine and alternative therapies prevalent in some MENA countries may have also contributed to COVID-19 vaccine hesitancy (Hassen et al. 2022; Elbarazi et al. 2021). Lastly, hurdles to easy access to vaccines may have been a barrier for some individuals, particularly those in remote or marginalized MENA communities (Kaadan et al. 2021). Issues, such as vaccine availability, reminders to get vaccination, appointments, transportation, and affordability may have contributed to vaccine hesitancy in the MENA region (Abu-Farha et al. 2021). Based on the previous points, addressing the common themes that have contributed to COVID-19 vaccine hesitancy in the MENA region is crucial to promoting public health and preventing the spread of COVID-19 in the region. Therefore, effective strategies to address vaccine hesitancy must consider the complex factors and common themes contributing to vaccination hesitancy and be tailored to specific countries' needs. Moreover, promoting education, building trust, and addressing the root causes of COVID-19 vaccine hesitancy are essential steps to achieving COVID-19 vaccine acceptance and uptake in the MENA countries.

17.6.1 Issues Related to Lack of Confidence in COVID-19 Vaccines in the MENA Region

Since the early days of the COVID-19 vaccine development, safety concerns were raised and have been considered as a significant factor driving COVID-19 vaccine hesitancy (Piltch-Loeb et al. 2021; Forman et al. 2021). The swift development of the vaccines with authorization for emergency use could have led to concerns regarding the long-term safety of COVID-19 vaccines (Kashte et al. 2021). Subsequently, this resulted in debates regarding the long-term efficacy and safety of these vaccines particularly in specific sub-populations (immunocompromised patients, pregnant women, children, etc.), as well as doubts raised regarding the efficacy of the early-developed vaccines toward the emerging virus variants, and the accountability issues if adverse events take place (Beretta and Marelli 2023; Lipsitch et al. 2022; Duly et al. 2022; Tian et al. 2022; Tormen et al. 2023). These issues were also found in the countries of the MENA region. For example, a recent study in Egypt indicated that despite a satisfactory level of vaccine acceptance, concerns were expressed in the context of inadequate clinical trials and fear of its side effects (Elgendy and Abdelrahim 2021). Consistent with this finding, the overwhelming majority of the health professionals in another study from Egypt who were hesitant to get COVID-19 vaccination cited inadequate clinical trials and fear of side effects as the major factors (Fares et al. 2021).

In a similar fashion, a study among the general public in Saudi Arabia reported that the major reasons for COVID-19 vaccine hesitancy were the notion that vaccines were insufficiently tested, that pharmaceutical corporations prioritized sales of the COVID-19 vaccine, and the fear of adverse effects associated with COVID-19 vaccines (Al-Mansour et al. 2021). Concerns regarding COVID-19 vaccination adverse effects, long-lasting immunity, or even the fear of getting COVID-19 from the vaccine were reported among health professionals in Palestine (Maraqa et al. 2021). Safety concerns were also reported as a major reason for COVID-19 vaccine hesitancy for booster doses in a large multinational MENA study (Abouzid et al. 2022). Additionally, a large study from Kuwait highlighted the role of confidence in COVID-19 vaccine hesitancy (Alibrahim and Awad 2021). The participants who did not have high confidence in the health system's ability to control the COVID-19 pandemic were hesitant to get COVID-19 vaccination (Alibrahim and Awad 2021). Thus, several studies in the MENA region highlighted the need to address safety concerns as a key determinant to promote COVID-19 vaccine acceptance and uptake. Achieving this relevant aim can be done by providing accurate and transparent information about the safety and efficacy of COVID-19 vaccines, and by addressing concerns and misconceptions about vaccine development and testing (Kerr et al. 2021).

17.6.2 Complacency Issues Driving COVID-19 Vaccine Hesitancy in the MENA Region

Complacency denoting the low perceived risk from the disease itself or the acquisition of the virus can be another major factor contributing to COVID-19 vaccine hesitancy (Patterson et al. 2022). To be more specific, complacency toward the risk of COVID-19 involves the feeling that the risk of contracting SARS-CoV-2 is low or that the consequences of infection are not severe enough to justify the decision to get vaccinated (Bennett et al. 2022). Complacency in the context of COVID-19 can be fueled by misinformation or misconceptions about the severity of the pandemic, which may have led some individuals to underestimate the risks of COVID-19 and, therefore, to doubt the benefits of vaccination and reinforce distrust in vaccination (El-Elimat et al. 2021; Zimmerman et al. 2023; Lalot et al. 2022). The role of complacency as an important determinant driving COVID-19 vaccination hesitancy has appeared in several studies from the MENA region, including a multinational study to validate the 5C scale for psychological determinants of vaccination in Arabic (Abdou et al. 2021; El-Elimat et al. 2021). This pattern was also revealed among health professionals in Kuwait where complacency toward the disease was associated with higher odds of COVID-19 vaccine hesitancy (Al-Sanafi and Sallam 2021).

Addressing the complacency issue as a determinant of COVID-19 vaccine hesitancy/rejection requires promoting education and awareness about the risks of COVID-19 and the importance of vaccination as a safe approach to reaching immunity (Peters 2022). This includes providing accurate information about the potential consequences of infection and the benefits of vaccination, as well as addressing misconceptions or misinformation that may contribute to complacency (Hammad et al. 2022). Moreover, effective communication strategies should consider the diversity of populations (e.g., educational level, professional level, etc.) and the varying unique perceptions and beliefs about the pandemic and vaccination that could result in low perceived threat and subsequently vaccination hesitancy (Yang et al. 2022).

17.6.3 Constraints Resulting in COVID-19 Vaccine Hesitancy in the MENA Region

Access to COVID-19 vaccines could be regarded as a significant constraint for some individuals, particularly those in remote or marginalized communities. Thus, issues such as vaccine availability free-of-charge, transportation issues, registration for vaccination, and affordability in terms of convenient time and place may culminate to contribute to COVID-19 vaccine hesitancy (Ali et al. 2021). In several countries in the MENA region, there may be limited vaccine supplies due to the absence of domestic production and distribution challenges in other countries (Wouters et al. 2021). This issue could have resulted in delays in vaccine rollouts or limited access to vaccines in several regions or communities (Duroseau et al. 2022). Additionally, several MENA countries took a similar global approach of prioritizing certain population groups for COVID-19 vaccination, such as healthcare workers and vulnerable populations, which may have resulted in delays in vaccine access for the general population (Paloyo et al. 2022).

Affordability is another significant factor contributing to vaccine hesitancy in the MENA region. While COVID-19 vaccines are provided free of charge in a majority of MENA countries, some individuals faced financial barriers to accessing COVID-19 vaccines, such as travel expenses or lost wages from taking time off work to receive the vaccine (Galal et al. 2022). Additionally, individuals may face indirect costs associated with vaccination, such as the cost of transportation or childcare, which may make it more challenging to access vaccines. A study from the UAE showed the importance of free vaccination and reducing constraints to promote confidence and achieve high vaccination rates against COVID-19 (Suliman et al. 2021). Moreover, the importance of providing the vaccine for free was also evident in a study from Libya where the vast majority of participants (93%) believed that the vaccine should be provided freely while only 48% were willing to pay for the vaccine (Elhadi et al. 2021). Thus, the policymakers and healthcare providers in the MENA region are encouraged to work collaboratively to improve vaccine distribution networks and to ensure that vaccines are available conveniently in all regions and communities to promote COVID-19 vaccine equity. This may involve developing mobile vaccination clinics and partnering with local organizations and volunteers to provide COVID-19 vaccines in areas where transportation infrastructure is limited (Logue et al. 2023).

17.6.4 Conspiracy Beliefs and Misinformation Resulting in COVID-19 Vaccine Hesitancy in the MENA Region

Conspiracy beliefs and misinformation have been identified as important contributors to COVID-19 vaccine hesitancy in the MENA region (Sallam et al. 2020, 2021b). Within this context, several studies from the MENA region have reported a consistent finding of the widespread prevalence of endorsement of conspiratorial ideas toward various emerging infections and toward COVID-19 with the spread of false information and conspiracy theories about COVID-19 vaccines. These were linked to the reluctance to receive COVID-19 vaccines among various populations including health professionals and educated groups (Alsanafi et al. 2023; Sallam et al. 2021c, 2022b, c). A range of false claims and conspiracy theories regarding SARS-CoV-2 and COVID-19 vaccines propagated swiftly in the MENA region, including the idea that vaccines contain harmful substances that might lead to infertility, being a biological warfare plot or are part of a large conspiracy plot to control the population size as reported in different studies from Jordan and Kuwait (Sallam et al. 2021b, c; Al-Qerem et al. 2022a; Hammad et al. 2022). These claims were found more commonly among individuals who rely on obtaining COVID-19 vaccine information through social media platforms, which may have contributed to mistrust and skepticism toward COVID-19 vaccines (Sallam et al. 2021b, c). Thus, the finding that exposure to misinformation and conspiracy ideas about COVID-19 vaccines could reduce vaccine acceptance rates and contribute to vaccine hesitancy was also replicated in the studies from the MENA region highlighting the relevance of this theme in vaccination hesitancy in the region. To address this important challenge, it is critical to develop effective communication strategies for countering misinformation and specific conspiracy theories by ensuring the delivery of accurate information about COVID-19 vaccines (Whitehead et al. 2023). This can involve targeted education and outreach campaigns that concentrate on building public confidence in COVID-19 vaccines. These efforts should also involve social media and digital platforms to promote the spread of accurate vaccine information (Steffens et al. 2020).

17.7 Strategies to Address COVID-19 Vaccine Hesitancy in the MENA Region

Addressing the root causes of COVID-19 vaccine hesitancy and developing effective communication strategies to promote COVID-19 vaccine acceptance and uptake in the MENA region can be considered a necessary step to prevent the spread of COVID-19 in the region. An important approach to addressing COVID-19 vaccine hesitancy in the MENA region can be the development of culturally tailored evidence-based communication strategies that address the safety and efficacy concerns as well as the misconceptions about COVID-19 vaccines (Kassianos et al. 2022). These strategies should be constructed considering the evidence provided by the growing literature on the prevalence of COVID-19 vaccine hesitancy and its associated determinants in the region (Sallam et al. 2022a; Ghazy et al. 2022). Additionally, the role of collaborative work between healthcare providers and policymakers cannot be overstated to build trust and promote education and awareness about the benefits of vaccination in the community (Chirico and Teixeira da Silva 2023).

Effective communication strategies can play a critical role in addressing COVID-19 vaccine hesitancy. One effective strategy is to work with community leaders and influencers to build trust and promote accurate information about vaccines as reported in a study from Oman (Al-Marshoudi et al. 2021). Healthcare providers can use a variety of channels, including the various social media platforms, as well as the traditional media channel such as television, newspapers, and radio, to reach a broad audience delivering messages to counter misinformation and conspiracy theories about COVID-19 vaccines (Islam et al. 2020; Romer and Jamieson 2020; Southwell et al. 2020; Jacquellyn Nambi et al. 2022). Additionally, an important step to building public confidence in COVID-19 vaccination can be accomplished by providing clear and accurate information about vaccine safety and efficacy (Hong 2023). This includes providing concise information about the vaccine development process, clinical trials, and authorization steps. Thus, prioritizing transparency and promoting individual autonomy are essential to build trust in COVID-19 vaccination (Thorpe et al. 2022).

Misinformation and conspiracy ideas regarding COVID-19 have been identified as important contributors to COVID-19 vaccine hesitancy in the MENA region. To address this challenge, it is critical to develop effective strategies for countering these unsubstantiated claims and to promote accurate information about COVID-19 vaccines (Chirico and Teixeira da Silva 2023). In this respect, healthcare providers have a key role together with community leaders and influencers to provide accurate information about COVID-19 vaccines highlighting their public health benefits (Seale et al. 2022; Kuehn 2022). Promoting convenient access to COVID-19 vaccines and their affordability can also help overcome COVID-19 vaccine hesitancy in the MENA region (Wouters et al. 2021). The governments in the MENA together with support from international organizations are recommended to improve COVID-19 vaccine distribution networks and to ensure that vaccines are affordable without constraints for all individuals in the region (Yarlagadda et al. 2022). This includes ensuring that COVID-19 vaccines are available freely and accessible to individuals of different socioeconomic status and various locations (Ye et al. 2022).

Finally, imposing COVID-19 vaccine mandates can be considered the last resort to address the issue of COVID-19 vaccine hesitancy in the MENA region. The controversial status of the attitude toward compulsory vaccination has been shown recently reflecting the need for meticulous consideration of this strategy (Mahameed et al. 2023; Giannakou et al. 2022). Vaccine mandates may exacerbate existing mistrust and conspiracy beliefs, leading to resistance to vaccination (Kevin et al. 2022). This issue can be particularly problematic if these mandates are implemented without transparency or public engagement and feedback. Therefore, it is essential to ensure that COVID-19 vaccine mandates are implemented cautiously concerning transparency, public engagement, and respect for individual autonomy and rights (Law et al. 2022).

17.8 Conclusions and Future Perspectives

In the MENA region, various studies showed that COVID-19 vaccine hesitancy was widely prevalent even before the approval of the vaccines; this hesitancy remains a substantial barrier to the efforts aiming to control the pandemic. Investigation of the root causes and associated factors of this phenomenon is essential for devising and tailoring effective strategies that could help to promote acceptance and uptake of COVID-19 vaccines in the region. Additionally, continuous monitoring of COVID-19 vaccine hesitancy in the MENA region is a relevant aim considering the continuous emergence of novel SARS-CoV-2 variants and the short-term immunity following natural infection and vaccination. Thus, COVID-19 booster doses are deemed necessary to prevent forward transmission of the virus and to reduce the devastating health effects of the disease that became endemic in the human population.

Effective strategies to address COVID-19 vaccine hesitancy should consider the role of healthcare providers and policymakers in building trust and promoting education and awareness about the benefits of vaccination. To effectively communicate the importance of COVID-19 vaccination, healthcare providers are recommended to take their role in community education and deliver accurate information about vaccines. This role also involves highlighting the disease risks to reduce complacency and dismissing misinformation and conspiracy ideas about COVID-19 vaccines. Lastly, promoting vaccine affordability can also help to overcome COVID-19 vaccine hesitancy in the MENA region. This includes ensuring that COVID-19 vaccines are available and easily accessible free of charge, to all community members regardless of their socioeconomic status or geographic location. Additionally, collaboration between the governments in the MENA and international organizations is recommended to improve vaccine distribution to ensure that COVID-19 vaccines are affordable for all individuals in the region.

Implementing COVID-19 vaccine mandates in the MENA region is a potential strategy to address vaccine hesitancy; however, this strategy should be considered carefully, since it may exacerbate the existing mistrust and conspiracy beliefs, leading to vaccine resistance.

Addressing COVID-19 vaccine hesitancy in the MENA region remains critical to promoting public health in the region and reducing the negative impacts of COVID-19. In this respect, effective strategies could benefit from the growing literature from the MENA region that points to the complex factors contributing to vaccination hesitancy. Community education can help to build confidence in COVID-19 vaccination and can be valuable to address misinformation and conspiracy theories that may result in vaccination hesitancy. Finally, it is important to highlight the need for promoting COVID-19 vaccine accessibility and affordability, which can result in higher rates of COVID-19 vaccine acceptance and uptake in the MENA region.

Abbreviations

COVID-19	Coronavirus Disease 2019
HBM	The Health Belief Model
MENA	Middle East and North Africa
PACV	Parent Attitudes About Childhood Vaccines
SAGE	Strategic Advisory Group of Experts on Immunization
SARS-CoV-2	Severe Acute Respiratory Syndrome Coronavirus 2
UAE	The United Arabic Emirates
VAX	Vaccination Attitudes Examination
VCBS	Vaccine Conspiracy Beliefs Scale
VPDs	Vaccine-Preventable Diseases
VHS	Vaccine Hesitancy Scale
WHO	World Health Organization

References

Abdelmoneim SA, Sallam M, Hafez DM, Elrewany E, Mousli HM, Hammad EM, Elkhadry SW, Adam MF, Ghobashy AA, Naguib M, Nour El-Deen AE, Aji N, Ghazy RM (2022) COVID-19 vaccine booster dose acceptance: systematic review and meta-analysis. Trop Med Infect Dis 7(10). https://doi.org/10.3390/tropicalmed7100298

Abdou MS, Kheirallah KA, Aly MO, Ramadan A, Elhadi YAM, Elbarazi I, Deghidy EA, El Saeh HM, Salem KM, Ghazy RM (2021) The coronavirus disease 2019 (COVID-19) vaccination psychological antecedent assessment using the Arabic 5c validated tool: an online survey in 13 Arab countries. PLoS ONE 16(11):e0260321. https://doi.org/10.1371/journal.pone.0260321

Abouzid M, Ahmed AA, El-Sherif DM, Alonazi WB, Eatmann AI, Alshehri MM, Saleh RN, Ahmed MH, Aziz IA, Abdelslam AE, Omran AA, Omar AA, Ghorab MA, Islam SMS (2022) Attitudes toward receiving COVID-19 booster dose in the Middle East and North Africa (MENA) region: a cross-sectional study of 3041 fully vaccinated participants. Vaccines (Basel) 10(8). https://doi.org/10.3390/vaccines10081270

Abu-Farha R, Mukattash T, Itani R, Karout S, Khojah HMJ, Abed Al-Mahmood A, Alzoubi KH (2021) Willingness of Middle Eastern public to receive COVID-19 vaccines. Saudi Pharm J 29(7):734–739. https://doi.org/10.1016/j.jsps.2021.05.005

Ackah BBB, Woo M, Stallwood L, Fazal ZA, Okpani A, Ukah UV, Adu PA (2022) COVID-19 vaccine hesitancy in Africa: a scoping review. Glob Health Res Policy 7(1):21. https://doi.org/10.1186/s41256-022-00255-1

Adhikari B, Yeong Cheah P, von Seidlein L (2022) Trust is the common denominator for COVID-19 vaccine acceptance: a literature review. Vaccine X 12:100213. https://doi.org/10.1016/j.jvacx.2022.100213

Al-Ghuraibi M, Dighriri IM, Elrggal ME, Obaid NA (2022) The socio-cultural factors behind the Saudi attitude toward COVID-19 vaccination: a survey-based study. Front Public Health 10:1026252. https://doi.org/10.3389/fpubh.2022.1026252

Al-Mansour K, Alyahya S, AbuGazalah F, Alabdulkareem K (2021) Factors affecting COVID-19 vaccination among the general population in Saudi Arabia. Healthcare (Basel) 9(9). https://doi.org/10.3390/healthcare9091218

Al-Marshoudi S, Al-Balushi H, Al-Wahaibi A, Al-Khalili S, Al-Maani A, Al-Farsi N, Al-Jahwari A, Al-Habsi Z, Al-Shaibi M, Al-Msharfi M, Al-Ismaili A, Al-Buloshi H, Al-Rawahi B, Al-Barwani K, Al-Abri S (2021) Knowledge, attitudes, and practices (KAP) toward the COVID-19 vaccine in Oman: a pre-campaign cross-sectional study. Vaccines (Basel) 9(6). https://doi.org/10.3390/vaccines9060602

Al-Qerem W, Al Bawab AQ, Hammad A, Ling J, Alasmari F (2022a) Willingness of the Jordanian population to receive a COVID-19 booster dose: a cross-sectional study. Vaccines (Basel) 10(3). https://doi.org/10.3390/vaccines10030410

Al-Qerem W, Hammad A, Alsajri AH, Al-Hishma SW, Ling J, Mosleh R (2022b) COVID-19 vaccination acceptance and its associated factors among the Iraqi population: a cross sectional study. Patient Prefer Adherence 16:307–319. https://doi.org/10.2147/ppa.S350917

Al-Sanafi M, Sallam M (2021) Psychological determinants of COVID-19 vaccine acceptance among healthcare workers in Kuwait: a cross-sectional study using the 5C and vaccine conspiracy beliefs scales. Vaccines (Basel) 9(7). https://doi.org/10.3390/vaccines9070701

Albahri AH, Alnaqbi SA, Alshaali AO, Alnaqbi SA, Shahdoor SM (2021) COVID-19 vaccine acceptance in a sample from the United Arab Emirates general adult population: a cross-sectional survey, 2020. Front Public Health 9:614499. https://doi.org/10.3389/fpubh.2021.614499

Albrecht D (2022) Vaccination, politics and COVID-19 impacts. BMC Public Health 22(1):96. https://doi.org/10.1186/s12889-021-12432-x

Ali I, Ali S, Iqbal S (2021) COVID-19 vaccination: concerns about its accessibility, affordability, and acceptability. Front Med (Lausanne) 8:647294. https://doi.org/10.3389/fmed.2021.647294

Alibrahim J, Awad A (2021) COVID-19 vaccine hesitancy among the public in Kuwait: a cross-sectional survey. Int J Environ Res Public Health 18(16). https://doi.org/10.3390/ijerph18168836

Alsanafi M, Salim NA, Sallam M (2023) Willingness to get HPV vaccination among female university students in Kuwait and its relation to vaccine conspiracy beliefs. Hum Vaccin Immunother 19(1):2194772. https://doi.org/10.1080/21645515.2023.2194772

Alsuwaidi AR, Hammad HAA, Elbarazi I, Sheek-Hussein M (2023) Vaccine hesitancy within the Muslim community: Islamic faith and public health perspectives. Hum Vaccin Immunother 19(1):2190716. https://doi.org/10.1080/21645515.2023.2190716

Amanna IJ, Slifka MK (2020) Successful vaccines. Curr Top Microbiol Immunol 428:1–30. https://doi.org/10.1007/82_2018_102

Anderson MG, Ballinger EA, Benjamin D, Frenkel LD, Hinnant CW Jr, Zucker KW (2020) A clinical perspective of the U.S. anti-vaccination epidemic: considering marginal costs and benefits, CDC best practices guidelines, free riders, and herd immunity. Vaccine 38(50):7877–7879. https://doi.org/10.1016/j.vaccine.2020.10.068

Andre FE, Booy R, Bock HL, Clemens J, Datta SK, John TJ, Lee BW, Lolekha S, Peltola H, Ruff TA, Santosham M, Schmitt HJ (2008) Vaccination greatly reduces disease, disability, death and inequity worldwide. Bull World Health Organ 86(2):140–146. https://doi.org/10.2471/blt.07.040089

Athamneh RY, Arıkan A, Sayan M, Mahafzah A, Sallam M (2021) Variable proportions of phylogenetic clustering and low levels of antiviral drug resistance among the major HBV sub-genotypes in the Middle East and North Africa. Pathogens 10(10). https://doi.org/10.3390/pathogens10101333

Banerjee P, Seth R, Dhaliwal BK, Sullivan A, Qiayum Y, Thankachen B, Closser S, Shet A (2022) Vaccine acceptance in rural India: engaging faith leaders as vaccine ambassadors. Front Public Health 10:979424. https://doi.org/10.3389/fpubh.2022.979424

Bennett MM, Douglas M, da Graca B, Sanchez K, Powers MB, Warren AM (2022) Attitudes and personal beliefs about the COVID-19 vaccine among people with COVID-19: a mixed-methods analysis. BMC Public Health 22(1):1936. https://doi.org/10.1186/s12889-022-14335-x

Beretta G, Marelli L (2023) Fast-tracking development and regulatory approval of COVID-19 vaccines in the EU: a review of ethical implications. Bioethics. https://doi.org/10.1111/bioe.13151

Betsch C, Schmid P, Heinemeier D, Korn L, Holtmann C, Böhm R (2018) Beyond confidence: development of a measure assessing the 5C psychological antecedents of vaccination. PLoS ONE 13(12):e0208601. https://doi.org/10.1371/journal.pone.0208601

Betsch C, Schmid P, Verger P, Lewandowsky S, Soveri A, Hertwig R, Fasce A, Holford D, De Raeve P, Gagneur A, Vuolanto P, Correia T, Tavoschi L, Declich S, Marceca M, Linos A, Karnaki P, Karlsson L, Garrison A (2022) A call for immediate action to increase COVID-19 vaccination uptake to prepare for the third pandemic winter. Nat Commun 13(1):7511. https://doi.org/10.1038/s41467-022-34995-y

Bitar AN, Zawiah M, Al-Ashwal FY, Kubas M, Saeed RM, Abduljabbar R, Jaber AAS, Sulaiman SAS, Khan AH (2021) Misinformation, perceptions towards COVID-19 and willingness to be vaccinated: a population-based survey in Yemen. PLoS ONE 16(10):e0248325. https://doi.org/10.1371/journal.pone.0248325

Bizri NA, Alam W, Mobayed T, Tamim H, Makki M, Mushrrafieh U (2021) COVID-19 in conflict region: the Arab Levant response. BMC Public Health 21(1):1590. https://doi.org/10.1186/s12889-021-11580-4

Brynne G, Nina G, Claudia Abreu L, Alpha AD, Sanghita B, de Vergil C, Rawlance N, Elizabeth Nyamupachitu M, Adalbert T (2022) Community engagement to support COVID-19 vaccine uptake: a living systematic review protocol. BMJ Open 12(9):e063057. https://doi.org/10.1136/bmjopen-2022-063057

Burgess RA, Osborne RH, Yongabi KA, Greenhalgh T, Gurdasani D, Kang G, Falade AG, Odone A, Busse R, Martin-Moreno JM, Reicher S, McKee M (2021) The COVID-19 vaccines rush: participatory community engagement matters more than ever. Lancet 397(10268):8–10. https://doi.org/10.1016/s0140-6736(20)32642-8

Bussink-Voorend D, Hautvast JLA, Vandeberg L, Visser O, Hulscher MEJL (2022) A systematic literature review to clarify the concept of vaccine hesitancy. Nat Hum Behav 6(12):1634–1648. https://doi.org/10.1038/s41562-022-01431-6

Cascini F, Pantovic A, Al-Ajlouni YA, Failla G, Puleo V, Melnyk A, Lontano A, Ricciardi W (2022) Social media and attitudes towards a COVID-19 vaccination: a systematic review of the literature. EClinicalMedicine 48:101454. https://doi.org/10.1016/j.eclinm.2022.101454

Cerda AA, García LY (2021) Hesitation and refusal factors in individuals' decision-making processes regarding a coronavirus disease 2019 vaccination. Front

Public Health 9:626852. https://doi.org/10.3389/fpubh.2021.626852
Charura D, Hill AP, Etherson ME (2022) COVID-19 vaccine hesitancy, medical mistrust, and mattering in ethnically diverse communities. J Racial Ethn Health Disparities 1–8. https://doi.org/10.1007/s40615-022-01337-z
Chirico F, Teixeira da Silva JA (2023) Evidence-based policies in public health to address COVID-19 vaccine hesitancy. Future Virol. https://doi.org/10.2217/fvl-2022-0028
Cole A, Webster P, Van Liew D, Salas M, Aimer O, Malikova MA (2022) Safety surveillance and challenges in accelerated COVID-19 vaccine development. Therap Adv Drug Saf 13:20420986221116452. https://doi.org/10.1177/20420986221116452
Cornelia B, Katrine Bach H, Sergei D, Dorothee H, Nikolay B, Natalia K, Janusz K, Robert B, Ingo Z, Charles Shey W, Ève D, Arnaud G, Elisabeth B-N, Amandine G-B, Jonas S (2020) Sample study protocol for adapting and translating the 5C scale to assess the psychological antecedents of vaccination. BMJ Open 10(3):e034869. https://doi.org/10.1136/bmjopen-2019-034869
Cucinotta D, Vanelli M (2020) WHO declares COVID-19 a pandemic. Acta Biomed 91(1):157–160. https://doi.org/10.23750/abm.v91i1.9397
Dada D, Djiometio JN, McFadden SM, Demeke J, Vlahov D, Wilton L, Wang M, Nelson LE (2022) Strategies that promote equity in COVID-19 vaccine uptake for black communities: a review. J Urban Health 99(1):15–27. https://doi.org/10.1007/s11524-021-00594-3
Debus M, Tosun J (2021) Political ideology and vaccination willingness: implications for policy design. Policy Sci 54(3):477–491. https://doi.org/10.1007/s11077-021-09428-0
Depoux A, Martin S, Karafillakis E, Preet R, Wilder-Smith A, Larson H (2020) The pandemic of social media panic travels faster than the COVID-19 outbreak. J Travel Med 27(3). https://doi.org/10.1093/jtm/taaa031
Dhama K, Khan S, Tiwari R, Sircar S, Bhat S, Malik YS, Singh KP, Chaicumpa W, Bonilla-Aldana DK, Rodriguez-Morales AJ (2020) Coronavirus disease 2019-COVID-19. Clin Microbiol Rev 33(4). https://doi.org/10.1128/cmr.00028-20
Dobson GP (2021) Wired to doubt: why people fear vaccines and climate change and mistrust science. Front Med (Lausanne) 8:809395. https://doi.org/10.3389/fmed.2021.809395
Dubé E, Gagnon D, Nickels E, Jeram S, Schuster M (2014) Mapping vaccine hesitancy—country-specific characteristics of a global phenomenon. Vaccine 32(49):6649–6654. https://doi.org/10.1016/j.vaccine.2014.09.039
Duly K, Farraye FA, Bhat S (2022) COVID-19 vaccine use in immunocompromised patients: a commentary on evidence and recommendations. Am J Health Syst Pharm 79(2):63–71. https://doi.org/10.1093/ajhp/zxab344
Duroseau B, Kipshidze N, Limaye RJ (2022) The impact of delayed access to COVID-19 vaccines in low- and lower-middle-income countries. Front Public Health 10:1087138. https://doi.org/10.3389/fpubh.2022.1087138
El-Elimat T, AbuAlSamen MM, Almomani BA, Al-Sawalha NA, Alali FQ (2021) Acceptance and attitudes toward COVID-19 vaccines: a cross-sectional study from Jordan. PLoS ONE 16(4):e0250555. https://doi.org/10.1371/journal.pone.0250555
Elbarazi I, Al-Hamad S, Alfalasi S, Aldhaheri R, Dubé E, Alsuwaidi AR (2021) Exploring vaccine hesitancy among healthcare providers in the United Arab Emirates: a qualitative study. Hum Vaccin Immunother 17(7):2018–2025. https://doi.org/10.1080/21645515.2020.1855953
Elgendy MO, Abdelrahim MEA (2021) Public awareness about coronavirus vaccine, vaccine acceptance, and hesitancy. J Med Virol 93(12):6535–6543. https://doi.org/10.1002/jmv.27199
Elhadi M, Alsoufi A, Alhadi A, Hmeida A, Alshareea E, Dokali M, Abodabos S, Alsadiq O, Abdelkabir M, Ashini A et al (2021) Knowledge, attitude, and acceptance of healthcare workers and the public regarding the COVID-19 vaccine: a cross-sectional study. BMC Public Health 21(1):955. https://doi.org/10.1186/s12889-021-10987-3
Fajar JK, Sallam M, Soegiarto G, Sugiri YJ, Anshory M, Wulandari L, Kosasih SAP, Ilmawan M, Kusnaeni K, Fikri M et al (2022) Global prevalence and potential influencing factors of COVID-19 vaccination hesitancy: a meta-analysis. Vaccines (Basel) 10(8). https://doi.org/10.3390/vaccines10081356
Fares S, Elmnyer MM, Mohamed SS, Elsayed R (2021) COVID-19 vaccination perception and attitude among healthcare workers in Egypt. J Prim Care Community Health 12:21501327211013304. https://doi.org/10.1177/21501327211013303
Ferreira Caceres MM, Sosa JP, Lawrence JA, Sestacovschi C, Tidd-Johnson A, Rasool MHU, Gadamidi VK, Ozair S, Pandav K, Cuevas-Lou C, Parrish M, Rodriguez I, Fernandez JP (2022) The impact of misinformation on the COVID-19 pandemic. AIMS Public Health 9(2):262–277. https://doi.org/10.3934/publichealth.2022018
Fieselmann J, Annac K, Erdsiek F, Yilmaz-Aslan Y, Brzoska P (2022) What are the reasons for refusing a COVID-19 vaccine? A qualitative analysis of social media in Germany. BMC Public Health 22(1):846. https://doi.org/10.1186/s12889-022-13265-y
Fiske A, Schönweitz F, Eichinger J, Zimmermann B, Hangel N, Sierawska A, McLennan S, Buyx A (2022) The COVID-19 vaccine: trust, doubt, and hope for a future beyond the pandemic in Germany. PLoS ONE 17(4):e0266659. https://doi.org/10.1371/journal.pone.0266659

Forman R, Shah S, Jeurissen P, Jit M, Mossialos E (2021) COVID-19 vaccine challenges: what have we learned so far and what remains to be done? Health Policy 125(5):553–567. https://doi.org/10.1016/j.healthpol.2021.03.013

Freeman D, Waite F, Rosebrock L, Petit A, Causier C, East A, Jenner L, Teale AL, Carr L, Mulhall S, Bold E, Lambe S (2022) Coronavirus conspiracy beliefs, mistrust, and compliance with government guidelines in England. Psychol Med 52(2):251–263. https://doi.org/10.1017/s0033291720001890

Frenkel LD (2021) The global burden of vaccine-preventable infectious diseases in children less than 5 years of age: implications for COVID-19 vaccination. How can we do better? Allergy Asthma Proc 42(5):378–385. https://doi.org/10.2500/aap.2021.42.210065

Galagali PM, Kinikar AA, Kumar VS (2022) Vaccine hesitancy: obstacles and challenges. Curr Pediatr Rep 10(4):241–248. https://doi.org/10.1007/s40124-022-00278-9

Galal B, Lazieh S, Al-Ali S, Khoshnood K (2022) Assessing vaccine hesitancy in Arab countries in the Middle East and North Africa (MENA) region: a scoping review protocol. BMJ Open 12(2):e045348. https://doi.org/10.1136/bmjopen-2020-045348

Garett R, Young SD (2021) Online misinformation and vaccine hesitancy. Transl Behav Med 11(12):2194–2199. https://doi.org/10.1093/tbm/ibab128

Gaweł A, Mańdziuk M, Żmudziński M, Gosek M, Krawczyk-Suszek M, Pisarski M, Adamski A, Cyganik W (2021) Effects of Pope Francis' religious authority and media coverage on Twitter user's attitudes toward COVID-19 vaccination. Vaccines (Basel) 9(12). https://doi.org/10.3390/vaccines9121487

Geiger M, Rees F, Lilleholt L, Santana AP, Zettler I, Wilhelm O, Betsch C, Böhm R (2021) Measuring the 7Cs of vaccination readiness. Eur J Psychol Assess 38(4):261–269. https://doi.org/10.1027/1015-5759/a000663

Germani F, Pattison AB, Reinfelde M (2022) WHO and digital agencies: how to effectively tackle COVID-19 misinformation online. BMJ Glob Health 7(8). https://doi.org/10.1136/bmjgh-2022-009483

Gerretsen P, Kim J, Quilty L, Wells S, Brown EE, Agic B, Pollock BG, Graff-Guerrero A (2021) Vaccine hesitancy is a barrier to achieving equitable herd immunity among racial minorities. Front Med (Lausanne) 8:668299. https://doi.org/10.3389/fmed.2021.668299

Ghare F, Meckawy R, Moore M, Lomazzi M (2023) Determinants of acceptance of COVID-19 vaccination in healthcare and public health professionals: a review. Vaccines (Basel) 11(2). https://doi.org/10.3390/vaccines11020311

Ghazy RM, Abdou MS, Awaidy S, Sallam M, Elbarazi I, Youssef N, Fiidow OA, Mehdad S, Hussein MF, Adam MF, Abdullah FSA, Rebai WK, Raad EB, Hussein M, Shehata SF, Ismail, II, Salam AA, Samhouri D (2022) Acceptance of COVID-19 vaccine booster doses using the health belief model: a cross-sectional study in low-middle- and high-income countries of the east Mediterranean region. Int J Environ Res Public Health 19(19). https://doi.org/10.3390/ijerph191912136

Giannakou K, Kyprianidou M, Heraclides A (2022) Attitudes and determinants of mandatory vaccination against COVID-19 among the general population of Cyprus: a nationwide cross-sectional study. Vaccines (Basel) 10(3). https://doi.org/10.3390/vaccines10030438

Gisondi MA, Barber R, Faust JS, Raja A, Strehlow MC, Westafer LM, Gottlieb M (2022) A deadly infodemic: social media and the power of COVID-19 misinformation. J Med Internet Res 24(2):e35552. https://doi.org/10.2196/35552

Hammad AM, Al-Qerem W, Abu Zaid A, Khdair SI, Hall FS (2022) Misconceptions related to COVID 19 vaccines among the Jordanian population: myth and public health. Disaster Med Public Health Prep 17:e207. https://doi.org/10.1017/dmp.2022.143

Hassen HD, Welde M, Menebo MM (2022) Understanding determinants of COVID-19 vaccine hesitancy; an emphasis on the role of religious affiliation and individual's reliance on traditional remedy. BMC Public Health 22(1):1142. https://doi.org/10.1186/s12889-022-13485-2

Holzmann-Littig C, Frank T, Schmaderer C, Braunisch MC, Renders L, Kranke P, Popp M, Seeber C, Fichtner F, Littig B, Carbajo-Lozoya J, Meerpohl JJ, Haller B, Allwang C, On Behalf of the CC (2022) COVID-19 vaccines: fear of side effects among German health care workers. Vaccines (Basel) 10(5). https://doi.org/10.3390/vaccines10050689

Hong S-A (2023) COVID-19 vaccine communication and advocacy strategy: a social marketing campaign for increasing COVID-19 vaccine uptake in South Korea. Humanit Soc Sci Commun 10(1):109. https://doi.org/10.1057/s41599-023-01593-2

Hunter CM, Chou WS, Webb Hooper M (2021) Behavioral and social science in support of SARS-CoV-2 vaccination: National Institutes of Health initiatives. Transl Behav Med 11(7):1354–1358. https://doi.org/10.1093/tbm/ibab067

Islam MS, Sarkar T, Khan SH, Mostofa Kamal AH, Hasan SMM, Kabir A, Yeasmin D, Islam MA, Amin Chowdhury KI, Anwar KS, Chughtai AA, Seale H (2020) COVID-19-related infodemic and its impact on public health: a global social media analysis. Am J Trop Med Hyg 103(4):1621–1629. https://doi.org/10.4269/ajtmh.20-0812

Islam MS, Kamal AM, Kabir A, Southern DL, Khan SH, Hasan SMM, Sarkar T, Sharmin S, Das S, Roy T, Harun MGD, Chughtai AA, Homaira N, Seale H (2021) COVID-19 vaccine rumors and conspiracy theories: the need for cognitive inoculation against misinformation to improve vaccine adherence. PLoS ONE 16(5):e0251605. https://doi.org/10.1371/journal.pone.0251605

Jacquellyn Nambi S, Ronald K, Ronald O, Juliet K, Freddy Eric K (2022) Using community influencer groups to address COVID-19 misinformation and vaccine hesitancy in Uganda: a protocol for a prospective quasi-experimental study. BMJ Open 12(8):e057994. https://doi.org/10.1136/bmjopen-2021-057994

Johns Hopkins University & Medicine (2023) Vaccination progress across the world. https://coronavirus.jhu.edu/vaccines/international. Accessed 13 May 2023

Jones CL, Jensen JD, Scherr CL, Brown NR, Christy K, Weaver J (2015) The health belief model as an explanatory framework in communication research: exploring parallel, serial, and moderated mediation. Health Commun 30(6):566–576. https://doi.org/10.1080/10410236.2013.873363

Joseph AM, Fernandez V, Kritzman S, Eaddy I, Cook OM, Lambros S, Jara Silva CE, Arguelles D, Abraham C, Dorgham N, Gilbert ZA, Chacko L, Hirpara RJ, Mayi BS, Jacobs RJ (2022) COVID-19 misinformation on social media: a scoping review. Cureus 14(4):e24601. https://doi.org/10.7759/cureus.24601

Kaadan MI, Abdulkarim J, Chaar M, Zayegh O, Keblawi MA (2021) Determinants of COVID-19 vaccine acceptance in the Arab world: a cross-sectional study. Glob Health Res Policy 6(1):23. https://doi.org/10.1186/s41256-021-00202-6

Kacimi SEO, Klouche-Djedid SN, Riffi O, Belaouni HA, Yasmin F, Essar MY, Taouza FA, Belakhdar Y, Fellah SC, Benmelouka AY, Ahmed S, Aloulou M, Bendelhoum A, Merzouk H, Ghozy S, Shah J, Haireche MA (2022) Determinants of COVID-19 vaccine engagement in Algeria: a population-based study with systematic review of studies from Arab countries of the MENA region. Front Public Health 10:843449. https://doi.org/10.3389/fpubh.2022.843449

Karafillakis E, Van Damme P, Hendrickx G, Larson HJ (2022) COVID-19 in Europe: new challenges for addressing vaccine hesitancy. Lancet 399(10326):699–701. https://doi.org/10.1016/s0140-6736(22)00150-7

Kashte S, Gulbake A, El-Amin Iii SF, Gupta A (2021) COVID-19 vaccines: rapid development, implications, challenges and future prospects. Hum Cell 34(3):711–733. https://doi.org/10.1007/s13577-021-00512-4

Kasrine Al Halabi C, Obeid S, Sacre H, Akel M, Hallit R, Salameh P, Hallit S (2021) Attitudes of Lebanese adults regarding COVID-19 vaccination. BMC Public Health 21(1):998. https://doi.org/10.1186/s12889-021-10902-w

Kassianos G, Puig-Barberà J, Dinse H, Teufel M, Türeci Ö, Pather S (2022) Addressing COVID-19 vaccine hesitancy. Drugs Context 11. https://doi.org/10.7573/dic.2021-12-3

Kerr JR, Freeman ALJ, Marteau TM, van der Linden S (2021) Effect of information about COVID-19 vaccine effectiveness and side effects on behavioural intentions: two online experiments. Vaccines (Basel) 9(4). https://doi.org/10.3390/vaccines9040379

Kevin B, de Alex F, Rachel G-A, Euzebiusz J, James D, Trudo L, Salmaan K, Janice EG, Stefan B (2022) The unintended consequences of COVID-19 vaccine policy: why mandates, passports and restrictions may cause more harm than good. BMJ Glob Health 7(5):e008684. https://doi.org/10.1136/bmjgh-2022-008684

Khaled SM, Petcu C, Bader L, Amro I, Al-Hamadi A, Al Assi M, Ali AAM, Le Trung K, Diop A, Bellaj T, Al-Thani MH, Woodruff PW, Alabdulla M, Haddad PM (2021) Prevalence and potential determinants of COVID-19 vaccine hesitancy and resistance in Qatar: results from a nationally representative survey of Qatari nationals and migrants between December 2020 and January 2021. Vaccines (Basel) 9(5). https://doi.org/10.3390/vaccines9050471

Kibongani Volet A, Scavone C, Catalán-Matamoros D, Capuano A (2022) Vaccine hesitancy among religious groups: reasons underlying this phenomenon and communication strategies to rebuild trust. Front Public Health 10:824560. https://doi.org/10.3389/fpubh.2022.824560

Knudsen J, Perlman-Gabel M, Uccelli IG, Jeavons J, Chokshi DA (2023) Combating misinformation as a core function of public health. NEJM Catalyst 4(2):CAT.22.0198. https://doi.org/10.1056/CAT.22.0198

Kricorian K, Civen R, Equils O (2022) COVID-19 vaccine hesitancy: misinformation and perceptions of vaccine safety. Hum Vaccin Immunother 18(1):1950504. https://doi.org/10.1080/21645515.2021.1950504

Kruk ME, Gage AD, Arsenault C, Jordan K, Leslie HH, Roder-DeWan S, Adeyi O, Barker P, Daelmans B, Doubova SV et al (2018) High-quality health systems in the sustainable development goals era: time for a revolution. Lancet Glob Health 6(11):e1196–e1252. https://doi.org/10.1016/s2214-109x(18)30386-3

Kuehn BM (2022) Health care professionals' advice can increase COVID-19 vaccination. JAMA 327(6):518. https://doi.org/10.1001/jama.2022.0338

Kumar D, Chandra R, Mathur M, Samdariya S, Kapoor N (2016) Vaccine hesitancy: understanding better to address better. Israel J Health Policy Res 5(1):2. https://doi.org/10.1186/s13584-016-0062-y

Kumari M, Lu R-M, Li M-C, Huang J-L, Hsu F-F, Ko S-H, Ke F-Y, Su S-C, Liang K-H, Yuan JP-Y, Chiang H-L, Sun C-P, Lee IJ, Li W-S, Hsieh H-P, Tao M-H, Wu H-C (2022) A critical overview of current progress for COVID-19: development of vaccines, antiviral drugs, and therapeutic antibodies. J Biomed Sci 29(1):68. https://doi.org/10.1186/s12929-022-00852-9

Kyriakidis NC, López-Cortés A, González EV, Grimaldos AB, Prado EO (2021) SARS-CoV-2 vaccines strategies: a comprehensive review of phase 3 candidates. npj Vaccines 6(1):28. https://doi.org/10.1038/s41541-021-00292-w

Lalot F, Abrams D, Heering MS, Babaian J, Ozkececi H, Peitz L, Davies Hayon K, Broadwood J (2022) Distrustful complacency and the COVID-19 vaccine: how concern and political trust interact to affect vaccine hesitancy. Polit Psychol. https://doi.org/10.1111/pops.12871

Larson HJ (2022) Defining and measuring vaccine hesitancy. Nat Hum Behav 6(12):1609–1610. https://doi.org/10.1038/s41562-022-01484-7

Larson HJ, Jarrett C, Schulz WS, Chaudhuri M, Zhou Y, Dube E, Schuster M, MacDonald NE, Wilson R (2015) Measuring vaccine hesitancy: the development of a survey tool. Vaccine 33(34):4165–4175. https://doi.org/10.1016/j.vaccine.2015.04.037

Larson HJ, Gakidou E, Murray CJL (2022) The vaccine-hesitant moment. N Engl J Med 387(1):58–65. https://doi.org/10.1056/NEJMra2106441

Law C, Skapetis T, Rodricks R (2022) A scoping review of ethical considerations of mandatory COVID-19 vaccination of healthcare workers. Asian Bioeth Rev 14(4):397–408. https://doi.org/10.1007/s41649-022-00214-5

Lazarus JV, Ratzan SC, Palayew A, Gostin LO, Larson HJ, Rabin K, Kimball S, El-Mohandes A (2021) A global survey of potential acceptance of a COVID-19 vaccine. Nat Med 27(2):225–228. https://doi.org/10.1038/s41591-020-1124-9

Lazarus JV, Romero D, Kopka CJ, Karim SA, Abu-Raddad LJ, Almeida G, Baptista-Leite R, Barocas JA, Barreto ML, Bar-Yam Y et al (2022a) A multinational Delphi consensus to end the COVID-19 public health threat. Nature 611(7935):332–345. https://doi.org/10.1038/s41586-022-05398-2

Lazarus JV, Wyka K, White TM, Picchio CA, Rabin K, Ratzan SC, Parsons Leigh J, Hu J, El-Mohandes A (2022b) Revisiting COVID-19 vaccine hesitancy around the world using data from 23 countries in 2021. Nat Commun 13(1):3801. https://doi.org/10.1038/s41467-022-31441-x

Lazarus JV, Wyka K, White TM, Picchio CA, Gostin LO, Larson HJ, Rabin K, Ratzan SC, Kamarulzaman A, El-Mohandes A (2023) A survey of COVID-19 vaccine acceptance across 23 countries in 2022. Nat Med 29(2):366–375. https://doi.org/10.1038/s41591-022-02185-4

Lee J, Huang Y (2022) COVID-19 vaccine hesitancy: the role of socioeconomic factors and spatial effects. Vaccines (Basel) 10(3). https://doi.org/10.3390/vaccines10030352

Lee SK, Sun J, Jang S, Connelly S (2022) Misinformation of COVID-19 vaccines and vaccine hesitancy. Sci Rep 12(1):13681. https://doi.org/10.1038/s41598-022-17430-6

Lin C, Tu P, Beitsch LM (2020) Confidence and receptivity for COVID-19 vaccines: a rapid systematic review. Vaccines (Basel) 9(1). https://doi.org/10.3390/vaccines9010016

Lipsitch M, Krammer F, Regev-Yochay G, Lustig Y, Balicer RD (2022) SARS-CoV-2 breakthrough infections in vaccinated individuals: measurement, causes and impact. Nat Rev Immunol 22(1):57–65. https://doi.org/10.1038/s41577-021-00662-4

Logue M, Haylock C, Scarborough C, Mackenzie J (2023) "If people are hesitant at all, you just want a really big front door": a rapid qualitative interview study on the Luton COVID-19 vaccination outreach clinics. BMC Public Health 23(1):440. https://doi.org/10.1186/s12889-023-15016-z

Luyten J, Bruyneel L, van Hoek AJ (2019) Assessing vaccine hesitancy in the UK population using a generalized vaccine hesitancy survey instrument. Vaccine 37(18):2494–2501. https://doi.org/10.1016/j.vaccine.2019.03.041

MacDonald NE (2015) Vaccine hesitancy: definition, scope and determinants. Vaccine 33(34):4161–4164. https://doi.org/10.1016/j.vaccine.2015.04.036

Magadmi RM, Kamel FO (2021) Beliefs and barriers associated with COVID-19 vaccination among the general population in Saudi Arabia. BMC Public Health 21(1):1438. https://doi.org/10.1186/s12889-021-11501-5

Mahameed H, Al-Mahzoum K, AlRaie LA, Aburumman R, Al-Naimat H, Alhiary S, Barakat M, Al-Tammemi AB, Salim NA, Sallam M (2023) Previous vaccination history and psychological factors as significant predictors of willingness to receive mpox vaccination and a favorable attitude towards compulsory vaccination. Vaccines 11(5). https://doi.org/10.3390/vaccines11050897

Mangla S, Zohra Makkia FT, Pathak AK, Robinson R, Sultana N, Koonisetty KS, Karamehic-Muratovic A, Nguyen UDT, Rodriguez-Morales AJ, Sanchez-Duque JA, Zamba PT, Aghamohammadi N, Cs F, Haque U (2021) COVID-19 vaccine hesitancy and emerging variants: evidence from six countries. Behav Sci (Basel) 11(11). https://doi.org/10.3390/bs11110148

Maraqa B, Nazzal Z, Rabi R, Sarhan N, Al-Shakhra K, Al-Kaila M (2021) COVID-19 vaccine hesitancy among health care workers in Palestine: a call for action. Prev Med 149:106618. https://doi.org/10.1016/j.ypmed.2021.106618

Marco-Franco JE, Pita-Barros P, Vivas-Orts D, González-de-Julián S, Vivas-Consuelo D (2021) COVID-19, fake news, and vaccines: should regulation be implemented? Int J Environ Res Public Health 18(2). https://doi.org/10.3390/ijerph18020744

Mardian Y, Shaw-Shaliba K, Karyana M, Lau C-Y (2021) Sharia (Islamic law) perspectives of COVID-19 vaccines. Front Trop Dis 2. https://doi.org/10.3389/fitd.2021.788188

Martin LR, Petrie KJ (2017) Understanding the dimensions of anti-vaccination attitudes: the vaccination attitudes examination (VAX) scale. Ann Behav Med 51(5):652–660. https://doi.org/10.1007/s12160-017-9888-y

Marzo RR, Sami W, Alam MZ, Acharya S, Jermsittiparsert K, Songwathana K, Pham NT, Respati T, Faller EM, Baldonado AM, Aung Y, Borkar SM, Essar MY, Shrestha S, Yi S (2022) Hesitancy in COVID-19 vaccine uptake and its associated factors among the general adult population: a cross-sectional study in six Southeast Asian countries. Trop Med Health 50(1):4. https://doi.org/10.1186/s41182-021-00393-1

Na L, Banks S, Wang PP (2023) Racial and ethnic disparities in COVID-19 vaccine uptake: a mediation framework. Vaccine 41(14):2404–2411. https://doi.org/10.1016/j.vaccine.2023.02.079

Nature Editorial (2023) There's no room for COVID complacency in 2023. Nature 613(7942):7. https://doi.org/10.1038/d41586-022-04476-9

Ngai CSB, Singh RG, Yao L (2022) Impact of COVID-19 vaccine misinformation on social media virality: content analysis of message themes and writing strategies. J Med Internet Res 24(7):e37806. https://doi.org/10.2196/37806

Nguyen KH, Srivastav A, Lindley MC, Fisher A, Kim D, Greby SM, Lee J, Singleton JA (2022) Parental vaccine hesitancy and association with childhood diphtheria, tetanus toxoid, and acellular pertussis; measles, mumps, and rubella; rotavirus; and combined 7-series vaccination. Am J Prev Med 62(3):367–376. https://doi.org/10.1016/j.amepre.2021.08.015

Norhayati MN, Che Yusof R, Azman YM (2021) Systematic review and meta-analysis of COVID-19 vaccination acceptance. Front Med (Lausanne) 8:783982. https://doi.org/10.3389/fmed.2021.783982

Nuwarda RF, Ramzan I, Weekes L, Kayser V (2022) Vaccine hesitancy: contemporary issues and historical background. Vaccines (Basel) 10(10). https://doi.org/10.3390/vaccines10101595

Oduwole EO, Pienaar ED, Mahomed H, Wiysonge CS (2022) Overview of tools and measures investigating vaccine hesitancy in a ten year period: a scoping review. Vaccines (Basel) 10(8). https://doi.org/10.3390/vaccines10081198

Omar DI, Hani BM (2021) Attitudes and intentions towards COVID-19 vaccines and associated factors among Egyptian adults. J Infect Public Health 14(10):1481–1488. https://doi.org/10.1016/j.jiph.2021.06.019

Omer SB, Benjamin RM, Brewer NT, Buttenheim AM, Callaghan T, Caplan A, Carpiano RM, Clinton C, DiResta R, Elharake JA, Flowers LC, Galvani AP, Lakshmanan R, Maldonado YA, McFadden SM, Mello MM, Opel DJ, Reiss DR, Salmon DA, Schwartz JL, Sharfstein JM, Hotez PJ (2021) Promoting COVID-19 vaccine acceptance: recommendations from the Lancet Commission on vaccine refusal, acceptance, and demand in the USA. Lancet 398(10317):2186–2192. https://doi.org/10.1016/s0140-6736(21)02507-1

Opel DJ, Mangione-Smith R, Taylor JA, Korfiatis C, Wiese C, Catz S, Martin DP (2011) Development of a survey to identify vaccine-hesitant parents: the parent attitudes about childhood vaccines survey. Hum Vaccin 7(4):419–425. https://doi.org/10.4161/hv.7.4.14120

Padmawati RS, Heywood A, Sitaresmi MN, Atthobari J, MacIntyre CR, Soenarto Y, Seale H (2019) Religious and community leaders' acceptance of rotavirus vaccine introduction in Yogyakarta, Indonesia: a qualitative study. BMC Public Health 19(1):368. https://doi.org/10.1186/s12889-019-6706-4

Paloyo SR, Caballes AB, Hilvano-Cabungcal AM, De Castro L (2022) Prioritizing the vulnerable over the susceptible for COVID-19 vaccination. Dev World Bioeth 22(3):162–169. https://doi.org/10.1111/dewb.12327

Pandey A, Galvani AP (2023) Exacerbation of measles mortality by vaccine hesitancy worldwide. Lancet Glob Health 11(4):e478–e479. https://doi.org/10.1016/s2214-109x(23)00063-3

Pascual-Iglesias A, Canton J, Ortega-Prieto AM, Jimenez-Guardeño JM, Regla-Nava JA (2021) An overview of vaccines against SARS-CoV-2 in the COVID-19 pandemic era. Pathogens 10(8). https://doi.org/10.3390/pathogens10081030

Patterson NJ, Paz-Soldan VA, Oberhelman R, Moses L, Madkour A, Miles TT (2022) Exploring perceived risk for COVID-19 and its role in protective behavior and COVID-19 vaccine hesitancy: a qualitative study after the first wave. BMC Public Health 22(1):503. https://doi.org/10.1186/s12889-022-12900-y

Patwary MM, Bardhan M, Disha AS, Hasan M, Haque MZ, Sultana R, Hossain MR, Browning M, Alam MA, Sallam M (2021) Determinants of COVID-19 vaccine acceptance among the adult population of Bangladesh using the health belief model and the theory of planned behavior model. Vaccines (Basel) 9(12). https://doi.org/10.3390/vaccines9121393

Peretti-Watel P, Larson HJ, Ward JK, Schulz WS, Verger P (2015) Vaccine hesitancy: clarifying a theoretical framework for an ambiguous notion. PLoS Curr 7. https://doi.org/10.1371/currents.outbreaks.6844c80ff9f5b273f34c91f71b7fc289

Perrone C, Fiabane E, Maffoni M, Pierobon A, Setti I, Sommovigo V, Gabanelli P (2023) Vaccination hesitancy: to be vaccinated, or not to be vaccinated, that is the question in the era of COVID-19. Public Health Nurs 40(1):90–96. https://doi.org/10.1111/phn.13134

Pertwee E, Simas C, Larson HJ (2022) An epidemic of uncertainty: rumors, conspiracy theories and vaccine hesitancy. Nat Med 28(3):456–459. https://doi.org/10.1038/s41591-022-01728-z

Peters MDJ (2022) Addressing vaccine hesitancy and resistance for COVID-19 vaccines. Int J

Nurs Stud 131:104241. https://doi.org/10.1016/j.ijnurstu.2022.104241

Piltch-Loeb R, Harriman NW, Healey J, Bonetti M, Toffolutti V, Testa MA, Su M, Savoia E (2021) COVID-19 vaccine concerns about safety, effectiveness, and policies in the United States, Canada, Sweden, and Italy among unvaccinated individuals. Vaccines (Basel) 9(10). https://doi.org/10.3390/vaccines9101138

Piltch-Loeb R, Silver DR, Kim Y, Norris H, McNeill E, Abramson DM (2022) Determinants of the COVID-19 vaccine hesitancy spectrum. PLoS ONE 17(6):e0267734. https://doi.org/10.1371/journal.pone.0267734

Poland CM, Ratishvili T (2022) Vaccine hesitancy and health care providers: using the preferred cognitive styles and decision-making model and empathy tool to make progress. Vaccine X 11:100174. https://doi.org/10.1016/j.jvacx.2022.100174

Qunaibi E, Basheti I, Soudy M, Sultan I (2021a) Hesitancy of Arab healthcare workers towards COVID-19 vaccination: a large-scale multinational study. Vaccines (Basel) 9(5). https://doi.org/10.3390/vaccines9050446

Qunaibi EA, Helmy M, Basheti I, Sultan I (2021b) A high rate of COVID-19 vaccine hesitancy in a large-scale survey on Arabs. Elife 10. https://doi.org/10.7554/eLife.68038

Rawlings L, Looi JCL, Robson SJ (2022) Economic considerations in COVID-19 vaccine hesitancy and refusal: a survey of the literature*. Econ Rec 98(321):214–229. https://doi.org/10.1111/1475-4932.12667

Rodrigues F, Block S, Sood S (2022) What determines vaccine hesitancy: recommendations from childhood vaccine hesitancy to address COVID-19 vaccine hesitancy. Vaccines (Basel) 10(1). https://doi.org/10.3390/vaccines10010080

Romer D, Jamieson KH (2020) Conspiracy theories as barriers to controlling the spread of COVID-19 in the U.S. Soc Sci Med 263:113356. https://doi.org/10.1016/j.socscimed.2020.113356

Sallam M (2021) COVID-19 vaccine hesitancy worldwide: a concise systematic review of vaccine acceptance rates. Vaccines (Basel) 9(2). https://doi.org/10.3390/vaccines9020160

Sallam M, Şahin G, Ingman M, Widell A, Esbjörnsson J, Medstrand P (2017) Genetic characterization of human immunodeficiency virus type 1 transmission in the Middle East and North Africa. Heliyon 3(7):e00352. https://doi.org/10.1016/j.heliyon.2017.e00352

Sallam M, Dababseh D, Yaseen A, Al-Haidar A, Taim D, Eid H, Ababneh NA, Bakri FG, Mahafzah A (2020) COVID-19 misinformation: mere harmless delusions or much more? A knowledge and attitude cross-sectional study among the general public residing in Jordan. PLoS ONE 15(12):e0243264. https://doi.org/10.1371/journal.pone.0243264

Sallam M, Ababneh NA, Dababseh D, Bakri FG, Mahafzah A (2021a) Temporal increase in D614G mutation of SARS-CoV-2 in the Middle East and North Africa. Heliyon 7(1):e06035. https://doi.org/10.1016/j.heliyon.2021.e06035

Sallam M, Dababseh D, Eid H, Al-Mahzoum K, Al-Haidar A, Taim D, Yaseen A, Ababneh NA, Bakri FG, Mahafzah A (2021b) High rates of COVID-19 vaccine hesitancy and its association with conspiracy beliefs: a study in Jordan and Kuwait among other Arab countries. Vaccines (Basel) 9(1). https://doi.org/10.3390/vaccines9010042

Sallam M, Dababseh D, Eid H, Hasan H, Taim D, Al-Mahzoum K, Al-Haidar A, Yaseen A, Ababneh NA, Assaf A, Bakri FG, Matar S, Mahafzah A (2021c) Low COVID-19 vaccine acceptance is correlated with conspiracy beliefs among university students in Jordan. Int J Environ Res Public Health 18(5). https://doi.org/10.3390/ijerph18052407

Sallam M, Al-Sanafi M, Sallam M (2022a) A global map of COVID-19 vaccine acceptance rates per country: an updated concise narrative review. J Multidiscip Healthc 15:21–45. https://doi.org/10.2147/jmdh.S347669

Sallam M, Eid H, Awamleh N, Al-Tammemi AB, Barakat M, Athamneh RY, Hallit S, Harapan H, Mahafzah A (2022b) Conspiratorial attitude of the general public in Jordan towards emerging virus infections: a cross-sectional study amid the 2022 monkeypox outbreak. Trop Med Infect Dis 7(12). https://doi.org/10.3390/tropicalmed7120411

Sallam M, Ghazy RM, Al-Salahat K, Al-Mahzoum K, AlHadidi NM, Eid H, Kareem N, Al-Ajlouni E, Batarseh R, Ababneh NA, Sallam M, Alsanafi M, Umakanthan S, Al-Tammemi AB, Bakri FG, Harapan H, Mahafzah A, Al Awaidy ST (2022c) The role of psychological factors and vaccine conspiracy beliefs in influenza vaccine hesitancy and uptake among Jordanian healthcare workers during the COVID-19 pandemic. Vaccines (Basel) 10(8). https://doi.org/10.3390/vaccines10081355

Sapienza A, Falcone R (2022) The role of trust in COVID-19 vaccine acceptance: considerations from a systematic review. Int J Environ Res Public Health 20(1). https://doi.org/10.3390/ijerph20010665

Schmid P, Rauber D, Betsch C, Lidolt G, Denker M-L (2017) Barriers of influenza vaccination intention and behavior—a systematic review of influenza vaccine hesitancy, 2005–2016. PLoS ONE 12(1):e0170550. https://doi.org/10.1371/journal.pone.0170550

Schoch-Spana M, Brunson EK, Long R, Ruth A, Ravi SJ, Trotochaud M, Borio L, Brewer J, Buccina J, Connell N, Hall LL, Kass N, Kirkland A, Koonin L, Larson H, Lu BF, Omer SB, Orenstein WA, Poland GA, Privor-Dumm L, Quinn SC, Salmon D, White A (2021) The public's role in COVID-19 vaccination: human-centered recommendations to enhance pandemic vaccine awareness, access, and acceptance in the United States. Vaccine 39(40):6004–6012. https://doi.org/10.1016/j.vaccine.2020.10.059

Schwartz JL (2012) New media, old messages: themes in the history of vaccine hesitancy and refusal. Virtual Mentor 14(1):50–55. https://doi.org/10.1001/virtualmentor.2012.14.1.mhst1-1201

Seale H, Harris-Roxas B, Heywood A, Abdi I, Mahimbo A, Chauhan A, Woodland L (2022) The role of community leaders and other information intermediaries during the COVID-19 pandemic: insights from the multicultural sector in Australia. Humanit Soc Sci Commun 9(1):174. https://doi.org/10.1057/s41599-022-01196-3

Shapiro GK, Holding A, Perez S, Amsel R, Rosberger Z (2016) Validation of the vaccine conspiracy beliefs scale. Papillomavirus Res 2:167–172. https://doi.org/10.1016/j.pvr.2016.09.001

Shearn C, Krockow EM (2023) Reasons for COVID-19 vaccine hesitancy in ethnic minority groups: a systematic review and thematic synthesis of initial attitudes in qualitative research. SSM Qual Res Health 3:100210. https://doi.org/10.1016/j.ssmqr.2022.100210

Shen SC, Dubey V (2019) Addressing vaccine hesitancy: clinical guidance for primary care physicians working with parents. Can Fam Phys 65(3):175–181

Singh DD, Parveen A, Yadav DK (2021) SARS-CoV-2: emergence of new variants and effectiveness of vaccines. Front Cell Infect Microbiol 11:777212. https://doi.org/10.3389/fcimb.2021.777212

Singh P, Dhalaria P, Kashyap S, Soni GK, Nandi P, Ghosh S, Mohapatra MK, Rastogi A, Prakash D (2022) Strategies to overcome vaccine hesitancy: a systematic review. Syst Rev 11(1):78. https://doi.org/10.1186/s13643-022-01941-4

Siu JY, Cao Y, Shum DHK (2022) Perceptions of and hesitancy toward COVID-19 vaccination in older Chinese adults in Hong Kong: a qualitative study. BMC Geriatr 22(1):288. https://doi.org/10.1186/s12877-022-03000-y

Skafle I, Nordahl-Hansen A, Quintana DS, Wynn R, Gabarron E (2022) Misinformation about COVID-19 vaccines on social media: rapid review. J Med Internet Res 24(8):e37367. https://doi.org/10.2196/37367

Soares P, Rocha JV, Moniz M, Gama A, Laires PA, Pedro AR, Dias S, Leite A, Nunes C (2021) Factors associated with COVID-19 vaccine hesitancy. Vaccines (Basel) 9(3). https://doi.org/10.3390/vaccines9030300

Southwell BG, Wood JL, Navar AM (2020) Roles for health care professionals in addressing patient-held misinformation beyond fact correction. Am J Public Health 110(S3):S288-s289. https://doi.org/10.2105/ajph.2020.305729

Steffens MS, Dunn AG, Leask J, Wiley KE (2020) Using social media for vaccination promotion: practices and challenges. Digit Health 6:2055207620970785. https://doi.org/10.1177/2055207620970785

Strully KW, Harrison TM, Pardo TA, Carleo-Evangelist J (2021) Strategies to address COVID-19 vaccine hesitancy and mitigate health disparities in minority populations. Front Public Health 9:645268. https://doi.org/10.3389/fpubh.2021.645268

Suliman DM, Nawaz FA, Mohanan P, Modber M, Musa MK, Musa MB, El Chbib D, Elhadi YAM, Essar MY, Isa MA, Lucero-Prisno DE 3rd, Moonesar IA (2021) UAE efforts in promoting COVID-19 vaccination and building vaccine confidence. Vaccine 39(43):6341–6345. https://doi.org/10.1016/j.vaccine.2021.09.015

Tang JW, Caniza MA, Dinn M, Dwyer DE, Heraud J-M, Jennings LC, Kok J, Kwok KO, Li Y, Loh TP, Marr LC, Nara EM, Perera N, Saito R, Santillan-Salas C, Sullivan S, Warner M, Watanabe A, Zaidi SK (2022) An exploration of the political, social, economic and cultural factors affecting how different global regions initially reacted to the COVID-19 pandemic. Interface Focus 12(2):20210079. https://doi.org/10.1098/rsfs.2021.0079

The New York Times (2023) Tracking coronavirus vaccinations around the world. https://www.nytimes.com/interactive/2021/world/covid-vaccinations-tracker.html. Accessed 30 Apr 2023

Thorpe A, Fagerlin A, Drews FA, Butler J, Stevens V, Riddoch MS, Scherer LD (2022) Communications to promote interest and confidence in COVID-19 vaccines. Am J Health Promot 36(6):976–986. https://doi.org/10.1177/08901171221082904

Tian F, Yang R, Chen Z (2022) Safety and efficacy of COVID-19 vaccines in children and adolescents: a systematic review of randomized controlled trials. J Med Virol 94(10):4644–4653. https://doi.org/10.1002/jmv.27940

Tormen M, Taliento C, Salvioli S, Piccolotti I, Scutiero G, Cappadona R, Greco P (2023) Effectiveness and safety of COVID-19 vaccine in pregnant women: a systematic review with meta-analysis. BJOG 130(4):348–357. https://doi.org/10.1111/1471-0528.17354

Tregoning JS, Flight KE, Higham SL, Wang Z, Pierce BF (2021) Progress of the COVID-19 vaccine effort: viruses, vaccines and variants versus efficacy, effectiveness and escape. Nat Rev Immunol 21(10):626–636. https://doi.org/10.1038/s41577-021-00592-1

Troiano G, Nardi A (2021) Vaccine hesitancy in the era of COVID-19. Public Health 194:245–251. https://doi.org/10.1016/j.puhe.2021.02.025

Tuckerman J, Kaufman J, Danchin M (2022) Effective approaches to combat vaccine hesitancy. Pediatr Infect Dis J 41(5):e243–e245. https://doi.org/10.1097/inf.0000000000003499

Ullah I, Khan KS, Tahir MJ, Ahmed A, Harapan H (2021) Myths and conspiracy theories on vaccines and COVID-19: potential effect on global vaccine refusals. Vacunas 22(2):93–97. https://doi.org/10.1016/j.vacun.2021.01.001

Urrunaga-Pastor D, Bendezu-Quispe G, Herrera-Añazco P, Uyen-Cateriano A, Toro-Huamanchumo CJ, Rodriguez-Morales AJ, Hernandez AV, Benites-Zapata VA (2021) Cross-sectional analysis of COVID-19 vaccine intention, perceptions and

hesitancy across Latin America and the Caribbean. Travel Med Infect Dis 41:102059. https://doi.org/10.1016/j.tmaid.2021.102059
Wang B, Nolan R, Marshall H (2021) COVID-19 immunisation, willingness to be vaccinated and vaccination strategies to improve vaccine uptake in Australia. Vaccines (Basel) 9(12). https://doi.org/10.3390/vaccines9121467
Wang D, Zhou Y, Ma F (2022) Opinion leaders and structural hole spanners influencing echo chambers in discussions about COVID-19 vaccines on social media in China: network analysis. J Med Internet Res 24(11):e40701. https://doi.org/10.2196/40701
Wang G, Yao Y, Wang Y, Gong J, Meng Q, Wang H, Wang W, Chen X, Zhao Y (2023) Determinants of COVID-19 vaccination status and hesitancy among older adults in China. Nat Med 29(3):623–631. https://doi.org/10.1038/s41591-023-02241-7
Wehbe S, Fahme SA, Rizk A, Mumtaz GR, DeJong J, Sibai AM (2021) COVID-19 in the Middle East and North Africa region: an urgent call for reliable, disaggregated and openly shared data. BMJ Glob Health 6(2). https://doi.org/10.1136/bmjgh-2021-005175
Whitehead HS, French CE, Caldwell DM, Letley L, Mounier-Jack S (2023) A systematic review of communication interventions for countering vaccine misinformation. Vaccine 41(5):1018–1034. https://doi.org/10.1016/j.vaccine.2022.12.059
Wiysonge CS, Ndwandwe D, Ryan J, Jaca A, Batouré O, Anya BM, Cooper S (2022) Vaccine hesitancy in the era of COVID-19: could lessons from the past help in divining the future? Hum Vaccin Immunother 18(1):1–3. https://doi.org/10.1080/21645515.2021.1893062
Wonodi C, Obi-Jeff C, Adewumi F, Keluo-Udeke SC, Gur-Arie R, Krubiner C, Jaffe EF, Bamiduro T, Karron R, Faden R (2022) Conspiracy theories and misinformation about COVID-19 in Nigeria: implications for vaccine demand generation communications. Vaccine 40(13):2114–2121. https://doi.org/10.1016/j.vaccine.2022.02.005
Wouters OJ, Shadlen KC, Salcher-Konrad M, Pollard AJ, Larson HJ, Teerawattananon Y, Jit M (2021) Challenges in ensuring global access to COVID-19 vaccines: production, affordability, allocation, and deployment. Lancet 397(10278):1023–1034. https://doi.org/10.1016/s0140-6736(21)00306-8
Yang X, Wei L, Liu Z (2022) Promoting COVID-19 vaccination using the health belief model: does information acquisition from divergent sources make a difference? Int J Environ Res Public Health 19(7). https://doi.org/10.3390/ijerph19073887
Yarlagadda H, Patel MA, Gupta V, Bansal T, Upadhyay S, Shaheen N, Jain R (2022) COVID-19 vaccine challenges in developing and developed countries. Cureus 14(4):e23951. https://doi.org/10.7759/cureus.23951
Ye Y, Zhang Q, Wei X, Cao Z, Yuan H-Y, Zeng DD (2022) Equitable access to COVID-19 vaccines makes a life-saving difference to all countries. Nat Hum Behav 6(2):207–216. https://doi.org/10.1038/s41562-022-01289-8
Zimmerman T, Shiroma K, Fleischmann KR, Xie B, Jia C, Verma N, Lee MK (2023) Misinformation and COVID-19 vaccine hesitancy. Vaccine 41(1):136–144. https://doi.org/10.1016/j.vaccine.2022.11.014

Interest Groups and COVID-19 in Africa (2019–2022)

18

Samuel Adu-Gyamfi, Razak Mohammed Gyasi, Kabila Abass, and Derrick Sarpong Owusu

Abstract

The study highlights the crucial roles played by interest groups in shaping the definition and redefinition of policy issues related to the COVID-19 pandemic in Africa. The study focused on selected countries to discuss strategies that were deployed to combat the COVID quandary. The selected countries are Nigeria, Ghana, Kenya, Tanzania, South Africa, Zimbabwe, Egypt, and Tunisia. A purely qualitative research, it relied on existing literature sources to produce thematic analysis which explore the various strategies adopted by interest groups in addressing the COVID-19 pandemic in Africa. The study revealed the challenges posed by the COVID-19 pandemic to the already fragile healthcare system in Africa. Again, it revealed the innovative approaches such as herbal medicine and local production of ventilators developed by African countries to tackle COVID-19 including efforts of other interests groups like government agencies, private institutions and international organizations in the fight against COVID-19 in Africa.

Keywords

Interest groups · World Health Organization · West-African Health Organization · African Health Organization · ECSA Health Community · Africa · Ghana · COVID-19

18.1 Summary

COVID-19 claimed millions of lives globally and resulted in unprecedented societal and economic disruptions. Before and during the early stages of the COVID-19 pandemic in Africa, the continent due to the general fragility of its healthcare system, was projected to record the highest mortality rate from COVID-19 as compared to the mortality rate of other continents. The popular philanthropist giant Bill Gates, earlier in 2020, predicted that the devastating impacts of COVID-19 in Africa could be far greater than the country where the virus originated from; China (Wasserman and Moynihan 2020). While there is no doubt about the fragile

S. Adu-Gyamfi (✉) · D. S. Owusu
Department of History and Political Studies, Kwame Nkrumah University of Science and Technology, Kumasi, Ghana
e-mail: mcgyamfi@yahoo.com

R. M. Gyasi
African Population and Health Research Center, Nairobi, Kenya

K. Abass
Department of Geography and Rural Development, Kwame Nkrumah University of Science and Technology, Kumasi, Ghana

N. Rezaei (ed.), *The COVID-19 Aftermath*, Advances in Experimental Medicine and Biology 1457,
https://doi.org/10.1007/978-3-031-61939-7_18

state of Africa's healthcare system, the belief that the mortality rate was projected to be higher than the rest of the other continents was one of the numerous stereotypes on the COVID-19 situation in Africa. As of November 18, 2022, the overall deaths due to COVID-19 in the whole African continent were 257,984 deaths compared to many countries that recorded higher rates (Galal 2022b). This was a surprising development to many people, especially those outside the African continent. Many African countries came up with innovative approaches for the treatment of COVID-19 infections, testing, and contact tracing. For instance, Africans and their beliefs in traditional and herbal medicine resulted in the local production of medicines in Africa for COVID-19 treatment. COVID-organics (CVO) produced from the plant artemisia by the Malagasy Institute of Applied Research are among such herbal medicines that Africans used for COVID-19 treatment. Also from Ghana, herbal medicines such as Nibima, Immunim, and Ampoforte, and other medicinal plants like neem trees are some of the innovative ways deployed by Africans for COVID-19 treatment. As seen in the case of Ghana, there was also the local production of low-cost ventilators by Professor Fred McBagonluri, a Ghanaian engineer, to address the issue of inadequate ventilators in the country during the pandemic. Irrespective of these and other laudable innovative approaches, little attention has been paid to these efforts, which can be seen as a way of downplaying the continent's mitigation efforts against COVID-19.

Due to the novel nature of the virus, there was limited knowledge about it and as such, the various health organizations in Africa similarly adopted training and capacity building as a response to the COVID-19 pandemic. Indeed, in Africa, the COVID-19 pandemic exposed the inadequate preparedness of the healthcare system. Some scholars have argued that bad governance largely accounts for this. Obi-Ani et al. (2021) throw more light on this with their argument that Nigeria's healthcare problems have been in existence for decades, yet, government after government always neglect the poor state of the healthcare system in the country. The authors disclosed how the poor leadership style of the various Nigerian governments over the years has resulted in major adverse impacts in the health sector of Nigeria. Some of these adverse impacts from the poor governance include: lack of basic equipment in hospitals, epileptic power supply, poor remuneration of medical personnel, and lack of essential drugs on the shelves, among others.

Almost all of the governments in Africa encountered several challenges in their mitigation measures against COVID-19. Nonetheless, the governments worked together with their respective health organizations, civil society organizations, non-governmental organizations, chiefs, and other concerned parties to adopt measures that would ensure the safety of their citizens. Arguably, it can be concluded that Africa had a successful response to the COVID-19 pandemic.

Though the organizations listed or discussed as major players are interests groups, they did not embark on actions in the sense of demonstrations or petitioning, rather they augmented and developed strategies to fight the COVID-19 dilemma. However, in some jurisdictions in Africa, we found political opponents or opposition parties, also as interest groups proffering alternative solutions to the COVID question and equally criticizing governments for their shortfalls.

18.2 Introduction

The WHO declared COVID-19 as a Public Health Emergency of International Concern (PHEIC) and a pandemic on March 2020 (Adu-Gyamfi et al. 2022). As a result of this global pandemic, different countries adopted strategies largely emanating from the WHO and other sister organizations on the global, continental, and sub-continental levels.

In all manner of spheres, the outbreak of COVID-19 required political leaders, especially presidents, heads of states and governments of respective countries to take national and institutional actions, which further required a proper definition of the COVID-19 challenge. Essentially, global, continental, and national level policies were seen as requirements to ensure proper disease governance. The proper interpretation of the COVID question or quandary (issue definition) by the political leaders on the continent of Africa would have had a tremendous impact on the fight against the COVID pandemic. Their ability to do same would not only emanate from indiscrimination but also from broader interest groups.

Interest groups as argued by Asare (2009) refer to "Associations or Movements" that exist in a political system with an aim of influencing policies toward their interest (Asare 2009). Concerning COVID-19, the major interest groups on the global level included the WHO and its sister organizations on the continental and sub-continental levels. For example, in the West African block, the WHO's counterpart is the West African Health Organization (WAHO). WAHO contributed tremendously toward the fight against Ebola in countries like Guinea, Liberia, Sierra Leone, and Nigeria in 2014. Within the scheme of things, interest groups like doctors, local traditional, and medical practitioners (herbalists) among others, chiefs or native heads including the broader citizenry who were affected, put up a fight against Ebola.

For the sake of this contribution concerning COVID-19, interest groups include the WHO, the WHO African office in Africa, and sister organizations at the sub-continental level like West African Health Organizations (WAHO), the East, Central, and Southern African Health communities (ECSA-HC). The core mandate of their organization has been to promote and encourage efficiency and relevance in the provision of healthcare services in the region.

The others include the African Centre for Disease Control and Prevention-Africa CDC under the African Union, the respective governments in Africa, their Ministries of Health, Medical service, doctors, nurses, epidemiologists, public health workers, opinion leaders and leaders of civil society groups, members of respective local economic and social groups, and opposition parties among others.

These interest groups either used means such as demonstrations, petitions and venue shopping to air their grievances. For example, venue shopping refers to the activities of advocacy groups and policy makers who seek out a decision setting where they can air their grievances with current policy and present alternative policy proposals (Pralle 2003). Concerning the COVID dilemma, these regional and state as well as para-state organizations and non-state actors or interest groups played a significant role in ensuring some success in the fight against the novel pandemic.

Plein (1991) referencing Schattschneider (1960, 68) reported that "the definition of alternatives is the supreme instrument of power, and the antagonists can rarely agree on what the issues are because power is involved in the definition" (Plein 1991; Schattschneider 1960, 68). Again, referencing Cobb and Elder (1972, 101), Plein (1991) quoted that "Issue definition is by no means an a priori given." The conflicting parties will not necessarily agree on how the issues are to be defined (Cobb and Elder 1972; Plein 1991). It is within this wider concept that the current discourse seeks to argue concerning how these listed interest groups have either used demonstrations, petitions, venue shopping, and cooperation among other things to deal with the challenge of COVID-19 within the global disease governance discourse. It highlights examples from some jurisdictions in Africa to make a case and draws lessons for the future.

Indeed, infectious diseases do not respect political boundaries, yet public health policy decisions are often made at national or subnational levels of governance framework. The following questions are critical in deciphering the COVID quandary. How did the African region with different layers of disease governance manage the COVID-19 quandary? How did the interest groups use demonstrations, petitions, venue shopping, and cooperation among other things to proffer alternative solutions, and whether they were successful in doing same?

The other question worthy of deciphering, concerns how the global-level initiatives from the WHO and global disease governance policies clashed with local interest groups who were helping with the definition and redefinition of issues at the local level in Africa? We argue that this discussion has wider ramifications on broader disease governance issues on the African continent and the world at large. It is important to state that the discussion has also focused on country-specific issues to reflect on the events on the continental level. By way of organization, the rest of the article pays attention to the approach of the research, efforts made by interest groups at the continental and sub-continental levels including actions of the local level interest groups, and challenges with overarching WHO policies against local action.

The current research relies purely on sources from the websites of respective global, continental organizations listed elsewhere in the introduction. Some of them include the website of the WHO-Africa Region. It has also relied on sources from JSTOR, BMC Public Health, among others. Themes were generated and discussed analytically to suit the objectives of the research.

18.3 Response to COVID-19 by Some Health Organizations in Africa Serving as Interest Groups

COVID-19 is not the first pandemic recorded in the world and likely won't be the last. There have been several pandemics recorded globally throughout mankind's history such as the White Plague, Spanish Flu, and Black Death, among many others. In Africa, various health organizations were mandated to bring up measures against the COVID-19 pandemic. These organizations in many respects could be referred to as interest groups. This section shows some of the health organizations and how they responded to the novel coronavirus as interest groups.

The first case of COVID-19 in Africa was recorded on February 4, 2020, in Egypt (Frimpong et al. 2020). Since then, there was a significant increase in the reported cases in Africa. For instance, according to a WHO Africa Report, from February 2020 to March 2020, over 200 new confirmed cases were reported daily with 1079 deaths and 29,463 confirmed cases as of May 3, 2020 (Strategic Response to COVID-19 in the WHO African Region: February to December 2020, 2023). Key drivers for the transmission of COVID in Africa were identified by the WHO-Africa Region to include demographics, risk of exposure, health systems, and disease burden. The WHO-Africa Region, in their response strategy against COVID, grouped their approaches into containment, mixed containment/mitigation, and mitigation scenarios (Strategic Response to COVID-19 in the WHO African Region: February to December 2020, 2023).

The containment scenario was adopted for countries where containment measures were possible. Countries with a limited number of cases and few imported cases fell under this category, and as of May 2020, six African countries were under this category (Strategic Response to COVID-19 in the WHO African Region: February to December 2020, 2023). The measure adopted by the WHO-African Region for the containment scenario was to provide technical support through virtual training and initial supplies during the lockdown.

The mixed containment/mitigation was applicable in such countries which had some areas where containment was possible and other areas with sustained transmission leading to challenges in implementing control measures (Strategic Response to COVID-19 in the WHO African Region: February to December 2020, 2023). For this category, the WHO-Africa Region sought to give maximum technical support through national surging capacity and virtual training in addition to essential supplies.

Countries with widespread and intense transmission in the community were grouped under the mitigation scenario (Strategic Response to COVID-19 in the WHO African Region: February to December 2020, 2023). This scenario was adopted in places where containment failed to achieve its goal. The WHO-Africa Report revealed that 29 countries were under this scenario as of May 3, 2020. Countries under this category were to be provided with major surge, guidance, training, and logistics supplies for critical gap filling with partners by the WHO-Africa Region as the response strategy (Strategic Response to COVID-19 in the WHO African Region: February to December 2020, 2023).

Other health organizations such as the West African Health Organization (WAHO), Africa Centres for Disease Control and Prevention, and East, Central, and Southern Africa Health Community (ECSA-HC) all joined in the fight against COVID-19 in Africa as interest groups.

As a specialized institution of ECOWAS, the West African Health Organization was formed in 1987 with the core mandate of dealing with health issues. As a result of this, all their efforts since the spread of COVID-19 into Africa were to contribute significantly to the fight against the novel virus, especially in its member states. According to Sombié et al. (2020), one of the important measures that the WAHO adopted against COVID-19 was the development of epidemiologic models to enlighten decision-makers in the quest for selection and implementation of response interventions (Sombié et al. 2020). In addition to this, WAHO also made deliveries of personal protective equipment (PPEs) to the various member states, raised funds for COVID-19 regional response, and conducted online seminars to train health practitioners and frontline workers.

Established in 2016 but officially launched in January 2017 as a specialized technical institution of the African Union, Africa Centres for Disease Control and Prevention (Africa CDC) was to provide the necessary support to public health initiatives of member states. One of its core objectives is to address all health threats and emergencies in a timely and effective manner. With COVID making its entry into Africa, the organization was faced with the responsibility of providing measures against the virus. The Africa CDC brought forward several measures which can be grouped into such themes as training, surveillance, and disease intelligence. The organization performed several tasks under these response themes. For instance, under the Africa CDC Institute for Pathogen Genomics, a major scientific report revealed a collaboration on genomics surveillance to show the power of African science and how the majority of COVID-19 variants were introduced in Africa (Africa CDC 2022).

Another health organization that served as an interest group in the mitigation against COVID-19 was the East, Central and Southern Africa Health Community (ECSA-HC). The ECSA-HC is made up of countries such as Eswatini, Kenya, Uganda, Tanzania, Malawi, Mauritius, Lesotho, Zambia, and Zimbabwe. The organization was established in 1974 with the aim of addressing the health needs of its member states and as such was faced with the task of bringing responses or measures against COVID. Much similar to other health organizations, the ECSA-HC conducted webinars on how to manage the COVID-19 pandemic and virtual training on COVID-19 workplace response. Again, in February 2020, the highest governing body of the organization, the ECSA Conference of Health Ministers proposed a resolution to its member states to "strengthen surveillance systems for prevention, early detection, and appropriate response to the threats of disease outbreak including COVID-19" (East, Central, and Southern Africa Health Community 2020).

18.4 Other Interest Groups

In addition to the above measures taken by the WHO-Africa Region, the WAHO, Africa CDC, and the ECSA-HC, several measures were taken by the various governments in Africa against COVID. Some of these were: enforced lockdowns of schools, workplaces, and places of worship, contact tracing, public bans, and other measures like the mandatory wearing of nose masks and social distancing. This next section pays attention to the specific activities of the interest groups and country-specific examples in Africa. In particular, attention is paid to the activities of governments and their ministries of health, opposition parties, chiefs, and other civil society organizations and non-governmental organizations among others. In Sub-Saharan Africa, attention is paid to countries like Nigeria and Ghana, representing the West African region, Kenya and Tanzania representing the East-African block, and South Africa and Zimbabwe representing the Southern African block.

In Northern Africa, attention is paid to Egypt and Tunisia. These countries have been selected because they invariably mimicked the quintessential African situation and paradox concerning the COVID-19 fight within the period 2020–2022.

18.5 COVID-19 in Nigeria and Ghana: Efforts of Interest Groups

18.5.1 Nigeria

Before Nigeria recorded her first case of the novel coronavirus, the Nigerian Federal Government had already put in place measures to intensify surveillance at selected international airports with the aim of preventing the virus from entering Nigeria. Mr. Osagie Ehanire, the Minister of Health in Nigeria, talking about these measures on January 28, 2020, disclosed that:

> … Attention will be given to five international airports to ensure that the point of entries are monitored and the ministry is working to forestall the importation of coronavirus to the country … The risk of importation is possible in all countries. However, the ministry wishes to assure all Nigerians that the capacity to detect, access and respond to this and other public health challenges are put in place. (Agency Report 2020)

Nonetheless, the country was one of the first Sub-Saharan African countries to confirm a case of COVID-19, and as of March 28, 2021, Nigeria was the 5th most affected African country and 77th affected country globally (Uchechi Okoroiwu et al. 2021). The first case of COVID-19 in Nigeria was recorded on February 27, 2020, when an affected 44-year-old Italian citizen arrived at the Murtala Muhammed International Airport, Lagos. The virus was recorded in all 36 states of Nigeria with Lagos State being the state with the highest number of cases. The Federal Government put in place a number of measures in line with WHO guidelines to prevent the spread of the virus. Some of these measures were screening passengers at various international airports, training health workers on such issues as proper use of personal protective equipment (PPEs), case definition guide, how to inform the Federal Ministry of Health on suspected cases, and closure of worship centers, schools, organizations, and businesses for 14 days in Lagos, Ogun State, and the Federal Capital Territory. This closure was extended for another 14 days. Also, the highest emergency level in Nigeria, Emergency Operations Centre (EOC) at level 3 was activated by the Federal Government as a response to the pandemic. Further, on March 17, 2020, Nigeria's President Muhammadu Buhari launched the Presidential Task Force (PTF). Initially planned to last for six months, the PTF's mandate was to coordinate and manage the Federal Government's efforts in the quest to identify, contain and prevent the spread of the virus (Bolu et al. 2022).

An effective COVID-19 testing system is crucial in mitigating the spread of the virus, especially in densely populated countries. Nigeria being a densely populated country fell short of her testing output during the onset of the virus. The country had slow-paced testing for the virus. As compared to Ghana which as of April 27, 2020, had completed over 100,000 tests, Nigeria only had around 12,000 tests completed that time (Simons 2020). A major reason for this was the shortage of testing kits and reagents in Nigeria (Onyeji 2020). This shortage hindered Nigeria's ability to conduct large-scale testing. Another challenge the Federal Government encountered in COVID-19's control was from Nigerians. There were reports of some COVID-19 patients leaving isolation centers to protest against government negligence, inadequate food, and medical attention (Sahara Reporters 2020). Those engaged in such protests defied the government's COVID-19 mitigation measures. According to Akinyemi et al. (2020), some Nigerians, because of their distrust of the

Federal Government, wrongly assumed that the pandemic was a ploy by the government to embezzle public funds. In addition to the belief that COVID-19 was a hoax, those group of Nigerians also believed that they had immunity to the virus and this resulted in the non-compliance of the Federal Government of Nigeria's mitigation measures against COVID-19.

In addition to the Federal Government's responses, non-governmental organizations (NGOs) came up with measures to respond to the COVID-19 outbreak. In Nigeria, the Women's Rights Advancement and Protection Alternative (WRAPA) was one such NGO. WRAPA was formed in 1999 with the core mission of "actualizing the legal rights of women, in private and in public through mobilization, sensitization, and the provision of legal and support services" (Wrapanigeria 2023). With a $50,000 grant support from MacArthur Foundation, the organization came up with a policy advisory that contained recommendations on how to declare sexual gender-based violence (SGBV) response and women's health needs as essential services within COVID-19 and other pandemics. Also, Participatory Communication for Gender Development Initiative, an NGO in Nigeria partnered with the National Orientation Agency Nigeria to combat false news on Nigeria's COVID-19 situation and also informed the citizens on how best to protect themselves from getting the virus.

Traditional rulers or chiefs in many African traditional societies are regarded as a link between the ancestors and the living community. As such, these rulers who are non-state actors are highly respected and admired. With the advent of COVID-19 in Nigeria, these traditional rulers played a significant role in helping the Federal Government's efforts to mitigate the virus. One of the significant roles that these rulers played was the dissemination of information about the virus to create awareness among their subjects on COVID-19. In a study concerning the roles of traditional rulers in mitigating COVID-19 in Enugu State; South-Eastern Nigeria, Ekoh et al. (2021) reported in an interview with a traditional ruler that:

> We create the awareness in the market square, we create awareness in the churches, our town criers move around early in the morning spreading the information. They hit the Ogene [gong] and tell them: 'the Igwe [traditional ruler] said that anyone who is to have a funeral or wedding or anything that brings people together, before going there you should make sure you cover your face with facemasks'. Luckily for us, the local government chairman distributed facemasks to us and we shared it to our people. Everyone should wear it and if you are coming you should wash your hand and when you get there, also wash your hand. That's what we do, we use the town criers to spread the information.

Again, it was made known from the study that traditional rulers in Enugu State aided the Federal Government's measures by stopping all ceremonial events that attracted a large crowd in their state. Traditional rulers in Nigeria through their efforts led to people presenting themselves voluntarily for COVID-19 testing. Mr. Mustapha Katibi, WHO State Coordinator for Nigeria, throws more light on this when he revealed that:

> Shortly after our sensitization of traditional leaders in Kwara state during epidemiological week 46 in 2020, testing increased from zero to 875 by week 52 … So far this year, weekly tests have remained higher than 300 samples, thanks to the ongoing sensitization efforts in the communities by community and religious leaders. (WHO Africa 2021)

18.5.2 Ghana

Just like Nigeria, Ghana's government put in place measures as a response strategy before confirming her first case of COVID-19. On March 11, 2020, the Government of Ghana dedicated $100 million to enhance the country's readiness against the virus (Ghanaweb 2020a). A day after this, on March 12, 2020, Kweku-Agyeman-Manu, the Health Minister confirmed Ghana's first case of COVID-19 (Duncan 2020). According to the minister, these cases have been tagged as imported cases because the affected individuals came to Ghana with the virus from Norway and Turkey. Immediately after the case confirmation, the government instituted measures to detect, contain and prevent the spread of the virus. The first of these measures ensued on March 15 when Ghana's President Akufo-Addo announced a ban on all public gatherings, and the closure of schools, and places of worship. Other measures were implemented by the government including a ban on travelers from a country with more than 200 cases and mandatory quarantine of all travelers who came to the country 48 hours before the closure of the country's borders. A partial lockdown of Accra, Kumasi, and Kasoa was announced by the president and took effect on March 30, 2020, and on April 26, the usage of face masks was made mandatory. Case search and contact tracing strategies were also implemented as a response by the government to enhance surveillance to ensure early detection of COVID-19 affected persons, and isolation of same. The government also embarked on a nationwide disinfection and fumigation exercise which led to the disinfection of about 266 markets in the Northern, North East, and Savannah Regions of Ghana (Duodu and Fugu 2020). The response strategies of the Federal Government of Nigeria and the Government of Ghana were similar; however, Ghana had the upper hand when it came to testing capabilities. With the adoption of pooled sampling (testing), Ghana catapulted to be the leading African country in testing its population as of April 14, 2020 (Daily Guide 2020). One commendable measure adopted by the Government of Ghana was the Coronavirus Alleviation Programme Business Support Scheme which was launched by the President of Ghana on May 19, 2020, to help micro, small, and medium enterprises (MSMEs) affected by the pandemic (Myjoyonline 2020).

Measures adopted by Ghana's government to deal with the pandemic were laudable. However, there were some challenges that the government faced in its mitigation response to COVID-19. For instance, the government faced the major challenge of securing isolation and quarantine centers at the onset of the pandemic. As an alternative, the government resorted to temporarily using hotels and churches as isolation and quarantine centers. Again, Amponsah et al. (2021) reveal that there was another challenge of a low physician-to-patient ratio, lack of intensive care unit (ICU) beds and ventilators. In April 2020, the health minister, Kwaku Agyeman Manu, revealed that Ghana had only 67 ventilators. The minister further revealed that:

> An extra order of 50 has been made out of which 20 has arrived whilst we await the remaining 30. That together with what we have in the private gives us about 200 ventilators in Ghana. (Arhinful 2020)

This was discussed into details in a later study by Siaw-Frimpong et al. (2021). In a study concerning the capacity of intensive care units in Ghana during the COVID-19 pandemic, the authors disclosed that there were 113 adult and 36 pediatric ICU beds in the country with over 30 million population. It further reported that the country had only 67 ventilators, and no non-invasive ventilation machines. There was also the issue of regional disparities in the distribution of the ICU beds, with majority of them being distributed in the large urban areas of the country. The authors also reported that there was inadequate ICU staff in the country who could operate the ventilators. Essentially, in an instance where new ventilators were brought into the country, there would have been an underutilization of the same. Ghanaian health workers became overwhelmed with the increasing number of confirmed COVID-19 cases. In addition to this, the health workers also expressed concerns about contracting the virus due to inadequate PPE's including N-95 face masks, and the poor or sordid state of isolation centers. This even triggered threats of strike by some health workers in Ghana. For instance, part of a memo to the head of Department of the Accident and Emergency unit of the Korle-Bu Teaching Hospital, written by disgruntled doctors read that:

> We write to express our general dissatisfaction with the preparedness or lack thereof of the department to combat the current COVID-19 pandemic. We also wish to express our displeasure with the events surrounding the hospital's first confirmed case. Firstly, the pandemic, an existential threat to us, you will agree has to be handled with decisiveness and transparency. It is based on this that, we are against the actions or inactions taken before, during and after the case had been confirmed … We will like to remind management

> of the department that, it has failed woefully in providing adequate protection for the staff and as such, we're left with little choice than to resort to protecting ourselves in the face of clear and present danger. (Ghanaweb 2020b)

The ruling government in Ghana during the pandemic was the New Patriotic Party (NPP). The NPP Government's approach to handling the COVID-19 pandemic received lots of criticism from the main opposition party, the National Democratic Congress (NDC). Sammy Gyamfi, the National Communications Officer of NDC on May 12, 2020, claimed that:

> To aver for starters, the President's speech was full of deliberate distortions and doesn't reflect Ghana's true COVID situation. Contrary to what the President would have us believe, Ghana's current COVID-19 situation is very bleak. We have recorded a total of 4,700 positive cases with 22 deaths, making us the number one country with more COVID-19 cases in the whole of West Africa. This fact becomes even more glaring when Ghana's situation is juxtaposed with that of developing countries like Rwanda (284 cases with no deaths) and Vietnam (288 cases with no deaths, even though Vietnam shares a vast land border with China, with a population of 97 million). The President's claim that Ghana has done better in the COVID fight because we have tested more persons is untenable. It must be understood that a higher number of tests based on contacts traced, is no proof that a country is doing better than a country with a lower number of tests. For instance, the UK has conducted more coronavirus tests than Taiwan. However, it cannot be said that the UK has performed better in the COVID fight than Taiwan who are internationally acclaimed as a shining light in the containment of pandemic. It is worth emphasizing that several countries were proactive in adopting aggressive preventive and containment measures early, thereby reducing community infections. Hence, they have lower numbers of contacts to trace and test. Ghana was sloppy from the outset, as we delayed in closing our borders and airport and allowed an influx of people from abroad into the country. This is why we have more contacts to trace and test than most African countries. Simply put, we are in this mess because President Akufo-Addo failed to adopt aggressive preventive and containment measures on time. This feat cannot be an achievement but rather, a monumental failure on the part of President Akufo-Addo to protect Ghanaians in the first place. (Yeboah 2020)

Just like how traditional rulers played a significant role in COVID-19 response strategies in Nigeria, traditional rulers/chiefs in Ghana also supported the government massively in the fight against COVID-19. These rulers served as enforcers of the government's policies. For instance, the Asantehene through the Otumfuo Osei Tutu II Charity Foundation partnered with Mastercard Foundation to enforce strict adherence to all COVID-19 protocols in Kumasi markets (Kasapafmonline 2020). Also, the Asantehene, after the partial lockdown, aided the government in its economic relief bid when he took the initiative of providing food and other essential commodities to the needy and vulnerable in Kumasi. These included the distribution of bags of rice, salt, vegetables, boxes of hand sanitizers, and other items worth GHC 1 million (Awuah 2020).

Civil societies and NGOs in Ghana also complemented the government's efforts in the fight against COVID-19. Star Ghana Foundation, an NGO based in Ghana received support from UKAid and European Union and on April 28, 2020, launched COVID-19 Response Project (CRP). This was launched to support the government's efforts and one key objective of this project was to improve "citizen's access to accurate, timely, inclusive and relevant information on COVID-19 and response actions" (Kankam Kusi 2021). Good Neighbors Ghana, another NGO, provided items such as antiseptic soaps, nose masks, bags of rice, and other items to 600 Kayayei in Accra during the pandemic. Also, the Medical and Surgical Skills Institute, a health training center in Ghana, partnered with the University of Minnesota's Center for Global Health and Social Responsibility to deliver online COVID-19 training (Vang 2020).

18.6 COVID-19 in Kenya and Tanzania: Efforts of Interest Groups

18.6.1 Kenya

In the Eastern region of the African continent, Kenya was the second country which was most affected by COVID-19. As of June 1, 2022, the country had recorded over 5600 deaths caused by COVID-19 (Kamer 2022a). Kenya's first case was confirmed on March 12, 2020. The infected person was a Kenyan citizen who had returned to Nairobi from the United States (Ministry of Health 2020). AS steps toward cushioning Kenyans against COVID-19, the government of Kenya put measures in place to prevent the spread of the virus. On March 25, 2020, President Uhuru Kenyatta outlined these measures to Kenyans in a state address. Before these strategies were made public, there was already a committee instituted on February 28, 2020, to coordinate Kenya's mitigation measures. These mitigation measures included mandated increases in testing capacity, supplies, and human resources, training of health workers, travel ban, closures of workplaces, schools, and worship centers, and nationwide curfew, among others. Also, as a measure to contain the spread of the virus, 4800 prisoners who had been imprisoned for petty offenses were released on April 2, 2020 (McDade et al. 2020).

Kenya's government encountered a number of challenges in its pandemic response. Enforcement of the curfew measures resulted in police brutalities that led to the deaths of about 15 people with several others injured (Anadolu Agency 2020). These police brutalities resulted in protests which were contrary to the government's policy of social distancing. Mwangangi (2020), the Chief Administrative Secretary of Kenya's Ministry of Health, outlined some other challenges the government faced in its response to COVID-19 including such challenges as shortage of PPEs, difficulties in contact tracing, and shortage of reagents.

Village chiefs in Kenya as seen in the case of other African countries assisted the government in its mitigation measures. Magwanda (2022) reveals that village chiefs were educated on COVID-19 and its dangers and the chiefs' responsibility was to disseminate information regarding COVID-19 to their communities through their local dialects. Food aid was part of the measures implemented by the government of Kenya to support struggling citizens as a result of the economic constraints caused by COVID-19. Village chiefs were instrumental in the identification of vulnerable households deserving of this policy of food aid. Again, during the pandemic, it was reported that some rural communities took advantage of the closure of schools and performed Female Genital Mutilation on girls. As a result of this, some rural chiefs took it upon themselves to make checks in every house in their communities. This was done to prevent such rural communities from performing Female Genital Mutilation (Bhalla 2020).

NGOs in Kenya did not exempt themselves from the fight against COVID-19. Shining Hope for Communities, an NGO in Nairobi engaged in distributing water, hygiene supplies, food, supporting women survivors of gender-based violence, and cash transfers to the vulnerable in Kenya (OCHA 2020). This NGO also partnered with Kenya's Health Ministry to administer COVID-19 vaccination. There was an increase in the number of gender-based violence during the pandemic, and NGOs such as Wangu Kanja Foundation, in tandem with Kenya's government bodies, collaborated to fight against gender-based violence.

18.6.2 Tanzania

In about two months period after announcing its first case, the United Republic of Tanzania's late president John Magufuli declared that there were no more active cases of COVID-19 in Tanzania, and as such, the country was COVID-19 free (Dahir 2020). John Magufuli's government of Tanzania's mitigation measures against COVID-19 triggered a lot of criticisms and concerns globally. The first COVID-19 case in Tanzania was recorded on March 16, 2020. The victim was a Tanzanian woman who had returned to the country from Belgium. Shortly after, the government took some measures against the virus which included: the closure of learning institutions, supplies of PPEs, suspension of international flights and 14 days of mandatory quarantine for international travelers, among others. However, it is reported that in May 2020, John Magufuli's government began to downplay the severity of the impacts of COVID-19 and encouraged new measures which many have criticized as bungled measures against COVID-19.

The government suspended releasing statistics on COVID-19 to the public and WHO in May 2020 when the case counts from the last official update on April 29, 2020, stood at 509 active cases with 21 deaths (AFP 2020). Magufuli on May 22, 2020, declared Tanzania a COVID-19 free territory. All COVID-19 protocols were no longer enforced; learning institutions were ordered to reopen, isolation centers were closed and international flights were expected to resume. Further, the government publicly announced that it would not accept any COVID-19 vaccine. The president in a speech justified this stance by saying:

> "The Ministry of Health should be careful, they should not hurry to try these vaccines without doing research, not every vaccine is important to us, we should be careful. We should not be used as 'guinea pigs … Vaccinations are dangerous. If the white man was able to come up with vaccinations, he should have found a vaccination for AIDS, cancer, and TB by now" (BBC 2021). As a devoted Christian, the late president described COVID-19 as "the devil" and that it "cannot survive in the body of Christ … It will burn instantly." (BBC 2021)

The president admonished Tanzanians to pray in churches and mosques as he believed COVID-19 could only be cured by divine interventions. Lack of transparency, misinformation, and denial of COVID-19 can be seen as the main challenges which were intertwined in Magufuli's government responses to COVID-19 in Tanzania. Predictably, these measures generated a lot of attention; mainly criticisms. For instance, Tedros Adhanom Ghebreyesus, Director General of the World Health Organization, in a statement on Tanzania's COVID-19 situation expressed his concerns by saying:

> This situation remains very concerning. I renew my call for Tanzania to start reporting COVID-19 cases and share data. I also call on Tanzania to implement the public health measures that we know work in breaking the chains of transmission, and to prepare for vaccination. (Jerving 2021)

Also, the Director of Africa CDC called out the government of Tanzania on COVID by saying:

> We continue to hope and plead that Tanzania could come forth and report the situation as it is so that we can work collaboratively to stem this virus out of the continent … we continue to reach out [to Tanzania] but we are not having the response that we expect. (Mitumba 2020)

The COVID-19 pandemic in Tanzania took a new turn after President John Magufuli's death on March 17, 2021. Opposition leaders in parliament argued that the cause of his death was COVID-19, which would have made it ironic, but heart complications have been cited as the cause of his death. Samia Suluhu ascended as the new president, and under her leadership, a massive change in measures against COVID-19 was introduced by her new government. The Suluhu-led government became transparent about the virus and acknowledged the presence of COVID-19 in the country. The new government resorted to WHO's recommended guidelines against COVID. The government through its appeal to Tanzanians on COVID-19 vaccination resulted in about 8.8 million doses of COVID vaccine being administered as of June 19, 2022 (Kamer 2022b). In the efforts to increase its mitigation measures, the government of Tanzania received technical support from international organizations such as UICEF, USAID, Centers for Disease Control and Prevention, and British Council, among others. With this approach from the Suluhu-led government, NGOs in Tanzania aided the government in the fight against COVID. Seven (7) isolation centers in Dar es Salaam, Dodoma, Shinyanga Geita, and Mara were refurbished by the Benjamin Mkapa Foundation (BMF), an NGO in Dar es Salaam, Tanzania. The BMF also collaborated with UNICEF to reach out to over 5 million Tanzanians on COVID-19 prevention messages and other essential services. Other NGOs such as the Community Concern of Orphans and Development Association received financial support from The Rockefeller Foundation to start house-to-house vaccination/immunization outreach in Njombe District and Makambo Town, Tanzania.

18.7 COVID-19 in South Africa and Zimbabwe: Efforts of Interest Groups

18.7.1 South Africa

With about 8078 cases recorded in a single day on March 6, 2021, South Africa is the country in Africa, which was deeply or highly affected by COVID-19 (Galal 2022a). In July 2020, South Africa was ranked globally as the fifth country in terms of the high cumulative number of COVID-19 cases. The country's first case of COVID-19 was confirmed on March 5, 2020, which involved a 38-year-old KwaZulu-Natal man who had contracted the virus in Italy and returned to South Africa on March 1, 2020. The WHO-recommended measures were swiftly implemented by South Africa's government as interventions to curb the spread of COVID-19. In line with the Disaster Management Act of South Africa, a national state of disaster was declared by the president of South Africa, Cyril Ramaphosa on March 15, 2020. The declaration of the national state of disaster was necessary as the president revealed that it empowered the government to take measures that were effective in mitigating the spread of the virus. A full restrictive lockdown followed shortly after the declaration of the national state of disaster. Some other interventions by the government included: the closure of schools, travel bans, revoked visas, closure of 35 out of the country's 53 land ports, and a ban on large gatherings. According to De Villiers et al. (2020), South Africa's response strategy mainly focused on four factors which were: preparation, primary detection, lockdown, and enhanced surveillance.

One major challenge that jeopardized the South African government's COVID-19 response was the violent unrest and protests that happened in the country during the pandemic. Tardzenyuy Thomas (2021) revealed that these protests were organized by NGOs, health workers, and minibus drivers on the allegations of corruption and mismanagement of funds linked to COVID-19 mitigation. Lizette Lancaster and Godfrey Mulaudzi, both from the Institute for Security Studies-Africa, expressed their opinion on the protests:

> The country is currently caught in myriad corruption scandals linked to politically connected people stealing funds meant to help tackle the COVID-19 pandemic. So, we can expect to see increasing numbers of angry South Africans taking to the streets to express their frustration and disgust at the political elite believed to be out of touch with the plight of the ordinary people. (Lancaster and Mulaudzi 2020)

A major protest accompanied by looting ensued after the arrest of South Africa's former president Jacob Zuma. The looting and riots were also fueled by the job layoffs and economic inequalities that emerged from the government's COVID-19 policies (Bauer 2021). These protests and unrest happened against the government's policies on crowd gatherings and significantly imperiled the government's COVID-19 response. Also, the government and its healthcare system encountered the challenge of capacity shortfalls in their response to COVID-19. There was an acute shortage of PPEs for healthcare professionals in South Africa. There were also reports of some Mfuleni nurses refusing to work because of their colleagues testing positive for the virus (Hlati 2020).

NGOs in South Africa augmented the government's efforts in the fight against COVID-19. South Africa Solitary Fund was among such NGOs. The fund distributed about 20 million units of PPEs to healthcare workers in South Africa and also made food donations to vulnerable households in the country (Eribo 2021). The Solitary Fund also received support from other organizations such as Direct Relief and a $2.5 million grant from the UK to support the organization's efforts against COVID-19. FoodForward SA, Gift of the Givers, and Peace Parks Foundation are other NGOs that played prominent roles in the mitigation of COVID-19 in South Africa. Traditional rulers in South Africa also aided the government's efforts through the suspention of cultural activities in their communities and supporting vaccination campaigns. Mr. Chupu Stanley Mathabatha, the Premier of Limpopo Province of South Africa, heaped praises on traditional leaders for the support they have given the government in the fight against COVID-19 (South African Government 2021).

18.7.2 Zimbabwe

As of July 8, 2020, the deadly COVID-19 had affected 213 countries and territories globally (Dao and Nguyen 2020). Zimbabwe was not spared by COVID-19. According to the WHO, as of March 21, 2023, the total number of confirmed cases was 264,391 with 5672 deaths in Zimbabwe (WHO 2023b). The country's first confirmed case was reported on March 1, 2020. The government of Zimbabwe quickly introduced stringent mitigation strategies to contain and prevent the spread of the virus. The pandemic was pronounced as a state of national disaster by H. E. Emmerson Mnangagwa, president of Zimbabwe. This pronouncement was accompanied by several mitigation measures such as expanding testing strategies, ban of public gatherings of more than 100 people, closure of schools and borders, cancelation of Independence Day celebrations, the mandatory wearing of face masks, and restriction of hospital visits, among others.

Challenges in the healthcare system disrupted the efficiency of the mitigation measures implemented by the government of Zimbabwe. There were shortages of PPEs, testing kits, among other important medical equipment. This lack of medical equipment and COVID-19 protective equipment resulted in a court case. The Zimbabwe Association of Doctors for Human Rights (ZADHR) sued the government of Zimbabwe on this issue (Chingono 2020). Murewanhema (2021) also disclosed patients running away from isolation centers due to the bad conditions of the quarantine centers.

Various Civil Societies in Zimbabwe adopted measures as a way of addressing both the impact of the pandemic and the impact of the government's mitigation responses. These civil society groups represented the interests of such sectors as the media, health sector, human rights, gender, and disability rights. For instance, in the health sector, the ZADHR sued the Zimbabwe government over its failure to provide personal protective equipment and other medical equipment that were highly needed by health workers to fight against COVID-19. Information for Development Trust, a media NGO in Zimbabwe through their investigative journalism, compelled the government of Zimbabwe to provide access to accurate information regarding the COVID-19 pandemic in the country (Masuku 2020). Also, with the announcement of the lockdown in Zimbabwe, the National Association of Non-Governmental Organizations (NANGO), the officially recognized coordinating body of all NGOs in Zimbabwe, recommended some measures to the government which ensured the general well-being of the vulnerable and disenfranchised groups in Zimbabwe (GCAP 2020).

The institution of traditional leadership is highly respected in Zimbabwe. Traditional leaders in Zimbabwe are made up of chiefs, headmen, and village heads who play a pivotal role in local governance and development. At the onset of COVID-19 in Zimbabwe, traditional rulers supported the government in its mitigation measures by enforcing various measures from the government. For example, the headman of Chipiro village in Zimbabwe, always incorporated village health workers in meetings he chaired. These village health workers in turn enforced COVID-19 safety measures such as social distancing and mask-wearing, among others (UNICEF 2022). Also, in his study of the prospects of Zimbabwean traditional leaders against COVID-19, Dziva (2020) reported how chiefs through their networks of headmen and village heads could provide factual information regarding COVID-19 in Zimbabwe.

18.8 COVID-19 in Egypt and Tunisia: Efforts of Interest Groups

18.8.1 Egypt

Egypt holds the unpleasant record as the first African country to report the first case of COVID-19. The first case of COVID-19 in Egypt was confirmed on February 14, 2020. As of March 29, 2023, the total number of confirmed cases in Egypt stood at 515,882 cases with 110,505,666 administered doses of vaccines (WHO 2023a). The Egyptian government quickly took measures to control and slow down the spread of the virus. These measures included: a nationwide lockdown, which preceded a public health emergency, closure of learning institutions and places of worship, suspension of flights and festivals, a nationwide disinfection programme with a focus on tourist locations, restaurants, workplaces, and mandatory quarantine of people from overseas. The Ministry of Health and Population in Egypt was pivotal in overseeing the government's health responses to COVID-19, and other state organizations were mobilized to enforce the government's mitigation startegies.

The government of Egypt's measures to deal with the outbreak of COVID-19 encountered several challenges. First of all, the virus exacerbated the existing health system's challenges. According to the World Bank, Egypt is among the countries with the lowest level of government health expenditures in the Middle East and Northern Africa (MENA) Region (Elshalakani 2020). During the pandemic, frontline health workers complained about negligence from the government and a shortage of medical supplies and PPEs against COVID-19. In a situation where these important medical items were needed in the country, a surprising development ensued when Hala Zayed, Egyptian Health Minister with instructions from President Abel Fattah el-Sisi, visited Italy with a shipment of medical supplies. This led to criticism and outrage as Egyptians pondered over why the government did this, especially when these medical supplies were not enough in Egypt and were sold at higher prices even when they were available (Farouk 2020). This supports Maged Mandour's assertion that the government of Egypt's mitigation measures against COVID became "a case study in repression, propaganda, and misinformation" (Mandour 2020). Maged Mandour, an Egyptian political analyst, reported how doctors who criticized the government's mishandling of the virus were arrested on grounds of spreading false news. The issue of the medical shipment to Italy has been described by Mandour as organized propaganda by the government to portray false claims about the government's prowess and strength in its healthcare system.

As a response to the crisis that came with COVID-19, civil society groups in Egypt took some initiatives in the general fight against COVID. Plan International Egypt, an NGO in Egypt for instance, implemented policies that sought to prevent the spread of COVID and also addressed the economic, social, and emotional impact of COVID. The organization collaborated with the Egyptian government, and with their efforts mainly focused on the most vulnerable girls and young women, the organization provided cash grants to 7800 households. The organization also used different communication channels including social media to provide accurate information on COVID-19. This created awareness cocnerning the virus in Egypt. Also, Cairo Foundation for Development and Law, a non-profit organization in Giza, Egypt, addressed the challenges that women faced as a result of the COVID-19 pandemic. Entisar Elsaeed, founder and director of the organization, revealed that:

> Women practically have become the first responders directly responsible for protecting family members, and the ones who bear responsibility for whether or not they will get infected … That role has become one of the most important challenges that women face, especially those who belong to lower social, economic, and cultural groups. (OHCHR 2021)

Also, the foundation's objective at the onset of the COVID pandemic was to address the impact of the Egyptian's government COVID responses on women. Elsaeed disclosed that:

> We also thought about how best to inform women on precautionary measures and physical distancing. We produced awareness-raising papers to prevent infection with the coronavirus, using Egyptian colloquial language, we also set up a hotline and held online group meetings to provide psychological support and advice to women survivors of sexual violence. (OHCHR 2021)

18.8.2 Tunisia

As of November 22, 2021, Tunisia was noted as the second-ranked country in North Africa with the highest cumulative number of COVID-19 cases (Galal 2021). During this same period, Tunisia reported the highest number of COVID-19 deaths in North Africa with the death toll at 25,347. On March 2, 2020, Tunisia's Health Minister Abdelatif el-Maki disclosed the first case of COVID in Tunisia which involved a 40-year-old man who had returned from Italy. Soon afterward, stringent measures in line with WHO guidelines were implemented in the country. On March 13, 2020, the government cancelled all sporting activities and cultural activities, closed movie theaters, schools, and mosques, and mandated closure by 4:00 p.m.; restaurants, cafés, and night clubs. The government closed the country's borders on March 16, 2020, and suspended international flights. A 12-hour night curfew was implemented on March 18, 2020, and in a few days' time, on March 22, the government implemented a national lockdown till May 3, 2020. The National Coronavirus Response Authority (NCRA) was created on March 25, 2020, by the Tunisian government as part of the pandemic response. The NCRA was created mainly to coordinate the government's mitigation responses against COVID-19 (Abouzzohour and Mimoune 2020).

As seen in other African countries, the Tunisian government was not exempted from the challenges associated with its mitigation against COVID. One of the challenges encountered at the onset of the virus was the exacerbation of the imbalances in Tunisia's healthcare system. Abouzzohour and Mimoune (2020) disclosed that the geographic disparity in Tunisia's healthcare system where the distribution of resources is always in favor of the bigger cities and coastal towns exacerbated the quandary. During the pandemic, there were reports of uneven distribution of ICU beds and other vital medical equipment. The coastal regions were the major beneficiaries of these ICU beds. This led to complaints from inhabitants and medical professionals in areas where there were inadequate facilities and medical supplies. For instance, in the Gabes Governorate, it was reported that the main regional hospital only had two respirators for COVID-19 and two intensive care doctors for the whole hospital. According to the report, these doctors only treated non-COVID-19 cases (The Arabweekly 2020). Even with this uneven distribution of medical supplies, the healthcare system in Tunisia became overwhelmed by the surge in COVID-19 cases and had to rely on external help to address the crisis. This was evident when Ahmed Ghoul, a nurse at Charles Nicolle Hospital in Tunis disclosed that:

> Doctors are exhausted, as the number of patients exceeded the capacity of the hospital … Even the mortuary was full and we could no longer find a place for them. (France24 2021)

Further, Dr. Pym Hamed, head of the Emergency Department at Charles Nicolle, revealed that "*We are suffering, we urgently need oxygen, it (demand) has exceeded the stocks that we have*" (France24 2021). Also, there were some Tunisians who disregarded the government's measures of mandatory wearing of face masks. Mohammed Habib Kechaou, the acting Tunisian Health Minister, revealed that:

> the emergence of new circles of coronavirus contamination is due to non-compliance with sectoral health protocols and health recommendations, essentially not wearing the mask in public places and spaces.

The minister further added that:

> wearing a mask is now compulsory and any offender will be punished with a fine of 1000 (370 dollars) to 5000 dinars (1850 dollars) in accordance with Article 5 of Decree-Law No. 2020-9 of April 17, 2020, relating to the repression of the violation of the curfew, the limitation of movement, total confinement and the measures taken with regard to persons affected or suspected of being affected by the Coronavirus … according to article 312 of the penal code, is punished by 6 months' imprisonment and a fine of 120 dinars, anyone who contravenes the prohibitions and prophylactic or control measures ordered in times of epidemic. (Saanouni 2020)

Tunisia's governmental efforts of mitigating the transmission of COVID-19 were supported by some civil society groups. Reiffenstuel (2020) has highlighted the contribution of the Red Crescent Society of Hammam Sousse in Tunisia's fight against COVID-19. This organization disinfected four main post offices, market, significant streets, and courtyards. Again, the Red Crescent Society raised awareness of the virus by communicating the symptoms, preventive methods, andeducated the people to follow the implementation of government's measures to fight the pandemic. Sustainable Food Bank, another NGO in Tunisia, set up a food bank that supported about 300 families that were adversely affected by the Tunisian government's COVID-19 mitigation policies.

18.9 Conclusion

COVID-19 claimed millions of lives globally and resulted in unprecedented societal and economic disruptions. Before and during the early stages of the COVID-19 pandemic in Africa, the continent was projected to record the highest mortality rate from COVID-19 as compared to the mortality rate of other continents. The popular philanthropis, Bill Gates, earlier in 2020, predicted that the devastating impacts of COVID-19 in Africa could be far greater than the country where the virus originated from; China (Wasserman and Moynihan 2020). While there is no doubt about the fragile state of Africa's healthcare system, the belief that the mortality rate would be higher than the rest of the other continents was one of the numerous stereotypes cocnerning the COVID-19 situation in Africa. As of November 18, 2022, the overall deaths due to COVID-19 in the whole African continent were 257,984 compared to many countries that recorded higher rates (Galal 2022b). This was a surprising development to many people, especially those outside the African continent. Many African countries came up with innovative approaches for the treatment of COVID-19 infections including testing and contact tracing. For instance, Africans and their beliefs in traditional and herbal medicine resulted in the local production of medicines in Africa for the treatment of COVID-19. COVID-organics (CVO) produced from the plant artemisia by the Malagasy Institute of Applied Research is among such herbal medicines that Africans used for COVID-19 treatment. Also, from Ghana, herbal medicines such as *Nibima*, *Immunim*, and *Ampoforte*, and other medicinal plants like neem trees were some of the innovative ways by Africans to fight the COVID pandemic. As seen in the case of Ghana, there was also the local production of low-cost ventilators by Professor Fred McBagonluri, a Ghanaian engineer, to address the issue of inadequate ventilators in the country during the pandemic. Irrespective of these and other laudable innovative approaches, some organizations on the continent of Africa and some world bodies have found disningenuous ways of downplaying the continent's mitigation efforts against COVID-19.

It has been argued that due to the novel nature of the virus, there was limited knowledge about it. As a result of this, the various health organizations in Africa similarly adopted training and capacity building as a response. Indeed, in Africa, the COVID-19 pandemic exposed the inadequate and fragile healthcare system. Some scholars have argued that bad governance largely accounts for this. Obi-Ani et al. (2021) throw more light on this with their argument that Nigeria's healthcare problems have been in existence for decades now, yet, government after government have neglected the poor state of the healthcare system in the country. The authors disclosed how the poor leadership style of the various Nigerian governments over the years resulted in major adverse impacts in the health sector of the country. Some of these adverse impacts from the poor governance include; lack of basic equipment in hospitals, epileptic power supply, poor remuneration of medical personnel, and lack of essential drugs in the shelves, among others.

Almost all the governments in Africa encountered several challenges in their mitigation measures against COVID-19. Nonetheless, the governments worked together with their respective health organizations, civil society organizations, non-governmental organizations, chiefs, and other concerned parties to adopt measures that would ensure the safety of their citizens. Arguably, it can be concluded that Africa had a successful response to the COVID-19 pandemic.

Though the organizations listed or discussed as major players are interests groups, they did not embark on actions in the sense of demonstrations or petitioning, rather; they augmented and developed strategies to fight the COVID-19 dilemma. However, in some jurisdictions in Africa, we found political opponents or opposition parties, serving as interest groups; proffering alternative solutions to the COVID question and equally criticizing governments for their shortfalls.

References

Abouzzohour Y, Mimoune NB (2020) Policy and institutional responses to COVID-19 in the Middle East and North Africa: Tunisia. Brookings Reports. https://www.brookings.edu/research/policy-and-institutional-responses-to-covid-19-in-the-middle-east-and-north-africa-tunisia/

Adu-Gyamfi S, Mohammed A, Ago Obeng J, Osei-Poku S, Tettey Yartey H (2022) COVID-19 in Africa: an economic and social interpretation (2019–2022). Hist Pogl 5(8):388–415. https://doi.org/10.52259/historijskipogledi.2022.5.8.388

AFP (2020) Tanzania opposition angry over no coronavirus update in two weeks. https://www.aljazeera.com/news/2020/5/13/tanzania-opposition-angry-over-no-coronavirus-update-in-two-weeks

Africa CDC (2022) A continent-wide collaboration on genomics surveillance show the power of African science and how the majority of COVID-19 variants were introduced into Africa. https://africacdc.org/news-item/a-continent-wide-collaboration-on-genomics-surveillance-show-the-power-of-african-science-and-how-the-majority-of-covid-19-variants-were-introduced-into-africa/

Agency Report (2020) Coronavirus: Nigeria 'strengthens' surveillance at five international airports. https://www.premiumtimesng.com/news/top-news/374865-coronavirus-nigeria-strengthens-surveillance-at-five-international-airports.html?tztc=1

Akinyemi K, Fakorede C, Anjorin A, Abegunrin R, Adunmo O, Ajoseh S, Akinkunmi F (2020) Intrigues and challenges associated with COVID-19 pandemic in Nigeria. Health 12:954–971. https://doi.org/10.4236/health.2020.128072

Amponsah SK, Tagoe B, Afriyie DK (2021) One year after first case of COVID-19 in Ghana: epidemiology, challenges and accomplishments. Pan Afr Med J 39:226. https://doi.org/10.11604/pamj.2021.39.226.29069

Anadolu Agency (2020) 15 killed by Kenyan police enforcing COVID-19 curfew. https://www.aa.com.tr/en/africa/15-killed-by-kenyan-police-enforcing-covid-19-curfew/1862803

Arhinful E (2020) COVID-19: Ghana has 67 ventilators—health minister. https://citinewsroom.com/2020/04/covid-19-ghana-has-67-ventilators-health-minister/

Asare BE (2009) Tobacco regulation in South Africa: interest groups and public policy. Afr J Polit Sci Int Relat 3(3):99–106

Awuah J (2020) Otumfuo distributes Ghc1m worth food…to help needy. https://dailyguidenetwork.com/otumfuo-distributes-ghc1m-worth-foodto-help-needy/

Bauer N (2021) 'Little to lose': poverty and despair fuel South Africa's unrest. https://www.aljazeera.com/news/2021/7/14/little-to-lose-poverty-and-despair-fuel-south-africas-unrest

BBC (2021) John Magufuli: Tanzania's late president in his own words. https://www.bbc.com/news/world-africa-56441421

Bhalla N (2020) Kenyan chiefs go door-to-door to stop female genital cutting amid coronavirus. https://news.trust.org/item/20200421133410-40f0a

Bolu O, Mustapha B, Ihekweazu C, Muhammad M, Hassan A, Abdulwahab A, Asekun AA, Nsirim R, Okechukwu E, Attah I, Swaminathan M, Greby S, Adebiyi A, Alex-Okoh M, Okwor T, Ilori E, Mba N, Mutah J, Akujobi J, Battah N, Haggai W, Okatubo G, Okigbo A, Castle E, Abubakar I, Akataobi C, Adekunle O, Aliyu SH (2022) Effect of Nigeria Presidential Task Force on COVID-19 pandemic, Nigeria. Emerg Infect Dis 28(13):S168–S176. https://doi.org/10.3201/eid2813.220254

Chingono N (2020) Doctors sue Zimbabwe government over lack of COVID-19 protective equipment. https://www.theguardian.com/global-development/2020/apr/09/doctors-sue-zimbabwe-government-over-lack-of-covid-19-protective-equipment

Cobb R, Elder CD (1972) Participation in American politics: the dynamics of agenda building. Johns Hopkins University Press, Baltimore

Dahir AL (2020) Tanzania's president says country is virus free. Others warn of disaster. https://www.nytimes.com/2020/08/04/world/africa/tanzanias-coronavirus-president.html

Daily Guide (2020) COVID-19: Ghana highest testing country in Africa. https://www.modernghana.com/news/995432/covid-19-ghana-highest-testing-country-in-africa.html

Dao TL, Nguyen TD (2020) Controlling the COVID-19 pandemic: useful lessons from Vietnam. Travel Med Infect Dis 37:101822. https://doi.org/10.1016/j.tmaid.2020.101822

De Villiers C, Cerbone D, Van Zijl W (2020) The South African government's response to COVID-19. J Public Budg Account Financ Manag 32(5):797–811. https://doi.org/10.1108/JPBAFM-07-2020-0120

Duncan J (2020) Two cases of coronavirus confirmed in Ghana. https://citinewsroom.com/2020/03/two-cases-of-coronavirus-confirmed-in-ghana/

Duodu S, Fugu M (2020) Northern, North East, Savannah regions disinfect markets. https://www.graphic.com.gh/news/general-news/northern-north-east-savannah-regions-disinfect-markets.html

Dziva C (2020) Prospects and challenges for traditional leaders in combating COVID-19 pandemic in rural Zimbabwe. https://doi.org/10.2139/ssrn.3599770

East, Central and Southern Africa Health Community (2020) Resolutions of the health ministers' conferences 1974–2020, p 176

Ekoh PC, Ejimkaraonye C, George EO, Agbawodikeizu PU (2021) 'Better to die of disease than die of hunger': the experience of Igwes (traditional rulers) in the fight against the COVID-19 pandemic in rural South-East Nigeria. Rural Remote Health 21(4). https://doi.org/10.22605/RRH6691

Elshalakani A (2020) Project information document—Egypt COVID-19 emergency response—P173912 (English). World Bank Group, Washington, DC. http://documents.worldbank.org/curated/en/906831588109454283/Project-Information-Document-Egypt-COVID-19-Emergency-Response-P173912

Eribo S (2021) COVID-19 and African civil society organizations: impact and responses. Alliance Afr Partnersh Perspect 1:147–155. https://doi.org/10.1353/aap.2021.0017

Farouk M (2020) Egyptians outraged over medical aid to Italy as supplies dwindle at home. https://www.al-monitor.com/originals/2020/04/egypt-regeni-italy-supplies-covid19-coronavirus.html

France24 (2021) Tunisia struggles to cope with surge in COVID-19 cases. https://www.france24.com/en/africa/20210715-tunisia-struggles-to-cope-with-surge-in-covid-19-cases

Frimpong OB, Bihuzo RM, Commodore R (2020) The COVID-19 pandemic in Africa: impact, responses and lessons from Ghana, The Democratic Republic of the Congo, and Rwanda. Afr Program Occas Pap. https://www.wilsoncenter.org/publication/covid-19-pandemic-africa-impact-responses-and-lessons-ghana-democratic-republic-congo

Galal S (2021) Coronavirus cases in Northern Africa as of November 22, 2021, by country. https://www.statista.com/statistics/1189031/coronavirus-cases-by-country-in-north-africa/

Galal S (2022a) Coronavirus (COVID-19) daily cases in South Africa as of March 6, 2022. https://www.statista.com/statistics/1107993/coronavirus-cases-in-south-africa/

Galal S (2022b) Coronavirus deaths in Africa 2022, by country. https://www.statista.com/statistics/1170530/coronavirus-deaths-in-africa/

GCAP (2020) COVID-19: promoting support for diverse communities in Zimbabwe. https://gcap.global/news/promoting-support-for-diverse-communities/

Ghanaweb (2020a) Panic 'hits' social media as Ghana confirms first cases of coronavirus. https://www.ghanaweb.com/GhanaHomePage/NewsArchive/Panic-hits-social-media-as-Ghana-confirms-first-cases-of-coronavirus-893413

Ghanaweb (2020b) Korle-Bu doctors threaten to strike over poor coronavirus preparedness. https://www.ghanaweb.com/GhanaHomePage/NewsArchive/Korle-Bu-doctors-threaten-tostrike-over-poor-coronavirus-preparedness-912370

Hlati O (2020) City of Cape Town acquiring stand-in staff as Mfuleni nurses refuse to work. https://www.iol.co.za/capetimes/news/city-of-cape-town-acquiring-stand-in-staff-as-mfuleni-nurses-refuse-to-work-47554674

Jerving S (2021) In brief: WHO calls out Tanzania for its lack of COVID-19 response. https://www.devex.com/news/in-brief-who-calls-out-tanzania-for-its-lack-of-covid-19-response-99217

Kamer L (2022a) Coronavirus deaths in East Africa 2022, by country. https://www.statista.com/statistics/1175313/coronavirus-deaths-by-country-in-east-africa/

Kamer L (2022b) Cumulative number of COVID-19 vaccination doses in Tanzania 2022. https://www.statista.com/statistics/1258567/total-number-of-covid-19-vaccination-doses-in-tanzania/

Kankam Kusi N (2021) Civil society's response to COVID-19 in Ghana: actions, results and lessons learned. WACSeries Issue Pap 7(5). West Africa Civil Society Institute, Accra, Ghana

Kasapafmonline (2020) Manhyia partners Master Card Foundation to enforce COVID 19 protocols in Kumasi markets. https://kasapafmonline.com/?p=205141

Lancaster L, Mulaudzi G (2020) Rising protests are a warning sign for South Africa's government. https://issafrica.org/iss-today/rising-protests-are-a-warning-sign-for-south-africas-government

Magwanda J (2022) COVID 19 restrictions and the impact on criminal justice and human rights. Africa Criminal Justice Reform. https://acjr.org.za/

acjr-publications/kenya-covid-19-socio-economic-impact-j-mangwanda.pdf/view
Mandour M (2020) Repression and coronavirus response in Egypt. https://carnegieendowment.org/sada/82304
Masuku J (2020) How a media NGO changed the COVID-19 narrative in Zimbabwe. https://cmds.ceu.edu/article/2020-05-07/how-media-ngo-changed-covid-19-narrative-zimbabwe
McDade KK, Ogira D, Onyango J, Ojal J, Kokwaro G, Mao W, Yamey G (2020) Kenya's policy response to COVID-19. Policy report. The Center for Policy Impact in Global Health. http://centerforpolicyimpact.org/our-work/the-4ds/kenya-policy-response-to-covid-19/
Ministry of Health (2020) First case of coronavirus disease confirmed in Kenya. https://www.health.go.ke/first-case-of-coronavirus-disease-confirmed-in-kenya/
Mitumba F (2020) Africa CDC concerned over Tanzania's virus response. https://www.faceofmalawi.com/2020/08/07/africa-cdc-concerned-over-tanzanias-virus-response/
Murewanhema G (2021) COVID-19 control pitfalls and challenges and drivers of SARS-CoV-2 transmission in Zimbabwe. Pan Afr Med J 38:28. https://doi.org/10.11604/pamj.2021.38.28.25758
Mwangangi M (2020) COVID-19 pandemic: Kenya's experience. https://apps.who.int/gb/COVID-19/pdf_files/30_04/Kenya.pdf
Myjoyonline (2020) Coronavirus alleviation programme business support scheme (CAP BuSS): win-win for all! https://www.myjoyonline.com/coronavirus-alleviation-programme-business-support-scheme-cap-buss-win-win-for-all/
Obi-Ani NA, Ezeaku DO, Ikem O, Isiani MC, Obi-Ani P, Chisolum OJ, Adu-Gyamfi S (2021) COVID-19 pandemic and the Nigerian primary healthcare system: the leadership question. Cogent Arts Humanit 8:1. https://doi.org/10.1080/23311983.2020.1859075
OCHA (2020) 'It is in me to work to support my community': how a local NGO in Kenya is helping vulnerable people during COVID-19. https://shorturl.at/nwCH2
OHCHR (2021) Responding to the needs of women, 'first responders' to Egypt's COVID-19 crisis. https://www.ohchr.org/en/stories/2021/02/responding-needs-women-first-responders-egypts-covid-19-crisis
Onyeji E (2020) Coronavirus: Nigeria in 'desperate need' of testing kits—NCDC. https://www.premiumtimesng.com/news/headlines/390058-coronavirus-nigeria-in-desperate-need-of-testing-kits-ncdc.html?tztc=1
Plein LC (1991) Popularizing biotechnology: the influence of issue definition. Sci Technol Hum Values 16(4):474–490
Pralle S (2003) Venue shopping, political strategy, and policy change: the internationalization of Canadian forest advocacy. J Publ Policy 23(3):233–260. https://doi.org/10.1017/S0143814X03003118
Reiffenstuel A (2020) Tunisia's civil society supports governmental efforts of slowing COVID-19 transmission in Tunisia. https://www.researchgate.net/publication/340453824_Tunisia%27s_Civil_Society_Supports_Governmental_Efforts_of_Slowing_COVID-19_Transmission_in_Tunisia
Saanouni M (2020) Covid-19/Tunisia: sanctions going as far as prison for not wearing a mask. https://www.aa.com.tr/fr/afrique/covid-19-tunisie-des-sanctions-allant-jusqu%C3%A0-la-prison-pour-non-port-du-masque/1954133
Sahara Reporters New York (2020) BREAKING: COVID-19 patients protest over inadequate medical attention, block highway, destroy properties in Gombe. https://saharareporters.com/2020/05/05/breaking-covid-19-patients-protest-over-inadequate-medical-attention-block-highway
Schattschneider EE (1960) The semi-sovereign people. Holt, Rinehart & Winston, New York
Siaw-Frimpong M, Touray S, Sefa N (2021) Capacity of intensive care units in Ghana. J Crit Care 61:76–81. https://doi.org/10.1016/j.jcrc.2020.10.009
Simons B (2020) Should African countries embrace Ghana's COVID-19 'pooled testing'? https://www.myjoyonline.com/should-african-countries-embrace-ghanas-covid-19-pooled-testing-2/
Sombié I, Johnson E, Lokossou V, Amouh T, Sow A, Ogbureke N, Okolo S (2020) How does the West African Health Organisation (WAHO) contribute to the evidence based decision-making and practice during COVID-19 pandemic in ECOWAS region? Pan Afr Med J 37:2. https://doi.org/10.11604/pamj.supp.2020.37.1.25625
South African Government (2021) Premier Stanley Mathabatha on the support of traditional leaders in the fight against the coronavirus COVID-19 pandemic. https://www.gov.za/speeches/limpopo-premier-thanks-traditional-leaders-support-fight-against-covid-19-pandemic-20-aug
Tardzenyuy Thomas M (2021) Protests during the 2020 COVID-19 lockdown in South Africa. In: MoLab inventory of mobilities and socioeconomic changes. Max Planck Institute for Social Anthropology, Halle/Saale. https://doi.org/10.48509/MoLab.8077
The Arabweekly (2020) Tunisia's pandemic outbreak reflects health infrastructure imbalances. https://thearabweekly.com/tunisias-pandemic-outbreak-reflects-health-infrastructure-imbalances
Uchechi Okoroiwu H, Ogar Ogar C, Mbe Egom Nja G, Kongfe Abunimye D, Idu Ejemot-Nwadiaro R (2021) COVID-19 in Nigeria: account of epidemiological events, response, management, preventions and lessons learned. Germs 11(3):391–402. https://doi.org/10.18683/germs.2021.1276
UNICEF (2022) A traditional leader turns to village health workers to dismantle cultural and religious practices. https://www.unicef.org/zimbabwe/stories/traditional-leader-turns-village-health-workers-dismantle-cultural-and-religious-practices
Vang G (2020) The University of Minnesota partners with Ghana NGO to deliver online COVID-19

training. https://globalhealthcenter.umn.edu/news/university-minnesota-partners-ghana-ngo-deliver-online-covid-19-training
Wasserman H, Moynihan Q (2020) Bill Gates warned the coronavirus could hit Africa worse than China. https://www.businessinsider.com/bill-gates-warns-coronavirus-could-hit-africa-with-dire-consequences-2020-2?IR=T
WHO (2023a) Egypt situation. https://covid19.who.int/region/emro/country/eg
WHO (2023b) Zimbabwe situation. https://covid19.who.int/region/afro/country/zw
WHO Africa (2021) Nigeria rallies over 7000 traditional leaders against COVID-19. https://www.afro.who.int/news/nigeria-rallies-over-7000-traditional-leaders-against-covid-19
Wrapanigeria (2023) What is WRAPA. https://wrapanigeria.org/whatiswrapa/
Yeboah I (2020) Ghana's COVID-19 situation bleak, not what Prez Akufo-Addo wants us believe-Sammy Gyamfi. https://www.graphic.com.gh/news/politics/ghana-s-covid-19-situation-bleak-not-what-prez-akufo-addo-wants-us-believe-sammy-gyamfi.html

Lessons Learned About Trauma Related to Racial Discrimination During COVID-19 in the United States

19

Michelle N. Strong, Mallory Constantine,
Allyson Donovan, and Gloria Wong-Padoongpatt

Abstract

Race scholars have discussed how the pandemic has disproportionately burdened marginalized communities and exacerbated pre-existing inequities, particularly for Black Indigenous People of Color (BIPOC) in the United States (U.S.). One glaring social determinant during the time of the COVID-19 is racial discrimination. This chapter will discuss lessons learned regarding the negative impact of discrimination on BIPOC, especially as it pertains to their experiences of trauma. Some of these lessons include (1) the need for clinical psychologist to improve access to treatment through increased research on culturally adaptive interventions, (2) increased research on the effects of race-based trauma on mental health symptomatology, (3) policy and institutional changes that reduce disparities in access to care, and (4) increased education for psychologists around billing procedures for individuals with race-based stress.

Keywords

Racism · Discrimination · BIPOC · Disparities · COVID-19 pandemic · Trauma · Collective trauma · Mental health care

19.1 Introduction

Racism has been an undercurrent in the American fabric since the nation's inception, and during the COVID-19 pandemic, the tensions across racial communities and racial inequities seemed to exacerbate. Racism-related hate crimes against Asian Americans increased 300% during the critical months of the pandemic, with 1 in 5 Asian Americans reporting one blatant encounter with daily racism (Lee and Waters 2021; Wong-Padoongpatt et al. 2022). Several attention-grabbing violences also happened against the Black and LatinX American communities that further spotlighted macro and systemic level concerns of racism; and there were nation-wide—and even world-wide—demands for critical institutional changes. Ibram X. Kendi highlighted that these experiences of racism and "racist abuse" negatively affect the physical and

M. N. Strong (✉) · M. Constantine · A. Donovan · G. Wong-Padoongpatt
University of Nevada, Las Vegas, USA
e-mail: stronm1@unlv.nevada.edu

M. Constantine
e-mail: harrim11@unlv.nevada.edu

A. Donovan
e-mail: donova4@unlv.nevada.edu

G. Wong-Padoongpatt
e-mail: gloria.wong@unlv.edu

N. Rezaei (ed.), *The COVID-19 Aftermath*, Advances in Experimental Medicine and Biology 1457,
https://doi.org/10.1007/978-3-031-61939-7_19

psychological health of targets (Fix et al. 2022; Kendi 2019). There are innumerable examples of racism and discrimination in the United States, and while the examples in this book chapter are not an extensive list of harm perpetrated against marginalized groups during the pandemic, it is important to highlight poignant examples. The COVID-19 pandemic, however, served as a crucible for American culture that amplified the need for improved race relations, reduction of economic and health disparities, and careful analysis of their impact.

One aspect of the ethno-racial experience in diverse Americans is not only the insidious effects of daily racism and discrimination but the compounding negative effects that trauma has on the well-being of targets. Studies have shown that trauma decreases well-being and increases negative mental health symptomatology, including depression, anxiety, suicide, and addictions (Stampfer 1990). Other research has shown how racism as microaggressions is traumatizing and has been particularly damaging because of its cumulative effect (Nadal et al. 2019). Hankerson et al. (2022) reviewed the effect that cumulative trauma affects mental health in Black Americans. The scholars found Black Americans were more likely to experience depression because of repeated exposure to race-based trauma. They also pointed to the impact of structural and historical factors on the intergenerational transmission of depression in this group. With the COVID-19 pandemic, minoritized groups, along with majority groups, experienced a collective trauma with the COVID-19 pandemic and shutdown. The literature suggests that rates of depression, loneliness, anxiety, and isolation increased during the pandemic (Palgi et al. 2020). Other researchers suggest that negative mental health symptomatology during this time mirrored individuals who have suffered from significant traumas (Jeftić et al. 2021). These negative mental health effects were further exacerbated in marginalized groups. The disparities highlighted during the COVID-19 pandemic are likely because of the compounded effect of daily racism, and the added effect that many minoritized groups were on the "front lines" and were more likely to be infected with and die from COVID-19 (Phiri et al. 2021). These compounding effects can include the experience of discrimination due to ethno-racial identity in LatinX Americans and their decreased access to health care services because of systemic discrimination and immigration status (Olayo-Méndez et al. 2021).

However, despite these cumulative and negative experiences, researchers have failed to address this issue properly, especially as it pertains to clinical diagnosis and treatment of these ethno-racial groups. Indeed, the American Psychological Association (APA) Multicultural Guidelines (2017) argues that psychologists should adapt clinical interventions and services to meet the needs of diverse groups, however as emphasized through the low levels of trust and treatment engagement in clinical care by minoritized groups, clinicians' abilities, and skills to treat diverse groups warrants improvement. Policy makers impede access to mental health care for minoritized groups by insurance, state, and federal policy. Therefore, this chapter will discuss lessons learned regarding the negative impact of discrimination on BIPOC, especially as it pertains to their experiences of trauma during the COVID-19 pandemic.

19.2 COVID-19 Pandemic as a Collective Trauma

Traumatic experiences occur across individual, group, communal/cultural, and collective levels. The magnitude and societal impact of any event differentiates each of the categories. Reactions to traumatic experiences also differ across these categories. Careful analysis of traumatic events and preceding reactions is essential to classifying, understanding, and learning from historical and modern events. Beginning analysis of individual experiences and responses to trauma informs a greater understanding of how an entire population responds to a collective trauma, such as the COVID-19 pandemic.

A significant proportion of the population will experience a trauma in their lifetime. In a nationally representative study ($N=2953$) conducted by Kilpatrick et al. (2013) the researchers found that 89.7% of the United States adults reported experiencing a significant trauma in their lifetime, with exposure to multiple traumatic events occurring the most often. Types of trauma include interpersonal violence (58.6; 47.1%), sexual assault (42.4; 15.8%), and physical assault (44.9; 42.4%) for women and men, respectively. The researchers also surveyed the rates of Post-Traumatic Stress Disorder (PTSD) prevalence rates according to the Diagnostic and Statistical Manual Fifth Version (DSM-5). Survey results suggest that 8.3% of U.S. adults will meet criteria for PTSD and 3.8% of them met criteria for PTSD within the 6-month period prior to the survey. These results are staggering and underline the ubiquitous experience of trauma in the United States.

U.S. adults often erroneously interchange the terms "PTSD" and "trauma." Trauma refers to, "... an emotional response to a terrible event like an accident, rape, or natural disaster" (American Psychological Association [APA] 2022). PTSD refers to a cluster of symptoms that may arise following an exposure to specific types of trauma, including "exposure to actual or threatened death, serious injury, or sexual violence" (American Psychiatric Association 2022). According to the DSM-5-TR, these clusters of symptoms can include intrusive memories, avoidance, negative changes in mood and cognition, and hyperarousal (American Psychiatric Association 2022). To diagnose an individual with PTSD, clinicians will see if the symptoms cause clinically significant distress (e.g., impairs an individual's ability to live a meaningful life) and occur at least 3 months after the traumatic event (American Psychiatric Association 2022). These differences are important to note because many forms of trauma do not qualify for a PTSD diagnosis; however, their effects on the emotional and general well-being of those affected are noteworthy.

Individual trauma refers to an event (e.g., a mugging or physical attack) or series of events that happen to one person (e.g., Multiple sexual assaults; Center for Substance Abuse Treatment 2014). As a function of individual trauma, an individual can experience a change in cognition, often for the person to "make sense" of the traumatic event. In the instance of sexual assault, for example, a person may have believed that the world was "safe" and a "good place" prior to their experience of trauma; however, after the traumatic event, they may think that the world is "unsafe" and "bad." They may even think that they are "bad" or "dirty" because of thoughts that they are "no longer clean" because of the traumatic event happened to them. These lead to negative psychological outcomes, including depression, anxiety, and symptoms of PTSD (Muran and Motta 1993; Hirschberger 2018).

Group trauma refers to a trauma that happens to a group of people that share a type of similar identity or history—such as first responders, medical personnel, or military service members. For survivors of group trauma, their reactions post-trauma differ from individual trauma. This is because there are group norms that can affect the healing process, including only sharing about the trauma with group members, pressure to suppress emotions, and pressure to not seek help to prevent feelings of personal and group shame. Those who have experienced group trauma are more likely to experience multiple traumas, which can complicate the healing and recovery process (Center for Substance Abuse Treatment 2014; Hirschberger 2018).

Last, c*ommunal and cultural trauma* refers to a wide range of stressful events that affect large communities, including hate crimes, workplace violence, shootings, physical or sexual assaults. Cultural traumas often erode the sense of connection and break down the sense of cultural heritage. This can cause significant negative effects, including medical, physical, and psychological difficulties. Types of cultural traumas can include both historical and mass trauma. Historical trauma includes trauma that affects generations of a cultural group, an example of this includes the enslavement and institutional oppression of African Americans (Phipps and Thorne 2019). While mass trauma refers to events that either directly or indirectly affect a large group of people, an example of this includes large-scale natural disasters. People can easily identify these traumas, facilitating a sense of togetherness and mobilizing support for those affected. Identifying various types of trauma is important, as different trauma can affect individuals and societies differently. Considering a trauma such as the COVID-19 pandemic, its widespread impact negatively affected arguably the entire world, creating a collective sense of trauma experience (Hirschberger 2018).

Collective trauma refers to a traumatic event that occurs at multiple levels that impacts a "population's social, cultural, and physical ecologies" (Bava and Saul 2013). While collective traumas typically threaten collective identities, often an integration of the traumatic experiences is assimilated into a population's historical narrative. Often groups who experience collective traumas make meaning from those traumatic experiences and incorporate those meanings into their collective identity (Hirschberger 2018). For example, collective feelings of resiliency, solidarity, and perseverance were felt in tandem with feelings of sadness, anger, and confusion during historical collective traumas including the Holocaust (Ekman 2016). During the COVID-19 pandemic, meaning making could be seen as advertisements and social media campaigns celebrating front-line workers.

As collective memory shapes collective identity, societal rules shift to assimilate the collective trauma. The COVID-19 pandemic brought about many shifts to societal rules including altering ways to engage in social gatherings (Guay et al. 2021), re-evaluating the necessity for on-site employment (Agba et al. 2020), and how many seek physical and mental health treatment (Feijt et al. 2020). Technology including conferencing platforms, remote log-in, and online purchasing propelled these shifts in societal rules and how American society functions daily (Khan et al. 2021).

While technology allowed for some semblance in maintaining societal norms, the public policy and societal changes impacted nearly every aspect of life. These abrupt changes left little room for vulnerability and disparately impacted low socioeconomic status groups. A systematic review conducted by Khanijahani et al. (2021) looked at the relationship between socioeconomic factors and race on health outcomes and access to care during the pandemic. The researchers found that factors such as education, poverty, language ability, and poor housing conditions increased risk for death and confirmed infection of COVID-19, and these factors were particularly prevalent in ethno-racially diverse individuals highlighting the ways that individuals from lower socioeconomic and racially diverse populations were more vulnerable to the virus during the pandemic. Further, telework was simply not possible for many low-income jobs or jobs that do not require a college degree (Geranios et al. 2022), ultimately resulting in increased job loss for these individuals. Many factors contributed to the significant decline in unemployment during the COVID-19 pandemic; however, low-income (i.e., earning $25,000 or less annually) workers were most impacted by involuntary layoffs (Geranios et al. 2022). Additionally, individuals of low SES had the most serious and longer-lasting financial impact from COVID-19 (Geranios et al. 2022).

Further societal shifts were seen in the sudden changes in social interactions and expected social norms. Increased regulations and high adherence to health promoting behaviors, required people to actively go against previously normative behaviors (i.e., greeting others through physical touch, standing in crowds, touching public property like grocery carts; Neville et al. 2021). Many social interactions were lost or significantly reduced due to social distancing. During the most stringent restrictions of COVID-19, many people maintained core social networks with family members; however, many people no longer maintained relationships with acquaintances in the periphery of their social networks (Long et al. 2022). Additionally, daily social interactions were lost when no longer working in close physical proximity to others. These alterations in social norms have continued to be maintained for many people despite the reduction in government regulations and are a hallmark indicator of collective traumas.

Collective traumatic experiences often elicit a feeling of helplessness or hopelessness across the population (Taylor 2020; Hirschberger 2018; Saricali et al. 2022). Levels of suicidal ideation, which often overlap with constructs of hopelessness, increased across the general population during the COVID-19 pandemic (Saricali et al. 2022). Additionally, many medical personnel reported higher rates of burn-out, compassion fatigue, and hopelessness as death rates rose (Lluch et al. 2022). Hospitals world-wide were not equipped to treat the number of COVID-19 patients nor properly handle the deaths. Many hospitals resorted to using cooling trucks as makeshift morgues due to the overwhelming deaths (Bumsted and Sisak 2021; Villarreal 2020). Social media and constant news coverage of the COVID-19 pandemic brought many of these medical traumas and deaths into our hands through news feeds inducing panic (Venegas-Vera et al. 2020). Both images of overwhelmed hospitals and videos of people infected with COVID-19 struggling to breath and being unresponsive to medical treatment flooded the internet. People world-wide constantly had access to timely, graphic, and traumatic updates of the impacts COVID-19 pandemic (Zhao and Zhou 2020).

The many significant stressors stemming from the COVID-19 pandemic increased mental illness across the nation. Stressors directly resulting from the COVID-19 pandemic ranged from financial concerns, physical and mental health concerns, loss of family and friends, reduced access to child care, supply chain issues, lack of access to mental and physical health care, and many more (Fegert et al. 2020). Increased stressors frequently result in increases in rates of mental illness such as anxiety and depressive disorders, which held true throughout the pandemic (Pfeifer et al. 2021; Shah et al. 2021).

Of note, anxiety, depressive, and trauma/stressor disorders include overlapping constructs within the diagnostic criteria (American Psychiatric Association 2022), meaning that people with these disorders may experience similar responses and types of distress. Conceptualizing the COVID-19 pandemic as a collective trauma may provide greater understanding of the increases of mental illness in the subsequent years. Many have experienced residual levels of vigilance relating to social distancing, cleanliness, and financial instability that were adaptive at the height of the pandemic. Ultimately, the COVID-19 altered the world populations' understanding of their social, cultural, and physical ecologies resulting in new collective identities. As such, the COVID-19 pandemic has impacted the population in ways that suggest qualifying as a collective trauma.

19.3 Race-Based Trauma

Minoritized groups have historically experienced significant communal and cultural traumas. Perhaps one of the most salient examples includes the horrific implementation of United States Federal policy designed to assimilate Native and Indigenous youth into European American cultural norms through American Indian Residential Schooling (Woolford 2015). The United States Federal government selected youth and children because of their race, removed them from their homes to attend boarding schools, stripped them of Indigenous cultural signifiers, forced them to only speak English, and frequently exploited and abused them (Woolford 2015). The horrific implementation of United States Federal policy caused race-based trauma.

Experiences of discrimination and racism are important to conceptualize and address considering the literature on minority stress theory and race-based traumatic stress. Meyer (2003) originally introduced the Minority Stress Theory to understand how discrimination negatively affects the mental health of minoritized groups and sexual minorities. Minority Stress Theory works to understand how discrimination, stigma, and prejudice negatively impact mental health because of the extremely negative and "hostile environment" it creates over the sociopolitical climate that minoritized groups are living in (Binion and Gray 2020). Carter (2007) explored the concept of stress because of discrimination in ethno-racially diverse individuals in a concept termed race-based trauma. The author explored how discrimination as threats of harm, demeaning events, or witnessing harm to other individuals from minoritized backgrounds can cause symptoms that align with acute stress disorder and PTSD. Notably, Carter argued these experiences can lead to a state of "racial battle fatigue" because of the disproportionate use of psychological resources to cope with distress caused by racism.

In his book, Microaggressions and Traumatic Stress: Theory, Research, and Clinical Treatment, Nadal (2019) synthesized the current research and literature surrounding microaggressions and traumatic stress. Specifically, Nadal discusses how the current Diagnostic and Statistical Manual of Mental Disorders, Fifth Edition, Text Revision (DSM-5-TR) fails to integrate discrimination trauma in their conceptualization and diagnoses of mental disorders. Nadal describes four types of discrimination trauma (1) traumatic discrimination, (2) microaggressive trauma, (3) insidious trauma, and (4) race-based traumatic stress. First, traumatic discrimination refers to experiences of trauma that are like current criterion A trauma in the DSM-5-TR. Specifically, traumatic discrimination refers to a target's experience of a discrimination event that is a threat to their sense of safety, is intense and extensive. The experience of traumatic discrimination can lead to sequelae similar to Post-Traumatic Stress Disorder (PTSD) such as avoidance and dissociation. Examples of traumatic discrimination include racial hate crimes, sexual assault and harassment, or non-violent and overt forms of discrimination, such as bullying or racist policy. Second, microaggressive trauma refers to the cumulative distress that results as a function of the unremitting experience of subtle forms of discrimination. This refers to the cumulative and negative effect that microaggressions have on targets. Last, insidious trauma points to the targeted experience of daily discrimination. This form of trauma can be harmful and unseen. Insidious trauma can refer to historical forms of oppression through racist policy (Root 1992; Nadal et al. 2019). Significant examples of this in the United States include the lingering negative effects of slavery on the Black community. Epigenetic studies conducted by Costa et al. (2018) have suggested that trauma can be passed down through one's genome. Although the current research in this area wants clarification, Fraiberg et al. (1975) argued that even without direct exposure to traumatic mistreatment, parents' unconscious reenactment of their own experiences of trauma can pass significant forms of race-based trauma to their children intergenerationally. This can lead to harmful psychological effects and negative mental health outcomes (Bernard et al. 2021).

Other scholars, such as Bryant-Davis (2007), refer to the harm caused by racism as "race-based traumatic stress." In his article, Bryan-Davis defines race-based trauma as (1) harm that is racially motivated and directed toward a target that causes emotional harm, (2) a stressor that impairs an individual's ability to cope, (3) a stressor perpetrated on a victim that causes physical harm and is racially motivated, and (4) a stressor that is race-based, systemic, and interpersonal. To understand the clinical presentation of race-based stress, Carter and Sant-Barket (2015) conducted a study using the Race-Based Traumatic Stress Scale (RBTSS). The results of the study were impactful. Indeed, the researchers found that individuals who understood their experience of racism as trauma were more likely to experience significant personality change, and researchers correlated experiences of trauma with decreased daily functioning and general well-being in targets.

Further, discrimination and race-based trauma do not happen as isolated events; however, they occur as multilevel and institutional oppression. People often misunderstand the concept of macroaggressions, resulting in misuse of the term, as Pérez Huber and Solorzano (2015) noted regarding institutional forms of racism. In their book, Microintervetion Strategies: What You Can Do To Disarm and Dismantle Individual and Systemic Racism and Bias, Sue et al. (2020) defined macroaggressions as there, "active manifestation of systemic or institutional biases that live in the philosophy, policies, programs, practices, and structures of institutions and communities" (Sue et al. 2020 p 10; Pérez Huber and Solorzano 2015; Sue et al. 2019). In the most basic sense, macroaggressions refer to the institutional, multilevel, and cultural expressions of racism that oppress marginalized groups. During the pandemic, these instances of institutional racism were poignant, especially with marginalized groups disproportionately representing essential workers such as medical staff (Johnson-Agbakwu et al. 2022), store attendants (Lopez et al. 2021), and agriculture and food production (Yearby and Mohapatra 2020). Further, marginalized groups are more likely to live in multi-generational homes, because of cultural and socioeconomic factors, which increase their probability of exposure to the virus (Woods 2018; Ghosh et al. 2021). Indeed, the research suggests that marginalized groups showed higher rates of exposure and mortality because of the virus (Rose et al. 2020) and were less likely to seek medical care because of fears of discrimination (Zhang et al. 2022).

These factors are important to conceptualize, especially in the increased experiences and awareness around race and racism during COVID-19 in marginalized groups (Tan et al. 2021). It highlights the complexity of treating and diagnosing BIPOC, by emphasizing the negative and deleterious effects racism has on general and mental well-being in targets while also holding the multilevel effects that racism has on BIPOC well-being. Studies suggest that racism inhibits access to food security (Bowen et al. 2021), housing (Shavers and Shavers 2006), and quality education (Shavers and Shavers 2006)—all factors that are instrumental in quality of life. Further, during the pandemic not only were experiences of racism elevated in BIPOC groups, but the inherently racist American institutions lead to marginalized groups to be disproportionately and negatively affected resulting in higher mortality rates, decreased access to food resources, and decreased access to health care.

19.4 Racial Discrimination During COVID-19

During the pandemic, instances of racism and discrimination were brought to the forefront of the American consciousness through the various marches and protests against police brutality (Jean 2020), abuse of native nations rights (Panda 2021), Asian hate (NW et al. 2015), and anti-immigration rhetoric targeting LatinX Americans (Cerda et al. 2023). Researchers conducted a study looking at rates of discrimination and found that of the 5,500 BIPOC individuals sampled 22.1% experienced discrimination related to COVID-19 and 42.7% reported people acting as though they were afraid of them. BIPOC were also more likely to experience COVID-related discrimination compared to Whites (Strassle et al. 2022). The research suggests that communities of color were disproportionately affected by the pandemic as evidenced by higher rates of unemployment, infection rates, and death (Rodriguez-Diaz et al. 2020; Zelner et al. 2021). Another form of collective trauma experienced during COVID-19 includes the systemic racism that BIPOC experienced. In the forms of both overt and covert forms of racism, overt forms of racism can include hate crimes that target a specific group; while covert forms of racism can include "subtle remarks" or "slight injustices." Covert forms of racism can be better categorized as microaggressions, while overt forms of racism are better thought of as macroaggressions as categorized by Derald Wing Sue.

In the seminal article published by Sue and Constantine (2007) the researchers define microaggressions. Specifically, Sue and his colleagues point out three types of microaggressions: microassault, microinsult, and microinvalidation. The microassault refers to non-verbal and verbal attacks that are meant to hurt the target. Microassaults can include name calling and derogatory terms meant to insult and hurt targets; they can include racist slurs, ableist remarks, or degrading comments toward gender and sexual minoritized individuals. The microinsult demeans an individual's personhood through snide remarks. An example of a microinsult can include that a minoritized individual obtained employment as a function of policy, such as affirmative action. These types of remarks suggest that the individual is not qualified for the position (Hinton 2003). The microinvalidation invalidates a target's experience of discrimination through actions or words, this can include the "All lives matter" and "I don't see color" rhetoric. Microinvalidations are particularly harmful as by invalidating persons of color's experience of historic systemic and institutional oppression (Helms 1992). Microaggressions have often been described as "small" or "subtle" remarks; however, their effect is insidious. Indeed, the research suggests that the harmful effect of microaggressions is not in their individual harm, but the cumulative effect and longitudinal experience of degradation on targets. Negative effects on targets can include anger, distrust, anxiety, fear, and depression (Spanierman and Heppner 2004; Thompson and Neville 1999).

During the COVID-19 pandemic, instances of microaggressions were prevalent across minoritized groups, especially with the rise of social media platforms such as Facebook, Twitter, Instagram, and Redditt (Chong and Chen 2021; Gee et al. 2022; Jang et al. 2021). One particular group that was impacted was LatinX and Asian Americans. The research suggests that race-related discrimination against Asian Americans increased by over 300% during the pandemic (Wong-Padoongpatt et al. 2022). Indeed, Nguyen et al. (2020) analyzed the prevalence of negative tweets toward Asian Americans, and found that during the pandemic such negative tweets increased by 68.4%. Themes that emerged in the study included: (1) racism and blame, (2) anti-racism, and (3) impact on daily living. In their article, Wong-Padoongpatt et al. (2022) completed a two-part study that investigated into (1) ratings of overall well-being and race-based psychosocial factors in a sample of Asian and LatinX Americans, and (2) differences in stress due to COVID-19 victimization in Asian and LatinX Americans after the lockdown mandate was lifted. The researchers found that (1) Asian Americans experienced more internalized racism during the pandemic and significant differences in well-being due to COVID-19 fear by racial groups, and (2) Asian Americans reported more distress due to the COVID-19 pandemic-related discrimination compared to LatinX Americans. Other research found that during the pandemic, Asian Americans experienced a significant increase in experience of everyday racism during the pandemic and significantly more so than other groups such as LatinX Americans and White Americans (Wong-Padoongpatt et al. 2022).

The types of microaggressions and the settings they were experienced in varied. Kim et al. (2022) found that Asian American nursing students reported experiencing microaggressions during the completion of their schooling and clinical work. These microaggressions were perpetrated on the part of patients, other nurses, and instructors. The researchers noted that the students aimed to remain professional during these interactions; however, these experiences during an already stressful time likely compounded the existing stressors. Indeed, the research suggests that experiences of microaggressions decrease mental well-being in targets (Nadal et al. 2017). In their article, Wu et al. (2021) found that Asian Americans and Asian immigrants experienced higher rates of discrimination and significantly worse mental health during the pandemic compared to Whites. Specifically, participants reported the following anxiety and depression scores (1) Whites 1.98, (2) Asian Americans 2.96, and (3) Asian immigrants 2.16. The researchers also found higher rates of "no mental health" symptomatology in Whites (47%) compared to Asian Americans (33%) and Asian immigrants (46%).

During the COVID-19 pandemic, overt Anti-Asian hate was pertinent (NW et al. 2015). An example of this includes Yunhan Zhang, an Asian-American tea shop owner, working in his shop one morning in Washington, DC. An individual came into his shop, pepper sprayed him in his eyes, and yelled, "Chinese" and "COVID-19" (Hermann 2020). These types of hate crimes can be placed in the context of racist commentary such as "Chinese virus" and "Kung flu" by individuals in positions of power, including former President Trump. These types of hate crimes have often been placed within the context of the racist remarks of former President Trump referring to the coronavirus (Real 2020). Indeed, research suggests that the rise in anti-Asian hate led to decreased mental well-being in Asian Americans, including elevated anxiety and depression scores during the pandemic, above and beyond their white counterparts (Wu et al. 2021). Further, the research suggests that rates of trauma symptoms in Asian Americans increased significantly during the pandemic (Litam 2020).

Further, the notable instances of racism against the Black community including the unarmed murder of 17-year-old Trayvon Martin in Florida (Blow 2012; Botelho 2012), Sandra Bland's death by suicide after a false traffic stop (Klein 2018), and the murder by asphyxiation of George Floyd and his cry for help saying, "I can't breathe, Sir. Please!" (Alang et al. 2017), point to the unfortunately common instances of police brutality and racism in America. Following events like these, feelings of anger, hopelessness, and grief plague the community (Alang et al. 2017). Further, the harm perpetuated by police brutality further impairs the mental well-being of the Black community due to harm resulting from historic and systemic racism. These instances include *racist policy,* a term coined by Kendi (2019) to describe institutional forms of racism such as (1) the Fugitive Slave Clause that required individuals to return escaped slaves to their master (Hall et al. 2005), (2) the utilization of property tax to fund schools, which disproportionately affects BIPOC students due to their overrepresentation in low-income neighborhoods (Boustan et al. 2013), and (3) the "War on Drugs" initiative started by the Nixon administration that led to the unequal drug sentencing of Black V. White men, with Black men being twice as likely to receive the maximum drug sentence due to drug charges (Sirin 2011). During the COVID-19 pandemic, these instances of discrimination became more apparent, especially through the Black Lives Matter protests (Buchanan et al. 2020; Jean 2020). Further, instances of racism and discrimination against Black Americans were at the forefront of the collective American consciousness due to media coverage of police brutality (Bryant-Davis 2007; Williams et al. 2021). It is important to note that this type of repeated media exposure is a type of trauma (Bryant-Davis 2007; Williams et al. 2021) and has been shown to increase levels of anxiety, depression, and hopelessness in targets (Alang et al. 2017).

During the pandemic, social factors also impacted marginalized, specifically LatinX individuals, general well-being (Garcini et al. 2020). For example, food insecurity increased from 16 to 46% in LatinXs (Cadenas et al. 2022) during the pandemic. Further, research suggests that documentation status impacts the psychosocial well-being of LatinXs by restricting individuals from health and social services (Chang 2019; Martinez et al. 2015; Cadenas et al. 2022). Further, during the pandemic, the psychosocial disparities between whites and LatinXs increased. Indeed, a nationally representative sample indicated that LatinX 37% of individuals reported in 2017 experiencing discrimination and fear of discrimination led 20% of LatinXs to avoid help seeking behavior (e.g., seeking medical help, food resources; Findling et al., 2019) which magnified the negative effects of the pandemic. Further, due to documentation status, access to jobs and resources is limited, further restricting access to resources (Moyce and Schenker 2018). These reasons may emphasize some of the reasons LatinX individuals were disproportionately affected by the pandemic. In their study, Cadenas et al. 2022 looked at the relationship between everyday discrimination, food insecurity, immigration status, and negative health outcomes during the pandemic. The researchers found that immigration status significantly impacted the relationship between discrimination and negative health outcomes. Further, the relationship between discrimination, immigration status, and negative health outcomes was mediated by food insecurity (Cadenas et al. 2022). These results are significant as it emphasizes the complexity of the negative impact of psychosocial factors on mental health and general well-being of LatinXs. These negative effects are further compounded by experiences of discrimination trauma, which was prevalent during the pandemic. For example, during the 2020, United States Presidential election candidates emphasized immigration policy and rhetoric such as, "build the wall" became common (Castañeda

2019). This type of rhetoric is a form of macroaggression that degrades the personhood of LatinXs by insinuating that this group of people should be "kept out" of the country.

Certainly, media circulation and exposure of race-based discrimination and violence is not a recent phenomenon. Historical examples include, but are certainly not limited to, reenactments of white European settlers attacking Indigenous peoples in Buffalo Bill's Wild West show from 1870 to 1920 (Hund et al. 2013), lynching postcards widely circulated in the United States from the 1880s to the 1970s (Wood 2005), wide distribution of Anti-Japanese World War II propaganda posters encouraging race-based violence (Sapre 2004), political platforms encouraging anti-immigration laws against LatinX populations starting as early as the late 1800s and continuing throughout American history (Androff et al. 2011) dispersion of video footage of the beating of Rodney King in 1991 (Adams 2016), and Facebook live streams of harsh police tactics on Indigenous peoples protesting the Dakota Access pipeline near Standing Rock (Hunt and Gruszczynski 2021). While these types of media exposures are not new, access to details of traumatic events and intensity of graphic images has increased due to the number of streaming platforms and increased individual content creators (Mahamid and Berte 2020). Additionally, these technological advances and media coverage of events allow for traumatic events to be widely broadcasted, watched, and re-watched. During the COVID-19 pandemic, many individuals reported being inundated with repeated exposure to graphic and violent images of the murder of Mr. George Floyd when checking news sources and social media outlets (Williams et al. 2021). These sudden and often unexpected repeated exposures to traumatic content impact individuals across community groups. As such, repeated exposure to race-based violence through social media may also be best understood as an additional collective trauma during the COVID-19 pandemic.

The APA multicultural guidelines state that clinicians should strive to use culturally adaptive interventions in their work with patients (APA 2017). Although psychologists have made strides to improve resources for treating marginalized groups (Lefforge et al. 2020), disparities in access to resources remain staggering. Research suggests that BIPOC are less likely to seek out treatment due to fears of discrimination, based on past experiences of discrimination by health care professionals (Becerra et al. 2013; Mann-Jackson et al. 2018). These disparities were highlighted during the pandemic with BIPOC reporting less treatment seeking behaviors due to experiences of discrimination and fear of discriminatory treatment (Bruening et al. 2017). One way that clinicians can improve this gap is by understanding the clinical presentation of mental health symptomatology in BIPOC and integrating cultural considerations into clinical treatment of mental health symptomatology.

As mentioned previously, BIPOC are particularly vulnerable to experiencing decreased mental health well-being due to experiences of racism and discrimination (Williams and Etkins 2021). These experiences of discrimination can lead to BIPOC to experience symptoms similar to individuals diagnosed with PTSD including anxiety, avoidance, and distrust of others (Kirkinis et al. 2021). However, these experiences of racism are often below the clinical threshold for a formal PTSD diagnosis. To be diagnosed with PTSD individuals must meet the criterion A qualifier for PTSD which states that an individual must be directly or indirectly exposed to "death, actual or threatened serious injury, or actual or threatened sexual violence" (American Psychological Association 2022). Many instances of discrimination in the form of microaggressions lead to negative harm and can have an insidious negative effect on their mental well-being in the form of increased anxiety, depression, and hopelessness (Kirkinis et al. 2021). Although the authors are not arguing that race-based trauma be considered a criterion A qualifier for PTSD. However, it is imperative for clinicians and researchers to begin to discuss where to place race-based trauma in our clinical conceptualizations and treatment of BIPOC.

This need was particularly apparent during the pandemic with more recent research beginning to highlight the ways that racism and discrimination complicate the clinical presentation of BIPOC mental health. For example, LatinXs who experienced discrimination experienced lower mental health well-being that was further impacted by food scarcity, and access to food was also limited due to systemic inequality that limited access to resources based on immigration status (Kirkinis et al. 2021). As the APA multicultural guidelines highlight, individuals exist within a multilayered interpersonal environment including the communal, institutional, and international social worlds they interact with. Integrating this complexity in the conceptualization, diagnosis, and treatment of BIPOC may prove efficacious in ameliorating treatment outcomes.

With that being said, psychologists strive to not pathologize normative responses to extraordinary experiences. As such, recognition that marginalized persons' experiences of actual harm through discrimination are not pathological is warranted. Indeed, traumatic responses to discrimination are well founded and there is a dire need to increase access to care within the existing insurance medical structure. Indeed, within the existing structure, most insurance policies require a formal diagnosis prior to authorizing reimbursement or coverage for mental health treatment. As such, many who need effective trauma treatment are excluded from utilizing their insurance to receive such care due to an incomplete definition of trauma. Typical courses of treatment would then require "out-of-pocket" payment, which many do not have access to such resources, ultimately leaving them untreated (Roll et al. 2013).

The authors offer the following considerations for the clinicians to consider with regards to the categorization and treatment of BIPOC. First, it is important to consider experiences of discrimination and race-based trauma in the clinical conceptualization of BIPOC, this is due to its deleterious effect on mental health. Second, clinicians should also work to understand the multiple and intersecting experiences (e.g., food scarcity and immigration status) of minoritized groups to improve clinical understanding of presenting problems. Third, psychologists should work to advocate for changes in funding, to increase access to mental health services.

19.5 Considerations for Treatment

Just as the pandemic highlighted clinicians' need to conceptualize race-based trauma in clinical presentation, the pandemic also highlighted limitations in culturally adapted interventions for marginalized groups. Indeed, during the pandemic, the APA set out several calls for training and research studies to be conducted because of the limitations of current treatment modalities (Causadias et al. 2023; Heidelburg and Leverett 2023). During this time, presentations on the colonization of treatment and the non-diverse samples of psychology research on treatment gained traction (Gergen 2007; Singh et al. 2020). These limitations and disparities are salient regarding racism-related traumas. With limited psychological research focusing on the implementation interventions focused on healing racism-related psychological wounds, clinicians must work to integrate an understanding of the impacts of race-based trauma for individual clients (Carter and Pieterse 2020).

Psychologists center current treatments for trauma on a PTSD diagnosis. Some common treatments for PTSD include Cognitive Processing Therapy (CPT; Galovski et al. 2015), Prolonged Exposure Therapy (PE; Peterson et al. 2019), and Eye Movement Desensitization and Reprocessing (EMDR; Stickgold 2002). The research suggests that PTSD treatments are effective in decreasing symptoms of distress caused by the index trauma (Lenz et al. 2014). Indeed, individuals who completed a PE (Eftekhari et al. 2013), CPT (Galovski et al. 2015, or EMDR (Stickgold 2002) protocol significantly reduced symptoms of PTSD, anxiety, and depression. The ability to administer evidence-based psychotherapy interventions effectively is critical for clinicians if they wish to assist clients in their journey toward health.

Despite the efficacy of current treatment protocols, access to treatment is often limited because of funding. Many insurance agencies often limit funding for psychological services. The research suggests that limits in funding deter individuals from receiving psychological interventions. A study conducted by Walker et al. (2015) found that in a nationally representative sample ($N=36{,}647$), Americans were less likely to seek mental health services if they did not have insurance, and even among those with access to insurance funding 46.6% experienced structural barriers to receive services, and 12.2% noted that cost was still a deterrent to seeking mental health services. Other funding agencies, such as the United States Office of Veterans Affairs (VA) require individuals to be diagnosed with an official PTSD diagnosis to receive benefits for trauma symptoms (Veterans 2023). Thus, if someone does not meet criteria for PTSD, it may exclude them from receiving financial help or limit access to specialized services. These factors are important to consider, especially in the pandemic's context. Many Americans experienced financial hardship during 2020 (Lueck 2021), and BIPOC groups were disproportionately affected economically (Soria et al. 2020). With experiences of discrimination by medical institutions, and economic barriers, deterring BIPOC from receiving psychological resources (Hingwe 2021), it highlights the importance of improving access to care either through changing the current funding systems and/or improving our clinical conceptualization and treatment of minoritized groups.

Another factor to consider is the way current manualized treatments manage psychological distress. Current treatments for trauma, such as CPT (Resick et al. 2016) and PE (Foa et al. 2009), focus on one individual trauma. Graduate programs teach clinicians to ask their clients to pick the worst trauma experience to process during treatment. The theoretical framework for this approach suggests that the skills learned from the processing and treatment of one trauma will generalize to the other experiences of trauma. Often, clinicians note that this can prove invalidating to the patient's experiences of trauma, with each type of trauma feeling like the "worst" for the individual. However, as these treatment modalities are effective within research settings, it is important to note that, within clinical practice, individuals often experience more than one trauma (Kilpatrick et al. 2013). Beyond this, these types of treatment modalities can often perpetuate harm if clinicians are not trained in the cultural implementation of the intervention.

For example, a Black male may present for treatment because of an experience of police brutality. The individual may state to the clinician that the police targeted him because he is Black, and that he cannot trust the police. Within the modality of CPT, clinicians are directed to help patients restructure unhelpful or distorted thoughts. Therefore, the clinician may feel prompted to restructure the way he conceptualizes his experience of trauma, to "improve" well-being. However, this may act as a type of microinvalidation by invalidating the patient's experience as a Black man in America. Brian Stevenson's book, *Just Mercy: A Story of Redemption*, highlights the structural inequalities within the American judicial system, including the disproportionate incarceration of Black men on death row. Stevenson also details his own experience with the police. He emphasized to the reader that despite being a lawyer and a Harvard graduate in his own car in his own neighborhood, his neighbors called the police on him for "suspicious behavior." These thoughts and feelings are not unfounded, especially while considering the targeting of Black bodies in America from its inception, from slavery, lynchings, and racist policies such as "separate but equal" (Schneider and Schneider 2014; Groves 1951).

This experience has prima facie been a part of the collective unconscious for decades, as highlighted through Harper Lee's (1982) book, *To Kill a Mockingbird*, which is a fictional story that details the unjust prosecution and sentencing of a Black man based on false rape charges. Indeed, during the pandemic, scholars found that facial recognition algorithms misidentified three Black men, which led to thousands of dollars spent in legal fees and time in jail by the Black men victimized by the error (Perkowitz 2021). This is particularly startling as research conducted at MIT and Microsoft Research and at the US National Institute of Standards and Technology (NIST) shows that facial recognition technology shows high levels of inaccuracies when applied to BIPOC (Buolamwini and Gebru 2018). The targeting and unequal treatment of Black men is a real and unfortunately common experience within the Black community. Communities affected by police brutality experience elevated levels of anxiety, depression, and hopelessness (Alang et al. 2017). Therefore, if a clinician were to "restructure" this "unhelpful thought," they would negate the actual reality that members of the Black community face, which can cause further harm. Studies have shown that microaggressions perpetuated by clinicians cause further detriment and diminish trust in psychology. This is because of the inherent role of the therapist as healer and their representation of a person in a position of power.

The pandemic has highlighted how clinicians need to conceptualize their patients dynamically, in the context of their communities, cultures, and historical experiences. Further, it highlighted clinicians need to improve training and cultural considerations of evidence-based treatments for minoritized groups to not perpetuate harm. Last, it showed clinicians the need to integrate and conceptualize race-based trauma in the treatment of BIPOC, as race-based trauma can have a significant and deleterious effect on mental well-being.

19.6 Conclusion

Perhaps the most significant lesson learned during the COVID-19 pandemic is the significant amount of work that remains toward reducing racism and racial disparities in the United States. The collective traumatic experiences throughout the United States of America during the COVID-19 pandemic disproportionately impacted BIPOC groups (Andrasik et al. 2022). The data and research that emerged throughout the pandemic were clear; America is still in the midst of a racism pandemic (Mills 2020). As mental health providers search for meaning from the COVID-19 pandemic and ways to address the disparities in access to mental health care, reflecting on failures of the mental health care system may prove most beneficial.

America has a significant issue in access to mental health care with more than half of adults with a mental illness not receiving treatment (Reinert et al. 2021). Access to mental health care decreases for BIPOC individuals due to lack of culturally adaptive psychological research and treatments (Bernal et al. 2009), lack of culturally competent mental health providers (Bernal et al. 2009), and lack of effective insurance billing policies that reflect presenting concerns that may be unique to BIPOC individuals (Thomas and Snowden 2001).

Increased funding and dedication toward culturally inclusive and adaptive research is needed to gain a more accurate understanding of impacts of systemic and individual racism on BIPOC mental health (Friis-Healy et al. 2021). To fill in these gaps, researchers should focus on (1) the unique impact of micro and macroaggressions on mental health symptomatology, (2) the role race-based trauma plays in mental health diagnosis and treatment, (3) include diverse populations in treatment outcome studies, and (4) run factor analysis with diverse populations to see if diagnostic criterion remains significant. Additionally, increased focus on researching and developing effective treatments with BIPOC individuals across presenting concerns and mental illness would assist in reduction of mental health disparities by increasing the likelihood of therapeutic benefit.

The benefits of culturally adaptive research and treatment are more likely to be realized through purposeful cultural training to mental health providers across all training levels. One of the most predictive influences in BIPOC clients early termination of mental health services is lack of perceived cultural competence of the therapist (Anderson et al. 2019). Cultural training requirements are different in every training program and widely differ across states (Triandis 1989). Unification of training expectations and requirements to include diversity training would increase mental health trainees cultural competency. Additionally, instating state legislature to require continuing education credits for diversity and multiculturalism across mental health professions would assist in the continuous learning and development for mental health providers.

Policy and institutional changes are likely to have a greater impact on the overall reduction in mental health disparities, than individual efforts. The role of diagnoses for insurance billing policy needs to be re-evaluated focusing on increasing access to care for BIPOC individuals and presenting concerns resulting from experiences of racism and discrimination. Current methods for billing insurance companies rely on clients meeting specific diagnostic criteria. Within the current DSM, clients seeking treatment for traumatic responses to racist and discriminatory experiences may not meet full diagnostic criteria for a diagnosable mental health disorder. As such, they are limited to cash-pay services which are frequently financially unfeasible for clients.

Minimally, the American Psychological Association (APA) is encouraged to provide guidance for therapists in navigating billing procedures for client's presenting with race-based stress. Ideally, American Psychiatric Association would develop or alter existing diagnostic criteria to increase access to billing insurance for BIPOC clients. In doing so, the American Psychiatric Association should seek to ensure not to pathologize BIPOC individuals' reactions to racist or discriminatory experiences but to instead provide a functional means for clinicians to increase access to care.

Lastly, the need for mental health professionals who are educated in navigating the U.S. legislative process has never been more apparent. In order to truly address the underlying systemic disparities, the abolishment of existing policy and creation of new policy to unravel and rebuild systems of care to ensure equal access. Mental health professionals are needed at every step in the legislative process, from drafting policy, to lobbying, and to advocating for impactful policy change. In doing so, mental health professionals will truly make meaning and learn from mistakes made during the COVID-19 pandemic.

References

Adams C (2016) March 3, 1991: Rodney King beating caught on video

Agba AMO, Ocheni SI, Agba MS (2020) COVID-19 and the world of work dynamics: a critical review. JESR 10(5):119. https://doi.org/10.36941/jesr-2020-0093

Alang S, McAlpine D, McCreedy E, Hardeman R (2017) Police brutality and black health: setting the Agenda for public health scholars. Am J Public Health 107(5):662–665. https://doi.org/10.2105/AJPH.2017.303691

American Psychiatric Association (ed) (2022) Diagnostic and statistical manual of mental disorders: DSM-5-TR, 5th edn, text revision. Am Psychiatr Assoc Publ

American Psychological Association (2017) Multicultural guidelines: an ecological approach to context, identity, and intersectionality. http://www.apa.org/about/policy/mulitcultural-guidelines.pdf

American Psychological Association (2022) Trauma. https://www.apa.org/topics/trauma

Anderson KN, Bautista CL, Hope DA (2019) Therapeutic alliance, cultural competence and minority status in premature termination of psychotherapy. Am J Orthopsychiatry 89(1):104–114. https://doi.org/10.1037/ort0000342

Andrasik MP, Maunakea AK, Oseso L, Rodriguez-Diaz CE, Wallace S, Walters K, Yukawa M (2022) Awakening: the unveiling of historically unaddressed social inequities during the COVID-19 pandemic in the United States. Infect Dis Clin North Am 36(2):295–308. https://doi.org/10.1016/j.idc.2022.01.009

Androff D, Ayon C, Becerra D, Gurrola M, Salas L, Krysik J, Gerdes K, Segal E (2011) U.S. immigration policy and immigrant children's well-being: the impact of policy shifts. J Sociol Soc Welf 38(1)

Bava S, Saul J (2013) Implementing collective approaches to massive trauma/loss in western contexts, pp 21–40

Becerra D, Androff D, Cimino A, Alex Wagaman M, Blanchard KN (2013) The impact of perceived discrimination and immigration policies upon perceptions of quality of life among latinos in the United States. Race Soc Probl 5(1):65–78. https://doi.org/10.1007/s12552-012-9084-4

Bernal G, Jiménez-Chafey MI, Domenech Rodríguez MM (2009) Cultural adaptation of treatments: a resource for considering culture in evidence-based practice. Prof Psychol Res Pract 40(4):361–368. https://doi.org/10.1037/a0016401

Bernard DL, Calhoun CD, Banks DE, Halliday CA, Hughes-Halbert C, Danielson CK (2021) Making the "C-ACE" for a culturally-informed adverse childhood experiences framework to understand the pervasive mental health impact of racism on Black Youth. Journ Child Adol Trauma 14(2):233–247. https://doi.org/10.1007/s40653-020-00319-9

Binion K, Gray MJ (2020) Minority stress theory and internalized homophobia among LGB sexual assault survivors: implications for posttraumatic adjustment. J Loss Trauma 25(5):454–471. https://doi.org/10.1080/15325024.2019.1707987

Blow CM (2012) The curious case of Trayvon Martin. The New York Times, 17 Mar 2012

Botelho G (2012) What happened the night Trayvon Martin died. CNN

Boustan L, Ferreira F, Winkler H, Zolt EM (2013) The effect of rising income inequality on taxation and public expenditures: evidence from U.S. municipalities and school districts, 1970–2000. Rev Econ Stat 95(4):1291–1302. https://doi.org/10.1162/REST_a_00332

Bowen S, Elliott S, Hardison-Moody A (2021) The structural roots of food insecurity: how racism is a fundamental cause of food insecurity. Sociol Compass 15(7). https://doi.org/10.1111/soc4.12846

Bruening M, Dinour LM, Chavez JBR (2017) Food insecurity and emotional health in the USA: a systematic narrative review of longitudinal research. Public Health Nutr 20(17):3200–3208. https://doi.org/10.1017/S1368980017002221

Bryant-Davis T (2007) Healing requires recognition: the case for race-based traumatic stress. Couns Psychol 35(1):135–143. https://doi.org/10.1177/0011000006295152

Buchanan L, Bui Q, Patel JK (2020) Black lives matter may be the largest movement in U.S. History. N Y Times. https://www.nytimes.com/interactive/2020/07/03/us/george-floyd-protests-crowd-size.html

Bumsted R, Sisak MR (2021) Hospitals overflowing with bodies in US epicenter of virus. AP NEWS

Buolamwini J, Gebru T (2018) Gender shades: intersectional accuracy disparities in commercial gender classification. In: Conference on fairness, accountability and transparency, 21 Jan 2018. PMLR, pp 77–91

Cadenas GA, Cerezo A, Carlos Chavez FL, Capielo Rosario C, Torres L, Suro B, Fuentes M, Sanchez D (2022) The citizenship shield: mediated and moderated links between immigration status, discrimination, food insecurity, and negative health outcomes for latinx immigrants during the COVID-19 pandemic. J Commun Psychol jcop.22831. https://doi.org/10.1002/jcop.22831

Carter RT (2007) Racism and psychological and emotional injury: recognizing and assessing race-based traumatic stress. Couns Psychol 35(1):13–105. https://doi.org/10.1177/0011000006292033

Carter RT, Pieterse AL (2020) Measuring the effects of racism: guidelines for the assessment and treatment of race-based traumatic stress injury, 1st edn. Columbia University Press, New York

Carter RT, Sant-Barket SM (2015) Assessment of the impact of racial discrimination and racism: how to use the race-based traumatic stress symptom scale in practice. Traumatology 21(1):32–39. https://doi.org/10.1037/trm0000018

Castañeda E (2019) Building walls: excluding Latin people in the United States. Lexington Books, Lanham

Causadias JM, Korous KM, Cahill KM, Rea-Sandin G (2023) The importance of research about research on culture: a call for meta-research on culture. Cultur Divers Ethnic Minor Psychol 29(1):85–95. https://doi.org/10.1037/cdp0000516

Center for Substance Abuse Treatment (US) (2014) Trauma-informed care in behavioral health services. Subst Abuse Ment Health Serv Adm (US). http://www.ncbi.nlm.nih.gov/books/NBK207201/

Cerda IH, Macaranas AR, Liu CH, Chen JA (2023) Strategies for naming and addressing structural racism in immigrant mental health. Am J Public Health 113(S1):S72–S79. https://doi.org/10.2105/AJPH.2022.307165

Chang CD (2019) Social determinants of health and health disparities among immigrants and their children. Curr Probl Pediatr Adolesc Health Care 49(1):23–30. https://doi.org/10.1016/j.cppeds.2018.11.009

Chong M, Chen H (2021) Racist framing through stigmatized naming: a topical and geo-locational analysis of #Chinavirus and #Chinesevirus on Twitter. Proc Assoc Inf Sci Technol 58(1):70–79. https://doi.org/10.1002/pra2.437

Costa DL, Yetter N, DeSomer H (2018) Intergenerational transmission of paternal trauma among US Civil War ex-POWs. Proc Natl Acad Sci USA 115(44):11215–11220. https://doi.org/10.1073/pnas.1803630115

Eftekhari A, Ruzek JI, Crowley JJ, Rosen CS, Greenbaum MA, Karlin BE (2013) Effectiveness of national implementation of prolonged exposure therapy in Veterans Affairs Care. JAMA Psychiat 70(9):949. https://doi.org/10.1001/jamapsychiatry.2013.36

Ekman K (2016) The ties that bind: Holocaust narrative, Jewish identity, and resilience

Fegert JM, Vitiello B, Plener PL, Clemens V (2020) Challenges and burden of the Coronavirus 2019 (COVID-19) pandemic for child and adolescent mental health: a narrative review to highlight clinical and research needs in the acute phase and the long return to normality. Child Adolesc Psychiatry Ment Health 14(1):20. https://doi.org/10.1186/s13034-020-00329-3

Feijt M, De Kort Y, Bongers I, Bierbooms J, Westerink J, Ijsselsteijn W (2020) Mental health care goes online: practitioners' experiences of providing mental health care during the COVID-19 pandemic. Cyberpsychol Behav Soc Netw 23(12):860–864. https://doi.org/10.1089/cyber.2020.0370

Findling MG, Bleich SN, Casey LS, Blendon RJ, Benson JM, Sayde JM, Miller C (2019) Discrimination in the United States: experiences of Latinos. Health Serv Res 54(S2):1409–1418. https://doi.org/10.1111/1475-6773.13216

Fix RL, Risco CM, Fix ST, Bernat EM (2022) How racial identity and worry about discrimination impact coping responses to racial discrimination among

black American community members. J Racial Ethn Health Disparities 9(2):641–654. https://doi.org/10.1007/s40615-021-00996-8
Foa EB, Chrestman KR, Gilboa-Schechtman E (2009) Prolonged exposure therapy for adolescents with PTSD: emotional processing of traumatic experiences: therapist guide. Programs that work. Oxford University Press, Oxford, New York
Fraiberg S, Adelson E, Shapiro V (1975) Ghosts in the nursery. J Am Acad Child Psychiatry 14(3):387–421. https://doi.org/10.1016/S0002-7138(09)61442-4
Friis-Healy EA, Nagy GA, Kollins SH (2021) It is time to REACT: opportunities for digital mental health apps to reduce mental health disparities in racially and ethnically minoritized groups. JMIR Ment Health 8(1):e25456. https://doi.org/10.2196/25456
Galovski TE, Wachen JS, Chard KM, Monson CM, Resick PA (2015) Cognitive processing therapy. In: Schnyder U, Cloitre M (eds) Evidence based treatments for Trauma-related psychological disorders. Springer International Publishing, Cham, pp 189–203
Garcini LM, Domenech Rodríguez MM, Mercado A, Paris M (2020) A tale of two crises: the compounded effect of COVID-19 and anti-immigration policy in the United States. Psychol Trauma Theory Res Pract Policy 12(S1):S230–S232. https://doi.org/10.1037/tra0000775
Gee GC, Morey BN, Bacong AM, Doan TT, Penaia CS (2022) Considerations of racism and data equity among Asian Americans, Native Hawaiians, and Pacific Islanders in the context of COVID-19. Curr Epidemiol Rep 9(2):77–86. https://doi.org/10.1007/s40471-022-00283-y
Geranios K, Kagabo R, Kim J (2022) Impact of COVID-19 and socioeconomic status on delayed care and unemployment. Health Equity 6(1):91–97. https://doi.org/10.1089/heq.2021.0115
Gergen K (2007) The self: colonization in psychology and society. In: Psychology's territories: historical and contemporary perspectives from different disciplines, pp 149–167
Ghosh AK, Venkatraman S, Soroka O, Reshetnyak E, Rajan M, An A, Chae JK, Gonzalez C, Prince J, DiMaggio C, Ibrahim S, Safford MM, Hupert N (2021) Association between overcrowded households, multigenerational households, and COVID-19: a cohort study. Public Health 198:273–279. https://doi.org/10.1016/j.puhe.2021.07.039
Groves HE (1951) Separate but equal—the doctrine of Plessy v. Ferguson. Phylon (1940–1956) 12(1):66. https://doi.org/10.2307/272323
Guay F, Bureau JS, Boulet J, Bradet R (2021) COVID-19 illegal social gatherings: Predicting rule compliance from autonomous and controlled forms of motivation. Motiv Sci 7(3):356–362. https://doi.org/10.1037/mot0000242
Hall KL, Ely JW, Grossman JB (2005) The Oxford companion to the Supreme Court of the United States, 2nd edn. Oxford University Press, Oxford, New York
Hankerson SH, Moise N, Wilson D, Waller BY, Arnold KT, Duarte C, Lugo-Candelas C, Weissman MM, Wainberg M, Yehuda R, Shim R (2022) The intergenerational impact of structural racism and cumulative trauma on depression. AJP 179(6):434–440. https://doi.org/10.1176/appi.ajp.21101000
Heidelburg K, Leverett P (2023) Call for papers: cultural adaptations to evidence-based interventions and practices in psychology. https://www.apa.org
Helms JE (1992) Why is there no study of cultural equivalence in standardized cognitive ability testing? Am Psychol 47(9):1083–1101. https://doi.org/10.1037/0003-066X.47.9.1083
Hermann P (2020) Police investigating attack on Chinese tea shop owner as hate crime. Washington Post, 11 Dec 2020
Hingwe S (2021) Mental health considerations for black, indigenous, and people of color: trends, barriers, and recommendations for collegiate mental health. In: Riba MB, Menon M (eds) College psychiatry, vol 1. Springer International Publishing, Cham, pp 85–96
Hinton E (2003) Microinequities: when small slights lead to huge problems in the workplace. Divers Inc., Mar/April 2003, pp 79–82
Hirschberger G (2018) Collective trauma and the social construction of meaning. Front Psychol 9:1441. https://doi.org/10.3389/fpsyg.2018.01441
Hund WD, Pickering M, Ramamurthy A (2013) Colonial advertising & commodity racism. In: Racism analysis. Series B, yearbook, vol 4. Lit, Zürich
Hunt K, Gruszczynski M (2021) The influence of new and traditional media coverage on public attention to social movements: the case of the Dakota access pipeline protests. Inf Commun Soc 24(7):1024–1040. https://doi.org/10.1080/1369118X.2019.1670228
Jang H, Rempel E, Roth D, Carenini G, Janjua NZ (2021) Tracking COVID-19 discourse on Twitter in North America: infodemiology study using topic modeling and aspect-based sentiment analysis. J Med Internet Res 23(2):e25431. https://doi.org/10.2196/25431
Jean T (2020) Black lives matter: police brutality in the era of COVID-19. Population health research brief series
Jeftić A, Ikizer G, Tuominen J, Chrona S, Kumaga R (2021) Connection between the COVID-19 pandemic, war trauma reminders, perceived stress, loneliness, and PTSD in Bosnia and Herzegovina. Curr Psychol. https://doi.org/10.1007/s12144-021-02407-x
Johnson-Agbakwu CE, Ali NS, Oxford CM, Wingo S, Manin E, Coonrod DV (2022) Racism, COVID-19, and health inequity in the USA: a call to action. J Racial Ethn Health Disparities 9(1):52–58. https://doi.org/10.1007/s40615-020-00928-y
Kendi IX (2019) How to be an antiracist. First edn. One World, New York
Khan MM, Rahman MH, Barua E (2021) Development of web-based system for essential services during the COVID-19 pandemic. In: 2021 24th

international conference on computer and information technology (ICCIT), 18 Dec 2021. IEEE, Dhaka, Bangladesh, pp 1–6. https://doi.org/10.1109/ICCIT54785.2021.9689799

Khanijahani A, Iezadi S, Gholipour K, Azami-Aghdash S, Naghibi D (2021) A systematic review of racial/ethnic and socioeconomic disparities in COVID-19. Int J Equity Health 20(1):248. https://doi.org/10.1186/s12939-021-01582-4

Kilpatrick DG, Resnick HS, Milanak ME, Miller MW, Keyes KM, Friedman MJ (2013) National estimates of exposure to traumatic events and PTSD prevalence using DSM-IV and DSM-5 criteria: DSM-5 PTSD prevalence. J Trauma Stress 26(5):537–547. https://doi.org/10.1002/jts.21848

Kim SS, Roberts TME, Khusbo JE, Watriboonruang W, Parks A, Lewczyk J (2022) Asian American nursing students' experiences of racial microaggressions amid the COVID-19 pandemic: focus group discussions. Nurse Educ Pract 64:103459. https://doi.org/10.1016/j.nepr.2022.103459

Kirkinis K, Pieterse AL, Martin C, Agiliga A, Brownell A (2021) Racism, racial discrimination, and trauma: a systematic review of the social science literature. Ethn Health 26(3):392–412. https://doi.org/10.1080/13557858.2018.1514453

Klein GC (2018) On the death of Sandra Bland: a case of anger and indifference. SAGE Open 8(1):215824401875493. https://doi.org/10.1177/2158244018754936

Lee H (1982) To kill a mockingbird. Warner Books edn. Warner Books, New York

Lee S, Waters SF (2021) Asians and Asian Americans' experiences of racial discrimination during the COVID-19 pandemic: impacts on health outcomes and the buffering role of social support. Stigma Health 6(1):70–78. https://doi.org/10.1037/sah0000275

Lefforge NL, McLaughlin S, Goates-Jones M, Mejia C (2020) A training model for addressing microaggressions in group psychotherapy. Int J Group Psychother 70(1):1–28. https://doi.org/10.1080/00207284.2019.1680989

Lenz S, Bruijn B, Serman N, Bailey L (2014) Effectiveness of cognitive processing therapy for treating posttraumatic stress disorder. J Ment Health Couns 36(4):360–376. https://doi.org/10.17744/mehc.36.4.1360805271967kvq

Litam SDA (2020) "Take your kung-flu back to Wuhan": counseling Asians, Asian Americans, and Pacific Islanders with race-based trauma related to COVID-19. TPC 10(2):144–156. https://doi.org/10.15241/sdal.10.2.144

Lluch C, Galiana L, Doménech P, Sansó N (2022) The impact of the COVID-19 pandemic on burnout, compassion fatigue, and compassion satisfaction in healthcare personnel: a systematic review of the literature published during the first year of the pandemic. Healthcare 10(2):364. https://doi.org/10.3390/healthcare10020364

Long E, Patterson S, Maxwell K, Blake C, Bosó Pérez R, Lewis R, McCann M, Riddell J, Skivington K, Wilson-Lowe R, Mitchell KR (2022) COVID-19 pandemic and its impact on social relationships and health. J Epidemiol Community Health 76(2):128–132. https://doi.org/10.1136/jech-2021-216690

Lopez L, Hart LH, Katz MH (2021) Racial and ethnic health disparities related to COVID-19. JAMA 325(8):719. https://doi.org/10.1001/jama.2020.26443

Lueck JA (2021) Help-seeking intentions in the U.S. population during the COVID-19 pandemic: examining the role of COVID-19 financial hardship, suicide risk, and stigma. Psychiatry Res 303:114069. https://doi.org/10.1016/j.psychres.2021.114069

Mahamid FA, Berte DZ (2020) Portrayals of violence and at-risk populations: symptoms of trauma in adolescents with high utilization of social media. Int J Ment Health Addiction 18(4):980–992. https://doi.org/10.1007/s11469-018-9999-0

Mann-Jackson L, Song EY, Tanner AE, Alonzo J, Linton JM, Rhodes SD (2018) The health impact of experiences of discrimination, violence, and immigration enforcement among Latino Men in a New Settlement State. Am J Mens Health 12(6):1937–1947. https://doi.org/10.1177/1557988318785091

Martinez O, Wu E, Sandfort T, Dodge B, Carballo-Dieguez A, Pinto R, Rhodes S, Moya E, Chavez-Baray S (2015) Evaluating the impact of immigration policies on health status among undocumented immigrants: a systematic review. J Immigrant Minority Health 17(3):947–970. https://doi.org/10.1007/s10903-013-9968-4

Meyer IH (2003) Prejudice, social stress, and mental health in lesbian, gay, and bisexual populations: conceptual issues and research evidence. Psychol Bull 129(5):674–697. https://doi.org/10.1037/0033-2909.129.5.674

Mills KI (2020) 'We are living in a racism pandemic,' says APA President. https://www.apa.org

Moyce SC, Schenker M (2018) Migrant workers and their occupational health and safety. Annu Rev Public Health 39(1):351–365. https://doi.org/10.1146/annurev-publhealth-040617-013714

Muran EM, Motta RW (1993) Cognitive distortions and irrational beliefs in post-traumatic stress, anxiety, and depressive disorders. J Clin Psychol 49(2):166–176. https://doi.org/10.1002/1097-4679(199303)49:2<166::AID-JCLP2270490207>3.0.CO;2-6

Nadal KL, Griffin KE, Wong Y, Davidoff KC, Davis LS (2017) The injurious relationship between racial microaggressions and physical health: implications for social work. J Ethn Cult Divers Soc Work 26(1–2):6–17. https://doi.org/10.1080/15313204.2016.1263813

Nadal KL, Erazo T, King R (2019) Challenging definitions of psychological Trauma: connecting

racial microaggressions and traumatic stress. JSACP 11(2):2–16. https://doi.org/10.33043/JSACP.11.2.2-16
Neville FG, Templeton A, Smith JR, Louis WR (2021) Social norms, social identities and the COVID-19 pandemic: theory and recommendations. Soc Personal Psychol Compass 15(5). https://doi.org/10.1111/spc3.12596
Nguyen TT, Criss S, Dwivedi P, Huang D, Keralis J, Hsu E, Phan L, Nguyen LH, Yardi I, Glymour MM, Allen AM, Chae DH, Gee GC, Nguyen QC (2020) Exploring U.S. shifts in anti-Asian sentiment with the emergence of COVID-19. Int J Environ Res Publ Health 17(19):7032. https://doi.org/10.3390/ijerph17197032
NW, 1615 L. St, Washington, S. 800, & Inquiries, D. 20036 U.-419-4300 | M.-419-4349 | F.-419-4372 | M (2015, November 4) How working parents share parenting and household responsibilities. Pew Research Center's Social & Demographic Trends Project. https://www.pewsocialtrends.org/2015/11/04/raising-kids-and-running-a-household-how-working-parents-share-the-load/
Olayo-Méndez A, Vidal De Haymes M, García M, Cornelius LJ (2021) Essential, disposable, and excluded: the experience of Latino immigrant workers in the US during COVID-19. J Poverty 25(7):612–628. https://doi.org/10.1080/10875549.2021.1985034
Palgi Y, Shrira A, Ring L, Bodner E, Avidor S, Bergman Y, Cohen-Fridel S, Keisari S, Hoffman Y (2020) The loneliness pandemic: Loneliness and other concomitants of depression, anxiety and their comorbidity during the COVID-19 outbreak. J Affect Disord 275:109–111
Panda CK (2021) Re-imagining the Native Americans in the Wake of 'Black Lives Matter' movement through Louise Erdrich's The Round House
Pérez Huber L, Solorzano DG (2015) Racial microaggressions as a tool for critical race research. Race Ethn Educ 18(3):297–320. https://doi.org/10.1080/13613324.2014.994173
Perkowitz S (2021) The bias in the machine: facial recognition technology and racial disparities. MIT Case Stud Soc Ethical Responsib Comput 5:15. https://doi.org/10.21428/2c646de562272586
Peterson AL, Foa EB, Riggs DS (2019) Prolonged exposure therapy
Pfeifer LS, Heyers K, Ocklenburg S, Wolf OT (2021) Stress research during the COVID-19 pandemic and beyond. Neurosci Biobehav Rev 131:581–596
Phipps R, Thorne S (2019) Utilizing Trauma-focused cognitive behavioral therapy as a framework for addressing cultural Trauma in African American children and adolescents: a proposal. Profess Counselor 9(1):35–50
Phiri P, Delanerolle G, Al-Sudani A, Rathod S (2021) COVID-19 and black, Asian, and minority ethnic communities: a complex relationship without just cause. JMIR Public Health Surveill 7(2):e22581
Real JAD (2020) With 'kung flu,' 'thugs,' and 'our heritage,' Trump leans on racial grievance as he reaches for a campaign reset. Washington Post
Reinert M, Fritze D, Nguyen T (2021) The state of mental health in America 2022
Resick PA, Monson CM, Chard KM (2016) Cognitive processing therapy for PTSD: a comprehensive manual. Guilford Publications
Rodriguez-Diaz CE, Guilamo-Ramos V, Mena L, Hall E, Honermann B, Crowley JS, Baral S, Prado GJ, Marzan-Rodriguez M, Beyrer C (2020) Risk for COVID-19 infection and death among Latinos in the United States: examining heterogeneity in transmission dynamics. Ann Epidemiol 52:46–53
Roll JM, Kennedy J, Tran M, Howell D (2013) Disparities in unmet need for mental health services in the United States, 1997–2010. Psychiatr Serv 64(1):80–82
Root MPP (1992) Reconstructing the impact of trauma on personality
Rose TC, Mason K, Pennington A, McHale P, Buchan I, Taylor-Robinson DC, Barr B (2020) Inequalities in COVID19 mortality related to ethnicity and socioeconomic deprivation. MedRxiv: 2020–2004
Sapre EE (2004) Wartime propaganda: enemies defined by race. W V Univ Philol Pap 51:91–104
Saricali M, Satici SA, Satici B, Gocet-Tekin E, Griffiths MD (2022) Fear of COVID-19, mindfulness, humor, and hopelessness: a multiple mediation analysis. Int J Ment Health Addiction 20(4):2151–2164
Schneider D, Schneider CJ (2014) Slavery in America. Infobase Publishing
Shah SMA, Mohammad D, Qureshi MFH, Abbas MZ, Aleem S (2021) Prevalence, psychological responses and associated correlates of depression, anxiety and stress in a global population, during the coronavirus disease (COVID-19) pandemic. Community Ment Health J 57:101–110
Shavers VL, Shavers BS (2006) Racism and health inequity among Americans. J Natl Med Assoc 98(3):386
Singh AA, Parker B, Aqil AR, Thacker F (2020) Liberation psychology and LGBTQ+ communities: naming colonization, uplifting resilience, and reclaiming ancient his-stories, her-stories, and t-stories
Sirin CV (2011) From Nixon's war on drugs to Obama's drug policies today: presidential progress in addressing racial injustices and disparities. Race Gend Class 82–99
Soria KM, Roberts BJ, Horgos B, Hallahan K (2020) Undergraduates' experiences during the COVID-19 pandemic: disparities by race and ethnicity
Spanierman LB, Heppner MJ (2004) Psychosocial costs of racism to whites scale (PCRW): construction and initial validation. J Couns Psychol 51(2):249
Stampfer HG (1990) "Negative symptoms": a Cu Mu Latlve Trauma stress disorder? Aust N Z J Psychiatry 24(4):516–528
Stickgold R (2002) EMDR: a putative neurobiological mechanism of action. J Clin Psychol 58(1):61–75

Strassle PD, Stewart AL, Quintero SM, Bonilla J, Alhomsi A, Santana-Ufret V, Maldonado AI, Forde AT, Nápoles AM (2022) COVID-19—related discrimination among racial/ethnic minorities and other marginalized communities in the United States. Am J Public Health 112(3):453–466

Sue DW, Constantine MG (2007) Racial microaggressions as instigators of difficult dialogues on race: implications for student affairs educators and students. Coll Stud Aff J 26(2):136–143

Sue DW, Alsaidi S, Awad MN, Glaeser E, Calle CZ, Mendez N (2019) Disarming racial microaggressions: microintervention strategies for targets, white allies, and bystanders. Am Psychol 74(1):128. https://doi.org/10.1037/amp0000296

Sue DW, Calle CZ, Mendez N, Alsaidi S, Glaeser E (2020) Microintervention strategies: what you can do to disarm and dismantle individual and systemic racism and bias. Wiley

Tan S, Bin de Souza P, Raifman M (2021) Structural racism and COVID-19 in the USA: a county-level empirical analysis. J Racial Ethn Health Disparities

Taylor M (2020) Collective trauma and the relational field. Human Psychol 48(4):382

Thomas KC, Snowden LR (2001) Minority response to health insurance coverage for mental health services. J Ment Health Policy Econ 4(1):35–42

Thompson CE, Neville HA (1999) Racism, mental health, and mental health practice. Couns Psychol 27(2):155–223

Triandis HC (1989) Intercultural education and training. Understanding the USA: a cross-cultural perspective, pp 305–322

Venegas-Vera AV, Colbert GB, Lerma EV (2020) Positive and negative impact of social media in the COVID-19 era. Rev Cardiovasc Med 21(4):561–564

Veterans A (2023) PTSD: national center for PTSD

Villarreal D (2020) Here are the states to request refrigerated trucks as COVID-19 spikes

Walker ER, Cummings JR, Hockenberry JM, Druss BG (2015) Insurance status, use of mental health services, and unmet need for mental health care in the United States. Psychiatr Serv 66(6):578–584. https://doi.org/10.1176/appi.ps.201400248

Williams DR, Etkins OS (2021) Racism and mental health. World Psychiatry 20(2):194

Williams MT, Osman M, Gran-Ruaz S, Lopez J (2021) Intersection of racism and PTSD: assessment and treatment of racial stress and Trauma. Curr Treat Options Psychiatry 8(4):167–185. https://doi.org/10.1007/s40501-021-00250-2

Wong-Padoongpatt G, Barrita A, King A, Strong M (2022) The slow violence of racism on Asian Americans during the COVID-19 pandemic. Front Publ Health 3800

Wood AL (2005) Lynching photography and the visual reproduction of white supremacy. Am Nineteenth Century Hist 6(3):373–399. https://doi.org/10.1080/14664650500381090

Woods LL (2018) The inevitable products of racial segregation: multigenerational consequences of exclusionary housing policies on African Americans, 1910–1960. Am J Econ Sociol 77(3–4):967–1012

Woolford A (2015) This benevolent experiment: indigenous boarding schools, genocide, and redress in Canada and the United States. University of Nebraska press

Wu C, Qian Y, Wilkes R (2021) Anti-Asian discriination and the Asian-white mental health gap during COVID-19. In: Race and ethnicity in pandemic times, Routledge Taylor & Francis Ltd (p 17)

Yearby R, Mohapatra S (2020) Law, structural racism, and the COVID-19 pandemic. J Law Biosci 7(1):lsaa036. https://doi.org/10.1093/jlb/lsaa036

Zelner J, Trangucci R, Naraharisetti R, Cao A, Malosh R, Broen K, Masters N, Delamater P (2021) Racial disparities in Coronavirus Disease 2019 (COVID-19) mortality are driven by unequal infection risks. Clin Infect Dis 72(5):e88–e95. https://doi.org/10.1093/cid/ciaa1723

Zhang D, Li G, Shi L, Martin E, Chen Z, Li J, Chen L, Li Y, Wen M, Chen B (2022) Association between racial discrimination and delayed or forgone care amid the COVID-19 pandemic. Prev Med 162:107153

Zhao N, Zhou G (2020) Social media use and mental health during the COVID-19 pandemic: moderator role of disaster stressor and mediator role of negative affect. Appl Psych Health Well 12(4):1019–1038. https://doi.org/10.1111/aphw.12226

Media and Interpersonal Channels Uses and Preferences Among International Students in the U.S. During the COVID-19 Pandemic

20

Pavica Sheldon

Abstract

This chapter provides an overview of how international students in the United States used media and interpersonal channels to communicate with family, friends, and each other during the COVID-19 pandemic. The chapter starts by discussing the challenges of international students during the health pandemic. It then provides a review of the role of media during a crisis. Factors influencing media use during the COVID-19 crisis include culture, personality, and time spent in a foreign country. Finally, the chapter ends by providing suggestions for how universities can better equip their international offices to deal with international students during challenging times such as the COVID-19 pandemic.

Keywords

COVID-19 · Health crisis · Media use · International students · Cultural differences

P. Sheldon (✉)
Department of Communication, University of South Alabama, 1747 Gwin Court, Mobile, AL 36695, USA
e-mail: psheldon@soouthalabama.edu

20.1 Introduction

The term "international students" refers to the category of students that require student visas (either an F-1 or J-1) to study in the United States. These students contribute billions to the US economy. They also enhance the intellectual and cultural environment for American peers. While typically more than 1 million international students are enrolled in US institutions of higher education every year, that number dropped by 15% during the 2020/21 academic year, reflecting the ongoing effects of the COVID-19 pandemic (Institute of International Education [IIE] 2022). The drop in the number of international students was due to border closings, flight cancelations, political decisions, and a variety of challenges caused by the COVID-19 pandemic (Pew Research Center 2021). The largest number of international students in the United States (approximately 35%) come from China, followed by India (18%), South Korea (4%), and Canada (3%). 17% of international students live in California, most of any state.

20.2 Challenges of Studying in a Foreign Country

International students often face multiple transitions. Not only do they have to adapt to the new educational system and values, but there are

N. Rezaei (ed.), *The COVID-19 Aftermath*, Advances in Experimental Medicine and Biology 1457,
https://doi.org/10.1007/978-3-031-61939-7_20

often other stressors such as financial vulnerabilities, discrimination, and language barriers (Lee and Rice 2007; Sawir et al. 2012). Culture shock is one of the most reported challenges experienced by international students studying in the United States (Mbous et al. 2022). Research (e.g., Zhang and Brunton 2007), also shows that many international students have limited contact with host nationals due to having a heavy academic load. On the other side, American students are often uninterested in their international peers (Ward 2001). While some international students adjust quickly, for others acculturation can take a long time (Chen 1992). According to Brislin (1993), three factors are needed for successful adjustment: (1) productive relationships with people from other cultures; (2) feeling that interpersonal interactions are warm, cordial, and cooperative; and (3) accomplishing tasks in an effective manner. Forming successful relationships with people from other cultures is what we often call connectedness. Connectedness has been defined as the number of friends and frequency of contact (Hendrickson et al. 2011), as well as feelings of belongingness or social support (Ong and Ward 2005). Being connected helps with loneliness and a more successful cross-cultural transition (Sümer et al. 2008). Although connectedness brings a lot of benefits to international students, research has shown that international students find it difficult to form friendships with local students (Zheng and Berry 1991; Ward and Masgoret 2004). This is due to both individual and contextual factors. Individual factors include language proficiency while contextual factors include perceived cultural distance and perceived cultural inclusiveness in the classroom (Bethel et al. 2020). When the culture of international students is vastly different from those of local students, international students are less willing to bond (Bethel et al 2016). Khanal and Gaulee (2019) noted that international students suffer from culture shock upon their arrival to the U.S. Their support system is typically in a different country and many pop culture references are not easily understood. In addition, they must learn how to make friends in the new environment. As research confirms, all these factors are leaving international students feeling left out and longing for connectedness (Charoensap-Kelly and Punyanunt-Carter 2022; Tran and Gomes 2017).

20.2.1 Challenges of International Students Due to the COVID-19 Pandemic

In 2020, international students experienced another unexpected challenge. COVID-19 caused challenging life circumstances. The pandemic was officially declared as such in March 2020 by the World Health Organization (Yang et al. 2020a, b). The contagious disease caused by a virus originated in Wuhan, China in December 2019 (Page et al. 2021). The pandemic impacted international student enrollment in US institutions, but also their acculturation process. The closure of schools and the shift to e-learning kept international students at a disadvantage. Many students had on-campus jobs that no longer existed leading to financial strains and struggles to provide for basic needs (Fischer and Whatley 2020). Many employers started prioritizing domestic students over international (Mbous et al. 2022). In addition, the restricted travel prevented international students from going back home even though they were not attending school (Alaklabi et al. 2021). Due to the origin of the coronavirus, international students of Asian descent reported increased incidents of hate and discrimination (Yang et al. 2020a, b). Charoensap-Kelly and Punyanunt-Carter (2022) noted that international students are among the most vulnerable groups of students impacted by the pandemic. This was particularly true for students who have just arrived in the country and did not have time to acculturate. With COVID-19 safety measures and masks, there were very few or no events ongoing for students when the pandemic started. This has furthermore exacerbated the opportunities to engage in activities with host nationals. In addition, both domestic and international students experienced negative emotions such as increased stress, anxiety, and hopelessness, as well as difficulty concentrating, disruptions to sleeping and eating patterns, loneliness, and increased concerns about academic performance (Aristovnik et al. 2020; Lynch et al. 2023). While mental health is important, many international students come from cultures where it is seldom addressed and students might feel reluctant in seeking out treatment or mental health resources (Forbes-Mewett and Sawyer 2016). Many of these resources were also not available face-to-face during the pandemic, which furthermore had the potential to increase distress in students (Lynch et al. 2023). Research has also shown that international students are more likely to seek help from informal sources such as parents, partners, and friends, as well as the Internet (Nguyen et al. 2019; Yeh et al. 2001).

In addition, international students did not qualify for pandemic-specific government support including the Coronavirus Aid, Relief, and Economic Security (CARES ACT) (Murakami 2020).

Martirosyan et al. (2022) examined the academic and nonacademic challenges of international students studying in a rural southwest Texas community during the COVID-19 pandemic. While only 59% of the participants indicated that they faced academic challenges during the pandemic, most of the participants experienced nonacademic challenges. The top five academic challenges were (a) difficulties in engaging with classmates, having group discussions, or completing group work, (b) difficulties in reaching out to instructors, asking questions, or receiving clarification/guidance when needed, (c) difficulties understanding course expectations and completing course assignments, (d) challenges to learning in a fully online environment due to English language barriers, and (e) difficulties in understanding hybrid/blended course schedules and requirements.

The top five nonacademic challenges reported were (a) general financial hardship, (b) mental health issues, (c) travel-related issues, (d) difficulty paying tuition, and (e) food insecurity. The findings of Martirosyan et al. (2022) study confirmed the need to take Maslow's (1943) Theory of Human Motivation when planning and responding to international students' needs during crises. Food insecurity and financial difficulties are basic needs, and they take precedence over other academic challenges. Other authors (Banaszak 2020; Fayazpour 2021) have also argued for Maslow's hierarchy of needs to be used when studying the impact of the COVID-19 pandemic on students. Not only did international students face food and housing insecurity due to lost work, but the government lockdown and online schooling contributed to isolation, or a deficiency in social and belonging needs (Martirosyan et al. 2022). Online learning dampened class work and the exchange of ideas, which are another fundamental element in Maslow's hierarchy (Chirikov and Soria 2020). For international students, however, not having basic needs fulfilled was a more important concern than academics (Martirosyan et al. 2022).

20.3 International Students' Media Use During the Pandemic

According to Kim's (2001) cross-cultural adaptation theory, the best way to adjust to a host country is to participate in communication with natives. This also includes mass communication such as watching American TV, reading American newspapers, and listening to the American radio. However, for the students already in the United States, the early days of the COVID-19 pandemic presented unique challenges as they had to quarantine and stay in apartments and dorms miles away from their family and friends back home but also isolated from their American peers. This furthermore hindered their adjustment process as they had fewer opportunities to interact with the natives and learn the new language and social norms. To satisfy their social needs, international students started relying on social media. Media were often the only source of information and the only way to socialize during the pandemic as schools were closed and face-to-face interactions with friends were limited. According to the media systems dependency theory

(Ball-Rokeach 1985; Ball-Rokeach and DeFleur 1976), when individuals spend more time on a specific media platform, they will become more dependent on it. During crises such as the COVID-19 pandemic, individuals are more inclined to use media they are already dependent on to access information and also socialize and relax. The higher instability and ambiguity of a society, the greater the dependency (Ball-Rokeach and DeFleur 1976).

Sheldon et al. (2023) looked at how international students studying in the United States used media and interpersonal channels during this global crisis. They surveyed 194 US-based international students from four continents, representing 54 countries to answer this question. As Fig. 20.1 shows, African students reported spending the highest number of hours ($n=6$) with media per day. European students reported spending the lowest number of hours ($n=4.7$) compared to students from other continents.

Of all types of media, Sheldon et al. (2023) found that social media were the primary source of information about COVID-19 for international students from all four continents included in their study (Europe, Asia, Africa, and South America) (Fig. 20.2). This is not surprising as younger generations use social media for information, entertainment, and social interaction during normal times. International students were spending a lot more time on social media during the pandemic, either reading news and updates from their home country or interacting with their friends and family (Seo et al. 2023). Chao et al. (2020) also found that during the pandemic, social media helped individuals learn safety measures and helped reduce loneliness due to isolation. Television was the least utilized channel among international students during the COVID-19 pandemic (Sheldon et al. 2023). Past studies have found that people who seek companionship via Facebook and those who use social media for relationship purposes were more likely to be honest, intentional, and positive than users with different motivations (Hollenbaugh and Ferris 2015). However, studies also showed that international students who had more frequent interactions with local people via social media tend to perceive a higher level of support from the host country though these contacts are not sufficient to adapt to the host country (Cao et al. 2018).

While there are many studies indicating the benefits of social media in helping international students make connections in new environments, the medium can create barriers to interaction with domestic students (Hodis and Hodis 2012; Olding 2013). This dual experience of being physically present in a foreign country but digitally connected to home friends has been named "translocal" (Martin and Rizvi 2014). Other studies also indicated that when international students seek out information online, they tend to stay in their comfort zones (Alzougool et al. 2013; Gomes 2015). One of the reasons mentioned is a lack of interest in the politics

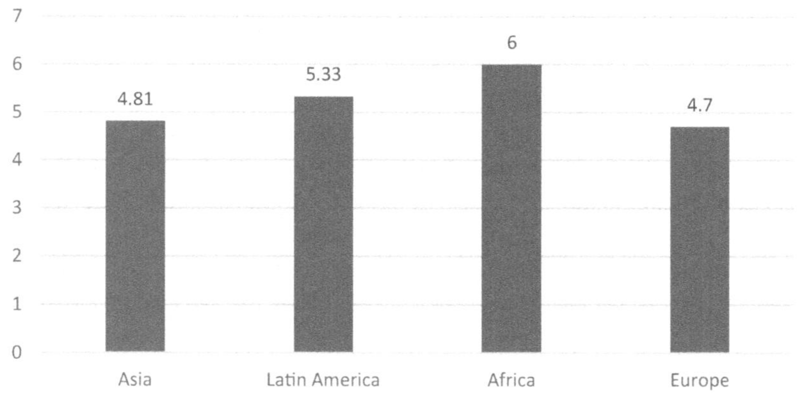

Fig. 20.1 Average media use (hour/day) of international students studying in the US during the COVID-19

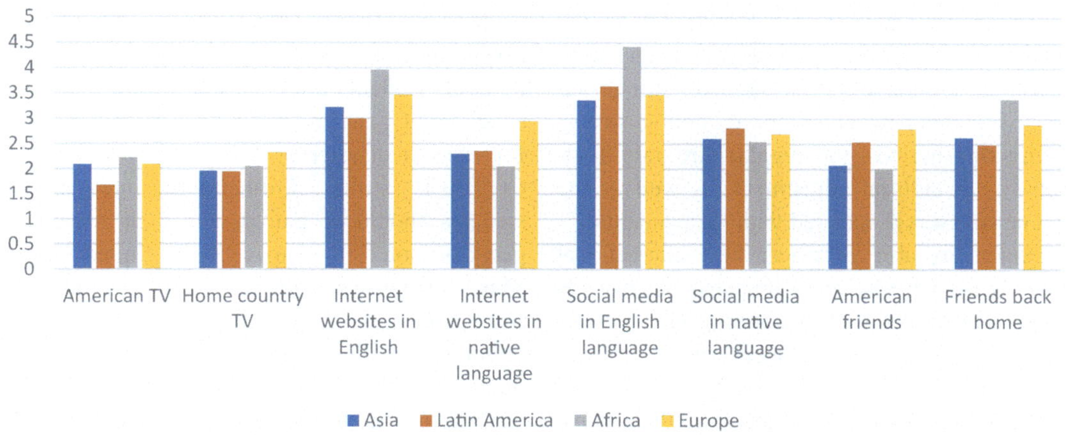

Fig. 20.2 International students' channel choices during the COVID-19 pandemic

and news of the host country unless the news directly affected them (Gomes 2015). Research has also found that spending long hours on social media may lead to addiction, depression, anxiety, and mood swings (Fung et al. 2014; Balakrishnan and Shamim 2013). Escobar-Viera et al. (2018) revealed that there are differences in how passive versus active users react to social media content. Thus, passive social media use was significantly related to depression. Gaitán-Aguilar et al. (2022) found that international students need support from other international students to achieve cross-cultural adaptation. International students share similar experiences being new to the host country and facing similar academic and life challenges.

However, not all international students act the same. Sheldon et al. (2023) found some differences among international students from different continents. Thus, students from African countries spent more time on social media than students from Europe, Latin America, or Asia. They also reported more contact with friends back home. This may be explained by cultural differences. Most African cultures are considered collectivist (Triandis 1988). As media systems dependency theory explains, during times of crisis, people rely on media that they are familiar with (Cheng et al. 2015). Similarly, other research (e.g., Evivie 2009) found that African international students see Americans as less interested in long-term friendships. Chen (2000) also found that international students coming from Europe are more satisfied in their interactions with a host society. Their culture is more similar and therefore the less social difficulty they experience (Olaniran 1996). Other studies revealed that non-white and non-European students experience the greatest difficulties when studying in the United States, including poor mental and physical health and higher acculturative stress (Poyrazli and Lopez 2007). International students who are not of Western and Eastern-European descent have attitudes and behavior that differ from those in the United States and are therefore at more risk for alienation (Dadfar and Friendlander 1982; Schram and Lauver 1988). Asian students are most isolated, followed by African students (Constantine et al. 2004).

The amount of time spent in the United States was also found to be a predictor of using American media and friends during the COVID-19 pandemic (Sheldon et al. 2023). The more time the students spent in the United States, the higher their host communication competence (Sheldon 2010) and acculturation. In other words, students had time to acquire the languages, get used to food, value and belief systems, and rituals and holidays observed (Lum 2004). For students who arrived in the country right before the pandemic, this was challenging as they did not have many opportunities, either in class or on campus, to interact with their

American peers and other host nationals as the universities were closed or operated under safety measures. Tseu et al. (2023) argued that international students have experienced "double stress" during this period—the regular stressors associated with cultural adjustment, as well as the unique stressors associated with the pandemic.

Personality also influenced how international students used media and interpersonal channels during the COVID-19 pandemic. Thus, extroverted and conscientious students interacted with American friends more frequently. They were also less likely to use social media in their native language (Sheldon et al. 2023). This is also true during normal times (Allehyani 2020; Swagler and Jome 2005). Extroverted individuals are more open and creative and have stable self-esteem, social maturity, assertiveness, persistence, and instrumental competence (Hayes and Lin 1994; Gareis 2000). On the other hand, shy students and those that exhibit low self-esteem, pessimism, and intergroup anxiety experience less success in maintaining relationships (Hayes and Lin 1994; Gareis 2000). Female international students also face fewer challenges than their male counterparts (Hayes and Lin 1994). This may be because women have been brought up to be warmer and more compassionate (Hayes and Lin 1994).

20.4 Conclusion

The number of international students studying in the United States is declining. The decline started following the COVID-19 lockdown in 2020 when the US government implemented measures to restrict travel from countries that traditionally sent most of its students to study in the U.S. While international offices cannot control the government measures, they can implement programs to help with both academic and nonacademic needs of international students. Among nonacademic needs were food insecurity and financial difficulties (Martirosyan et al. 2022). To overcome these challenges, international offices should work to open food pantries and closets with free items for international students. They should also promote these items as students might not be aware that they exist. Food pantries could be particularly useful when a disaster or a health pandemic like COVID-19 strikes. Furthermore, international students should be paired with their American peers as soon as they arrive in the United States. This may be especially important for students from Africa, as they had the least amount of interaction with American peers and rely on their native media to communicate with friends back home during the pandemic (Sheldon et al. 2023). New international students should also be provided information about the on-campus clubs where they can meet other students from the same or different countries. This could help with feeling connected to the campus and American culture. As Koo et al. (2021) found, international students with more social connectedness had fewer adjustment challenges. Even before their arrival on campus, international students should be provided extra support, both by their faculty and international offices. Faculty could also organize happy hours to provide students with an informal environment to interact with their professors and peers.

As Martirosyan et al. (2022) noted, international students should also be encouraged to use counseling services on campus. Due to cultural differences and social norms, many international students refrain from using these services. University officials should make extra efforts to commit to understanding the cultural differences that international students bring and work with them to overcome obstacles that they face upon their arrival. Because international students lack familiar support systems that are geographically close, they rely on institutions of higher education as ad hoc families, providing support in acculturation to both the new country and the new educational system (Liu et al. 2022). Of course, international students are not all the same. They have unique experiences and understanding of digital environments including online learning platforms.

References

Alaklabi M, Alaklabi J, Almuhlafi A (2021) Impacts of COVID-19 on international students in the U.S. High Educ Stud 11:37. https://doi.org/10.5539/hes.v11n3p37

Allehyani FS (2020) Personality traits as indicators of international students' intercultural communication competence. Doctoral Dissertation, Anglia Ruskin University

Alzougool B, Chang S, Gomes C, Berry M (2013) Finding their way around: international students' use of information sources. J Adv Manag Sci 1:43–49. https://doi.org/10.12720/joams.1.1.43-49

Aristovnik A, Keržič D, Ravšelj D et al (2020) Impacts of the COVID-19 pandemic on life of higher education students: a global perspective. Sustainability 12:8438. https://doi.org/10.3390/su12208438

Balakrishnan V, Shamim A (2013) Malaysian facebookers: motives and addictive behaviours unraveled. Comput Hum Behav 29:1342–1349. https://doi.org/10.1016/j.chb.2013.01.010

Ball-Rokeach SJ (1985) The origins of individual media-system dependency. Commun Res 12:485–510. https://doi.org/10.1177/009365085012004003

Ball-Rokeach SJ, DeFleur ML (1976) A dependency model of mass-media effects. Commun Res 3:3–21. https://doi.org/10.1177/009365027600300101

Banaszak B (2020) International students and COVID-19 needs. https://www.franklin.edu/current-students/blog/international-students-covid-19-needs. Accessed 8 Mar 2023

Bethel A, Szabo A, Ward C (2016) Parallel lives: predicting and enhancing connections between international and domestic students. In: Jindal-Snape RB (eds) Multi-dimensional transitions of international students to higher education. Taylor & Francis, London, pp 21–36

Bethel A, Ward C, Fetvadjiev VH (2020) Cross-cultural transition and psychological adaptation of international students: the mediating role of host national connectedness. Front Educ. https://doi.org/10.3389/feduc.2020.539950

Brislin RW (1993) Understanding culture's influence on behavior. Wadsworth, Belmont, CA

Cao C, Meng Q, Shang L (2018) How can Chinese international students' host-national contact contribute to social connectedness, social support and reduced prejudice in the mainstream society? Testing a moderated mediation model. Int J Intercult Relat 63:43–52. https://doi.org/10.1016/j.ijintrel.2017.12.002

Chao M, Xue D, Liu T, Yang H, Hall BJ (2020) Media use and acute psychological outcomes during COVID-19 outbreak in China. J Anxiety Disord 72. https://doi.org/10.1016/j.janxdis.2020.102248

Charoensap-Kelly P, Punyanunt-Carter N (2022) Renewing pedagogical research and practices: helping international students succeed post-COVID-19. Int J Commun 17:602–616. https://ijoc.org/index.php/ijoc/article/view/20107

Chen GM (1992) A test of intercultural communication competence. Intercult Commun Stud 2:23–42. https://web.uri.edu/iaics/files/04-Guo-Ming-Chen.pdf

Chen ZJ (2000) International students' preparation for and adaptation to the American higher educational system: a study of cross-cultural communication. World Commun 29:25–48

Cheng JW, Mitomo H, Otsuka T, Jeon SY (2015) The effects of ICT and mass media in postdisaster recovery—a two model case study of the Great East Japan earthquake. Telecommun Policy 39:515–532. https://doi.org/10.1016/j.telpol.2015.03.006

Chirikov I, Soria KM (2020) International students' experiences and concerns during the pandemic. SERU consortium. University of California-Berkeley and University of Minnesota. https://cshe.berkeley.edu/seru/about-seru/seru-surveys/seru-covid-survey-reports. Accessed 8 Mar 2023

Constantine MG, Okazaki S, Utsey SO (2004) Self-concealment, social self-efficacy, acculturative stress, and depression in African, Asian, and Latin American international college students. Am J Orthopsychiatry 74:230–241. https://doi.org/10.1037/0002-9432.74.3.230

Dadfar S, Friedlander ML (1982) Differential attitudes of international students toward seeking professional psychological help. J Counsel Psychol 29:335–338. https://psycnet.apa.org/doi/10.1037/0022-0167.29.3.335

Escobar-Viera CG, Shensa A, Bowman ND, Sidani JE, Knight J, James AE, Primack BA (2018) Passive and active social media use and depressive symptoms among United States adults. Cyberpsycholol Behav Soc Netw 21:437–443. https://doi.org/10.1089/cyber.2017.0668

Evivie LG (2009) Challenges faced by African international students at a metropolitan research university: a phenomenological case study. Master's thesis. University of North Carolina at Charlotte

Fayazpour S (2021) COVID-19, distance education, and students' needs. UNLV best teaching practices expo. https://digitalscholarship.unlv.edu/btp_expo/153. Accessed 8 Mar 2023

Fischer MA, Whatley M (2020) COVID-19 impact research brief: international students at community colleges. https://www.nafsa.org/professional-resources/browse-by-interest/covid-19-impact-research-brief-international-students. Accessed 8 Mar 2023

Forbes-Mewett H, Sawyer AM (2016) International students and mental health. J Int Stud 6:661–677. https://doi.org/10.32674/jis.v6i3.348

Fung ICH, Tse ZTH, Cheung CN, Miu AS, Fu KW (2014) Ebola and the social media. Lancet 384:2207. https://doi.org/10.1016/S0140-6736(14)62418-1

Gaitán-Aguilar L, Hofhuis J, Bierwiaczonek K, Carmona C (2022) Social media use, social identification and cross-cultural adaptation of international students: a longitudinal examination. Front Psychol 13. https://doi-org.libproxy.usouthal.edu/10.3389/fpsyg.2022.1013375

Gareis E (2000) Intercultural friendship: five case studies of German students in the USA. J Intercult Stud 21:67–91. https://doi.org/10.1080/17513057.2012.691525
Gomes C (2015) Footloose transients: International students in Australia and their aspirations for transnational mobility after graduation. Cross J Migr Cult 6:41–57. https://doi.org/10.1386/cjmc.6.1.41_1
Hayes RL, Lin HR (1994) Coming to America: developing social support systems for international students. J Multicult Counsel Dev 22:7–16. https://doi.org/10.1002/j.2161-1912.1994.tb00238.x
Hendrickson B, Rosen D, Aune RK (2011) An analysis of friendship networks, social connectedness, homesickness, and satisfaction levels of international students. Int J Intercult Relat 35:281–295. https://doi.org/10.1016/j.ijintrel.2010.08.001
Hodis GM, Hodis FA (2012) A mediation analysis of international students' patterns of computer-mediated communication. Int J Commun 6:2846–2869. https://ijoc.org/index.php/ijoc/article/viewFile/1328/827
Hollenbaugh EE, Ferris AL (2015) Predictors of honesty, intent, and valence of Facebook self-disclosure. Comput Hum Behav 50:456–464. https://doi.org/10.1016/j.chb.2015.04.030
Institute of International Education (2022) International students' enrollment trends, 1948/49–2021/2022. Open Doors Report on International Education Exchange. http://www.opendoorsdata.org/. Accessed 8 Mar 2023
Khanal J, Gaulee U (2019) Challenges of international students from pre-departure to post-study: a literature review. J Int Stud 9:560–581. https://doi.org/10.32674/jis.v9i2.673
Kim YY (2001) Becoming intercultural: an integrative theory of communication and cross-cultural adaptation. Sage Publications, Thousand Oaks, CA
Koo K, Baker I, Yoon J (2021) The first year of acculturation: a longitudinal study on acculturative stress and adjustment among first-year international college students. J Int Stud 11:278–298. https://doi.org/10.32674/jis.v11i2.1726
Lee JJ, Rice C (2007) Welcome to America? International students' perceptions of discrimination. High Educ 53:381–409. https://doi.org/10.1007/s10734-005-4508-3
Liu CH, Pinder-Amaker S, Hahm HC, Chen JA (2022) Priorities for addressing the impact of the COVID-19 pandemic on college student mental health. J Am Coll Health 70:1356–1358. https://doi.org/10.1080/07448481.2020.1803882
Lum D (2004) Social work practice and people of color. Brooks/Cole, Belmont, CA
Lynch J, Gesing P, Cruz N (2023) International student trauma during COVID-19: relationships among mental health, visa status, and institutional support. J Am Coll Health. https://doi.org/10.1080/07448481.2023.2166350
Martin F, Rizvi F (2014) Making Melbourne: digital connectivity and international students' experience of locality. Media Cult Soc 36:1016–1031. https://doi.org/10.1177/0163443714541223
Martirosyan NM, Van De Walker D, Saxon DP (2022) The impact of the COVID-19 pandemic on international students in a public university in the United States: academic and non-academic challenges. J Compar Int High Educ 14:90–102. https://doi.org/10.32674/jcihe.v14i4.4429
Maslow AH (1943) A theory of human motivation. Psychol Rev 50:370–396. https://doi.org/10.1037/h0054346
Mbous YPV, Mohamed R, Rudisill TM (2022) International students challenges during the COVID-19 pandemic in a university in the United States: a focus group study. Curr Psychol. https://doi.org/10.1007/s12144-022-02776-x
Murakami K (2020) Will push to reopen threaten aid? Inside Higher Education. https://www.insidehighered.com/news/2020/07/16/uncertainty-over-aid-higher-education-possible-next-coronavirus-relief-bill. Accessed 8 Mar 2023
Nguyen MH, Serik M, Vuong TT, Ho MT (2019) Internationalization and its discontents: help-seeking behaviors of students in a multicultural environment regarding acculturative stress and depression. Sustainability 11:1865. https://doi.org/10.3390/su11071865
Olaniran BA (1996) Social skill acquisition: a close look at foreign students on college campuses and factors influencing their level of social difficulty in social situations. Commun Stud 47:72–88. https://doi.org/10.1080/10510979609368465
Olding LA (2013) An investigation of the social relationships and social interactions amongst international students studying in Australia: a case study using Facebook. Doctoral Dissertation, University of Tasmania
Ong ASJ, Ward C (2005) The construction and validation of a social support measure for sojourners: the index of sojourner social support (ISSS) scale. J Cross Cult Psychol 36:637–661. https://doi.org/10.1177/0022022105280508
Page J, Hinshaw D, McKay B (2021) In Hunt for Covid-19 origin, patient zero points to second Wuhan market—The Man with the first confirmed infection of the New Coronavirus Told the WHO team that his parents had shopped there. Wall Street J. https://www.wsj.com/articles/in-hunt-for-covid-19-origin-patient-zero-points-to-second-wuhan-market-11614335404
Pew Research Center (2021) Amid pandemic. International Student Enrollment at U.S. Universities Fell 15% in the 2020-21 School Year. https://www.pewresearch.org/fact-tank/2021/12/06/amid-pandemic-international-student-enrollment-at-u-s-universities-fell-15-in-the-2020-21-school-year/

Poyrazli S, Lopez MD (2007) An exploratory study of perceived discrimination and homesickness: a comparison of international students and American students. J Psychol 141:263–280. https://doi.org/10.3200/JRLP.141.3.263-280

Sawir E, Marginson S, Forbes-Mewett H, Nyland C, Ramia G (2012) International student security and English language proficiency. J Stud Int Educ 16:434–454. https://doi.org/10.1177/1028315311435418

Schram JL, Lauver PJ (1988) Alienation in international students. J Coll Stud Dev 29:146–150

Seo H, Liu Y, Ebrahim MI, Chung D (2023) The COVID-19 pandemic and international students: a mixed-methods approach to relationships between social media use, social support, and mental health. First Monday 28. https://doi.org/10.5210/fm.v28i2.11516

Sheldon P (2010) Host communication competence and locus of control of international students in the United States. J Int Commun 24. https://immi.se/old-website/nr24/sheldon.htm

Sheldon P, Arif D, Charoensap-Kelly P, Mushtarin N, Choi J, Zhang Z (2023) Media use of international students studying in the United States during the COVID-19 pandemic: the role of dependency, acculturation, and personality. Presented at the Southern States communication Association conference. St. Petersburg, FL

Swagler MA, Jome LM (2005) The effects of personality and acculturation on the adjustment of North American sojourners in Taiwan. J Counsel Psychol 52:527–536. https://psycnet.apa.org/doi/10.1037/0022-0167.52.4.527

Sümer S, Poyrazli S, Grahame K (2008) Predictors of depression and anxiety among international students. J Couns Dev 86:429–437. https://doi.org/10.1002/j.1556-6678.2008.tb00531.x

Tran LT, Gomes C (2017) Student mobility, connectedness, and identity. In: Tran L, Gomes C (eds) International student connectedness and identity: cultural studies and transdisciplinarity in education. Springer, Singapore, pp 1–13. https://doi.org/10.1007/978-981-10-2601-0_1

Triandis HC (1988) Collectivism and individualism: a reconceptualization of a basic concept in cross-cultural social psychology. In: Verma GK, Bagley C (eds) Cross-cultural studies of personality, attitudes, and cognition. Palgrave Macmillan, London, pp 60–95. https://doi.org/10.1007/978-1-349-08120-2_3

Tseu A, Ganesh K, Wirtz D (2023) Adjusting in a pandemic: experiences of incoming international students. Front Educ 8. https://www.frontiersin.org/articles/10.3389/feduc.2023.1003051/full

Ward C (2001) The impact of international students on domestic students and host institutions. Ministry of Education. Wellington, New Zealand

Ward C, Masgoret AM (2004) The experiences of international students in New Zealand: report on the results of the national survey. Ministry of Education. Wellington, New Zealand

Yang Y, Liu K, Li S, Shu M (2020a) Social media activities, emotion regulation strategies, and their interactions on people's mental health in COVID-19 pandemic. Int J Environ Res Public Health 17:8931. https://doi.org/10.3390/ijerph17238931

Yang CC, Tsai JY, Pan S (2020b) Discrimination and well-being among Asians/Asian Americans during COVID-19: the role of social media. Cyberpsychol Behav Soc Netw 23:865–870. https://doi.org/10.1089/cyber.2020.0394

Yeh C, Inose M, Kobori A, Chang T (2001) Self and coping among college students in Japan. J Coll Stud Dev 42:242–256. https://repository.usfca.edu/cgi/viewcontent.cgi?article=1006&context=soe_fac

Zhang Z, Brunton M (2007) Differences in living and learning: Chinese international students in New Zealand. J Stud Int Educ 11:124–140

Zheng X, Berry JW (1991) Psychological adaptation of Chinese sojourners. Int J Psychol 26:451–470. https://doi.org/10.1016/j.ijintrel.2010.11.004

21

Estimating the Real Impact of the COVID-19 Pandemic in Kazakhstan: Factors Associated with Detection of the "True Infections"

Antonio Sarria-Santamera, Nurlan Abdukadyrov, Usman Abubakar Haruna, Natalya Glushkova, Yuliya Semenova, Zhandos Salpynov, Paolo Colet, and Zaule Zhumambayeva

Abstract

The COVID-19 pandemic is ongoing worldwide, and various case and death numbers are being reported to track its spread. However, the number of actual cases is uncertain due to under-reporting. Using mortality data as a more reliable indicator, this study in Kazakhstan evaluated the extent of under-reporting and under-detection of COVID-19 cases from March 2020 to September 2022 using back-casting and capture-recapture methods. The results indicate that official case reporting in Kazakhstan significantly underestimates the number of infections by at least 50%. The study also suggests that improved testing capabilities may have led to a decrease in the percentage of unreported cases, however, early in the pandemic, Kazakhstan faced significant testing shortages. The study presents a mathematical model based on mortality data that highlights the severe under-reporting of COVID-19 cases in Kazakhstan and argues

A. Sarria-Santamera (✉) · Y. Semenova ·
Z. Salpynov · P. Colet
Department of Medicine Nazarbayev University School of Medicine, Nazarbayev University, Astana 020000, Kazakhstan
e-mail: antonio.sarria@nu.edu.kz

Y. Semenova
e-mail: yuliya.semenova@nu.edu.kz

Z. Salpynov
e-mail: zhandos.salpynov@nu.edu.kz

P. Colet
e-mail: paolo.colet@nu.edu.kz

N. Abdukadyrov
Department of Mathematics, Statistics and Computer Sciences, University of Illinois at Chicago, Chicago, IL 60607, USA
e-mail: nabduk2@uic.edu

U. A. Haruna
Department of Biomedical Sciences, Nazarbayev University School of Medicine, Astana 020000, Kazakhstan
e-mail: usman.haruna@nu.edu.kz

N. Glushkova
Department of Epidemiology, Biostatistics and Evidence-Based Medicine, Al-Farabi Kazakh National University, Almaty 050040, Kazakhstan

Z. Zhumambayeva
Department of Children Diseases, Astana Medical University, Beibitshilik St 49/A, Astana 010000, Kazakhstan

N. Rezaei (ed.), *The COVID-19 Aftermath*, Advances in Experimental Medicine and Biology 1457,
https://doi.org/10.1007/978-3-031-61939-7_21

that understanding the true estimate of actual cases could aid in making informed decisions to end the pandemic.

Keywords

COVID-19 · Pandemic · Under-reporting · Estimate · True infections · Kazakhstan

21.1 Introduction

The novel Coronavirus Disease (COVID-19) pandemic broke out in December 2019 in China and spread rapidly around the world in 2020. Globally, the number of confirmed COVID-19 cases has grown to 639 million, and 6.6 million deaths have been reported as of November 29, 2022 (WHO n.d). The rapid spread of COVID-19 cases was troubling and had a significant negative impact on the population, national healthcare systems, and societies across the world. Most nations had to implement successive measures aimed at social distancing, including lockdowns and mandatory use of facemasks, to prevent SARS-CoV-2 from spreading further in areas or communities where the virus has already established a foothold. The successful rollout of effective vaccines was critical to controlling the escalating impact. However, as with all viruses, the genome of coronaviruses, including SARS-CoV-2, can change through mutations, recombination, insertions, and deletions, resulting in new variants. Since the beginning of the pandemic, various variants of SARS-CoV-2 have emerged and spread globally. By the end of 2021, multiple new variants with increased transmission and infectivity potential have been reported. These changes could have direct implications for disease pathogenesis, transmission potential, and severity (Sharun et al. 2021).

Kazakhstan, like most countries, has experienced different waves while implementing measures with variable levels of stringency (Haruna et al. 2022). The government declared a state of emergency and implemented the first lockdown three days after the initial cases were identified on March 13th, 2020, and extended it until May 11th, 2020; the second lockdown was subsequently implemented on July 5, 2020 (UNSDG n.d). Several waves and periods of non-pharmaceutical interventions (NPIs) have since been implemented, each with varying stringency. As in all countries, the public health system began to implement a reporting of daily cases as diagnosed by RT-PCR or rapid antigen tests. Vaccination in the country started initially with Sputnik V and later with other vaccines in February 2021.

However, since the early phases of the pandemic, there has been a widespread perception in Kazakhstan and elsewhere that, despite massive efforts by public health systems to identify, diagnose, and report daily cases to monitor pandemic dynamics, identifying cases and tracing contacts is difficult and difficult to perform (Sarría-Santamera et al. 2022). The dwindling number of reported cases suggested that these figures were probably only a portion of the real number of cases (Gaipov et al. 2021; Griette and Magal 2021). The weakness of reporting systems has been one of the most critical findings associated with the overall public health failure in addressing the COVID-19 pandemic (Sachs et al. 2020). Many countries could not implement necessary diagnostic testing strategies (Haruna et al. 2022), especially in the early stages but also in cases of significant incidence surges caused by highly transmissible variants. A second issue is SARS-CoV-2's unique biological characteristics, which include the fact that the majority of transmission occurs 24–48 h before the onset of symptoms and that a significant number of patients are mild or nearly asymptomatic (Yegorov et al. 2021). However, this is crucial given the ongoing increase in incidence; asymptomatic or mild cases are just as infectious as severe cases (Yegorov et al. 2021).

Earlier forecasting models suggested that fewer case identification and contact tracing, and increased transmission before symptoms, would decrease the chance of controlling the spread of COVID-19 (Hellewell et al. 2020). The Diamond Prince Cruise ship case documented the main problems that surveillance systems will be facing in controlling SARS-CoV-2: 66.7% covert infections among 0–19-year-olds, 30.8% among 20–58-year-olds, and 52.8% among 60 years and older were reported (Mizumoto and Chowell 2020). These findings provided evidence that the SARS-CoV-2 virus would spread rapidly across the globe and be extremely difficult to contain. In fact, very early in the pandemic, there was evidence that many cases went undocumented, and those undocumented infections were the source of many secondary documented cases (Li et al. 2020). Comprehensive epidemiological data representing the disease's true picture is important to guide resource allocation, implementation, monitoring, and evaluation of the diverse NPIs, such as lockdowns, quarantines, isolation, wearing masks, and contact tracing, as well as vaccination. Without precise estimates, it is challenging to accurately quantify the dynamics of the pandemic (Sharun et al. 2021).

As the world prepares for a post-COVID-19 period, it is important to reflect on what may have been the discrepancies between reported and actual infections, as a scenario in which circulating variants may display a high transmissibility rate and potentially evade the immune response induced from either natural immunity or vaccination is possible, leading to new waves of undetected transmission (Sharun et al. 2021). COVID-19 epidemiological models have been developed by several research groups to help in the understanding of this situation. These models predict actual infections and other critical metrics using existing data (confirmed cases and fatalities, testing rates, and so on) as well as a variety of hypotheses and epidemiological expertise.

As of November 30, 2022, Kazakhstan had reported 1.49 million confirmed cases and 19,053 deaths (Mathieu et al. 2020). A previous work conducted by our group revealed that official case reporting undercounts infections by at least 60% from March 2020–2021 (Sarría-Santamera et al. 2022). In this paper, we aim to update estimates of the dynamics of true infections in Kazakhstan from March 2020, when the first cases were reported, through September 2022, providing a general overview of the pandemic in Kazakhstan.

21.2 Methods

We used back-casting and capture-recapture approaches to compute the true number of infections in Kazakhstan from the first cases detected through September 2022.

21.2.1 Back-Casting Method

The back-casting technique is statistical modeling for the prediction of a real number of infections in a given population (Sarría-Santamera et al. 2022). It is assumed that the period lag after infection to death follows a gamma distribution, where $\alpha = (\mu/s)^2$ and $\beta = (s^2/\mu)$ and μ represents the mean and s standard deviation (Phipps et al. 2020), and has been used in several studies (Böhning et al. 2020; Niwitpong et al. 2012; Sarría-Santamera et al. 2022). By using mortality data count and gamma distribution, the model enables a backward forecast from the period of the initial infection. To compute the number of infectious diseases present on a given day that caused deaths on day t, the following formula can be used:

$$n_i(t', t) = \frac{N_f(t) x f(t - t'; \alpha, \beta)}{\text{IFR}} \quad (21.1)$$

$N_f(t)$	Number of fatalities that occurred on the day (t).
$f(x; \alpha, \beta)$	Probabilistic density of gamma distribution.
IFR (Infection Fatality Rate)	Proportion of deaths to cases for both covert and non-covert cases (5).

Therefore, the cumulative number of new cases on day $t\prime$ can be calculated by adding the figures of $N_i(t', t)$ for all possible values of $t > t'$ This is adjusted because not all cases that happened on day t' will result in mortality (Phipps et al. 2020). To compute the total sum of new cases, we then use the cumulative gamma distribution function and the day on which fatality data are available.

$$N_i(t^i) = \frac{1}{F(t_0 - t^i; \alpha, \beta)} \sum_{t=t'+1}^{t_0} n_i(t', t) \tag{21.2}$$

21.2.2 Capture-Recapture Method

The capture-recapture method is used to estimate the number of people infected with COVID-19 by analyzing the number of reported deaths from the disease. Unlike the back-casting method, the capture-recapture approach enables the estimation of the lower bound of the number of infected individuals and the overall estimate.

The application of this method necessitates the use of a geometric distribution for the frequency of unit identification during sampling. To estimate a lower end, the Cauchy–Schwarz inequality formula must be used. The following equation is used to estimate unreported or concealed cases:

$$H_{t_0} = \sum_{t=t_0+1}^{t_m} \frac{\triangle N(t)[\triangle N(t) - 1]}{1 + (\triangle N(t-1) - \triangle D(t), 0)} \tag{21.3}$$

$\triangle$N (t) Number of new cases at day t.

$\triangle$D (t) Number of new mortalities at day t.

t_0 Beginning day (day of the first reported COVID-19 case).

t_m Final day for which data are accessible. The following formula can be used to determine the overall number of infections: $N(t_m) + H_{t_0}$.

21.2.3 Data

This project covers the time frame that corresponds from March 2020 through September 2022. Data from Our World in Data was used to calculate the total number of tests conducted per capita in the country (https://ourworldindata.org/coronavirus).

21.3 Results

We applied the back-casting and capture-recapture approaches to Kazakhstan previously. It is necessary to estimate μ, s, and IFR for the back-casting approach, because IFR varies with age (Levin et al. 2020). Based on Levin et al., we calculated an IFR of 0.05 for Kazakhstan. Philipp et al. describe estimates of 21.55 and 8.64, respectively (Phipps et al. 2020). To obtain a more accurate result, it is recommended that exponential smoothing combined with back-casting be used. In our situation, t was September 1, 2022, we back-cast the data using exponential smoothing up to July 13, 2022. The gap of 50 days allows us to calculate the number of cases almost entirely based on the death rate. Table 21.1 and Figs. 21.1 and 21.2 summarize the findings obtained using the methods described above.

21.4 Discussion

The findings of this study not only reflect the presence of a substantial fraction of undiagnosed SARS-CoV-2 infections in Kazakhstan but also show certain variations in different periods, with a reduced differential between documented and undocumented cases during 2022. In any event, and generally, these findings are not surprising and confirm the occurrence of a large number of undocumented cases. Official figures may undercount overall cases by 55–60%. In this work, we harnessed the use of the back-casting and capture-recapture approaches leveraging the use of statistical modeling frameworks based on mortality data to estimate the extent of under-reporting. This study does not aim to establish a causal link between unreported cases and the spread of the pandemic or to analyze possible causes of the under-reporting, but it provides a clear picture of the challenges faced by public health systems in managing the

COVID-19 pandemic and points out the impact under-reporting may have on the virus' transmission (Böhning et al. 2016).

The accurate forecast of COVID-19 cases is critical for the implementation of interventions as well as for assessing their effectiveness, to better preparedness and resource management. However, a body of evidence suggests that the current reporting system may seem to be under-reporting the number of cases, thereby jeopardizing having a complete understanding of the real dynamic of the pandemic and limiting the implementation of public health measures and their evaluation (Srivastava and Prasanna 2020).

Why is this important? A precise assessment of how widespread the disease is within

Table 21.1 Estimation of the proportion of true infections using capture-recapture and back-casting methods

Date	Reported cases	Capture-recapture		Back-casting	
		Estimated true infections	% Reported	Estimated true infections	% Reported
01.07.2020	20,734	47,010	4411	77,685	2669
01.10.2020	139,337	285,789	4876	377,225	3694
01.01.2021	196,471	400,672	4904	511,058	3844
01.04.2021	289,046	589,654	4902	1,092,612	2645
01.07.2021	472,776	959,634	4927	1,460,774	3236
01.10.2021	1,009,613	2,049,068	4927	2,955,543	3416
01.01.2022	1,129,300	2,288,974	4934	3,263,.772	3460
01.04.2022	1,452,936	2,940,378	4941	3,384,446	4293
01.07.2022	1,454,788	2,943,966	4942	3,385,620	4297
01.09.2022	1,525,983	3,107,728	4910	3,390,719	4500

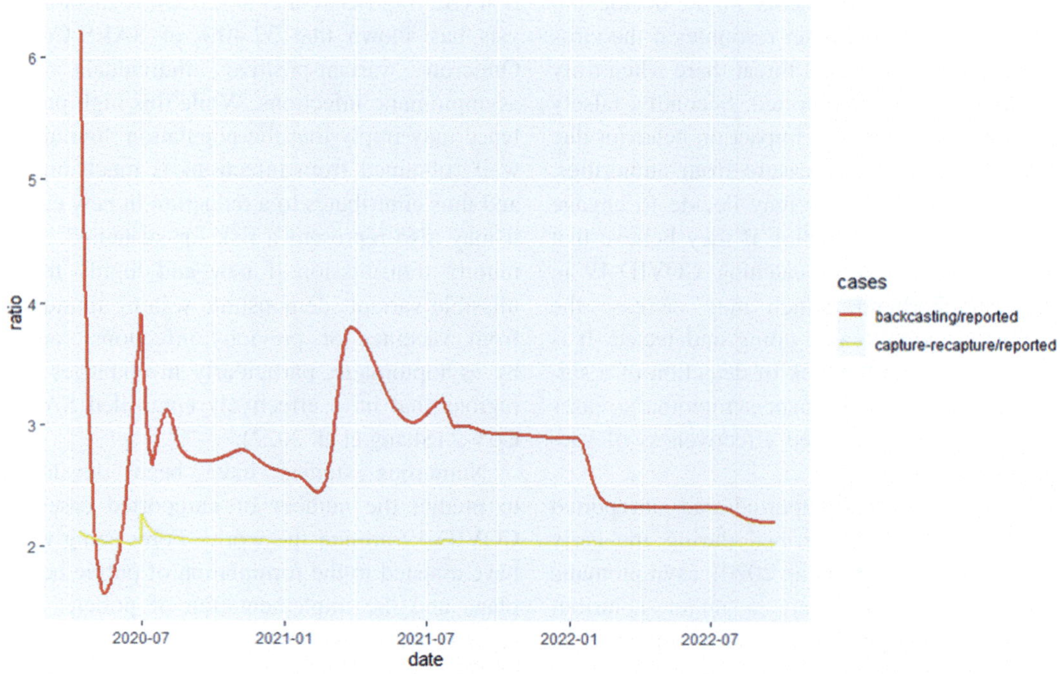

Fig. 21.1 Ratio between the number of reported and estimated COVID-19 cases

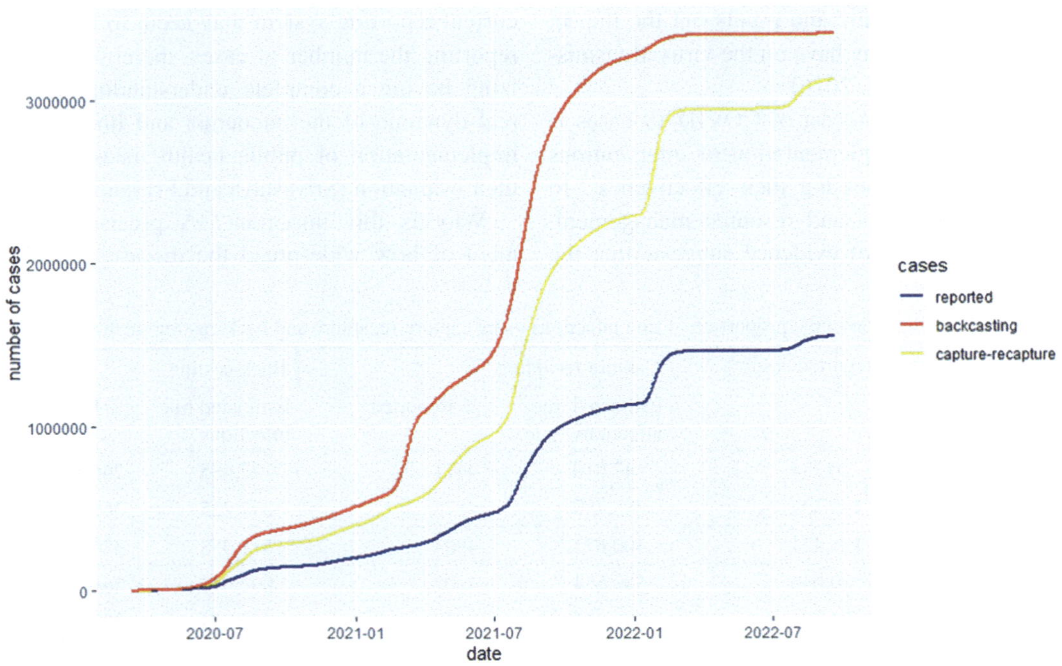

Fig. 21.2 Cumulative number of estimated COVID-19 cases

communities is critical in understanding and controlling the pandemic. Under-reporting has a profound impact on how nations respond to the pandemic. Governments might decide not to allocate funds and other resources if the virus seems to pose a minimal threat there when truly the cases were underreported. Secondly, falsely low counts may have an impact on behavior due to a lack of testing, pressure from authorities, or other factors. People may decide to engage in greater social interaction if they believe that there is a low risk of catching COVID-19 in their neighborhood, which may increase the number of overall cases going undetected. It is also possible that the lack of detection of a significant percentage of paucisymptomatic cases may reduce the perceived effectiveness of vaccination strategies.

The gap between reported and unreported cases was already described during the early outbreak in China (Li et al. 2020): asymptomatic SARS-CoV-2 cases were a major contributor to the widespread community transmission and highlight the required commitment and necessary resources for testing to mitigate and contain the COVID-19 pandemic (Liu et al. 2020). Asymptomatic cases may range from 14 to 50% of people with SARS-CoV-2 infection (Buitrago-Garcia et al. 2022). A meta-analysis has shown that 32.40% of SARS-CoV-2 Omicron variant-positive individuals had asymptomatic infections. While this high prevalence may imply that the population "immunity wall" obtained from infections is much higher and thus contributes to a reduction in new cases, it may also represent a risk for sustained community transmission if new and highly transmissible variants demonstrate waning immunity from vaccines or previous infections caused by asymptomatic, particularly in countries and regions that have effectively controlled SARS-CoV-2 (Shang et al. 2022).

Numerous studies have been developed to predict the number of unreported cases of COVID-19 around the world. Forecasting tools have assisted in the formulation of public health plans and the implementation of policies that support COVID-19 mitigation (de Lima et al. 2020). Countries have also conducted seroprevalence studies to investigate the extension of the pandemic. But a critical not-resolved question is what the actual number of infections is.

Studies in various countries with diverse social, cultural, ethnic, and healthcare characteristics have shown significant under-reporting of COVID-19 cases. Research in the USA suggests that the total number of COVID-19 cases may be 30–40 times higher than reported cases (Srivastava and Prasanna 2020). In Brazil, estimates indicate that around 90% of positive cases are not recorded (Aronna et al. n.d). Analysis using different methods have revealed that the number of infected individuals in Brazil could reach 30%, which is six times more than the number observed in the USA (Albani et al. 2021). Djilali et al. in North African countries (Algeria, Egypt, and Morocco) reported that the number of reported cases in these countries is just a fraction of the actual number of infections (Djilali et al. 2020). Krantz et al. revealed significant differences ranging between 2 and 25% in the capacity of various countries to detect COVID-19 cases during the first peak in March 2022 (Krantz and Rao 2020). These authors, using the same methodology, predicted that from December 2019 to May 6, 2022, the real number of global COVID-19 infections would range between 601 million and 2.4 billion. Interestingly, by mid-May 2022, 518 million cases had been reported (Maciej Serda et al. 2022). The overall reflection of all authors: increasing testing capacity is a determinant factor to control the pandemic.

Our previous estimate of COVID-19 cases in Kazakhstan revealed significant under-reporting (Sarría-Santamera et al. 2022), therefore we intend to present an updated estimate to reflect the virus's current dynamics and the level of under-reporting. These findings are consistent with our prior analysis, albeit with a certain upturn in the gap during the last year.

Available data investigating the epidemiological characteristics of SARS-CoV-2 in Kazakhstan are very limited. Data collected from 5116 laboratory-confirmed COVID-19 cases from all regions of Kazakhstan between March 13 and June 6, 2020 shows that a high proportion (40%) of these cases were asymptomatic (Zhussupov et al. 2021). Although no nationwide seroprevalence studies have been conducted in Kazakhstan, local and small-scale published data indicate that the number of infections is substantially higher than what is reported through public health monitoring systems. Kadyrova reported that in the period April–May 2021, before the rollout of vaccination in the country, 64% of their cohort tested positive for at least one antibody (Kadyrova et al. 2022). A study conducted in October 2020 and January 2021 in several Kazakhstani cities reported a prevalence of positive IgM/IgG ranging between 39 and 60% (Smagul et al. 2022). Semenova analyzed data from individual self-referrals for antibody testing between July 15, 2020 and July 10, 2021 and reported positive IgG in 63% of them. The study's major limitation is the fact that the data were collected from test centers, rather than from persons who took the test. Thus, a single person may appear multiple times in this analysis (Semenova et al. 2022).

In October 2021, Kazakhstan was reporting a cumulative of 1.02 million cases. This represents a 4.3% prevalence of the infected population, even considering that in June 2021 the Government of Kazakhstan confirmed the circulation of the Delta variant in the country. The surge in cases during that summer led to the then-highest number of daily reported cases during the pandemic, peaking at almost 8000 new daily cases. By the end of January 2022, after the Omicron wave in the country, which was associated with the highest number of daily reported cases (more than 15,000), the overall cumulative total of cases was 1.33 million, representing a prevalence of 6.6%. In October 2022, when this work was elaborated, the total number of reported cases was 1.48 million.

Data on severity of cases is also very limited but some it has been estimated that of the total diagnoses cases, mild-moderate cases may represent 85% of total, while only 2.5% could have been extremely severe cases requiring mechanical ventilation (Zhandosova et al. 2021).

Information on the sequencing of variants of the virus in Kazakhstan is scarce. A study of hospitalized patients between February and April 2020 found that 87% of the isolates were from early Asian lineages, while the rest were related to lineages from Europe and the Americas (Yegorov et al. 2021). More recent data, covering the period from August 2021 to May 2022 and limited to people diagnosed with COVID-19 in Astana, showed that the majority of cases were the Omicron and Delta variants (Kairov et al. 2022).

Zhalmagambetov et al. asserted that the relatively low number of cases reported during the first wave in Kazakhstan was due to the NPI implemented by the government (Zhalmagambetov et al. 2020). Ibrayeva also suggests in their work the effectiveness of social mobility restrictive measures implemented in the country also during the first wave (Ibrayeva et al. 2022). Kazakhstan experienced an increase in illness transmission after the initial lockdown was lifted. In June 2020, a second lockdown was necessarily implemented to take control of the problem. This led to a decrease in the number of infected people (Alekesheva et al. 2021), and highlights the impact of NPI to control number of cases through a decrease in reproductive number.

During the first wave, the positivity rate of PCR was even higher than 20%, while it was just over 4% during April–May 2021. After June 2021, data on tests was unavailable (Fig. 21.3). Several factors may be attributable to the increase in under-reporting. First, these policies had a negative socioeconomic impact on vulnerable groups as well as enterprises that provide income, jobs, and taxes that support public services, and were difficult for people to comply with (Haruna et al. 2022). Secondly, under-testing. Government testing centers are free, but they may be overwhelmed, and the expenses of private testing are much beyond the means of most individuals. Additionally, testing centers tend to be in larger cities, which limits the number of tests being carried out in rural areas.

Furthermore, another specific unresolved issue in Kazakhstan is the emergence of

Fig. 21.3 COVID-19 daily reported cases and daily PCR tests performed

pneumonia with COVID-19-like symptoms but negative PCR (Gaipov et al. 2021). Negative PCR COVID-19-like pneumonia has been a non-resolved problem in Kazakhstan. The Ministry of Health reported more than 90,000 cases in October. An analysis of those cases indicated that PCR-positive cases had a two-fold mortality risk than PCR-negative cases (Gaipov et al. 2021).

The main limitation of our work is that we use reported deaths to determine IFR. It is still unclear what the true IFR is for COVID-19, which significantly impacts the accuracy of the results (Zhalmagambetov et al. 2020), and there has also been significant concern regarding the uncertainty of deaths being correctly reported (Haruna et al. 2022). The number of deaths caused by COVID-19 might have been underestimated in less developed countries, particularly during the outbreak's first wave, but this phenomenon has affected all countries. Countries such as the United States, the United Kingdom, and Spain are also likely to have underreported the number of deaths due to COVID-19 (Lau et al. 2021). Previous studies suggest that the death toll of COVID-19 may have been underestimated by 35% (Krantz and Rao 2020). Depending on the assumption of IFR, Unnikrishnan estimates that in India, COVID-19 cases reported may be as low as 3–18% (Unnikrishnan et al. 2021). While globally it was reported that 5.94 million individuals had died as a result of COVID-19 as of December 2021, they estimated that 18.2 million people died from the disease. It has been estimated that excess mortality in Kazakhstan may be 3.8 times greater than the reported COVID-19 mortality (Sachs et al. 2022). IFR would ideally be estimated objectively by randomized general testing. However, randomized testing, particularly in the early phases of the outbreak, may not be practical due to the limited testing supply and the increased demand from those with symptoms.

The excess mortality rate captures not only the confirmed deaths associated with the pandemic but also those that were not detected and reported. In addition, deaths from other causes resulted from the overall crisis, which makes it a more comprehensive measure of the pandemic's impact on deaths. Under-reporting of deaths may be due to a lack of testing and the difficulty of tracing COVID-19 deaths, which may affect the more underprivileged, who may also be more affected by COVID-19 (Aronna et al. n.d.). Despite the inability to determine the COVID-19 proportion that contributes to this excess, it may indicate that the actual infection may even be higher than estimated.

Deaths have been, nevertheless, a better indicator than cases (Lau et al. 2021), not because deaths are not underreported, but because they may be more stable. The conceptual basis for the methods we describe is that COVID-19-related mortalities typically occur a few days after infection. Some authors have proposed using hospitalizations to estimate underreported COVID-19 cases. The problem is that hospitalizations are probably affected by the specific local availability of healthcare resources. This local availability of hospital resources may vary both across countries and within the same country at different moments of the pandemic, as has been the case in Kazakhstan. Those changes in the availability of medical care services, including intensive care beds and mechanical ventilation, may have also contributed to possible differences in CFR (KBTUHW 2021).

The study's results suggest that Kazakhstan's official reporting of cases continues to underreport infections by at least 55% despite an increase in a testing capacity. The country faced a significant testing deficit during the early stages of the pandemic. However, as testing became more widely available, it became easier to identify previously missed infections. The study used two methods that are easy to apply in practice, both of which use a series of cumulative data from official sources to identify cases that may have gone undetected. By combining these methods, the true scale of infection can be more accurately estimated. This estimation can help understand the pandemic's future development and inform forecasting models.

The Lancet Commission has determined that countries were slow to respond to the pandemic.

The commission recommends that countries need to enhance their health systems by expanding their plans to prevent and respond to new infectious diseases, including improved surveillance and monitoring. A significant obstacle was the high number of asymptomatic cases, which coupled with slow case identification, contact tracing, and isolation, led to a rapid increase in COVID-19 infections, overwhelming public health and healthcare systems within a few weeks, as surveillance systems were unable to trace the contacts of a large number of infected individuals (Sachs et al. 2022).

21.5 Conclusion

This paper presents methods to estimate the number of undiagnosed COVID-19 cases in Kazakhstan and highlights the limited capacity of public health system to detect COVID-19 infections. The challenges posed by a shortage of testing resources and the unique characteristics of SARS-CoV-2 such as its ability to spread before symptoms appear and the high proportion of asymptomatic cases suggest that reported cases represent only a fraction of the "true infections" in Kazakhstan. The models presented using mortality data show that the true number of infections is likely much higher than reported. The authors estimate that no more than 45% of cases have been reported. Mortality may have also been underestimated and if this is the case, the proportion of reported cases may be even much lower. The limited capacity of public health systems to detect cases could make it difficult to effectively implement public health measures such as isolation and movement restrictions, and may prolong the pandemic. The authors argue that a more accurate estimate of the true number of infections would aid in making informed decisions to bring the pandemic to an end in Kazakhstan.

Funding This work has been financed by Grants NU 280720FD1901 and NU 021220CRP0822 from Nazarbayev University.

Compliance with Ethical Standards This work did not involve human research and no ethical approval was necessary. The authors declare no conflict of interest regarding this work.

References

Albani V, Loria J, Massad E, Zubelli J (2021) COVID-19 underreporting and its impact on vaccination strategies. BMC Infect Dis 21(1):1–13. https://doi.org/10.1186/S12879-021-06780-7/TABLES/4

Alekesheva LZh, Toguzbaeva KK, Dzhusupov KOLZh. Addressing COVID-19 in Kazakhstan: what are theefforts and challenges? УДК 616.24-002:578.834.1(574). https://doi.org/10.53065/kaznmu.2021.41.35.060

Alekesheva Z, Toguzbaeva KK, Dzhusupov KO (2021) Addressing COVID-19 in Kazakhstan: what are the efforts and challenges? Vestnik KazNMU. Available at: https://kaznmu.edu.kz/press/2021/08/05/addressing-covid-19-in-kazakhstan-what-are-the-efforts-and-challenges/. Accessed 18 Dec 2022

Aronna MS, Guglielmi R, Moschen LM (2022) Estimate of the rate of unreported COVID-19 cases during the first outbreak in Rio de Janeiro. Infect Dis Model. Sep 7(3):317–332. https://doi.org/10.1101/2021.10.08.21264741

Böhning D, Rocchetti I, Alfó M, Holling H (2016) A flexible ratio regression approach for zero-truncated capture–recapture counts. Biometrics 72(3):697–706. https://doi.org/10.1111/BIOM.12485

Böhning D, Superiore C, Magistratura D, Maruotti IA, Holling H (2020) Estimating the undetected infections in the Covid-19 outbreak by harnessing capture-recapture methods. MedRxiv, 2020.04.20.20072629. https://doi.org/10.1101/2020.04.20.20072629

Buitrago-Garcia D, Ipekci AM, Heron L, Imeri H, Araujo-Chaveron L, Arevalo-Rodriguez I, Ciapponi A, Cevik M, Hauser A, Alam MI, Meili K, Meyerowitz EA, Prajapati N, Qiu X, Richterman A, Robles-Rodriguez WG, Thapa S, Zhelyazkov I, Salanti G, Low N (2022) Occurrence and transmission potential of asymptomatic and presymptomatic SARS-CoV-2 infections: update of a living systematic review and meta-analysis. PLoS Med 19(5):e1003987. https://doi.org/10.1371/JOURNAL.PMED.1003987

de Lima CL, da Silva CC, da Silva ACG, Luiz Silva E, Marques GS, de Araújo LJB, Albuquerque Júnior LA, de Souza SBJ, de Santana MA, Gomes JC, de Freitas Barbosa VA, Musah A, Kostkova P, dos Santos WP, da Silva Filho AG (2020) COVID-SGIS: a smart tool for dynamic monitoring and temporal forecasting of Covid-19. Front Publ Health 8:761. https://doi.org/10.3389/FPUBH.2020.580815/BIBTEX

Djilali S, Benahmadi L, Tridane A, Niri K (2020) Modeling the impact of unreported cases of the COVID-19 in the North African Countries.

Biology 9(11):373. https://doi.org/10.3390/BIOLOGY9110373

Gaipov A, Gusmanov A, Abbay A, Sakko Y, Issanov A, Kadyrzhanuly K, Yermakhanova Z, Aliyeva L, Kashkynbayev A, Moldaliyev I, Crape B, Sarria-Santamera A (2021) SARS-CoV-2 PCR-positive and PCR-negative cases of pneumonia admitted to the hospital during the peak of COVID-19 pandemic: analysis of in-hospital and post-hospital mortality. BMC Infectious Diseases 21(1). https://doi.org/10.1186/S12879-021-06154-Z

Griette Q, Magal P (2021) Clarifying predictions for COVID-19 from testing data: the example of New York State. Infect Dis Model 6:273–283. https://doi.org/10.1016/J.IDM.2020.12.011

Haruna UA, Amos OA, Gyeltshen D, Colet P, Almazan J, Ahmadi A, Sarria-Santamera A (2022) Towards a post-COVID world: challenges and progress of recovery in Kazakhstan. Pub Health Chall 1(3):e17. https://doi.org/10.1002/PUH2.17

Hellewell J, Abbott S, Gimma A, Bosse NI, Jarvis CI, Russell TW, Munday JD, Kucharski AJ, Edmunds WJ, Sun F, Flasche S, Quilty BJ, Davies N, Liu Y, Clifford S, Klepac P, Jit M, Diamond C, Gibbs H, Eggo RM et al (2020) Feasibility of controlling COVID-19 outbreaks by isolation of cases and contacts. Lancet Glob Health 8(4):e488–e496. https://doi.org/10.1016/S2214-109X(20)30074-7

Ibrayeva LK, Rybalkina DK, Minbayeva LS, Minbayev SK, Bacheva IV, Zharylgapova AM (2022) Development of the COVID-2019 pandemic in Kazakhstan. Med News N Caucasus 17(1):27–31. https://doi.org/10.14300/MNNC.2022.17008

Kadyrova I, Yegorov S, Negmetzhanov B, Kolesnikova Y, Kolesnichenko S, Korshukov I, Akhmaltdinova L, Vazenmiller D, Stupina Y, Kabildina N, Ashimova A, Raimbekova A, Turmukhambetova A, Miller MS, Hortelano G, Babenko D (2022) High SARS-CoV-2 seroprevalence in Karaganda, Kazakhstan before the launch of COVID-19 vaccination. PLoS ONE 17(7):e0272008. https://doi.org/10.1371/JOURNAL.PONE.0272008

Kairov U, Amanzhanova A, Karabayev D, Rakhimova S, Aitkulova A, Samatkyzy D, Kalendar R, Kozhamkulov U, Molkenov A, Gabdulkayum A, Sarbassov D, Akilzhanova A (2022) A high scale SARS-CoV-2 profiling by its whole-genome sequencing using Oxford Nanopore Technology in Kazakhstan. Frontiers Genet 13. https://doi.org/10.3389/FGENE.2022.906318

Kazakhstan Branch Trade Union of Healthcare Workers (KBTUHW). Responding to the Challenges of COVID-19 through Analysis, Capacity Building, and Stimulating Social Dialogue in Kazakhstan (2021). Available at: https://zdravunion.kz/wp-content/uploads/2021/06/Doklad-rus-okonch.pdf

Krantz SG, Rao ASRS (2020) Level of underreporting including underdiagnosis before the first peak of COVID-19 in various countries: preliminary retrospective results based on wavelets and deterministic modeling. Infect Control Hosp Epidemiol 41(7):1. https://doi.org/10.1017/ICE.2020.116

Lau H, Khosrawipour T, Kocbach P, Ichii H, Bania J, Khosrawipour V (2021) Evaluating the massive underreporting and undertesting of COVID-19 cases in multiple global epicenters. Pulmonology 27(2):110–115. https://doi.org/10.1016/J.PULMOE.2020.05.015

Levin AT, Hanage WP, Owusu-Boaitey N, Cochran KB, Walsh SP, Meyerowitz-Katz G (2020) Assessing the age specificity of infection fatality rates for COVID-19: systematic review, meta-analysis, and public policy implications. Eur J Epidemiol 35(12):1123–1138. https://doi.org/10.1007/S10654-020-00698-1

Li R, Pei S, Chen B, Song Y, Zhang T, Yang W, Shaman J (2020) Substantial undocumented infection facilitates the rapid dissemination of novel coronavirus (SARS-CoV-2). Science (New York, N.Y.) 368(6490):489–493. https://doi.org/10.1126/SCIENCE.ABB3221

Liu J, Huang J, Xiang D (2020) Large SARS-CoV-2 outbreak caused by asymptomatic traveler, China. Emerg Infect Dis 26(9):2260. https://doi.org/10.3201/EID2609.201798

Maciej Serda, Becker FG, Cleary M, Team R M, Holtermann H, The D, Agenda N, Science P, Sk SK, Hinnebusch R, Hinnebusch AR, Rabinovich I, Olmert Y, Uld DQGLQ, Ri WKHU, Lq V, Frxqwu WKH, Zklfk E, Edvhg LV, ىمطاف, ح et al (2022) Global underreporting of COVID-19 cases during 1 January 2020 to 6 May 2022. Curr Sci, 123(6):741–742. 10.2/JQUERY.MIN.JS

Mathieu E, Ritchie H, Rodés-Guirao L, Appel C, Giattino C, Hasell J, Macdonald B, Dattani S, Beltekian D, Ortiz-Ospina E, Roser M (2020) Coronavirus pandemic (COVID-19). Our World Data. https://ourworldindata.org/coronavirus

Mizumoto K, Chowell G (2020) Transmission potential of the novel coronavirus (COVID-19) onboard the diamond Princess Cruises Ship, 2020. Infect Dis Model 5:264–270. https://doi.org/10.1016/J.IDM.2020.02.003

Niwitpong S, Böhning D, van der Heijden PGM, Holling H (2012) Capture–recapture estimation based upon the geometric distribution allowing for heterogeneity. Metrika 2012 76:4, 76(4):495–519. https://doi.org/10.1007/S00184-012-0401-0

Phipps SJ, Grafton RQ, Kompas T (2020) Robust estimates of the true (population) infection rate for COVID-19: a backcasting approach. R Soc Open Sci 7(11). https://doi.org/10.1098/RSOS.200909

Sachs JD, Horton R, Bagenal J, Ben Amor Y, Karadag Caman O, Lafortune G (2020) The lancet COVID-19 commission. Lancet (london, England) 396(10249):454. https://doi.org/10.1016/S0140-6736(20)31494-X

Sachs JD, Karim SSA, Aknin L, Allen J, Brosbøl K, Colombo F, Barron GC, Espinosa MF, Gaspar V, Gaviria A, Haines A, Hotez PJ, Koundouri

P, Bascuñán FL, Lee JK, Pate MA, Ramos G, Reddy KS, Serageldin I, Michie S et al (2022) The Lancet Commission on lessons for the future from the COVID-19 pandemic. The Lancet 400(10359):1224–1280. https://doi.org/10.1016/S0140-6736(22)01585-9

Sarría-Santamera A, Abdukadyrov N, Glushkova N, Peck DR, Colet P, Yeskendir A, Asúnsolo A, Ortega MA (2022) Towards an accurate estimation of COVID-19 cases in Kazakhstan: back-casting and capture-recapture approaches. Medicina (Kaunas, Lithuania) 58(2). https://doi.org/10.3390/MEDICINA58020253

Semenova Y, Kalmatayeva Z, Oshibayeva A, Mamyrbekova S, Kudirbekova A, Nurbakyt A, Baizhaxynova A, Colet P, Glushkova N, Ivankov A, Sarria-Santamera A (2022) Seropositivity of SARS-CoV-2 in the Population of Kazakhstan: a nation-wide laboratory-based surveillance. Int J Environ Res Pub Health 19(4). https://doi.org/10.3390/IJERPH19042263

Shang W, Kang L, Cao G, Wang Y, Gao P, Liu J, Liu M (2022) Percentage of asymptomatic infections among SARS-CoV-2 omicron variant-positive individuals: a systematic review and meta-analysis. Vaccines 10(7):1049. https://doi.org/10.3390/VACCINES10071049/S1

Sharun K, Tiwari R, Dhama K, Emran TB, Rabaan AA, Al Mutair A (2021) Emerging SARS-CoV-2 variants: impact on vaccine efficacy and neutralizing antibodies. Hum Vaccin Immunother 17(10):3491–3494. https://doi.org/10.1080/21645515.2021.1923350

Smagul M, Esmagambetova A, Nusupbaeva G, Kirpicheva U, Kasabekova L, Nukenova G, Saliev T, Fakhradiyev I, Tanabayeva S, Zhussupov B (2022) Sero-prevalence of SARS-CoV-2 in certain cities of Kazakhstan. Health Sci R 5(2). https://doi.org/10.1002/HSR2.562

Srivastava A, Prasanna VK (2020) Data-driven identification of number of unreported cases for COVID-19: bounds and limitations. Undefined

Unnikrishnan J, Mangalathu S, Kutty RV (2021) Estimating under-reporting of COVID-19 cases in Indian states: an approach using a delay-adjusted case fatality ratio. BMJ Open 11(1):e042584. https://doi.org/10.1136/BMJOPEN-2020-042584

Yegorov S, Goremykina M, Ivanova R, Good SV, Babenko D, Shevtsov A, MacDonald KS, Zhunussov Y (2021) Epidemiology, clinical characteristics, and virologic features of COVID-19 patients in Kazakhstan: a nation-wide retrospective cohort study. The Lancet Regional Health. Europe 4. https://doi.org/10.1016/J.LANEPE.2021.100096

Zhalmagambetov B, Madikenova M, Paizullayeva S, Abbay A, Gaipov A (2020) COVID-19 Outbreak in Kazakhstan: current status and challenges. J Clin Med Kaz 1(55):6–8. https://doi.org/10.23950/1812-2892-JCMK-00763

Zhandosova JS, Sharipbaeva AE, Baitugelova NY, Smagulova SK, Kudasheva TV, Dzhandosova FS (2021) COVID-19 in Kazakhstan: problems of scale, evaluation of services health care and social protection. Nur-Sultan, 2021. p 180

Zhussupov B, Saliev T, Sarybayeva G, Altynbekov K, Tanabayeva S, Altynbekov S, Tuleshova G, Pavalkis D, Fakhradiyev I (2021) Analysis of COVID-19 pandemics in Kazakhstan. J Res Health Sci 21(2). https://doi.org/10.34172/JRHS.2021.52

The Impact of COVID-19 on Children and Adolescents with Chronic Illness

22

Finiki Nearchou and Clodagh Flinn

Abstract

Children and adolescents living with chronic illness may already be at a disadvantage when compared to their healthy peers in achieving developmental milestones as they are required to manage healthcare aspects of their condition in addition to the typical transitions associated with this developmental stage. The COVID-19 pandemic has significantly impacted on the physical and mental well-being of children and adolescents including those young people living with a chronic illness. Disruptions and delays of healthcare service provision, prolonged lockdowns, strict social distancing measures, and school closures are some of the pandemic implications that have affected daily routines imposing strains on young people themselves, but also on their caregivers. This chapter presents a critical elaboration on the available evidence documenting the unique impact of the pandemic at an individual-, family-, and system-level on children and adolescents aged up to 18 years old living with a non-communicable disease (e.g., cancer, diabetes, asthma).

Keywords

Children · Adolescent · Non-communicable disease · Developmental stage · Puberty · Healthcare management · Cancer · Diabetes · Asthma · Pediatric patient

F. Nearchou (✉) · C. Flinn
School of Psychology, University College Dublin, Belfield Campus, Newman Building, Dublin 4, Ireland
e-mail: foiniki.nearchou@ucd.ie

C. Flinn
e-mail: Clodagh.flinn@ucdconnect.ie

22.1 Introduction

Chronic illness refers to health conditions that persist for a long period of time, usually over one year, require medical attention, and/or limit daily activities (Centers for Disease Control and Prevention 2022; World Health Organization 2014). The present chapter will focus on those chronic illnesses that are classified, according to the World Health Organization (WHO), as non-communicable diseases (NCDs). These types of chronic illnesses are not transmitted from individual to individual and are rather caused by the interaction of genetic, environmental, physiological, and behavioral factors. The main types of chronic illnesses are cardiovascular diseases (such as strokes and heart attacks), diabetes, cancers, and respiratory chronic diseases (such as chronic obstructive pulmonary disease and asthma). Chronic illness is the cause of death

N. Rezaei (ed.), *The COVID-19 Aftermath*, Advances in Experimental Medicine and Biology 1457,
https://doi.org/10.1007/978-3-031-61939-7_22

for 41 million people annually which accounts for 74% of all global deaths, while it may affect people from any region and at any age including children and adolescents (World Health Organization 2022).

Despite the global decrease in mortality of children and adolescents, non-fatal ill-health remains high with the morbidity burden due to chronic illness in these cohorts globally accounting for one-third of disability adjusted life-years (DALY). Most non-communicable conditions have their onset in childhood with advancements in medicine and technology resulting in treating and managing effectively these conditions, which has led to increasing numbers of young people surviving into adulthood. For instance, recent reports indicate that both type 1 and type 2 diabetes affect significant numbers of children and adolescents worldwide, and type 1 diabetes typically persists over the course of their lifespan (Patterson et al. 2019). Annual estimates show that 400,000 children and adolescents up to the age of 19 years old are diagnosed with cancers and in high-income countries 80% of children with cancers are cured (World Health Organization 2021). Childhood asthma is the most common respiratory NCD accounting for the highest disability burden in the United States (2013), while recent evidence based on the Global Burden Disease 2019 data showed that childhood asthma accounted for 5.1 million DALY rates globally (Zhang and Zheng 2019). Dermatological diseases are also quite common in young people. Acne vulgaris (acne) affects approximately 85% of those aged 12–25 years old (Seattle 2013), and eczema affects 12% of children usually between 2 and 8 years old (Shi et al. 2021), often peaking at 17% in two-year-old children (Lusignan et al. 2021).

Human development is marked with navigating from one stage to another such as from childhood to adulthood wherein the individual is expected to complete a series of developmental milestones (Pinquart 2014). Adolescence is an important stage of human development, usually considered part of childhood, during which the young individual undergoes considerable developmental changes at behavioral, social, and cognitive level, while at the same time puberty signifies immense hormonal and physical changes. Young people experience their bodies rapidly transforming as they enter a stage of sexual exploration and experimentation. As young people grow and mature, they develop their identity, start to form romantic relationships, try to balance parental and peer influences, establish their autonomy, and prepare for career and family life. Achieving these developmental milestones is deemed important by young people and they are required to obtain and consolidate skills that will enable them to navigate through this transitional period of rapid changes. During this journey to adulthood young people may encounter challenges while some of them struggle to navigate through this period of their lives. Considering that achieving developmental milestones has been associated with better subjective well-being (Pinquart et al. 2004) it is not surprising that mental health difficulties are highly prevalent in this age group with recent evidence showing that globally 14% of children and adolescents are diagnosed with a mental disorder (Polanczyk et al. 2015).

The burden of living with chronic illness as youth grow into adulthood may have repercussions on physical, cognitive, social, and emotional development for the affected individuals, as well as impact on the immediate family/caregivers of the young patient (Yeo and Sawyer 2005). Children and adolescents living with chronic illness may already be at a disadvantage when compared to their healthy peers in achieving developmental milestones as they are required to manage healthcare aspects of their condition in addition to the typical transitions associated with this developmental stage (Yeo and Sawyer 2005). Recent meta-analytic evidence indicated that children with chronic illness (<18 years old) tend to have poorer health-related quality of life including domains of psychosocial functioning than their peers without a chronic illness (Pinquart 2020). Meta-analytic evidence also showed that adolescents with chronic illness tend to be less successful in completing developmental milestones such as being socially integrated in peer groups, forming romantic relationships, obtaining autonomy from parents, or preparing for career and family (Pinquart and Pfeiffer 2015).

Being socially active as well as forming and sustaining romantic relationships are very important developmental milestones for adolescents (Jordan et al. 2020). However young people with chronic illness face significant challenges on that front due to the constraints imposed by their poor health status as well as due to low self-esteem and body confidence especially in those living with a visible chronic illness (Jordan et al. 2020). Thus, it is not surprising that young people with chronic illness tend to experience higher levels of loneliness than their healthy peers. Notably, young people with or without chronic illness who experience the painful emotion of loneliness are more susceptible to exhibiting poorer physical health as well as increased levels of depression and anxiety.

The COVID-19 pandemic altered people's lives and daily routines abruptly and imminently with governments imposing strict and prolonged lockdowns resulting in social distancing measures, school closures, healthcare service disruptions, job losses, social event prohibitions, and many more consequences that changed the way of life in unprecedented manners. Very early in the pandemic concerns had been raised about its potential impacts on the mental and physical health of children and adolescents (Ghosh et al. 2020; Wang et al. 2020). In November 2020, we published the findings of a systematic review that provided a synthesis of the empirical evidence generated during the first months of pandemic up to June 21st, 2020 on the youth mental health impact of this global public health crisis. Although this was one of the first systematic reviews on this topic, the available evidence at the time strongly indicated that the mental health of young people would be negatively affected including those youth living with a chronic illness (Nearchou et al. 2020).

22.2 Young People with Chronic Illness at Risk for Severe COVID-19-Related Consequences

The COVID-19 pandemic had significant impacts on physical and mental health, particularly among subgroups such as healthcare workers (Pablo et al. 2020), young people (Nearchou et al. 2020; Iacobucci 2022), and people with chronic illness (Singh et al. 2021). COVID-19 had indirect consequences on children and adolescents globally: for example, school closures and home quarantine measures have been identified as causes of anxiety and loneliness in children and adolescents (Giorgio et al. 2021). However, it had even more severe, direct, and immediate consequences for children and adolescents with chronic illness, in comparison with their healthy counterparts. The pandemic presented unique difficulties for young people living with chronic illness carrying implications at an individual-, family-, and system-level (Logan 2022).

22.2.1 Individual Level

At an individual level, even under normal circumstances (i.e., without the presence of a global pandemic), young people with chronic illness can be at an increased risk for experiencing emotional, behavioral, and psychiatric symptoms (Knapp and Harris 1998; Witt et al. 2003). This risk, depending on the population, can be a result of the severity, progression, and treatment of their chronic illness, as well as stress related to their disease burden. They may be psychologically affected, or even traumatized, by medical treatments associated with their illness (Stuber et al. 2006). Children and adolescents with chronic illnesses have experienced greater anxiety about their morbidity and mortality associated with COVID-19 infection (Logan 2022). During the pandemic children and adolescents with chronic illness were more vulnerable in terms of developing a severe COVID-19 infection compared to their healthy peers. Meta-analytic evidence suggests that among children who had COVID-19, 5.1% of children with a wide range of underlying comorbidities developed a severe infection, in comparison with only 0.2% of children without comorbidities (Tsankov et al. 2021). The most common comorbidities included were obesity, chronic respiratory diseases, and cardiovascular diseases. Children with comorbidities were at a higher risk of developing severe COVID-19 than healthy children, with a relative risk ratio of 1.79. In addition, children with underlying comorbidities also had a higher risk of mortality associated with COVID-19 than healthy children did, with a relative risk ratio of 2.81. A study of pediatric patients in tertiary medical care in the Netherlands reported that out of all patients with acute COVID-19, 66% of them had a pre-existing comorbidity, most commonly neurologic, cardiovascular, and endocrine system disorders (Biharie et al. 2022). Significantly more cases of severe acute COVID-19 were observed in the patient group with comorbidities, than in the group without comorbidities. Furthermore, all patients with severe or critical infection had pre-existing comorbidities.

22.2.2 Family Level

At a family level, COVID-19 has resulted in parental stress and concern in relation to their child's individual health risks, disruptions to work and educational opportunities, and childcare interruptions (Nearchou et al. 2020; Tse et al. 2021; Lee et al. 2021). A qualitative study showed that parents caring for children with chronic illness experienced the pandemic as an extremely stressful period where they encountered cumulative challenges, which impacted on their mental health (McLoone et al. 2022). A study of parents with children with chronic kidney disease reported that parents were fearful of the virus, in addition to being concerned that about catching the virus themselves and then passing it onto their child (Tse et al. 2021). Many families have also experienced financial stress partly due to job losses, which can subsequently affect their ability to afford their child's medications and other healthcare costs. For families with children who are immunocompromised or at a higher risk of severe illness from infection with COVID-19, family-level decisions about wearing protective face masks, attending school, and interacting with others outside the family may have been confounded by disease-specific factors (Logan 2022). However, it should be noted that there is some evidence suggesting that children with chronic illness and their parents reported lower stress levels than parents of healthy children during the COVID-19 pandemic (Krijger et al. 2022). Although seemingly surprising, parents who care long-term for children with chronic illness may have developed coping strategies and may also be accustomed to undergoing periods of isolation due their child's health condition.

22.2.3 System-Level

At a system-level, health regimes and adherence to daily medications in children and adolescents are often promoted by a network of individuals who provide social support within home, school, or healthcare settings (Plevinsky et al. 2020). With schools moving to online learning measures and healthcare settings implementing telemedicine for a large proportion of 2020, this social support network was interrupted for many youth living with chronic illness. Schools in particular provide essential support for children with chronic illness including mental health support, peer socialization, learning and nursing support, daily routine or structure, as well as respite for parents. Within healthcare systems, there have been fewer physical (in-person) appointments and increased provision of routine care through video or telephone calls (i.e., telemedicine) (Baum et al. 2021). Further, supply chain disruptions have impacted the provision of medications for some illness groups. A qualitative study found that pharmacists reported facing shortages of medications used for chronic illness including diabetes, hypertension, human immunodeficiency virus (HIV), and oncology drugs during the pandemic (Ramakrishnan et al. 2023). Drug shortages may lead to serious consequences in terms of patient outcomes. For example, alternative prescribed drugs may be less effective for some patients or put them at a higher risk of developing adverse effects (Badreldin and Atallah 2021). Furthermore, extended drug shortage periods may increase patient deterioration and mortality rates (Walker et al. 2017), as well as increasing patient's levels of distress, frustration, and confusion (Shah et al. 2015). Such distress and uncertainty, in combination with increased morbidity and mortality risks, may have all contributed to an increased vulnerability of chronically ill children during the global COVID-19 pandemic.

22.3 Impact of COVID-19 on Physical Health and the Disruption of Health Services

22.3.1 Impact of COVID-19 on Physical Health of Children with Chronic Illness

Children with underlying illnesses were at a higher risk of developing severe COVID-19 infections and COVID-19-associated mortality, than healthy children (Tsankov et al. 2021). Serious infection with COVID-19 can manifest itself in children as severe pneumonia, autoinflammatory shock, and respiratory distress or even failure (Sankar et al. 2020). Such cases can often require mechanical ventilation, and antiviral and immune system modulating treatments (Sankar et al. 2020; Cronin et al. 2022; Alijotas-Reig et al. 2020). Hence, being exposed to COVID-19 infection may have serious complications for children with a wide range of chronic illness including among others diabetes and asthma.

The presence of diabetes may increase the risk of severe morbidity and mortality associated with COVID-19 through the disruption of immunity, or as a result of diabetes-related complications including hypertension, cardiovascular diseases, obesity, and nephropathy (Remuzzi and Remuzzi 2020; Richardson et al. 2020). In adult diabetes patients, meta-analytic evidence suggests that the presence of diabetes increases mortality (Wu et al. 2021). Given that hypertension and obesity, which are both robustly linked to type 2 diabetes in youth, are also associated with more severe cases of COVID-19 (Tsankov et al. 2021; Nogueira-de-Almeida 2020), children with type 2 diabetes may be at higher risk for severe COVID-19 (DiMeglio et al. 2020). This aligns with findings from adult patient studies wherein patients with higher body mass index values (greater than or equal to 35 kg/m^2) required more invasive COVID-19-related care (mechanical ventilation) more often than patients with lower body mass index values (Simonnet et al. 2020). The high visceral adiposity in obese individuals is known to increase levels of systemic and local inflammatory cytokines, including C-reactive protein and Interleukin-6 (Fontana et al. 2007). These cytokines have been positively associated with more severe COVID-19 infection (Zeng et al. 2020). There is some evidence showing that COVID-19 is linked with new diabetes onset in children (Prosperi and Chiarelli 2022). Specifically, COVID-19 in children may lead to diabetes through direct destruction of pancreatic β-cells. β-cell dysfunction is a critical determinant of type 2 diabetes (Ashcroft and Rorsman 2012). Thus after infection with COVID-19, the risk of incident diabetes may increase in children and adults (Prosperi and Chiarelli 2022). A study conducted in the United States covering the period between March 2020 and June 2021 showed that young people aged <18 years old who have been infected by COVID-19 and recovered are at increased risk to develop type 1 diabetes or type 2 diabetes (Barrett et al. 2022). Although Rabbone, Schiaffini (Rabbone et al. 2020) reported a 23% reduction in new diabetes cases in 2020 compared to 2019; the authors also reported significant increases of diabetic ketoacidosis cases during 2020 in comparison with 2019 levels. These figures may be attributed to underlying causes such as fear about accessing healthcare systems and reductions in medical services during COVID-19 (Kamrath et al. 2020). Thus, the true rate of incident diabetes may have increased during the pandemic, but the number of diagnosed cases declined due to issues with healthcare systems and delivery. Lastly, COVID-19 may have indirectly increased type 2 diabetes risks through aspects such as increased body weight and obesity (Prosperi and Chiarelli 2022). During lockdown periods physical activity settings for children were often shut down, and combined with a more sedentary lifestyle this resulted in increases to children's body mass index and prevalence of obesity (Chaabane et al. 2021), which are known to greatly increase risk of developing type 2 diabetes.

Pediatric asthma, a chronic respiratory illness associated with high healthcare utilization, has been suggested to be one of the pre-existing conditions which increases risk of developing severe COVID-19 (Logan 2022). Some researchers have reported decreases in the average daily visits to emergency departments for asthma-related reasons (Kenyon et al. 2020), and decreases in pediatric intensive care unit (ICU) admissions for asthma and other respiratory illnesses (Zee-Cheng et al. 2021). These supposed improvements in asthma symptoms have been posited to be a result of environmental changes which occurred during the pandemic such as improvements in air quality or less exposure to allergens and seasonal respiratory illnesses, which have been indirect results of social distancing and mask-wearing initiatives (Logan 2022). It must also be considered that some parents may have avoided taking their child to the emergency department for fear or anxiety about the child contracting COVID-19 and developing a severe illness. Regarding health outcomes, a study of Israeli children with respiratory diseases found that one-third of pediatric patients improved, including a reduced need for reliever medications (e.g., inhaled bronchodilators), fewer respiratory symptoms, and exacerbation frequencies (Cahal et al. 2021). The clinical condition of roughly 10% of the patients worsened, with these patients more likely to be over five years old, have increased screen time, reduced physical activity, and shorter sleep duration. The majority of caregivers in the study reported increased anxiety levels and increased perceptions of their child's elevated health risk, irrespective of the child's actual clinical status. Caregivers also reported increased use of healthcare by alternative methods (e.g., phone and online consultations, email) during COVID-19, especially among those with increased anxiety (Cahal et al. 2021).

Children and adolescents with a range of illnesses may receive immunosuppressant therapies as part of their treatment regimes. For example, immunosuppressants can be used in the treatment of dermatological diseases such as hidradenitis suppurativa, psoriasis, and eczema; gastrointestinal diseases like inflammatory bowel disease; organ or bone marrow transplant settings; cancers; and others. Weakened immune systems may put people at a higher risk of becoming infected with COVID-19, and of infections becoming more severe and even fatal (Calder 2021). Weakened immune systems may put people at a higher risk of developing a severe COVID-19 infection. A review on the impact of COVID-19 in immunocompromised patients reported an increased risk of severe infection and death in cancer patients and solid organ transplant recipients (Fung and Babik 2020). However, evidence for patients with other types of immunocompromise such as people with HIV, and patients taking biologic or targeted disease-modifying antirheumatic drugs, were less clear.

Pediatric organ transplant recipients have been characterized as being at high-risk during the pandemic primarily as a result of immunosuppression, which is secondary to medications required in order to prevent graft rejection (i.e., when a recipient's immune system attacks the donated graft and begins destroying the transplanted organ). Organ recipients are at an increased risk of developing common respiratory illnesses, and other comorbidities associated with their underlying conditions can predispose them to further adverse health outcomes (Logan 2022). Children and adolescents awaiting organ transplants experienced higher waiting times during COVID-19, which placed them at an increased risk of waitlist mortality. A US study examining the pediatric transplantation over the initial six months of the pandemic found that the number of liver transplants was 11% lower in comparison with 2019 (Lobritto et al. 2020). In Canada, many pre-emptive kidney transplants were suspended, resulting in patients having to start dialysis before their transplant (Teoh et al. 2020). Further, biopsies to assess graft rejection were also suspended. A study of families of pediatric solid organ transplant recipients and those of healthy children reported that caregivers of transplant recipients perceived a higher risk of their children developing a severe illness or dying from a COVID-19 infection, than caregivers of healthy children (Forner-Puntonet et al. 2021). Exposure to COVID-19 was higher in families of healthy children, as measured by self-reported COVID-19 diagnosis or close contact with others diagnosed with COVID-19. Further, families of transplant recipients perceived isolation and hygiene measures as protective and de-stigmatizing. A study of pediatric liver transplant patients found that approximately three-quarters (76%) of patients reported fear of severe COVID-19 infection as a result of transplant or transplant-associated medications, and half of the patients reported that they required emotional support (Dorfman et al. 2022).

It has been suggested that approximately 4–9% of pediatric oncology patients experienced severe or critical COVID-19 infection in comparison with 55% of adult oncology patients. However, although these pediatric patients have generally better outcomes than adult patients, their risk of hospitalization, ICU admission, and mortality are increased in comparison with the general pediatric population (Kahn et al. 2022). A global registry from 51 countries reported on 1814 cases of COVID-19 in pediatric oncology patients (as of 03/04/2023) (St. Jude Global 2023). Among these patients, 8.7% had a moderate infection, 10.7% had a severe infection, and 7.5% were in critical condition. Further, 8.5% were admitted to ICU and 3.6% died from COVID-19. Positive COVID-19 infection in pediatric oncology patients can also lead to modifications and delays to their cancer-related treatment, subsequently affecting cancer-related disease outcomes. For example, 38.9% of patients in the above global registry had chemotherapy withheld, while 5.2% experienced reductions in chemotherapy, 2.8% experienced surgery delays, and 2.0% experienced radiation therapy delays. Unfortunately, associations between cancer treatment delays and increased mortality have been established for numerous cancer treatments including surgery, systematic treatments, and radiotherapy (Hanna et al. 2020). This is yet another example of how children and adolescents with chronic illnesses have been indirectly affected by the COVID-19 pandemic. Table 22.1 summarizes the negative health consequences of the pandemic on children and adolescents reported for each type of chronic illness.

22.3.2 The Disruption of Regular Health Services and the Move to Telehealth

Children with chronic illnesses may require both planned and urgent care over time. The quality of this care may depend on therapeutic relationships and the communication or integration of different care services, for example inpatient and outpatient health and social services (Royal College of Paediatrics and Child Health 2019). COVID-19-related restrictions had direct effects on the provision and delivery of health services worldwide (Kinoshita et al. 2022). For example, in some countries (e.g., the U.K.) the pandemic brought additional demands onto already stretched workforces (Kinoshita et al. 2022). Issues such as low pre-pandemic staffing levels and staff burnout during the pandemic (Academy of Medical Sciences 2021), in addition to shortages in medication and supplies (Diskin et al. 2022) and the redeployment of pediatric staff to adult services (Royal College

Table 22.1 Overview of findings discussed in the chapter relating to the impact of COVID-19 on children's physical health, by chronic illness

Illness	COVID-19-associated finding	Illness-specific finding
Asthma	Possibility of increased risk of developing a severe COVID-19 infection.	• Decreases in emergency department visits and ICU admissions for asthma-related reasons. • Reduced use of reliever medications, fewer respiratory symptoms and exacerbations frequencies.
Cancers	Increased risk of hospitalization, ICU admission, severe COVID-19 infection, and mortality Higher risk of contracting COVID-19 due to immunosuppressing medications.	• Modifications or delays to, and withholding of, cancer treatments.
Dermatological diseases	Immunosuppressant therapies can be used in the treatment of some dermatological conditions (e.g., Hidradenitis suppurativa, psoriasis, eczema). Immunosuppression increases the risk of contracting respiratory illnesses like COVID-19, and the risk of the infection becoming severe.	
Diabetes	Increased risk of COVID-19-associated severe morbidity and mortality, through disruption of immunity.	• Hypertension, cardiovascular disease, obesity, nephropathy.
Type 1 diabetes	Reductions in the number of incident T1D cases from 2019 to 2020, possibly reflecting reduced provision of medical services during the pandemic.	• Increased episodes of diabetic ketoacidosis from 2019 to 2020, possibly due to delayed patient presentation to healthcare settings. • Possible increased risk of developing T1D following a COVID-19 infection.
Type 2 diabetes	Increased risk of severe COVID-19 infection.	• Hypertension and obesity linked to T2D and severe COVID-19. • COVID-19 indirectly increased T2D risks through increased body weight and obesity. • COVID-19 can lead to pancreatic β-cell dysfunction (a determinant of T2D), thus increasing the risk of incident T2D following infection with COVID-19.
Gastrointestinal diseases	Immunosuppressant medications are used in the treatment of gastrointestinal diseases, including inflammatory bowel disorders. Immunosuppression increases the risk of contracting respiratory illnesses like COVID-19, and the risk of the infection becoming severe.	
Organ transplant recipients	Organ transplant recipients are at high-risk because of immunosuppressant therapies, which are needed to prevent graft rejection. Immunosuppression increases risk of contracting COVID-19 and the risk of the infection becoming severe, even fatal. Higher family-wide exposure to COVID-19 compared with families of healthy children.	• Fewer transplants performed during COVID-19. Associated higher waiting times during COVID-19 increased risk of waitlist mortality.

ICU, Intensive Care Unit; T1D, Type 1 Diabetes; T2D, Type 2 Diabetes of Paediatrics and Child Health 2020), may all have contributed to the disruption of pediatric healthcare services during COVID-19. A review of changes in healthcare provision during COVID-19 and the subsequent impact on children with chronic illness in the U.K. found that there were reductions in the use of primary care facilities and difficulties accessing medications (Lignou et al. 2022). Furthermore, there were delayed presentations to emergency departments and pediatric assessment units, reduced surgical activity, and delays in diagnoses.

Impacts of the increased use of telemedicine and telehealth during the pandemic on children's mental and physical health is still unfolding. The reliance on remote healthcare could have resulted in missed diagnoses for common mental health problems, cerebrovascular and cardiovascular disease, diabetes, and cancers (Williams et al. 2020). Delayed diagnosis for these chronic illnesses has been associated with increased rates of mortality, particularly for patients with depression (Prosperi and Chiarelli 2022; Williams et al. 2020).

It has been suggested that over one-quarter (26%) of young people in the U.K. with existing mental health problems were unable to access psychological support through phone or video calls during the pandemic (Lignou et al. 2022; Young 2020). Emergency and routine care was provided to people with diabetes through phone or video consultations, including psychological support (Hartmann-Boyce et al. 2020). Although the impact of this delivery on diabetes outcomes is not yet clear, the majority of participants in a pre-COVID-19 survey reported that virtual appointments were useful (Lignou et al. 2022). Research with oncology patients suggests a larger proportion of missed diagnoses (Maringe et al. 2020). For children with epilepsy, the use of telemedicine and virtual diagnosis without the use of electroencephalography could have resulted in reduced accuracy and misdiagnosis (Wirrell et al. 2020). A study of young people (0–25 years old) with epilepsy and their caregivers in the U.K. found that 59% of participants reported the same level of satisfaction and 14% reported more satisfaction than usual with telehealth services during the pandemic (Reilly et al. 2021). However, a qualitative study of caregivers of children accessing acute pediatric services in London during the pandemic found that many respondents felt that the challenges experienced during telehealth consultations with their general practitioner rendered the care inadequate to meet their child's needs (Watson et al. 2021). A study of Canadian pediatricians found that 12% of participants reported adverse patient health outcomes, including hospital or ICU admissions, unplanned surgery, and death as a result of COVID-19-related disruption in healthcare delivery (Diskin et al. 2022). Furthermore, 12% reported disruptions of medications or other supplies, while 48% reported disruptions in family caregiving and 40% in homecare. These disruptions to healthcare delivery and the associated effects are expected to dominate delivery in coming years in the wake of COVID-19 (Academy of Medical Sciences 2021).

22.4 Impact of the COVID-19 Pandemic on Parents/Caregivers Living with a Chronically Ill Child

The COVID-19 pandemic has had negative impacts on parental mental health and family functioning. A US study of parents and children found that 27% of parents reported worsening mental health and 14% reported worsening behavioral health of their children (Patrick et al. 2020). Furthermore, the proportion of families with moderate-to-severe food insecurity increased from 6 to 8% (from before March 2020 to after), and almost one-quarter of families reported a loss in regular childcare. In addition to these difficulties, parents/guardians of children with chronic illness had the added burden of caring for and managing their child's health condition during a global pandemic.

Although the pandemic had negative impacts on parental and family mental health for many people, the added burden of coping with a chronic illness within the household could make the stress and worry so often associated with COVID-19 even more difficult for parents of chronically ill children. Across pediatric chronic illness, parents and caregivers can fill gaps in disjointed or uncoordinated healthcare systems, for example acting as patients' advocates or medical record keepers. The COVID-19 pandemic required parents to quickly navigate their children's new medical pathways while healthcare systems became overwhelmed and the delivery of care moved to remote or virtual services (Edge et al. 2021; Garfan et al. 2021). Learning new work platforms, procedures, and home-learning responsibilities, all while caring for a chronically ill child, is likely to have put parents/caregivers under immense and unprecedented pressure. Psychological stress in parents is concerning for a number of reasons including the negative impacts on the individuals themselves, as well as consequences to family functioning as a whole (Casaña-Granell et al. 2018; Shudy et al 2006).

Parents of children with chronic illnesses have experienced greater levels of stress, than parents of healthy children. A study of Greek parents found that parents with chronically ill children had significantly higher stress levels than those of healthy children, especially those caring for children with cardiovascular, respiratory, or immunodeficiency disorders (Kostopoulou et al. 2021). A study of parents of children with multiple chronic illnesses found that parents demonstrated high stress levels which were likely associated with both the COVID-19 pandemic, and their child's medical complexities (Gallegos et al. 2022). Parents of children who were diagnosed more than ten years ago reported higher parenting stress than those whose children were diagnosed less than two years ago.

A study of caregivers of childhood cancer survivors found that although caregivers expressed resiliency and commonalities between caring for a child with cancer and COVID-19-related preventative measures, they also reported higher anxiety, poorer sleep, and less social support (Wimberly et al. 2021). Half of the caregivers also reported delayed or canceled medical appointments and 19% reported delayed or canceled medical imaging, which can impede diagnosis and negatively impact patient health and caregiver distress (Yan et al. 2020; McLoone et al. 2020). A qualitative study with parents of children receiving tertiary-level healthcare (i.e., highly specialized and often involving advanced and complex diagnostics, procedures, and treatments) found that parents reported intense fears about their child's additional vulnerabilities (McLoone et al. 2022). These fears included their child's risk of developing a severe COVID-19 infection, as well as pandemic-related disruptions to accessing medications, clinical care, and allied health support. Parents also reported fears regarding what would happen to their child's daily care protocols in the event they became unwell, and concerns about postponing their child's treatments if someone in the household contracted COVID-19 and had to self-isolate or quarantine.

Camporesi and colleagues (Camporesi 2022) sought to compare psychological distress in parents of children admitted to ICU with no changes to visitation policies, with those with restricted visitation access (no access up to one hour per day). Parents who were separated from their child in ICU due to visitation restrictions showed higher depression, anxiety, and psychological distress, in comparison with parents who were not. The experience of having a child admitted to ICU is extremely stressful, often having long-term consequences such as traumatic stress in parents and impaired family functioning (Balluffi et al. 2004; Murphy et al. 2020). Additional to this experience, not being allowed to be at their child's bedside likely adds additional distress and experiences of isolation for parents.

For parents of chronically ill children, the pandemic was an additional stress borne to multiple existing stressors. The accumulation of additional stressors and responsibilities during COVID-19, experienced in combination with fewer social interactions and limited access to normal support networks has been damaging to the mental health of parents with chronically ill children (McLoone et al. 2022). Parental stress has been reported to be related to experiences of isolation, a lack of resources and support, and concerns about the mental health of other children in the household (Kostopoulou et al. 2021). Many families were separated during the pandemic as a result of COVID-19-related restrictions. On top of those separations experienced by the general population (such as school closures and social distancing measures), families with chronically ill children may also have experienced hospital-based visitation restrictions. It has been suggested that hospital-based visitation restrictions led to decreased emotional support, coping, and resilience for parents and children alike, sometimes leading to marital discord, increased distress of siblings, and financial loss (McLoone et al. 2022). For example, to limit virus transmission in Australia some hospitals implemented a 'one parent, no siblings' rule for a number of months at the beginning of the pandemic (McLoone et al. 2022). Although parents could understand this policy, they felt it put families at risk of mental health impacts as a result of the reduced emotional support they were able to provide to one and other during critical and distressing times (McLoone et al. 2022). Parents highlighted the negative impacts of siblings not being able to see one and other during hospital admission periods. Regarding the impact of COVID-19 on support for parents and families of chronically ill children, parents reported that they usually would seek support from their extended family. However, with the restrictions and risks associated with the pandemic, parents reported that they often could not access this support (McLoone et al. 2022). Furthermore, the study found that community organizations were reported as essential support networks for parents, however many of them had limited capacity during the pandemic. Moreover, support experienced from families in similar situations within hospital settings was also interrupted, as many families tried to maintain social distancing and communal areas in the hospital were closed (McLoone et al. 2022).

Importantly, parents also reported some supportive factors during the pandemic including spending more time together at home and improved hygiene practices at school, which consequently reduced the spread of communicable non-COVID-19-related illnesses (e.g., flu) in their children (McLoone et al. 2022). Additionally, parents suggested that their local community (made up of families of children without a chronic illness) quickly became more understanding of their situation and what is it like to live with ongoing uncertainty and fear for their child's health.

22.5 Conclusion and Directions for Future Research

Living with underlying chronic illness imposes psychological, social, and financial burden to individuals and their families. Children and adolescents living with chronic illness have to manage the healthcare aspects of their condition in addition to achieving developmental milestones associated with their journey toward adulthood. Especially adolescence, usually spanning from 12 to 18 years old, can be a turmoil for young people characterized by rapid physiological, emotional, behavioral, and social changes in the lives of young people. The unpredicted and immense changes in the daily lives of young people with chronic illness introduced during the COVID-19 pandemic exposed them to unprecedented situations. The disruption of regular healthcare service provision, the prolonged lockdowns, the social isolation, the school closures, and other implications associated with the pandemic disrupted daily routine for many young people putting at risk their physical and psychological health. The increasing risk of a COVID-19 infection especially before the vaccine roll-out was accompanied with significant concerns about health and fear of death.

Overall, available evidence suggests that the COVID-19 pandemic has been linked to negative physical health outcomes in children and adolescents with chronic illness related to the COVID-19 per se and/or to the specific clinical manifestations of each type of chronic illness. We are now in the post-pandemic era, however the consequences of the pandemic on youth health are still unraveling. Considering that young people living with chronic illness are already at increased risk for experiencing poorer quality of life and mental health difficulties, researchers, policy makers, and healthcare professionals should continue monitoring the impact of the pandemic on those cohorts of young people. Generating robust research evidence on the physical and psychological impact of the pandemic can inform the design of interventions or decision-making with the view of counteracting negative outcomes that are not yet apparent. Young people even as young as 6 or 7 years old can be excellent informants on their health and well-being, while achieving data triangulation through collecting data from multiple sources and systems (such as the young individual, their parents, and healthcare professionals) can offer valuable and holistic insight into the impact of the pandemic on young people with chronic illness.

Evidence presented in this chapter suggests that the COVID-19 pandemic may have impacted aspects of youth's lives that are universal for all types of chronic illness as well as may have introduced unique challenges tailored to a specific disease. Future research employing epidemiological designs including clinical cohorts of participants diagnosed with different chronic illnesses can offer robust evidence on the wider impact of pandemic on subgroups of young patients (e.g., children with diabetes or immunosuppressed adolescents). In the past three years, starting only a few months since the WHO declared the COVID-19 outbreak (March 11, 2020) a global pandemic, an enormous amount of studies have been published in several outlets including scientific journals. However, because the situation was constantly and unpredictably developing, especially during 2020, it was of time essence to obtain as much research evidence as possible investigating this unprecedented global health crisis. This has resulted in several studies that did not necessarily meet scientific rigor benchmarks. Systematic reviews and meta-analyses when properly conducted comprise a very robust research design in evaluating but also in synthesizing as well as updating existing evidence. Considering that research in COVID-19 is still ongoing, systematic reviews and meta-analyses can be a powerful tool in synthesizing evidence that offers insight into post-pandemic developments for young people living with chronic illness and other stakeholders such as governments, policymakers, researchers, and healthcare professionals.

References

Academy of Medical Sciences (2021) COVID-19: preparing for the future. Looking ahead to winter 2021/22 and beyond

Alijotas-Reig J et al (2020) Immunomodulatory therapy for the management of severe COVID-19. Beyond the anti-viral therapy: a comprehensive review. Autoimmun Rev 19(7):102569

Ashcroft FM, Rorsman P (2012) Diabetes mellitus and the β cell: the last ten years. Cell 148(6):1160–1171

Badreldin HA, Atallah B (2021) Global drug shortages due to COVID-19: impact on patient care and mitigation strategies. Res Social Adm Pharm 17(1):1946–1949

Balluffi A et al (2004) Traumatic stress in parents of children admitted to the pediatric intensive care unit. Pediatr Crit Care Med 5(6):547–553

Barrett CE et al (2022) Risk for newly diagnosed diabetes >30 days after SARS-CoV-2 infection among persons aged <18 Years - United States, March 1, 2020-June 28, 2021. MMWR Morb Mortal Wkly Rep 71(2):59–65

Baum A, Kaboli PJ, Schwartz MD (2021) Reduced in-person and increased telehealth outpatient visits during the COVID-19 pandemic. Ann Intern Med 174(1):129–131

Biharie A et al (2022) Comorbidities, clinical characteristics and outcomes of COVID-19 in pediatric patients in a tertiary medical center in the Netherlands. World J Pediatr 18(8):558–563

Cahal M et al (2021) Real-time effects of COVID-19 pandemic lockdown on pediatric respiratory patients. Pediatr Pulmonol 56(6):1401–1408

Calder PC (2021) Nutrition and immunity: lessons for COVID-19. Nutr Diabetes 11(1):19

Camporesi A et al (2022) The parental psychological distress caused by separation from their critically ill child during the COVID-19 pandemic: a tale of two cities. Front Pediatr 10

Casaña-Granell S et al (2018) A brief version of the pediatric inventory for parents (PIP) in Spanish population: stress of main family carers of chronic paediatric patients. PLoS ONE 13(7):e0201390

Centers for Disease Control and Prevention (2022) About chronic diseases, N.C.f.C.D.P.a.H. Promotion, Editor. 2022: Georgia, United States. p. https://www.cdc.gov/chronicdisease/about/index.htm

Chaabane S et al (2021) The impact of COVID-19 school closure on child and adolescent health: a rapid systematic review. Children 8(5):415

Cronin JN, Camporota L, Formenti F (2022) Mechanical ventilation in COVID-19: a physiological perspective. Exp Physiol 107(7):683–693

de Lusignan S et al (2021) The epidemiology of eczema in children and adults in England: a population-based study using primary care data. Clin Exp Allergy 51(3):471–482

de Pablo GS et al (2020) Impact of coronavirus syndromes on physical and mental health of health care workers: Systematic review and meta-analysis. J Affect Disord 275:48–57

Di Giorgio E et al (2021) The interplay between mothers' and children behavioral and psychological factors during COVID-19: an Italian study. Eur Child Adolesc Psychiatry 30(9):1401–1412

DiMeglio LA et al (2020) COVID-19 and children with diabetes—Updates, unknowns, and next steps: first. Do No Extrapolation. Diabetes Care 43(11):2631–2634

Diskin C et al (2022) The impact of the COVID-19 pandemic on children with medical complexity. BMC Pediatr 22(1):496

Dorfman L et al (2022) Treatment adherence and behavior of pediatric liver transplant recipients during the COVID-19 pandemic. Pediatr Transplant 26(4):e14250

Edge R et al (2021) Cancer care disruption and reorganisation during the COVID-19 pandemic in Australia: a patient, carer and healthcare worker perspective. PLoS ONE 16(9):e0257420

Fontana L et al (2007) Visceral fat Adipokine secretion is associated with systemic inflammation in obese humans. Diabetes 56(4):1010–1013

Forner-Puntonet M et al (2021) Impact of COVID-19 on families of pediatric solid organ transplant recipients. J Pediatr Psychol 46(8):927–938

Fung M, Babik JM (2020) COVID-19 in immunocompromised hosts: what we know so far. Clin Infect Dis 72(2):340–350

Gallegos C et al (2022) Parenting a child with a chronic illness during a pandemic. J Pediatr Nurs 66:64–69

Garfan S et al (2021) Telehealth utilization during the Covid-19 pandemic: a systematic review. Comput Biol Med 138:104878

Ghosh R et al (2020) Impact of COVID -19 on children: special focus on the psychosocial aspect. Minerva Pediatr 72(3):226–235

Hanna TP et al (2020) Mortality due to cancer treatment delay: systematic review and meta-analysis. BMJ 371:m4087

Hartmann-Boyce J et al (2020) Diabetes and COVID-19: risks, management, and learnings from other national disasters. Diab Care 43(8):1695–1703

Iacobucci G (2022) Covid-19: Pandemic has disproportionately harmed children's mental health, report finds. BMJ 376:o430

Jordan A et al (2020) Romantic relationships in young people with long-term health conditions: a scoping review. J Pediatr Psychol 46(3):264–279

Kahn AR et al (2022) COVID-19 in children with cancer. Curr Oncol Rep 24(3):295–302

Kamrath C et al (2020) Ketoacidosis in children and adolescents with newly diagnosed type 1 diabetes during the COVID-19 pandemic in Germany. JAMA 324(8):801–804

Kenyon CC et al. (2020) Initial effects of the COVID-19 pandemic on pediatric asthma emergency department utilization. J Allergy Clin Immunol Pract 8(8):2774–2776. e1

Kinoshita S et al (2022) Changes in telepsychiatry regulations during the COVID-19 pandemic: 17 countries and regions' approaches to an evolving healthcare landscape. Psychol Med 52(13):2606–2613

Knapp PK, Harris ES (1998) Consultation-liaison in child psychiatry: a review of the past 10 years. Part I: Clinical findings. J Am Acad Child Adolesc Psychiatry 37(1):17–25

Kostopoulou E et al (2021) Parents under siege: the psychological impact of COVID-19 outbreak on children's caregivers. Swiss Med Wkly 151(3132):w30012–w30012

Krijger A et al (2022) Perceived stress, family impact, and changes in physical and social daily life activities of children with chronic somatic conditions during the COVID-19 pandemic. BMC Pub Health 22(1):1106

Lee SJ et al (2021) Parenting activities and the transition to home-based education during the COVID-19 pandemic. Child Youth Serv Rev 122:105585

Lignou S et al (2022) Changes in healthcare provision during covid-19 and their impact on children with chronic illness: a scoping review. Inquiry 59:469580221081445

Lobritto S et al (2020) Impact of COVID-19 pandemic on pediatrics and pediatric transplantation programs. Front Pediatr 8:612627

Logan BA (2022) The impact of the COVID-19 pandemic on pediatric chronic illness groups. Brown Univ Child and Adolesc Behav Lett 38(5):1–4

Maringe C et al (2020) The impact of the COVID-19 pandemic on cancer deaths due to delays in diagnosis in England, UK: a national, population-based, modelling study. Lancet Oncol 21(8):1023–1034

McLoone J et al (2020) The COVID-19 pandemic: distance-delivered care for childhood cancer survivors. Pediatr Blood Cancer 67(12):e28715

McLoone J et al (2022) It's made a really hard situation even more difficult: The impact of COVID-19 on families of children with chronic illness. PLoS ONE 17(9):e0273622

Murphy LK et al (2020) Longitudinal trajectories of caregiver distress and family functioning after community-acquired pediatric septic shock. Pediatr Crit Care Med J Soc Crit Care Med World Federation Pediatr Intensive Crit Care Soc 21(9):787

Nearchou F et al (2020) Exploring the impact of COVID-19 on mental health outcomes in children and adolescents: a systematic review. Int J Environ Res Public Health 17(22)

Nogueira-de-Almeida CA et al (2020) COVID-19 and obesity in childhood and adolescence: a clinical review. J de Pediatr 96

Patrick SW et al (2020) Well-being of parents and children during the COVID-19 pandemic: a national survey. Pediatrics 146(4)

Patterson CC et al (2019) Worldwide estimates of incidence, prevalence and mortality of type 1 diabetes in children and adolescents: Results from the International Diabetes Federation Diabetes Atlas, 9th edition. Diabetes Research and Clinical Practice. vol 157, p 107842

Pinquart M (2014) Achievement of developmental milestones in emerging and young adults with and without pediatric chronic illness–a meta-analysis. J Pediatr Psychol 39(6):577–587

Pinquart M (2020) Health-related quality of life of young people with and without chronic conditions. J Pediatr Psychol 45(7):780–792

Pinquart M, Pfeiffer JP (2015) Solving developmental tasks in adolescents with a chronic physical illness or physical/sensory disability: a meta-analysis. Int J Disabil Dev Educ 62:249–264

Pinquart M, Silbereisen RK, Wiesner M (2004) Changes in discrepancies between desired and present states of developmental tasks in adolescence: a 4-process model. J Youth Adolesc 33(6):467–477

Plevinsky JM et al (2020) The impact of COVID-19 on pediatric adherence and self-management. J Pediatr Psychol 45(9):977–982

Polanczyk GV et al (2015) Annual research review: a meta-analysis of the worldwide prevalence of mental disorders in children and adolescents. J Child Psychol Psychiatry 56(3):345–365

Prosperi S, Chiarelli F (2022) COVID-19 and diabetes in children. Ann Pediatr Endocrinol Metab 27(3):157–168

Rabbone I et al (2020) Has COVID-19 delayed the diagnosis and worsened the presentation of type 1 diabetes in children? Diab Care 43(11):2870–2872

Ramakrishnan M et al (2023) Impact of COVID-19 pandemic on medicine supply chain for patients with chronic diseases: experiences of the community pharmacists. Clin Epidemiol Glob Health 20:101243

Reilly C, Muggeridge A, Cross JH (2021) The perceived impact of COVID-19 and associated restrictions on young people with epilepsy in the UK: young people and caregiver survey. Seizure 85:111–114

Remuzzi A, Remuzzi G (2020) COVID-19 and Italy: what next? The Lancet 395(10231):1225–1228

Richardson S et al (2020) Presenting characteristics, comorbidities, and outcomes among 5700 patients hospitalized with COVID-19 in the New York City area. JAMA 323(20):2052–2059

Royal College of Paediatrics and Child Health (2019) The case for investing in integrated child health services for the acutely unwell child and long term conditions. RCPCH London, UK

Royal College of Paediatrics and Child Health (2020) The impact of COVID-19 on child health services-report

Sankar J et al (2020) COVID-19 in children: clinical approach and management. Indian J Pediatr 87(6):433–442

Seattle W (2013) GBD compare. University of Washington, Seattle

Shah S et al (2015) Impact of supply problems of preservative-free glaucoma medications on patients and hospital staff. Ophthalmic Physiol Opt 35(2):236–241

Shi H et al (2021) Prevalence and influencing risk factors of eczema among preschool children in Urumqi city: a cross-sectional survey. BMC Pediatr 21(1):347

Shudy M et al (2006) Impact of pediatric critical illness and injury on families: a systematic literature review. Pediatrics 118(Supplement_3):S203–S218

Simonnet A et al (2020) High prevalence of obesity in severe acute respiratory syndrome coronavirus-2 (SARS-CoV-2) requiring invasive mechanical ventilation. Obesity 28(7):1195–1199

Singh K et al (2021) Health, psychosocial, and economic impacts of the COVID-19 pandemic on people with chronic conditions in India: a mixed methods study. BMC Public Health 21(1):685

St. Jude Global (2023) Global registry of COVID-19 in childhood cancer. 2023, St. Jude Children's Research Hospital International Society of Paediatric Oncology

Stuber ML et al (2006) The medical traumatic stress toolkit. CNS Spectr 11(2):137–142

Teoh CW et al (2020) Management of pediatric kidney transplant patients during the COVID-19 pandemic: guidance from the canadian society of transplantation pediatric group. Can J Kidney Health Dis 7:2054358120967845

Tsankov BK et al (2021) Severe COVID-19 infection and pediatric comorbidities: a systematic review and meta-analysis. Int J Infect Dis 103:246–256

Tse Y et al (2021) COVID-19: experiences of lockdown and support needs in children and young adults with kidney conditions. Pediatr Nephrol 36(9):2797–2810

Walker J et al (2017) Medicine shortages in Fiji: a qualitative exploration of stakeholders' views. PLoS ONE 12(6):e0178429

Wang G et al (2020) Mitigate the effects of home confinement on children during the COVID-19 outbreak. The Lancet 395(10228):945–947

Watson G et al (2021) 'Do I, don't I?' A qualitative study addressing parental perceptions about seeking healthcare during the COVID-19 pandemic. Arch Dis Child 106(11):1118–1124

Williams R et al (2020) Diagnosis of physical and mental health conditions in primary care during the COVID-19 pandemic: a retrospective cohort study. Lancet Pub Health 5(10):e543–e550

Wimberly CE et al (2021) Impacts of COVID-19 on caregivers of childhood cancer survivors. Pediatr Blood Cancer 68(4):e28943

Wirrell EC et al (2020) Care delivery for children with epilepsy during the COVID-19 pandemic: an international survey of clinicians. J Child Neurol 35(13):924–933

Witt WP, Riley AW, Coiro MJ (2003) Childhood functional status, family stressors, and psychosocial adjustment among school-aged children with disabilities in the United States. Arch Pediatr Adolesc Med 157(7):687–695

World Health Organization (2014) Global status report on noncommunicable diseases 2014. World Health Organization, Geneva

World Health Organization (2021) Childhood cancer, World Health Organization, Editor. Geneva, Switzerland. https://www.who.int/news-room/fact-sheets/detail/cancer-in-children

World Health Organization (2022) Noncommunicable diseases, World Health Organization, Editor. Geneva, Switzerland. https://www.who.int/news-room/fact-sheets/detail/noncommunicable-diseases#:~:text=Key%20facts%2D%20and%20middle%2Dincome%20countries

Wu ZH, Tang Y, Cheng Q (2021) Diabetes increases the mortality of patients with COVID-19: a meta-analysis. Acta Diabetol 58(2):139–144

Yan AP et al (2020) Adherence to surveillance for second malignant neoplasms and cardiac dysfunction in childhood cancer survivors: a childhood cancer survivor study. J Clin Oncol 38(15):1711

Yeo M, Sawyer S (2005) Chronic illness and disability. BMJ (clinical Research Ed.) 330(7493):721–723

Young M (2020) Coronavirus: impact on young people with mental health needs

Zee-Cheng JE et al (2021) Changes in pediatric ICU utilization and clinical trends during the coronavirus pandemic. Chest 160(2):529–537

Zeng F et al (2020) Association of inflammatory markers with the severity of COVID-19: a meta-analysis. Int J Infect Dis 96:467–474

Zhang D, Zheng J (2022) The burden of childhood Asthma by age group, 1990–2019: a systematic analysis of global burden of disease 2019 data. Frontiers in Pediatrics, 10

23 Management of COVID-19 in Ophthalmology

Maria Letizia Salvetat and Marco Zeppieri

Abstract

From its outbreak in early 2020, the new SARS-CoV-2 infection has deeply affected the entire eye care system for several reasons. Since the beginning of the COVID-19 pandemic, ophthalmologists were among the "high-risk category" for contracting the SARS-CoV-2 infection based on the notion that the eye was suspected to be a site of inoculation, infection, and transmission. Clinical ophthalmologists have been required to learn and promptly recognize the ocular manifestations associated with the COVID-19 disease, with its treatments and vaccinations. Restriction measures, lockdown periods, guidelines to prevent SARS-CoV-2 infection transmission, and the use of telemedicine and artificial intelligence modalities have induced profound modifications. These changes, which are most likely to be irreversible, influence routine clinical practice, education, and research, thus giving rise to a "new ophthalmology in the COVID era". This book chapter aims to provide several notions regarding COVID-19 in ophthalmology, including the SARS-CoV-2 virus infection and transmission modalities; the ocular manifestation associated with the COVID-19 disease; the drugs and vaccines used for COVID-19; the precautions adopted in the ophthalmic practice to limit the spread of the virus; the consequences of the pandemic on the ophthalmic patients, clinicians, and the entire eye care system; and, the future of ophthalmology in the era of "COVID new normal".

Keywords

SARS-CoV-2 · COVID-19 · Pandemic · Lockdown · COVID-19 vaccination · COVID-19-associated eye manifestations

Abbreviations

SARS-CoV-2	Severe Acute Respiratory Syndrome Coronavirus 2
COVID-19	Coronavirus Disease 2019
WHO	World Health Organization
RNA	Ribonucleic Acid
ACE2	Angiotensin-Converting Enzyme 2
RT-PCR	Reverse Transcriptase-Polymerase Chain Reaction
VEGF	Vascular Endothelial Growth Factor

M. L. Salvetat
Department of Ophthalmology, Azienda Sanitaria Friuli Occidentale, Pordenone, Italy

M. Zeppieri (✉)
Department of Ophthalmology, University Hospital of Udine, P.le S. Maria della Misericordia 15, 33100 Udine, Italy
e-mail: markzeppieri@hotmail.com

N. Rezaei (ed.), *The COVID-19 Aftermath*, Advances in Experimental Medicine and Biology 1457,
https://doi.org/10.1007/978-3-031-61939-7_23

23.1 Introduction

Severe acute respiratory syndrome coronavirus 2 (SARS-CoV-2), which can cause the coronavirus disease 2019 (COVID-19), was first detected in the city of Wuhan (China), in December 2019 (Khalili et al. 2020). The infection spread rapidly across the globe and was declared a pandemic by the World Health Organization (WHO) on March 11, 2020 (WHO 2020; Yesudhas et al. 2021). Since December 2019, the COVID-19 disease has devastated the lives of millions of people worldwide, with an incredible socio-economic impact. By January 2023, more than 670 million confirmed cases of SARS-CoV-2 infection and 6,750,000 deaths have been reported by the WHO (Worldometer 2020).

Although not completely understood, the COVID-19 disease can present with a wide range of signs and symptoms of different severity, ranging from completely asymptomatic to critically ill, with viral pneumonia, acute respiratory distress syndrome (ARDS), and multiple organ failure that can lead to death (Pascarella et al. 2020). From its first appearance in late 2019, the SARS-CoV-2 virus has shown several mutations, two of which have determined important pandemic effects. The two variants include Delta, which was first discovered in October 2020, and the variant Omicron, which was reported in November 2021 (Wang and Cheng 2022).

An extremely tight relationship between ophthalmology and COVID-19 disease was evident from the beginning of the pandemic, with several subsequent important repercussions. The first physician to sound the alarm about the new SARS-CoV-2 infection causing human disease in late 2019 was a Chinese ophthalmologist, Dr. Lin Wenliang. He was working in the Wuhan Center Hospital, in China, and most probably contracted the virus from patients with COVID-19-related conjunctivitis. He died soon after due to respiratory complications at the age of 33 (Yesudhas et al. 2021). Since then, numerous other ophthalmologists across the world have been infected by the virus with severe COVID-19 disease complications and death. Eye care professionals have thus been regarded as a "high-risk category" for contracting the SARS-CoV-2 infection, probably because of the near-distance conditions when dealing with patients during routine examinations and surgical procedures (Danesh-Meyer and McGhee 2021; Li et al. 2020; Romano et al. 2020; Shah et al. 2021; Shetty et al. 2020a; Wan et al. 2020).

The SARS-CoV-2 ribonucleic acid (RNA) has been isolated from the ocular surface of both COVID-19 patients with ocular manifestations and those without ocular signs and symptoms (Al-Sharif et al. 2021; Sopp and Sharda 2021; Xia et al. 2020; Zhou et al. 2020b). Studies have reported that the eye could be a possible route for infection and transmission of the disease (La Distia Nora et al. 2020; Sopp and Sharda 2021; Zhu et al. 2022). The ocular manifestations associated with the COVID-19 disease tend to be rare and usually show a good prognosis in most patients (Baj et al. 2022; Luo et al. 2022; Nasiri et al. 2021; Shetty et al. 2020a). Patients with ocular signs and symptoms should be promptly recognized, especially considering that these acute ocular manifestations could be the first signs (Casalino et al. 2020) or even the only sign (Ozturker 2022) of the SARS-CoV-2 infection.

The ocular adverse events associated with the COVID-19 vaccination are extremely uncommon and normally transient (Ichhpujani et al. 2022; Ng et al. 2021). It is important to note that post-vaccination ocular manifestations seem to target predisposed or genetically susceptible individuals, especially patients affected by autoimmune inflammatory diseases, including Graves' disease, non-infective uveitis, or subjects with undiagnosed autoimmune conditions (Furer et al. 2021; Ng et al. 2022; Sen and Honavar 2021; Wang et al. 2022).

The pandemic has induced distancing measures and lockdown periods to prevent the spread of the disease. New guidelines and recommendations by the major ophthalmological societies worldwide are constantly being updated to prevent SARS-CoV-2 infection transmission (Chodosh et al. 2022; Li et al. 2020). This has led to profound modifications in ophthalmic care, education, and research (Danesh-Meyer and McGhee 2021; Mack and Fraser-Bell 2021), most likely in an irreversible way, giving rise to a "new ophthalmology in the COVID era".

This book chapter aims to provide a brief overview of several subtopics regarding coronavirus, which include the SARS-CoV-2 virus infection and transmission modalities; the ocular manifestations associated with the COVID-19 disease; the drugs and vaccines used for the virus; the precautions adopted in an ophthalmic clinical setting to limit the spread of the virus; the consequences of the pandemic on ophthalmic patients, the ophthalmologists, and the entire eye care system; and, the future of ophthalmology in the "COVID new normal".

23.2 The SARS-CoV-2 Virus Infection

23.2.1 The Coronaviridae Family

According to the WHO, more than 23 viral infections can currently be considered active or prone to becoming pandemics (Dhiman et al. 2022). The novel SARS-CoV-2 virus is one of the most recent significant addition to the list of viral pandemics, which also includes Ebola and the new Monkeypox viruses (Dhiman et al. 2022). The SARS-CoV-2 virus, which was isolated and identified in January 2020, belongs to the Coronaviridae family and can be classified in the subgroup of the Beta-coronaviruses (Yesudhas et al. 2021). Six other kinds of Coronaviruses are able to infect the human body, two of which include the Severe Acute Respiratory Syndrome Coronavirus (SARS-CoV) and the Middle East Respiratory Syndrome Coronavirus (MERS-CoV). These Coronaviruses caused a pandemic in 2002 and 2015, respectively, with a mortality rate of 10% for SARS-CoV and 37% for MERS-CoV (der Hoek 2007).

All seven human Coronaviruses are zoonotic in origin, infecting birds and mammals, including camels, cattle, cats, and bats (Al-Sharif et al. 2021). For reasons yet to be explained, these viruses can cross species barriers, cause illness in humans and produce secondary transmission from human to human being. Genomic characterization has shown that bats and rodents are probably the gene sources of alpha and beta-coronaviruses (Fung and Liu 2019). The SARS-CoV-2 virus is likely to have originated in bats, which are the natural reservoir for the virus. The potential amplifying mammalian host intermediate between bats and humans, however, has yet to be known. Studies have reported that it is not even certain that this intermediate exists (Fung and Liu 2019).

23.2.2 SARS-CoV-2 Virus Characteristics and Host Cell Receptors

The SARS-CoV-2 virus is an enveloped, single-chain, positive-stranded ribonucleic acid (RNA) virus. The RNA genome of the SARS-CoV-2 is about 30 kb in length and encodes for four proteins: S, M, E, and N proteins. The spike protein S, located on the viral surface, is responsible for the attachment to the host receptors and represents the main antigen that stimulates neutralizing antibodies and cytotoxic lymphocytes. When seen on microscopy, it has a crown-like surface appearance, which gives the name to the Coronaviridae family. The membrane protein M helps to shape the virus particles and their binding to the nucleocapsid. The envelope protein E promotes the packaging and reproduction of the virus particles. The nucleocapsid protein N helps the genome with binding and replication (Seah and Agrawal 2020).

The SARS-CoV-2 virus enters the host cells by binding its spike protein S to a host receptor situated on the surface of the host cells. The presence of the host receptor is thus necessary for viral infection, replication, and transmission. Two human host SARS-CoV-2 binding receptors have been identified. The angiotensin-converting enzyme 2 (ACE2) is considered the main virus receptor (Abid et al. 2022; Hoffmann et al. 2020; Lan et al. 2020). The cluster of differentiation 147 (CD147) receptor is considered an alternative receptor, especially in ACE2-deficient cells (Wang et al. 2020a). After the interaction between the virus and its membrane receptor (ACE2 or CD147), the transmembrane protease serine 2 (TMPRSS2), located on the host cell membrane, is necessary to induce receptor protein priming, which permits the virus entry into the host cell through endocytosis or membrane fusion. Cathepsin L is then needed to cleave the virus so that it can fuse the endosome membrane, release its RNA into the cytoplasm, and use the ribosomes of the host cell for replication, thereby producing new virus particles that infect other cells (Hirano and Murakami 2020; Perrotta et al. 2020; Sun et al. 2006). The infection normally stimulates the innate immune system, which causes the release of inflammatory mediators, especially cytokines. Numerous studies have shown that inflammation, abnormal cytokines release, and tissue damage are typically found in the infected cells (Guo et al. 2020).

The ACE2 receptor is part of the renin-angiotensin system. The binding of these receptors by the SARS-CoV-2 could be dangerous for the tissues because it can induce a local accumulation of Angiotensin II, which has a pro-inflammatory effect. This mechanism has been already demonstrated for SARS-CoV infection (Kuba et al. 2005). The ACE2 receptors are mainly expressed in alveolar, bronchial, and tracheal epithelial cells. Studies have reported that these receptors can also be found in other organs and tissues, such as the heart, brain, vascular endothelial cells, kidneys, liver, intestine, lymphocytes, and eye (Abid et al. 2022; Guo et al. 2020), which may explain the wide range of clinical manifestations of the COVID-19 disease. Coagulopathies tend to be frequently associated with the COVID-19 disease, in which there seems to be a direct effect of the SARS-CoV-2 virus on the ACE2 receptors present on the endothelial cells, leading to endothelial dysfunction, arterial and venous thromboembolic disease, and pulmonary embolism (Abou-Ismail et al. 2020).

The genetic variation of ACE2 receptor expression levels and binding affinities could explain the different infection rates of the SARS-CoV-2 and COVID-19 disease severity found among different ethnicities, age groups, and gender. The ACE2 receptors expression has been found to be higher in Black and Asian ethnicities, older individuals, and males, which can explain why Caucasians, females, and children tend to show less disease severity and morbidity (Abid et al. 2022).

Both ACE2 and CD147 receptors are expressed in the conjunctiva, limbus, cornea, aqueous humor, iris, retina, retinal pigment epithelium, and blood vessel endothelial cells (Abid et al. 2022; Holappa 2015; Ma et al. 2020; Senanayake et al. 2007; Zhou et al. 2020a). TMPRSS2 and Cathepsin L have been isolated in cornea and conjunctival cells (Ma et al. 2020). Previous studies have demonstrated that ACE2 and TMPRSS2, which are both necessary for the SARS-CoV-2 infection, show a significantly higher expression in human corneal cells when compared to conjunctival ones. It thus seems more likely that the cornea could potentially be more susceptible to the SARS-CoV-2 infection, as opposed to the conjunctiva surface (Ma et al. 2020).

23.2.3 SARS-CoV-2 Virus Human-to-Human Transmission Routes and COVID-19 Disease Systemic Clinical Manifestations

The principal human-to-human routes of transmission of the SARS-CoV-2 virus include direct contact with virus-contaminated respiratory secretions (droplets, aerosols), exchange of body fluids (i.e., saliva, urine, semen, tears, etc.) and, transmission from mother-to-child. Indirect transmission modalities are due to contact with virus-contaminated fomites, objects, and surfaces (Karia et al. 2020). Both symptomatic and asymptomatic COVID-19 patients can spread the virus (Luo et al. 2020). The asymptomatic carrier rate is still difficult to calculate and represents the most important obstacle to infection control.

The incubation period of the SARS-CoV-2 infection ranges between 2 and 14 days (Cao et al. 2020). The clinical and systemic manifestations of COVID-19 disease have a wide range of severity, ranging from being completely asymptomatic to life-threatening conditions due to severe viral pneumonia, acute respiratory distress syndrome (ARDS), and multiple organ failure (Pascarella et al. 2020; Wang and Ma 2008). The mean proportion of asymptomatic COVID-19 patients was reported to be 30.8%, with a 95% confidence interval (CI) of 7.7–53.8% (Day 2020).

Studies have demonstrated that human Coronaviridae are responsible for about 5–10% of acute respiratory infections and that 2% of the population are healthy carriers (Al-Sharif et al. 2021). The respiratory system is mainly compromised in patients with SARS-CoV-2 infection. Several other organs, however, can also be affected by an infection. There are over 70 different clinical manifestations associated with the COVID-19 disease that have been reported in the current literature (Luo et al. 2022). The typical, yet not specific, clinical manifestations include fever, dry cough, headache, muscle weakness, fatigue, dyspnea, and loss of smell and taste. Several atypical manifestations have been described, including neurological, renal, dermatological, cardiovascular, and ophthalmological signs and symptoms (Guan et al. 2020; Kouhsari et al. 2021; Luo et al. 2022). Hypertension and diabetes mellitus are the most common comorbidities associated with severe complications of the COVID-19 disease (Madjid et al. 2020).

23.2.4 COVID-19 Disease Diagnosis and Detection of the SARS-CoV-2 RNA in the Ocular Tissues

The diagnosis of COVID-19 disease is based on the research of the viral RNA with the reverse transcriptase-polymerase chain reaction (RT-PCR) test in specimens such as nasopharyngeal or oropharyngeal swabs, bronchoalveolar lavage fluid, sputum, bronchial aspirate, and blood (Rai et al. 2021). The highest detection rates have been reported in the bronchoalveolar lavage fluid, sputum, and nasal swabs (Wang et al. 2020b). Although the gold standard for the diagnosis of SARS-CoV-2 infection is the RT-PCR test, the combination of IgM and IgG antibodies has demonstrated good diagnostic accuracy (Rai et al. 2021).

Studies have reported the presence of SARS-CoV-2 RNA in tears (collected with Schirmer's strips), conjunctival swabs (Arora et al. 2020; Seah et al. 2020; Sopp and Sharda 2021; Zhou et al. 2020b), in aqueous samples (Koo et al. 2021) and in vitreous samples (Hada et al. 2021) taken form laboratory-confirmed COVID-19 patients (i.e., having a positive RT-PCR test in either respiratory or blood specimens) with and without ocular manifestations.

The mean percentage of positive RT-PCR tests in tears or conjunctival swabs of laboratory-confirmed COVID-19 patients reported by different authors varies from 0 to 57.1%, with higher rates in the presence of conjunctivitis and in moderate to severe COVID-19 disease cases (Al-Sharif et al. 2021; Seah et al. 2020; Sopp and Sharda 2021; Xia et al. 2020; Zhang et al. 2020; Zhou et al. 2020b). This wide range of positivity can be explained by the different levels of disease severity, the variety in sampling time and methods used, the different sensitivities of the RT-PCR kits, the mutation of the virus genome, etc. (Forouhari et al. 2022). A recent meta-analysis has calculated that the mean rate of detection of the SARS-CoV-2 RNA in tears and conjunctival swabs was 2.7% (95% CI of 1.4–4.8%) (Sopp and Sharda 2021).

When the research of the viral RNA in the nasopharyngeal swab is taken as the gold standard for the diagnosis, the positivity of the RT-PCR test in ocular fluid and tissues has been demonstrated to have a very high specificity (100%) but poor sensitivity (0.6%) (Ulhaq and Soraya 2020). In COVID-19 patients (confirmed by a positive RT-PCT test in the bronchoalveolar lavage fluid or bronchial aspirate) with ocular surface manifestations, the rate of positivity of the RT-PCR test in the conjunctival and nasopharyngeal swabs has been demonstrated to be

16.7% and 91.7%, respectively (Wu et al. 2020). The use of tears or conjunctival swabs can thus not be considered a reliable diagnostic method for the SARS-CoV-2 infection.

The lower sensitivity of the RT-PCR test in conjunctival swabs when compared with nasopharyngeal samples, even in the presence of ocular manifestations, could be related to several factors, which include poor collection, transportation, and handling of the conjunctival samples, inadequate viral loading for RT-PCR detection, inopportune window periods during which the virus is released on the ocular surface, etc. (Zhu et al. 2022).

The RT-PCR test can be positive in the conjunctival swabs of COVID-19 patients without any ocular manifestation (Arora et al. 2020; Sopp and Sharda 2021) and, surprisingly, even in tears of patients after complete clinical recovery (Colavita et al. 2020). The absence of ocular signs and symptoms thus does not exclude the possibility of viral shedding from tears and conjunctival secretions. Moreover, it is important to note that the SARS-CoV-2 RNA can also be found in the aqueous humor of asymptomatic subjects with negative nasopharyngeal swab tests (Koo et al. 2021), suggesting that the virus might survive in immune-privileged spaces despite the lack of symptoms. It is certainly important to underline that a negative RT-PCR test does not exclude the presence of the virus and that the presence of the viral RNA in ocular samples does not imply an infection.

With regard to the other seven human Coronaviruses, no viral RNA has ever been found on the human ocular surface except for the SARS-CoV-1 virus, in which no ocular symptoms were associated with the SARS-CoV-1 infection (Al-Sharif et al. 2021).

23.2.5 Possibility of SARS-CoV-2 Virus Ocular Inoculation and Transmission

The means of SARS-CoV-2 virus ocular inoculation or transmission are not known and still debatable (La Distia Nora et al. 2020; Deng et al. 2020a; Willcox et al. 2020). The ocular surface, however, is regarded as a possible inoculation site, in which the virus can enter the respiratory system through the nasolacrimal duct and reach other parts of the body through hematogenous metastasis, and infected conjunctival secretions or tears (Al-Sharif et al. 2021; Eissa et al. 2022; La Distia Nora et al. 2020; Lu et al. 2020; Kharel Sitaula et al. 2020; Sopp and Sharda 2021; Willcox et al. 2020; Zhu et al. 2022). This hypothesis is supported by several data, which include:

a. The components necessary for the SARS-CoV-2 virus to infect the cells, i.e., the binding receptors (ACE2 and CD147) and the mechanisms for cell entry (TPS2 and cathepsin L) are all present on the ocular surface (Hirano and Murakami 2020; Perrotta et al. 2020);
b. The SARS-CoV-2 RNA has been detected in tears and conjunctival swabs of laboratory-confirmed COVID-patients (Al-Sharif et al. 2021; Sopp and Sharda 2021; Xia et al. 2020; Zang et al. 2020; Zhou et al. 2020b), even if in a low percentage of cases (Sopp and Sharda 2021; Willcox et al. 2020). Moreover, it is still not clear whether these viral particles are infectious (Seah et al. 2020);
c. The SARS-CoV-2 virus has demonstrated conjunctival tropism. Studies have shown that the virus isolated from a nasopharyngeal aspirate can replicate in ex vivo cultures of human conjunctiva (Hui et al. 2020);
d. The SARS-CoV-2 virus isolated from ocular swabs of laboratory-confirmed COVID-19 patients has shown to be capable of replication in vitro, thus supporting the possibility of conjunctival infectivity (Colavita et al. 2020). An active SARS-CoV-2 virus replication in the ocular tissues has yet to be demonstrated (Ali 2020);
e. Conjunctivitis is the most commonly reported ophthalmic manifestation associated with the COVID-19 disease, with a mean prevalence of 5–9% of cases (Al-Namaeh 2021; Baj et al. 2022; Loffredo et al. 2020; Nasiri et al. 2021; Sen et al. 2021; Shetty et al. 2020a;

Taha et al. 2022; Willcox et al. 2020). It can be the first sign (Casalino et al. 2020), or even a unique manifestation of the SARS-CoV-2 infection (Ozturker 2022; Scalinci and Trovato Battagliola 2020). The presence of ocular manifestations before any other systemic symptom can suggest that the ocular surface can be the portal of SARS-CoV-2 entry into the human body. Furthermore, patients with conjunctivitis associated with the COVID-19 disease can be a source of SARS-CoV-2 transmission. In the presence of a living virus in conjunctival secretions, an eye-hand-nose or eye transmission can most likely occur;

f. Previous authors have detected SARS-CoV-2 RNA in nasal and throat swabs after conjunctival inoculation of rhesus monkeys (Deng et al. 2020b);
g. Some case reports suggested a possible SARS-CoV-2 infection transmission with droplets, infected conjunctival secretions, or infected tears through the ocular surface when no eye protection was worn (Dockery et al. 2020; Lu et al. 2020; Zhang et al. 2020);
h. Several studies have demonstrated that wearing eyeglasses for more than 8 h a day has a protective effect against the SARS-CoV-2 infection (Zeng et al. 2020).

After the entry and replication on the ocular surface, it is speculated that the SARS-CoV-2 virus can reach the respiratory tract or other parts of the body in different ways: through the nasolacrimal duct, as already demonstrated in animal models (Deng et al. 2020b); or by hematogenous metastasis after reaching the lacrimal gland or the choroideal plexus. The different possible ways of entry of the SARS-CoV-2 virus through the eyes include direct inoculation of the conjunctiva from infected droplets; migration of upper respiratory tract infection through the nasolacrimal duct; hematogenous infection of the lacrimal gland or the choroideal plexus; and exudation of the virus through the conjunctival vessels (Amesty et al. 2020).

Several studies, however, have demonstrated that patients with confirmed COVID-19 disease and symptoms of the upper respiratory tract did not show any viral shedding in tears or conjunctival swabs (Sopp and Sharda 2021; Willcox et al. 2020), suggesting that the hypothesis of the lacrimal duct being the conduct of the virus from the respiratory tract into the eye surface may not be true. Further studies are needed to elucidate the role of the eye in SARS-CoV-2 infection. Although the virus ocular transmission in humans remains unproven and the current literature suggests that there is a very low risk of transmission through the ocular surface (Sen et al. 2021), shreds of evidence support this possible hypothesis (Davis et al. 2022; Zhang et al. 2020).

Several national ophthalmological societies have recommended appropriate precautions to prevent viral transmission, including the use of eye protection with goggles or face shields by all healthcare providers during contact with the patients, to protect themselves against the infection; the generalized invitation to avoid touching the eyes, nose, and mouth, regardless the presence of ocular signs and symptoms; the suggestion for patients to consider switching from contact lenses to glasses, to reduce the times the patient touch the eyes and because the glasses provide a physical barrier between the respiratory droplets and the ocular surface, etc. (Chodosh et al. 2022; Chu et al. 2020; Danesh-Meyer and McGhee 2021; Lam et al. 2020b; Li et al. 2020).

It is important to note that during the COVID-19 pandemic, guidelines for corneal donation and eye banking have been changing in some countries, in which suspected or confirmed COVID-19 cases have become part of the exclusion criteria (Sharma et al. 2020). Although the viral RNA has been detected in the corneas of deceased COVID-19 patients, there has not been any reported case of COVID-19 transmission through cornea transplantation (Ballouz and Mian 2020) and further studies are required to determine the possible transmission of the SARS-CoV-2 infection after a corneal transplant.

23.3 The Ocular Manifestations of the COVID-19 Disease

The mean prevalence of ocular manifestations among COVID-19 patients is estimated to be approximately 5–10%, ranging from 0.8 to 64% in the various studies (Aggarwal et al. 2020; Akbari and Dourandeesh 2022; Al-Namaeh 2021; Al-Sharif et al. 2021; Alnahdi and Alkharashi 2023; Amesty et al. 2020; Baj et al. 2022; Eissa et al. 2022; Farvardin et al. 2021; Gold and Galetta 2021; Loffredo et al. 2020; Lin et al. 2021; Nasiri et al. 2021; Shetty et al. 2020a; Sen et al. 2021; Sopp and Sharda 2021; Taha et al. 2022; Willcox et al. 2020; Wu et al. 2020). These data could be overestimated considering that most studies were conducted in hospital settings.

The lower occurrence of the ocular manifestations associated with the COVID-19 disease when compared with the respiratory ones could have several explanations, including the lower concentration of SARS-CoV-2 ACE2 receptors in the eye than in the respiratory tissues (Liu et al. 2004), and the protective effect of the tears against pathogens due to the constant flow and production and the presence of antimicrobial agents in tears, which include lactoferrin and immunoglobulin A (Liu and Sun 2020). Furthermore, the continuous flow of tears across the eye's surface into the nasolacrimal duct could also explain the lower ocular and the higher respiratory involvement (Liu and Sun 2020).

Approximately 25% of cases of symptomatic COVID-19 patients show ocular manifestations, which can be observed at any stage of the disease, and even several weeks after the recovery (Baj et al. 2022; La Distia Nora et al. 2020; Nasiri et al. 2021; Sen et al. 2021) In some cases, the ocular signs and symptoms can be the only presentation of the COVID-19 disease. Studies have reported that the ocular manifestation could be linked (Loffredo et al. 2020; Wu et al. 2020) or not linked (La Distia Nora et al. 2020) to the COVID-19 disease clinical course severity. The risk factors for ocular involvement include advanced age, high fever, increased neutrophil-to-lymphocyte ratio, and elevated acute-phase reactant levels (Bostanci Ceran and Ozates 2020). The ocular pathologies associated with the COVID-19 disease are listed in Table 23.1.

The most commonly reported ocular symptoms in COVID-19 patients were dry eye, foreign body sensation, blurring of vision, chemosis, conjunctival hyperemia, tearing, discharge, itching, photophobia, and eye pain (Bostanci Ceran and Ozates 2020; Wu et al. 2020). The mean occurrence of symptoms in COVID-19 patients has been reported to be 6.9% in severe cases, and 4.1% in mild to moderate cases (Aggarwal et al. 2020).

Among all ophthalmic manifestations associated with the COVID-19 disease, conjunctivitis has been the most commonly reported, representing approximately 85–90% of ocular symptomatic cases (Al-Namaeh 2021; Baj et al. 2022; Loffredo et al. 2020; Nasiri et al. 2021; Shetty et al. 2020a; Sopp and Sharda 2021; Taha et al. 2022, 61; Willcox et al. 2020). A previous meta-analysis reported that conjunctivitis can be considered an uncommon manifestation of COVID-19, with a mean prevalence estimation of 6.2% (95% CI of 4.1–9%) (Sopp and Sharda 2021). The first described case of conjunctivitis associated with the COVID-19 disease was unilateral follicular conjunctivitis preceding pneumonia (Casalino et al. 2020).

In most cases, conjunctivitis associated with the SARS-CoV-2 infection is mild, uni- or bilateral, follicular type, without corneal complications, typically lasting for 2–24 days (6 days on average) (Güemes-Villahoz et al. 2020). Conjunctivitis tends to be rarely associated with tarsal hemorrhage, pseudomembranes, mucous filaments, corneal epithelial defects, sub-epithelial infiltrations, and pre-auricular lymphadenopathy (Güemes-Villahoz et al. 2020).

Conjunctivitis can develop at any stage of the COVID-19 disease, presenting as a first sign (Casalino et al. 2020), during the disease, and even after the disease (Akbari and Dourandeesh 2022). Conjunctivitis has also been reported in patients as the first and unique manifestation of the SARS-CoV-2 infection, without any other systemic features (Ozturker 2022; Scalinci and

Table 23.1 Ocular manifestations associated with the COVID-19 disease

Anterior segment
Blepharitis
Dry eye disease
Conjunctival chemosis
Conjunctivitis
Keratoconjunctivitis
Keratitis
Herpes simplex virus (HSV) keratitis
Acute corneal graft rejection
Xerophtalmia
Episcleritis
Angle-closure glaucoma
Posterior segment
Vitritis
Retinal microvascular changes (cotton-wool spots, micro-hemorrhages, hard exudates, perivascular infiltrates, tortuous vessels, hyperreflective lesions at the ganglion cell and internal plexiform macular)
Central retinal vein occlusion
Central retinal artery occlusion
Acute macular neuroretinopathy and paracentral acute middle maculopathy
Outer retinal abnormalities
HSV- or varicella zoster virus-related acute retinal necrosis
Chorioretinitis
Serpiginous choroiditis
Uveitis
Central serous chorioretinopathy
Endogenous bacterial or fungal endophthalmitis
Neuro-ophthalmological manifestations
Adie's tonic pupil
Optic neuritis
Papillophlebitis
Ischemic optic neuropathy
Neurogenic ptosis
Abducens nerve palsy
Oculomotor nerve palsy
Horner's syndrome
Miller–Fisher syndrome
Myastenia gravis
Cerebrovascular accidents with vision loss
Cerebral venous sinus thromboses
Pseudotumor cerebry
Thromboencephalitis
Adnexa and orbital region
Dacryoadenitis
Retro-orbital pain
Orbital inflammatory disease
Bacterial and fungal orbital cellulitis and sinusitis
Orbital mucormycosis
Others
Kawasaki-like syndrome

Trovato Battagliola 2020), suggesting that the COVID-19 disease should be included in the differential diagnosis for patients presenting with conjunctivitis. Thorough attention to ocular manifestations can enhance and anticipate the detection of COVID-19 infection (Shetty et al. 2020a).

Although no specific antivirals for SARS-CoV-2-related ocular surface manifestations are currently available, several topical medications with antiviral properties have been suggested to prevent the ocular infection, even after exposure to the viral particles, including chloroquine, trehalose, antihistamines, interferons, zinc, and azithromycin (Shetty et al. 2020b). Studies have shown that the majority of patients with conjunctivitis associated with the COVID-19 disease resolved with conservative management, without ocular or systemic complications (Güemes-Villahoz et al. 2020).

The other ocular pathologies associated with SARS-CoV-2 infections appear to be rare and with good prognosis with or without appropriate therapy, with the exception to uveitis and vascular occlusions (Baj et al. 2022; Nasiri et al. 2021; Sen et al. 2021; Sopp and Sharda 2021; Taha et al. 2022).

The ocular manifestations of COVID-19 in the pediatric age group can be divided into two groups (Alnahdi and Alkharashi 2023): isolated events associated with the COVID-19 disease, and manifestations occurring in new multisystem inflammatory syndrome in children (MIS-C), a novel entity associated with SARS-CoV-2 infection (Lo and Chen 2021). In both cases, and as seen in adults, conjunctivitis was the most common ocular manifestation (Alnahdi and Alkharashi 2023), present in nearly 50% of patients with MIS-C (Lo and Chen 2021). The ocular manifestations associated with the COVID-19 disease in children tend to show a benign course in the majority of cases, with complete visual recovery (Alnahdi and Alkharashi 2023).

Although the SARS-CoV-2 RNA has been identified in different parts of the eye (Arora et al. 2020; Hada et al. 2021; Koo et al. 2021;

Seah et al. 2020; Sopp and Sharda 2021; Zhou et al. 2020b), its replication and infectivity in the ocular tissues have yet to be established (Sopp and Sharda 2021). The causal or coincidental relationship between numerous reported ocular manifestations and the SARS-CoV-2 virus remains unclear and needs to be clarified. Several hypotheses on the pathophysiology of eye involvement during the COVID-19 disease have been proposed, including the direct effect of the virus on the eye; the tissue damage caused by the hypercoagulability, inflammation, and immune response induced by the virus; the manifestation of underlying diseases; the aggravation of pre-existing systemic conditions; the adverse effect of the medications prescribed against the virus; the result of systemic abnormalities induced by the COVID-disease, including hypoxia and severe hypertension; etc. (Alnahdi and Alkharashi 2023; Baj et al. 2022; Gold and Galetta 2021; Lin et al. 2021; Sen et al. 2021).

Recent studies have demonstrated that the SARS-CoV-2 virus can cause a strong production of pro-inflammatory cytokines, the so-called cytokine storm syndrome or hypercytokinemia, which is an important cause of ARDS and multiple organ failure (Wang and Ma 2008). Moreover, studies have shown an important decrease in the T cells (CD4 and CD8) in patients infected with SARS-CoV-2, suggesting a subsequent reduction of the immune response that can allow for secondary opportunistic infections (Diao et al. 2020).

With regard to physiopathological mechanisms involved during infection, a direct effect of the virus on ocular tissues has been proposed for COVID-19-associated conjunctivitis (5–9% of cases), episcleritis (2.2% of cases), keratitis (0.5% of cases), and blepharitis or lid margin hyperemia (5% of cases) (Baj et al. 2022; Sen et al. 2021). Further studies, however, are needed to confirm this hypothesis. Although the SARS-CoV-2 RNA has been isolated in tears and conjunctival secretions (Arora et al. 2020; Seah et al. 2020; Sopp and Sharda 2021; Zhou et al. 2020b), no cases of SARS-CoV-2-confirmed specific ocular surface manifestations have been reported to date (Sopp and Sharda 2021). This could reflect the direct viral activity or the inflammatory or immune response to the virus or the ocular surface disturbance related to the prevention and therapeutic strategies. Misdiagnoses of an opportunistic conjunctival infection or the presence of a conjunctival reaction as part of a general immune response to the respiratory infection are possible confounding factors that could lead to an overestimated prevalence of COVID-19 ocular surface manifestations.

As underlined by several recent studies, dry eye symptoms have dramatically increased during the SARS-CoV-2 pandemic (Giannaccare et al. 2020; Saldanha et al. 2021; Tatti et al. 2022). Possible explanations of this phenomenon include face mask-wearing, in which the expiratory air is directed toward the eyes, thus accelerating tears evaporation; limited use of lubricating agents due to fear of hand contamination; increased screen and computer time in smart working during the pandemic and lockdown periods, with associated reduction of rate and intensity of eye blinks; loss of follow-up visits in patients with known dry eye conditions; etc.

All ocular findings with underlying thromboembolic pathogenesis, including retinal microvascular changes, retinal vein or artery occlusions, acute macular neuropathy, central acute middle maculopathy, papillophlebitis, cerebrovascular accidents with vision loss, cerebral venous sinus thromboses, are thought to be related to the direct effect of SARS-CoV-2 on the ACE2 receptors present on the endothelial cells, which can induce endothelial dysfunction, hypercoagulability and arterial and venous thromboembolic disease (Abou-Ismail et al. 2020; Sarkar et al. 2021).

In COVID-19 patients, neurological manifestations have been documented in almost 36% of cases, especially in males and young adults. These signs and symptoms mostly include headache, dizziness, anosmia, hypogeusia, Guillain-Barrè syndrome, and ischemic stroke (Gold and Galetta 2021; Tisdale and Chwalisz 2020). The neuro-ophthalmic manifestations, such as Horner's syndrome, Adie's tonic pupil, optic neuritis, neurogenic ptosis, abducens, and oculomotor nerve palsy are attributed to the demonstrated neuro-invasive and neurotrophic abilities of the SARS-CoV-2 virus (Aghagoli et al. 2021) or the inflammation or immune response induced by the virus (Sen et al. 2021).

COVID-19 patients are more susceptible to opportunistic infections for several reasons, such as the compromised immune system with decreased $CD4^{+}$ and $CD8^{+}$ lymphocytes induced by the virus; the reduced pulmonary function; the use of immunosuppressive therapy used to manage severe cases (corticosteroids and immunosuppressant agents); etc. Associated comorbidities, such as diabetes mellitus and hypertension, can act as co-factors. All these conditions can explain the activation or reactivation of the following: herpes simplex virus (HSV) keratitis or kerato-conjunctivitis; varicella-zoster virus (VZV)-related acute retinal necrosis; infection-related vitritis and uveitis; bacterial and fungal orbital cellulitis and sinusitis; and of the potentially life-threatening orbital mucormycosis (Sen et al. 2021).

Disproportionate immune responses or autoimmunity activated by the SARS-CoV-2 virus are believed to play a role in several ocular manifestations, such as some immune-mediated-like kerato-conjunctivitis, acute corneal graft rejection, delayed retinal vascular occlusions that develop few weeks after the onset of the COVID-19 systemic symptoms, serpiginous choroiditis, and other uveitis or retinochoroiditis, optic neuritis, myasthenia gravis, immune-mediated thrombo-encephalitis, Kawasaki-like disease in children, and post-COVID Miller–Fisher Syndrome (Gold and Galetta 2021; Sen et al. 2021).

Kawasaki-like disease is a specific vasculitis of the small and medium vessels that has been associated with COVID-19 infection in children (Verdoni et al. 2020). The typical ocular manifestations include conjunctivitis, subconjunctival hemorrhage, punctate keratitis, iridocyclitis, vitreous opacities, and papilledema. Most patients have a complete clinical recovery in a few days, but 0.5–5% can present toxic shock (Verdoni et al. 2020).

The post-COVID Miller–Fisher Syndrome, characterized by neurological symptoms usually observed 5–20 days after the COVID-19 diagnosis, is probably related to an immune-mediated injury. The most common symptoms include paresthesia, ophthalmoplegia, blurred vision, ataxia, and areflexia (Reyes-Bueno et al. 2020).

The local and systemic inflammation induced by the SARS-CoV-2 infection, with the well-known "cytokine storm", and the compromised blood-ocular barrier, especially the blood-retinal barrier, can exacerbate all previous clinical conditions, and explain other manifestations such as the conjunctival chemosis, described in almost 15% of the COVID-patients, dacryoadenitis, retro-orbital pain, orbital inflammatory disease in children, papillophlebitis, retinochoroiditis, and uveitis. It is also important to note that prevention and therapeutic methods used in COVID-19 patients may explain some ocular manifestations.

23.4 Ocular Manifestations from the COVID-19 Treatment Measures and Vaccines

23.4.1 Ocular Side Effects After Prevention/Treatment Measures

Since no definite consensus about the treatment of the COVID-19 disease has been reached by the medical community, several different drugs have been used during the pandemic for clinical and research settings to prevent and/or treat the disease manifestations, geared especially at attenuating the cytokine storm induced by the SARS-CoV-2 virus (Ferreira-da-Silva et al. 2022). Chloroquine/hydroxychloroquine, systemic corticosteroids, intravenous immunoglobulin, and antiviral agents have been the most commonly used drugs during the pandemic (Ferreira-da-Silva et al. 2022). Although several adverse effects have been reported for drugs used in COVID-19 patients, there is limited data on causality assessment (Ferreira-da-Silva et al. 2022). There are currently seven drugs that have been approved in Europe for COVID-19 patients, which include: remdesivir, dexamethasone, anakinra, regdanvimab, tocilizumab, casirivimab/imdevimab, and sotrovimab (Asili et al. 2021).

As summarized in Table 23.2, both preventative measures (with the exclusion of the vaccines) and drugs used in managing the COVID-19 disease may have ocular side effects. The routine use of face masks, during which the expiratory air is directed toward the eyes and the periocular area, may increase the risk of infections, which is also possible during procedures and intravitreal injections (Raevis et al. 2021). Using masks can also accelerate tear evaporation, thereby inducing or increasing dry eye symptoms (Giannaccare et al. 2020; Pandey and Sharma 2021). Furthermore, during the pandemic, several studies have reported an increase in accidental ocular injuries in the pediatric population due to contact with alcohol-based hand sanitizers, some of which have been reported to be severe and required surgical intervention (Yangzes et al. 2021).

With regard to medications used in COVID-19 patients, only a few reports of drug-induced ophthalmic side effects have been published, which include: transient blurred vision associated with the use of hydroxychloroquine and irvemectin (Ferreira-da-Silva et al. 2022); and central serous chorioretinopathy (Mohd-Alif et al. 2022), orbital mucormycosis and endogenous endophthalmitis (Khatwani et al. 2021)

Table 23.2 Possible ocular adverse events related to preventive measures/drugs used in COVID-19 patients

Preventive measure/drug	Ocular adverse events
Face mask	Dry eye syndrome, periocular infections
Alcohol-based hand sanitizers	Blepharitis, conjunctivitis, keratitis, corneal erosion
Systemic corticosteroids (dexamethasone)	Early-onset nuclear and posterior subcapsular cataract, ocular hypertension, open-angle glaucoma, opportunistic infections, central serous chorioretinopathy, new cases of diabetes mellitus and progression in the established ones, with a consequently increase of diabetes mellitus ocular complications, such as cataracts and diabetic retinopathy
Chloroquine/hydroxychloroquine	Retinal toxicity with bull's eye maculopathy; Vogt–Koyanagi–Harada (VKH) disease, conjunctivitis, uveitis, optic neuropathy, Sjoegren syndrome, posterior subcapsular lens opacity, ciliary body dysfunction, and whorl-like corneal intraepithelial deposits
Anti-retroviral medications (Lopinavir/Ritonavir combination)	Bull's eye or granular pattern maculopathy, crystalline intra-retinal deposits, pigment alterations in the mid-peripheral retina, macular atrophy
Ribavirin	Retinopathy, retinal vein occlusion, serous retinal detachment, non-arteritic ischemic optic neuropathy, VKH disease
Interferon-beta-1	Retinopathy, VKH disease, conjunctivitis, uveitis, optic neuropathy, corneal ulcers, Sjoegren syndrome
Interleukine-1 inhibitors (such as anakinra)	Nystagmus
Interleukine-6 inhibitors (such as sarilumab, siltuximab, tocilizumab)	Bilateral papilledema, uveitis, ophthalmic herpes zoster infection, retinopathy with cotton-wool spots and retinal hemorrhages
IVIG immunoglobuline	Central retinal vein occlusion
Positive end-expiratory pressure ventilation	Orbital and periocular emphysema
Prone position	Orbital compartment syndrome, exposure keratopathy, angle-closure glaucoma

associated with high doses of intravenous corticosteroids treatments.

Several other ocular manifestations, such as dry eye, pain, conjunctival hyperemia, chemosis exposure, and microbial keratitis have been associated with critically ill patients. Studies have reported that more than 60% of patients admitted to the intensive care units (ICU) for causes other than SARS-CoV-2 virus (i.e., long periods of ICU admission, use of sedation, renal and cardiopulmonary failure, mechanical respiratory support, carbon dioxide retention, etc.) can show ocular signs and symptoms (Ferreira-da-Silva et al. 2022; Sen et al. 2021). Several conditions used in ICU patients have been reported to favor the onset of acute angle-closure glaucoma, which include treatment with salbutamol, noradrenaline, and other drugs used in COVID-19-related pneumonia, and darkroom conditions and prone position (used to improve oxygenation in critically ill COVID-19 patients) (Ferreira-da-Silva et al. 2022).

23.4.2 COVID-19 Vaccine-Associated Ocular Adverse Effects

The publication of the SARS-CoV-2 virus genome sequence in January 2020 allowed for the development of a preventive vaccine to fight the SARS-CoV-2 pandemic (Poland et al. 2020). The process culminated in two recombinant mRNA vaccines (Pfizer/BioNTech BNT162b2 and Moderna mRNA-1273), approved in December 2020 by the Food and Drug Administration and by the WHO with Emergency Use Authorizations (EUA) (Creech et al. 2021). Since then, several COVID-19 vaccines have been developed in different countries, including Germany, the Czech Republic, Belgium, Spain, India, China, the Republic of Korea, etc.

The most important vaccine antigen of the SARS-CoV-2 virus is represented by the spike S glycoprotein (Seah et al. 2020). A total of 11 COVID-19 vaccines have been approved to date by the EUA, which can be subdivided into four types: mRNA vaccines (BNT162b2, Pfizer-BioNTech14; mRNA-1273, Moderna15); whole inactivated virus vaccines (PiCoVacc, Sinovac19; BBIBP-CorV, Sinopharm20); protein subunit vaccines (NXV-CoV2373, Novavax16); and adenovirus vector vaccines (Ad26.COV2, Janssen Johnson and Johnson17; ChAdOx1 nCoV-19/AZD1222, Oxford-Astra-Zeneca18) (Forchette et al. 2021).

As of January 2023, more than 13 billion vaccine doses have been administered worldwide (WHO Coronavirus (COVID-19) Dashboard 2022), with over 5 billion individuals fully vaccinated. COVID-19 vaccines have been demonstrated to be effective in protecting individuals from hospitalization, serious illness, and death (Soiza et al. 2021).

Since the widespread administration of COVID-19 vaccinations, several adverse effects have been reported (Sharif et al. 2021). The most commonly reported COVID-19 vaccination side effect is the injection site's local reaction followed by several non-specific flu-like symptoms that usually start within 24–48 h of the inoculation and last for 1–2 days (Sharif et al. 2021). Numerous other systemic and organ-specific adverse manifestations, however, have been also described in the literature, which include rare yet severe manifestations such as cerebral venous thrombosis, immune thrombocytopenia, and acute myocarditis (Hippisley-Cox et al. 2021).

As summarized in Table 23.3, different ocular adverse effects associated with vaccinations and busters have been published in the literature (Ichhpujani et al. 2022; Lee and Huang 2021; Lin et al. 2021; Ng et al. 2021, 2022; Nyankerh et al. 2022; Park et al. 2022; Sen and Honavar 2021; Testi et al. 2022; Yasaka et al. 2022; Wang et al. 2022). The types of vaccines associated with ocular adverse effects to date have been the BNT162b2 (Pfizer-BioNTecn14), mRNA-1273 (Moderna), AZD1222 (Astra-Zeneca), and Ad26.COV2.S (Johnson and Johnson) (Lin et al. 2021).

Similar ocular adverse events have been reported after other vaccines, such as vaccinations against influenza, yellow fever, hepatitis B,

Table 23.3 Ocular manifestations associated with the COVID-19 vaccination

Anterior segment
Activation or reactivation of herpes simplex keratitis
Reactivation of herpes simplex keratitis on corneal graft
Acute corneal allograft rejection
Corneal melting
Episcleritis and anterior scleritis
Anterior uveitis
Angle-closure glaucoma
Posterior segment
Central and branch retinal vein occlusion
Acute zonal occult outer retinopathy (AZOOR)
Acute macular neuroretinopathy
Paracentral acute middle maculopathy
Acute retinal necrosis for VZV reactivation
Multifocal choroiditis
Central serous chorioretinopathy
Vogt–Koyanagi–Harada disease onset or reactivation
Sarcoidosis
Toxoplasma retinochoroiditis
Intermediate uveitis
Other posterior uveitis
Panuveitis
Neuro-ophthalmological manifestations
Optic neuritis
Anterior ischemic optic neuropathy
Abducens nerve palsy
Oculomotor nerve palsy
Trochlear palsy
Bell's palsy
Ramsay–Hunt syndrome
Vertical gaze palsy
Cerebral venous sinus thrombosis
Pseudotumor cerebry
Superior ophthalmic vein thrombosis
Miller–Fisher syndrome
Myastenia gravis
Adnexa and orbital region
Blepharitis, transient eyelid edema or purpuric eyelid lesion
Herpes zoster ophthalmicus
Graeves' disease
Tolosa–Hunt syndrome
Orbital apex syndrome
Myositis

Neisseria meningitides, tetanus, and human papillomavirus (Marinho et al. 2019).

In general, the ocular manifestations begin, on average, about 4–9 days after the COVID-19 vaccination (ranging from 1 to 60 days), with the majority of cases occurring normally within 3 weeks after the vaccine (Ichhpujani et al. 2022; Ng et al. 2021; Park et al. 2022; Sen and Honavar 2021). The signs and symptoms tend to be mild and transient, with or without treatment (Ichhpujani et al. 2022; Sen and Honavar 2021).

The most frequent ocular symptoms associated with the vaccines are blurred vision, eye pain, and flashes, with an incidence ranging from 0.25 to 3.5% of cases (Sen and Honavar 2021). Retina and uvea seem to be the most frequently affected eye tissues (Ichhpujani et al. 2022; Lin et al. 2021; Mack and Fraser-Bell 2021; Ng et al. 2021; Sen and Honavar 2021; Yasaka et al. 2022; Wang et al. 2022). Corneal graft rejection is the most commonly reported anterior segment manifestation (Ichhpujani et al. 2022; Ng et al. 2022; Sen and Honavar 2021).

Ocular adverse events post-COVID-19 vaccination have a remarkably low incidence (Ichhpujani et al. 2022; Ng et al. 2022; Sen and Honavar 2021). A recent systematic review collected 158 cases among a total of 12,640,866,343 vaccine doses administered up till September 19, 2022 (Ichhpujani et al. 2022). The reported prevalence rate of post-vaccination adverse events per million doses included: ≤ 0.1 cases of corneal graft rejection; ≤ 0.05 cases for Herpes Simplex Virus activation, ≤ 0.5 cases for Herpes Zoster Virus activation, ≤ 0.2 cases for scleritis, ≤ 0.3 cases for episcleritis, ≤ 0.3 cases for retinopathies, ≤ 1 cases of optic neuropathy, and ≤ 0.9 for uveitis (Ichhpujani et al. 2022; Wang et al. 2022).

The causal relationship between ocular manifestations and the COVID-19 vaccines has not been established and it is strongly debated, which requires long-term and large multicenter data. Although several cases of retinal vein and artery occlusion have been reported in the literature (Ichhpujani et al. 2022; Ng et al. 2022; Sen and Honavar 2021), a recent retrospective multicenter study did not find any evidence of a causal association between SARS-CoV-2 vaccination and a higher risk of retinal vascular occlusion disease (Feltgen et al. 2022).

Several mechanisms have been suggested to explain the causal association between ocular manifestations and COVID-19 vaccinations

(Ichhpujani et al. 2022; Forchette et al. 2021; Yasaka et al. 2022; Wang et al. 2022; Watad et al. 2021), which include:

- upregulation of pro-inflammatory molecules like cytokines, chemokines, and complement system;
- immune dysregulation with decreases of lymphocytes, monocytes, eosinophils, and T helper cells $CD4^+$ and $CD8^+$, which could explain for example the reactivation of the HSV (Richardson-May et al. 2021) or VZV;
- hyperstimulation of the host immune system;
- molecular mimicry due to the similarity between vaccine fragments, including adjuvants and excipients, and ocular peptides with the induction of an autoimmune response. Ex-Novo Vogt–Koyanagi–Harada cases have been explained as the result of the molecular mimicry between peptide fragments of the vaccine and the uveal tissue (Wang et al. 2022). Fundamental importance seems to be the production of auto-antibodies against the platelets and the platelet factor 4, with consequent vaccine-induced immune thrombotic thrombocytopenia (VITT), which may explain the vascular occlusions (Lee et al. 2021);
- delayed hypersensitivity with deposition of Ag-Ab immune complexes;
- failure of the immune privilege related to the hemato-ophthalmic barrier;
- anemia, hypertension, hypotension, hypoxia, and other systemic comorbidities induced by the vaccines.

The current literature suggests a substantial overlap between the ocular manifestation associated with the COVID-19 infection and COVID-19 vaccination, suggesting that virus- and vaccine-mediated immune responses in humans follow a common pathway (Ng et al. 2022). This is evident when comparing data from Tables 23.1 and 23.4. Bilateral uveitis, for example, has been described to begin acutely within 2 weeks in both infected patients and post-vaccination (Mazzotta and Giancipoli 2020). Moreover, post-vaccination ocular manifestations seem to target predisposed or genetically susceptible individuals or subjects having undiagnosed autoimmune conditions (Ng et al. 2021, 2022; Sen and Honavar 2021; Wang et al. 2022). A previous multicenter observational study showed that uveitis post-COVID-19 vaccination occurred in none of the healthy subjects, however, was reported in 0.3% of patients affected by an autoimmune inflammatory rheumatic disease (Furer et al. 2021). These data suggest that it could be extremely useful to screen the characteristics of the patients at high risk of developing ocular adverse effects in response to COVID-19 infection or vaccination. Patients with autoimmune diseases, including Graves' disease and non-infective uveitis, are known to be in a pro-inflammatory state with a predisposition of ocular inflammation; thus require more careful clinical attention for possible ocular and/or systemic adverse manifestations (Furer et al. 2021; Ng et al. 2022).

Corneal graft rejection has been reported in several patients after COVID-19 vaccination (Moura-Coelho et al. 2022). It has thus been suggested in the literature that elective corneal transplants should be postponed till at least after 3–6 months after the final dose of the vaccine. Moreover, increasing immune suppressants in the peri-vaccination period in patients with corneal grafts may decrease the risk of immune rejection, especially in those at high-risk (Moura-Coelho et al. 2022; Sen and Honavar 2021). The low prevalence rates of post-vaccination ophthalmic HSV and HZV exclude the need for prophylactic antiviral therapy for patients with previous herpetic eye diseases in the peri-vaccination period (Wang et al. 2022).

The majority of the COVID-19 vaccine-related adverse ocular manifestations reported in the literature have been treated conservatively or with steroidal or non-steroidal immunosuppressants, which tend to show a good to fair prognosis, except for cases of retinal and ophthalmic vascular occlusions and graft rejections (Sen and Honavar 2021).

The data from the pharmacovigilance surveillance systems suggest that the prevalence of vaccination-associated ocular adverse events is very rare, and patients with ocular signs and symptoms tend to have a good prognosis. There is currently no evidence to avoid the COVID-19 vaccination, even in patients with pre-existing ophthalmic conditions (Ichhpujani et al. 2022; Ng et al. 2021). Patients having a corneal graft, previous uveitis, other inflammatory eye diseases, or autoimmune diseases need to be educated about the most common signs and symptoms of possible ocular adverse events after vaccination (blurring of vision, decreased contrast sensitivity, photopsia, scotomas, and floaters) and should seek immediate medical attention for acute disorders (Sen and Honavar 2021).

23.5 Ophthalmic Care, Education, and Research During the COVID-19 Pandemic and in the Post-COVID Era

The outbreak of the new COVID-19 pandemic in early 2020 had a great social and economic impact worldwide. Extraordinary restriction measures for the population were decided by many governments, and a complete reorganization of the different national health systems was necessary for numerous countries. To prevent the spreading of the infection and to reduce the pressure on the national health services, social distancing and other preventative measures, as well as national lockdown periods, were imposed in several ways in different periods in numerous countries. The restrictions during the lockdown periods included strict home isolation unless strictly necessary; travel restrictions between different municipalities except for emergencies, work, or health reasons; and, closure of all commercial and productive activities, with the exception of those strictly necessary for survival. Hospitals were forced to halt all nonessential activities, limit the access of patients, and dedicate most resources to treating symptomatic COVID-19 patients.

Recommendations on general precautionary measures to limit virus transmission were proposed by several international health organizations, including the WHO (2020). Guidelines to prevent COVID-19 transmission in both patients and healthcare workers in the ophthalmic practice were immediately elaborated by the major ophthalmological societies worldwide (Chodosh et al. 2022; Li et al. 2020; Wong et al. 2020) because, since the beginning of the pandemic, eye care providers were shown to be at high risk of contracting the SARS-CoV-2 infection (Danesh-Meyer and McGhee 2021; Li et al. 2020; Romano et al. 2020; Shah et al. 2021; Shetty et al. 2020a; Wan et al. 2020). The factors for the high-risk label include: close proximity to the patient, especially during slit lamp examination and direct ophthalmoscopy, with increased exposure to respiratory droplets; possible contact with the infected nose, mouth, and ocular secretions; daily contact with large numbers of patients; prevalent interaction with a particular population subgroup, i.e., aging people, at high risk of comorbidities and of COVID-19 severe disease, etc. (Li et al. 2020; Romano et al. 2020; Shah et al. 2021; Shetty et al. 2020a; Wan et al. 2020). A summary of the prevention guidelines in ophthalmic practice is provided in Table 23.4.

In response to the SARS-CoV-2 pandemic, the ophthalmic community needed to deeply modify clinical practice modalities, due to the increasing pressure on the healthcare system, the restrictions imposed by the governments, and the new ocular manifestations related to the COVID-19 infection, treatments, and vaccination, so that a "new ophthalmology in the Covid-19 era" took form.

The most important consequences induced by the COVID-19 pandemic on the eye care system include:

a. Many ophthalmological departments worldwide were invited to defer or drastically reduce elective visits, procedures, and surgeries, and to provide only mandatory assistance for acute and chronic sight-threatening conditions and emergency eye care service, which precluded many patients from attending appointments for visits or surgery. A high number of routine out-patient visits, ranging approximately from 75 to 95%, were

Table 23.4 Recommendations to prevent COVID-19 infection in the ophthalmic practice

Outpatient practice
To test and isolate healthcare workers showing any signs of affection
To limit the outpatient visits, to postpone non-urgent visits and to visit physically only emergencies such as trauma, ocular infections, acute angle-closure glaucoma, retinal arterial or venous occlusions, retinal detachment, uveitis, or other entities that can lead to irreversible vision loss
To screen patients for SARS-CoV-2 infection before the visit, in order to select those with high body temperature and those who had a history of recent international travel, contact with confirmed cases, and any disease symptom, and to postpone the visits of confirmed or highly suspected patients if not strictly necessary
To keep triaged patients in air-conditioning waiting rooms and maximum distance between them and to limit accompanying persons
To decontaminate waiting and examination rooms after visits
To warn patient about the increased risk of infection through the hand-to-face contact
To disinfect surfaces and medical devices following every visit with disinfectant solutions containing 70% alcohol or other solutions according to the manufacturer's instructions
To invite patients to frequent hands disinfection and facial masks wear
To invite patients to switch from contact lenses to glasses, when possible, to reduce the times the patient touch the eyes and because the glasses provide a physical barrier between the respiratory droplets and the ocular surface
To invite ophthalmologists and other eye care professionals of wearing face-mask, face-shield and eye protective gears during clinical examinations
In case of urgent assessment of confirmed or highly suspected patients, or patients that are unable to wear a mask or young children, to invite ophthalmologists to use personal protective equipment (PPE), including respiratory masks (FFP2/N95), waterproof long-sleeved gowns, face shields and gloves
To use protective slit-lamp shields and routinely disinfect them
To use disposable single-use tools, like tonometer (Goldmann applanation tonometer, iCare, TonoPen) tips
To disinfect to devices that come into contact with the patient (i.e., trial frame, trial lenses, Goldmann applanation tonometer tips, diagnostic contact lenses, ultrasound probe) with opportune solutions (i.e., 70% isopropyl alcohol for at least 5 min)
To avoid the use of air-puff tonometers due to the creation of micro-aerosols that can disperse the virus
To prefer indirect ophthalmoscopy, optical coherence tomography, or fundus photography rather than direct ophthalmoscopy
To prefer perimetry with enclosed screen rather than bowl types
To adopt the telemedicine model of clinical care when possible (teleconsultations, online prescription programs, store-and-forward asynchronous models, home-monitoring devices, etc.)
Surgical procedures
To reduce or stop and postpone non-urgent procedures like refraction, cataract, refractive, and squint surgery
To use the minimum number of staff in the operation room
To perform pre-operative RT-PCR test in asymptomatic patients for ophthalmic surgical procedure that may generate aerosol particles, such as phacoemulsification
Prefer local anesthesia when possible, in order to avoid the generation of aerosols during the general anesthesia procedures
In case of confirmed COVID-19 patients that need a surgical procedure, full PPE, use airborne isolation in negative pressure isolation room, experienced surgeon, and familiar procedure

suspended or postponed (Babu et al. 2020; Borrelli et al. 2020; Legrottaglie et al. 2020; Reddy 2021). Meanwhile, patients' insecurity and fear of the infection produced an increase in missed ophthalmological visits worldwide (Brant et al. 2021). The balance between SARS-CoV-2 infection control and adequate ophthalmology care service was critical, considering that some ocular pathologies such as glaucoma, wet age-related macular degeneration, and retinal detachment can cause irreversible loss of visual function if treatment is delayed. Previous studies have demonstrated that the postponement of the in-person follow-up visits and/or their replacement with telematic visits increased the delayed or wrong diagnoses, leading to insufficient therapy, with consequent exacerbation of eye conditions and complications (Attzs and Lakhani 2021; Berkenstock et al. 2021).

b. The Emergency Eye Department (EED) attendance was reduced by more than 50% when compared with previous periods (Pellegrini et al. 2020; Poyser et al. 2020; Salvetat et al. 2022; Wickham et al. 2020). Moreover, the ocular pathologies presented at the EED during the pandemic lockdown registered a significantly higher percentage of retinal detachment and a lower number of retinal tears. The data suggested the possibility of the patient's delayed presentation, and a lower percentage of ocular traumas and conjunctivitis, likely related to hand hygiene practices, social distancing, working activities, and school closures (Poyser et al. 2020; Salvetat et al. 2022).

c. Numerous elective surgical procedures were suspended or postponed (Domínguez-Varela et al. 2021; Tognetto et al. 2020). For example, a 97% reduction in elective, non-urgent cataract surgery was recorded in several European countries (Tognetto et al. 2020; Toro et al. 2021; Tsironi et al. 2022). Cataract surgery performed during the "COVID era" included a significantly high number of advanced white-mature/hypermature and/or bilateral cataracts (Tsironi et al. 2022). Bilateral same-day cataract surgery or telemedicine for the follow-up of patients after cataract surgery was also introduced by some clinics (Domínguez-Varela et al. 2021; Tognetto et al. 2020). The postponement of the major part of the elective cataract surgeries has been demonstrated to significantly decrease the patient's quality of life, and to lead to an aggravation of the cataract clinical presentation, increasing surgery duration and possible complications (Tsironi et al. 2022).

d. Numerous urgent surgical procedures were also suspended or postponed. In Europe, a mean reduction of 95%, 91%, and 76% was reported, respectively, for keratoplasty, glaucoma surgery, and pars plana vitrectomy (Toro et al. 2021).

e. Anti-VEGF and steroids intravitreal injections were reduced by 50–70% in several countries worldwide (Borrelli et al. 2020; dell'Omo et al. 2022). International retinal disease expert guidelines for intravitreal injections during the COVID-19 pandemic included the following recommendations: to prioritize treatments for patients at greater risk of irreversible vision loss (recent wet age-related macular degeneration (AMD), new patients with significant vision loss from recent diabetic macular edema (DME), proliferative diabetic retinopathy, acute-phase retinal vein occlusion (RVO), ischemic RVO, neovascular glaucoma, and monocular or quasi-monocular patients); to simplify intravitreal regimens (to prefer fixed dosing regimen in patients with AMD; to prefer the dexamethasone intravitreal implant in patients with DME/RVO in absence of contraindications); and to consider pan-retinal photocoagulation instead of intravitreal anti-VEGF for patients with severe proliferative diabetic retinopathy (PDR) (Korobelnik et al. 2021). As demonstrated by several studies, the delay of the anti-VEGF therapy induced a significantly negative effect on patients' visual acuity (Stone et al. 2021).

f. Stay-home recommendations, school closures, online school classes, and home-working enormously increased the time of interaction with digital screens, videos, and near-vision tasks, especially among school-age children. Digital eye strain (DES) is the most common eye problem associated with prolonged digital device use (laptops, desktops, smartphones), characterized by several symptoms including foreign body sensation, itching, dry eyes, watering, blurring vision, and headache (Mohan et al. 2021). A significant increase in the prevalence of DES among the general population, especially among children and young adults, has been described during the lockdown periods

(Mohan et al. 2021). A significant increase in the prevalence of myopia in the pediatric population, especially in 6–8-year-old children, has been documented during the COVID-19 pandemic (Li et al. 2022). The causes of the myopic shift of the young populations during the pandemic have been attributed to increased screen time and decreased outdoor activity (Wong et al. 2021). Ophthalmology education and training programs for medical students were suspended or virtually delivered during the pandemic (Mack and Fraser-Bell 2021; Wendt et al. 2021), with a consequent negative impact on the surgical training of the medical students. New virtual approaches to learning ophthalmology were proposed, including virtual teaching conferences; instructional videos or virtual experiences with simulator tools (allowing for example the direct ophthalmoscope examination without direct contact with the patient); e-learning resources for medical students, such as EyeWiki, Eyeguru.org, etc. (Rose 2020; Wendt et al. 2021; Wong and Bandello 2020). Moreover, research and publication about ophthalmological topics slowed worldwide (Mack and Fraser-Bell 2021). The COVID-19 pandemic has temporarily suspended medical student involvement in clinical rotations, resulting in the need to develop virtual approaches to learning ophthalmology.

g. Telemedicine and artificial intelligence gained power in ophthalmology as in several other medical areas (Mack and Fraser-Bell 2021; Saleem et al. 2020; Sommer and Blumenthal 2020; Tham et al. 2022). Technological advances in recent years have allowed for the development and implementation of various modalities and techniques enabling the ophthalmologist, in some contexts, to remotely provide patients eye care and ophthalmology education and training (Mack and Fraser-Bell 2021; Tham et al. 2022; Wendt et al. 2021). Several teleophthalmology models have been described, ranging from triage, screening of eye diseases, consultative services, remote supervision, educational purposes, and emergency services (Table 23.5). Since the start of the COVID-19 pandemic, some countries, such as the USA, UK, India, Hong Kong, and Singapore have promoted telemedicine in ophthalmology (Jayadev et al. 2020). Studies have shown that telemedicine was useful in decreasing the number of clinical visits during the COVID-19 lockdown (Lam et al. 2020a). With the most used teleophthalmology model, the traditional ocular examinations are replaced with investigations conducted outside of the eye clinic by primary care physicians or trained ophthalmic technicians that can perform visual acuity measurement, intraocular pressure measurement, anterior and posterior segment photography, optical coherence tomography (OCT), and perimetry. The clinician can then review the examination results remotely, in real-time or later, thus reducing physical contact between patient and physician. Artificial intelligence may provide a step further. The development of algorithms able to identify lesions in ophthalmic examinations without human intervention could allow the analysis of a large amount of data without the employment of eye care professionals.

h. At present, teleophthalmology is mostly utilized in the screening of retinal conditions such as retinopathy of prematurity, diabetic retinopathy, and age-related macular degeneration; in diagnosing anterior segment and neuro-ophthalmological conditions; and, in managing glaucoma (Sommer and Blumenthal 2020; Tham et al. 2022). These new digital solutions require information technology infrastructures, Internet connectivity, and cybersecurity which are still lacking in some settings (Sommer and Blumenthal 2020). The possible applications of telemedicine and artificial intelligence for ophthalmic care are listed in Table 23.5.

i. Ethical questions about eye care became imperative in the pandemic era, which include: What is the potential visual cost of postponed appointments and elective surgery? In presence of a COVID-positive

Table 23.5 Possible telemedicine and artificial intelligence applications to the ophthalmic care

Video consultation and tele-health platforms allow for anamnesis; patients triage; external adnexal, pupil, ocular motility, alignment, anterior segment, iris, corneal light reflexes examination; and follow-up of patients after cataract surgery
Store-and-forward asynchronous models for anterior and posterior segment images
Remotely controlled slit-lamp devices, non-mydriatic fundus cameras, and optical coherence tomography (OCT) devices situated in public areas
Use of algorithms able to identify lesions in ophthalmic examinations or images without human intervention
Amsler grids or visual acuity applications via printed chats or mobile app
Smartphone anterior and posterior segment imaging
Home OCT devices
Home intraocular pressure measuring devices (iCare tonometer and Sensimed Triggerfish contact lens)
Home computerized perimetry: Rarebit Perimetry (); on-line app-based perimetry Peristat (Keep Your Sight Foundation, Los Angeles, California) (22–24 di 19); preferential hyperacuity perimeter ForSee device (Notal Vision) to monitor the central macular 10°
Home monitoring of the central macular 10° with smartphone applications: mVT (Vital Art and Science), Alleye TM hyperacuity testing, MyVisionTrack TM shape discrimination hyperacuity testing to identify metamorphopsia
Ophthalmology virtual teaching (with conferences, clinical experiences, and simulators) and e-learning resources for ophthalmology students (EyeWiki, Eyeguru.org, Retinoscopy Simulator, etc.)

patient needing an urgent surgical intervention, what are the factors to consider to balance the risk of surgical delay with that of exacerbation of the patient illness, and the risk of infecting the healthcare team? What are the benefits of the redeployment of care personnel, in addition to the ophthalmologists, in the emergency rooms or ICU? (Prager et al. 2021).

23.6 Conclusions

The COVID-19 pandemic has changed all aspects of the eye care system. Some of these effects are still present and will likely modify the healthcare system in the future. The normal "physical" eye care model is not suitable for the COVID-19 pandemic. Considering that COVID-19 is gradually becoming an endemic disease and future contagions may appear in the world scenario, a post-pandemic "new normal" needs to be envisioned and structured (Webster 2020). It is thus likely that several of the measures taken during the COVID-19 infection will continue long after the pandemic, depicting a "new normal" after the COVID-19 pandemic. Although the technology for a remote ophthalmic eye examination or ophthalmology teaching is not available yet in many settings, it is likely that, after the COVID-19 pandemic, teleophthalmology and artificial intelligence will find a more important and permanent role in the healthcare system across the globe.

Financial Disclosures/Support and Conflict of Interests None.

References

Abid A, Khan MA, Lee B, White A, Carnt N, Arshad S, Samarawickrama C (2022) Ocular distribution of the renin-angiotensin-aldosterone system in the context of the SARS-CoV-2 pandemic. J Renin Angiotensin Aldosterone Syst 2022:9970922. https://doi.org/10.1155/2022/9970922. PMID: 35173799; PMCID: PMC8831072

Abou-Ismail MY, Diamond A, Kapoor S, Arafah Y, Nayak L (2020) The hypercoagulable state in COVID-19: incidence, pathophysiology, and management. Thromb Res 194:101–115. https://doi.org/10.1016/j.thromres.2020.06.029. Epub 20 June 2020. Erratum in: Thromb Res 2020. PMID: 32788101; PMCID: PMC7305763

Aggarwal K, Agarwal A, Jaiswal N, Dahiya N, Ahuja A, Mahajan S, Tong L, Duggal M, Singh M, Agrawal R, Gupta V (2020) Ocular surface manifestations of coronavirus disease 2019 (COVID-19): a systematic review and meta-analysis. PLoS ONE 15(11):e0241661. https://doi.org/10.1371/journal.pone.0241661. PMID: 33151999; PMCID: PMC7643964

Aghagoli G, Gallo Marin B, Katchur NJ, Chaves-Sell F, Asaad WF, Murphy SA (2021) Neurological involvement in COVID-19 and potential mechanisms: a review. Neurocrit Care 34(3):1062–1071. https://doi.org/10.1007/s12028-020-01049-4. PMID: 32661794; PMCID: PMC7358290

Akbari M, Dourandeesh M (2022) Update on overview of ocular manifestations of COVID-19. Front Med (Lausanne) 9:877023. https://doi.org/10.3389/fmed.2022.877023. PMID: 36177323; PMCID: PMC9513125

Al-Namaeh M (2021) COVID-19 and conjunctivitis: a meta-analysis. Ther Adv Ophthalmol 13:25158414211003368. https://doi.org/10.1177/25158414211003368. PMID: 34192224; PMCID: PMC7994922

Al-Sharif E, Strianese D, AlMadhi NH, D'Aponte A, dell'Omo R, Di Benedetto R, Costagliola C (2021) Ocular tropism of coronavirus (CoVs): a comparison of the interaction between the animal-to-human transmitted coronaviruses (SARS-CoV-1, SARS-CoV-2, MERS-CoV, CoV-229E, NL63, OC43, HKU1) and the eye. Int Ophthalmol 41(1):349–362. https://doi.org/10.1007/s10792-020-01575-2. Epub 3 Sept 2020. PMID: 32880786; PMCID: PMC7471531

Ali MJ (2020) The SARS-CoV-2, tears, and ocular surface debate: what we know and what we need to know. Indian J Ophthalmol 68(7):1245–1246. https://doi.org/10.4103/ijo.IJO_1881_20. PMID: 32587141; PMCID: PMC7574068

Alnahdi MA, Alkharashi M (2023) Ocular manifestations of COVID-19 in the pediatric age group. Eur J Ophthalmol 33(1):21–28. https://doi.org/10.1177/11206721221116210. Epub 27 July 2022. PMID: 35898182; PMCID: PMC9339424

Amesty MA, Alió Del Barrio JL, Alió JL (2020) COVID-19 disease and ophthalmology: an update. Ophthalmol Ther 9(3):1–12. https://doi.org/10.1007/s40123-020-00260-y. Epub 22 May 2020. PMID: 32445134; PMCID: PMC7243216

Arora R, Goel R, Kumar S, Chhabra M, Saxena S, Manchanda V, Pumma P (2020) Evaluation of SARS-CoV-2 in tears of patients with moderate to severe COVID-19. Ophthalmology 128(4):494–503. https://doi.org/10.1016/j.ophtha.2020.08.029. Epub 31 Aug 2020. PMID: 32882309; PMCID: PMC7458068

Asili P, Mirahmad M, Tabatabaei-Malazy O, Manayi A, Haghighat E, Mahdavi M, Larijani B (2021) Characteristics of published/registered clinical trials on COVID-19 treatment: a systematic review. Daru 29(2):449–467. https://doi.org/10.1007/s40199-021-00422-8. Epub 11 Nov 2021. PMID: 34762250; PMCID: PMC8581284

Attzs MS, Lakhani BK (2021) COVID-19 and its effect on the provision of ophthalmic care in the United Kingdom. Int J Clin Pract 75(7):e14052. https://doi.org/10.1111/ijcp.14052. Epub 1 Feb 2021. PMID: 33527547; PMCID: PMC7994995

Babu N, Kohli P, Mishra C, Sen S, Arthur D, Chhablani D, Baliga G, Ramasamy K (2020) To evaluate the effect of COVID-19 pandemic and national lockdown on patient care at a tertiary-care ophthalmology institute. Indian J Ophthalmol 68:1540–1544

Baj J, Forma A, Teresińska B, Tyczyńska M, Zembala J, Januszewski J, Flieger J, Buszewicz G, Teresiński G (2022) How does SARS-CoV-2 affect our eyes-what have we learnt so far about the ophthalmic manifestations of COVID-19? J Clin Med 11(12):3379. https://doi.org/10.3390/jcm11123379. PMID: 35743449; PMCID: PMC9225256

Ballouz D, Mian SI (2020) Eye banking in the coronavirus disease 2019 era. Curr Opin Ophthalmol 31(5):389–395. https://doi.org/10.1097/ICU.0000000000000684. PMID: 32740064

Berkenstock MK, Liberman P, McDonnell PJ, Chaon BC (2021) Changes in patient visits and diagnoses in a large academic center during the COVID-19 pandemic. BMC Ophthalmol 21(1):139. https://doi.org/10.1186/s12886-021-01886-7. PMID: 33743634; PMCID: PMC7980730

Borrelli E, Grosso D, Vella G, Sacconi R, Querques L, Zucchiatti I, Prascina F, Bandello F, Querques G (2020) Impact of COVID-19 on outpatient visits and intravitreal treatments in a referral retina unit: let's be ready for a plausible "rebound effect." Graefes Arch Clin Exp Ophthalmol 258:2655–2660

Bostanci Ceran B, Ozates S (2020) Ocular manifestations of coronavirus disease 2019. Graefes Arch Clin Exp Ophthalmol 258(9):1959–1963. https://doi.org/10.1007/s00417-020-04777-7. Epub 6 Jun 2020. PMID: 32504100; PMCID: PMC7274940

Brant AR, Pershing S, Hess O, Rooney D, Goldberg J, Tabin G, Wang SY (2021) The impact of COVID-19 on missed ophthalmology clinic visits. Clin Ophthalmol 15:4645–4657. https://doi.org/10.2147/OPTH.S341739. PMID: 34916776; PMCID: PMC8667753

Cao K, Kline B, Han Y, Ying GS, Wang NL (2020) Current evidence of 2019 novel coronavirus disease (COVID-19) ocular transmission: a systematic review and meta-analysis. Biomed Res Int 2020:7605453. https://doi.org/10.1155/2020/7605453. PMID: 33134387; PMCID: PMC7588829

Casalino G, Monaco G, Di Sarro PP et al (2020) Coronavirus disease 2019 presenting with conjunctivitis as the first symptom. Eye 34:1235–1236

Chodosh J, Holland GN, Yeh S (2022) Important coronavirus updates for ophthalmologists. American Academy of Ophthalmology. Available at: https://www.org/headline/alert-important-coronavirus-context

Chu DK, Akl EA, Duda S, Solo K, Yaacoub S, Schünemann HJ (2020) COVID-19 Systematic Urgent Review Group Effort (SURGE) study authors. Physical distancing, face masks, and eye protection to prevent person-to-person transmission of SARS-CoV-2 and COVID-19: a systematic review and meta-analysis. Lancet 395(10242):1973–1987. https://doi.org/10.1016/S0140-6736(20)31142-9. Epub 1 June 2020. PMID: 32497510; PMCID: PMC7263814

Colavita F, Lapa D, Carletti F, Lalle E, Bordi L, Marsella P, Nicastri E, Bevilacqua N, Giancola ML, Corpolongo A, Ippolito G, Capobianchi MR, Castilletti C (2020) SARS-CoV-2 isolation from ocular secretions of a patient with COVID-19 in Italy with prolonged viral RNA detection. Ann Intern Med 173(3):242–243. https://doi.org/10.7326/M20-1176. Epub 17 Apr 2020. PMID: 32302380; PMCID: PMC7175424

Creech CB, Walker SC, Samuels RJ (2021) SARS-CoV-2 vaccines. JAMA 325:1318–1320

Danesh-Meyer HV, McGhee CN (2021) Implications of coronavirus disease 2019 for ophthalmologists. Am J Ophthalmol 223:108–118

Davis G, Li K, Thankam FG, Wilson DR, Agrawal DK (2022) Ocular transmissibility of COVID-19: possibilities and perspectives. Mol Cell Biochem 477(3):849–864. https://doi.org/10.1007/s11010-021-04336-6. Epub 23 Jan 2022. PMID: 35066705; PMCID: PMC8783769

Day M (2020) Covid-19: identifying and isolating asymptomatic people helped eliminate virus in Italian village. BMJ 368:m1165. https://doi.org/10.1136/bmj.m1165. PMID: 32205334

dell'Omo R, Filippelli M, Virgili G, Bandello F, Querques G, Lanzetta P, Avitabile T, Viola F, Reibaldi M, Semeraro F, Quaranta L, Rizzo S, Midena E, Campagna G, Costagliola C (2022) Eyecare in Italy during COVID-19 pandemic (EICO) study group. Effect of COVID-19-related lockdown on ophthalmic practice in Italy: a report from 39 institutional centers. Eur J Ophthalmol 32(1):695–703. https://doi.org/10.1177/11206721211002442. Epub 16 Mar 2021. PMID: 33724078

Deng C, Yang Y, Chen H, Chen W, Chen Z, Ma K, Wang J (2020a) Low risk of SARS-CoV-2 transmission through the ocular surface. Acta Ophthalmol 98(7):e926–e927. https://doi.org/10.1111/aos.14471. Epub 21 May 2020. PMID: 32436625; PMCID: PMC7280640

Deng W, Bao L, Gao H, Xiang Z, Qu Y, Song Z, Gong S, Liu J, Liu J, Yu P, Qi F, Xu Y, Li F, Xiao C, Lv Q, Xue J, Wei Q, Liu M, Wang G, Wang S, Yu H, Chen T, Liu X, Zhao W, Han Y, Qin C (2020b) Ocular conjunctival inoculation of SARS-CoV-2 can cause mild COVID-19 in rhesus macaques. Nat Commun 11(1):4400. https://doi.org/10.1038/s41467-020-18149-6. PMID: 32879306; PMCID: PMC7467924

der Hoek LV (2007) Human coronaviruses: what do they cause? Anitivir Ther 12(4Pt B):651–658

Dhiman R, Rakheja V, Saxena R (2022) An ophthalmologist's insight into the viral pandemics. J Optom 15(1):35–43. https://doi.org/10.1016/j.optom.2020.10.005. Epub 6 Jan 2021. PMID: 33422421; PMCID: PMC8045749

Diao B, Wang C, Tan Y, Chen X, Liu Y, Ning L, Chen L, Li M, Liu Y, Wang G, Yuan Z, Feng Z, Zhang Y, Wu Y, Chen Y (2020) Reduction and functional exhaustion of T cells in patients with coronavirus disease 2019 (COVID-19). Front Immunol 11:827. https://doi.org/10.3389/fimmu.2020.00827. PMID: 32425950; PMCID: PMC7205903

Dockery DM, Rowe SG, Murphy MA, Krzystolik MG (2020) The ocular manifestations and transmission of COVID-19: recommendations for prevention. J Emerg Med 59(1):137–140. https://doi.org/10.1016/j.jemermed.2020.04.060. Epub 8 May 2020. PMID: 32456959; PMCID: PMC7205711

Domínguez-Varela IA, Ortiz-Morales G, Gómez-Elizondo DE, Nava-García JA, Valdez-García JE (2021) Ophthalmological care management of cataract patients during the COVID-19 times: perspectives from a Mexican eye health care system. Asia Pac J Ophthalmol (Phila) 10(6):598. https://doi.org/10.1097/APO.0000000000000448. PMID: 34789675; PMCID: PMC8673855

Eissa M, Abdelrazek NA, Saady M (2022) Covid-19 and its relation to the human eye: transmission, infection, and ocular manifestations. Graefes Arch Clin Exp Ophthalmol 1–10. https://doi.org/10.1007/s00417-022-05954-6. Epub ahead of print. PMID: 36585987; PMCID: PMC9803899

Farvardin M, Johari M, Tahamtan M, Najafi MT, Farvardin R, Rao KSJ, Nami M (2021) Ophthalmic manifestations of COVID-19; a less-appreciated yet significant challenge. Int Ophthalmol 41(3):1141–1147. https://doi.org/10.1007/s10792-020-01664-2. Epub 2 Jan 2021. PMID: 33389366; PMCID: PMC7778483

Feltgen N, Ach T, Ziemssen F, Quante CS, Gross O, Abdin AD, Aisenbrey S, Bartram MC, Blum M, Brockmann C, Dithmar S, Friedrichs W, Guthoff R, Hattenbach LO, Herrlinger KR, Kaskel-Paul S, Khoramnia R, Klaas JE, Krohne TU, Lommatzsch A, Lueken S, Maier M, Nassri L, Nguyen-Dang TA, Radeck V, Rau S, Roider J, Sandner D, Schmalenberger L, Schmidtmann I, Schubert F, Siegel H, Spitzer MS, Stahl A, Stingl JV, Treumer F, Viestenz A, Wachtlin J, Wolf A, Zimmermann J, Schargus M, Schuster AK (2022) Retinal vascular occlusion after COVID-19 vaccination: more coincidence than causal relationship? Data from a retrospective multicentre study. J Clin Med 11(17):5101. https://doi.org/10.3390/jcm11175101. PMID: 36079030; PMCID: PMC9457026

Ferreira-da-Silva R, Ribeiro-Vaz I, Morato M, Junqueira Polónia J (2022) A comprehensive review of adverse events to drugs used in COVID-19 patients: recent clinical evidence. Eur J Clin Invest 52(7):e13763. https://doi.org/10.1111/eci.13763. Epub 27 Feb 2022. PMID: 35224719; PMCID: PMC9111855

Forchette L, Sebastian W, Liu T (2021) A comprehensive review of COVID-19 virology, vaccines, variants, and therapeutics. Curr Med Sci 41(6):1037–1051. https://doi.org/10.1007/s11596-021-2395-1. Epub 9 July 2021. PMID: 34241776; PMCID: PMC8267225

Forouhari A, Mansouri V, Safi S, Ahmadieh H, Ghaffari Jolfayi A (2022) A systematic literature review and bibliometric analysis of ophthalmology and COVID-19 research. J Ophthalmol 2022:8195228. https://doi.org/10.1155/2022/8195228. PMID: 35646394; PMCID: PMC9133895

Fung TS, Liu DX (2019) Human coronavirus: host-pathogen interaction. Annu Rev Microbiol 73:529–557

Furer V, Eviatar T, Zisman D, Peleg H, Paran D, Levartovsky D, Zisapel M, Elalouf O, Kaufman I, Meidan R, Broyde A, Polachek A, Wollman J, Litinsky I, Meridor K, Nochomovitz H, Silberman A, Rosenberg D, Feld J, Haddad A, Gazzit T, Elias M, Higazi N, Kharouf F, Shefer G, Sharon O, Pel S, Nevo S, Elkayam O (2021) Immunogenicity and safety of the BNT162b2 mRNA COVID-19 vaccine in adult patients with autoimmune inflammatory rheumatic diseases and in the general population: a multicentre study. Ann Rheum Dis 80(10):1330–1338. https://doi.org/10.1136/annrheumdis-2021-220647. Epub 14 June 2021. Erratum in: Ann Rheum Dis 81(7):e133. PMID: 34127481; PMCID: PMC8206170

Giannaccare G, Vaccaro S, Mancini A, Scorcia V (2020) Dry eye in the COVID-19 era: how the measures for controlling pandemic might harm ocular surface. Graefes Arch Clin Exp Ophthalmol 258(11):2567–2568. https://doi.org/10.1007/s00417-020-04808-3. Epub 19 June 2020. PMID: 32561978; PMCID: PMC7304378

Gold DM, Galetta SL (2021) Neuro-ophthalmologic complications of coronavirus disease 2019 (COVID-19). Neurosci Lett 742:135531. https://doi.org/10.1016/j.neulet.2020.135531. Epub 25 Nov 2020. PMID: 33248158; PMCID: PMC7687583

Guan WJ et al (2020) Clinical characteristic of coronavirus disease 2019 in China. N Engl J Med 18:1708–1720

Güemes-Villahoz N, Burgos-Blasco B, García-Feijoó J, Sáenz-Francés F, Arriola-Villalobos P, Martinez-de-la-Casa JM, Benítez-Del-Castillo JM, Herrera de la Muela M (2020) Conjunctivitis in COVID-19 patients: frequency and clinical presentation. Graefes Arch Clin Exp Ophthalmol 258(11):2501–2507. https://doi.org/10.1007/s00417-020-04916-0. Epub 29 Aug 2020. PMID: 32860573; PMCID: PMC7455778

Guo YR, Cao QD, Hong ZS, Tan YY, Chen SD, Jin HJ, Tan KS, Wang DY, Yan Y (2020) The origin, transmission and clinical therapies on coronavirus disease 2019 (COVID-19) outbreak—an update on the status. Mil Med Res 7(1):11. https://doi.org/10.1186/s40779-020-00240-0. PMID: 32169119; PMCID: PMC7068984

Hada M, Khilnani K, Vyas N, Chouhan JK, Dharawat KS, Bhandari S, Tripathy K (2021) Evaluating the presence of SARS-CoV-2 in the intraocular fluid of COVID-19 patients. Indian J Ophthalmol 69(9):2503–2506. https://doi.org/10.4103/ijo.IJO_820_21. PMID: 34427253; PMCID: PMC8544035

Hippisley-Cox J, Patone M, Mei XW, Saatci D, Dixon S, Khunti K, Zaccardi F, Watkinson P, Shankar-Hari M, Doidge J, Harrison DA, Griffin SJ, Sheikh A, Coupland CAC (2021) Risk of thrombocytopenia and thromboembolism after COVID-19 vaccination and SARS-CoV-2 positive testing: self-controlled case series study. BMJ 374:n1931. https://doi.org/10.1136/bmj.n1931. PMID: 34446426; PMCID: PMC8388189

Hirano T, Murakami M (2020) COVID-19: a new virus, but a familiar receptor and cytokine release syndrome. Immunity 52(5):731–733. https://doi.org/10.1016/j.immuni.2020.04.003. Epub 22 Apr 2020. PMID: 32325025; PMCID: PMC7175868

Hoffmann M, Kleine-Weber H, Schroeder S, Krüger N, Herrler T, Erichsen S, Schiergens TS, Herrler G, Wu NH, Nitsche A, Müller MA, Drosten C, Pöhlmann S (2020) SARS-CoV-2 cell entry depends on ACE2 and TMPRSS2 and is blocked by a clinically proven protease inhibitor. Cell 181(2):271–280.e8. https://doi.org/10.1016/j.cell.2020.02.052. Epub 5 Mar 2020. PMID: 32142651; PMCID: PMC7102627

Holappa M (2015) Ocular renin-angiotensin system with special reference in the anterior part of the eye. WJO 5:110

Hui KPY, Cheung MC, Perera RAPM, Ng KC, Bui CHT, Ho JCW, Ng MMT, Kuok DIT, Shih KC, Tsao SW, Poon LLM, Peiris M, Nicholls JM, Chan MCW (2020) Tropism, replication competence, and innate immune responses of the coronavirus SARS-CoV-2 in human respiratory tract and conjunctiva: an analysis in ex-vivo and in-vitro cultures. Lancet Respir Med 8(7):687–695. https://doi.org/10.1016/S2213-2600(20)30193-4. Epub 7 May 2020. PMID: 32386571; PMCID: PMC7252187

Ichhpujani P, Parmar UPS, Duggal S, Kumar S (2022) COVID-19 vaccine-associated ocular adverse effects: an overview. Vaccines (Basel) 10(11):1879. https://doi.org/10.3390/vaccines10111879. PMID: 36366386; PMCID: PMC9697513

Jayadev C, Mahendradas P, Vinekar A, Kemmanu V, Gupta R, Pradhan ZS, D'Souza S, Aroor CD, Kaweri L, Shetty R, Honavar SG, Shetty B (2020) Tele-consultations in the wake of COVID-19—suggested guidelines for clinical ophthalmology. Indian J Ophthalmol 68(7):1316–1327. https://doi.org/10.4103/ijo.IJO_1509_20. PMID: 32587157; PMCID: PMC7574118

Karia R, Gupta I, Khandait H, Yadav A, Yadav A (2020) COVID-19 and its modes of transmission. SN Compr Clin Med 2(10):1798–1801. https://doi.org/10.1007/s42399-020-00498-4. Epub 1 Sept 2020. PMID: 32904860; PMCID: PMC7461745

Khalili M, Karamouzian M, Nasiri N, Javadi S, Mirzazadeh A, Sharifi H (2020) Epidemiological characteristics of COVID-19: a systematic review and meta-analysis. Epidemiol Infect 148:e130. https://doi.org/10.1017/S0950268820001430. PMID: 32594937; PMCID: PMC7343974

Kharel Sitaula R, Khatri A, Janani MK, Mandage R, Sadhu S, Madhavan HN, Upadhyay MP, Biswas J

(2020) Unfolding COVID-19: lessons-in-learning in ophthalmology. Clin Ophthalmol 14:2807–2820. https://doi.org/10.2147/OPTH.S259857. PMID: 33061265; PMCID: PMC7532306
Khatwani PR, Goel NP, Trivedi KY, Aggarwal SV (2021) Unveiling endophthalmitis post COVID-19—a case series. Indian J Ophthalmol 69(10):2869–2871. https://doi.org/10.4103/ijo.IJO_1616_21. PMID: 34571655; PMCID: PMC8597453
Koo EH, Eghrari AO, Dzhaber D, Shah A, Fout E, Dubovy S, Maestre-Mesa J, Miller D (2021) Presence of SARS-CoV-2 viral RNA in aqueous humor of asymptomatic individuals. Am J Ophthalmol 230:151–155. https://doi.org/10.1016/j.ajo.2021.05.008. Epub 19 May 2021. PMID: 34022227; PMCID: PMC8132502
Korobelnik JF, Loewenstein A, Eldem B, Joussen AM, Koh A, Lambrou GN, Lanzetta P, Li X, Lövestam-Adrian M, Navarro R, Okada AA, Pearce I, Rodríguez FJ, Wong DT, Wu L (2021) Anti-VEGF intravitreal injections in the era of COVID-19: responding to different levels of epidemic pressure. Graefes Arch Clin Exp Ophthalmol 259(3):567–574. https://doi.org/10.1007/s00417-021-05097-0. Epub 2 Feb 2021. PMID: 33528647; PMCID: PMC7852054
Kouhsari E, Azizian K, Sholeh M, Shayestehpour M, Hashemian M, Karamollahi S, Yaghoubi S, Sadeghiifard N (2021) Clinical, epidemiological, laboratory, and radiological characteristics of novel coronavirus (2019-nCoV) in retrospective studies: a systemic review and meta-analysis. Indian J Med Microbiol 39(1):104–115. https://doi.org/10.1016/j.ijmmb.2020.10.004. Epub 4 Nov 2020. PMID: 33610239; PMCID: PMC7667392
Kuba K, Imai Y, Rao S (2005) A crucial role of angiotensin converting enzyme 2 (ACE2) in SARS coronavirus-induced lung injury. Nat Med 22:875–879
La Distia Nora R, Putera I, Khalisha DF, Septiana I, Ridwan AS, Sitompul R (2020) Are eyes the windows to COVID-19? Systematic review and meta-analysis. BMJ Open Ophthalmol 5(1):e000563. https://doi.org/10.1136/bmjophth-2020-000563. PMID: 34192152; PMCID: PMC7476422
Lam D, Downie C, Cornish E, McCluskey P, Zagora SL (2020a) Telehealth in an acute ophthalmic setting during COVID-19 lock-down. Clin Exp Ophthalmol 48(9):1312–1315. https://doi.org/10.1111/ceo.13876. Epub 3 Nov 2020. PMID: 33118272
Lam DSC, Wong RLM, Lai KHW, Ko CN, Leung HY, Lee VYW, Lau JYN, Huang SS (2020b) COVID-19: special precautions in ophthalmic practice and FAQs on personal protection and mask selection. Asia Pac J Ophthalmol (Phila) 9(2):67–77. https://doi.org/10.1097/APO.0000000000000280. PMID: 32349113; PMCID: PMC7227209
Lan J, Ge J, Yu J, Shan S, Zhou H, Fan S, Zhang Q, Shi X, Wang Q, Zhang L, Wang X (2020) Structure of the SARS-CoV-2 spike receptor-binding domain bound to the ACE2 receptor. Nature 581(7807):215–220. https://doi.org/10.1038/s41586-020-2180-5. Epub 30 Mar 2020. PMID: 32225176
Lee YK, Huang YH (2021) Ocular manifestations after receiving COVID-19 vaccine: a systematic review. Vaccines (Basel) 9(12):1404. https://doi.org/10.3390/vaccines9121404. PMID: 34960150; PMCID: PMC8709261
Lee EJ, Cines DB, Gernsheimer T, Kessler C, Michel M, Tarantino MD, Semple JW, Arnold DM, Godeau B, Lambert MP, Bussel JB (2021) Thrombocytopenia following Pfizer and Moderna SARS-CoV-2 vaccination. Am J Hematol 96(5):534–537. https://doi.org/10.1002/ajh.26132. Epub 9 Mar 2021. PMID: 33606296; PMCID: PMC8014568
Legrottaglie EF, Balia L, Camesasca FI, Vallejo-Garcia JL, Fossati G, Vinciguerra R, Rosetta P, Vinciguerra P (2020) Management of an ophthalmology department during COVID-19 pandemic in Milan, Italy. Eur J Ophthalmol 22:1120672120960334
Li JPO, Shantha J, Wong TY et al (2020) Preparedness amongst ophthalmologists: during and beyond the COVID-19 pandemic. Ophthalmology 127(5):56
Li M, Xu L, Tan CS, Lanca C, Foo LL, Sabanayagam C, Saw SM (2022) Systematic review and meta-analysis on the impact of COVID-19 pandemic-related lifestyle on myopia. Asia Pac J Ophthalmol (Phila) 11(5):470–480. https://doi.org/10.1097/APO.0000000000000559. PMID: 36179338
Lin TPH, Ko CN, Zheng K, Lai KHW, Wong RLM, Lee A, Zhang S, Huang SS, Wan KH, Lam DSC (2021) COVID-19: update on its ocular involvements, and complications from its treatments and vaccinations. Asia Pac J Ophthalmol (Phila) 10(6):521–529. https://doi.org/10.1097/APO.0000000000000453. PMID: 34839344; PMCID: PMC8673850
Liu Z, Sun CB (2020) Conjunctiva is not a preferred gateway of entry for SARS-CoV-2 to infect respiratory tract. J Med Virol 92(9):1410–1412. https://doi.org/10.1002/jmv.25859. Epub 3 Jun 2020. PMID: 32275082; PMCID: PMC7262346
Liu L, Sun Y, Pan X (2004) Expression of SARS coronavirus S protein functional receptor-angiotensin-converting enzyme 2 in human cornea and conjunctiva. Chin Ophthal Res 22:561–564
Lo TC, Chen YY (2021) Ocular and systemic manifestations in paediatric multisystem inflammatory syndrome associated with COVID-19. J Clin Med 10(13):2953. https://doi.org/10.3390/jcm10132953. PMID: 34209269; PMCID: PMC8269386
Loffredo L, Pacella F, Pacella E, Tiscione G, Oliva A, Violi F (2020) Conjunctivitis and COVID-19: a meta-analysis. J Med Virol 92(9):1413–1414. https://doi.org/10.1002/jmv.25938. Epub 22 May 2020. PMID: 32330304; PMCID: PMC7264785.9-572
Lu CW, Liu XF, Jia ZF (2020) 2019-nCoV transmission through the ocular surface must not be ignored. Lancet 395(10224):e39. https://doi.org/10.1016/S0140-6736(20)30313-5. Epub 6 Feb 2020. PMID: 32035510; PMCID: PMC7133551

Luo SH, Liu W, Liu ZJ, Zheng XY, Hong CX, Liu ZR, Liu J, Weng JP (2020) A confirmed asymptomatic carrier of 2019 novel coronavirus. Chin Med J (Engl) 133(9):1123–1125. https://doi.org/10.1097/CM9.0000000000000798. PMID: 32149768; PMCID: PMC7213620

Luo X, Lv M, Zhang X, Estill J, Yang B, Lei R, Ren M, Liu Y, Wang L, Liu X, Wang Q, Meng M, Chen Y (2022) COVID-19 evidence and recommendations working group. Clinical manifestations of COVID-19: an overview of 102 systematic reviews with evidence mapping. J Evid Based Med 15(3):201–215. https://doi.org/10.1111/jebm.12483. Epub 31 July 2022. PMID: 35909298; PMCID: PMC9353366

Ma D, Chen CB, Jhanji V, Xu C, Yuan XL, Liang JJ, Huang Y, Cen LP, Ng TK (2020) Expression of SARS-CoV-2 receptor ACE2 and TMPRSS2 in human primary conjunctival and pterygium cell lines and in mouse cornea. Eye (Lond) 34(7):1212–1219. https://doi.org/10.1038/s41433-020-0939-4. Epub 7 May 2020. PMID: 32382146; PMCID: PMC7205026

Mack HG, Fraser-Bell S (2021) "COVID new normal" in ophthalmology: implications for ophthalmologists, eye care, ophthalmic education and research. Clin Exp Ophthalmol 49(1):9–11. https://doi.org/10.1111/ceo.13898. Epub 18 Jan 2021. PMID: 33462979; PMCID: PMC8013672

Madjid M, Safavi-Naeini P, Solomon SD, Vardeny O (2020) Potential effects of coronaviruses on the cardiovascular system: a review. JAMA Cardiol 5(7):831–840. https://doi.org/10.1001/jamacardio.2020.1286. PMID: 32219363

Marinho PM, Nascimento H, Romano A, Muccioli C, Belfort R Jr (2019) Diffuse uveitis and chorioretinal changes after yellow fever vaccination: a re-emerging epidemic. Int J Retina Vitreous 5:30. https://doi.org/10.1186/s40942-019-0180-0. PMID: 31608161; PMCID: PMC6778973

Mazzotta C, Giancipoli E (2020) Anterior acute uveitis report in a SARS-CoV-2 patient managed with adjunctive topical antiseptic prophylaxis preventing 2019-nCoV spread through the ocular surface route. Int Med Case Rep J 13:513–520. https://doi.org/10.2147/IMCRJ.S260252. PMID: 33116943; PMCID: PMC7568617

Mohan A, Sen P, Shah C, Jain E, Jain S (2021) Prevalence and risk factor assessment of digital eye strain among children using online e-learning during the COVID-19 pandemic: digital eye strain among kids (DESK study-1). Indian J Ophthalmol 69(1):140–144. https://doi.org/10.4103/ijo.IJO_2535_20. PMID: 33323599; PMCID: PMC7926141

Mohd-Alif WM, Nur-Athirah A, Hanapi MS, Tuan Jaffar TN, Shatriah I (2022) Bilateral and multiple central serous chorioretinopathy following COVID-19 infection: a case report and literature review. Cureus 14(3):e23246. https://doi.org/10.7759/cureus.23246. PMID: 35449667; PMCID: PMC9012563

Moura-Coelho N, Cunha JP, Papa-Vettorazzi R, Gris Ó, Güell JL (2022) Acute corneal allograft rejection following SARS-CoV-2 vaccination—a systematic review. Acta Ophthalmol https://doi.org/10.1111/aos.15211. Epub ahead of print. PMID: 35781792; PMCID: PMC9349413

Nasiri N, Sharifi H, Bazrafshan A, Noori A, Karamouzian M, Sharifi A (2021) Ocular manifestations of COVID-19: a systematic review and meta-analysis. J Ophthalmic Vis Res 16(1):103–112. https://doi.org/10.18502/jovr.v16i1.8256. PMID: 33520133; PMCID: PMC7841281

Ng XL, Betzler BK, Testi I, Ho SL, Tien M, Ngo WK, Zierhut M, Chee SP, Gupta V, Pavesio CE, de Smet MD, Agrawal R (2021) Ocular adverse events after COVID-19 vaccination. Ocul Immunol Inflamm 29(6):1216–1224. https://doi.org/10.1080/09273948.2021.1976221. Epub 24 Sept 2021. PMID: 34559576; PMCID: PMC8477588

Ng XL, Betzler BK, Ng S, Chee SP, Rajamani L, Singhal A, Rousselot A, Pavesio CE, Gupta V, de Smet MD, Agrawal R (2022) The eye of the storm: COVID-19 vaccination and the eye. Ophthalmol Ther 11(1):81–100. https://doi.org/10.1007/s40123-021-00415-5. Epub 16 Dec 2021. PMID: 34914035; PMCID: PMC8675299

Nyankerh CNA, Boateng AK, Appah M (2022) Ocular complications after COVID-19 vaccination, vaccine adverse event reporting system. Vaccines (Basel) 10(6):941. https://doi.org/10.3390/vaccines10060941. PMID: 35746549; PMCID: PMC9230098

Ozturker ZK (2022) Conjunctivitis as sole symptom of COVID-19: a case report and review of literature. Eur J Ophthalmol 31(2):NP161–NP166. https://doi.org/10.1177/1120672120946287. Epub 24 July 2020. Erratum in: Eur J Ophthalmol 2020:1120672120956486. PMID: 32703010; PMCID: PMC7383095

Pandey SK, Sharma V (2021) Mask-associated dry eye disease and dry eye due to prolonged screen time: are we heading towards a new dry eye epidemic during the COVID-19 era? Indian J Ophthalmol 69(2):448–449. https://doi.org/10.4103/ijo.IJO_3250_20. PMID: 33380621; PMCID: PMC7933894

Park KA, Jeon H, Choi DG, Jung JH, Shin HJ, Lee BJ, Moon Y, Lee SY, Lee DC, Cho SY, Kim SJ, Oh SY, Moon S, Oh SY, Choi DD, Choi MY, Kim WJ, Kim US, Lee HJ, Kim Y (2022) Ocular motility disorders following coronavirus disease-19 vaccination. Graefes Arch Clin Exp Ophthalmol 1–13. https://doi.org/10.1007/s00417-022-05888-z. Epub ahead of print. PMID: 36383278; PMCID: PMC9667443

Pascarella G, Strumia A, Piliego C, Bruno F, Del Buono R, Costa F, Scarlata S, Agrò FE (2020) COVID-19 diagnosis and management: a comprehensive review. J Intern Med 288(2):192–206. https://doi.org/10.1111/joim.13091. Epub 13 May 2020. PMID: 32348588; PMCID: PMC7267177

Pellegrini M, Roda M, Lupardi E, Di Geronimo N, Giannaccare G, Schiavi C (2020) The impact of COVID-19 pandemic on ophthalmological emergency department visits. Acta Ophthalmol 98:e1058–e1059
Perrotta F, Matera MG, Cazzola M, Bianco A (2020) Severe respiratory SARS-CoV2 infection: does ACE2 receptor matter? Respir Med 168:105996. https://doi.org/10.1016/j.rmed.2020.105996. Epub 25 Apr 2020. PMID: 32364961; PMCID: PMC7194970
Poland GA, Ovsyannikova IG, Kennedy RB (2020) SARS-CoV-2 immunity: review and applications to phase 3 vaccine candidates. Lancet 396(10262):1595–1606. https://doi.org/10.1016/S0140-6736(20)32137-1. Epub 13 Oct 2020. PMID: 33065034; PMCID: PMC7553736
Poyser A, Deol SS, Osman L, Kuht HJ, Sivagnanasithiyar T, Manrique R, Okafor LO, DeSilva I, Sharpe D, Savant V et al (2020) Impact of COVID-19 pandemic and lockdown on eye emergencies. Eur J Ophthalmol 19:1120672120974944
Prager KM, Dagi Glass LR, Wang M, Chen RWS, Liebmann JM, Cioffi GA (2021) Ophthalmology and ethics in the COVID-19 era. Am J Ophthalmol 224:158–162. https://doi.org/10.1016/j.ajo.2020.11.008. Epub 26 Nov 2020. PMID: 33249054; PMCID: PMC7690940
Raevis JJ, Gjyzeli G, Mititelu M, Rogers J, Lasarev M, Chang JS (2021) Face masks and bacterial dispersion toward the periocular area. Ophthalmology 128(8):1236–1238. https://doi.org/10.1016/j.ophtha.2021.01.007. Epub 12 Jan 2021. PMID: 33444660; PMCID: PMC7801188
Rai P, Kumar BK, Deekshit VK, Karunasagar I, Karunasagar I (2021) Detection technologies and recent developments in the diagnosis of COVID-19 infection. Appl Microbiol Biotechnol 105(2):441–455. https://doi.org/10.1007/s00253-020-11061-5. Epub 4 Jan 2021. PMID: 33394144; PMCID: PMC7780074
Reddy PA (2021) Commentary: COVID-19 lockdown-I and rural eye centers. Indian J Ophthalmol 69(1):156. https://doi.org/10.4103/ijo.IJO_3350_20. PMID: 33323603; PMCID: PMC7926110
Reyes-Bueno JA, García-Trujillo L, Urbaneja P, Ciano-Petersen NL, Postigo-Pozo MJ, Martínez-Tomás C, Serrano-Castro PJ (2020) Miller–Fisher syndrome after SARS-CoV-2 infection. Eur J Neurol 27(9):1759–1761. https://doi.org/10.1111/ene.14383. PMID: 32503084; PMCID: PMC7300794
Richardson-May J, Rothwell A, Rashid M (2021) Reactivation of herpes simplex keratitis following vaccination for COVID-19. BMJ Case Rep 14(9):e245792. https://doi.org/10.1136/bcr-2021-245792. PMID: 34493563; PMCID: PMC8424829
Romano MR, Montericcio A, Montalbano C, Raimondi R, Allegrini D, Ricciardelli G, Angi M, Pagano L, Romano V (2020) Facing COVID-19 in ophthalmology department. Curr Eye Res 45:653–658
Rose S (2020) Medical student education in the time of COVID-19. JAMA 323(21):2131–2132. https://doi.org/10.1001/jama.2020.5227. PMID: 32232420
Saldanha IJ, Petris R, Makara M, Channa P, Akpek EK (2021) Impact of the COVID-19 pandemic on eye strain and dry eye symptoms. Ocul Surf 22:38–46. https://doi.org/10.1016/j.jtos.2021.06.004. Epub 13 Jun 2021. PMID: 34133976; PMCID: PMC8462938
Saleem SM, Pasquale LR, Sidoti PA, Tsai JC (2020) Virtual ophthalmology: telemedicine in a COVID-19 era. Am J Ophthalmol 216:237–242. https://doi.org/10.1016/j.ajo.2020.04.029. Epub 30 Apr 2020. PMID: 32360862; PMCID: PMC7191296
Salvetat ML, Salati C, Busatto P, Zeppieri M (2022) The impact of COVID-19 related national lockdown on ophthalmic emergency in Italy: a multicenter study. Eur J Ophthalmol 32(3):1782–1794. https://doi.org/10.1177/11206721211028046. Epub 3 July 2021. PMID: 34219482; PMCID: PMC9111919
Sarkar M, Madabhavi IV, Quy PN et al (2021) COVID-19 and coagulopathy. Clin Respir J 15:1259–1274
Scalinci SZ, Trovato Battagliola E (2020) Conjunctivitis can be the only presenting sign and symptom of COVID-19. IDCases 20:e00774. https://doi.org/10.1016/j.idcr.2020.e00774. Epub 24 Apr 2020. PMID: 32373467; PMCID: PMC7195291
Seah I, Agrawal R (2020) Can the coronavirus disease 2019 (COVID-19) affect the eyes? A review of coronaviruses and ocular implications in humans and animals. Ocul Immunol Inflamm 28(3):391–395. https://doi.org/10.1080/09273948.2020.1738501. Epub 16 Mar 2020. PMID: 32175797; PMCID: PMC7103678
Seah IYJ, Anderson DE, Kang AEZ, Wang L, Rao P, Young BE, Lye DC, Agrawal R (2020) Assessing viral shedding and infectivity of tears in coronavirus disease 2019 (COVID-19) patients. Ophthalmology 127(7):977–979. https://doi.org/10.1016/j.ophtha.2020.03.026. Epub 24 Mar 2020. PMID: 32291098; PMCID: PMC7151491
Sen M, Honavar SG (2021) After the storm: ophthalmic manifestations of COVID-19 vaccines. Indian J Ophthalmol 69(12):3398–3420. https://doi.org/10.4103/ijo.IJO_2824_21. PMID: 34826968; PMCID: PMC8837328
Sen M, Honavar SG, Sharma N, Sachdev MS (2021) COVID-19 and eye: a review of ophthalmic manifestations of COVID-19. Indian J Ophthalmol 69(3):488–509. https://doi.org/10.4103/ijo.IJO_297_21. PMID: 33595463; PMCID: PMC7942063
Senanayake PD, Drazba J, Shadrak K et al (2007) Angiotensin II and its receptor subtypes in the human retina. Invest Ophthalmol Vis Sci 48:3301
Shah SH, Garg AK, Patel S, Yim W, Jokerst JV, Chao DL (2021) Assessment of respiratory droplet transmission during the ophthalmic slit-lamp exam: a particle tracking analysis. Am J Ophthalmol 222:76–81. https://doi.org/10.1016/j.ajo.2020.08.046. Epub 25 Sept 2020. PMID: 32980331; PMCID: PMC7516395

Sharif N, Alzahrani KJ, Ahmed SN, Dey SK (2021) Efficacy, immunogenicity and safety of COVID-19 vaccines: a systematic review and meta-analysis. Front Immunol 12:714170. https://doi.org/10.3389/fimmu.2021.714170. PMID: 34707602; PMCID: PMC8542872

Sharma N, D'Souza S, Nathawat R, Sinha R, Gokhale NS, Fogla R, Titiyal JS, Maskati QB, Mukherjee G, Sachdev Writing Committee MS; Composition of the All India Ophthalmological Society (AIOS)-Eye Bank Association of India (EBAI) Expert Group for COVID-19 Practice Guidelines for Cornea and Eyebanking Includes the Writing Committee. Prepared in Association with the AIOS and EBAI Expert Group (2020) All India Ophthalmological Society—Eye Bank Association of India consensus statement on guidelines for cornea and eyebanking during COVID-19 era. Indian J Ophthalmol 68(7):1258–1262. https://doi.org/10.4103/ijo.IJO_1554_20. PMID: 32587148; PMCID: PMC7574107

Shetty R, D'Souza S, Lalgudi VG (2020a) What opthalmologists should know about conjunctivitis in the COVID-19 pandemic? Indian J Ophthlmol 68(5):683–687

Shetty R, Lalgudi VG, Khamar P, Gupta K, Sethu S, Nair A, Honavar SG, Ghosh A, D'Souza S (2020b) Potential ocular and systemic COVID-19 prophylaxis approaches for healthcare professionals. Indian J Ophthalmol 68(7):1349–1356. https://doi.org/10.4103/ijo.IJO_1589_20. PMID: 32587162; PMCID: PMC7574070

Soiza RL, Scicluna C, Thomson EC (2021) Efficacy and safety of COVID-19 vaccines in older people. Age Ageing 50(2):279–283. https://doi.org/10.1093/ageing/afaa274. PMID: 33320183; PMCID: PMC7799251

Sommer AC, Blumenthal EZ (2020) Telemedicine in ophthalmology in view of the emerging COVID-19 outbreak. Graefes Arch Clin Exp Ophthalmol 258(11):2341–2352. https://doi.org/10.1007/s00417-020-04879-2. Epub 19 Aug 2020. PMID: 32813110; PMCID: PMC7436071

Sopp NM, Sharda V (2021) An eye on COVID-19: a meta-analysis of positive conjunctival reverse transcriptase-polymerase chain reaction and SARS-CoV-2 conjunctivitis prevalence. Optom Vis Sci 98(5):429–436. https://doi.org/10.1097/OPX.0000000000001687. PMID: 33973912; PMCID: PMC8136458

Stone LG, Grinton ME, Talks JS (2021) Delayed follow-up of medical retina patients due to COVID-19: impact on disease activity and visual acuity. Graefes Arch Clin Exp Ophthalmol 259(7):1773–1780. https://doi.org/10.1007/s00417-021-05174-4. Epub 11 May 2021. PMID: 33977317; PMCID: PMC8112469

Sun Y, Liu L, Pan X, Jing M (2006) Mechanism of the action between the SARS-CoV S240 protein and the ACE2 receptor in the eye. Int Ophthlmol 6:783–786

Taha MJJ, Abuawwad MT, Alrubasy WA, Sameer SK, Alsafi T, Al-Bustanji Y, Abu-Ismail L, Nashwan AJ (2022) Ocular manifestations of recent viral pandemics: a literature review. Front Med (Lausanne) 9:1011335. https://doi.org/10.3389/fmed.2022.1011335. PMID: 36213628; PMCID: PMC9537761

Tatti F, Mangoni L, Pirodda S, Demarinis G, Iovino C, Siotto Pintor E, Orrù G, Lecca LI, Campagna M, Denotti G, Peiretti E (2022) Ocular surface changes associated with face masks in healthcare personnel during COVID-19 pandemic. Life (Basel) 12(10):1491. https://doi.org/10.3390/life12101491. PMID: 36294925; PMCID: PMC9604707

Testi I, Brandão-de-Resende C, Agrawal R, Pavesio C (2022) COVID-19 vaccination ocular inflammatory events study group. Ocular inflammatory events following COVID-19 vaccination: a multinational case series. J Ophthal Inflamm Infect 12(1):4. https://doi.org/10.1186/s12348-021-00275-x. PMID: 34982290; PMCID: PMC8725430

Tham YC, Husain R, Teo KYC, Tan ACS, Chew ACY, Ting DS, Cheng CY, Tan GSW, Wong TY (2022) New digital models of care in ophthalmology, during and beyond the COVID-19 pandemic. Br J Ophthalmol 106(4):452–457. https://doi.org/10.1136/bjophthalmol-2020-317683. Epub 22 Mar 2021. PMID: 33753407; PMCID: PMC8961770

Tisdale AK, Chwalisz BK (2020) Neuro-ophthalmic manifestations of coronavirus disease 19. Curr Opin Ophthalmol 31(6):489–494. https://doi.org/10.1097/ICU.0000000000000707. PMID: 33009081

Tognetto D, Brézin AP, Cummings AB, Malyugin BE, Evren Kemer O, Prieto I, Rejdak R, Teus MA, Törnblom R, Toro MD, Vinciguerra AL, Giglio R, De Giacinto C (2020) Rethinking elective cataract surgery diagnostics, assessments, and tools after the COVID-19 pandemic experience and beyond: insights from the EUROCOVCAT group. Diagnostics (Basel) 10(12):1035. https://doi.org/10.3390/diagnostics10121035. PMID: 33276612; PMCID: PMC7761628

Toro MD, Brézin AP, Burdon M, Cummings AB, Evren Kemer O, Malyugin BE, Prieto I, Teus MA, Tognetto D, Törnblom R et al (2021) Early impact of COVID-19 outbreak on eye care: insights from EUROCOVCAT group. Eur J Ophthalmol 31:5–9

Tsironi S, Kavvadas D, Delis G, Bekiaridou A, Kapourani V, Loizou F, Apostolidou PS, Misiou K, Theofrastou E, Panakleridou T, Psimenidou E, Sarafi A, Fadel E, Karachrysafi S (2022) Cataract surgery during the COVID-19 pandemic: insights from a Greek Tertiary Hospital. Geriatrics (Basel) 7(4):77. https://doi.org/10.3390/geriatrics7040077. PMID: 35893324; PMCID: PMC9394252

Ulhaq ZS, Soraya GV (2020) The prevalence of ophthalmic manifestations in COVID-19 and the diagnostic value of ocular tissue/fluid. Graefes Arch Clin Exp Ophthalmol 258(6):1351–1352. https://doi.org/10.1007/s00417-020-04695-8. Epub 23 Apr 2020. PMID: 32328758; PMCID: PMC7180670

Verdoni L, Mazza A, Gervasoni A, Martelli L, Ruggeri M, Ciuffreda M, Bonanomi E, D'Antiga L (2020) An outbreak of severe Kawasaki-like disease at the Italian epicentre of the SARS-CoV-2 epidemic: an observational cohort study. Lancet 395(10239):1771–1778. https://doi.org/10.1016/S0140-6736(20)31103-X. Epub 13 May 2020. PMID: 32410760; PMCID: PMC7220177

Wan KH, Huang SS, Young A, Lam DSC (2020) Precautionary measures need for ophthalmologists during pandemic of the coronavirus disease 2019 (COVID-19). Acta Ophthalmol 98:221–222

Wang LL, Cheng GH (2022) Sequence analysis of the emerging SARS-CoV-2 variant Omicron in South Africa. J Med Virol 94(4):1728–1733

Wang H, Ma S (2008) The cytokine storm and factors determining the sequence and severity of organ dysfunction in multiple organ dysfunction syndrome. Am J Emerg Med 26(6):711–715. https://doi.org/10.1016/j.ajem.2007.10.031. PMID: 18606328

Wang K, Chen W, Zhang Z, Deng Y, Lian JQ, Du P, Wei D, Zhang Y, Sun XX, Gong L, Yang X, He L, Zhang L, Yang Z, Geng JJ, Chen R, Zhang H, Wang B, Zhu YM, Nan G, Jiang JL, Li L, Wu J, Lin P, Huang W, Xie L, Zheng ZH, Zhang K, Miao JL, Cui HY, Huang M, Zhang J, Fu L, Yang XM, Zhao Z, Sun S, Gu H, Wang Z, Wang CF, Lu Y, Liu YY, Wang QY, Bian H, Zhu P, Chen ZN (2020a) CD147-spike protein is a novel route for SARS-CoV-2 infection to host cells. Signal Transduct Target Ther 5(1):283. https://doi.org/10.1038/s41392-020-00426-x. PMID: 33277466; PMCID: PMC7714896

Wang W, Xu Y, Gao R, Lu R, Han K, Wu G, Tan W (2020b) Detection of SARS-CoV-2 in different types of clinical specimens. JAMA 323(18):1843–1844. https://doi.org/10.1001/jama.2020.3786. PMID: 32159775; PMCID: PMC7066521

Wang MTM, Niederer RL, McGhee CNJ, Danesh-Meyer HV (2022) COVID-19 vaccination and the eye. Am J Ophthalmol 240:79–98. https://doi.org/10.1016/j.ajo.2022.02.011. Epub 25 Feb 2022. PMID: 35227700; PMCID: PMC8875854

Watad A, De Marco G, Mahajna H, Druyan A, Eltity M, Hijazi N, Haddad A, Elias M, Zisman D, Naffaa ME, Brodavka M, Cohen Y, Abu-Much A, Abu Elhija M, Bridgewood C, Langevitz P, McLorinan J, Bragazzi NL, Marzo-Ortega H, Lidar M, Calabrese C, Calabrese L, Vital E, Shoenfeld Y, Amital H, McGonagle D (2021) Immune-mediated disease flares or new-onset disease in 27 subjects following mRNA/DNA SARS-CoV-2 vaccination. Vaccines (Basel) 9(5):435. https://doi.org/10.3390/vaccines9050435. PMID: 33946748; PMCID: PMC8146571

Webster P (2020) Virtual health care in the era of COVID-19. Lancet 395(10231):1180–1181. https://doi.org/10.1016/S0140-6736(20)30818-7. PMID: 32278374; PMCID: PMC7146660

Wendt S, Abdullah Z, Barrett S, Daruwalla C, Go JA, Le B, Li E, Livingston C, Miller M, Nakhleh L, Pecha J, Pothula S, Pradhan S, Sathappan V, Shah A, Sonuyi AM, Ugoh P, Wang Q, Weber N, Succar T, Blieden L, Mortensen P, Elkin Z, Sun G, Lee AG (2021) A virtual COVID-19 ophthalmology rotation. Surv Ophthalmol 66(2):354–361. https://doi.org/10.1016/j.survophthal.2020.10.001. Epub 12 Oct 2020. PMID: 33058927; PMCID: PMC7550053

WHO (2020) Statement on the second meeting of the international health regulations (2005) emergency committee regarding the outbreak of the novel coronavirus (2019-nCoV). Accessed 2023. https://www.who.int/news-room/detail/statement-on-the-second-meeting-of-the-international-health-regulation-(2005)-emergency-commitee-regarding-the-outbreak-of-the-novel-coronavirus-(2019-ncov)

WHO Coronavirus (COVID-19) Dashboard/WHO Coronavirus (COVID-19) Dashboard with vaccination data (2022). Available online: https://covid-19.who.int/

Wickham L, Hay G, Hamilton R, Wooding J, Tossounis H, da Cruz L, Siriwardena D, Strouthidis N (2020) The impact of COVID policies on acute ophthalmology services—experiences from Moorfields Eye Hospital NSH Foundation Trust. Eye 34:1189–1192

Willcox MD, Walsh K, Nichols JJ, Morgan PB, Jones LW (2020) The ocular surface, coronaviruses and COVID-19. Clin Exp Optom 103(4):418–424. https://doi.org/10.1111/cxo.13088. Epub 13 May 2020. PMID: 32406140; PMCID: PMC7272971

Wong TY, Bandello F (2020) Academic ophthalmology during and after the COVID-19 pandemic. Ophthalmology 127(8):e51–e52. https://doi.org/10.1016/j.ophtha.2020.04.029. Epub 21 Apr 2020. PMID: 32359842; PMCID: PMC7194607

Wong RLM, Ting DSW, Wan KH, Lai KHW, Ko CN, Ruamviboonsuk P, Huang SS, Lam DSC, Tham CCY (2020) COVID-19: ocular manifestations and the APAO prevention guidelines for ophthalmic practices. Asia Pac J Ophthalmol (Phila) 9(4):281–284. https://doi.org/10.1097/APO.0000000000000308. PMID: 32739937; PMCID: PMC7480794

Wong CW, Tsai A, Jonas JB, Ohno-Matsui K, Chen J, Ang M, Ting DSW (2021) Digital screen time during the COVID-19 pandemic: risk for a further myopia boom? Am J Ophthalmol 223:333–337. https://doi.org/10.1016/j.ajo.2020.07.034. Epub 30 July 2020. PMID: 32738229; PMCID: PMC7390728

Worldometer (2020) Covid-19 coronavirus pandemic (Internet). Worldometer. Available from: https://www.worldometers.info/coronavirus/

Wu P, Duan F, Luo C, Liu Q, Qu X, Liang L, Wu K (2020) Characteristics of ocular findings of patients with coronavirus disease 2019 (COVID-19) in Hubei Province, China. JAMA Ophthalmol 138(5):575–578. https://doi.org/10.1001/jamaophthalmol.2020.1291. PMID: 32232433; PMCID: PMC7110919

Xia J, Tong J, Liu M, Shen Y, Guo D (2020) Evaluation of coronavirus in tears and conjunctival secretions of patients with SARS-CoV-2 infection. J Med Virol 92(6):589–594. https://doi.org/10.1002/jmv.25725. Epub 12 Mar 2020. PMID: 32100876; PMCID: PMC7228294

Yangzes S, Grewal S, Gailson T, Grewal SPS (2021) Hand sanitizer-induced ocular injury. JAMA Ophthalmology 139:362

Yasaka Y, Hasegawa E, Keino H, Usui Y, Maruyama K, Yamamoto Y, Kaburaki T, Iwata D, Takeuchi M, Kusuhara S, Takase H, Nagata K, Yanai R, Kaneko Y, Iwahashi C, Fukushima A, Ohguro N, Sonoda KH, JOIS Uveitis Survey Working Group (2022) A multicenter study of ocular inflammation after COVID-19 vaccination. Jpn J Ophthalmol 1–8. https://doi.org/10.1007/s10384-022-00962-9. Epub ahead of print. PMID: 36417027; PMCID: PMC9684958

Yesudhas D, Srivastava A, Gromiha MM (2021) COVID-19 outbreak: history, mechanism, transmission, structural studies and therapeutics. Infection 49(2):199–213. https://doi.org/10.1007/s15010-020-01516-2. Epub 4 Sept 2020. PMID: 32886331; PMCID: PMC7472674

Zeng W, Wang X, Li J, Yang Y, Qiu X, Song P, Xu J, Wei Y (2020) Association of daily wear of eyeglasses with susceptibility to coronavirus disease 2019 infection. JAMA Ophthalmol 138(11):1196–1199. https://doi.org/10.1001/jamaophthalmol.2020.3906. PMID: 32936214; PMCID: PMC7495310

Zhang X, Chen X, Chen L, Deng C, Zou X, Liu W, Yu H, Chen B, Sun X (2020) The evidence of SARS-CoV-2 infection on ocular surface. Ocul Surf 18(3):360–362. https://doi.org/10.1016/j.jtos.2020.03.010. Epub 11 Apr 2020. PMID: 32289466; PMCID: PMC7194535

Zhou L, Xu Z, Castiglione GM, Soiberman US, Eberhart CG, Duh EJ (2020a) ACE2 and TMPRSS2 are expressed on the human ocular surface, suggesting susceptibility to SARS-CoV-2 infection. Ocul Surf 18(4):537–544. https://doi.org/10.1016/j.jtos.2020.06.007. Epub 13 June 2020. PMID: 32544566; PMCID: PMC7293510

Zhou Y, Duan C, Zeng Y, Tong Y, Nie Y, Yang Y, Chen Z, Chen C (2020b) Ocular findings and proportion with conjunctival SARS-COV-2 in COVID-19 patients. Ophthalmology 127(7):982–983. https://doi.org/10.1016/j.ophtha.2020.04.028. Epub 21 Apr 2020. PMID: 32359840; PMCID: PMC7194804

Zhu R, Yu ZY, Han L (2022) Insights on the possibility of SARS-CoV-2 transmission through the eyes. Int J Ophthalmol 15(11):1857–1863. https://doi.org/10.18240/ijo.2022.11.18. PMID: 36404966; PMCID: PMC9631193

Obesity, Physical Activity, and COVID-19

24

Zachary Zeigler and Anthony Acevedo

Abstract

Estimates suggest that over 80% of people with COVID-19 may experience mild to no symptoms. The Centers for Disease Control and Prevention (CDC) reported that around 7–8% of positive COVID-19 cases were hospitalized in the first year of the pandemic. A substantial body of evidence points to obesity, sedentary behavior, and low physical activity levels as risk factors for susceptibility to the virus and the severity of symptoms. Obese individuals are 46% more likely to contract the virus than those of normal weight, and those who are physically active have an 11% reduced risk of contracting the virus. Medical experts have also identified obesity, sedentary behavior, and physical inactivity as risk factors for severe COVID-19 symptoms and long COVID. Evidence from bariatric surgery suggests that weight loss decreases the susceptibility and severity of COVID-19. The mechanisms linking obesity, sedentary behavior, and physical activity to COVID-19 are somewhat similar. Obesity, sedentary behavior, and low physical activity are all linked to inflammation, immune dysfunction, and comorbidities such as diabetes and cardiovascular disease that increase COVID-19 risk. Additionally, achieving the recommended 150 min of moderate-intensity exercise seems to be the desired dose for protection from COVID-19.

Keywords

Obesity · Sedentary behavior · Physical activity · Upper body obesity · COVID-19 · Immunity

24.1 Introduction

Estimates of the percentage of people who get COVID-19 and are asymptomatic vary depending on the population studied and the testing strategies used. This broad estimate varies widely depending on age, overall health, and other factors. However, according to the World Health Organization (WHO), over 80% of people with COVID-19 may experience mild to no symptoms (World Health Organization 2020). Studies conducted in various settings have reported different rates of asymptomatic cases. For example, a recent systematic review found that the proportion of asymptomatic cases ranged from 6 to 96%, with an average of 40% (Buitrago-Garcia et al. 2020).

The severity of COVID-19 among those who are symptomatic also varies widely. Hospitalization rates, for example, differ

Z. Zeigler (✉) · A. Acevedo
Grand Canyon University, 19013 W Solano Dr. Litchfield Park, Phoenix, AZ 85340, USA
e-mail: Zachary.zeigler@gcu.edu

N. Rezaei (ed.), *The COVID-19 Aftermath*, Advances in Experimental Medicine and Biology 1457,
https://doi.org/10.1007/978-3-031-61939-7_24

depending on several factors, including age, underlying health conditions, and the availability of healthcare resources. The Centers for Disease Control and Prevention (CDC) reported that around 7–8% of people with COVID-19 were hospitalized in the first year of the pandemic (Centers for Disease Control and Prevention 2021). Additionally, one data report from more than 2.7 million people in the USA who tested positive for COVID-19 found that approximately 14% were hospitalized (Garg et al. 2020).

As research started to catch up to the new COVID-19 virus, risk factors for acquiring and predicting disease severity have emerged. In addition to the most pronounced risk factors, such as advanced age and a compromised immune system, obesity has been highlighted as a decisive risk factor for COVID-19 susceptibility and severity. In addition, an often understudied but influential risk factor for COVID-19 susceptibility and severity is physical activity status.

24.2 Obesity and COVID-19 Susceptibility and Severity

24.2.1 Obesity and COVID-19 Susceptibility

A substantial body of evidence points to obesity being a risk factor for both susceptibilities to the virus and the severity of symptoms. Figure 24.1 outlines BMI categories according to the World Health Organization. Researchers have found that obesity is associated with a significantly increased risk of COVID-19 infection, with obese individuals being 46% more likely to contract the virus than those of normal weight (Caussy et al. 2020). According to the CDC, people with a body mass index (BMI) of 30 kg/m^2 or higher are at increased risk of severe illness from COVID-19. Across the globe, researchers have demonstrated a correlation between obesity and severe COVID-19. A meta-analysis that included 25 studies conducted in the USA, six in China, four in Italy, four in the UK, four in Mexico, three in France, and one in Brazil, Singapore, South Korea, and Bolivia found that obese subjects had a 39% higher risk of SARS-CoV2 infection as compared to those without obesity (Yang et al. 2021).

24.2.2 Obesity and COVID-19 Severity

Medical experts have also identified obesity as a risk factor for severe or extreme COVID-19 symptoms. Studies have found that people with obesity are at increased risk of hospitalization, intensive care unit (ICU) admission, mechanical ventilation, and death from COVID-19.

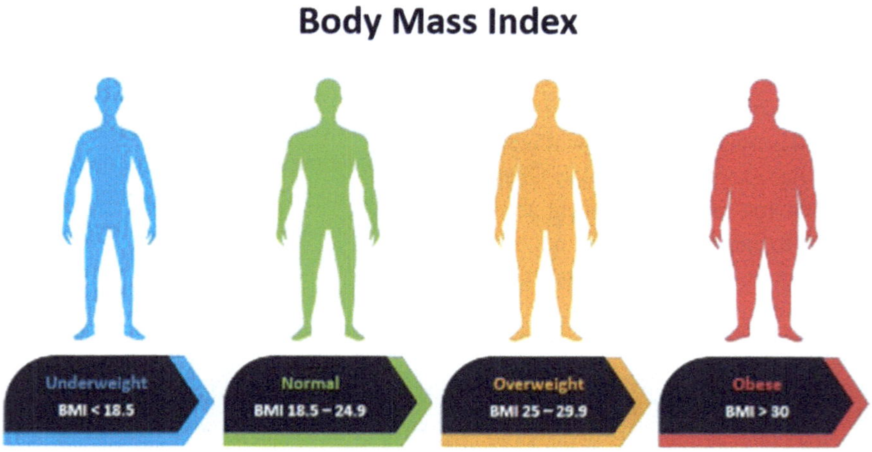

Fig. 24.1 BMI categories ranging from underweight to obese. Expressed as kg/m^2

One review found that individuals with obesity had a 113% higher risk for hospitalization, 74% higher risk for ICU admission, and 48% increased risk of death (Barry et al. 2020). Some data also suggest a dose response relationship between BMI and mortality risk from COVID-19. For instance, those with class I obesity (30–34.9 kg/m^2) had 27%, class II (35–39.9 kg/m^2) had a 56%, and class III (>40 kg m^2) had 92% increased risk of COVID-19 mortality when compared to a BMI < 30 kg/m^2 (Poly et al. 2021). Not all data supports this dose response relationship between BMI level and COVID-19 severity. As noted in a later section, some data support a U-shaped, as opposed to a linear, relationship between BMI category and COVID-19 severity.

24.3 Obesity and Long COVID

Long COVID, also known as post-acute sequelae of SARS-CoV-2 infection or chronic COVID-19 syndrome, refers to a range of symptoms (see Fig. 24.2). These symptoms persist for weeks or months after the acute phase of COVID-19 (Nalbandian et al. 2021). These symptoms can vary widely and may include fatigue, shortness of breath, chest pain, cough, joint pain, muscle aches, headaches, cognitive difficulties, insomnia, depression, and anxiety. Some individuals with long COVID may also experience new or ongoing symptoms, such as heart palpitations, gastrointestinal issues, and skin rashes. The exact cause of long COVID is not yet fully understood, but experts believe it to be related to a persistent immune response or inflammation in the body. Long COVID is a relatively new area of research, and there is still much to learn about its prevalence, risk factors, and optimal management. However, long COVID can significantly impact a person's quality of life.

The impact of obesity on long COVID is not yet fully understood. However, research has identified obesity as a possible risk factor for developing long-term symptoms from COVID-19. One study involved data from a smartphone app-based survey of individuals who had self-reported a positive COVID-19 test and were experiencing symptoms. The authors found that obesity was a risk factor for developing long COVID-19 symptoms, such as breathlessness, fatigue, and chest pain, after controlling for age, sex, and other comorbidities (Carole et al. 2021). Other research studies have also found that obesity was associated with an increased risk of persistent COVID-19 symptoms, even after controlling for age, sex, and other comorbidities (Carfì et al. 2020; Fernández-de-Las-Peñas et al. 2021).

Obesity may contribute to the development of long COVID by causing or exacerbating

Mechanisms Linking Obesity to Increased Susceptibility and Severity of COVID-19

- Inflammation: The combination of pre-existing inflammation in obese individuals and the inflammatory response to COVID-19 may exacerbate the severity of the disease.
- Immune dysfunction: Obesity can lead to immune system dysregulation, allowing the virus to replicate more easily and cause more severe COVID-19 disease.
- Comorbidities: Obesity is often associated with other health conditions, such as type 2 diabetes, hypertension, and cardiovascular disease, which are also risk factors for severe COVID-19 symptoms.
- Angiotensin-converting enzyme 2 (ACE2) receptor expression: The virus that causes COVID-19, SARS-CoV-2, enters human cells through the ACE2 receptor. ACE2 expression is upregulated in adipose tissue in obesity.
- Adipokines: It is not yet clear exactly how adipokines are involved in the pathogenesis of COVID-19, however, it is thought that they may contribute to the inflammatory response and immune dysregulation that is observed in severe cases of the disease.
- Brown adipose tissue (BAT): Thermogenic tissue that can help regulate energy balance and glucose metabolism. Dysfunction of BAT may contribute to metabolic and immune dysfunction in COVID-19 and has a link to long COVID. Shown to be less prevalent in individuals that exhibit Obesity.

Mechanisms Linking Physical Activity to Decreased Susceptibility and Severity of COVID-19

- Inflammation: Chronic inflammation can weaken the immune system, making it less effective at fighting off COVID-19. Physical activity has been shown to reduce levels of inflammation in the body.
- Immune Function: Physical activity can increase blood flow and circulation of immune cells throughout the body. This can help distribute immune cells to areas of the body where they are needed to detect and destroy COVID-19 cells
- Comorbidities: Physical Activity protects against health conditions such as Obesity, type 2 diabetes, hypertension, and cardiovascular disease, which are all risk factors for severe COVID-19 symptoms.
- Cytokines: Cytokines are signaling molecules that play an important role in the immune response to COVID-19. Exercise can stimulate the production of cytokines, which can help activate and mobilize immune cells.
- T and B cells: White blood cells that play an integral role in the detection and neutralization of COVID-19. Studies have shown that physical activity can increase the production and function of both T and Be cells, which may enhance the immune response and reduce COVID-19 severity
- Neutrophils: White blood cells that act as part of the early innate immune response. Studies have shown that exercise can increase the number of circulating neutrophils in the blood, which may help the body fight off COVID-19 more effectively.

Fig. 24.2 Mechanisms linking obesity and physical activity to COVID-19 susceptibility and severity

underlying health conditions, such as cardiovascular disease, diabetes, and sleep apnea, which can increase the risk of long-term complications after COVID-19. Obesity may also impair immune function and increase inflammation in the body, contributing to the persistence of symptoms.

24.4 Mechanisms Linking Obesity to Increased Susceptibility and Severity of COVID-19

The mechanisms behind why obesity is a risk factor for COVID-19 and why an obese person may be at increased risk for severe complications from the virus are not fully understood (Belančić et al. 2020). However, several hypotheses have been proposed based on the available evidence (Fig. 24.3).

- Inflammation: Obesity is associated with chronic low-grade inflammation, which may impair the immune response to viral infections (Trayhurn and Wood 2024). The combination of pre-existing inflammation in obese individuals and the inflammatory response to COVID-19 may exacerbate the severity of the disease (Sattar et al. 2020). COVID-19 can also trigger a robust inflammatory response, leading to a cytokine storm that can cause severe lung damage and other complications.
- Immune dysfunction: Obesity can lead to dysregulation of the immune system, including a decreased number and function of T cells and natural killer cells (O'Shea and Andrew 2019). These immune cells play an essential role in fighting viral infections, so a weakened immune response in obese individuals may allow the virus to replicate more easily and cause more severe COVID-19 disease (Stefan et al. 2020). Additionally, chronic inflammation associated with obesity can lead to increased oxidative stress and accumulated damaged proteins and

Fig. 24.3 Symptomology of long COVID

organelles, leading to cellular dysfunction and ultimately impairing immune function.

- Comorbidities: Obesity is often associated with other health conditions, such as type 2 diabetes, hypertension, and cardiovascular disease, which are also risk factors for severe COVID-19 symptoms. These comorbidities may increase the risk of severe disease in obese individuals. For example, researchers looked at risk factors for COVID-19 hospitalizations and, as previously noted, found that obesity was a strong risk factor. However, adjusting for biomarkers, such as hemoglobin A1C and HDL cholesterol, attenuated the association by 33 to 46%, suggesting the mechanisms may involve comorbid conditions such as impaired glucose and lipid metabolism (Hamer et al. 2020).
- Angiotensin-converting enzyme 2 (ACE2) receptor expression: The virus that causes COVID-19, SARS-CoV-2, enters human cells through the ACE2 receptor. Some studies suggest that the expression of ACE2 receptors is increased in obese individuals, which may make them more susceptible to infection with SARS-CoV-2. In obesity, adipose tissue upregulates the expression of ACE2. Additionally, dysregulation of the renin-angiotensin system, associated with obesity, may contribute to increased ACE2 expression and COVID-19 susceptibility.
- Adipokines: Research has shown that adipokines may play a role in the pathogenesis of COVID-19. Studies have found that levels of certain adipokines, such as adiponectin and leptin, are altered in patients with COVID-19 and may be associated with disease severity. Adiponectin, for example, is inversely related to mortality and respiratory failure in persons with COVID-19 (Hindsberger et al. 2023.). Also, leptin levels are increased in patients with COVID-19, which may be associated with a greater risk of developing severe disease (Stefan et al. 2020). It is not yet clear exactly how adipokines are involved in the pathogenesis of COVID-19. However, experts believe they may contribute to severe disease cases' inflammatory response and immune dysregulation.
- Brown adipose tissue (BAT): BAT is a thermogenic tissue that can help regulate energy balance and glucose metabolism. Dysfunction of BAT may contribute to metabolic and immune dysfunction in COVID-19 and also has a link to long COVID. BAT may protect against COVID-19 by producing specific cytokines and promoting anti-inflammatory responses (Cinti and Saverio 2022). Some research has suggested potential approaches for targeting BAT in COVID-19 and long COVID, including cold exposure, physical exercise, and pharmacological interventions (Muzyka et al. 2023). It is important to note that researchers have not yet conducted empirical studies demonstrating the effectiveness of such interventions. Further research is needed to fully understand the role of BAT in COVID-19 and its potential as a therapeutic target for long COVID.

24.5 Obesity Phenotype and COVID-19

Some evidence suggests that upper body obesity may be a risk factor for severe COVID-19 outcomes. Upper body obesity, or visceral obesity, refers to the accumulation of fat around the abdomen and organs, as opposed to subcutaneous fat, which accumulates under the skin. Experts have suggested that visceral adipose tissue's metabolic and inflammatory effects can produce inflammatory cytokines, possibly contributing to developing a cytokine storm and acute respiratory distress syndrome in severe COVID-19 cases. Upper body obesity is metabolically active and is associated with a higher risk of insulin resistance, type 2 diabetes, and cardiovascular disease, which are also risk factors for severe COVID-19 outcomes. Furthermore, some studies have shown that visceral adipose tissue expresses higher levels of ACE2 receptors, the primary receptor for

SARS-CoV-2, the virus that causes COVID-19. Higher levels of ACE2 receptors may increase the likelihood of viral entry into cells and subsequent infection. Researchers have also linked upper body fat with impaired lung function, which may further increase the risk of respiratory complications in COVID-19 patients (Gao et al. 2020).

A systematic review and meta-analysis of 21 studies, including 10,149 patients with COVID-19, found that patients with visceral adiposity were more likely to experience critical conditions, such as admission to the intensive care unit or the need for mechanical ventilation. Research investigating the relationship between different obesity phenotypes and the severity of COVID-19 as measured by chest X-ray severity scores found that patients with abdominal obesity phenotype had a 1.6 times higher chest X-ray severity score than patients without abdominal obesity. In contrast, there was no significant difference in chest X-ray severity scores between patients with BMI-based obesity and those without obesity. Lastly, researchers have also associated visceral adiposity with a higher mortality risk in patients with COVID-19 (Földi et al. 2021). Accordingly, abdominal obesity phenotype may better predict COVID-19 severity than BMI-based obesity (Malavazos et al. 2021).

24.6 Sarcopenia as a Risk Factor for COVID-19

Obesity, however, is not always a risk factor for COVID-19 susceptibility and severity. Many studies show a U-shaped relationship between BMI and COVID-19 outcomes (Klang et al. 2020). A meta-analysis of ten countries found that obese subjects had a 39% higher risk of SARS-CoV2 infection than those without obesity (Yang et al. 2021). Patients who had a BMI between 30 and 34.9 kg/m^2 did not show any significant increase in their hospital admission rates. Additionally, patients with a BMI > 25 kg/m^2 or BMI 30–34.9 kg/m^2 showed no significantly higher rate of ICU admission than those with a BMI < 25 kg/m^2 or BMI < 30 kg/m^2. Additional research examined 17 million adults in the UK and found that individuals with a BMI between 23 and 28 kg/m^2 had the lowest risk of severe COVID-19 outcomes, while those with a BMI below 20 kg/m^2 and above 35 kg/m^2 had the highest risk (Gao et al. 2020).

24.6.1 Sarcopenia and COVID-19

Individuals with a BMI below the normal range (< 18.5 kg/m^2) may be at increased risk of severe illness and death from COVID-19. High risk in low BMI categories is likely due to complications of frailty, sarcopenia, and possible nutritional deficiencies that could impact immune function. A meta-analysis investigated the association between sarcopenia and COVID-19 outcomes. The study included 11 studies with 4788 COVID-19-positive patients, of which 637 had sarcopenia. The study found, after adjusting for age, that patients with sarcopenia had a higher risk of severe COVID-19 outcomes, including a 2.08 increased odds of ICU admission, 2.13 increased odds of mechanical ventilation, and 1.95 increased odds of death. These findings suggest that sarcopenia may be a risk factor for severe COVID-19 outcomes (Siahaan et al. 2019).

24.6.2 Possible Obesity Paradox and COVID-19

Medical professionals use the term "obesity paradox" to describe a seemingly contradictory finding. Despite being a risk factor for numerous chronic health conditions, some research shows obesity is associated with improved survival in specific populations with certain diseases. While the concept of the obesity paradox has been somewhat controversial, some studies have suggested that being overweight or mildly obese may confer a survival advantage in patients with certain conditions, specifically those with heart disease and the elderly. Whether or not an obesity paradox exists concerning obesity and

COVID-19 has not been determined, but some data suggest it may.

Although the majority of evidence points to obesity as a risk factor for COVID-19 susceptibility and severity, some studies do show that specific obese populations, especially the older population, may have a reduced risk of mortality from COVID-19 (Vulturar et al. 2022). A large observational study was conducted at the San Matteo Hospital in Pavia, Italy, to investigate the relationship between obesity and outcomes in patients hospitalized with COVID-19, including mortality and the need for mechanical ventilation. The mortality rate among obese patients was 13.2%, compared to 16.9% among non-obese patients. Similarly, the need for mechanical ventilation was not significantly different between obese and non-obese patients. The rate of ICU admission among obese patients was 16.3%, compared to 31.7% among non-obese patients. This finding is consistent with the obesity paradox observed in other contexts, where being overweight or mildly obese has been associated with improved outcomes in specific patient populations (Biscarini et al. 2020).

24.6.3 Metabolically Healthy Obesity

Metabolically healthy obesity describes a condition in which an individual has a BMI classified as obese but has no or few metabolic abnormalities such as insulin resistance, high blood pressure, or dyslipidemia. Some data suggest that a metabolically healthy obese individual may not be at increased risk for COVID-19 severity and susceptibility. Researchers compared 135 COVID-19-positive patients, and 135 COVID-19-negative controls matched for age, sex, and BMI. The study found that metabolically healthy obesity was not associated with an increased risk of COVID-19. The metabolically healthy obese patients had a higher risk of severe COVID-19 outcomes than metabolically healthy normal weight individuals, which other research studies have corroborated (Li et al. 2021).

Research that included 7337 COVID-19 patients found that metabolically unhealthy individuals, regardless of obesity status, had a higher risk of critical COVID-19 outcomes than metabolically healthy individuals without obesity. These findings suggest that metabolic health status is an essential determinant of COVID-19 outcomes, and metabolically unhealthy individuals, either with obesity or not, are at a higher risk of critical outcomes (Kim et al. 2022).

24.7 Can Weight Loss Decrease COVID-19 Severity

Researchers have suggested that because obesity increases the risk for COVID-19 susceptibility and severity, weight loss would decrease these risks. The data available on this topic focus on weight loss from bariatric surgery. One study of 363 COVID-19-positive patients with obesity, of which 33 had prior bariatric surgery, found that patients with prior bariatric surgery had a lower risk of developing severe COVID-19, defined as the need for ICU admission, mechanical ventilation, or death. Specifically, none of the patients with prior bariatric surgery required ICU admission or mechanical ventilation, and none died from COVID-19, compared to 11% of patients without prior bariatric surgery who required ICU admission or mechanical ventilation and 3% who died (Aminian et al. 2019).

There may be a dose response between the degree of weight loss and COVID-19 protection. Researchers investigated the association between weight loss achieved through metabolic surgery and the risk and severity of COVID-19 infection. The study included 4251 patients who underwent metabolic surgery between 2006 and 2019, of which 27 tested positive for COVID-19. The study found that patients with more significant weight loss after metabolic surgery had a lower risk of testing positive for COVID-19. Specifically, patients who achieved more than 10% excess weight loss had a significantly lower risk of testing positive for COVID-19 than those who achieved less than 10% excess weight loss. The study also found that patients with more significant weight loss after metabolic surgery had a lower risk of developing severe COVID-19 outcomes. Specifically, patients who achieved more than 20% excess weight loss had a significantly lower risk of requiring hospitalization, ICU admission, or mechanical ventilation than patients who achieved less than 20% excess weight loss (Aminian et al. 2019).

24.8 The Interplay Between the COVID-19 Lockdown, Decreased Physical Activity, and Weight Gain

The COVID-19 pandemic and subsequent lockdown measures have significantly impacted worldwide physical activity. The closure of gyms and sports facilities and restrictions on outdoor activities significantly reduced people's ability to engage in physical activity. Decreased physical activity has increased the risk of COVID susceptibility and severity (Zeigler et al. 2021). Researchers have identified reduced physical activity levels as a risk factor for weight gain during the COVID-19 pandemic (Zachary 2021; Zachary et al. 2020). The evidence linking obesity and decreased physical activity to severe COVID and lessons learned from the COVID-19 pandemic should provide guidelines that government agencies should apply to future pandemics. For example, while it is important to employ measures to control the spread of the pathogen, it is equally vital to ensure enough opportunities and venues for recreation. Maintaining physical activity levels during a future pandemic could stave off weight gain. Decreasing weight gain from the pandemic could have decreased COVID-19 severity.

24.9 Physical Activity and COVID-19 Severity

The COVID-19 pandemic has had an inordinate impact on global health and welfare, causing a copious amount of illness and death in high-risk populations worldwide. As the world continues to navigate the pandemic, it has become apparent that certain lifestyle factors, like physical activity, will play a role in mitigating the risk of COVID-19 symptom presentation and severity. Physical activity can be defined as any bodily movement created by skeletal muscles resulting in energy expenditure, which includes but is not limited to walking, running, swimming, cycling, and resistance training. While the mechanisms through which physical activity protects individuals against this novel virus are still being examined, there is evidence to support that it is a valid method to prevent susceptibility to severe cases. Physical activity has been shown to reduce inflammation, improve lung health, and boost the immune system. It is also known to reduce the risk of obesity, type 2 diabetes, and cardiovascular disease, which are all comorbidities that have been found to be associated with a greater risk of complications from COVID-19. Emerging research suggests that physical activity plays a role in reducing the severity of COVID-19 symptoms, including hospitalizations, admission to the intensive care unit (ICU), and death (Sallis et al. 2021). Regular physical activity was associated with an 11% lower risk of contracting COVID-19 (Ezzatar et al. 2022), suggesting a stronger innate immune response in warding off the virus. While vaccines and other preventative measures, such as wearing masks, are important tools for controlling the spread of the virus, engaging in regular and appropriately prescribed physical activity is proving an effective way for individuals to reduce their risk of COVID-19 infection and severity.

24.10 How Physical Activity Affects Immunity

In the early 1900s, (Larrabee 1902) found that there were significant changes in the immune systems of marathon runners. Early immunology studies (1900–1979) mainly focused on exercise-induced changes, but it was not until the 1980s that new technology gave researchers the ability to conduct modern-era exercise immunology research. In recent decades, there has been conflicting evidence on the effects of exercise on immunity, showing that duration and intensity play a role in the outcomes. Research has shown that acute bouts of intense physical activity can lead to temporary immune suppression, but regular moderate-intensity physical activity has been shown to enhance immune function. The influence on the immune system is multifactorial, but the evidence is consistent in showing a J-shaped curve in the relationship between exercise workload and the risk for upper respiratory tract infection (Nieman 2020). (Fig. 24.1) This means that excessive exercise in duration and intensity could have deleterious effects on the immune system. However, the specific mechanisms establishing the relationship between physical activity and immune function are intricate and not fully understood.

24.10.1 Innate Immune Response to the SARS-CoV-2 Virus and Physical Activity

Upon exposure to the SARS-CoV-2 virus, it will enter the lungs through the respiratory tract to target alveolar epithelial type 2 cells (AT2). The virus contains spike proteins that have a high affinity for the angiotensin-converting enzyme 2 (ACE2) on the AT2 cell. Once inside the host cell, the SARS-CoV-2 virus will begin replicating to infect other cells. The immune system will respond by producing an inflammatory response. This initial reaction is the innate immune system, which is the primary defense against viruses, pathogens, bacteria, fungi, and parasites. This system is comprised of physical and chemical barriers, as well as various immune cells. The acute immune response from physical activity can enhance the innate immune system by increasing blood circulation. This vasodilation is accomplished through the release of chemicals such as nitric oxide, which assists in relaxing the smooth muscle in blood vessel walls, in turn, allowing immune cells to distribute through the body more efficiently. Physical activity has been shown to stimulate the production of cytokines, including interleukin-6 (IL-6) and tumor necrosis factor-alpha (TNF-alpha). These cytokines can activate other immune cells, such as macrophages, and promote an inflammatory response that can help to fight infection. Notably, acute physical activity is now considered an important addition to the immune system that enhances the circulation of leukocytes to the tissues (Adams et al., 2011). Furthermore, physical activity can ameliorate the function of physical barriers such as the skin and mucous membranes.

24.10.2 Adaptive Immune Response to the SARS-CoV-2 Virus and Physical Activity

The body's next line of defense is the adaptive immune system, which is a more targeted response that involves the activation of T cells and the production of antibodies by B cells. Once activated, these specific proteins proliferate and differentiate into effector cells. The B cells will produce specific antibodies that target the spike proteins of the SARS-CoV-2 virus, inhibiting their ability to infect host cells. Helper T cells will activate other immune cells, while killer T cells directly kill the virus. The adaptive immune system has a memory component, where T cells and B cells produce specific antibodies to help defend the body if the same antigen is encountered. This immunity response does not prevent you from contracting the virus but has the ability to reduce the severity. The challenge with the SARS-CoV-2 virus is its ability to mutate and present with new variants, which may be immune resistant. Regular physical activity has been shown to increase the circulation of T cells and B cells, increasing immune surveillance throughout the body. This creates the need for regular physical activity to reinforce the immune system.

24.11 Physical Activity and Long COVID

Long COVID, previously mentioned as PASC can deliver more than 200 symptoms with a minimum of 65 million people worldwide reported being impacted. We do not know the exact cause of long COVID, but it is assumed to affect multiple organ systems (Fig. 24.2). Studies have shown that even those with mild COVID-19 had alterations in their T cells and reduced CD4 and CD8 effector memory cell numbers for 13 months (Glynne et al., 2022). Another study showed that the innate immune cells were highly activated for long periods, naïve T and B cells were suppressed, and type I and type III interferons were elevated (Phetsouphanh et al., 2022). Physical activity is a suggested intervention to help with managing long COVID symptoms, but its efficacy is not yet fully understood. A New Zealand questionnaire short form (NZPAQ-SF) was used to collect pre- and post-COVID levels of activities of daily living and physical activity, which consisted of brisk walking, moderate physical activity, and vigorous physical activity. The study of 477 global participants, consisting of mostly white (92.7%) females (89.1%) with a mean age of 45.69, reported that symptoms worsened (78.84%), improved (0.84%), mixed effect (20.96%), or no effect (28.72%) by adding physical activity (Wright et al., 2022a, b). The main finding of this study was long COVID severely reduced physical activity and 64.4% of the participants reported that fatigue worsened. Similar findings were shown in a meta-analysis that reported 32% of participants from around the world had a higher level of fatigue, which could be a correlation with lower physical activity. A systematic review analyzing 6 studies consisting of 264 participants concluded that a physical activity program including a 5-to-30-min moderate-intensity aerobic program with a limited amount of resistance training could be an efficacious program for those with long COVID (Hekmatikar et al. 2022). Being that COVID is heterogenous and has shown multiple effects on people, it is only right to assess each person and develop an individualized comprehensive physical activity program.

24.12 Sedentary and COVID-19 Severity

Overall, there is a high level of importance in not only engaging in regular physical activity but also reducing sedentary behavior throughout the day. The COVID-19 pandemic brought more people to become less physically active. Most countries went into lockdown, which inhibited people from going to work, gym, and or participating in leisure activities. According to data provided by the WHO, it was estimated that physical inactivity played a role in an estimated 3.2 million deaths worldwide in 2018. It was found that a consistently inactive lifestyle had a 1.6% greater chance of death after contracting COVID-19 (Sallis et al., 2021). As the world normalizes, encouraging individuals to break up prolonged sitting with short bouts of physical activity, such as taking short walks or doing desk exercises, may help to reduce the risk of severe COVID-19 outcomes and improve overall health.

24.13 Physical Activity Prescription/Recommendation

The COVID-19 pandemic brought the world to take many protective measures, which led to a myriad of physical inactivity. While there is not any evidence to suggest that physical inactivity is a direct cause of death, it has been shown to increase the risk of contraction, COVID-19 severity, hospitalizations, ICU, and fatality. A meta-analysis found that those who met the WHO minimum standards of physical activity had reduced COVID-19 severity.

- The World Health Organization (WHO) has delineated physical activity guidelines that consist of adults reaching at least 150 min of moderate-intensity aerobic exercise per week, or at least 75 min of vigorous-intensity aerobic exercise per week. In addition, adults should do resistance training exercises at least twice a week (Fig. 24.4).

Children and Adolescents aged 5-17 years •Average of 60 minutes per day of moderate-to-vigorous intensity, mostly aerobic, physical activity, across the week. •Vigorous-intensity aerobic activities, as well as those that strengthen muscle and bone, at least 3 days a week. •Limit the amount of time spent being sedentary, particularly recreational screen time.	**Adults aged 18–64 years** •At least 150–300 minutes of moderate-intensity aerobic physical activity; or at least 75–150 minutes of vigorous-intensity aerobic physical activity; or an equivalent combination of both •Muscle-strengthening activities at moderate or greater intensity that involve all major muscle groups on 2 or more days a week •Limit the amount of time spent being sedentary.
Adults aged 65 years and above •Same as for adults; and •As part of their weekly physical activity, older adults should do varied multicomponent physical activity that emphasizes functional balance and strength training at moderate or greater intensity, on 3 or more days a week, to enhance functional capacity and to prevent falls.	**People with Chronic Conditions (Obesity, Hypertension, type 2 diabetes, HIV and cancer survivors)** •Same as adults; and •To reduce the detrimental effects of high levels of sedentary behavior, all adults should aim to do more than the recommended levels of moderate- to vigorous-intensity physical activity.

Fig. WHO guidelines for different age groups and specific population groups on how much physical activity is needed for good health

Fig. 24.4 WHO guidelines for different age groups and specific populations on how much physical activity is recommended

There is much literature to support the benefits of physical activity on lung function. This raises importance due to the previously mentioned effect that COVID-19 has on the lungs. Those who met the suggested standards of physical activity can improve respiratory muscle strength, lung volume, and cardiovascular fitness, and reduce inflammation. A meta-analysis showed that adults with lung disease had a greater than 30% chance of hospitalization with COVID-19. In fact, those who were hospitalized had a greater chance of requiring mechanical ventilation and an elevated mortality rate. The effectiveness of physical activity was shown in observational research indicating that those who met the weekly recommended physical activity had a lower risk of infection (11%), hospitalization (36%), severe COVID-19 illness (34%), and COVID-19-related death (43%) as compared with their inactive peers (Ezzatvar et al. 2022) (Fig. 24.5). A study conducted in 2020 included data from 3142 US counties that showed physical activity moderated COVID-19 cases and deaths (Cunningham, 2021). Although the results varied from state to state, there was a consistent negative relationship showing that less physical activity led to higher death rates. While the majority of the research shows the benefits of physical activity, there are limitations with the studies and the results need to be interpreted with caution due to the confounding variables of sex, age, underlying health conditions, and socioeconomic status.

24.13.1 Dose Response to Physical Activity

Dose response is shown to play a vital role in the association of COVID-19 severity, with research showing those who met the physical activity standards set by the WHO had their hospital length decrease by two days. The dose response was shown to have a non-linear relationship between physical activity and COVID-19 severity flattening out at the WHOs recommendations. Some activity is still shown to be better than none, which was recognized as 11–149 min/week of moderate intensity. A minimum dose of 10 min a day for a total of 50 min per week is shown to reduce COVID-19 severity. This suggests that more is not necessarily better and too much exercise has been shown to have deleterious effects on the immune system. Exercise intensity has been shown to play a role, signifying a J-shaped curve. Furthermore, numerous studies show that healthy people that conducted heavy exercise before or while

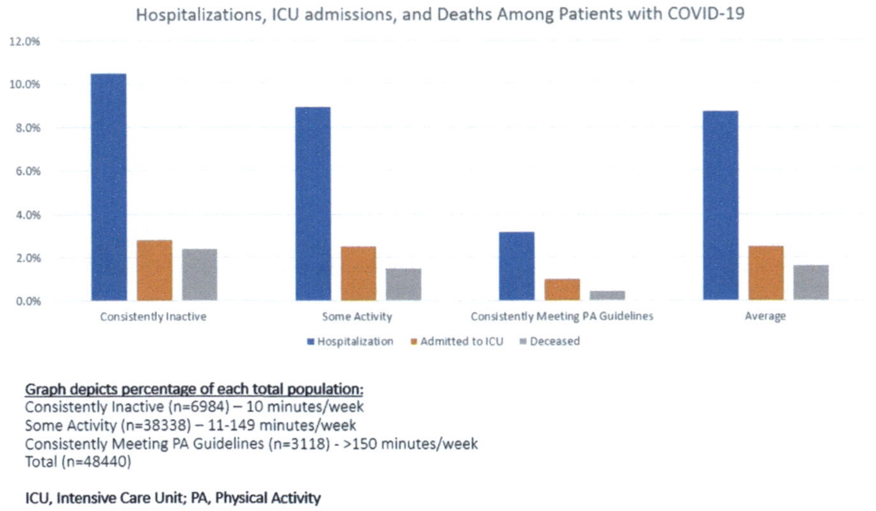

Fig. 24.5 Hospitalization, ICU, and deaths among patients with COVID-19, stratified by physical activity level

infected increased their COVID-19 severity (Nieman 2020). Additionally, engaging in outdoor activities or exercises that allow for social distancing, practicing good hygiene, and wearing a mask or face covering while exercising can also help reduce the risk of contracting COVID-19.

24.14 Physical Activity and Cardiovascular Disease

Cardiovascular disease (CVD) is responsible for the majority of the deaths in the world with studies showing those individuals with this preexisting disease to be at higher risk for COVID-19 severity. The Center for Disease Control and Prevention (CDC) reported approximately 30% of deaths in the USA due to COVID-19 had CVD. CVD is associated with diseases affecting the heart and blood vessels, which can be mitigated by adopting a healthy lifestyle factor like physical activity. There is an extensive amount of literature that shows those who meet the minimum physical activity recommendations suggested by the WHO had a significant reduction in CVD. This is important to understand as 8 in 10 US adults were shown to not meet the recommended guidelines for physical activity (Lobelo et al. 2018). Physical activity can strengthen the heart muscle due to it causing the heart to work harder, in turn improving the contractility of the heart. High blood pressure (HBP), or as it is also known as the silent killer, can be reduced with physical activity. HBP was found to be a common comorbidity among those with COVID-19, with the CDC reporting approximately 32% of deaths in the USA due to COVID-19 having HBP. It is important to mention that those with CVD often had other comorbidities, such as diabetes and obesity, which can also increase the risk of COVID-19 severity.

24.15 Physical Activity and Diabetes

Diabetes has three main types with type 1 being an autoimmune condition where the immune system attacks the pancreatic cells that produce insulin. A recent retrospective study showed that patients 55 and older with type 1 diabetes had a threefold risk of death from COVID-19 when compared to the general population (Rawshani et al. 2021). Gestational diabetes occurs during pregnancy and generally resolves after delivery. However, it can lead to an increased risk of developing type 2 diabetes. Type 2 diabetes is the most prevalent in middle-aged or older adults, accounting for 90–95% of cases in the world. Type 2 diabetes is caused by the body becoming resistant to insulin and or does not produce enough insulin to help keep glucose levels within normal ranges. A study conducted in Europe showed that adults 40 and older with type 2 diabetes were independently associated with a twofold increased risk of hospitalization, admission to ICU, and death for COVID-19. The CDC reported that of those who were hospitalized with COVID-19 severity, 39.7% had type 2 diabetes and the percentage increased for those older than 50 to 46.5%. The exact cause of type 2 diabetes is not fully understood; it is well understood that physical inactivity plays a significant role. The American Diabetes Association recommends that adults engage in at least 150 min of moderate-intensity aerobic exercise per week, plus two days of resistance training. Moderate-intensity exercise has many positive effects, which include improvement in insulin sensitivity, lowering blood sugar, temporarily increasing IL-6 that have direct anti-inflammatory effects and can reduce the rate of obesity.

24.16 Conclusion

The COVID-19 pandemic has resulted in an alarming number of confirmed cases and deaths globally, with over 447 million confirmed cases and over 6 million COVID-19-related deaths recorded to date. Given the gravity of the situation, there is a need for interventions targeting lifestyle factors to mitigate the severity of COVID-19. Research has shown that obese individuals are 46% more likely to contract the virus than those with normal weight, whereas physically active individuals have an 11% reduced risk of contracting the virus. Physical activity has emerged as a promising intervention in this regard. Regular physical activity has been linked to a range of health benefits, including improving immune function, and cardiovascular health, and reducing obesity, all of which can contribute to reducing the severity of COVID-19.

Older males are identified as the group most at risk of death from COVID-19, and they have been shown to be less physically active than younger males. Additionally, individuals with underlying health conditions such as cardiovascular disease, lung disease, metabolic disease, and obesity are among the most inactive populations and are therefore at the greatest risk of severe illness and death. Given the constant mutations of the SARS-CoV-2 virus resulting in different variants, maintaining a strong immune system through physical activity has become imperative. The pandemic has brought about significant restrictions, including stay-at-home orders, social distancing, and mask-wearing, which have impacted people's ability to be physically active. However, research has shown that there are various ways to engage in physical activity, even amidst such restrictions. Furthermore, mask-wearing during physical activity has been extensively studied and has been found to have no deleterious effects on healthy individuals. It is recommended to consult with a doctor for individuals with certain respiratory conditions. Moving forward, the COVID-19 pandemic has highlighted the significant value of physical activity and its benefits for overall health. It is crucial that individuals prioritize physical activity to maintain a healthy immune system and reduce the severity of COVID-19 (Fig. 24.6).

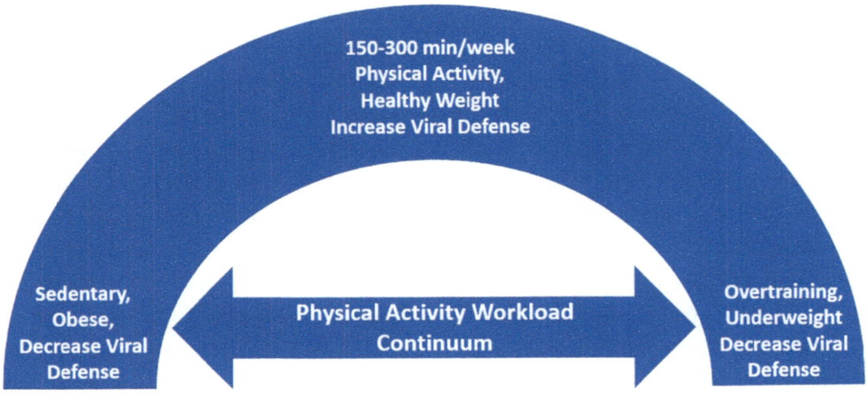

Fig. 24.6 U-shaped relationship between physical activity level and viral defense

References

Adams GR et al (2011) Exercise and leukocyte interchange among central circulation, lung, spleen, and muscle. Brain Behav Immun 25(4):658–66. Print

Aminian A, Fathalizadeh A, Tu C, Scott Butsch W, Pantalone KM, Griebeler ML, Kashyap SR, Rosenthal RJ, Burguera B, Nissen SE (2021) Association of prior metabolic and bariatric surgery with severity of coronavirus disease 2019 (COVID-19) in patients with obesity. Surg Obes Relat Dis 17(1):208–214. Print

Popkin BM, Du S, Green WD, Beck MA, Algaith T, Herbst CH, Alsukait RF, Alluhidan M, Alazemi N, Shekar M (2020) Individuals with obesity and COVID-19: a global perspective on the epidemiology and biological relationships. Obes Rev 21(11):e13128. Print

Belančić A, Kresović A, Valentino R (2020) Potential pathophysiological mechanisms leading to increased COVID-19 susceptibility and severity in obesity. Obes Med 19:100259. Print

Biscarini S, Colaneri M, Ludovisi S, Seminari E, Teresa CP, Valsecchi P, Gallazzi I, Giusti E, Cammà C, Zuccaro V (2020) The obesity paradox: analysis from the SMAtteo COvid-19 REgistry (SMACORE) cohort. Nutr Metab Cardiovasc Dis 30(11):1920–1925. Print

Buitrago-Garcia D, Egli-Gany D, Michel JC, Hossmann S, Imeri H, Ipekci AM, Salanti G, Low N (2020) Occurrence and transmission potential of asymptomatic and presymptomatic SARS-CoV-2 infections: a living systematic review and meta-analysis. PLoS Med 17(9):e1003346. Print

Carfì A, Bernabei R, Francesco L (2020) Persistent symptoms in patients after acute COVID-19. Jama 324(6):603–605. Print

Carole HS, Murray B, Varsavsky T, Graham MS, Penfold RS, Bowyer RC, Pujol JC, Klaser K, Antonelli M, Canas LS (2021) Attributes and predictors of long COVID. Nat Med 27(4):626–631. Print

Caussy C, Wallet F, Laville M, Emmanuel D (2020) Obesity is associated with severe forms of COVID-19. Obesity 28(7):1175. Print

Centers for Disease Control and, Prevention (2021) COVID-19 hospitalization and death by age (2021). 2023. Print

Cinti F, Saverio C (2022) The endocrine adipose organ: a system playing a central role in COVID-19. Cells 11(13):2109. Print

Cunningham GB (2021) Physical activity and its relationship with COVID-19 cases and deaths: Analysis of U.S. counties. J Sport Health Sci 10(5):570–76. Print

Ezzatvar Y et al (2022) Physical activity and risk of infection, severity and mortality of COVID-19: a systematic review and non-linear dose–response meta-analysis of data from 1 853 610 adults. Br J Sports Med 56(20):1188–1193. Print

Fernández-de-Las-Peñas C, Palacios-Ceña D, Gómez-Mayordomo V, Lidiane LF, Cuadrado ML, Plaza-Manzano G, Navarro-Santana M (2021) Prevalence of post-COVID-19 symptoms in hospitalized and non-hospitalized COVID-19 survivors: a systematic review and meta-analysis. Eur J Int Med 92:55–70. Print

Földi M, Farkas N, Kiss S, Dembrovszky F, Szakács Z, Balaskó M, Erőss B, Hegyi P, Andrea S (2021) Visceral adiposity elevates the risk of critical condition in COVID-19: a systematic review and meta-analysis. Obesity 29(3):521–528. Print

Gao F, Zheng KI, Wang X-B, Sun Q-F, Pan K-H, Wang T-Y, Chen Y-P, Targher G, Byrne CD, George J (2020) Obesity is a risk factor for greater COVID-19 severity. Diab Care 43(7):e72–e74. Print

Garg S, Kim L, Whitaker M, O'Halloran A, Cummings C, Holstein R, Prill M, Chai SJ, Kirley PD, Alden NB (2020) Hospitalization rates and characteristics of patients hospitalized with laboratory-confirmed coronavirus disease 2019—COVID-NET, 14 States, March 1–30, 2020. Morb Mortal Wkly Rep 69(15):458. Print

Glynne P et al (2022) Long COVID following mild SARS-CoV-2 infection: characteristic T cell alterations and response to antihistamines. J Invest Med 70(1):61–67. Print

Hamer M, Catharine RG, Kivimäki M, David Batty G (2020) Overweight, obesity, and risk of hospitalization for COVID-19: a community-based cohort study of adults in the United Kingdom. Proc Nat Acad Sci 117(35):21011–21013. Print

Hekmatikar A, Hossein A et al (2022) Functional and psychological changes after exercise training in post-COVID-19 patients discharged from the hospital: a PRISMA-compliant systematic review. Int J Environ Res Publ Heal 19(4):2290. Print

Hindsberger B, Lindegaard B, Andersen LR, Israelsen SB, Pedersen L, Szecsi PB, Benfield T (2023) Circulating adiponectin levels are inversely associated with mortality and respiratory failure in patients hospitalized with COVID-19. Int J Endocrinol 2023 (2023). Print

David J-P, Carbonell-Baeza A, Lavie CJ (2020) Physical exercise as therapy to fight against the mental and physical consequences of COVID-19 quarantine: Special focus in older people. Prog Cardiovasc Dis 63(3):386. Print

Kim NH, Kim KJ, Choi J, Sin Gon K (2022) Metabolically unhealthy individuals, either with obesity or not, have a higher risk of critical coronavirus disease 2019 outcomes than metabolically healthy individuals without obesity. Metabolism 128:154894. Print

Klang E, Kassim G, Soffer S, Freeman R, Matthew AL, Reich DL (2020) Severe obesity as an independent risk factor for COVID-19 mortality in hospitalized patients younger than 50. Obesity 28(9):1595–99. Print

Larrabee RC (1902) Leucocytosis after violent exercise. J Med Res 7(1):76. Print

Li S, Cao Z, Yang H, Zhang Y, Xu F, Wang Y (2021) Metabolic healthy obesity, vitamin D status, and risk of COVID-19. Aging Dis 12(1):61. Print

Lobelo F et al (2020) Routine assessment and promotion of physical activity in healthcare settings: a scientific statement from the American Heart Association. Circulation 137(18):e495–e522. Print

Malavazos AE, Secchi F, Basilico S, Capitanio G, Boveri S, Milani V, Dubini C, Schiaffino S, Morricone L, Foschini C (2021) Abdominal obesity phenotype is associated with COVID-19 chest X-ray severity score better than BMI-based obesity. Eat Weight Disord –Stud Anorexia Bulimia Obes (2021):1–15. Print

Muzyka I, Revenko O, Kovalchuk I, Savytska M, Bekesevych A, Oksana Z (2023) What is the role of brown adipose tissue in metabolic health: lessons learned and future perspectives in the long COVID? Inflammopharmacology 1–9. Print

Nalbandian A, Sehgal K, Gupta A, Madhavan MV, McGroder C, Stevens JS, Cook JR, Nordvig AS, Shalev D, Sehrawat TS (2021) Post-acute COVID-19 syndrome. Nat Med 27(4):601–615. Print

Nieman DC (2020) Coronavirus disease-2019: A tocsin to our aging, unfit, corpulent, and immunodeficient society. J Sport Health Sci 9(4):293–301. Print

O'Shea D, Hogan AE (2019) Dysregulation of natural killer cells in obesity. Cancers 11(4):573. Print

Phetsouphanh C et al (2022) Immunological dysfunction persists for 8 months following initial mild-to-moderate SARS-CoV-2 infection. Nat Immunol 23(2):210–216. Print

Poly TN, Islam MM, Yang HC, Lin MC, Jian W-S, Hsu M-H, Yu-Chuan Jack L (2021) Obesity and mortality among patients diagnosed with COVID-19: a systematic review and meta-analysis. Frontiers Med 8:620044. Print

Rawshani A et al (2021) Severe COVID-19 in people with type 1 and type 2 diabetes in Sweden: a nationwide retrospective cohort study. Lancet Reg Health-Eur 4:100105. Print

Sallis R et al (2021) Physical inactivity is associated with a higher risk for severe COVID-19 outcomes: a study in 48 440 adult patients. Brit J Sports Med 55(19):1099–1105. Print

Sattar N, McInnes IB, McMurray JJV (2020) Obesity is a risk factor for severe COVID-19 infection: multiple potential mechanisms. Circulation 142(1):4–6. Print

Siahaan YMT, Hartoyo V, Hariyanto TI, Andree K (2022) Coronavirus disease 2019 (Covid-19) outcomes in patients with sarcopenia: a meta-analysis and meta-regression. Clin Nutrition ESPEN 48 (2022):158–66. Print

Stefan N, Andreas LB, Schulze MB, Ludwig DS (2020) Obesity and impaired metabolic health in patients with COVID-19. Nat Rev Endocrinol 16(7):341–342. Print

Trayhurn P, Wood IS (2004) Adipokines: inflammation and the pleiotropic role of white adipose tissue. Br J Nutr 92(3):347–355. Print

Vulturar D-M, Crivii C-B, Orăsan OH, Palade E, Buzoianu A-D, Zehan IG, Doina Adina T (2022) Obesity impact on SARS-CoV-2 infection: pros and cons "Obesity Paradox"—a systematic review. J Clin Med 11(13):3844. Print

World Health, Organization (2020) World Health Organization coronavirus disease (COVID-19) dashboard. World Health Organization (2020). Print

Wright J, Astill SL, Sivan M (2022a) The relationship between physical activity and long COVID: a Cross-Sectional Study. Int J Environ Res Pub Heal 19(9):5093. Print

Wright J, Astill SL, Sivan M (2022b) The relationship between physical activity and long covid: a cross-sectional study. Int J Environ Res Publ Heal 19(9):5093. Print

Yang J, Ma ZY, Lei YC (2021)A meta-analysis of the association between obesity and COVID-19. Epidemiol Infect 149 (2021). Print

Zachary Z, Brianna F, Brianna L, Garrett P, Jade W, Alyssa D, Mikayla K (2020) Self-quarantine and weight gain related risk factors during the COVID-19 pandemic. Obes Res Clin Pract 14(3):210–216. Print

Zachary Z (2021) COVID-19 self-quarantine and weight gain risk factors in adults. Curr Obes Rep 10:423–433. Print

Zeigler Z, Lawrence J, Jamison A, Pete S (2021) Physical activity behaviors and COVID-19 symptoms. OBM Integr Complement Med 6(3):1–16. Print

Global Breakdowns and Transformations in the World of Sports After Covid-19

25

Özge Ercan

Abstract

The world has spent the first phase of the crisis caused by the Covid-19 pandemic, which is powerful and invisible, threatening the lives of every person without discrimination, with the struggle in the field of health. The second stage, which we are still in, is the period in which the economic crisis that occurred as a result of the measures taken is tried to be overcome. The third stage is a new order that Covid-19 has begun to shape. The-sports world has been affected by the crisis caused by the Covid-19 pandemic in the first two stages. It is clear, then, that it will be heavily affected by the third stage. The strategies it will have globally after the Covid-19 pandemic will determine the new role of sports in the global order. In this section, considering these three stages, the breaks, challenges and transformations that may take place in the world of sports after the Covid-19 pandemic are discussed and evaluated.

Keywords

Covid-19 · Future · Resilience · Sports · Transformation

Ö. Ercan (✉)
Faculty of Sport Sciences,
Sinop University, 57000 Sinop, Türkiye
e-mail: ozgercan@windowslive.com

25.1 Introduction

The Covid-19 pandemic has been one of the turning points in human history. First of all, it threatened human life. It tested people with emotions such as anxiety, fear and pain, and affected their inner peace. Then it damaged the social fabric of human beings, which is a social being. By creating the before and after of today's world, it has forced countries to make changes in their internal structures and in their relations with other countries. It profoundly affected the social and cultural systems of countries.

The spread of Covid-19 to every corner of the world at a speed far beyond the predictions and its transformation into a global pandemic has increased the size of the danger in terms of health as well as in economic terms. It has become an important determinant in the economies of countries and caused a contraction in the global economy. The first wave of the pandemic revealed how unprepared it is in many areas such as the production of goods and/or services, supply chain, etc., which form the roof of the global economy.

Quarantine and curfews to reduce the risk of transmission of the virus have been the trigger for the economic downturn. In the meantime, the closure of businesses has reduced the economy's capacity to produce goods and services (Brinca et al. 2021; Gupta et al. 2020). It disrupted consumer services (Bartik et al. 2020). This was

N. Rezaei (ed.), *The COVID-19 Aftermath*, Advances in Experimental Medicine and Biology 1457,
https://doi.org/10.1007/978-3-031-61939-7_25

the period when the most severe fluctuations in supply and demand shocks in financial markets were experienced since the Second World War. The economic contraction and financial market turbulence caused by Covid-19 have affected different lines of business from different sectors in a vulnerable and disproportionate way.

Sports has been one of the sectors affected by the Covid-19 pandemic and where these effects have been felt on a large scale. Considering that sports is a giant sector that has its own sub-markets and creates added value for other sectors, far beyond a match, it is understandable how severe the effects of this pandemic on the-sports industry are. For example, social distancing rules and measures taken to prevent the spread of the pandemic have led to the postponement or cancellation of sports events by governments all over the world. Clubs and organizations have lost billions of dollars as a result of cancellations due to pandemic. Clubs in professional leagues were only able to make up for some of their losses, thanks to the media deals they made. Towards the middle of 2020, a large number of sports events were held in halls and stadiums without spectators.

In addition to global events such as the Olympic Games, World Championships, etc., all amateur and local competitions have also had their share from postponements and cancellations. Due to the bans imposed on sports centres and fitness centres, individual exercises could not be done. In short, the pandemic has also affected the-sports sector, which has an important place for the economies of many countries, both individually and professionally.

It has drastically changed the-sports habits. Digitalization in sports, which started before the pandemic, has increased its impact in this period. It has brought the use of digital tools to the forefront. The pandemic has stopped some things in the-sports industry. But it also paved the way for new beginnings. Considering that the-sports economy means more than just a sporting event, it is possible to say that the new normals and new paradigms after Covid-19 will create a butterfly effect and give birth to a new sports industry. Therefore, "Chaos Theory" needs to be elaborated in order to identify the damages caused by Covid-19 in the-sports economy and to create a comprehensive rescue strategy for them.

25.2 A Theoretical Perspective on the Chaos Caused by Covid-19

The developments caused by the Covid-19 pandemic and its effects worldwide can best be explained by chaos theory. This theory is an important tool for interpreting the negative impact of the pandemic on systems in an individual, social or economic sense and the unpredictable situations resulting from these effects.

The normal course of a system, the situations in which one or more inputs make physical interventions, incorporate different orders into the system and human beings cannot control these inputs with their abilities are called "chaos". Chaos theory is an appropriate approach to facilitate the identification and understanding of the new actors and paradigms that chaos has revealed.

Chaos theory explains the elements that enable relationships and systems to organize, transform and renew themselves in unpredictable situations (Seeger 2002). Therefore, chaos theory must be well-comprehended in order for each system to successfully emerge from the chaos environment and to turn each crisis environment into an opportunity.

Chaos theory is a theory proposed by Lorenz (1963) which attempts to understand how simple systems can change in sudden, unexpected or irregular ways (Warren 2013). According to chaos theory, an event that happens in one place can trigger another event in another place and may cause disorder. In this way, the consequences of the triggering event affect some or all of the components of the system.

Lorenz (1972) argues that a small change in a place can create enormous turbulence in a long-distance space, and explains this theory with the "butterfly effect". More specifically, he states that the flapping of a butterfly's wings in Asia can cause a storm all over the world. Therefore, considering that the Covid-19 pandemic has affected many systems in the world and caused chaos, it would not be wrong to compare it to a storm.

The Covid-19 pandemic, like the storm in Lorenz's chaos theory, has come upon the world and evolved into a path individually, socially and economically that is currently unpredictable in many areas. Sports is one of the systems affected by this storm. The Covid-19 pandemic has changed every aspect of sports around the world. It has led to a reinterpretation of sports in every aspect.

Therefore, chaos theory is needed to propose a model that will improve the competitiveness of sports after the pandemic. Chaos theory is necessary to understand the dimensions of chaos in the-sports sector, the butterfly effect caused by Covid-19 in the-sports sector and the new paradigms emerging within sports. As a matter of fact, the relevant literature suggests that chaos theory can help professionals in creating crisis management strategies.

25.3 Covid-19 Pandemic and Global Breakdowns in Sports

Until almost a few years ago, no one could have predicted where the world is today. The Covid-19 virus has infected millions of people around the world without skipping a single country and has again caused millions of deaths. In this process, almost all domestic and international human interactions have been stopped in order to reduce transmission. The whole world has been turned upside down by the expanding impact of the pandemic. In this process, the-sports sector also suffered a great disruption. Sports, which have a great place in the economy of many countries, became one of the sectors affected by the pandemic with its sub-elements such as its organizations, athletes, sponsors, media, facilities, equipment, etc. (Abdolmaleki et al. 2022).

The first reflections of the pandemic on sports began with the postponements and cancellations of leagues, matches and competitions. These organizations included hundreds of sporting events such as the Olympics, football leagues, basketball and volleyball competitions, formula racing, tennis tournaments, etc. Many giant national and international events have been postponed or cancelled. These organizations, which were postponed or cancelled due to Covid-19 in 2020, were led by the "Olympic Games" (Parnell et al. 2020). After many countries announced that they would not send athletes to the games to be held in Tokyo in 2020, the IOC President and the Japanese Government announced that they postponed the 2020 Olympic and Paralympic Games to 2021, which were seen for the first time in Olympic history, (Olympics 2020). UEFA also announced that it has postponed "Euro 2020" to 2021 as a result of a meeting of 55 member states (UEFA 2020). In the "NBA", the most popular basketball league in the world, the virus was detected in the players of the Utah Jazz team and the games of the whole season were suspended after this incident (Reuters 2020). Due to the pandemic, the second grand slam tennis tournament of the season, the "French Open", hosted by Paris, was postponed for 4 months (NBC Sports 2021). The open tennis tournament was not the only event in France that was postponed. One of the most important and prestigious races, the France Cycle Tour, which could not be organized in 1946 only due to the Second World War, experienced a first after 74 years. With the announcement of the International Cycling Union (UCI), the race has also been postponed to a later date (The Guardian 2020). The Wimbledon Championships suffered the same fate as the France Cycle Tour. The tournament, which was last denied to be organized by the organizers in World War II, was cancelled by the All England Lawn Tennis Club (AELTC) in 2020 due to the pandemic (Reuters 2020). Formula 1, the racing organization with millions of followers in the world, was only one of the organizations that were not allowed to allow spectators to the stands until the competitions held in Azerbaijan in 2022 (Formula 1 2020).

The Covid-19 pandemic has put the-sports providers, managers and media in a very difficult financial situation (Keshkar et al. 2021). Hundreds of sponsorship deals, the-sport's main beneficiaries, have been terminated due to postponed or cancelled events. In this process, technological solutions were tried to be found in order not to lose sponsors and to keep the sector afloat. By converting the insides of the houses into virtual stadiums or fields, interaction with fans was provided and digital sponsorship applications were put into operation. Thus, some of the sponsoring companies decided to continue their agreements in this way.

The Covid-19 pandemic's increase in the use of technology in the-sports sector has not been limited to virtual environments. In the early days of the shutdowns, television channels began to engage audiences with old sports programmes. But that didn't last long. For this reason, sports providers have turned to e-sports, which is still a matter of scientific debate whether it should be considered a sport (Cunningham et al. 2018; Pizzo et al. 2018; Hallmann and Giel 2018; Heere 2018). E-sports, which was on the rise before the pandemic, has become the shining star of the sector during the pandemic.

Covid-19 has also affected athletes' performances. In this process, both material and moral concerns on athletes increased. Because one of the ways that the organizers and the clubs, who got into a bottleneck, was the cut in the payments of the athletes in order to survive. For example, the Barcelona Club has announced that it has reduced the payments of all professional athletes by 70% until the pandemic is over. Due to the pandemic, coaches, referees and other sports workers have also faced the risk of losing their income and jobs.

The-sports industry is much more than just entertainment for spectators, generating revenue for sports organizations and the media. Therefore, it was not only professional sports and manpower suffered during the Covid-19 process. Fitness rooms were among the first businesses to close after the Covid-19 pandemic. During the pandemic period, fitness centres, where people receive individual services, have also come to the point of closure due to the bans introduced. To survive in this process, many fitness centres have developed their own methods regarding the social distancing rules. Those who provided services to their customers with online access managed to survive. Social media, in particular, has played a key role in ensuring customer continuity in this process. Although technology has been in place in the fitness industry since 2016, the pandemic has sharply triggered this growth (Ananthakrishnan et al. 2020). In the process, fitness applications have increased the population of people in their concerns. There has been a shift from traditional fitness rooms and studios to virtual fitness. People have rapidly adopted these technology-driven alternatives around the world and provided an increase in downloads and use of fitness applications.

Covid-19 has also been effective in the production and sale of sports products. Due to the closure of gyms, the-sports product preferences of customers who do sports at home have changed. Brands selling sports products have changed their products in line with the new needs of their customers. They marketed and sold their products through virtual stores to customers who could not leave home due to the closures. They changed customer relations and marketing methods. They have increased the online shopping habits of customers.

25.4 Transforming Sports After Covid-19

Covid-19 has not completely disappeared from our lives. Although it is not easy to predict the future, in the post-Covid-19 era, we have some signs that will shape the world. Firstly, the Covid-19 has left deep and indelible scars on individuals and societies, reshaping their way of life with technology and ensuring that technology has a singular importance by taking on everything in this process. Second, it forced countries to take radical decisions related with other countries, especially in the network of economic and political relations. Therefore, the impact of managerial significance of the pandemic on society and the economy has gradually increased.

The impact of Covid-19 on sports has once again opened the managerial dimension of sports to discussion and showed us that there are lessons to be learned from this process. First of all, in the-sports sector, it has shown that the current strategies to cope with the pandemic crisis are incomplete and insufficient. Because to be able to manage the crisis caused by the pandemic in sports, many managers had to initially put in place their own recovery programmes. Each manager tried to find a solution to the crisis caused by the pandemic with his own methods. In the face of this sudden and unexpected situation, the experience that managers have gained only to protect their organizational structures in the first place has led to the emergence of very different services and products in the-sports industry. At the end of the day, it has become necessary to develop comprehensive strategies and recovery programmes that would better manage the effects of the pandemic on the-sports economy and to deal with these effects internationally, and to conduct coordinated micro-, medium- and macro-level analyses for each sub-sector.

All crises, including the crisis caused by the Covid-19 pandemic, pave the way for some opportunities to arise. By bridging this gap in future managerial needs and challenges and bringing together different perspectives, the post-Covid-19 sports sector can transform current crises into enormous opportunities. Therefore, the failures caused by the crises caused by the Covid-19 pandemic can be turned into opportunities with the experiences gained in this process and the-sports sector can successfully continue its way with new normals. The reflections of these two complicated concepts caused by the Covid-19 pandemic on the axis of sports can be listed as follows.

25.4.1 Digitalization

The measures taken to prevent the spread of the pandemic have been a major factor in the use of digital tools by sports. The dizzying pace of developments in technology has led to an increase in the number of digital devices and applications, while the Covid-19 pandemic has increased the use of these devices and applications (Heidenreich and Talke 2020). Sports providers who want to maintain interaction with fans have started using virtual innovations as an alternative to cancelled or postponed matches. In this way, they were able to transform fans limited to arenas or stadiums into an unlimited global market (Shields and Rein 2020).

Especially during this period, social media has become an important part of sports marketing. The rapid increase in the use of social media during the pandemic period has increased its presence on the-sports sector. Sponsors and the media, in particular, have used digital platforms to the fullest within sports (Wang and Zhou 2015; Waters et al. 2011; Westberg et al. 2018). In addition to sponsors and media, sports providers have also communicated with fans instantly via social media during the Covid-19 pandemic and ensured the continuity of brand loyalty (Annamalai et al. 2021).

Even if the Covid-19 pandemic is completely eliminated from our lives, it is clear that digital devices and applications are now permanent in the-sports sector. As a matter of fact, when people are allowed to enter the fields and stadiums, gyms, fitness centres, etc., again, it is seen that real-time communication and the use of digital technologies will continue in all areas of sports. After the pandemic, sports providers will follow fans online and steer the-sports industry by interpreting their digital footprints.

25.4.2 Digital Disruption and Digital Innovation

The rapidly increasing use of technology in the Covid-19 period has led to the questioning of the traditional ways of working of sports organizations, the erosion of their established patterns and the digital destruction with the addition of innovations. In the post-Covid-19 era, it seems that technology will continue to undertake many things in the-sports sector. Sports organizations with sufficient technological infrastructure will be in a more advantageous position than others.

Market offers, business processes and the creation of models arising from the use of technology lead the-sports sector to digital innovations (Nambisan et al. 2017). All sports organizations, large or small, professional or amateur, involved in sports are putting forward their strategic initiatives with the Covid-19 pandemic. Therefore, in the near future, the knowledge economy and innovation will become more important in sports and will determine the status of all national and international organizations related to sports. Digital innovations such as virtual and augmented reality will affect traditional business models in sports after the pandemic and lead to the emergence of new professions in sports. With 5G technology, professions such as sports data analyst, sports information director, sports performance analyst, sports statistician, which have replaced some established services in sports (Beckman et al. 2012; Zupic 2014), seem to have become even more popular with the Covid-19 pandemic.

25.4.3 E-sport

Sports providers have started to look for new content that will attract the attention of the masses instead of the events and competitions cancelled by the Covid-19 pandemic. During the pandemic, they turned to two alternatives that would grab the audience's attention and interact with them. The first of these alternatives is the re-screening of important interviews with the celebrities of the-sports world and the match that was previously broadcast on television channels. The continuous repetition of these matches and interviews has enabled the audience to interact with a certain point. The second is live e-sports events, where it was debatable whether it is sports or not before the Covid-19 pandemic.

During the quarantine period, when people could not go out of their homes, interest in e-sports increased all over the world. E-sports has replaced international and national sports organizations and competitions that have been cancelled or postponed, and it is thought that it will attract more people after the pandemic. This is a new source of motivation for sports providers and a new market where they can generate revenue. The ecosystem of e-sports includes seamless and interconnected players, spectators, matches, training processes, devices, social channels, data, and multiple technologies (Carrillo Vera and Aguado Terrón 2019). Hence, it is open to all sectors as well as sports. Global investors, brands and media outlets are turning to e-sports marketing and e-sports sponsorships to capitalize on the popularity of e-sports (Kim et al. 2020; Mangeloja 2019).

25.4.4 Fitness Applications on Smart Devices

Fitness centres were among the first businesses to close with the measures taken by governments after the Covid-19 pandemic. In this period when health protection is more important than ever, people have started to turn to online exercise trends at home. This has provided technology investors with tremendous opportunities to develop fitness applications where users can be both socially isolated and maintain their physical health, which has increased the number of fitness applications.

In this process where mobile devices are now used intensively, mobile fitness applications are seen as a new marketing channel and an alternative to fitness businesses by customers (Parker et al. 2021). It is estimated that this trend will continue even after the Covid-19 pandemic. In the wake of the pandemic, traditional gyms are less likely to deal with customers alone. Therefore, hybrid models will most probably be implemented by fitness entrepreneurs.

25.5 Conclusion

Covid-19 pandemic is the largest and most comprehensive event that affected the world in every field and led to changes after World War II. It has directly or indirectly affected all systems and structures in the world and led to a crisis.

Although it is partially controlled as a health problem, its effects will be in daily life for many years. Since the exact extent of the effects is not clear yet, the process of change in the world will continue. The world of sports will also take its share from this change.

The Covid-19 pandemic initially affected sports in two main areas, social and economic. During the pandemic, social changes have triggered economic changes in the-sports sector. In order to reduce the risk of transmission of Covid-19, international and national sports organizations have been cancelled or postponed. This situation has changed the climate of being together and the habits of consuming sports of the spectators, which are the cornerstone of the organizations. Therefore, there is no longer a collective but an individual target group in front of the-sports sector.

Individuality in sports is not limited to spectators. During the pandemic period, the contribution of sports to preventive health services has once again come to the agenda. While the importance of people's sports habits has increased once again in this process, collective activities intertwined with other people in sports halls have been replaced by individual activities at home or in the open area.

Although the consumption of sports has evolved into individuality during the pandemic process, the-sports sector is a part of the global system. Therefore, the attitude of the global system after Covid-19 will also determine the direction of the-sports sector. The main parameter of the-sports sector after the pandemic will be global cooperation. It is clear that a multidimensional global cooperation is needed for a multifaceted struggle. For this reason, joint studies can enable the-sports world to overcome the current crisis successfully after Covid-19. Because the real reason why Covid-19 has brought sports to a near standstill is that it has caused an outage on a global scale.

Of course, when Covid-19 is completely eradicated, a more realistic damage assessment for the-sports will be possible and a more realistic impact will be measured. However, it is important to determine the possible opportunities that the-sports sector will face in the future and to make some preliminary evaluations. Some of the following suggestions and lessons to be learned can be taken into consideration in order to stabilize the-sports world after Covid-19 on a sectoral basis:

- The roles and capacities assigned to sport within the global system will have a significant impact on its future position. Therefore, considering the crisis caused by Covid-19 in the-sports world and the possibility of encountering similar situations again, some reforms should be made and strategic plans should be prepared. The strategic plan should include a sustainable vision with global inclusive and broad participation, and focus on the uninterrupted and safe realization of sporting events. In addition, these reforms should be arranged in such a way as to maximize the benefits provided by sports and physical activity, the importance of which has once again been understood by the pandemic.
- It is very difficult to solve the post-pandemic problems of today's sports enterprises with traditional solutions which have a single discipline and a single point of view. In order for the-sports sector to get out of the economic impasse created by Covid-19, it must be dynamic and variable.
- During the pandemic, the-sports industry has started to interact with the audience in digital environments (digital media, social media, interactive software, etc.). Innovations realized through knowledge economies and R&D activities will determine the status of each business in the-sports sector. Considering that the-sports sector has become more dependent on technology with the pandemic, it will be advantageous to create an advanced technological infrastructure. For this reason, a competition will arise in the-sports industry to obtain the latest technologies. Therefore, having a self-sufficient and reliable technological infrastructure will be in the interest of every business in the-sports sector. However, after the pandemic, significant and need-based investments should be made in technological infrastructure.

- A balance should be maintained by controlling the interaction of sports and machinery. Although it is not expected that digital experiences will completely replace physical experiences in the near future after the pandemic, arrangements should be made in such a way that the activities carried out in the digital environment shall not precede traditional sports, and sports applications downloaded to smartphones or devices shall not precede fitness and gyms.
- Digital skills, digital competencies and digital literacy have increased their importance after the pandemic. New professions that have emerged with digitalization in the-sports sector also attribute new roles to sports workers. If the managers and employees who have a good command of digital technologies in the-sports sector succeed in implementing the practical applications of new solutions and processes, they can increase the existing resources, gain superiority over their competitors and offer immediate solutions to the problems encountered in the process. For this reason, sports workers should be provided with digital skills.
- Sports have many insurmountable problems such as betting, match fixing and corruption that existed before the pandemic. In addition to these problems, post-pandemic digitalization will also raise some new problems and challenges. For this reason, studies on ensuring security in digital environments, protecting personal data and determining ethical rules must be carried out in advance.

As a result, the world's largest social experiment will have taken place after Covid-19 and its implications for the-sports industry will be better understood. What is clearly seen from the point we are at now is that the-sports organizations that can manage the pandemic process are one step closer to being the winners of the future after the pandemic.

Declaration of Conflicting Interests The author(s) declared no potential conflicts of interest with respect to the research, authorship and/or publication of this article.

Funding The author(s) received no financial support for the research, authorship and/or publication of this article.

References

Abdolmaleki H, Mohammadi S, Alonso-Dos-Santos M, García-Pascual F (2022) Challenges and opportunities of sports industry during the covid-19 pandemic. J Sports Econ Manag 12(1):41–67

Ananthakrishnan UM, Chen J, Susarla A (2020) No pain, no gain: examining the digital resilience of the fitness sector during the COVID-19 pandemic. SSRN Electron J. https://doi.org/10.2139/ssrn.3740143

Annamalai B, Yoshida M, Varshney S, Pathak AA, Venugopal P (2021) Social media content strategy for sport clubs to drive fan engagement. J Retail Consum Serv 62:102648

Bartik AW, Bertrand M, Cullen Z, Glaeser EL, Luca M, Stanton C (2020) The impact of COVID-19 on small business outcomes and expectations. Proc Natl Acad Sci 117(30):17656–17666

Beckman C, Eisenhardt K, Kotha S, Meyer A, Rajagopalan N (2012) Technology entrepreneurship. Strateg Entrep J 6(2):89–93

Brinca P, Duarte JB, Faria-e-Castro M (2021) Measuring labor supply and demand shocks during COVID-19. Eur Econ Rev 139:103901

Carrillo Vera JA, Aguado Terrón JM (2019) The eSports ecosystem: Stakeholders and trends in a new show business. Catalan J Commun Cultural Stud 11(1):3–22

Cunningham GB, Fairley S, Ferkins L, Kerwin S, Lock D, Shaw S, Wicker P (2018) ESport: construct specifications and implications for sport management. Sport Manage Rev 21(1):1–6

Formula 1 (2020). Available at: https://www.formula1.com/en/latest/article.formula-1-launches-virtual-grand-prix-series-to-replace-postponed-races. [Accessed: 25 Jan 2023]

Gupta S, Simon KI, Wing C (2020) Mandated and voluntary social distancing during the covid-19 epidemic: a review

Hallmann K, Giel T (2018) ESports—Competitive-sports or recreational activity? Sport Manage Rev 21(1):14–20

Heere B (2018) Embracing the-sportification of society: defining e-sports through a polymorphic view on sport. Sport Manage Rev 21(1):21–24

Heidenreich S, Talke K (2020) Consequences of mandated usage of innovations in organizations: developing an innovation decision model of symbolic and forced adoption. AMS Rev 10(3–4):279–298

Keshkar S, Dickson G, Ahonen A, Swart K, Addesa F, Epstein A, Murray D (2021) The effects of Coronavirus pandemic on the-sports industry: an update. Ann Appl Sport Sci 9(1):0–0

Kim YH, Nauright J, Suveatwatanakul C (2020) The rise of e-sports and potential for Post-COVID continued growth. Sport Soc 23(11):1861–1871

Lorenz EN (1963) Deterministic nonperiodic flow. J Atmos Sci 20(2):130–141

Lorenz E (1972) Predictability: does the flap of a butterfly's wing in Brazil set off a tornado in Texas? pp 260–263

Mangeloja E (2019) Economics of esports. Electron J Bus Ethics Organ Stud 24(2):34–42

Nambisan S, Lyytinen K, Majchrzak A, Song M (2017) Digital Innovation Management: reinventing innovation management research in a digital world. MIS Q 41(1):223–238

NBC Sports (2021) Available at: https://sports.nbcsports.com/2021/04/10/french-open-postponed-due-to-pandemic. [Accessed: 21 Jan 2023]

Olympics (2020) Available at: https://olympics.com/en/news/tokyo-olympic-games-postponed-ioc. [Accessed: 21 Jan 2023]

Parker K, Uddin R, Ridgers ND, Brown H, Veitch J, Salmon J, Arundell L (2021) The use of digital platforms for adults' and adolescents' physical activity during the COVID-19 pandemic (Our life at home): Survey study. J Med Int Res 23(2):e23389

Parnell D, Widdop P, Bond A, Wilson R (2020) COVID-19, networks and sport. Managing Sport and Leisure 1–7

Pizzo AD, Na S, Baker BJ, Lee MA, Kim D, Funk DC (2018) eSport versus sport: a comparison of spectator motives. Sport Mark Q 27(2):108–123

Reuters (2020) Available at: https://www.reuters.com/article/us-basketball-nba-idUSKBN20Z07V [Accessed: 20 Jan 2023]

Seeger MW (2002) Chaos and crisis: propositions for a general theory of crisis communication. Pub Relat Rev 28(4):329–337

Shields B, Rein I (2020) Strategies to reimagine the stadium experience. In: 21st Century sports. Springer, Cham, pp 233–248

The Guardian (2020) Available at: https://www.theguardian.com/sport/2020/apr/15/tour-de-france-rescheduled-nice-29-august-coronavirus-cycling. [Accessed: 21 Jan 2023]

UEFA (2020) Available at: https://www.uefa.com/insideuefa/about-uefa/news/025b-0f8e76aef315-8506a9de10aa-1000--uefa-postpones-euro-2020-by-12-months/ [Accessed: 21 Jan 2023]

Wang Y, Zhou S (2015) How do sports organizations use social media to build relationships? A content analysis of NBA clubs' Twitter use. Int J Sport Commun 8(2):133–148

Warren K (2013) Chaos theory and complexity theory. In: Encyclopedia of social work

Waters RD, Burke KA, Jackson ZH, Buning JD (2011) Using stewardship to cultivate fandom online: comparing how national football league teams use their web sites and Facebook to engage their fans. Int J Sport Commun 4(2):163–177

Westberg K, Stavros C, Smith AC, Munro G, Argus K (2018) An examination of how alcohol brands use sport to engage consumers on social media. Drug Alcohol Rev 37(1):28–35

Zupic I (2014) The knowledge base of technology entrepreneurship. In: Handbook of Research on Techno-Entrepreneurship, Second Edition. Edward Elgar Publishing

GPSR Compliance

The European Union's (EU) General Product Safety Regulation (GPSR) is a set of rules that requires consumer products to be safe and our obligations to ensure this.

If you have any concerns about our products, you can contact us on ProductSafety@springernature.com

In case Publisher is established outside the EU, the EU authorized representative is:

Springer Nature Customer Service Center GmbH
Europaplatz 3
69115 Heidelberg, Germany

The manufacturer's authorised representative in the EU is Springer Nature Customer Service Centre GmbH, Europaplatz 3, 69115 Heidelberg, Germany. If you have any concerns regarding our products, please contact ProductSafety@springernature.com

Printed and bound by CPI Group (UK) Ltd, Croydon, CR0 4YY
01/10/2026
02225906-0020